Beginning Algebra

Beginning Algebra

Fifth Edition

John Tobey

North Shore Community College
Danvers, Massachusetts

Jeffrey Slater

North Shore Community College
Danvers, Massachusetts

Prentice
Hall

Prentice Hall
Upper Saddle River, NJ 07458

Library of Congress Cataloging-in-Publication Data

Tobey, John
  Beginning algebra/John Tobey, Jeffrey Slater.—5th ed.
    p. cm.
  Includes index.
  ISBN 0-13-093228-0 — ISBN 0-13-090951-3 (pbk.)
  1. Algebra.   I. Slater, Jeffrey – II. Title.
  QA152.3 .T63 2001
  512—dc21   2001021662

Executive Acquisition Editor: Karin E. Wagner
Project Manager: Mary Beckwith
Editor in Chief: Christine Hoag
Senior Managing Editor: Linda Mihatov Behrens
Executive Managing Editor: Kathleen Schiaparelli
Vice President/Director of Production and Manufacturing: David W. Riccardi
Production Management: Elm Street Publishing Services, Inc.
Manufacturing Buyer: Alan Fischer
Manufacturing Manager: Trudy Pisciotti
Executive Marketing Manager: Eilish Collins Main
Development Editor: Kathy Sessa Frederico
Editor in Chief, Development: Carol Trueheart
Media Project Manager, Developmental Math: Audra J. Walsh
Art Director: Maureen Eide
Assistant to the Art Director: John Christiana
Interior Designer: Studio Montage
Cover Designer: Studio Montage
Managing Editor, Audio/Video Assets: Grace Hazeldine
Creative Director: Carole Anson
Director of Creative Services: Paul Belfanti
Photo Researcher: Julie Tesser
Photo Editor: Beth Boyd
Cover Photo: Patrick Ingrand/Stone
Art Studio: Scientific Illustrators
Compositor: Preparé, Inc. / Emilcomp, Srl

© 2002, 1998, 1995, 1991, 1984 by Prentice-Hall, Inc.
Upper Saddle River, NJ 07458

Printed in the United States of America
10 9 8 7 6 5

Student ISBN (paperback) 0-13-090951-3
Student ISBN (case) 0-13-093228-0

Prentice-Hall International (UK) Limited, *London*
Prentice-Hall of Australia Pty. Limited, *Sydney*
Prentice-Hall Canada Inc., *Toronto*
Prentice-Hall Hispanoamericana, S.A., *Mexico*
Prentice-Hall of India Private Limited, *New Delhi*
Prentice-Hall of Japan, Inc., *Tokyo*
Pearson Education Asia Pte. Ltd.
Editora Prentice-Hall do Brasil, Ltda., *Rio de Janeiro*

This book is dedicated to the memory of
Lexie Tobey and John Tobey, Sr.
They have left a legacy of love, a memory
of four decades of faithful teaching, and a sense
of helping others that will influence generations to come.
For their grandchildren they have left
an inspiring model of a loving family,
true character, and service
to God and community.

Contents

Chapter 0

A Brief Review of Arithmetic Skills   1

Chapter 1

Real Numbers and Variables   65

Chapter **2**

Equations and Inequalities   127

Chapter **3**

Solving Applied Problems   183

Chapter **4**

Exponents and Polynomials   245

Chapter **5**

Factoring   301

Chapter 8

Systems of Equations 473

Chapter 9

Radicals 521

Chapter **10**

Quadratic Equations   575

Preface

To the Instructor

We share a partnership with you. For over thirty years we have taught mathematics courses at North Shore Community College. Each semester we join you in the daily task of sharing the knowledge of mathematics with students who often struggle with this subject. We enjoy teaching and helping students—and we are confident that you share these joys with us.

Mathematics instructors and students face many challenges today. *Beginning Algebra* was written with these needs in mind. This textbook explains mathematics slowly, clearly, and in a way that is relevant to everyday life for the college student. As with previous editions, special attention has been given to problem solving in the fifth edition. This text is written to help students organize the information in any problem-solving situation, to reduce anxiety, and to provide a guide that enables students to become confident problem solvers.

One of the hallmark characteristics of *Beginning Algebra* that makes the text easy to learn and teach from is the building-block organization. Each section is written to stand on its own, and each homework set is completely self-testing. Exercises are paired and graded and are of varying levels and types to ensure that all skills and concepts are covered. As a result, the text offers students an effective and proven learning program suitable for a variety of course formats—including lecture-based classes; discussion-oriented classes; distance learning centers; modular, self-paced courses; mathematics laboratories; and computer-supported centers.

Beginning Algebra is part of a series that includes the following:

Tobey/Slater, *Basic College Mathematics*, Fourth Edition

Blair/Tobey/Slater, *Prealgebra*, Second Edition

Tobey/Slater, *Beginning Algebra*, Fifth Edition

Tobey/Slater, *Intermediate Algebra*, Fourth Edition

Tobey/Slater, *Beginning and Intermediate Algebra*

We have visited and listened to teachers across the country and have incorporated a number of suggestions into this edition to help you with the particular learning delivery system at your school. The following pages describe the key continuing features and changes in the fifth edition.

Key Features and Changes in the Fifth Edition

Developing Problem-Solving Abilities

We are committed as authors to producing a textbook that emphasizes mathematical reasoning and problem-solving techniques as recommended by AMATYC, NCTM, AMS, NADE, MAA, and other bodies. To this end, the problem sets are built on a wealth of real-life and real-data applications. Unique problems have been developed and incorporated into the exercise sets that help train students in data interpretation, mental mathematics, estimation, geometry and graphing, number sense, critical thinking, and decision making.

More Applied Problems

The exercises and applications have been extensively revised. Numerous real-world and real-data application problems show students the relevance of the math they are learning. The applications relate to everyday life, global issues beyond the borders of the United States, and other academic disciplines. Many include source citations. The number of real-data applications has significantly increased. Roughly 30 percent of the applications have been contributed by actual students based on scenarios they have encountered in their home or work lives.

Math in the Media

New Math in the Media applications appear at the end of each chapter to offer students yet another opportunity to see why developing mastery of mathematical concepts enhances their understanding of the world around them. The applications are based on a brief clip, illustration, or information from familiar media sources—either online or print. The exercises may ask students to interpret or verify information, perform calculations, make decisions or predictions, or provide a rationale for their responses.

Putting Your Skills to Work Applications

This highly successful feature has been revised in the fifth edition. There are 10 new Putting Your Skills to Work applications in the new edition. These nonroutine application problems challenge students to synthesize the knowledge they have gained and apply it to a totally new area. Each problem is specifically arranged for independent and cooperative learning or group investigation of mathematical problems that pique student interest. Students are given the opportunity to help one another discover mathematical solutions to extended problems. The investigations feature open-ended questions and extrapolation of data to areas beyond what is normally covered in such a course.

Internet Connections

As an integral part of each Putting Your Skills to Work problem, students are exposed to an interesting application of the Internet and encouraged to continue their investigations. This use of technology inspires students to have confidence in their abilities to successfully use mathematics. In the fifth edition, the Internet Connections have been completely revised and updated.

 The companion Web site (`http://www.prenhall.com/tobey_beginning`) now features annotated links to help students navigate the sites more efficiently and to provide a more user-friendly experience.

Increased Integration and Emphasis on Geometry

Due to the emphasis on geometry on many statewide exams, geometry problems are integrated throughout the text. The new edition contains over approximately 35 percent more geometry problems. Additionally, examples and exercises that incorporate a principle of geometry are now marked with a triangle icon for easy identification.

Blueprint for Problem Solving

The successful Mathematics Blueprint for Problem Solving strengthens problem-solving skills by providing a consistent and interactive outline to help students organize their approach to problem solving. Once students fill in the blueprint, they can refer back to their plan as they do what is needed to solve the problem. Because of its flexibility, this feature can be used with single-step problems, multistep problems, applications, and nonroutine problems that require problem-solving strategies. Students will not need to use the blueprint to solve every problem. It is available for those faced with a problem with which they are not familiar, to alleviate anxiety, to show them where to begin, and to assist them in the steps of reasoning.

Developing Your Study Skills

This highly successful feature has been retained in the new edition. The boxed notes are integrated throughout the text to provide students with techniques for improving their study skills and succeeding in math courses.

Graphs, Charts, and Tables

When students encounter mathematics in real-world publications, they often encounter data represented in a graph, chart, or table and are asked to make a reasonable conclusion based on the data presented. This emphasis on graphical interpretation is a continuing trend with the expanding technology of our day. The number of mathematical problems based on charts, graphs, and tables has been significantly increased in this edition. Students are asked to make simple interpretations, to solve medium-level problems, and to investigate challenging applied problems based on the data shown in a chart, graph, or table.

New Design

The fifth edition has a new design that enhances the accessible, student-friendly writing style. This new design includes new chapter opening applications and an improved and enhanced art program. See the walkthrough of features in the preface.

Mastering Mathematical Concepts

Text features that develop the mastery of concepts include the following:

Learning Objectives

Concise learning objectives listed at the beginning of each section allow students to preview the goals of that section.

Examples and Exercises

The examples and exercises in this text have been carefully chosen to guide students through *Beginning Algebra*. We have incorporated several different types of exercises and examples to assist your students in retaining the content of this course.

Chapter Pretests

Each chapter opens with a concise pretest to familiarize the students with the learning objectives for that particular chapter. The problems are keyed to appropriate sections of the chapter. All answers appear in the back of the book.

Practice Problems

Practice problems are found throughout the chapter, after the examples, and are designed to provide your students with immediate practice of the skills presented. The complete worked-out solution of each practice problem appears in the back of the book.

To Think About

These critical thinking questions follow some of the examples in the text and also appear in the exercise sets. They extend the concept being taught, providing the opportunity for all students to stretch their minds, to look for patterns, and to make conclusions based on their previous experience. The new edition includes an increased number of these problems.

Exercise Sets

Exercise sets are paired and graded. This design helps ease the students into the problems, and the answers provide students with immediate feedback.

Cumulative Review Problems

Each exercise set concludes with a section of cumulative review problems. These problems review topics previously covered and are designed to assist students in retaining the material. Many additional applied problems have been added to the cumulative review sections.

Graphing and Scientific Calculator Problems

Calculator boxes are placed in the margin of the text to alert students to a scientific or graphing calculator application. In the exercise section, icons indicate problems that are designed for solving with a graphing or scientific calculator.

Reviewing Mathematical Concepts

At the end of each chapter we have included problems and tests to provide students with several different formats to help them review and reinforce the ideas that they have learned. This not only assists them with this chapter, it reviews previously covered topics as well.

Chapter Organizers

The concepts and mathematical procedures covered in each chapter are reviewed at the end of the chapter in a unique chapter organizer. This device has been extremely popular with faculty and students alike. It not only lists concepts and methods, but provides a completely worked-out example for each type of problem. Students find that preparing a similar chapter organizer on their own in higher-level math courses becomes an invaluable way to master the content of a chapter of material.

Verbal and Writing Skills

These exercises provide students with the opportunity to extend a mathematical concept by allowing them to use their own words, to clarify their thinking, and to become familiar with mathematical terms.

Chapter Review Problems

These problems are grouped by section as a quick refresher at the end of the chapter. These problems can also be used by the student as a quiz of the chapter material.

Tests

Found at the end of the chapter, the chapter test is a representative review of the material from that particular chapter that simulates an actual testing format. This provides the students with a gauge to their preparedness for the actual examination.

Cumulative Tests

At the end of each chapter is a cumulative test. One-half of the content of each cumulative test is based on the math skills learned in previous chapters. By completing these tests for each chapter, the students build confidence that they have mastered not only the contents of the present chapter but the contents of the previous chapters as well.

Additional Content Changes in the Fifth Edition

- Throughout the text, explanations, definitions, and procedures have been carefully revised for greater clarity and precision.
- In Chapter 1, the section on order of operations has been moved to Section 1.5 so that all of the material on operations with real numbers is now covered together before the introduction to algebra.
- The discussion of solving linear equations in Chapter 2 now includes coverage of equations with no solution and equations with infinitely many solutions.
- Section 4.3 now offers a more thorough introduction to polynomials with the addition of new terminology at the beginning of the section and a new lesson on evaluating polynomials at the end.

Expanded and Enhanced Supplements Resource Package

The fifth edition is supported by a wealth of new supplements designed for added effectiveness and efficiency. New items include the MathPro 4.0 Explorer tutorial software together with a unique video clip feature, MathPro 5—the new online version of the popular tutorial program—providing online access anytime/anywhere and enhanced course management; a new computerized testing system—TestGenEQ with QuizMaster-EQ; all new lecture videos; lecture videos digitized on CD-ROM; Prentice Hall Tutor Center; and options for online and distance learning courses. Please see the list of supplements and descriptions.

Options for Online and Distance Learning

For maximum convenience, Prentice Hall offers online interactivity and delivery options for a variety of distance learning needs. Instructors may access or adopt these in conjunction with this text, *Beginning Algebra*, Fifth Edition.

Companion Web Site

Visit `http://www.prenhall.com/tobey_beginning`
The companion Web site includes basic distance learning access to provide links to the text's Internet Connections activities. For the fifth edition the Internet Connections activities have been completely revised and updated. The Web site now features annotated links to facilitate student navigation of the sites associated with the exercises. Links to additional sites are also included.

This text-specific site offers students an online study guide via online self-quizzes. Questions are graded and students can e-mail their results. Syllabus Manager gives professors the option of creating their own online custom syllabus. Visit the Web site to learn more.

WebCT

Visit `http://www.prenhall.com/demo`
WebCT includes distance learning access to content found in the Tobey/Slater companion Web site plus more. WebCT provides tools to create, manage, and use online course materials. Save time and take advantage of items such as online help, communication tools, and access to instructor and student manuals. Your college may already have WebCT software installed on its server or you may choose to download it. Contact your local Prentice Hall sales representative for details.

BlackBoard

Visit `http://www.prenhall.com/demo`
For distance learning access to content and features from the Tobey/Slater companion Web site plus more. BlackBoard provides simple templates and tools to create, manage, and use online course materials. Take advantage of items such as online help, course management tools, communication tools, and access to instructor and student manuals. Contact your local Prentice Hall sales representative for details.

CourseCompass™ powered by BlackBoard

Visit `http://www.prenhall.com/demo`
For distance learning access to content and features from the Tobey/Slater companion Web site plus more. Prentice Hall content is preloaded in a customized version of BlackBoard 5. CourseCompass™ provides all of BlackBoard 5's powerful course management tools to create, manage, and use online course materials. Contact your local Prentice Hall sales representative for details.

Supplements for the Instructor

Printed Resources

Annotated Instructor's Edition (ISBN: 0-13-090953-X)
- Complete student text.
- Answers appear in place on the same text page as exercises.
- Teaching Tips placed in the margin at key points where students historically need extra help.
- Answers to all exercises in pretests, review problems, tests, cumulative tests, diagnostic pretest, and practice final.

Instructor's Solutions Manual (ISBN: 0-13-092423-7)
- Detailed step-by-step solutions to the even-numbered exercises.
- Solutions to every exercise (odd and even) in the diagnostic pretest, pretests, review problems, tests, cumulative tests, and practice final.
- Solution methods reflect those emphasized in the text.

Instructor's Resource Manual with Tests (ISBN: 0-13-092424-5)
- Nine test forms per chapter—6 free response, 3 multiple choice. Two of the free-response tests are cumulative in nature.
- Four forms of final examination.
- Answers to all items.

Media Resources

New TestGen-EQ with QuizMaster-EQ (Windows/Macintosh) (ISBN: 0-13-092524-1)
- Algorithmically driven, text-specific testing program.
- Networkable for administering tests and capturing grades online.
- The built-in Question Editor allows you to edit or add your own questions to create a nearly unlimited number of tests and worksheets.
- Use the Function Plotter to create graphs.
- Side-by-side "Testbank" window and "Test" window show your test as you build it and as it will be printed.
- Extensive symbol palettes and expression templates assist professors in writing questions that include specialized tables and notation.
- Tests can be easily exported to HTML so they can be posted to the Web for student practice.
- QuizMaster-EQ tests can be used for practice and graded tests. Instructors can set preferences to determine test availability, time limits, and number of tries.
- QuizMaster-EQ provides detailed exam reports for individual students, classes, or the course.

New MathPro Explorer 4.0
Network Version for Windows/Macintosh (ISBN 0-13-092518-7)
- Enables instructors to create either customized or algorithmically generated practice tests from any section of a chapter, or a test of random items.
- Includes an e-mail function for network users, enabling instructors to send a message to a specific student or an entire group.
- Network-based reports and summaries for a class or student and for cumulative or selected scores are available.

New MathPro 5 Anytime. Anywhere. With Assessment.
- The popular MathPro tutorial software available over the Internet.
- Online tutorial access—anytime/anywhere.
- Enhanced course management tools.

Companion Web Site

Visit `http://www.prenhall.com/tobey_beginning`

- Internet Connections activities have been completely revised and updated. Annotated links facilitate student navigation of the sites associated with the Internet Connections exercises.
- Additional links provided to sites of interest or resources.
- Provides an online study guide via self-quizzes. Questions are graded and students can e-mail their results.
- Syllabus Manager gives professors the option of creating their own online custom syllabus. Visit the Web site to learn more.

Supplements for Students

Printed Resources

Student Solutions Manual (ISBN: 0-13-092417-2)

- Solutions to all odd-numbered exercises.
- Solutions to every (odd and even) exercise found in pretests, chapter tests, reviews, and cumulative reviews.
- Solution methods reflect those emphasized in the textbook.
- Ask your bookstore about ordering.

Media Resources

New MathPro Explorer 4.0 CD-ROM (Student version: 0-13-092519-5)

- Keyed to each section of the text for text-specific tutorial exercises and instruction.
- Warm-up exercises and graded practice problems.
- Video clips, providing a problem similar to the one being attempted, explained and worked out on the board.
- Algorithmically generated exercises; includes bookmark, online help, glossary, and summary of scores for the exercises tried.
- Explorations enable students to explore concepts associated with each objective in more detail.

New MathPro 5 Anytime. Anywhere. With Assessment.

- The popular MathPro tutorial software available over the Internet.
- Online tutorial access—anytime/anywhere.

New Lecture Videos (ISBN: 0-13-092522-5)

- All-new videotapes accompany the fifth edition.
- Keyed to each section of the text.
- Key concepts are explained step-by-step.

New Digitized Lecture Videos on CD-ROM (ISBN: 0-13-092523-3)

- The entire set of *Beginning Algebra*, Fifth Edition lecture videotapes in digital form.
- Convenient access anytime to video tutorial support from a computer at home or on campus.
- Available shrink-wrapped with the text or stand-alone.

New Prentice Hall Tutor Center

- Staffed with developmental math instructors and open 5 days a week, 7 hours per day.
- Obtain help for examples and exercises in Tobey/Slater, *Beginning Algebra*, Fifth Edition via toll-free telephone, fax, or e-mail.

- The Prentice Hall Tutor Center is accessed through a registration number that may be bundled with a new text or purchased separately with a used book.
- Contact your Prentice Hall sales representative for details, or visit `http://www.prenhall.com/tutorcenter`.

Companion Web Site

Visit `http://www.prenhall.com/tobey_beginning`

- Internet Connections activities have been completely revised and updated. Annotated links facilitate navigation of the sites associated with the Internet Connections exercises.
- Additional links provided to sites of interest or resources.
- Provides an online study guide via self-quizzes. Questions are graded and students can e-mail their results to the instructor.

Additional Printed Material

Have your instructor contact the local Prentice Hall sales representative about the following resources:

- *How to Study Mathematics*
- *Math on the Internet: A Student's Guide*
- *Prentice Hall/New York Times, Theme of the Times Newspaper Supplement*

Acknowledgments

This book is the product of many years of work and many contributions from faculty and students across the country. We would like to thank the many reviewers and participants in focus groups and special meetings with the authors in preparation of previous editions.

Our deep appreciation to each of the following:

George J. Apostolopoulos, DeVry Institute of Technology

Katherine Barringer, Central Virginia Community College

Jamie Blair, Orange Coast College

Larry Blevins, Tyler Junior College

Robert Christie, Miami-Dade Community College

Mike Contino, California State University at Heyward

Judy Dechene, Fitchburg State University

Floyd L. Downs, Arizona State University

Barbara Edwards, Portland State University

Janice F. Gahan-Rech, University of Nebraska at Omaha

Colin Godfrey, University of Massachusetts, Boston

Carl Mancuso, William Paterson College

Janet McLaughlin, Montclair State College

Gloria Mills, Tarrant County Junior College

Norman Mittman, Northeastern Illinois University

Elizabeth A. Polen, County College of Morris

Ronald Ruemmler, Middlesex County College

Sally Search, Tallahassee Community College

Ara B. Sullenberger, Tarrant County Community College

Michael Trappuzanno, Arizona State University

Jerry Wisnieski, Des Moines Community College

In addition, we want to thank the following reviewers and focus group participants for providing splendid insight and suggestions for this new edition.

Mark Billiris, St. Petersburg Junior College

Connie Buller, Metropolitan Community College

Nelson Collins, Joliet Junior College

Robert Dubuc, Jr., New England Institute of Technology

Mary Beth Headlee, Manatee Community College

Doug Mace, Baker College

James Matovina, Community College of Southern Nevada

Beverly Meyers, Jefferson College

Nancy Meyers, University of Southern Indiana

Wayne L. Miller, Lee College

Sharon L. Morrison, St. Petersburg Junior College

Jim Osborn, Baker College

Linda Padilla, Joliet Junior College

Cathy Panik, Manatee Community College

Joel Rappaport, Miami-Dade Community College

Jose Rico, Laredo Community College

Dennis Runde, Manatee Community College

Carolyn Gigi Smith, Armstrong Atlantic State University

Lee Ann Spahr, Durham Technical Community College

Richard Sturgeon, University of Southern Maine

Margie Thrall, Manatee Community College

We have been greatly helped by a supportive group of colleagues who not only teach at North Shore Community College but who have provided a number of ideas as well as extensive help on all of our mathematics books. Also, a special word of thanks to Hank Harmeling, Tom Rourke, Wally Hersey, Bob McDonald, Judy Carter, Bob Campbell, Rick Ponticelli, Russ Sullivan, Kathy LeBlanc, Lora Connelly, Sharyn Sharaf, Donna Stefano, Elisabeth Lucas, Jenny Crawford, and Nancy Tufo. Joan Peabody has done an excellent job of typing various materials for the manuscript and her help is gratefully acknowledged. Suellen Robinson provided new problems, new ideas, new answer keys, and new perspective. Her excellent help was much appreciated. Sarah Street provided excellent help identifying geometry problems and incorporating accuracy suggestions.

We want to thank Louise Elton for providing several new applied problems and suggested applications. Error checking is a challenging task and few can do it well. So we especially want to thank Lauri Semarne and the staff of Laurel Technical Services for accuracy checking the content of the book at different stages of text preparation.

Additionally, Sherm Rosen researched the Internet Connections and provided splendid suggestions for improvements and helpful link annotations. Dave Nasby and others' work with Math in the Media is much appreciated.

Each textbook is a combination of ideas, writing, and revisions from the authors and wise editorial direction and assistance from the editors. We want to thank our Prentice Hall editor, Karin Wagner, for her helpful insight and perspective on each phase of the revision of the textbook. Her patience, her willingness to listen, and her flexibility to adapt to changing publishing decisions have been invaluable to the production of this book. Mary Beckwith, our project manager, provided daily support and encouragement as the book progressed. Her patient assistance with the art program and her attention to a variety of details was most appreciated. Kathy Sessa Frederico, our developmental editor, sifted through mountains of material and offered excellent suggestions for improvement and change. Gina Linko, our production director, kept things moving on schedule and cheerfully solved many crises.

Nancy Tobey retired from teaching and joined the team as our administrative assistant. Mailing, editing, photocopying, collating, and taping were cheerfully done each day. A special thanks to Nancy. We could not have finished the book without you.

Book writing is impossible for us without the loyal support of our families. Our deepest thanks and love to Nancy, Johnny, Melissa, Marcia, Shelley, Rusty, and Abby. Your understanding, your love and help, and your patience have been a source of great encouragement. Finally, we thank God for the strength and energy to write and the opportunity to help others through this textbook.

We have spent more than 30 years teaching mathematics. Each teaching day we find that our greatest joy is helping students learn. We take a personal interest in ensuring that each student has a good learning experience in taking this course. If you have some personal comments, suggestions, or ideas for future editions of this textbook, please write to us at:

> Prof. John Tobey and Prof. Jeffrey Slater
> Prentice Hall Publishing
> Office of the College Mathematics Editor
> One Lake Street
> Upper Saddle River, NJ 07458
> or e-mail us at
> jtobey@nscc.mass.edu.

We wish you success in this course and in your future life!

John Tobey
Jeffrey Slater

Enhanced, Student-Friendly Pedagogy

The Tobey/Slater series is a comprehensive learning system that features several pedagogical tools designed for ease of use and student success.

● **Chapter Organizer**

*The Chapter Organizer appears at the end of each chapter and summarizes key concepts and mathematical procedures. It lists concepts and methods **and** provides a completely worked-out example for each type of problem.*

Chapter 4 Organizer

Topic	Procedure	Examples
Multiplying monomials, p. 248.	$x^a \cdot x^b = x^{a+b}$ 1. Multiply the numerical coefficients. 2. Add the exponents of a given base.	$3^{12} \cdot 3^{15} = 3^{27}$ $x^3 \cdot x^4 = x^7$ $(-3x^2)(6x^3) = -18x^5$ $(2ab)(4a^2b^3) = 8a^3b^4$
Dividing monomials, p. 250.	$\dfrac{x^a}{x^b} = \begin{cases} x^{a-b} & \text{Use if } a \text{ is greater than } b. \\ \dfrac{1}{x^{b-a}} & \text{Use if } b \text{ is greater than } a. \end{cases}$ 1. Divide or reduce the fraction created by the quotient of the numerical coefficients. 2. Subtract the exponents of a given base.	$\dfrac{16x^7}{8x^3} = 2x^4$ $\dfrac{5x^3}{25x^5} = \dfrac{1}{5x^2}$ $\dfrac{-12x^5y^7}{18x^3y^{10}} = -\dfrac{2x^2}{3y^3}$
o,	$x^0 = 1 \quad \text{if } x \neq 0$	$5^0 = 1 \quad \dfrac{x^6}{x^6} = 1$ $w^0 = 1 \quad 3x^0y = 3y$

Page 293

Immigration to the U.S.

Source: U.S. Census Bureau

Page 385

● **Graphs, Charts, and Tables**

Problems based on charts, graphs, and tables have been significantly increased. Students make simple interpretations, solve medium-level problems, and investigate challenging applied problems based on presented data.

● **Developing Your Study Skills**

Sprinkled throughout the text, these boxed notes provide students with techniques for improving their study skills and succeeding in math.

Developing Your Study Skills

Getting Help

Getting the right kind of help at the right time can be a key ingredient in being successful in mathematics. When you have gone to class on a regular basis, taken careful notes, methodically read your textbook, and diligently done your homework—all of which means making every effort possible to learn the mathematics—you may find that you are still having difficulty. If this is the case, then you need to seek help. Make an appointment with your instructor to find out what help is available to you. The instructor, tutoring services, a mathematics lab, videotapes, and computer software may be among the resources you can draw on.

Once you discover the resources available in your school, you need to take advantage of them. Do not put it off, or you will find yourself getting behind. You cannot afford that. When studying mathematics, you must keep up with your work.

Page 140

Math in the Media

Making Gravy

Compiled by LOS ANGELES TIMES Staff Writers

© 2000 Los Angeles Times. Reprinted by permission.

Making gravy is simple. The secret to divine gravy is deglazing the pan and using all the browned bits stuck on the roasting pan because they hold the flavor. What frustrates most gravy makers is getting rid of the lumps. But if you make a roux, a gravy paste of flour and turkey fat stirred until smooth, that will help get rid of stubborn lumps.

Recipe

Neck and giblets from 1 turkey
6 cups water
1 onion, quartered
1 carrot, cut in pieces
1 celery stalk, cut in pieces
Black peppercorns
Butter, optional
3 tablespoons flour
Salt, pepper

Preparation

1. Remove neck and giblets from bird. Separate liver from other giblets and discard or save for another meal.

2. Place remaining giblets and neck in 2-quart saucepan. Add water, onion, carrot, celery and peppercorns, and bring to boil over high heat. Cover, reduce heat to low, and simmer until tender, about $1\frac{1}{2}$ hours. Strain and reserve broth. Discard neck. Chop giblets, cover, and refrigerate.

3. When turkey is roasted, remove from oven and transfer to platter. Lightly cover with foil and let stand about 20 minutes to allow juices to set before carving.

4. Meanwhile, add 1 cup reserved broth to drippings in roasting pan. Place pan over medium heat and scrape browned particles free from bottom with wooden spoon. Pour mixture into clear measuring cup and let fat rise to top. Skim fat off with spoon, or use specially designed measuring cup that separates the fat.

5. To make 2 cups gravy, place $\frac{1}{4}$ cup turkey fat in saucepan. (If necessary, add butter to make $\frac{1}{4}$ cup.) Add enough reserved giblet broth to skimmed drippings to make 2 cups.

6. Heat fat over medium heat. Stir in flour and cook, stirring, until bubbly. Remove from heat. Gradually pour in dripping-stock mixture, stirring constantly with wire whisk. Return pan to heat and cook, stirring, until gravy boils and thickens. For thinner gravy, add more broth. Stir in giblets. Season to taste with salt and pepper.

7. *2 cups gravy, each tablespoon:* 20 calories; 67 mg sodium; 3 mg cholesterol; 1 gram fat; 1 gram carbohydrates; 1 gram protein; 0.05 gram fiber.

EXERCISES

Sometimes it's necessary to adapt a recipe to reduce or increase the number of servings—depending on the number of guests expected. Do you think you could use your skills to adapt this recipe to produce a different number of servings? Could you calculate the nutritional values for a specific portion size? Try answering questions 1–3 to test your skills.

1. Adapt the recipe given in bold print in Step 5 to make 4 cups gravy.

2. Adapt the recipe given in bold print in Step 5 to make 3 cups gravy.

3. If two tablespoons of gravy were used, what would be the nutritional values?

🌑 Math in the Media

New Math in the Media exercises help students make connections between the real world and concepts learned in class. They are based on scenarios from the media and appear at the end of each chapter. Related questions follow that ask students to interpret the information, perform necessary calculations, or provide a rationale for their decisions.

Page 127

Page 57

🌑 Chapter-Opening Application

A real-world application opens each chapter and links a specific situation to a Putting Your Skills to Work application that appears in that chapter to enhance students' awareness of the relevance of math.

Chapter 2

Equations and Inequalities

The Environmental Protection Agency monitors the level of emissions into the air of a number of pollutants. Progress has been made in reducing some pollutants, while others are still at significantly high levels. Could you analyze the level of emissions over the last 50 years and use your math skills to predict the level of emissions in the future? Turn to the Putting Your Skills to Work problems on page 161 to find out.

Integrated Problem Solving

Page 156

🔹 Problem Solving

Problem Solving is thorough and easy to follow; key steps are highlighted with the pedagogical use of color. A clear problem-solving process is defined and reinforced throughout.

Procedure to Solve a Formula for a Specified Variable

1. Remove any parentheses.
2. If fractions exist, multiply all terms on both sides by the LCD of all the fractions.
3. Combine like terms on each side if possible.
4. Add or subtract terms on both sides of the equation to get all terms with the desired variable on one side of the equation.
5. Add or subtract the appropriate quantities to get all terms that do *not* have the desired variable on the other side of the equation.
6. Divide both sides of the equation by the coefficient of the desired variable.
7. Simplify if possible.

1. *Understand the problem.*
   (a) Read the word problem carefully to get an overview.
   (b) Determine what information you will need to solve the problem.
   (c) Draw a sketch. Label it with the known information. Determine what needs to be found.
   (d) Choose a variable to represent one unknown quantity.
   (e) If necessary, represent other unknown quantities in terms of that very same variable.
2. *Write an equation.*
   (a) Look for key words to help you to translate the words into algebraic symbols and expressions.
   (b) Use a given relationship in the problem or an appropriate formula to write an equation.
3. *Solve and state the answer.*
4. *Check.*
   (a) Check the solution in the original eq[uation]
   (b) Be sure the solution to the equation [answers the question in the word] problem. You may need to do some a[dditional work.]

Page 195

🔹 EXAMPLE 4   The mean annual snowfall in Juneau, Alaska, is 105.8 inches. This is 20.2 inches less than three times the annual snowfall in Boston. What is the annual snowfall in Boston?

Understand the problem and write an equation.

Page 192

🔹 Mathematics Blueprint for Problem Solving

Students begin the problem-solving process and plan the steps to be taken along the way using an outline to organize their approach to problem solving. Once students fill in the blueprint, they can refer back to their plan to solve the problem.

Mathematics Blueprint For Problem Solving

Gather the Facts	Assign the Variable	Basic Formula or Equation	Key Points to Remember
Snowfall in Juneau is 105.8 inches. This is 20.2 inches less than three times the snowfall in Boston.	We do not know the snowfall in Boston. Let b = annual snowfall in Boston. Then $3b - 20.2$ = annual snowfall in Juneau.	Set $3b - 20.2$ equal to 105.8, which is the snowfall in Juneau.	All measurements of snowfall are recorded in inches.

Juneau's snowfall is 20.2 less than three times Boston's snowfall.

$$105.8 = 3b - 20.2$$

Solve and state the answer.
You may want to rewrite the equation to make it easier to solve.

$$3b - 20.2 = 105.8$$
$$3b = 126 \quad \text{Add 20.2 to both sides.}$$
$$b = 42 \quad \text{Divide both sides by 3.}$$

The annual snowfall in Boston is 42 inches.

Check.
Reread the word problem. Work backward.

Three times 42 is 126.
126 less 20.2 is 105.8.

Is this the annual snowfall in Juneau? Yes. ✓

Practice Problem 4   The maximum recorded rainfall for a 24-hour period in the United States occurred in Alvin, Texas, on July 25–26, 1979. This amount was 24 inches more than the maximum recorded rainfall for a 24-hour period in Canada, which occurred in Ucluelet Brynnor Mines, British Columbia, on October 6, 1977. The total rainfall from these two occurrences was 62 inches. How much rainfall was recorded for each location? (*Source:* National Oceanic and Atmospheric Administration)   🔹

Putting Your Skills to Work

Controlling Emissions Levels

The U.S. Environmental Protection Agency monitors the emissions of a number of pollutants in the air in the United States. Each year, measurements are made to identify the number of tons of each type of pollutant. The following chart records in millions of tons the amount of emissions of nitrogen dioxide and volatile organic compounds.

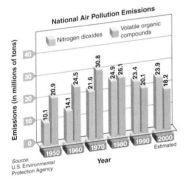

National Air Pollution Emissions

Nitrogen dioxides · Volatile organic compounds

Emissions (in millions of tons)

10.1, 20.9, 14.1, 24.5, 21.6, 30.8, 24.9, 26.1, 23.4, 20.1, 23.9, 18.2

1950, 1960, 1970, 1980, 1990, 2000 Estimated

Year

Source:
U.S. Environmental Protection Agency

Problems for Individual Study and Investigation

1. How many more tons of nitrogen dioxide were emitted in 1980 than in 1970?

2. How many fewer tons of volatile organic compounds were emitted in 2000 than in 1990?

Problems for Cooperative Study and Investigation

Scientists sometimes use the equation $n = 23.9 + 0.05x$, where x represents the number of years since the year 2000, to predict the number of millions of tons of nitrogen dioxide that will be emitted per year.

3. Use the formula to estimate how many tons of nitrogen dioxide will be emitted in 2005.

4. If the actual amount of nitrogen dioxide emitted in 2010 turns out to be 24.7 million tons, by how many tons will the formula be in error?

Scientists sometimes use the equation $v = 18.2 - 0.19x$, where x represents the number of years since the year 2000, to predict the number of millions of tons of volatile organic compounds that will be emitted per year.

5. Solve the formula for the variable x. Leave the answer in the form of a fractional expression. Do not divide out the decimals.

6. Use the result from question 5 to find the year when 16.68 million tons of volatile organic compounds will be emitted.

Internet Connections

 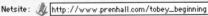 Netsite: http://www.prenhall.com/tobey_beginning

Site: U.S. Environmental Protection Agency

7. Examine the three separate tables for carbon monoxide, nitrogen oxides, and sulfur dioxide emissions estimates 1989–1998. For each of these, by what percent did total emissions increase or decrease in 1998 as compared with 1989?

8. In what year did the emissions of carbon monoxide from transportation sources have the greatest percent decrease from the previous year? What was that percent decrease?

Page 161

Interesting and Diverse Exercises and Applications

🔵 Real-World Applications

Numerous real-world and real-data applications relate topics to everyday life, global issues, and other academic disciplines. An abundance of new real-world application problems show students the relevance of math in their daily lives.

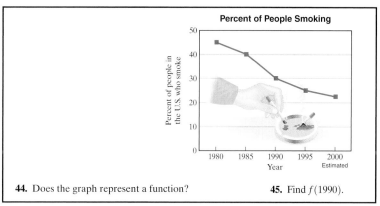

Percent of People Smoking

44. Does the graph represent a function?

45. Find $f(1990)$.

Page 460

🔵 **EXAMPLE 7** The number of motor vehicle accidents in millions is recorded in the following table for the years 1980 to 2000.

(a) Plot points that represent this data on the given coordinate system.

(b) What trends are apparent from the plotted data?

Number of Years Since 1980	Number of Motor Vehicle Accidents (in Millions)
0	18
5	19
10	12
15	11
20	15

Source: U.S. National Highway Traffic Safety Administration

(a)
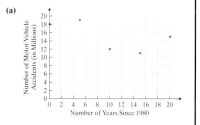

(b) From 1980 to 1985, there was a slight increase in the number of accidents. From 1985 to 1995, there was a significant decrease in the number of accidents. From 1995 to 2000, there was a moderate increase in the number of accidents.

Page 411

Page 173

🔵 Verbal and Writing Skills Exercises

These exercises ask students to use their own words. Writing helps students clarify their thinking and become familiar with mathematical terms.

Verbal and Writing Skills

15. Add −2 to both sides of the inequality $5 > 3$. What is the result? Why is the direction of the inequality not reversed?

16. Divide −3 into both sides of the inequality $-21 > -29$. What is the result? Why is the direction of the inequality reversed?

🔵 To Think About

Critical-thinking questions appear in the exercise sets to extend the concepts. They also give students the opportunity to look for patterns and draw conclusions.

To Think About

48. During the years from 1980 to 2000, the total income for the U.S. federal budget can be approximated by the equation $y = 14(4x + 35)$, where x is the number of years since 1980 and y is the amount of money in billions of dollars. (*Source:* U.S. Office of Management and Budget)
(a) Write the equation in slope–intercept form.

(b) Find the slope and the y-intercept.

(c) In this specific equation, what is the meaning of the slope? What does it indicate?

49. During the years from 1970 to 1990, the approximate number of civilians employed in the United States could be predicted by the equation $y = \frac{1}{10}(22x + 830)$, where x is the number of years since 1970 and y is the number of civilians employed, measured in millions. (*Source:* U.S. Bureau of Labor Statistics)
(a) Write the equation in slope–intercept form.

(b) Find the slope and the y-intercept.

(c) In this specific equation, what is the meaning of the slope? What does it indicate?

Page 436

Pretest Chapter 2

1. _____

2. _____

3. _____

4. _____

5. _____

6. _____

7. _____

8. _____

9. _____

If you are familiar with the topics in this chapter, take this test now. Check your answers with those in the back of the book. If an answer is wrong or you can't answer a question, study the appropriate section of the chapter.

If you are not familiar with the topics in this chapter, don't take this test now. Instead, study the examples, work the practice problems, and then take the test.

This test will help you identify which concepts you have mastered and which you need to study further.

Sections 2.1–2.3

Solve for x.

1. $x - 30 = -46$

2. $\frac{1}{4}x = 12$

3. $-5x = 20$

4. $7x + 2 = 30$

5. $3x - 8 = 7x + 6$

6. $5x + 3 - 6x = 4 - 8x - 2$

7. $5.2x + 0.9$

⬤ Chapter Pretest

Each chapter opens with a concise pretest to familiarize the students with the learning objectives for that particular chapter. A diagnostic pretest with chapter references appears at the beginning of the text. A final exam with chapter references appears at the back of the text. All answers are included.

⬤ Emphasis on Geometry

A new geometry icon in the examples and exercises marks an increased emphasis on geometry.

▲ ⬤ **EXAMPLE 4** A trapezoid is a four-sided figure with two parallel sides. If the parallel sides are a and b and the altitude is h, the area is given by

$$A = \frac{h}{2}(a + b).$$

Solve this equation for a.

$$A = \frac{h}{2}(a + b)$$

$$A = \frac{ha}{2} + \frac{hb}{2} \qquad \text{Remove the parentheses.}$$

$$2(A) = 2\left(\frac{ha}{2}\right) + 2\left(\frac{hb}{2}\right) \qquad \text{Multiply all terms by the LCD of 2.}$$

$$2A = ha + hb \qquad \text{Simplify.}$$

$$2A - hb = ha \qquad \text{We want to isolate the term containing } a. \text{ Therefore, we subtract } hb \text{ from both sides.}$$

$$\frac{2A - hb}{h} = \frac{ha}{h} \qquad \text{Divide both sides by } h \text{ (the coefficient of } a\text{).}$$

$$\frac{2A - hb}{h} = a \qquad \text{The solution is obtained.}$$

Note: Although the solution is in simple form, it could be written in an alternative way. Since

$$\frac{2A - hb}{h} = \frac{2A}{h} - \frac{hb}{h} = \frac{2A}{h} - b,$$

we could also have written $\frac{2A}{h} - b = a$.

▲ **Practice Problem 4** The relationship between the circumference C of a circle and the circle's diameter d is described by the equation $C = \pi d$. Solve it for d.

Practice Problem 8 Line h has a slope of $\frac{1}{4}$.

(a) If line j is parallel to line h, what is its slope?

(b) If line k is perpendicular to line h, what is its slope?

⬤ **EXAMPLE 9** The equation of line l is $y = -2x + 3$.

(a) What is the slope of a line that is parallel to line l?

(b) What is the slope of a line that is perpendicular to line l?

(a) Looking at the equation, we can see that the slope of line l is -2. The slope of a line that is parallel to line l is -2.

(b) Perpendicular lines have slopes whose product is -1.

$$m_1 m_2 = -1$$

$$(-2)m_2 = -1 \qquad \text{Substitute } -2 \text{ for } m_1.$$

$$m_2 = \frac{1}{2} \qquad \text{Because } (-2)\left(\frac{1}{2}\right) = -1.$$

The slope of a line that is perpendicular to line l is $\frac{1}{2}$.

Practice Problem 9 The equation of line n is $y = \frac{1}{4}x - 1$.

(a) What is the slope of a line that is parallel to line n?

(b) What is the slope of a line that is perpendicular to line n?

Graphing Calculator

Graphing Parallel Lines

If two equations are in the form $y = mx + b$, then it will be obvious that they are parallel because the slope will be the same. On a graphing calculator graph both of these equations:

$$y = -2x + 6$$

$$y = -2x - 4$$

Use the window of -10 to 10 for both x and y. Display:

⬤ Calculator Notes

Calculator notes in the margin alert students to calculator applications. Calculator icons in the exercise sets mark problems that can be solved using a scientific or graphing calculator.

To the Student

This book was written with your needs and interests in mind. The original manuscript and the first four editions of this book have been class-tested with students all across the country. Based on the suggestions of many students, the book has been refined and improved to maximize your learning while using this text.

We realize that students who enter college have sometimes never enjoyed math or never done well in a math course. You may find that you are anxious about taking this course. We want you to know that this book has been written to help you overcome those difficulties. Literally thousands of students across the country have found an amazing ability to learn math as they have used previous editions of this book. We have incorporated several learning tools and various types of exercises and examples to assist you in learning this material.

It helps to know that learning mathematics is going to help you in life. Perhaps you have entered this course feeling that little or no mathematics will be necessary for you in your future job. However, elementary school teachers, bus drivers, laboratory technicians, nurses, telephone operators, cable TV repair personnel, photographers, pharmacists, salespeople, doctors, architects, inspectors, counselors, and custodians who once believed that they needed little if any mathematics are finding that the mathematical skills presented in this course can help them. Mathematics, you will find, can help you too. In this book, a great number of examples and problems come from everyday life. You will be amazed at the number of ways mathematics is used in the world each day.

Our greatest wish is that you will find success and personal satisfaction in your mathematics course. We have written this book to help you accomplish that very goal.

Suggestions for Students

1. Be sure to take the time to read the boxes marked **Developing Your Study Skills**. These ideas come from faculty and students throughout the United States. They have found ways to succeed in mathematics and they want to share those ideas with you.

2. **Read** through each section of the book that is assigned by your instructor. You will be amazed at how much information you will learn as you read. It will "fill in the pieces" of mathematical knowledge. Reading a math book is a key part of learning.

3. Study carefully each **sample example**. Then work out the related practice problem. Direct involvement in doing problems and not just thinking about them is one of the greatest guarantees of success in this course.

4. Work out all assigned **homework problems**. Verify your answers in the back of the book. Ask questions when you don't understand or when you are not sure of an answer.

5. Be sure to take advantage of **end-of-chapter helps**. Study the chapter organizers. Do the chapter review problems and the practice test problems.

6. If you need help, remember that teachers and tutors can assist you. There is a **Student Solutions Manual** for this textbook that you will find most valuable. This manual shows worked-out solutions for all the odd-numbered exercises as well as diagnostic pretest, chapter review, chapter test, and cumulative test problems in this book. (If your college bookstore does not carry the Student Solutions Manual, ask them to order a copy for you. ISBN 0-13-092417-2)

7. Watch the **videotapes or digitized videos on CD-ROM**. Every section of this text is explained in detail on videotape. The work is solved using the same methods as explained in the text.

8. Use **MathPro Explorer 4.0**, a tutorial software package that allows you to be tutored in any section in the book. It allows you to test yourself on your mastery of any section of any chapter of the text. Practice problem solving using the resources available to you.

9. Remember, learning mathematics takes time. However, the time spent is well worth the effort. **Take the time** to study, do homework, review, ask questions, and just reflect over what you have learned. The time you invest in learning mathematics will reap dividends in your future courses and your future life.

We encourage you to look over your textbook carefully. Many important features have been designed into the book to make learning mathematics a more enjoyable activity. There are some special features, unique to this textbook, that students throughout the country have told us that they found especially helpful.

Four Key Textbook Features to Help You

1. **Practice problems with worked-out solutions.** Immediately following every sample example is a similar problem called the practice problem. If you can work it out correctly by following the sample example as a general point of reference, then you will likely be able to do the homework exercises. If you encounter some difficulty, then you will find helpful the completely worked-out solutions to the practice problems that appear at the end of the text.

2. **Student-friendly application problems.** You will find in every chapter of this book application problems that are realistic and interesting. Many of the problems were actually written or suggested by students. As you develop your problem-solving and reasoning skills in this course, you will encounter a number of real-world situations that help you see how very helpful mathematical skills are in today's complex world.

3. **Putting Your Skills to Work problems.** In each chapter are unique problem sets that ask you to analyze in depth some mathematical aspect of daily life. You will be asked to extend your knowledge, do some creative thinking, and to work in small groups in a cooperative learning situation. You will even have a chance to explore some Web sites and see how the Internet can assist you in learning. These sections will awaken some new interests and help you to develop the critical thinking skills so necessary both in college and after you graduate.

4. **The Chapter Organizer.** Everything you need to learn in any one chapter of this book is readily available at your fingertips in the chapter organizer. This very popular chart summarizes all methods covered in the chapter, gives page references, and shows a sample example completely worked out for every major topic covered. Students have found this tool a most helpful way to master the content of any chapter of the book.

Follow the directions for each problem. Simplify each answer.

Chapter 0

1. Add. $3\frac{1}{4} + 2\frac{3}{5}$

2. Multiply. $\left(1\frac{1}{6}\right)\left(2\frac{2}{3}\right)$

3. Divide. $\frac{15}{4} \div \frac{3}{8}$

4. Multiply. $(1.63)(3.05)$

5. Divide. $120 \div 0.0006$

6. Find 7% of 64,000.

Chapter 1

7. Add. $-3 + (-4) + (+12)$

8. Subtract. $-20 - (-23)$

9. Combine. $5x - 6xy - 12x - 8xy$

10. Evaluate. $2x^2 - 3x - 4$ when $x = -3$.

11. Remove the grouping symbols. $2 - 3\{5 + 2[x - 4(3 - x)]\}$

12. Evaluate. $-3(2 - 6)^2 + (-12) \div (-4)$

Chapter 2

In questions 13–16, solve each equation for x.

13. $40 + 2x = 60 - 3x$

14. $7(3x - 1) = 5 + 4(x - 3)$

15. $\frac{2}{3}x - \frac{3}{4} = \frac{1}{6}x + \frac{21}{4}$

16. $\frac{4}{5}(3x + 4) = 20$

17. Solve for p. $A = \frac{1}{2}(3p - 4f)$

18. Solve for x and graph the result. $42 - 18x < 48x - 24$

1. _____

2. _____

3. _____

4. _____

5. _____

6. _____

7. _____

8. _____

9. _____

10. _____

11. _____

12. _____

13. _____

14. _____

15. _____

16. _____

17. _____

18. _____

19. _____

20. _____

21. _____

22. _____

23. _____

24. _____

25. _____

26. _____

27. _____

28. _____

29. _____

30. _____

31. _____

32. _____

33. _____

34. _____

35. _____

36. _____

37. _____

38. _____

39. _____

40. _____

41. _____

42. _____

Chapter 3

19. The length of a rectangle is 7 meters longer than the width. The perimeter is 46 meters. Find the dimensions.

20. One side of a triangle is triple the second side. The third side is 3 meters longer than double the second side. Find each side of the triangle if the perimeter of the triangle is 63 meters.

21. Hector has four test scores of 80, 90, 83, and 92. What does he need to score on the fifth test to have an average of 86 on the five tests?

22. Marcia invested $6000 in two accounts. One earned 5% interest, while the other earned 7% interest. After one year she earned $394 in interest. How much did she invest in each account?

23. Melissa has three more dimes than nickels. She has twice as many quarters as nickels. The value of the coins is $4.20. How many of each coin does she have?

24. The drama club put on a play for Thursday, Friday, and Saturday nights. The total attendance for the three nights was 6210. Thursday night had 300 fewer people than Friday night. Saturday night had 510 more people than Friday night. How many people came each night?

Chapter 4

25. Multiply. $(-2xy^2)(-4x^3y^4)$

26. Divide. $\dfrac{36x^5y^6}{-18x^3y^{10}}$

27. Raise to the indicated power. $(-2x^3y^4)^5$

28. Evaluate. $(-3)^{-4}$

29. Multiply. $(3x^2 + 2x - 5)(4x - 1)$

30. Divide. $(x^3 + 6x^2 - x - 30) \div (x - 2)$

Chapter 5

Factor completely.

31. $5x^2 - 5$

32. $x^2 - 12x + 32$

33. $8x^2 - 2x - 3$

34. $3ax - 8b - 6a + 4bx$

Solve for x.

35. $x^3 + 7x^2 + 12x = 0$

36. $16x^2 - 24x + 9 = 0$

Chapter 6

37. Simplify. $\dfrac{x^2 + 3x - 18}{2x - 6}$

38. Multiply. $\dfrac{6x^2 - 14x - 12}{6x + 4} \cdot \dfrac{x + 3}{2x^2 - 2x - 12}$

39. Divide and simplify. $\dfrac{x^2}{x^2 - 4} \div \dfrac{x^2 - 3x}{x^2 - 5x + 6}$

40. Add. $\dfrac{3}{x^2 - 7x + 12} + \dfrac{4}{x^2 - 9x + 20}$

41. Solve for x. $2 - \dfrac{5}{2x} = \dfrac{2x}{x + 1}$

42. Simplify. $\dfrac{3 + \dfrac{1}{x}}{\dfrac{9}{x} + \dfrac{3}{x^2}}$

Chapter 7

43. Graph $y = 2x - 4$.

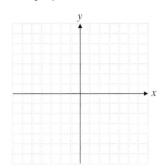

44. Graph $3x + 4y = -12$.

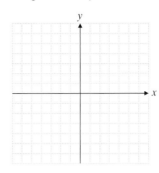

45. What is the slope of a line passing through $(6, -2)$ and $(-3, 4)$?

46. If $f(x) = 2x^2 - 3x + 1$, find $f(3)$.

47. Graph the region. $y \geq -\dfrac{1}{3}x + 2$

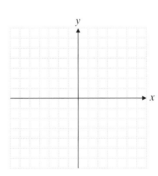

48. Find the equation of a line with a slope of $\dfrac{3}{5}$ that passes through the point $(-1, 3)$.

Chapter 8

Solve each system by the appropriate method.

49. Substitution method
$$x + y = 17$$
$$2x - y = -5$$

50. Addition method
$$-5x + 4y = 8$$
$$2x + 3y = 6$$

51. Any method
$$2(x - 2) = 3y$$
$$6x = -3(4 + y)$$

52. Any method
$$x + \dfrac{1}{3}y = \dfrac{10}{3}$$
$$\dfrac{3}{2}x + y = 8$$

53. Is $(2, -3)$ a solution for the system
$$3x + 5y = -9$$
$$2x - 3y = 13?$$

54. A man bought three pairs of gloves and four scarves for $53. A woman bought two pairs of the same-priced gloves and three scarves of the same-priced scarves for $38. How much did each item cost?

43. _____

44. _____

45. _____

46. _____

47. _____

48. _____

49. _____

50. _____

51. _____

52. _____

53. _____

54. _____

55. _____

56. _____

57. _____

58. _____

59. _____

60. _____

61. _____

62. _____

63. _____

64. _____

65. _____

66. _____

Chapter 9

55. Evaluate. $\sqrt{121}$

56. Simplify. $\sqrt{125x^3y^5}$

57. Multiply and simplify.

$$(\sqrt{2} + \sqrt{6})(2\sqrt{2} - 3\sqrt{6})$$

58. Rationalize the denominator.

$$\frac{\sqrt{5} - \sqrt{3}}{\sqrt{6}}$$

59. In the right triangle with sides a, b, and c, find side c if side $a = 4$ and side $b = 6$.

60. y varies directly with x. When $y = 56$, then $x = 8$. Find y when $x = 11$.

Chapter 10

In questions 61–64, solve for x.

61. $14x^2 + 21x = 0$

62. $2x^2 + 1 = 19$

63. $2x^2 - 4x - 5 = 0$

64. $x^2 - x + 8 = 5 + 6x$

65. Graph the equation. $y = x^2 + 8x + 15$

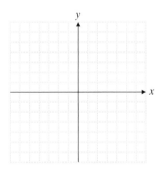

66. A rectangular box is 4 inches longer in length than in width. The area of the box is 96 square inches. Find the length and the width of the box.

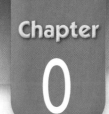

Chapter

0

A Brief Review of Arithmetic Skills

E very four years, the athletic competition at the summer Olympics captures the attention of the world. The records of the high jump competition are carefully measured with feet, inches, and fractions of an inch. Could you perform mathematical calculations with these records and make a prediction about future Olympic records? Turn to Putting Your Skills to Work on page 21 and find out.

Pretest Chapter 0

1. _____

2. _____

3. _____

4. _____

5. _____

6. _____

7. _____

8. _____

9. _____

10. _____

11. _____

12. _____

13. _____

14. _____

15. _____

16. _____

17. _____

Take this pretest for Chapter 0 and compare your answers to the solutions for the pretest listed in the Answer Key at the end of the book. If you obtain the correct answer for 25 or more problems, you are ready to begin Chapter 1. If you obtain more than one wrong answer in any one section, you should review the necessary sections of Chapter 0 to improve your arithmetic skills. If you take the time to master these arithmetic skills now, you will find Chapter 1 and the remaining chapters of this book easier to complete.

Section 0.1

In exercises 1 and 2, simplify each fraction.

1. $\dfrac{15}{55}$

2. $\dfrac{46}{115}$

3. Write $\dfrac{15}{4}$ as a mixed number.

4. Change $4\dfrac{5}{7}$ to an improper fraction.

Find the missing number.

5. $\dfrac{3}{7} = \dfrac{?}{14}$

6. $\dfrac{7}{4} = \dfrac{?}{20}$

Section 0.2

7. Find the LCD, but do not add. $\quad \dfrac{3}{8}, \dfrac{5}{6}$, and $\dfrac{7}{15}$

Perform the calculation indicated. Write the answer in simplest form.

8. $\dfrac{3}{7} + \dfrac{2}{7}$

9. $\dfrac{5}{14} + \dfrac{2}{21}$

10. $2\dfrac{3}{4} + 5\dfrac{2}{3}$

11. $3\dfrac{1}{5} - 1\dfrac{3}{8}$

Section 0.3

Perform the calculations indicated. Write the answer in simplest form.

12. $\dfrac{25}{7} \times \dfrac{14}{45}$

13. $2\dfrac{4}{5} \times 3\dfrac{3}{4}$

14. $4 \div \dfrac{8}{7}$

15. $2\dfrac{1}{3} \div 3\dfrac{1}{4}$

Section 0.4

Change to a decimal.

16. $\dfrac{7}{8}$

17. $\dfrac{5}{9}$

Perform the calculations indicated.

18. $15.23 + 3.6 + 0.821$

19. 3.28×0.63

20. $3.015 \div 6.7$

21. $12.13 - 9.884$

Section 0.5

Change to a percent.

22. 0.9

23. 0.007

Change to a decimal.

24. 327%

25. 2%

26. What is 25% of 1630?

27. What percent of 500 is 36?

Section 0.6

Estimate. Round each number so that there is one nonzero digit. Then perform the calculation.

28. $8456 + 1749$

29. 7386×2856

30. Estimate the number of miles per gallon achieved by Marcia's car if she drove for 235.8 miles and used 10.7 gallons of gas.

Section 0.7

Solve. You may want to use the Mathematics Blueprint for Problem Solving.

31. Susan drove 870 miles. She started and ended with a full tank of gas. During the trip, she made the following gasoline purchases: 3.7 gal, 10 gal, 12 gal, and 4.9 gal. How many miles per gallon did her car achieve? (Round to the nearest tenth.)

32. A pollster surveyed 450 registered voters in the city of Springfield. 198 of those surveyed said they had voted in the last election. What percent of the people interviewed said they voted in the last election?

18. _____

19. _____

20. _____

21. _____

22. _____

23. _____

24. _____

25. _____

26. _____

27. _____

28. _____

29. _____

30. _____

31. _____

32. _____

0.1 Simplifying and Finding Equivalent Fractions

Student Learning Objectives

After studying this section, you will be able to:

1 Understand and use some basic mathematical definitions.

2 Simplify fractions to lowest terms.

3 Change forms between improper fractions and mixed numbers.

4 Change a fraction into an equivalent fraction with a different denominator.

SSM   PH TUTOR CENTER   CD & VIDEO   MATH PRO   WEB

Chapter 0 is designed to give you a mental "warm-up." In this chapter you'll be able to step back a bit and tone up your math skills. This brief review of arithmetic will increase your math flexibility and give you a good running start into algebra.

1 Understanding and Using Some Basic Mathematical Definitions

Whole numbers are the set of numbers $0, 1, 2, 3, 4, 5, 6, 7, \ldots$. They are used to describe whole objects, or entire quantities.

    Fractions are a set of numbers that are used to describe parts of whole quantities. In the object shown in the figure there are four equal parts. The *three* of the *four* parts that are shaded are represented by the fraction $\frac{3}{4}$. In the fraction $\frac{3}{4}$ the number 3 is called the **numerator** and the number 4, the **denominator.**

$$\frac{3}{4}$$

$3 \leftarrow$ *Numerator* is on the top
$\overline{4} \leftarrow$ *Denominator* is on the bottom

The *denominator* of a fraction shows the number of equal parts in the whole and the *numerator* shows the number of these parts being talked about or being used.

    Numerals are symbols we use to name numbers. There are many different numerals that can be used to describe the same number. We know that $\frac{1}{2} = \frac{2}{4}$. The fractions $\frac{1}{2}$ and $\frac{2}{4}$ both describe the same number.

    Usually, we find it more useful to use fractions that are simplified. A fraction is considered to be in **simplest form** or **reduced form** when the numerator (top) and the denominator (bottom) can both be divided exactly by no number other than 1, and the denominator is greater than 1.

$$\frac{1}{2} \text{ is in simplest form.}$$

$$\frac{2}{4} \text{ is } not \text{ in simplest form, since the numerator and the denominator can both be divided by 2.}$$

    If you get the answer $\frac{2}{4}$ to a problem, you should state it in simplest form, $\frac{1}{2}$. The process of changing $\frac{2}{4}$ to $\frac{1}{2}$ is called **simplifying** or **reducing** the fraction.

2 Simplifying Fractions to Lowest Terms

Natural numbers or **counting numbers** are the set of whole numbers excluding 0. Thus the natural numbers are the numbers $1, 2, 3, 4, 5, 6, \ldots$.

    When two or more numbers are multiplied, each number that is multiplied is called a **factor.** For example, when we write $3 \times 7 \times 5$, each of the numbers 3, 7, and 5 is called a factor.

    Prime numbers are all natural numbers greater than 1 whose only natural number factors are 1 and itself. The number 5 is prime. The only natural number factors of 5 are 5 and 1.

$$5 = 5 \times 1$$

The number 6 is not prime. The natural number factors of 6 are 3 and 2 or 6 and 1.

$$6 = 3 \times 2 \qquad 6 = 6 \times 1$$

The first 15 prime numbers are

$$2, 3, 5, 7, 11, 13, 17, 19, 23, 29, 31, 37, 41, 43, 47.$$

Any natural number greater than 1 either is prime or can be written as the product of prime numbers. For example, we can take each of the numbers 12, 30, 14, 19, and 29 and either indicate that they are prime or, if they are not prime, write them as the product of prime numbers. We write as follows:

$$12 = 2 \times 2 \times 3 \qquad 30 = 2 \times 3 \times 5 \qquad 14 = 2 \times 7$$

19 is a prime number. 29 is a prime number.

To reduce a fraction, we use prime numbers to factor the numerator and the denominator. Write each part of the fraction (numerator and denominator) as a product of prime numbers. Note any *factors* that appear in both the *numerator* (top) and *denominator* (bottom) of the fraction. If we divide numerator and denominator by these values we will obtain an equivalent fraction in *simplest form*. When the new fraction is simplified, it is said to be in **lowest terms.** Throughout this text, to *simplify* a fraction will always mean to simplify the fraction to lowest terms.

EXAMPLE 1 Simplify each fraction. (a) $\dfrac{14}{21}$ (b) $\dfrac{15}{35}$ (c) $\dfrac{20}{70}$

(a) $\dfrac{14}{21} = \dfrac{\cancel{7} \times 2}{\cancel{7} \times 3} = \dfrac{2}{3}$
We factor 14 and factor 21. Then we divide numerator and denominator by 7.

(b) $\dfrac{15}{35} = \dfrac{\cancel{5} \times 3}{\cancel{5} \times 7} = \dfrac{3}{7}$
We factor 15 and factor 35. Then we divide numerator and denominator by 5.

(c) $\dfrac{20}{70} = \dfrac{2 \times \cancel{2} \times \cancel{5}}{7 \times \cancel{2} \times \cancel{5}} = \dfrac{2}{7}$
We factor 20 and factor 70. Then we divide numerator and denominator by both 2 and 5.

Practice Problem 1 Simplify.

(a) $\dfrac{10}{16}$ (b) $\dfrac{24}{36}$ (c) $\dfrac{42}{36}$

Sometimes when we simplify a fraction, all the prime factors in the top (numerator) are divided out. When this happens, we must remember that a 1 is left in the numerator.

EXAMPLE 2 Simplify each fraction. (a) $\dfrac{7}{21}$ (b) $\dfrac{15}{105}$

(a) $\dfrac{7}{21} = \dfrac{\cancel{7} \times 1}{\cancel{7} \times 3} = \dfrac{1}{3}$ (b) $\dfrac{15}{105} = \dfrac{\cancel{5} \times \cancel{3} \times 1}{7 \times \cancel{5} \times \cancel{3}} = \dfrac{1}{7}$

Practice Problem 2 Simplify.

(a) $\dfrac{4}{12}$ (b) $\dfrac{25}{125}$ (c) $\dfrac{73}{146}$

If all the prime numbers in the bottom (denominator) are divided out, we do not need to leave a 1 in the denominator, since we do not need to express the answer as a fraction. The answer is then a whole number and is not usually expressed as a fraction.

EXAMPLE 3 Simplify each fraction. (a) $\dfrac{35}{7}$ (b) $\dfrac{70}{10}$

(a) $\dfrac{35}{7} = \dfrac{5 \times \cancel{7}}{\cancel{7} \times 1} = 5$ (b) $\dfrac{70}{10} = \dfrac{7 \times \cancel{5} \times \cancel{2}}{\cancel{5} \times \cancel{2} \times 1} = 7$

Practice Problem 3 Simplify.

(a) $\dfrac{18}{6}$ (b) $\dfrac{146}{73}$ (c) $\dfrac{28}{7}$

Sometimes the fraction we use represents how many of a certain thing are successful. For example, if a baseball player was at bat 30 times and achieved 12 hits, we could say that he had a hit $\frac{12}{30}$ of the time. If we reduce the fraction, we could say he had a hit $\frac{2}{5}$ of the time.

EXAMPLE 4   Cindy got 48 out of 56 questions correct on a test. Write this as a fraction.

Express as a fraction in simplest form the number of correct responses out of the total number of questions on the test.

$$48 \text{ out of } 56 \rightarrow \frac{48}{56} = \frac{6 \times \cancel{8}}{7 \times \cancel{8}} = \frac{6}{7}$$

Cindy answered the questions correctly $\frac{6}{7}$ of the time.

Practice Problem 4   The major league pennant winner in 1917 won 56 games out of 154 games played. Express as a fraction in simplest form the number of games won in relation to the number of games played.

The number *one* can be expressed as $1, \frac{1}{1}, \frac{2}{2}, \frac{6}{6}, \frac{8}{8}$, and so on, since

$$1 = \frac{1}{1} = \frac{2}{2} = \frac{6}{6} = \frac{8}{8}.$$

We say that these numerals are *equivalent ways* of writing the number *one* because they all express the same quantity even though they appear to be different.

Sidelight

When we simplify fractions, we are actually using the fact that we can multiply any number by 1 without changing the value of that number. (Mathematicians call the number 1 the **multiplicative identity** because it leaves any number it multiplies with the same identical value as before.)

Let's look again at one of the previous examples.

$$\frac{14}{21} = \frac{7 \times 2}{7 \times 3} = \frac{7}{7} \times \frac{2}{3} = 1 \times \frac{2}{3} = \frac{2}{3}$$

So we see that

$$\frac{14}{21} = \frac{2}{3}$$

When we simplify fractions, we are using this property of multiplying by 1.

3   *Changing Forms Between Improper Fractions and Mixed Numbers*

If the numerator is less than the denominator, the fraction is a **proper fraction.** A proper fraction is used to describe a quantity smaller than a whole.

Fractions can also be used to describe quantities larger than a whole. The following figure shows two bars that are equal in size. Each bar is divided into 5 equal pieces. The first bar is shaded in completely. The second bar has 2 of the 5 pieces shaded in.

The shaded-in region can be represented by $\frac{7}{5}$ since 7 of the pieces (each of which is $\frac{1}{5}$ of a whole box) are shaded. The fraction $\frac{7}{5}$ is called an improper fraction. An **improper fraction** is one in which the numerator is larger than or equal to the denominator.

The shaded-in region can also be represented by 1 whole added to $\frac{2}{5}$ of a whole, or $1 + \frac{2}{5}$. This is written as $1\frac{2}{5}$. The fraction $1\frac{2}{5}$ is called a mixed number. A **mixed**

number consists of a whole number added to a proper fraction (the numerator is smaller than the denominator). The addition is understood but not written. When we write $1\frac{2}{5}$, it represents $1 + \frac{2}{5}$. The numbers $1\frac{7}{8}, 2\frac{3}{4}, 8\frac{1}{3}$, and $126\frac{1}{10}$ are all mixed numbers. From the preceding figure it seems clear that $\frac{7}{5} = 1\frac{2}{5}$. This suggests that we can change from one form to the other without changing the value of the fraction.

From a picture it is easy to see how to *change improper fractions to mixed numbers*. For example, suppose we start with the fraction $\frac{11}{3}$ and represent it by the following figure (where 11 of the pieces, each of which is $\frac{1}{3}$ of a box, are shaded). We see that $\frac{11}{3} = 3\frac{2}{3}$, since 3 whole boxes and $\frac{2}{3}$ of a box are shaded.

Changing Improper Fractions to Mixed Numbers

You can follow the same procedure without a picture. For example, to change $\frac{11}{3}$ to a mixed number, we can do the following:

$$\frac{11}{3} = \frac{3}{3} + \frac{3}{3} + \frac{3}{3} + \frac{2}{3} \qquad \text{By the rule for adding fractions (which is discussed in detail in Section 0.2)}$$

$$= 1 + 1 + 1 + \frac{2}{3} \qquad \text{Write 1 in place of } \frac{3}{3}, \text{ since } \frac{3}{3} = 1.$$

$$= 3 + \frac{2}{3} \qquad \text{Write 3 in place of } 1 + 1 + 1.$$

$$= 3\frac{2}{3} \qquad \text{Use the notation for mixed numbers.}$$

Now that you know how to change improper fractions to mixed numbers and why the procedure works, here is a shorter method.

To Change an Improper Fraction to a Mixed Number

1. Divide the denominator into the numerator.
2. The result is the whole-number part of the mixed number.
3. The remainder from the division will be the numerator of the fraction. The denominator of the fraction remains unchanged.

We can write the fraction as a division statement and divide. The arrows show how to write the mixed number.

$$\frac{7}{5} = \begin{array}{r} 1 \\ 5\overline{)7} \\ \underline{5} \\ 2 \end{array}$$

Whole-number part   Numerator of fraction

$\longrightarrow 1\frac{2}{5}$

2   Remainder

Thus, $\dfrac{7}{5} = 1\dfrac{2}{5}$.

$$\frac{11}{3} = \begin{array}{r} 3 \\ 3\overline{)11} \\ \underline{9} \\ 2 \end{array}$$

Whole-number part   Numerator of fraction

$\longrightarrow 3\frac{2}{3}$

2   Remainder

Thus, $\dfrac{11}{3} = 3\dfrac{2}{3}$.

Sometimes the remainder is 0. In this case, the improper fraction changes to a whole number.

EXAMPLE 5   Change to a mixed number or to a whole number.

(a) $\dfrac{7}{4}$ **(b)** $\dfrac{15}{3}$

(a) $\dfrac{7}{4} = 7 \div 4$ $\quad 4\overline{)7} \atop {4 \above 0pt \overline{3}}$ $\;$ Remainder

Thus $\dfrac{7}{4} = 1\dfrac{3}{4}$

(b) $\dfrac{15}{3} = 15 \div 3$ $\quad 3\overline{)15} \atop {15 \above 0pt \overline{0}}$ $\;$ Remainder

Thus $\dfrac{15}{3} = 5$

Practice Problem 5   Change to a mixed number or to a whole number.

(a) $\dfrac{12}{7}$ **(b)** $\dfrac{20}{5}$

Changing Mixed Numbers to Improper Fractions

It is not difficult to see how to change mixed numbers to improper fractions. Suppose that you wanted to write $2\frac{2}{3}$ as an improper fraction.

$$2\dfrac{2}{3} = 2 + \dfrac{2}{3} \qquad \text{The meaning of mixed number notation}$$

$$= 1 + 1 + \dfrac{2}{3} \quad \text{Since } 1 + 1 = 2$$

$$= \dfrac{3}{3} + \dfrac{3}{3} + \dfrac{2}{3} \quad \text{Since } 1 = \tfrac{3}{3}$$

When we draw a picture of $\frac{3}{3} + \frac{3}{3} + \frac{2}{3}$, we have this figure:

$$\dfrac{3}{3} \qquad\qquad \dfrac{3}{3} \qquad\qquad \dfrac{2}{3}$$

If we count the shaded parts, we see that

$$\dfrac{3}{3} + \dfrac{3}{3} + \dfrac{2}{3} = \dfrac{8}{3}. \quad \text{Thus} \quad 2\dfrac{2}{3} = \dfrac{8}{3}.$$

Now that you have seen how this change can be done, here is a shorter method.

To Change a Mixed Number to an Improper Fraction

1. Multiply the whole number by the denominator.
2. Add this to the numerator. The result is the new numerator. The denominator does not change.

EXAMPLE 6   Change to an improper fraction.   **(a)** $3\dfrac{1}{7}$ **(b)** $5\dfrac{4}{5}$

(a) $3\dfrac{1}{7} = \dfrac{(3 \times 7) + 1}{7} = \dfrac{21 + 1}{7} = \dfrac{22}{7}$ **(b)** $5\dfrac{4}{5} = \dfrac{(5 \times 5) + 4}{5} = \dfrac{25 + 4}{5} = \dfrac{29}{5}$

Practice Problem 6   Change to an improper fraction.

(a) $3\dfrac{2}{5}$ **(b)** $1\dfrac{3}{7}$ **(c)** $2\dfrac{6}{11}$ **(d)** $4\dfrac{2}{3}$

4 *Changing a Fraction to an Equivalent Fraction with a Different Denominator*

Fractions can be changed to an equivalent fraction with a different denominator by multiplying both numerator and denominator by the same number.

$$\frac{5}{6} = \frac{5}{6} \times 1 = \frac{5}{6} \times \frac{2}{2} = \frac{5 \times 2}{6 \times 2} = \frac{10}{12} \qquad \frac{3}{7} = \frac{3}{7} \times 1 = \frac{3}{7} \times \frac{3}{3} = \frac{3 \times 3}{7 \times 3} = \frac{9}{21}$$

$$\text{So } \frac{5}{6} \text{ is equivalent to } \frac{10}{12}. \qquad\qquad \frac{3}{7} \text{ is equivalent to } \frac{9}{21}.$$

We often multiply in this way to obtain an equivalent fraction with a *particular* denominator.

EXAMPLE 7   Find the missing number.

(a) $\dfrac{3}{5} = \dfrac{?}{25}$ **(b)** $\dfrac{3}{7} = \dfrac{?}{21}$ **(c)** $\dfrac{2}{9} = \dfrac{?}{36}$

(a) $\dfrac{3}{5} = \dfrac{?}{25}$ Observe that we need to multiply the denominator by 5 to obtain 25. So we multiply the numerator 3 by 5 also.

$\dfrac{3 \times 5}{5 \times 5} = \dfrac{15}{25}$ The desired numerator is 15.

(b) $\dfrac{3}{7} = \dfrac{?}{21}$ Observe that $7 \times 3 = 21$. We need to multiply the numerator by 3 to get the new numerator.

$\dfrac{3 \times 3}{7 \times 3} = \dfrac{9}{21}$ The desired numerator is 9.

(c) $\dfrac{2}{9} = \dfrac{?}{36}$ Observe that $9 \times 4 = 36$. We need to multiply the numerator by 4 to get the new numerator.

$\dfrac{2 \times 4}{9 \times 4} = \dfrac{8}{36}$ The desired numerator is 8.

Practice Problem 7   Find the missing number.

(a) $\dfrac{3}{8} = \dfrac{?}{24}$ **(b)** $\dfrac{5}{6} = \dfrac{?}{30}$ **(c)** $\dfrac{12}{13} = \dfrac{?}{26}$ **(d)** $\dfrac{2}{7} = \dfrac{?}{56}$

(e) $\dfrac{5}{9} = \dfrac{?}{27}$ **(f)** $\dfrac{3}{10} = \dfrac{?}{60}$ **(g)** $\dfrac{3}{4} = \dfrac{?}{28}$ **(h)** $\dfrac{8}{11} = \dfrac{?}{55}$

0.1 Exercises

Verbal and Writing Skills

1. In the fraction $\frac{12}{13}$, what number is the numerator?

2. In the fraction $\frac{13}{17}$, what is the denominator?

3. What is a factor? Give an example.

4. Give some examples of the number 1 written as a fraction.

5. Draw a diagram to illustrate $2\frac{2}{3}$.

Simplify each fraction.

6. $\frac{18}{24}$ **7.** $\frac{20}{35}$ **8.** $\frac{16}{48}$ **9.** $\frac{12}{60}$ **10.** $\frac{60}{12}$ **11.** $\frac{45}{15}$

Change to a mixed number.

12. $\frac{17}{6}$ **13.** $\frac{19}{5}$ **14.** $\frac{21}{9}$ **15.** $\frac{125}{4}$ **16.** $\frac{38}{7}$ **17.** $\frac{41}{6}$

Change to an improper fraction.

18. $3\frac{1}{5}$ **19.** $2\frac{6}{7}$ **20.** $6\frac{3}{5}$ **21.** $5\frac{3}{8}$ **22.** $8\frac{8}{9}$ **23.** $13\frac{1}{6}$

Find the missing numerator.

24. $\frac{3}{11} = \frac{?}{44}$ **25.** $\frac{5}{7} = \frac{?}{28}$ **26.** $\frac{3}{5} = \frac{?}{35}$ **27.** $\frac{2}{7} = \frac{?}{21}$ **28.** $\frac{9}{11} = \frac{?}{55}$ **29.** $\frac{13}{17} = \frac{?}{51}$

Applications

Solve.

30. Charles Barkley of the Phoenix Suns once scored 1560 points during 68 games played during the season. Express as a mixed number in simplified form how many points he averaged per game.

31. Alex Rodriguez, one of Major League Baseball's most successful hitters, got 215 hits out of 600 times at bat for the Seattle Mariners. Express as a fraction in simplified form how often he obtained a hit.

32. Last year, my parents had a combined income of $64,000. They paid $13,200 in federal income taxes. What simplified fraction shows how much my parents spent on their federal taxes?

33. R.E.M. is making a new CD. 7 out of 8 takes (tries) of their songs are acceptable. The recording engineer tells the band that there is time that day for only 136 takes. How many of the 136 takes will R.E.M. need to perform well to maintain the same fractional ratio of acceptable takes? (*Hint:* What fraction with a denominator of 136 will reduce to $\frac{7}{8}$?)

The following chart gives the recipe for a trail mix.

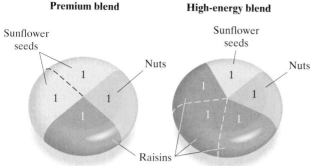

The Rocking ℞ trail mix

Premium blend   **High-energy blend**

34. What part of the premium blend is nuts?

35. What part of the high-energy blend is raisins?

36. What part of the premium blend is not sunflower seeds?

The following chart provides some statistics for the North Andover Knights.

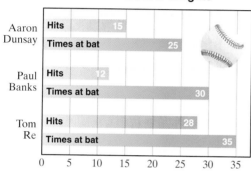

Baseball statistics for the North Andover Knights

37. Determine how often each player hit the ball based on the number of times at bat. Write each answer as a reduced fraction.

38. Which player was the best hitter?

The following chart provides some statistics for the Delphi Dunkers.

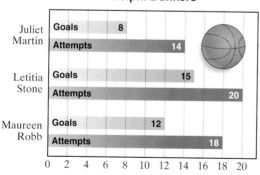

Basketball statistics for the Delphi Dunkers

39. Determine how often each player makes a basket based on the number of attempts at goal. Write each fraction as a reduced number.

40. Which player was the best scorer?

Developing Your Study Skills

Why Is Homework Necessary?

Mathematics is a set of skills that you learn by doing, not by watching. Your instructor may make solving a mathematics problem look very easy, but for you to learn the necessary skills, you must practice them over and over again, just as your instructor once had to do. There is no other way. Learning mathematics is like learning to play a musical instrument, to type, or to play a sport. No matter how much you watch someone else do it, how many books you may read on "how to" do it, or how easy it may seem to be, the key to success is to practice on a regular basis. Homework provides this practice.

1 *Adding or Subtracting Fractions with a Common Denominator*

If fractions have the same denominator, the numerators may be added or subtracted. The denominator remains the same.

To Add or Subtract Two Fractions with the Same Denominator

1. Add or subtract the numerators.
2. Keep the same (common) denominator.
3. Simplify the answer whenever possible.

EXAMPLE 1   Add the fractions. Simplify your answer whenever possible.

(a) $\dfrac{5}{7} + \dfrac{1}{7}$ **(b)** $\dfrac{2}{3} + \dfrac{1}{3}$ **(c)** $\dfrac{1}{8} + \dfrac{3}{8} + \dfrac{2}{8}$ **(d)** $\dfrac{3}{5} + \dfrac{4}{5}$

(a) $\dfrac{5}{7} + \dfrac{1}{7} = \dfrac{5+1}{7} = \dfrac{6}{7}$

(b) $\dfrac{2}{3} + \dfrac{1}{3} = \dfrac{2+1}{3} = \dfrac{3}{3} = 1$

(c) $\dfrac{1}{8} + \dfrac{3}{8} + \dfrac{2}{8} = \dfrac{1+3+2}{8} = \dfrac{6}{8} = \dfrac{3}{4}$

(d) $\dfrac{3}{5} + \dfrac{4}{5} = \dfrac{3+4}{5} = \dfrac{7}{5} = 1\dfrac{2}{5}$

Practice Problem 1   Add.

(a) $\dfrac{3}{6} + \dfrac{2}{6}$ **(b)** $\dfrac{3}{11} + \dfrac{2}{11} + \dfrac{6}{11}$ **(c)** $\dfrac{5}{8} + \dfrac{2}{8} + \dfrac{7}{8}$

EXAMPLE 2   Subtract the fractions. Simplify your answer whenever possible.

(a) $\dfrac{9}{11} - \dfrac{2}{11}$ **(b)** $\dfrac{5}{6} - \dfrac{1}{6}$

(a) $\dfrac{9}{11} - \dfrac{2}{11} = \dfrac{9-2}{11} = \dfrac{7}{11}$

(b) $\dfrac{5}{6} - \dfrac{1}{6} = \dfrac{5-1}{6} = \dfrac{4}{6} = \dfrac{2}{3}$

Practice Problem 2   Subtract.

(a) $\dfrac{11}{13} - \dfrac{6}{13}$ **(b)** $\dfrac{8}{9} - \dfrac{2}{9}$

Although adding and subtracting fractions with the same denominator is fairly simple, most problems involve fractions that do not have a common denominator. Fractions and mixed numbers such as halves, fourths, and eighths are often used. To add or subtract such fractions, we begin by finding a common denominator.

2 *Finding the Least Common Denominator of Two or More Fractions*

Before you can add or subtract fractions, they must have the same denominator. To save work, we select the smallest possible common denominator. This is called the **least common denominator** or LCD (also known as the *lowest common denominator*).

The LCD of two or more fractions is the smallest whole number that is exactly divisible by each denominator of the fractions.

● **EXAMPLE 3** Find the LCD. $\frac{2}{3}$ and $\frac{1}{4}$

The numbers are small enough to find the LCD by inspection. The LCD is 12, since 12 is exactly divisible by 4 and by 3. There is no smaller number that is exactly divisible by 4 and 3.

Practice Problem 3 Find the LCD. $\frac{1}{8}$ and $\frac{5}{12}$ ●

In some cases, the LCD cannot easily be determined by inspection. If we write each denominator as the product of prime factors, we will be able to find the LCD. We will use ($\cdot$) to indicate multiplication. For example, $30 = 2 \cdot 3 \cdot 5$. This means $30 = 2 \times 3 \times 5$.

Procedure to Find the LCD Using Prime Factors

1. Write each denominator as the product of prime factors.
2. The LCD is a product containing each different factor.
3. If a factor occurs more than once in any one denominator, the LCD will contain that factor repeated the greatest number of times that it occurs in any one denominator.

● **EXAMPLE 4** Find the LCD of $\frac{5}{6}$ and $\frac{1}{15}$ by this new procedure.

$$6 = 2 \cdot 3$$
$$15 = \;\;\;\; 3 \cdot 5$$
$$\text{LCD} = 2 \cdot 3 \cdot 5$$
$$\text{LCD} = 2 \cdot 3 \cdot 5 = 30$$

Write each denominator as the product of prime factors.

The LCD is a product containing each different prime factor. The different factors are 2, 3, and 5, and each factor appears at most once in any one denominator.

Practice Problem 4 Use prime factors to find the LCD of $\frac{8}{35}$ and $\frac{6}{15}$. ●

Great care should be used to determine the LCD in the case of repeated factors.

● **EXAMPLE 5** Find the LCD of $\frac{4}{27}$ and $\frac{5}{18}$.

$$27 = 3 \cdot 3 \cdot 3$$

Write each denominator as the product of prime factors. We observe that the factor 3 occurs three times in the factorization of 27.

$$18 = \;\;\;\; 3 \cdot 3 \cdot 2$$
$$\text{LCD} = 3 \cdot 3 \cdot 3 \cdot 2$$
$$\text{LCD} = 3 \cdot 3 \cdot 3 \cdot 2 = 54$$

The LCD is a product containing each different factor. The factor 3 *occurred most* in the factorization of 27, where it occurred *three* times. Thus the LCD will be the product of *three* 3's and *one* 2.

Practice Problem 5 Find the LCD of $\frac{5}{12}$ and $\frac{7}{30}$. ●

● **EXAMPLE 6** Find the LCD of $\frac{5}{12}, \frac{1}{15}$, and $\frac{7}{30}$.

$$12 = 2 \cdot 2 \cdot 3$$

Write each denominator as the product of prime factors. Notice that the only repeated factor is 2, which occurs twice in the factorization of 12.

$$15 = \;\;\;\; 3 \cdot 5$$
$$30 = \;\;\;\; 2 \cdot 3 \cdot 5$$
$$\text{LCD} = 2 \cdot 2 \cdot 3 \cdot 5$$
$$\text{LCD} = 2 \cdot 2 \cdot 3 \cdot 5 = 60$$

The LCD is the product of each different factor with the factor 2 appearing twice since it occurred twice in one denominator.

Practice Problem 6   Find the LCD of $\frac{2}{27}, \frac{1}{18}$, and $\frac{5}{12}$.

③ Adding or Subtracting Fractions That Do Not Have a Common Denominator

Before you can add or subtract them, fractions must have the same denominator. Using the LCD will make your work easier. First you must find the LCD. Then change each fraction to a fraction that has the LCD as the denominator. Sometimes one of the fractions will already have the LCD as the denominator. Once all the fractions have the same denominator, you can add or subtract. Be sure to simplify the fraction in your answer if this is possible.

To Add or Subtract Fractions That Do Not Have a Common Denominator

1. Find the LCD of the fractions.
2. Change each fraction to an equivalent fraction with the LCD for a denominator.
3. Add or subtract the fractions.
4. Simplify the answer whenever possible.

Let us return to the two fractions of Example 3. We have previously found that the LCD is 12.

EXAMPLE 7   Bob picked $\frac{2}{3}$ of a bushel of apples on Monday and $\frac{1}{4}$ of a bushel of apples on Tuesday. How much did he have in total?

To solve this problem we need to add $\frac{2}{3}$ and $\frac{1}{4}$, but before we can do so, we must change $\frac{2}{3}$ and $\frac{1}{4}$ to fractions with the same denominator. We change each fraction to an equivalent fraction with a common denominator of 12, the LCD.

$$\frac{2}{3} = \frac{?}{12} \qquad \frac{2}{3} \times \frac{4}{4} = \frac{8}{12} \quad \text{so} \quad \frac{2}{3} = \frac{8}{12}$$

$$\frac{1}{4} = \frac{?}{12} \qquad \frac{1}{4} \times \frac{3}{3} = \frac{3}{12} \quad \text{so} \quad \frac{1}{4} = \frac{3}{12}$$

Then we rewrite the problem with common denominators and add.

$$\frac{2}{3} + \frac{1}{4} = \frac{8}{12} + \frac{3}{12} = \frac{8+3}{12} = \frac{11}{12}$$

In total Bob picked $\frac{11}{12}$ of a bushel of apples.

Practice Problem 7   Add. $\dfrac{1}{8} + \dfrac{5}{12}$

Sometimes one of the denominators is the LCD. In such cases the fraction that has the LCD for the denominator will not need to be changed. If every other denominator divides into the largest denominator, the largest denominator is the LCD.

EXAMPLE 8 Find the LCD and then add. $\dfrac{3}{5} + \dfrac{7}{20} + \dfrac{1}{2}$

We can see by inspection that both 5 and 2 divide exactly into 20. Thus 20 is the LCD. Now add.

$$\frac{3}{5} + \frac{7}{20} + \frac{1}{2}$$

We change $\frac{3}{5}$ and $\frac{1}{2}$ to equivalent fractions with a common denominator of 20, the LCD.

$$\frac{3}{5} = \frac{?}{20} \qquad \frac{3}{5} \times \frac{4}{4} = \frac{12}{20} \quad \text{so} \quad \frac{3}{5} = \frac{12}{20}$$

$$\frac{1}{2} = \frac{?}{20} \qquad \frac{1}{2} \times \frac{10}{10} = \frac{10}{20} \quad \text{so} \quad \frac{1}{2} = \frac{10}{20}$$

Then we rewrite the problem with common denominators and add.

$$\frac{3}{5} + \frac{7}{20} + \frac{1}{2} = \frac{12}{20} + \frac{7}{20} + \frac{10}{20} = \frac{12 + 7 + 10}{20} = \frac{29}{20} \quad \text{or} \quad 1\frac{9}{20}$$

Practice Problem 8 Find the LCD and add. $\dfrac{3}{5} + \dfrac{4}{25} + \dfrac{1}{10}$

Now we turn to examples where the selection of the LCD is not so obvious. In Examples 9 through 11 we will use the prime factorization method to find the LCD.

EXAMPLE 9 Add. $\dfrac{7}{18} + \dfrac{5}{12}$

First we find the LCD.

$$18 = 3 \cdot 3 \cdot 2$$
$$12 = 3 \cdot 2 \cdot 2$$
$$\text{LCD} = 3 \cdot 3 \cdot 2 \cdot 2 = 9 \cdot 4 = 36$$

Now we change $\frac{7}{18}$ and $\frac{5}{12}$ to equivalent fractions that have the LCD.

$$\frac{7}{18} = \frac{?}{36} \qquad \frac{7}{18} \times \frac{2}{2} = \frac{14}{36}$$

$$\frac{5}{12} = \frac{?}{36} \qquad \frac{5}{12} \times \frac{3}{3} = \frac{15}{36}$$

Now we add the fractions.

$$\frac{7}{18} + \frac{5}{12} = \frac{14}{36} + \frac{15}{36} = \frac{29}{36} \quad \text{This fraction cannot be simplified.}$$

Practice Problem 9 Add. $\dfrac{1}{49} + \dfrac{3}{14}$

EXAMPLE 10   Subtract. $\dfrac{25}{48} - \dfrac{5}{36}$

First we find the LCD.

$$48 = 2 \cdot 2 \cdot 2 \cdot 2 \cdot 3$$
$$36 = \qquad 2 \cdot 2 \cdot 3 \cdot 3$$
$$\text{LCD} = 2 \cdot 2 \cdot 2 \cdot 2 \cdot 3 \cdot 3 = 16 \cdot 9 = 144$$

Now we change $\frac{25}{48}$ and $\frac{5}{36}$ to equivalent fractions that have the LCD.

$$\frac{25}{48} = \frac{?}{144} \qquad \frac{25}{48} \times \frac{3}{3} = \frac{75}{144}$$

$$\frac{5}{36} = \frac{?}{144} \qquad \frac{5}{36} \times \frac{4}{4} = \frac{20}{144}$$

Now we subtract the fractions.

$$\frac{25}{48} - \frac{5}{36} = \frac{75}{144} - \frac{20}{144} = \frac{55}{144} \qquad \text{This fraction cannot be simplified.}$$

Practice Problem 10   Subtract. $\dfrac{1}{12} - \dfrac{1}{30}$

EXAMPLE 11   Combine. $\dfrac{1}{5} + \dfrac{1}{6} - \dfrac{3}{10}$

First we find the LCD.

$$5 = 5$$
$$6 = \qquad 2 \cdot 3$$
$$10 = 5 \cdot 2$$
$$\text{LCD} = 5 \cdot 2 \cdot 3 = 10 \cdot 3 = 30$$

Now we change $\frac{1}{5}, \frac{1}{6}$, and $\frac{3}{10}$ to equivalent fractions that have the LCD for a denominator.

$$\frac{1}{5} = \frac{?}{30} \qquad \frac{1}{5} \times \frac{6}{6} = \frac{6}{30}$$

$$\frac{1}{6} = \frac{?}{30} \qquad \frac{1}{6} \times \frac{5}{5} = \frac{5}{30}$$

$$\frac{3}{10} = \frac{?}{30} \qquad \frac{3}{10} \times \frac{3}{3} = \frac{9}{30}$$

Now we combine the three fractions.

$$\frac{1}{5} + \frac{1}{6} - \frac{3}{10} = \frac{6}{30} + \frac{5}{30} - \frac{9}{30} = \frac{2}{30} = \frac{1}{15}$$

Note the important step of simplifying the fraction to obtain the final answer.

Practice Problem 11   Combine. $\dfrac{2}{3} + \dfrac{3}{4} - \dfrac{3}{8}$

4 *Adding or Subtracting Mixed Numbers*

If the problem you are adding or subtracting has mixed numbers, change them to improper fractions first and then combine (add or subtract). As a convention in this book, if the original problem contains mixed numbers, express the result as a mixed number rather than as an improper fraction.

EXAMPLE 12 Combine. Simplify your answer whenever possible.

(a) $5\frac{1}{2} + 2\frac{1}{3}$ **(b)** $2\frac{1}{5} - 1\frac{3}{4}$ **(c)** $1\frac{5}{12} + \frac{7}{30}$

(a) First we change the mixed numbers to improper fractions.

$$5\frac{1}{2} = \frac{5 \times 2 + 1}{2} = \frac{11}{2} \qquad 2\frac{1}{3} = \frac{2 \times 3 + 1}{3} = \frac{7}{3}$$

Next we change each fraction to an equivalent form with a common denominator of 6.

$$\frac{11}{2} = \frac{?}{6} \qquad \frac{11}{2} \times \frac{3}{3} = \frac{33}{6}$$

$$\frac{7}{3} = \frac{?}{6} \qquad \frac{7}{3} \times \frac{2}{2} = \frac{14}{6}$$

Finally, we add the two fractions and change our answer to a mixed number.

$$\frac{33}{6} + \frac{14}{6} = \frac{47}{6} = 7\frac{5}{6}$$

Thus $5\frac{1}{2} + 2\frac{1}{3} = 7\frac{5}{6}$.

(b) First we change the mixed numbers to improper fractions.

$$2\frac{1}{5} = \frac{2 \times 5 + 1}{5} = \frac{11}{5} \qquad 1\frac{3}{4} = \frac{1 \times 4 + 3}{4} = \frac{7}{4}$$

Next we change each fraction to an equivalent form with a common denominator of 20.

$$\frac{11}{5} = \frac{?}{20} \qquad \frac{11}{5} \times \frac{4}{4} = \frac{44}{20}$$

$$\frac{7}{4} = \frac{?}{20} \qquad \frac{7}{4} \times \frac{5}{5} = \frac{35}{20}$$

Now we subtract the two fractions.

$$\frac{44}{20} - \frac{35}{20} = \frac{9}{20}$$

Thus $2\frac{1}{5} - 1\frac{3}{4} = \frac{9}{20}$.

Note: It is not necessary to use these exact steps to add and subtract mixed numbers. If you know another method and can use it to obtain the correct answers, it is all right to continue to use that method throughout this chapter.

(c) Now we add $1\frac{5}{12} + \frac{7}{30}$.

The LCD of 12 and 30 is 60. Why? Change the mixed number to an improper fraction. Then change each fraction to an equivalent form with a common denominator.

$$1\frac{5}{12} = \frac{17}{12} \times \frac{5}{5} = \frac{85}{60} \qquad \frac{7}{30} \times \frac{2}{2} = \frac{14}{60}$$

Then add the fractions, simplify, and write the answer as a mixed number.

$$\frac{85}{60} + \frac{14}{60} = \frac{99}{60} = \frac{33}{20} = 1\frac{13}{20} \qquad \text{Thus } 1\frac{5}{12} + \frac{7}{30} = 1\frac{13}{20}.$$

Practice Problem 12   Combine.

(a) $1\dfrac{2}{3} + 2\dfrac{4}{5}$ **(b)** $5\dfrac{1}{4} - 2\dfrac{2}{3}$

▲ ● **EXAMPLE 13**   Manuel is enclosing a triangular-shaped exercise yard for his new dog. He wants to determine how many feet of fencing he will need. The sides of the yard measure $20\frac{3}{4}$ feet, $15\frac{1}{2}$ feet, and $18\frac{1}{8}$ feet. What is the perimeter of (total distance around) the triangle?

Understand the problem. Begin by drawing a picture.

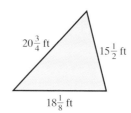

We want to add up the lengths of all three sides of the triangle. This distance around the triangle is called the **perimeter.**

$$20\frac{3}{4} + 15\frac{1}{2} + 18\frac{1}{8} = \frac{83}{4} + \frac{31}{2} + \frac{145}{8}$$

$$= \frac{166}{8} + \frac{124}{8} + \frac{145}{8} = \frac{435}{8} = 54\frac{3}{8} \text{ feet}$$

He will need $54\dfrac{3}{8}$ feet of fencing.

▲ **Practice Problem 13**   Find the perimeter of a rectangle with sides of $4\frac{1}{5}$ cm and $6\frac{1}{2}$ cm. Begin by drawing a picture. Label the picture by including the measure of *each* side.

Developing Your Study Skills

Class Attendance

A student of mathematics needs to get started in the right direction by choosing to attend class every day, beginning with the first day of class. Statistics show that class attendance and good grades go together. Classroom activities are designed to enhance learning and you must be in class to benefit from them. Vital information and explanations that can help you in understanding concepts are given each day. Do not be deceived into thinking that you can just find out from a friend what went on in class. There is no good substitute for firsthand experience. Give yourself a push in the right direction by developing the habit of going to class every day.

Class Participation

People learn mathematics through active participation, not through observation from the sidelines. If you want to do well in this course, be involved in classroom activities. Sit near the front where you can see and hear well and where your focus is on the instruction process and not on the students around you. Ask questions, be ready to contribute toward solutions, and take part in all classroom activities. Your contributions are valuable to the class and to yourself. Class participation requires an investment of yourself in the learning process, which you will find pays huge dividends.

Verbal and Writing Skills

1. Explain why the denominator 8 is the least common denominator of $\frac{3}{4}$ and $\frac{5}{8}$.

2. What must you do before you add or subtract fractions that do not have a common denominator?

Find the LCD (least common denominator) of each pair of fractions. Do not combine the fractions; only find the LCD.

3. $\frac{7}{15}$ and $\frac{11}{21}$

4. $\frac{13}{25}$ and $\frac{29}{40}$

5. $\frac{7}{10}$ and $\frac{1}{4}$

6. $\frac{3}{16}$ and $\frac{1}{24}$

7. $\frac{5}{63}$ and $\frac{5}{21}$

8. $\frac{7}{36}$ and $\frac{5}{54}$

9. $\frac{1}{2}, \frac{1}{18}$, and $\frac{13}{30}$

10. $\frac{5}{8}, \frac{3}{14}$, and $\frac{11}{16}$

Combine. Be sure to simplify your answer whenever possible.

11. $\frac{3}{8} + \frac{7}{12}$

12. $\frac{1}{12} + \frac{2}{9}$

13. $\frac{5}{14} - \frac{1}{4}$

14. $\frac{9}{20} - \frac{1}{15}$

15. $\frac{2}{9} + \frac{5}{6}$

16. $\frac{3}{4} + \frac{3}{10}$

17. $\frac{5}{7} - \frac{2}{9}$

18. $\frac{7}{8} - \frac{2}{3}$

19. $\frac{9}{8} + \frac{7}{12}$

20. $\frac{5}{6} + \frac{7}{15}$

21. $\frac{7}{18} + \frac{1}{12}$

22. $\frac{4}{7} + \frac{7}{9} + \frac{1}{3}$

23. $\frac{2}{3} + \frac{7}{12} + \frac{1}{4}$

24. $\frac{3}{35} + \frac{4}{7} - \frac{1}{5}$

25. $\frac{4}{15} + \frac{7}{12} - \frac{1}{3}$

26. $\frac{3}{10} - \frac{3}{25}$

27. $\frac{7}{12} - \frac{5}{18}$

28. $4\frac{1}{3} + 3\frac{2}{5}$

29. $3\frac{1}{8} + 2\frac{1}{6}$

30. $1\frac{5}{24} + \frac{5}{18}$

31. $6\frac{2}{3} + \frac{3}{4}$

32. $7\frac{1}{6} - 2\frac{1}{4}$

33. $7\frac{2}{5} - 3\frac{3}{4}$

34. $9\frac{4}{5} - 3\frac{1}{2}$

35. $4\frac{7}{12} - 1\frac{5}{6}$

36. $2\frac{1}{8} + 3\frac{2}{3}$

37. $3\frac{1}{7} + 4\frac{1}{3}$

38. $11\frac{1}{7} - 6\frac{5}{7}$

39. $17\frac{1}{5} - 10\frac{3}{5}$

40. $2\frac{1}{8} + 6\frac{3}{4}$

Applications

41. Jenny and Laura went inline skating. They skated $3\frac{1}{8}$ miles on Monday, $2\frac{2}{3}$ miles on Tuesday, and $4\frac{1}{2}$ miles on Wednesday. What was their total distance for those three days?

42. Clara is creating a fabric wall hanging. She plans to sew together a large rectangle with three different types of cloth to make the background of the wall hanging. The first piece is $11\frac{3}{8}$ inches long, the second is $21\frac{1}{2}$ inches long, and the third is $9\frac{3}{4}$ inches long. If she wishes to sew a single piece of trim along the edge over all three pieces, how long must the piece of trim be?

43. Sheryl has $8\frac{1}{2}$ hours this weekend to work on her new video. She estimates that it will take $2\frac{2}{3}$ hours to lip sync the new song and $1\frac{3}{4}$ hours to learn the new dance steps. How much time will she have left over for her MTV interview?

44. Mrs. Banks has an outdoor fountain that holds $12\frac{3}{4}$ gallons of water. While cleaning it, she drained $4\frac{5}{8}$ gallons of water from it. How many gallons of water are still in the fountain?

Carpenters use fractions in their work. The picture at the right is a diagram of a spice cabinet. The symbol " means inches. Use the picture to answer exercises 45 and 46.

45. Before you can determine where the cabinet will fit, you need to calculate the height, *A*, and the width, *B*. Don't forget to include the $\frac{1}{2}$-inch thickness of the wood where needed.

46. Look at the close-up of the drawer. The width is $4\frac{9}{16}''$. In the diagram, the width of the opening for the drawer is $4\frac{5}{8}''$. What is the difference? Why do you think the drawer is smaller than the opening?

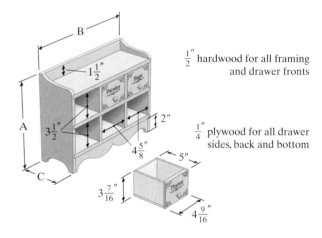

47. The Falmouth Country Club maintains the putting greens with a grass height of $\frac{7}{8}$ inch. The grass on the fairways is maintained at $2\frac{1}{2}$ inches. How much must the mower blade be lowered by a person mowing the fairways if that person will be using the same mowing machine on the putting greens?

48. The director of facilities maintenance at the club discovered that due to slippage in the adjustment lever that the lawn mower actually cuts the grass $\frac{1}{16}$ of an inch too long or too short on some days. What is the maximum height that the fairway grass could be after being mowed with this machine? What is the minimum height that the putting greens could be after being mowed with this machine?

Cumulative Review Problems

49. Simplify. $\dfrac{36}{44}$

50. Change to an improper fraction. $26\dfrac{3}{5}$

Putting Your Skills to Work

The High Jump: Raising the Bar

In the 1896 Olympics, Ellery Clark of the United States set the world record for the men's high jump by clearing a bar at the height of 5 ft $11\frac{1}{4}$ in. In the 1912 Olympics, Alma Richards of the United States set a new record by clearing a bar at the height of 6 ft 4 in. He had jumped $4\frac{3}{4}$ in. higher than Ellery Clark. An interesting pattern can be observed by examining the record heights achieved in the men's high jump in the Olympics over 20-year periods. The bar graph below shows the number of inches by which each jump exceeded the 1896 world record.

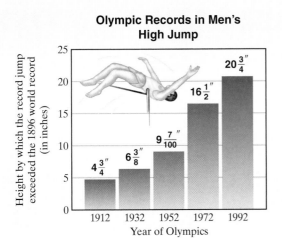

Olympic Records in Men's High Jump

Height by which the record jump exceeded the 1896 world record (in inches)

- 1912: $4\frac{3}{4}''$
- 1932: $6\frac{3}{8}''$
- 1952: $9\frac{7}{100}''$
- 1972: $16\frac{1}{2}''$
- 1992: $20\frac{3}{4}''$

Year of Olympics

Problems for Individual Investigation and Analysis

1. How much higher was the record jump in the 1972 Olympics than the record jump in the 1932 Olympics?

2. In what 20-year period did the men's record high jump measurement experience the greatest increase?

Problems for Group Investigation and Cooperative Learning

Together with some members of your class see if you can answer the following.

3. During the period from 1912 to 1992 the high jump record increased from $4\frac{3}{4}$ in. greater than the 1896 record to $20\frac{3}{4}$ in. greater than the 1896 record. What is the total increase from 1912 to 1992? What is the average increase expected for a 20-year period? What would you expect the average increase to be in a 4-year period?

4. A new men's high jump record was set in the 1996 Olympics in Atlanta. Using your results from exercise 3, what would you guess this jump was? The record jump was 7 ft 10 in. by Charles Austin of the United States. Was your guess too high or too low? How much was it in error?

Internet Connections

 www   Netsite: http://www.prenhall.com/tobey_beginning

Site: Olympic FAQ—Olympic Winter Games or a related site

This site gives basic information about the Winter Olympic Games for 1924 to 1994.

5. Make a table showing the number of male athletes, the number of female athletes, and the total number of athletes for these years. Find how many male athletes and female athletes competed in the 1998 Winter Olympic in Nagano, Japan. Add this information to the table.

6. In 1984, how many more men participated than women?

7. In what 12-year period did the number of female athletes increase the most?

8. Find the total increase in the number of athletes from 1952 to 1992. What is the average increase expected for an eight-year period? Using this result, what would you guess was the number of athletes in 2000? How does your guess compare with the actual number?

21

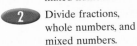

1 *Multiplying Fractions, Whole Numbers, and Mixed Numbers*

Multiplication of Fractions

During a recent snowstorm, the runway at Beverly Airport was plowed. However, the plow cleared only $\frac{3}{5}$ of the width and $\frac{2}{7}$ of the length. What fraction of the total runway area was cleared? To answer this question, we need to multiply $\frac{3}{5} \times \frac{2}{7}$.

The answer is that $\frac{6}{35}$ of the total runway area was cleared.

The multiplication rule for fractions states that to multiply two fractions, we multiply the two numerators and multiply the two denominators.

To Multiply Any Two Fractions

1. Multiply the numerators.

2. Multiply the denominators.

EXAMPLE 1 Multiply.

(a) $\frac{3}{5} \times \frac{2}{7}$ **(b)** $\frac{1}{3} \times \frac{5}{4}$ **(c)** $\frac{7}{3} \times \frac{1}{5}$ **(d)** $\frac{6}{5} \times \frac{2}{3}$

(a) $\frac{3}{5} \times \frac{2}{7} = \frac{3 \cdot 2}{5 \cdot 7} = \frac{6}{35}$ **(b)** $\frac{1}{3} \times \frac{5}{4} = \frac{1 \cdot 5}{3 \cdot 4} = \frac{5}{12}$

(c) $\frac{7}{3} \times \frac{1}{5} = \frac{7 \cdot 1}{3 \cdot 5} = \frac{7}{15}$ **(d)** $\frac{6}{5} \times \frac{2}{3} = \frac{6 \cdot 2}{5 \cdot 3} = \frac{12}{15} = \frac{4}{5}$

Note that we must simplify this fraction.

Practice Problem 1 Multiply.

(a) $\frac{2}{7} \times \frac{5}{11}$ **(b)** $\frac{8}{9} \times \frac{3}{10}$

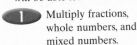

It is possible to avoid having to simplify a fraction at the last step. In many cases we can divide by a value that appears as a factor in both a numerator and a denominator. Often it is helpful to write a number as a product of prime factors in order to do this.

EXAMPLE 2 Multiply.

(a) $\frac{3}{5} \times \frac{5}{7}$ **(b)** $\frac{4}{11} \times \frac{5}{2}$ **(c)** $\frac{15}{8} \times \frac{10}{27}$

(a) $\frac{3}{5} \times \frac{5}{7} = \frac{3 \cdot 5}{5 \cdot 7} = \frac{3 \cdot \overset{1}{\cancel{5}}}{7 \cdot \cancel{5}} = \frac{3}{7}$ Note that here we divided numerator and denominator by 5.

If we factor each number, we can see the common factors.

(b) $\frac{4}{11} \times \frac{5}{2} = \frac{2 \cdot \overset{1}{\cancel{2}}}{11} \times \frac{5}{\underset{1}{\cancel{2}}} = \frac{10}{11}$ **(c)** $\frac{15}{8} \times \frac{10}{27} = \frac{\overset{1}{\cancel{3}} \cdot 5}{2 \cdot 2 \cdot 2} \times \frac{5 \cdot \overset{1}{\cancel{2}}}{\underset{1}{\cancel{3}} \cdot 3 \cdot 3} = \frac{25}{36}$

After dividing out common factors, the resulting multiplication problem involves smaller numbers and the answers are in simplified form.

Practice Problem 2 Multiply. **(a)** $\frac{3}{5} \times \frac{4}{3}$ **(b)** $\frac{9}{10} \times \frac{5}{12}$

Sidelight

Why does this method of dividing out a value that appears as a factor in both numerator and denominator work? Let's reexamine one of the examples we have solved previously.

$$\frac{3}{5} \times \frac{5}{7} = \frac{3}{\cancel{5}} \times \frac{\overset{1}{\cancel{5}}}{7} = \frac{3}{7}$$

Consider the following steps and reasons.

$$\frac{3}{5} \times \frac{5}{7} = \frac{3 \cdot 5}{5 \cdot 7} \qquad \text{Definition of multiplication of fractions.}$$

$$= \frac{5 \cdot 3}{5 \cdot 7} \qquad \text{Change the order of the factors in the numerator, since } 3 \cdot 5 = 5 \cdot 3. \text{ This is called the commutative property of multiplication.}$$

$$= \frac{5}{5} \cdot \frac{3}{7} \qquad \text{Definition of multiplication of fractions.}$$

$$= 1 \cdot \frac{3}{7} \qquad \text{Write 1 in place of } \frac{5}{5}, \text{ since 1 is another name for } \frac{5}{5}.$$

$$= \frac{3}{7} \qquad 1 \cdot \frac{3}{7} = \frac{3}{7}, \text{ since any number can be multiplied by 1 without changing the value of the number.}$$

Think about this concept. It is an important one that we will use again when we discuss rational expressions.

Multiplication of a Fraction by a Whole Number

Whole numbers can be named using fractional notation. $3, \frac{9}{3}, \frac{6}{2},$ and $\frac{3}{1}$ are ways of expressing the number *three*. Therefore,

$$3 = \frac{9}{3} = \frac{6}{2} = \frac{3}{1}.$$

When we multiply a fraction by a whole number, we merely express the whole number as a fraction whose denominator is 1 and follow the multiplication rule for fractions.

EXAMPLE 3   Multiply.   **(a)** $7 \times \frac{3}{5}$ **(b)** $\frac{3}{16} \times 4$

(a) $7 \times \frac{3}{5} = \frac{7}{1} \times \frac{3}{5} = \frac{21}{5} = 4\frac{1}{5}$ **(b)** $\frac{3}{16} \times 4 = \frac{3}{16} \times \frac{4}{1} = \frac{3}{4 \cdot \cancel{4}} \times \frac{\cancel{4}}{1} = \frac{3}{4}$

Notice that in (b) we did not use *prime* factors to factor 16. We recognized that $16 = 4 \cdot 4$. This is a more convenient factorization of 16 for this problem. Choose the factorization that works best for each problem. If you cannot decide what is best, factor into primes.

Practice Problem 3   Multiply.   **(a)** $4 \times \frac{2}{7}$ **(b)** $12 \times \frac{3}{4}$

Multiplication of Mixed Numbers

When multiplying mixed numbers, we first change them to improper fractions and then follow the multiplication rule for fractions.

EXAMPLE 4   How do we find the area of a rectangular field $3\frac{1}{3}$ miles long by $2\frac{1}{2}$ miles wide?

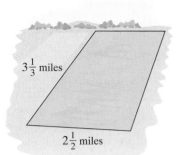

$3\frac{1}{3}$ miles

$2\frac{1}{2}$ miles

To find the area, we multiply length times width.

$$3\frac{1}{3} \times 2\frac{1}{2} = \frac{10}{3} \times \frac{5}{2} = \frac{\cancel{2} \cdot 5}{3} \times \frac{5}{\cancel{2}} = \frac{25}{3} = 8\frac{1}{3}$$

The area is $8\frac{1}{3}$ square miles.

Practice Problem 4   Multiply.   **(a)** $2\frac{1}{5} \times \frac{3}{7}$ **(b)** $3\frac{1}{3} \times 1\frac{2}{5}$

EXAMPLE 5   Multiply. $2\frac{2}{3} \times \frac{1}{4} \times 6$

$$2\frac{2}{3} \times \frac{1}{4} \times 6 = \frac{8}{3} \times \frac{1}{4} \times \frac{6}{1} = \frac{\cancel{4} \cdot 2}{\cancel{3}} \times \frac{1}{\cancel{4}} \times \frac{2 \cdot \cancel{3}}{1} = \frac{4}{1} = 4$$

Practice Problem 5   Multiply. $3\frac{1}{2} \times \frac{1}{14} \times 4$

2 *Dividing Fractions, Whole Numbers, and Mixed Numbers*

Division of Fractions

To divide two fractions, we invert the second fraction (that is, the divisor) and then multiply the two fractions.

To Divide Two Fractions

1. Invert the second fraction (that is, the divisor).
2. Now multiply the two fractions.

EXAMPLE 6   Divide.   **(a)** $\frac{1}{3} \div \frac{1}{2}$ **(b)** $\frac{2}{5} \div \frac{3}{10}$ **(c)** $\frac{2}{3} \div \frac{7}{5}$

(a) $\frac{1}{3} \div \frac{1}{2} = \frac{1}{3} \times \frac{2}{1} = \frac{2}{3}$ Note that we always invert the *second* fraction.

(b) $\frac{2}{5} \div \frac{3}{10} = \frac{2}{5} \times \frac{10}{3} = \frac{2}{\cancel{5}} \times \frac{\cancel{5} \cdot 2}{3} = \frac{4}{3} = 1\frac{1}{3}$ **(c)** $\frac{2}{3} \div \frac{7}{5} = \frac{2}{3} \times \frac{5}{7} = \frac{10}{21}$

Practice Problem 6   Divide.   **(a)** $\frac{2}{5} \div \frac{1}{3}$ **(b)** $\frac{12}{13} \div \frac{4}{3}$

Division of a Fraction and a Whole Number

The process of inverting the second fraction and then multiplying the two fractions should be done very carefully when one of the original values is a whole number. Remember, a whole number such as 2 is equivalent to $\frac{2}{1}$.

EXAMPLE 7   Divide.   **(a)** $\frac{1}{3} \div 2$ **(b)** $5 \div \frac{1}{3}$

(a) $\frac{1}{3} \div 2 = \frac{1}{3} \div \frac{2}{1} = \frac{1}{3} \times \frac{1}{2} = \frac{1}{6}$

(b) $5 \div \frac{1}{3} = \frac{5}{1} \div \frac{1}{3} = \frac{5}{1} \times \frac{3}{1} = \frac{15}{1} = 15$

Practice Problem 7   Divide.   **(a)** $\frac{3}{7} \div 6$ **(b)** $8 \div \frac{2}{3}$

Sidelight

Number Sense   Look at the answers to the problems in Example 7. In part (a), you will notice that $\frac{1}{6}$ is less than the original number $\frac{1}{3}$. Does this seem reasonable? Let's see.

If $\frac{1}{3}$ is divided by 2, it means that $\frac{1}{3}$ will be divided into two equal parts. We would expect that each part would be less than $\frac{1}{3}$. $\frac{1}{6}$ is a reasonable answer to this division problem.

In part (b), 15 is greater than the original number 5. Does this seem reasonable? Think of what $5 \div \frac{1}{3}$ means. It means that 5 will be divided into thirds. Let's think of an easier problem. What happens when we divide 1 into thirds? We get *three* thirds. We would expect, therefore, that when we divide 5 by thirds, we would get 5×3 or 15 thirds. 15 is a reasonable answer to this division problem.

Complex Fractions

Sometimes division is written in the form of a **complex fraction** with one fraction in the numerator and one fraction in the denominator. It is best to write this in standard division notation first; then complete the problem using the rule for division.

EXAMPLE 8 Divide.   **(a)** $\dfrac{\frac{3}{7}}{\frac{3}{5}}$ **(b)** $\dfrac{\frac{2}{9}}{\frac{5}{7}}$

(a) $\dfrac{\frac{3}{7}}{\frac{3}{5}} = \frac{3}{7} \div \frac{3}{5} = \frac{\cancel{3}}{7} \times \frac{5}{\cancel{3}} = \frac{5}{7}$

(b) $\dfrac{\frac{2}{9}}{\frac{5}{7}} = \frac{2}{9} \div \frac{5}{7} = \frac{2}{9} \times \frac{7}{5} = \frac{14}{45}$

Practice Problem 8 Divide.   **(a)** $\dfrac{\frac{3}{11}}{\frac{5}{7}}$ **(b)** $\dfrac{\frac{12}{5}}{\frac{8}{15}}$

Sidelight

Why does the method of "invert and multiply" work? The division rule really depends on the property that any number can be multiplied by 1 without changing the value of the number. Let's look carefully at an example of division of fractions:

$\dfrac{2}{5} \div \dfrac{3}{7} = \dfrac{\frac{2}{5}}{\frac{3}{7}}$ We can write the problem using a complex fraction.

$= \dfrac{\frac{2}{5}}{\frac{3}{7}} \times 1$ We can multiply by 1, since any number can be multiplied by 1 without changing the value of the number.

$= \dfrac{\frac{2}{5}}{\frac{3}{7}} \times \dfrac{\frac{7}{3}}{\frac{7}{3}}$ We write 1 in the form $\frac{\frac{7}{3}}{\frac{7}{3}}$, since any nonzero number divided by itself equals 1. We choose this value as a multiplier because it will help simplify the denominator.

$= \dfrac{\frac{2}{5} \times \frac{7}{3}}{\frac{3}{7} \times \frac{7}{3}}$ Definition of multiplication of fractions.

$= \dfrac{\frac{2}{5} \times \frac{7}{3}}{1}$ The product in the denominator equals 1.

$= \dfrac{2}{5} \times \dfrac{7}{3}$

Thus we have shown that $\frac{2}{5} \div \frac{3}{7}$ is equivalent to $\frac{2}{5} \times \frac{7}{3}$ and have shown some justification for the "invert and multiply rule."

Division of Mixed Numbers

This method for division of fractions can be used with mixed numbers. However, we first must change the mixed numbers to improper fractions and then use the rule for dividing fractions.

EXAMPLE 9   Divide.   **(a)** $2\frac{1}{3} \div 3\frac{2}{3}$ **(b)** $\dfrac{2}{3\frac{1}{2}}$

(a) $2\frac{1}{3} \div 3\frac{2}{3} = \frac{7}{3} \div \frac{11}{3} = \frac{7}{\cancel{3}} \times \frac{\cancel{3}}{11} = \frac{7}{11}$

(b) $\dfrac{2}{3\frac{1}{2}} = 2 \div 3\frac{1}{2} = \frac{2}{1} \div \frac{7}{2} = \frac{2}{1} \times \frac{2}{7} = \frac{4}{7}$

Practice Problem 9   Divide.

(a) $1\frac{2}{5} \div 2\frac{1}{3}$ **(b)** $4\frac{2}{3} \div 7$ **(c)** $\dfrac{1\frac{1}{5}}{1\frac{2}{7}}$

EXAMPLE 10   A chemist has 96 fluid ounces of a solution. She pours the solution into test tubes. Each test tube holds $\frac{3}{4}$ fluid ounce. How many test tubes can she fill?

We need to divide the total number of ounces, 96, by the number of ounces in each test tube, $\frac{3}{4}$.

$$96 \div \frac{3}{4} = \frac{96}{1} \div \frac{3}{4} = \frac{96}{1} \times \frac{4}{3} = \frac{\cancel{3} \cdot 32}{1} \times \frac{4}{\cancel{3}} = \frac{128}{1} = 128$$

She will be able to fill 128 test tubes.

Pause for a moment to think about the answer. Does 128 test tubes filled with solution seem like a reasonable answer? Did you perform the correct operation?

Practice Problem 10   A chemist has 64 fluid ounces of a solution. He wishes to fill several jars, each holding $5\frac{1}{3}$ fluid ounces. How many jars can he fill?

Sometimes when solving word problems involving fractions or mixed numbers, it is helpful to solve the problem using simpler numbers first. Once you understand what operation is involved, you can go back and solve using the original numbers in the word problem.

EXAMPLE 11   A car traveled 301 miles on $10\frac{3}{4}$ gallons of gas. How many miles per gallon did it get?

Use simpler numbers: 300 miles on 10 gallons of gas. We want to find out how many miles the car traveled on 1 gallon of gas. You may want to draw a picture.

10 gallons

Divide. $300 \div 10 = 30.$

300 miles

Now use the original numbers given in the problem.

$$301 \div 10\frac{3}{4} = \frac{301}{1} \div \frac{43}{4} = \frac{301}{1} \times \frac{4}{43} = \frac{1204}{43} = 28 \quad \text{This is 28 miles to the gallon.}$$

Practice Problem 11   A car can travel $25\frac{1}{2}$ miles on 1 gallon of gas. How many miles can a car travel on $5\frac{1}{4}$ gallons of gas? Check your answer to see if it is reasonable.

Verbal and Writing Skills

Multiply. Simplify your answer whenever possible.

1. Explain in your own words how to multiply two mixed numbers.

2. Explain in your own words how to divide two proper fractions.

3. $\dfrac{21}{5} \times \dfrac{10}{7}$

4. $\dfrac{7}{9} \times \dfrac{18}{5}$

5. $9 \times \dfrac{2}{5}$

6. $\dfrac{8}{11} \times 3$

Divide. Simplify your answer whenever possible.

7. $\dfrac{8}{5} \div \dfrac{8}{3}$

8. $\dfrac{13}{9} \div \dfrac{13}{7}$

9. $\dfrac{3}{7} \div 3$

10. $\dfrac{7}{8} \div 4$

11. $10 \div \dfrac{5}{7}$

12. $18 \div \dfrac{2}{9}$

13. $\dfrac{\frac{3}{4}}{\frac{4}{5}}$

14. $\dfrac{\frac{5}{7}}{\frac{5}{14}}$

15. $1\dfrac{3}{7} \div 6\dfrac{1}{4}$

16. $4\dfrac{1}{2} \div 3\dfrac{3}{8}$

17. $\dfrac{\frac{7}{9}}{1\frac{1}{3}}$

18. $\dfrac{\frac{5}{8}}{1\frac{3}{4}}$

Perform the proper calculations. Reduce your answer whenever possible.

19. $\dfrac{6}{5} \times \dfrac{10}{12}$

20. $\dfrac{5}{24} \times \dfrac{18}{15}$

21. $\dfrac{5}{16} \div \dfrac{1}{8}$

22. $\dfrac{2}{11} \div 4$

23. $10\dfrac{3}{7} \times 5\dfrac{1}{4}$

24. $10\dfrac{2}{9} \div 2\dfrac{1}{3}$

25. $2\dfrac{1}{8} \div \dfrac{1}{4}$

26. $4 \div 1\dfrac{7}{9}$

27. $8 \times 3\dfrac{1}{2}$

28. $\dfrac{3}{4} \div 6$

29. $2\dfrac{1}{2} \times \dfrac{1}{10} \times \dfrac{3}{4}$

30. $3\dfrac{1}{3} \times \dfrac{1}{5} \times \dfrac{2}{3}$

Applications

31. A denim shirt at the Gap requires $2\frac{3}{4}$ yards of material. How many yards would be needed to make 26 shirts?

32. A fleece pullover requires $1\frac{5}{8}$ yds of material. How many yards will it take to make 18 pullovers?

33. Jennifer rode her mountain bike for $4\frac{1}{5}$ miles after work. Two-thirds of the distance was over a mountain bike trail. How long is the mountain bike trail?

34. Phil ran a cross-country race that was $3\frac{3}{8}$ miles long. One-third of that race was on a hilly nature preserve trail. How long is the nature preserve trail?

Cumulative Review Problems

In exercises 35 and 36, simplify each fraction.

35. 116/124

36. 33/77

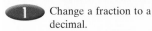 **Use of Decimals**

Student Learning Objectives

After studying this section, you will be able to:

1. Change a fraction to a decimal.

2. Change a decimal to a fraction.

3. Add and subtract decimals.

4. Multiply decimals.

5. Divide decimals.

6. Multiply or divide a number by a multiple of 10.

SSM
PH TUTOR CD & VIDEO MATH PRO WEB
CENTER

The Basic Concept of Decimals

We can express a part of a whole as a fraction or as a decimal. A **decimal** is another way of writing a fraction whose denominator is 10, 100, 1000, and so on.

$$\frac{3}{10} = 0.3 \qquad \frac{5}{100} = 0.05 \qquad \frac{172}{1000} = 0.172 \qquad \frac{58}{10,000} = 0.0058$$

The period in decimal notation is known as the **decimal point.** The number of digits in a number to the right of the decimal point is known as the number of **decimal places** of the number. The place value of decimals is shown in the following chart.

Hundred-thousands	Ten-thousands	Thousands	Hundreds	Tens	Ones	← Decimal point	Tenths	Hundredths	Thousandths	Ten-thousandths	Hundred-thousandths
100,000	10,000	1000	100	10	1	.	$\frac{1}{10}$	$\frac{1}{100}$	$\frac{1}{1000}$	$\frac{1}{10,000}$	$\frac{1}{100,000}$

EXAMPLE 1 Write each of the following decimals as a fraction. State the number of decimal places. Write out in words the way the number would be spoken.

(a) 0.6 **(b)** 0.29 **(c)** 0.527 **(d)** 1.38 **(e)** 0.00007

Decimal Form	Fraction Form	Number of Decimal Places	The Words Used to Describe the Number
(a) 0.6	$\frac{6}{10}$	one	six-tenths
(b) 0.29	$\frac{29}{100}$	two	twenty-nine hundredths
(c) 0.527	$\frac{527}{1000}$	three	five hundred twenty-seven thousandths
(d) 1.38	$1\frac{38}{100}$	two	one and thirty-eight hundredths
(e) 0.00007	$\frac{7}{100,000}$	five	seven hundred-thousandths

Practice Problem 1 Write each decimal as a fraction and in words.

(a) 1.371 **(b)** 0.09

You have seen that a given fraction can be written in several different but equivalent ways. There are also several different equivalent ways of writing the decimal form of fractions. The decimal 0.18 can be written in the following equivalent ways:

Fractional form: $\frac{18}{100} = \frac{180}{1000} = \frac{1800}{10,000} = \frac{18,000}{100,000}$

Decimal form: $0.18 = 0.180 = 0.1800 = 0.18000$

Thus we see that *any number of terminal zeros may be added to the right-hand side of a decimal* without changing its value.

$$0.13 = 0.1300 \qquad 0.162 = 0.162000$$

Similarly, *any number of terminal zeros may be removed from the right-hand side of a decimal* without changing its value.

1 Changing a Fraction to a Decimal

A fraction can be changed to a decimal by dividing the denominator into the numerator.

EXAMPLE 2   Write each of the following fractions as a decimal.

(a) $\dfrac{3}{4}$ **(b)** $\dfrac{21}{20}$ **(c)** $\dfrac{1}{8}$ **(d)** $\dfrac{3}{200}$

(a) $\dfrac{3}{4} = 0.75$ since

$$
\begin{array}{r}
0.75 \\
4\overline{)3.00} \\
28 \\
\hline
20 \\
20 \\
\hline
0
\end{array}
$$

(b) $\dfrac{21}{20} = 1.05$ since

$$
\begin{array}{r}
1.05 \\
20\overline{)21.00} \\
20 \\
\hline
100 \\
100 \\
\hline
0
\end{array}
$$

(c) $\dfrac{1}{8} = 0.125$ since

$$
\begin{array}{r}
0.125 \\
8\overline{)1.000} \\
8 \\
\hline
20 \\
16 \\
\hline
40 \\
40 \\
\hline
0
\end{array}
$$

(d) $\dfrac{3}{200} = 0.015$ since

$$
\begin{array}{r}
0.015 \\
200\overline{)3.000} \\
200 \\
\hline
1000 \\
1000 \\
\hline
0
\end{array}
$$

Calculator

Fraction to Decimal
You can use a calculator to change $\frac{3}{5}$ to a decimal. Enter:

$$3 \; \boxed{\div} \; 5 \; \boxed{=}$$

The display should read:

$$\boxed{\qquad 0.6}$$

Try the following.

(a) $\dfrac{17}{25}$ **(b)** $\dfrac{2}{9}$

(c) $\dfrac{13}{10}$ **(d)** $\dfrac{15}{19}$

Note: 0.7894737 is an approximation for $\frac{15}{19}$.

Practice Problem 2   Write as decimals.

(a) $\dfrac{3}{8}$ **(b)** $\dfrac{7}{200}$ **(c)** $\dfrac{33}{20}$

Sometimes division yields an infinite repeating decimal. We use three dots to indicate that the pattern continues forever. For example:

$$\frac{1}{3} = 0.3333\ldots \qquad
\begin{array}{r}
0.333 \\
3\overline{)1.000} \\
9 \\
\hline
10 \\
9 \\
\hline
10 \\
9 \\
\hline
1
\end{array}$$

An alternative notation is to place a bar over the repeating digit(s):

$$0.3333\ldots = 0.\overline{3} \qquad 0.575757\ldots = 0.\overline{57}$$

EXAMPLE 3 Write each fraction as a decimal.   **(a)** $\dfrac{2}{11}$ **(b)** $\dfrac{5}{6}$

(a) $\dfrac{2}{11} = 0.181818\ldots$ or $0.\overline{18}$

$$
\begin{array}{r}
0.1818 \\
11\overline{)2.0000} \\
\underline{11} \\
90 \\
\underline{88} \\
20 \\
\underline{11} \\
90 \\
\underline{88} \\
2
\end{array}
$$

(b) $\dfrac{5}{6} = 0.8333\ldots$ or $0.8\overline{3}$

$$
\begin{array}{r}
0.8333 \\
6\overline{)5.0000} \\
\underline{48} \\
20 \\
\underline{18} \\
20 \\
\underline{18} \\
20 \\
\underline{18} \\
2
\end{array}
$$

Note that the 8 does not repeat. Only the digit 3 is repeating.

Practice Problem 3 Write each fraction as a decimal.

(a) $\dfrac{1}{6}$ **(b)** $\dfrac{5}{11}$

Sometimes division must be carried out to many places in order to observe the repeating pattern. This is true in the following example:

$$\frac{2}{7} = 0.285714285714285714\ldots$$ This can also be written as $\dfrac{2}{7} = 0.\overline{285714}$.

It can be shown that the denominator determines the maximum number of decimal places that might repeat. So $\frac{2}{7}$ must repeat in the seventh decimal place or sooner.

2 *Changing a Decimal to a Fraction*

To convert from a decimal to a fraction, merely write the decimal as a fraction with a denominator of 10, 100, 1000, 10,000, and so on, and simplify the result when possible.

EXAMPLE 4 Write each decimal as a fraction.

(a) 0.2   **(b)** 0.35   **(c)** 0.516   **(d)** 0.74   **(e)** 0.138   **(f)** 0.008

(a) $0.2 = \dfrac{2}{10} = \dfrac{1}{5}$

(b) $0.35 = \dfrac{35}{100} = \dfrac{7}{20}$

(c) $0.516 = \dfrac{516}{1000} = \dfrac{129}{250}$

(d) $0.74 = \dfrac{74}{100} = \dfrac{37}{50}$

(e) $0.138 = \dfrac{138}{1000} = \dfrac{69}{500}$

(f) $0.008 = \dfrac{8}{1000} = \dfrac{1}{125}$

Practice Problem 4 Write each decimal as a fraction and simplify whenever possible.

(a) 0.8   **(b)** 0.88   **(c)** 0.45   **(d)** 0.148   **(e)** 0.612   **(f)** 0.016

All repeating decimals can also be converted to fractional form. In practice, however, repeating decimals are usually rounded to a few places. It will not be necessary, therefore, to learn how to convert $0.\overline{033}$ to $\frac{11}{333}$ for this course.

3 *Adding and Subtracting Decimals*

Last week Bob spent $19.83 on lunches purchased at the cafeteria at work. During this same period, Sally spent $24.76 on lunches. How much did the two of them spend on lunches last week?

Adding or subtracting decimals is similar to adding and subtracting whole numbers, except that it is necessary to line up decimal points. To perform the operation $19.83 + 24.76$, we line up the numbers in column form and add the digits:

$$\begin{array}{r} 19.83 \\ + 24.76 \\ \hline 44.59 \end{array}$$

Thus Bob and Sally spent $44.59 on lunches last week.

Addition and Subtraction of Decimals

1. Write in column form and line up decimal points.
2. Add or subtract the digits.

 **EXAMPLE 5** Perform the following operations.

(a) $3.6 + 2.3$ **(b)** $127.32 - 38.48$

(c) $3.1 + 42.36 + 9.034$ **(d)** $5.0006 - 3.1248$

(a) $\begin{array}{r} 3.6 \\ + 2.3 \\ \hline 5.9 \end{array}$
  (b) $\begin{array}{r} 127.32 \\ - 38.48 \\ \hline 88.84 \end{array}$
  (c) $\begin{array}{r} 3.1 \\ 42.36 \\ + 9.034 \\ \hline 54.494 \end{array}$
  (d) $\begin{array}{r} 5.0006 \\ - 3.1248 \\ \hline 1.8758 \end{array}$

Practice Problem 5   Add or subtract.

(a) $3.12 + 5.08 + 1.42$ **(b)** $152.003 - 136.118$

(c) $1.1 + 3.16 + 5.123$ **(d)** $1.0052 - 0.1234$

Sidelight

When we added fractions, we had to have common denominators. Since decimals are really fractions, why can we add them without having common denominators? Actually, we have to have common denominators to add any fractions, whether they are in decimal form or fraction form. However, sometimes the notation does not show this. Let's examine Example 5(c).

Original Problem   We are adding the three numbers:

$$\begin{array}{r} 3.1 \\ 42.36 \\ + 9.034 \\ \hline 54.494 \end{array}$$

$3\frac{1}{10} + 42\frac{36}{100} + 9\frac{34}{1000}$

$3\frac{100}{1000} + 42\frac{360}{1000} + 9\frac{34}{1000}$

$3.100 + 42.360 + 9.034$ This is the new problem.

New Problem   ***Original Problem***

$$\begin{array}{r} 3.100 \\ 42.360 \\ + 9.034 \\ \hline 54.494 \end{array} \qquad \begin{array}{r} 3.1 \\ 42.36 \\ + 9.034 \\ \hline 54.494 \end{array}$$

We notice that the results are the same. The only difference is the notation. We are using the property that any number of zeros may be added to the right-hand side of a decimal without changing its value.

This shows the convenience of adding and subtracting fractions in decimal form. Little work is needed to change the decimals so that they have a common denominator. All that is required is to add zeros to the right-hand side of the decimal (and we usually do not even write out that step except when subtracting).

As long as we line up the decimal points, we can add or subtract any decimal fractions.

In the following example we will find it useful to add zeros to the right-hand side of the decimal.

EXAMPLE 6 Perform the following operations.

(a) $1.0003 + 0.02 + 3.4$ **(b)** $12 - 0.057$

We will add zeros so that each number shows the same number of decimal places.

(a)
```
   1.0003
   0.0200
 + 3.4000
 ───────
   4.4203
```

(b)
```
   12.000
 −  0.057
 ───────
   11.943
```

Practice Problem 6 Perform the following operations.

(a) $0.061 + 5.0008 + 1.3$ **(b)** $18 - 0.126$

4 Multiplying Decimals

Multiplication of Decimals

To multiply decimals, you first multiply as with whole numbers. To determine the position of the decimal point, you count the total number of decimal places in the two numbers being multiplied. This will determine the number of decimal places that should appear in the answer.

EXAMPLE 7 Multiply. 0.8×0.4

```
        0.8   (one decimal place)
      × 0.4   (one decimal place)
      ─────
       0.32   (two decimal places)
```

Practice Problem 7 Multiply. 0.5×0.3

Note that you will often have to add zeros to the left of the digits obtained in the product so that you obtain the necessary number of decimal places.

EXAMPLE 8 Multiply. 0.123×0.5

```
      0.123   (three decimal places)
   ×    0.5   (one decimal place)
   ────────
     0.0615   (four decimal places)
```

Practice Problem 8 Multiply. 0.12×0.4

Here are some examples that involve more decimal places.

EXAMPLE 9 Multiply. **(a)** 2.56×0.003 **(b)** 0.0036×0.008

(a)
```
      2.56    (two decimal places)
   × 0.003    (three decimal places)
   ────────
   0.00768    (five decimal places)
```

(b)
```
      0.0036    (four decimal places)
    × 0.008    (three decimal places)
   ──────────
   0.0000288    (seven decimal places)
```

Practice Problem 9 Multiply.

(a) 1.23×0.005 **(b)** 0.003×0.00002

Sidelight

Why do we count the number of decimal places? The rule really comes from the properties of fractions. If we write the problem in Example 8 in fraction form, we have

$$0.123 \times 0.5 = \frac{123}{1000} \times \frac{5}{10} = \frac{615}{10{,}000} = 0.0615.$$

5 Dividing Decimals

When discussing division of decimals, we frequently refer to the three primary parts of a division problem. Be sure you know the meaning of each term.

The **divisor** is the number you divide into another.

The **dividend** is the number to be divided.

The **quotient** is the result of dividing one number by another.

In the problem $6 \div 2 = 3$ we represent each of these terms as follows:

When dividing two decimals, count *the number of decimal places* in the divisor. Then *move the decimal point to the right* that *same number of places* in both *the divisor* and *the dividend*. Mark that position with a caret ($_\wedge$). Finally, perform the division. Be sure to line up the decimal point in the quotient with the position indicated by the caret in the dividend.

EXAMPLE 10 Four friends went out for lunch. The total bill, including tax, was $32.68. How much did each person pay if they shared the cost equally?

To answer this question, we must calculate $32.68 \div 4$.

$$
\begin{array}{r}
8.17 \\
4\overline{)32.68} \\
\underline{32} \\
6 \\
\underline{4} \\
28 \\
\underline{28} \\
0
\end{array}
$$

Since there are no decimal places in the divisor, we do not need to move the decimal point. We must be careful, however, to place the decimal point in the quotient directly above the decimal point in the dividend.

Thus $32.68 \div 4 = 8.17$, and each friend paid $8.17.

Practice Problem 10 Divide. $0.1116 \div 5$

Note that sometimes we will need to place extra zeros in the dividend in order to move the decimal point the required number of places.

EXAMPLE 11 Divide. $16.2 \div 0.027$

$$0.027\overline{)16.200}$$

There are **three** decimal places in the divisor, so we move the decimal point **three places** to **the right** in the **divisor** and **dividend** and mark the new position by a caret. Note that we must add two zeros to 16.2 in order to do this.

three decimal places

$$
\begin{array}{r}
600. \\
0.027\overline{)16.200} \\
\underline{16\ 2} \\
000
\end{array}
$$

Now perform the division as with whole numbers. The decimal point in the answer is directly above the caret in the dividend.

Thus $16.2 \div 0.027 = 600$.

Practice Problem 11 Divide. $1800 \div 0.06$

Special care must be taken to line up the digits in the quotient. Note that sometimes we will need to place zeros in the quotient after the decimal point.

EXAMPLE 12 Divide. $0.04288 \div 3.2$

$$3.2\overline{)0.0\,4288}$$

There is **one** decimal place in the divisor, so we move the decimal point **one place** to **the right** in the **divisor** and **dividend** and mark the new position by a caret.

one decimal place

$$
\begin{array}{r}
0.0134 \\
3.2\overline{)0.0\,4288} \\
\underline{32} \\
108 \\
\underline{96} \\
128 \\
\underline{128} \\
0
\end{array}
$$

Now perform the division as for whole numbers. The decimal point in the answer is directly above the caret in the dividend. Note the need for the initial zero after the decimal point in the answer.

Thus $0.04288 \div 3.2 = 0.0134$.

Practice Problem 12 Divide. $0.01764 \div 4.9$

Sidelight

Why does this method of dividing decimals work? Essentially, we are using the steps we used in Section 0.1 to change a fraction to an equivalent fraction by multiplying both the numerator and denominator by the same number. Let's reexamine Example 12.

$$0.04288 \div 3.2 = \frac{0.04288}{3.2}$$

Write the original problem using fraction notation.

$$= \frac{0.04288 \times 10}{3.2 \quad \times 10}$$

Multiply the numerator and denominator by 10. Since this is the same as multiplying by 1, we are not changing the fraction.

$$= \frac{0.4288}{32}$$

Write the result of multiplication by 10.

$$= 0.4288 \div 32$$

Rewrite the fraction as an equivalent problem with division notation.

Notice that we have obtained a new problem that is the same as the problem in Example 12 when we moved the decimal one place to the right in the divisor and dividend. We see that the reason we can move the decimal point as many places as necessary to the right in divisor and dividend is that this is the same as multiplying the numerator and denominator of a fraction by a power of 10 to obtain an equivalent fraction.

6 *Multiplying and Dividing a Number by a Multiple of 10*

When multiplying by 10, 100, 1000, and so on, a simple rule may be used to obtain the answer. For every zero in the multiplier, move the decimal point one place to the right.

EXAMPLE 13   Multiply.   **(a)** 3.24×10 **(b)** 15.6×100 **(c)** 0.0026×1000

(a) $3.24 \times 10 = 32.4$ One zero—move decimal point one place to the right.

(b) $15.6 \times 100 = 1560$ Two zeros—move decimal point two places to the right.

(c) $0.0026 \times 1000 = 2.6$ Three zeros—move decimal point three places to the right.

Practice Problem 13   Multiply.

(a) 0.0016×100 **(b)** 2.34×1000 **(c)** $56.75 \times 10,000$

The reverse rule is true for division. When dividing by 10, 100, 1000, 10,000, and so on, move the decimal point one place to the left for every zero in the divisor.

EXAMPLE 14   Divide. **(a)** $52.6 \div 10$ **(b)** $0.0038 \div 100$ **(c)** $5936.2 \div 1000$

(a) $\dfrac{52.6}{10} = 5.26$ Move one place to the left.

(b) $\dfrac{0.0038}{100} = 0.000038$ Move two places to the left.

(c) $\dfrac{5936.2}{1000} = 5.9362$ Move three places to the left.

Practice Problem 14   Divide.

(a) $\dfrac{5.82}{10}$ **(b)** $123.4 \div 1000$ **(c)** $\dfrac{0.00614}{10,000}$

Verbal and Writing Skills

1. A decimal is another way of writing a fraction whose denominator is _____ .

2. We write 0.42 in words as _____ .

3. When dividing 7432.9 by 1000 we move the decimal point _____ places to the _____ .

Write each fraction as a decimal.

4. $\dfrac{15}{16}$　　**5.** $\dfrac{3}{20}$　　**6.** $\dfrac{7}{11}$　　**7.** $\dfrac{2}{3}$　　**8.** $\dfrac{3}{15}$　　**9.** $\dfrac{12}{15}$

Write each decimal as a fraction in simplified form.

10. 0.8　　**11.** 0.15　　**12.** 0.625　　**13.** 0.08　　**14.** 0.75　　**15.** 1.125

Add or subtract.

16. $1.0076 - 0.0982$　　**17.** $0.00381 - 0.00228$　　**18.** $5.3 + 6.8 + 3.78$　　**19.** $3.6 + 1.28 + 4.5$

20. $46.03 + 215.1 + 0.078$　　**21.** $33.01 + 0.38 + 175.401$　　**22.** $147.18 - 15.39$　　**23.** $131.43 - 86.95$

Multiply or divide.

24. 7.21×4.2　　**25.** 7.12×2.6　　**26.** 5.26×0.0015　　**27.** 2.18×0.013

28. $169{,}000 \times 0.0013$　　**29.** $368{,}000 \times 0.00021$　　**30.** $7.9728 \div 3.02$　　**31.** $6.519 \div 2.05$

32. $0.5230 \div 0.002$　　**33.** $0.031 \div 0.005$　　**34.** $186.16 \div 5.2$　　**35.** $46.62 \div 7.4$

Multiply or divide by moving the decimal point.

36. 1.36×1000　　**37.** $0.76 \div 100$　　**38.** $175{,}318 \div 1000$　　**39.** 3.45×1000

40. $3.52 \div 1000$　　**41.** $0.00243 \times 100{,}000$　　**42.** $7.36 \times 10{,}000$　　**43.** $73{,}892 \div 100{,}000$

Mixed Practice

Perform the indicated calculations.

44. 23.75×0.06　　**45.** 1.824×0.004　　**46.** $1.62 + 2.005 + 8.1007$　　**47.** $59 - 6.27$

48. $0.05724 \div 0.027$　　**49.** $77.136 \div 0.003$　　**50.** 0.7683×1000　　**51.** $34.72 \div 10{,}000$

Applications

52. Fred's car usually gets 32.5 miles per gallon when driven on the highway. The gas tank holds 14.4 gallons. How far could the car travel on the highway on one full tank of gas?

53. Melanie bought a compact car because of her limited budget. The compact averages 28.5 miles on the highway. The gas tank has a capacity of 23.4 gallons. How far can her car go on a full tank of gas?

54. Gillian decided that she would need 18.5 yards of material to make curtains. If the material she chose cost $11.50 per yard, what did the material for her curtains cost?

55. The Parkins family is going to order a turkey for Thanksgiving dinner from Ray's Turkey Farm. Mrs. Parkins estimates that she will need 12 pounds of meat to serve all her guests. The turkey farm owner claims that each turkey yields about 0.78 of its weight in meat. If Mrs. Parkins buys a 17.3-pound turkey, will there be enough meat for everyone?

56. The EPA standard for safe drinking water is a maximum of 1.3 milligrams of copper per liter of water. A water testing firm found 6.8 milligrams of copper in a 5-liter sample drawn from Jim and Sharon LeBlanc's house. Is the water safe or not? By how much does the amount of copper exceed or fall short of the maximum allowed?

57. Harry has a part-time job at Stop and Shop. He earns $8.50 an hour. Last week he worked 19 hours. He had hoped to earn at least $150. Did he reach his goal or not? By how much did he exceed or fall short of his goal?

Cumulative Review Problems

Perform each operation. Simplify all answers.

58. $3\frac{1}{2} \div 5\frac{1}{4}$

59. $\frac{3}{8} \cdot \frac{12}{27}$

60. $\frac{12}{25} + \frac{9}{20}$

61. $1\frac{3}{5} - \frac{1}{2}$

Developing Your Study Skills

Making a Friend in the Class

Attempt to make a friend in your class. You may find that you enjoy sitting together and drawing support and encouragement from one another. Exchange phone numbers so you can call each other whenever you get stuck in your study. Set up convenient times to study together on a regular basis, to do homework, and to review for exams.

You must not depend on a friend or fellow student to tutor you, do your work for you, or in any way be responsible for your learning. However, you will learn from one another as you seek to master the course. Studying with a friend and comparing notes, methods, and solutions can be very helpful. And it can make learning mathematics a lot more fun!

Putting Your Skills to Work

The Mathematics of Major World Languages

You probably know that English is the third most commonly spoken native language in the world. However, do you know that Chinese Mandarin is the most commonly spoken language? Are you aware that more people speak Bengali than Portuguese? Use the bar graph below to answer the following questions.

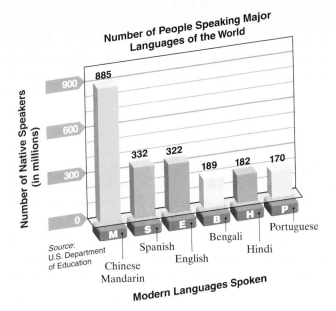

Number of People Speaking Major Languages of the World

Number of Native Speakers (in millions)

- M (Chinese Mandarin): 885
- S (Spanish): 332
- E (English): 322
- B (Bengali): 189
- H (Hindi): 182
- P (Portuguese): 170

Source: U.S. Department of Education

Modern Languages Spoken

Problems for Individual Study and Investigation

1. How many more people in the world speak Chinese Mandarin than English?

2. How many more people in the world speak Spanish than English?

Problems for Group Study and Investigation

Together with some other members of your class, see if you can determine the answers to the following questions. Round your answers to the nearest tenth.

3. If there were approximately 5,996,000,000 people in the world in 1999, what percent of the population of the world spoke English?
What percent of the population of the world spoke Chinese Mandarin?

4. In 1999, 562,800 college students in the United States took modern language courses in Spanish and 36,520 college students took modern language courses in Chinese Mandarin. How does this relate to the number of people in the world who speak these languages? What can you conclude?

Internet Connections

Netsite: http://www.prenhall.com/tobey_beginning

Site: Language Use Data (U.S. Census Bureau) or a related site

This site includes a table showing languages spoken at home by people age 5 and older, by state. Use the table to answer the following questions about the state where you live.

5. How many more people speak Greek than Yiddish in your state?

6. What percent of your state's population speaks only English at home?

7. What percent of your state's population speaks Italian at home?

8. Other than English, what is the most commonly spoken language in your state? What percent of your state's population speaks this language at home?

0.5 Use of Percent

The Basic Concept of Percents

A **percent** is a fraction that has a denominator of 100. When you say "sixty-seven percent" or write 67%, you are just expressing the fraction $\frac{67}{100}$ in another way. The word *percent* is a shortened form of the Latin words *per centum,* which means "by the hundred." In everyday use, percent means per one hundred.

Russell Camp owns 100 acres of land in Montana. 49 of the acres are covered with trees. The rest of the land is open fields. We say that 49% of his land is covered with trees.

It is important to see that 49% means 49 parts out of 100 parts. It can also be written as a fraction, $\frac{49}{100}$, or as a decimal 0.49. Understanding the meaning of the notation allows you to change from one form to another. For example,

$$49\% = 49 \text{ out of } 100 \text{ parts} = \frac{49}{100} = 0.49.$$

Similarly, you can express a fraction with denominator 100 as a percent or a decimal.

$\frac{11}{100}$ means 11 parts out of 100 or 11%.   So $\frac{11}{100} = 11\% = 0.11.$

1 Changing a Decimal to a Percent

Now that we understand the concept, we can use some quick procedures to change from decimals to percent, and vice versa.

Changing a Decimal to Percent

1. Move the decimal point two places to the right.
2. Add the % symbol.

EXAMPLE 1 Change to a percent.   **(a)** 0.23   **(b)** 0.461   **(c)** 0.4

We move the decimal point two places to the right and add the % symbol.

(a) $0.23 = 23\%$ **(b)** $0.461 = 46.1\%$ **(c)** $0.4 = 0.40 = 40\%$

Practice Problem 1 Change to a percent.   **(a)** 0.92   **(b)** 0.418   **(c)** 0.7

Be sure to follow the same procedure for percents that are less than 1%. Remember, 0.01 is 1%. Thus we would expect 0.001 to be less than 1%. $0.001 = 0.1\%$ or one-tenth (0.1) of a percent.

EXAMPLE 2 Change to a percent.   **(a)** 0.0364   **(b)** 0.0026   **(c)** 0.0008

We move the decimal point two places to the right and add the % symbol.

(a) $0.0364 = 3.64\%$ **(b)** $0.0026 = 0.26\%$ **(c)** $0.0008 = 0.08\%$

Practice Problem 2 Change to a percent.

(a) 0.0019   **(b)** 0.0736   **(c)** 0.0003

Be sure to follow the same procedure for percents that are greater than 100%. Remember that 1 is 100%. Thus we would expect 1.5 to be greater than 100%. In fact, $1.5 = 150\%$.

Calculator

Percent to Decimal
You can use a calculator to change 52% to a decimal. If your calculator has a $\boxed{\%}$ key, do the following:
Enter: 52 $\boxed{\%}$
The display should read:

$$\boxed{0.52}$$

If your calculator does not have a $\boxed{\%}$ key, divide the number by 100.
Enter:

$$52 \boxed{\div} 100 \boxed{=}$$

Try the following:
(a) 46%   **(b)** 137%
(c) 9.3%   **(d)** 6%

Note: The calculator divides by 100 when the percent key is pressed. If you do not have a $\boxed{\%}$ key, then you can divide by 100 instead.

⬮ **EXAMPLE 3**   Change to a percent.   **(a)** 1.48   **(b)** 2.938   **(c)** 4.5

We move the decimal point two places to the right and add the percent symbol.

(a) $1.48 = 148\%$ **(b)** $2.938 = 293.8\%$ **(c)** $4.5 = 4.50 = 450\%$

Practice Problem 3   Change to a percent.

(a) 3.04   **(b)** 5.186   **(c)** 2.1

②   *Changing a Percent to a Decimal*

In this procedure we move the decimal point to the left and remove the % symbol.

Changing a Percent to a Decimal

1. Move the decimal point two places to the left.
2. Remove the % symbol.

⬮ **EXAMPLE 4**   Change to a decimal.   **(a)** 4%   **(b)** 3.2%   **(c)** 0.6%

First we move the decimal point two places to the left. Then we remove the % symbol.

(a) $4\% = 4.\% = 0.04$
         ↑

The unwritten decimal point is understood to be here.

(b) $3.2\% = 0.032$ **(c)** $0.6\% = 0.006$

Practice Problem 4   Change to a decimal.   **(a)** 7%   **(b)** 9.3%   **(c)** 0.2%

⬮ **EXAMPLE 5**   Change to a decimal.   **(a)** 192%   **(b)** 254.8%   **(c)** 0.027%

First we move the decimal point two places to the left. Then we remove the % symbol.

(a) $192\% = 192.\% = 1.92$
           ↑

The unwritten decimal point is understood to be here.

(b) $254.8\% = 2.548$ **(c)** $0.027\% = 0.00027$

Practice Problem 5   Change to a decimal.

(a) 131%   **(b)** 301.6%   **(c)** 0.04%

③   *Finding the Percent of a Number*

How do we find 60% of 20? Let us relate it to a problem we did in Section 0.2. Consider the following problem.

$$\begin{array}{ccccc} \text{What} & \text{is} & \dfrac{3}{5} & \text{of} & 20? \\ \downarrow & \downarrow & \downarrow & \downarrow & \downarrow \\ \boxed{?} & = & \dfrac{3}{5} & \times & 20 \end{array}$$

$$\boxed{?} = \frac{3}{\cancel{5}} \times \cancel{20}^{4} = 12 \quad \text{The answer is 12.}$$

Since a percent is really a fraction, a percent problem is solved similarly to the way a fraction problem is solved. Since $\frac{3}{5} = \frac{6}{10} = 60\%$, we could write the problem as:

What   is   60%   of   20?
$\downarrow$ $\downarrow$ $\downarrow$ $\downarrow$ $\downarrow$

$\boxed{?}$ $=$ 60% $\times$ 20

$\boxed{?}$ $= 0.60 \times 20$

$\boxed{?}$ $= 12.0$ The answer is 12.

Thus we have developed the following rule.

Finding the Percent of a Number

To find the percent of a number, change the percent to a decimal and multiply the number by the decimal.

⬤ EXAMPLE 6   Find.

(a) 10% of 36   **(b)** 2% of 350   **(c)** 182% of 12   **(d)** 0.3% of 42

(a) 10% of 36 = 0.10 × 36 = 3.6   **(b)** 2% of 350 = 0.02 × 350 = 7
(c) 182% of 12 = 1.82 × 12 = 21.84   **(d)** 0.3% of 42 = 0.003 × 42 = 0.126

Practice Problem 6   Find.

(a) 18% of 50   **(b)** 4% of 64   **(c)** 156% of 35
(d) 0.8% of 60   **(e)** 1.3% of 82   **(f)** 0.002% of 564   ⬤

There are many real-life applications for finding the percent of a number. When you go shopping in a store, you may find sale merchandise marked 35% off. This means that the sale price is 35% off the regular price. That is, 35% of the regular price is subtracted from the regular price to get the sale price.

⬤ EXAMPLE 7   A store is having a sale of 35% off the retail price of all sofas. Melissa wants to buy a particular sofa that normally sells for $595.

(a) How much will Melissa save if she buys the sofa on sale?

(b) What will the purchase price be if Melissa buys the sofa on sale?

(a) To find 35% of $595 we will need to multiply 0.35 × 595.

$$\begin{array}{r} \$595 \\ \times\ \ 0.35 \\ \hline 2975 \\ 1785\ \ \ \\ \hline \$208.25 \end{array}$$

Thus Melissa will save $208.25 if she buys the sofa on sale.

(b) The purchase price is the difference between the original price and the amount saved.

$$\begin{array}{r} \$595.00 \\ -\ \ 208.25 \\ \hline \$386.75 \end{array}$$

If Melissa buys the sofa on sale, she will pay $386.75.

Practice Problem 7   John received a 4.2% pay raise at work this year. He had previously earned $18,000 per year.

(a) What was the amount of his pay raise in dollars?

(b) What is his new salary?   ⬤

Calculator

Using Percents
You can use a calculator to find 12% of 48.
Enter:

12 $\boxed{\%}$ $\boxed{\times}$ 48 $\boxed{=}$

The display should read:

$\boxed{\qquad 5.76}$

If your calculator does not have a $\boxed{\%}$ key, do the following:
Enter: 0.12 $\boxed{\times}$ 48 $\boxed{=}$
What is 54% of 450?

4 *Finding the Missing Percent When Given Two Numbers*

Recall that we can write $\frac{3}{4}$ as $\frac{75}{100}$ or 75%. If we were asked the question, "What percent is 3 of 4?" we would say 75%. This gives us a procedure for finding what percent one number is of a second number.

Finding the Missing Percent

1. Write a fraction with the two numbers. The number *after* the word "of" is always the denominator, and the other number is the numerator.
2. Simplify the fraction (if possible).
3. Change the fraction to a decimal.
4. Express the decimal as a percent.

EXAMPLE 8   What percent of 24 is 15?

This can be solved quickly as follows.

Step 1 $\dfrac{15}{24}$ Write the relationship as a fraction. The number after "of" is 24, so 24 is the denominator.

Step 2 $\dfrac{15}{24} = \dfrac{5}{8}$ Simplify the fraction (when possible).

Step 3 $= 0.625$ Change the fraction to a decimal.

Step 4 $= 62.5\%$ Change the decimal to percent.

Practice Problem 8   What percent of 148 is 37?

The question in Example 8 can also be written as "15 is what percent of 24?" To answer the question, we begin by writing the relationship as $\frac{15}{24}$. Remember that "of 24" means 24 will be the denominator.

EXAMPLE 9   **(a)** 82 is what percent of 200?   **(b)** What percent of 16 is 3.8?
(c) $150 is what percent of $120?

(a) 82 is what percent of 200?

$\dfrac{82}{200}$ Write the relationship as a fraction with 200 in the denominator.

$$\frac{82}{200} = \frac{41}{100} = 0.41 = 41\%$$

(b) What percent of 16 is 3.8?

$\dfrac{3.8}{16}$ Write the relationship as a fraction.

You can divide to change the fraction to a decimal and then change the decimal to a percent.

$$\begin{array}{r} 0.2375 \rightarrow 23.75\% \\ 16\overline{)3.8000} \end{array}$$

(c) $150 is what percent of $120?

$\dfrac{150}{120} = \dfrac{5}{4}$ Reduce the fraction whenever possible to make the division easier.

$$\begin{array}{r} 1.25 \rightarrow 125\% \\ 4\overline{)5.00} \end{array}$$

Practice Problem 9

(a) What percent of 48 is 24?

(b) 4 is what percent of 25?

EXAMPLE 10   Marcia made 29 shots on goal during the last high school field hockey season. She actually scored a goal eight times. What percent of her total shots were goals? Round your answer to the nearest whole percent.

   Marcia scored a goal eight times out of 29 tries. We want to know what percent of 29 is 8.

Step 1 $\dfrac{8}{29}$ Express the relationship as a fraction. The number after the word "of" is 29, so 29 appears in the denominator.

Step 2 $\dfrac{8}{29} = \dfrac{8}{29}$ Note that this fraction cannot be reduced.

Step 3 $= 0.2758\ldots$ The decimal equivalent of the fraction has many digits.

Step 4 $= 27.58\ldots\%$ We change the decimal to a percent, which we round to the nearest whole percent.

   $\approx 28\%$

Therefore, Marcia scored a goal approximately 28% of the time she made a shot on goal.

Practice Problem 10

Roberto scored a basket 430 times out of 1256 attempts during his high school basketball career. What percent of the time did he score a basket? Round your answer to the nearest whole percent.

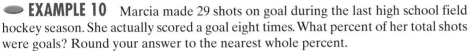

Calculator

Using Percents
You can use a calculator to find a missing percent. What percent of 95 is 19?

1. Enter as a fraction.

   19 $\div$ 95

2. Change to a percent.

   19 $\div$ 95 $\times$ 100 $=$

The display should read:

   20

This means 20%.
What percent of 625 is 250?

Verbal and Writing Skills

1. When you write 19%, what do you really mean? Describe the meaning in your own words.

2. When you try to solve a problem like "What percent of 80 is 30?" how do you know if you should write the fraction as $\frac{80}{30}$ or as $\frac{30}{80}$?

Change to a percent.

3. 0.568   **4.** 0.089   **5.** 0.409   **6.** 0.008   **7.** 2.39   **8.** 7.4

Change to a decimal.

9. 3%   **10.** 66%   **11.** 0.4%   **12.** 0.62%   **13.** 250%   **14.** 175%

Find the following.

15. What is 35.8% of 1000?

16. What is 7.7% of 450?

17. What is 0.8% of 65?

18. What is 7% of 69?

19. What is 112% of 65?

20. What is 154% of 270?

21. 36 is what percent of 24?

22. 49 is what percent of 28?

23. What percent of 340 is 17?

24. 48 is what percent of 600?

25. What percent of 16 is 22?

26. 45 is what percent of 18?

Applications

27. Dave took an exam with 80 questions. He had 12 wrong. What was his grade for the exam? Write the grade as a percent.

28. At one of the Lollapalooza concerts, there were 30,000 people in the crowd. 47% of the audience were women. How many women were at the concert?

29. Diana and Russ ate a meal costing $32.80 when they went out to dinner. If they want to leave the standard 15% tip for their server, how much will they tip and what will their total bill be?

30. In a local college survey, it was discovered that 137 out of 180 students had a grandparent not born in the United States. What percent of the students had a grandparent not born in the United States? (Round your answer to the nearest hundredth of a percent.)

31. The Gonzalez family has a combined monthly income of $1850. Their food budget is $380/month. What percentage of their monthly income is budgeted for food? (Round your answer to the nearest whole percent.)

32. Music CDs have a failure rate of 1.8%. (They skip or get stuck on a song.) If 36,000 CDs were manufactured last week, how many of them were defective?

33. Last Christmas season, the Jones Mill Outlet Store Chain calculated that they sold 36,000 gift items. If they assume that there will be a 1.5% return rate after the holidays, how many gifts can they expect to be exchanged?

34. The total cost of a downtown Boston apartment, including rent, heat, electricity, and water is $3690 per month. Sarah is renting this apartment with four friends. Based on a complicated formula, which takes into account the size of her bedroom and the furniture she brought to the apartment, she must pay 17% of the apartment cost each month. What is her monthly cost?

35. Eastern Shores charter airline had 2390 flights last year. Of these flights, 560 did not land on time. What percent of the flights landed on time? Round your answer to the nearest whole percent.

36. Jim is earning $12.50 per hour as a cook. He has earned an 8% pay raise this year. What will his raise be, and what will his new hourly rate be?

37. Bruce sells medical supplies on the road. He logged 18,600 miles in his car last year. He declared 65% of his mileage as business travel. (a) How many miles did he travel on business last year? (b) If his company reimburses him 31 cents for each mile traveled, how much should he be paid for travel expenses?

38. Abdul sells computers for a local computer outlet. He gets paid $450 per month plus a commission of 3.8% on all the computer hardware he sells. Last year he sold $780,000 worth of computer hardware. (a) What was his sales commission on the $780,000 worth of hardware he sold? (b) What was his annual salary (that is, his monthly pay and commission combined)?

To Think About

39. Fred and Nancy Adams go out once a week to a restaurant where their daughter is the chef. They are given a 33% discount on the cost of the meal. They always leave a tip of 20%. Fred thinks the tip should be based on the cost of the meal before the discount. Nancy thinks the tip should be based on the cost of the meal after the discount. If they normally purchase a meal each week that costs $45 for the two of them before the discount and use Nancy's method of calculating the tip, by how much will the tip increase if they switch to Fred's method?

40. Dave Bagley traveled 24,500 miles last year. He is a salesperson and 74% of his mileage is for business purposes. He was planning to deduct 31 cents per business mile on his income tax return. However, his accountant told him he can deduct 35 cents per mile. By how much will his deduction increase if he uses the new larger amount?

Cumulative Review Problems

41. Janeen had an opening balance of $45.50 in her checking account this month. She deposited her paycheck of $1189 and a refund check of $33.90. She was charged by the bank $1.50 for ATM fees. Then she made out checks for $98.00, $128.00, $56.89, and $445.88. What will be her final balance at the end of the month?

42. Tally is buying a car and has a bank loan. He has to make 33 more monthly payments of $188.50. How much money will he pay to the bank over the next 33 months?

43. Dan took a trip in his Ford Taurus. At the start of the trip his car odometer read 68,459.5 miles. At the end of the trip his car odometer read 69,229.5 miles. He used 35 gallons of gas on the trip. How many miles per gallon did his car achieve?

44. In Hilo, Hawaii, Brad spent three months working in a restaurant for a summer job. In June he observed that 4.6 inches of rain fell. In July the rainfall was 4.5 inches and in August it was 2.9 inches. What was the average monthly rainfall that summer for those three months in Hilo?

Student Learning Objectives

After studying this section, you will be able to:

 Use rounding to estimate.

SSM
PH TUTOR CENTER   CD & VIDEO   MATH PRO   WEB

1 Using Rounding to Estimate

As we begin this section, we will take some time to be sure you understand the idea of rounding a number. You will probably recall the following simple rule from your previous mathematics courses.

Rounding a Number

If the first digit to the right of the round-off place is:

1. less than 5, we make no change to the digit in the round-off place.
2. 5 or more, we increase the digit in the round-off place by 1.

To illustrate, 4689 rounded to the nearest hundred is 4700. Rounding 233,987 to the nearest ten thousand, we obtain 230,000. We will now use our experience in rounding as we discuss the general area of estimation.

Estimation is the process of finding an approximate answer. It is not designed to provide an exact answer. Estimation will give you a rough idea of what the answer might be. For any given problem, you may choose to estimate in many different ways.

Estimation by Rounding

1. Round each number to an appropriate place.
2. Perform the calculation with the rounded numbers.

EXAMPLE 1 Find an estimate of the product 5368 × 2864.

Step 1 Round 5368 to 5000.
Round 2864 to 3000.

Step 2 Multiply.

$$5000 \times 3000 = 15,000,000$$

An estimate of the product is 15,000,000.

Practice Problem 1 Find an estimate of the product 128,621 × 378.

EXAMPLE 2 The four walls of a college classroom are $22\frac{1}{4}$ feet long and $8\frac{3}{4}$ feet high. A painter needs to know the area of these four walls in square feet. Since paints is sold in gallons, an estimate will do. Estimate the area of the four walls.

Step 1 Round $22\frac{1}{4}$ feet to 20 feet.
Round $8\frac{3}{4}$ feet to 9 feet.

Step 2 Multiply 20 × 9 to obtain an estimate of the area of one wall.
Multiply 20 × 9 × 4 to obtain an estimate of the area of all four walls.

$$20 \times 9 \times 4 = 720 \text{ square feet}$$

Our estimate for the painter is 720 square feet of wall space.

Practice Problem 2 Mr. and Mrs. Ramirez need to carpet two rooms of their house. One room measures $12\frac{1}{2}$ feet by $9\frac{3}{4}$ feet. The other room measures $14\frac{1}{4}$ feet by $18\frac{1}{2}$ feet. Estimate the number of square feet (square footage) in these two rooms.

EXAMPLE 3 Won Lin has a small compact car. He drove 396.8 miles in his car and used 8.4 gallons of gas.

(a) Estimate the number of miles he gets per gallon.

(b) Estimate how much it will cost him for fuel to drive on a cross-country trip of 2764 miles if gasoline usually costs 1.59\frac{9}{10}$ per gallon.

(a) Round 396.8 miles to 400 miles. Round 8.4 gallons to 8 gallons. Now divide.

$$\frac{50}{8\overline{)400}}$$

Won Lin's car gets about 50 miles per gallon.

(b) We will need to use the information we found in part (a) to determine how many gallons of gasoline Won Lin will use on his trip. Round 2764 miles to 3000 miles and divide 3000 miles by 50 gallons.

$$\frac{60}{50\overline{)3000}}$$

Won Lin will use about 60 gallons of gas for his cross-country trip.

To estimate the cost, we need to ask ourselves, "What kind of an estimate are we looking for?" It may be sufficient to round $\$1.59\frac{9}{10}$ to $\$2.00$ and multiply.

$$60 \times \$2.00 = \$120.00$$

Keep in mind that this is a broad estimate. You may want an estimate that will be closer to the exact answer. In that case round $\$1.59\frac{9}{10}$ to $\$1.60$ and multiply.

$$60 \times \$1.60 = \$96.00$$

⊘ **CAUTION** An estimate is only a rough guess. If we are estimating the cost of something, we may want to round the unit cost to a value that is closer to the actual amount so that our estimate is more accurate. When we estimate costs, it is a good idea to round to a value above the actual unit cost to make sure that we will have enough money for the expenditure. The actual cost to the nearest penny of the cross-country trip in Example 3 is $93.56.

Practice Problem 3 Roberta drove 422.8 miles in her truck and used 19.3 gallons of gas. Assume that gasoline costs $\$1.69\frac{9}{10}$ per gallon.

(a) Estimate the number of miles she gets per gallon.

(b) Estimate how much it will cost her to drive to Chicago and back, a distance of 3862 miles.

Other words that indicate estimation in word problems are *about* and *approximate*.

EXAMPLE 4 A local manufacturing company releases 1684 pounds of sulfur dioxide per day from its plant. The company operates 294 days per year and has been operating at this level for 32 years. Approximately how many pounds of sulfur dioxide have been released into the air by this company over the last 32 years?

The word *approximately* means we are looking for an estimate. We begin by rounding the given data so that there is only one nonzero digit in each number.

Round 1684 tons to 2000 tons.

Round 294 days to 300 days.

Round 32 years to 30 years.

Now determine the calculation you would perform to find the answer and calculate using the rounded numbers. We multiply

amount released in one day × the number of days in an operating year × years.

$$2000 \times 300 \times 30 = 18,000,000$$

Thus 18,000,000 tons is an approximation of the sulfur dioxide released by the company during its 32 years of operation.

Practice Problem 4 A space probe is sent from Earth at 43,300 miles per hour toward the planet Pluto, which has an average distance from Earth of 3,580,000,000 miles.

(a) About how many hours will it take for the space probe to travel from Earth to Pluto?

(b) About how many days will it take for the space probe to travel from Earth to Pluto?

EXAMPLE 5 Find an estimate for 2.68% of $54,361.92.

Begin by writing 2.68% as a decimal.

$$2.68\% = 0.0268$$

Next round 0.0268 to the nearest hundredth.

0.0268 is 0.03 rounded to the nearest hundredth.

Finally, round $54,361.92 to $50,000.
Now multiply.

$0.03 \times 50,000 = 1500$ Remember that to find a percent of a number you multiply.

Thus we estimate the answer to be $1500.

To get a better estimate of 2.68% of $54,361.92, you may wish to round $54,361.92 to the nearest thousand. Try it and compare this estimate to the previous one. Remember, an estimate is only an approximation. How close you should try to get to the exact answer depends on what you need the estimate for.

Practice Problem 5 Find an estimate for 56.93% of $293,567.12.

In exercises 1–10, follow the principles of estimation to find an approximate value. Do not find the exact value. There are many acceptable answers, depending on your rounding method.

1. 693×307

2. 437×892

3. $28,362 \times 5986$

4. $386,215 \times 29$

5. $14 + 73 + 80 + 21 + 56$

6. $318 + 494 + 613 + 243$

7. $23 \overline{)578,962}$

8. $82 \overline{)4,320,760}$

9. $\dfrac{0.002714}{0.0315}$

10. $\dfrac{0.5361}{0.00786}$

11. Find 17% of $21,365.85.

12. Find 4.9% of $9321.88.

Applications

In exercises 13–22, determine an estimate of the exact answer. Use estimation by rounding. Do not find the exact value.

▲ **13.** Estimate the floor area of a regulation NBA basketball court measuring 94 feet long and 50 feet wide.

▲ **14.** Estimate the floor area of a ballet studio that measures $36\frac{3}{4}$ feet long and $24\frac{1}{8}$ feet wide.

15. A typical customer at the local Piggly Wiggly supermarket spends approximately $82 at the checkout register. The store keeps four registers open, and each handles 22 customers per hour. Estimate the amount of money the store receives in one hour.

16. The Westerly Community Credit Union in Rhode Island has found that on Fridays, the average customer withdraws $85 from the ATM for weekend spending money. Each ATM averages 19 customers per hour. There are five machines. Estimate the amount of money withdrawn by customers in one hour.

17. Rod's trip from Salt Lake City to his cabin in the mountains was 117.7 miles. If his car used 3.8 gallons of gas for the trip, estimate the number of miles his car gets to the gallon.

18. The Paltrows are transferring their stock of books from one store to another. To save money, they are moving the books themselves. If their car can transport 430 books per trip, and they have 11,900 books to move, estimate the number of trips it will take.

19. Riva took a holiday job at a kiosk in the mall. She was paid strictly on a commission basis, making 4% of the value of her sales. If her total sales for the week were $12,051.27, estimate her salary for the week.

20. The local McDonald's registered $437,690 in sales last month. They know from experience that 15% of their sales are Happy Meals. Estimate the amount of money spent on Happy Meals last month.

21. The governor of a Midwestern state estimates that only 8.3% of her budget of $19,364,282,153.18 goes toward education. Estimate the amount of money spent on education.

22. A recent survey of local hospitals by a state medical board estimates that 6.23% of all operations performed in hospitals are unnecessary. If 28,364,122 operations were performed last year, estimate the number of unnecessary operations.

To Think About

23. Laura's boyfriend, James, was transferred to San Diego. He averages 43 long-distance calls to her per month. The average cost of his long-distance calls is $3.24 per call. A new long-distance telephone company has promised James a 23% reduction in the cost of his phone bills. If that is true, estimate how much he will save in one year.

24. Eddie usually drives his car to his job in Denver. His car costs 8 cents per mile to drive. He commutes 22 miles round-trip every day, and drives to work 280 days per year. He can buy an unlimited-use bus pass for one year for $300. Estimate how much he would save by taking the bus to work.

Cumulative Review Problems

25. What is 0.6% of 350?

26. If Samuel plays basketball and scores 9 baskets out of 16 throws, what percent of the time does he score a basket?

27. Michael had breakfast at the Agawam Diner. The cost of his breakfast listed on the menu was $5.85. If he leaves a 15% tip, what will be the total amount he paid for his meal?

28. In a recent tire inspection for Ford Explorers, local dealers found that 12% of the treads were defective. For each Explorer that was defective, the dealer offered immediate replacement of all five tires (including the spare). If the local dealers inspected the tires on 450 Explorers, how many new tires did they need?

Developing Your Study Skills

Reading the Textbook

Your homework time each day should begin with the careful reading of the section(s) assigned in your textbook. Usually, much time and effort have gone into the selection of a particular text, and your instructor has decided that this is the book that will help you to become successful in this mathematics class. Textbooks are expensive, but they can be a wise investment if you take advantage of them by reading them.

Reading a mathematics textbook is unlike reading many other types of books that you may find in your literature, history, psychology, or sociology courses. Mathematics texts are technical books that provide you with exercises for practice. Reading a mathematics text requires slow and careful reading of each word, which takes time and effort.

Begin reading your textbook with a paper and pencil in hand. As you come across a new definition or concept, underline it in the text and/or write it down in your notebook. Whenever you encounter an unfamiliar term, look it up and make a note of it. When you come to an example, work through it step-by-step. Be sure to read each word and to follow directions carefully.

Notice the helpful hints that the author provides. They guide you to correct solutions and prevent you from making errors. Take advantage of these pieces of expert advice.

Be sure that you understand what you are reading. Make a note of any of those things that you do not understand and ask your instructor about them. Do not hurry through the material. Learning mathematics takes time.

0.7 Mathematics Blueprint for Problem Solving

1 Using the Mathematics Blueprint to Solve Real-life Problems

Student Learning Objectives

After studying this section, you will be able to:

1 Use the Mathematics Blueprint to solve real-life problems.

SSM   PH TUTOR   CD & VIDEO   MATH PRO   WEB
CENTER

When a builder constructs a new home or office building, he or she often has a blueprint. This accurate drawing shows the basic form of the building. It also shows the dimensions of the structure to be built. This blueprint serves as a useful reference throughout the construction process.

Similarly, when solving real-life problems, it is helpful to have a "mathematics blueprint." This is a simple way to organize the information provided in the word problem, in a chart, or in a graph. You can record the facts you need to use. You can determine what it is you are trying to find and how you can go about actually finding it. You can record other information that you think will be helpful as you work through the problem.

As we solve real-life problems, we will use three steps.

Step 1 Understand the problem.
Here we will read through the problem. Draw a picture if it will help, and use the Mathematics Blueprint as a guide to assist us in thinking through the steps needed to solve the problem.

Step 2 Solve and state the answer.
We will use arithmetic or algebraic procedures along with problem-solving strategies to find a solution.

Step 3 Check.
We will use a variety of techniques to see if the answer in step 2 is the solution to the word problem. This will include estimating to see if the answer is reasonable, repeating our calculation, and working backward from the answer to see if we arrive at the original conditions of the problem.

▲ ● **EXAMPLE 1** Nancy and John want to install wall-to-wall carpeting in their living room. The floor of the rectangular living room is $11\frac{2}{3}$ feet wide and $19\frac{1}{2}$ feet long. How much will it cost if the carpet is $18.00 per square yard?

1. Understand the problem.
First, read the problem carefully. Drawing a sketch of the living room may help you see what is required. The carpet will cover the floor of the living room, so we need to find the area. Now we fill in the Mathematics Blueprint.

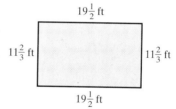

Mathematics Blueprint For Problem Solving

Gather the Facts	What Am I Solving For?	What Must I Calculate?	Key Points to Remember
The living room measures $11\frac{2}{3}$ ft by $19\frac{1}{2}$ ft. The carpet costs $18.00 per square yard.	**(a)** the area of the room in square feet **(b)** the area of the room in square yards **(c)** the cost of the carpet	**(a)** Multiply $11\frac{2}{3}$ ft by $19\frac{1}{2}$ ft to get area in square feet. **(b)** Divide the number of square feet by 9 to get the number of square yards. **(c)** Multiply the number of square yards by $18.00.	There are 9 square feet, 3 feet × 3 feet, in 1 square yard; therefore, we must divide the number of square feet by 9 to obtain square yards.

2. *Solve and state the answer.*

(a) To find the area of a rectangle, we multiply the length times the width.

$$11\frac{2}{3} \times 19\frac{1}{2} = \frac{35}{3} \times \frac{39}{2}$$

$$= \frac{455}{2} = 227\frac{1}{2} \text{ sq ft}$$

A minimum of $227\frac{1}{2}$ square feet of carpet will be needed. We say a minimum because some carpet may be wasted in cutting. Carpet is sold by the square yard. We will want to know the amount of carpet needed in square yards.

(b) To determine the area in square yards, we divide $227\frac{1}{2}$ by 9. (9 sq ft = 1 sq yd.)

$$227\frac{1}{2} \div 9 = \frac{455}{2} \div \frac{9}{1}$$

$$= \frac{455}{2} \times \frac{1}{9} = \frac{455}{18} = 25\frac{5}{18} \text{ sq yd}$$

A minimum of $25\frac{5}{18}$ square yards of carpet will be needed.

(c) Since the carpet costs $18.00 per square yard, we will multiply the number of square yards needed by $18.00.

$$25\frac{5}{18} \times 18 = \frac{455}{18} \times \frac{18}{1} = \$455$$

The carpet will cost a minimum of $455.00 for this room.

3. *Check.*

We will estimate to see if our answers are reasonable.

(a) We will estimate by rounding each number to the nearest 10.

$$11\frac{2}{3} \times 19\frac{1}{2} \longrightarrow 10 \times 20 = 200 \text{ sq ft}$$

This is close to our answer of $227\frac{1}{2}$ sq ft. Our answer is reasonble. ✓

(b) We will estimate by rounding to one significant digit.

$$227\frac{1}{2} \div 9 \longrightarrow 200 \div 10 = 20 \text{ sq yd}$$

This is close to our answer of $25\frac{5}{18}$ sq yd. Our answer is reasonable. ✓

(c) We will estimate by rounding each number to the nearest 10.

$$25\frac{5}{18} \times 18 \longrightarrow 30 \times 20 = \$600$$

This is close to our answer of $455. Our answer seems reasonable. ✓

"Remember to estimate. It will save you time and money!"

▲ **Practice Problem 1**   Jeff went to help Abby pick out wall-to-wall carpet for her new house. Her rectangular living room measures $16\frac{1}{2}$ feet by $10\frac{1}{2}$ feet. How much will it cost to carpet the room if the carpet costs $20 per square yard?

Mathematics Blueprint For Problem Solving

Gather the Facts	What Am I Solving For?	What Must I Calculate?	Key Points to Remember

To Think About   Assume that the carpet in Example 1 comes in a standard width of 12 feet. How much carpet will be wasted if it is laid out on the living room floor in one strip that is $19\frac{1}{2}$ feet long? How much carpet will be wasted if it is laid in two sections side by side that are each $11\frac{2}{3}$ feet long? Assuming you have to pay for wasted carpet, what is the minimum cost to carpet the room?

⬭ EXAMPLE 2   The following chart shows the 2000 sales of Micropower Computer Software for each of the four regions of the United States. Use the chart to answer the following questions (round all answers to the nearest whole percent):

(a) What percent of the sales personnel are assigned to the Northeast?

(b) What percent of the volume of sales is attributed to the Northeast?

(c) What percent of the sales personnel are assigned to the Southeast?

(d) What percent of the volume of sales is attributed to the Southeast?

(e) Which of these two regions of the country has sales personnel that appear to be more effective in terms of the volume of sales?

Region of the U.S.	Number of Sales Personnel	Dollar Volume of Sales
Northeast	12	1,560,000
Southeast	18	4,300,000
Northwest	10	3,660,000
Southwest	15	3,720,000
Total	55	13,240,000

1. *Understand the problem.*
We will only need to deal with figures from the Northeast region and the Southeast region.

Mathematics Blueprint For Problem Solving

Gather the Facts	What Am I Solving For?	What Must I Calculate?	Key Points to Remember
Personnel: 12 Northeast 18 Southeast 55 total Sales Volume: $1,560,000 NE $4,300,000 SE $13,240,000 Total	**(a)** the percent of the total personnel that is in the Northeast **(b)** the percent of the total sales made in the Northeast **(c)** the percent of the total personnel that is in the Southeast **(d)** the percent of the total sales made in the Southeast **(e)** compare the percentages from the two regions	**(a)** 12 of 55 is what percent? Divide. $12 \div 55$ **(b)** 1,560,000 of 13,240,000 is what percent? 1,560,000 $\div$ 13,240,000 **(c)** $18 \div 55$ **(d)** 4,300,000 $\div$ 13,240,000	We do not need to use the numbers relating to the Northwest or the Southwest in this problem.

2. *Solve and state the answer.*

(a) $\dfrac{12}{55} = 0.21818\ldots$

$\approx 22\%$

(b) $\dfrac{1,560,000}{13,240,000} = \dfrac{156}{1324} \approx 0.1178$

$\approx 12\%$

(c) $\dfrac{18}{55} = 0.32727\ldots$

$\approx 33\%$

(d) $\dfrac{4,300,000}{13,240,000} = \dfrac{430}{1324} \approx 0.3248$

$\approx 32\%$

(e) We notice that 22% of the sales force in the Northeast made 12% of the sales. The percent of the sales compared to the percent of the sales force is about half (12% of 24% would be half) or 50%. 33% of the sales force in the Southeast made 32% of the sales. The percent of sales compared to the percent of the sales force is close to 100%. We must be cautious here. *If there are no other significant factors,* it would appear that the Southeast sales force is more effective. (There may be other significant factors affecting sales, such as a recession in the Northeast, new and inexperienced sales personnel, or fewer competing companies in the Southeast.)

3. *Check.*

You may want to use a calculator to check the division in step 2, or you may use estimation.

(a) $\dfrac{12}{55} \rightarrow \dfrac{10}{60} \approx 0.17$

$= 17\%$ ✓

(b) $\dfrac{1,560,000}{13,240,000} \rightarrow \dfrac{1,600,000}{13,000,000} \approx 0.12$

$= 12\%$ ✓

(c) $\dfrac{18}{55} \rightarrow \dfrac{20}{60} \approx 0.33$

$= 33\%$ ✓

(d) $\dfrac{4,300,000}{13,240,000} \rightarrow \dfrac{4,300,000}{13,000,000} \approx 0.33$

$= 33\%$ ✓

Practice Problem 2 Using the chart for Example 2, answer the following questions. (Round all answers to the nearest whole percent.)

(a) What percent of the sales personnel are assigned to the Northwest?

(b) What percent of the sales volume is attributed to the Northwest?

(c) What percent of the sales personnel are assigned to the Southwest?

(d) What percent of the sales volume is attributed to the Southwest?

(e) Which of these two regions of the country has sales personnel that appear to be more effective in terms of volume of sales?

Mathematics Blueprint For Problem Solving

Gather the Facts	What Am I Solving For?	What Must I Calculate?	Key Points to Remember

To Think About Suppose in 2001 the number of sales personnel (55) increases by 60%. What would the new number of sales personnel be? Suppose in 2002 that the number of sales personnel decreases by 60% from the number of sales personnel in 2001. What would the new number be? Why is this number not 55, since we have increased the number by 60% and then decreased the result by 60%? Explain.

Use the Mathematics Blueprint for Problem Solving to help you solve each of the following exercises.

▲ **1.** Jocelyn wants to put new vinyl flooring in her kitchen. The kitchen measures $12\frac{3}{4}$ feet long by $9\frac{1}{2}$ feet wide. If the vinyl flooring she chose costs $20.00 per square yard, how much will the new flooring cost her? (Round your answer to the nearest cent.)

▲ **2.** The Carters need to replace the deck floor off their kitchen door. The deck is $11\frac{1}{2}$ feet by $20\frac{1}{2}$ feet. If the new decking costs $4.50 per square foot, how much will it cost them to replace the deck? (Round your answer to the nearest cent.)

▲ **3.** In order to put in a new lawn, the landscaper told Mr. Lopez to add new loam to a depth of $\frac{1}{2}$ foot, Mr. Lopez's lawn is $85\frac{1}{2}$ feet by 60 feet. How many cubic yards of loam does he need? (There are 27 cubic feet in 1 cubic yard.)

▲ **4.** The Brock family removed a built-in swimming pool and have decided to fill in the hole with dirt and seed the area. The pool hole is 30 feet by 12 feet by 9 feet deep. How many cubic yards of dirt are needed to fill in the hole? (There are 27 cubic feet in 1 cubic yard.)

The following directions are posted on the wall at the gym.

Beginning exercise training schedule

On day 1, each athlete will begin the morning as follows:

Jog................. $1\frac{1}{2}$ miles

Walk............. $1\frac{3}{4}$ miles

Rest.............. $2\frac{1}{2}$ minutes

Walk............. 1 mile

5. Betty's athletic trainer told her to follow the beginning exercise training schedule on day 1. On day 2, she is to increase all distances and times by $\frac{1}{3}$ that of day 1. On day 3, she is to increase all distances and times by $\frac{1}{3}$ that of day 2. What will be her training schedule on day 3?

6. Melinda's athletic trainer told her to follow the beginning exercise training schedule on day 1. On day 2, she is to increase all distances and times by $\frac{1}{3}$ that of day 1. On day 3, she is to once again increase all distances and times by $\frac{1}{3}$ that of day 1. What will be her training schedule on day 3?

To Think About

Refer to exercises 5 and 6 in working exercises 7–10.

7. Who will have a more demanding schedule on day 3, Betty or Melinda? Why?

8. If Betty kept up the same type of increase day after day, how many miles would she be jogging on day 5?

9. If Melinda kept up the same type of increase day after day, how many miles would she be jogging on day 7?

10. Which athletic trainer would appear to have the best plan for training athletes if they used this plan for 14 days? Why?

11. In 1985, the average selling price of an existing single-family home in Atlanta, Georgia, was $66,200. Between 1985 and 1990, the average price increased by 30%. Between 1990 and 2000, the average price increased again, this time by 15%. What was the median house price in Atlanta in 2000?

12. Chicken eggs are classified by weight per dozen eggs. Large eggs weigh 24 ounces per dozen and medium eggs weigh 21 ounces per dozen.
  (a) If you do not include the shell, which is 12% of the total weight of an egg, how many ounces of eggs do you get from a dozen large eggs? From a dozen medium eggs?

  (b) At a local market, large eggs sell for $1.79 a dozen, and medium eggs for $1.39 a dozen. If you do not include the shell, which is a better buy, large or medium eggs?

North Shore Community College students recently conducted a survey of 1000 passengers at Logan Airport. The passengers were classified as shown in the following table.

	Men age 21 or older	Women age 21 or older	Young people age 20 or younger	Total
Business	380	210	30	620
Pleasure	110	150	120	380
Total	490	360	150	1000

13. What percent of the total travelers were men age 21 or older? What percent of the business travelers were men age 21 or older? Which percentage is greater? Why do you suppose this is the case?

14. What percent of the business travelers were men age 21 or older? What percent of the business travelers were women age 21 or older? What percent of the business travelers were young people age 20 or younger? Why do you suppose this percentage is so small?

Use the following information from a paycheck stub to solve exercises 15–18.

TOBEY & SLATER INC. 5000 Stillwell Avenue Queens, NY 10001		Check Number	Payroll Period		Pay Date
			From Date	To Date	
		495885	10-30-99	11-30-99	12-01-99

Name	Social Security No.	I.D. Number	File Number	Rate/Salary	Department	MS	DEP	Res
Fred J. Gilliani	012-34-5678	01	1379	1150.00	0100	M	5	NY

	Current	Year to Date		Current	Year to Date
GROSS	1,150.00	6,670.00	STATE	67.76	388.45
FEDERAL	138.97	781.07	LOCAL	5.18	30.04
FICA	87.98	510.28	DIS-SUI	.00	.00
W-2 GROSS		6,670.00	NET	790.47	4,960.16

Earnings					Deductions/Specials		
No.	Type	Hours	Rate	Amount	Dept/Job No.	No.	Description	Amount
96	REGULAR			1,150.00	0100	82	Retirement	12.56
						75	Medical	36.28
						56	Union Dues	10.80

Gross pay is the pay an employee receives for his or her services before deductions. Net pay is the pay the employee actually gets to take home. You may round each amount to the nearest whole percent for exercises 15–18.

15. What percent of Fred's gross pay is deducted for federal, state, and local taxes?

16. What percent of Fred's gross pay is deducted for retirement and medical?

17. What percent of Fred's gross pay does he actually get to take home?

18. What percent of Fred's deductions are special deductions?

Math in the Media

Making Gravy

Compiled by LOS ANGELES TIMES Staff Writers

© 2000 Los Angeles Times. Reprinted by permission.

Making gravy is simple. The secret to divine gravy is deglazing the pan and using all the browned bits stuck on the roasting pan because they hold the flavor. What frustrates most gravy makers is getting rid of the lumps. But if you make a roux, a gravy paste of flour and turkey fat stirred until smooth, that will help get rid of stubborn lumps.

Recipe

Neck and giblets from 1 turkey

6 cups water

1 onion, quartered

1 carrot, cut in pieces

1 celery stalk, cut in pieces

Black peppercorns

Butter, optional

3 tablespoons flour

Salt, pepper

Preparation

1. Remove neck and giblets from bird. Separate liver from other giblets and discard or save for another meal.

2. Place remaining giblets and neck in 2-quart saucepan. Add water, onion, carrot, celery, and peppercorns, and bring to boil over high heat. Cover, reduce heat to low, and simmer until tender, about $1\frac{1}{2}$ hours. Strain and reserve broth. Discard neck. Chop giblets, cover, and refrigerate.

3. When turkey is roasted, remove from oven and transfer to platter. Lightly cover with foil and let stand about 20 minutes to allow juices to set before carving.

4. Meanwhile, add 1 cup reserved broth to drippings in roasting pan. Place pan over medium heat and scrape browned particles free from bottom with wooden spoon. Pour mixture into clear measuring cup and let fat rise to top. Skim fat off with spoon, or use specially designed measuring cup that separates the fat.

5. To make 2 cups gravy, place $\frac{1}{4}$ cup turkey fat in saucepan. (If necessary, add butter to make $\frac{1}{4}$ cup.) Add enough reserved giblet broth to skimmed drippings to make 2 cups.

6. Heat fat over medium heat. Stir in flour and cook, stirring, until bubbly. Remove from heat. Gradually pour in dripping-stock mixture, stirring constantly with wire whisk. Return pan to heat and cook, stirring, until gravy boils and thickens. For thinner gravy, add more broth. Stir in giblets. Season to taste with salt and pepper.

7. *2 cups gravy, each tablespoon:* 20 calories; 67 mg sodium; 3 mg cholesterol; 1 gram fat; 1 gram carbohydrates; 1 gram protein; 0.05 gram fiber.

EXERCISES

Sometimes it's necessary to adapt a recipe to reduce or increase the number of servings—depending on the number of guests expected. Do you think you could use your skills to adapt this recipe to produce a different number of servings? Could you calculate the nutritional values for a specific portion size? Try answering questions 1–3 to test your skills.

1. Adapt the directions given in Step 5 to make 4 cups gravy.

2. Adapt the directions given in Step 5 to make 3 cups gravy.

3. If two tablespoons of gravy were used, what would be the nutritional values?

Chapter 0 Organizer

Topic	Procedure	Examples
Simplifying fractions, p. 5.	1. Write the **numerator** and **denominator** as a product of prime factors. 2. Use the basic rule of fractions that $$\frac{a \times c}{b \times c} = \frac{a}{b}$$ for any factor that appears in both the numerator and the denominator. 3. Multiply the remaining factors for the numerator and separately for the denominator.	$$\frac{15}{25} = \frac{\cancel{5} \cdot 3}{\cancel{5} \cdot 5} = \frac{3}{5}$$ $$\frac{36}{48} = \frac{\cancel{2} \cdot \cancel{2} \cdot 3 \cdot \cancel{3}}{\cancel{2} \cdot \cancel{2} \cdot 2 \cdot 2 \cdot \cancel{3}} = \frac{3}{4}$$ $$\frac{26}{39} = \frac{2 \cdot \cancel{13}}{3 \cdot \cancel{13}} = \frac{2}{3}$$
Changing improper fractions to mixed numbers, p. 7.	1. Divide the denominator into the numerator to obtain the whole-number part of the mixed fraction. 2. The remainder from the division will be the numerator of the fraction. 3. The denominator remains unchanged.	$$\frac{14}{3} = 4\frac{2}{3} \qquad \frac{19}{8} = 2\frac{3}{8}$$ since $3\overline{)14}$: $\frac{4}{\underline{12}}\ 2$ since $8\overline{)19}$: $\frac{2}{\underline{16}}\ 3$
Changing mixed numbers to improper fractions, p. 8.	1. Multiply the whole number by the denominator and add the result to the numerator. This will yield the new numerator. 2. The denominator does not change.	$$4\frac{5}{6} = \frac{(4 \times 6) + 5}{6} = \frac{24 + 5}{6} = \frac{29}{6}$$ $$3\frac{1}{7} = \frac{(3 \times 7) + 1}{7} = \frac{21 + 1}{7} = \frac{22}{7}$$
Changing fractions to equivalent fractions with a given denominator, p. 9.	1. Divide the original denominator into the new denominator. This result is the value that we use for multiplication. 2. Multiply the numerator and the denominator of the original fraction by that value.	$$\frac{4}{7} = \frac{?}{21}$$ $\frac{3}{7\overline{)21}} \leftarrow$ Use this to multiply $\frac{4 \times 3}{7 \times 3} = \frac{12}{21}$
Addition and subtraction of fractions with the same denominator, p. 12.	1. Add or subtract the numerators. 2. Leave the denominator unchanged. 3. Simplify the answer if possible.	$$\frac{1}{12} + \frac{5}{12} = \frac{6}{12} = \frac{1}{2}$$ $$\frac{13}{17} - \frac{5}{17} = \frac{8}{17}$$
Finding the LCD (least common denominator) of two or more fractions, p. 13.	1. Write each denominator as the product of prime factors. 2. The LCD is a product containing each different factor. 3. If a factor occurs more than once in any one denominator, the LCD will contain that factor repeated the greatest number of times that it occurs in any one denominator.	Find the LCD of $\frac{4}{15}$ and $\frac{3}{35}$. $$15 = 5 \cdot 3$$ $$35 = 5 \cdot 7$$ $$\text{LCD} = 3 \cdot 5 \cdot 7 = 105$$ Find the LCD of $\frac{11}{18}$ and $\frac{7}{45}$. $18 = 3 \cdot 3 \cdot 2$ (factor 3 appears twice) $45 = 3 \cdot 3 \cdot 5$ (factor 3 appears twice) $\text{LCD} = 2 \cdot 3 \cdot 3 \cdot 5 = 90$
Adding and subtracting fractions that do not have a common denominator, p. 14.	1. Find the LCD. 2. Change each fraction to an equivalent fraction with the LCD for a denominator. 3. Add or subtract the fractions and simplify the answer if possible.	$$\frac{3}{8} + \frac{1}{3} = \frac{3 \cdot 3}{8 \cdot 3} + \frac{1 \cdot 8}{3 \cdot 8}$$ $$= \frac{9}{24} + \frac{8}{24} = \frac{17}{24}$$ $$\frac{11}{12} - \frac{1}{4} = \frac{11}{12} - \frac{1 \cdot 3}{4 \cdot 3}$$ $$= \frac{11}{12} - \frac{3}{12} = \frac{8}{12} = \frac{2}{3}$$

Topic	Procedure	Examples
Adding and subtracting mixed numbers, p. 17.	1. Change the mixed numbers to improper fractions. 2. Follow the rules for adding and subtracting fractions. 3. If necessary, change your answer to a mixed number.	$1\dfrac{2}{3} + 1\dfrac{3}{4} = \dfrac{5}{3} + \dfrac{7}{4} = \dfrac{5 \cdot 4}{3 \cdot 4} + \dfrac{7 \cdot 3}{4 \cdot 3}$ $= \dfrac{20}{12} + \dfrac{21}{12} = \dfrac{41}{12} = 3\dfrac{5}{12}$ $2\dfrac{1}{4} - 1\dfrac{3}{4} = \dfrac{9}{4} - \dfrac{7}{4} = \dfrac{2}{4} = \dfrac{1}{2}$
Multiplying fractions, p. 22.	1. If there are no common factors, multiply the numerators. Then multiply the denominators. 2. If possible, write the numerators and denominators as the product of prime factors. Use the basic rule of fractions to divide out any value that appears in both a numerator and a denominator. Multiply the remaining factors in the numerator. Multiply the remaining factors in the denominator.	$\dfrac{3}{7} \times \dfrac{2}{13} = \dfrac{6}{91}$ $\dfrac{6}{15} \times \dfrac{35}{91} = \dfrac{2 \cdot \cancel{3}}{\cancel{3} \cdot \cancel{5}} \times \dfrac{\cancel{5} \cdot \cancel{7}}{\cancel{7} \cdot 13} = \dfrac{2}{13}$ $3 \times \dfrac{5}{8} = \dfrac{3}{1} \times \dfrac{5}{8} = \dfrac{15}{8}$ or $1\dfrac{7}{8}$
Dividing fractions, p. 24.	1. Change the division sign to multiplication. 2. Invert the second fraction. 3. Multiply the fractions.	$\dfrac{4}{7} \div \dfrac{11}{3} = \dfrac{4}{7} \times \dfrac{3}{11} = \dfrac{12}{77}$ $\dfrac{5}{9} \div \dfrac{5}{7} = \dfrac{\cancel{5}}{9} \times \dfrac{7}{\cancel{5}} = \dfrac{7}{9}$
Multiplying and dividing mixed numbers, p. 23 and 26.	1. Change each mixed number to an improper fraction. 2. Use the rules for multiplying or dividing fractions. 3. Change your answer to a mixed number.	$2\dfrac{1}{4} \times 3\dfrac{3}{5} = \dfrac{9}{4} \times \dfrac{18}{5}$ $= \dfrac{3 \cdot 3}{2 \cdot \cancel{2}} \times \dfrac{\cancel{2} \cdot 3 \cdot 3}{5}$ $= \dfrac{3 \cdot 3 \cdot 3 \cdot 3}{2 \cdot 5} = \dfrac{81}{10} = 8\dfrac{1}{10}$ $1\dfrac{1}{4} \div 1\dfrac{1}{2} = \dfrac{5}{4} \div \dfrac{3}{2}$ $= \dfrac{5}{4} \times \dfrac{2}{3} = \dfrac{5}{2 \cdot \cancel{2}} \times \dfrac{\cancel{2}}{3} = \dfrac{5}{6}$
Changing fractional form to decimal form, p. 29.	Divide the denominator into the numerator.	$\dfrac{5}{8} = 0.625$ since $8\overline{)5.000}$ $\underline{48}$ 20 $\underline{16}$ 40 $\underline{40}$ 0
Changing decimal form to fractional form, p. 30.	1. Write the decimal as a fraction with a denominator of 10, 100, 1000, and so on. 2. Simplify the fraction, if possible.	$0.37 = \dfrac{37}{100}$ $\qquad$ $0.375 = \dfrac{375}{1000} = \dfrac{3}{8}$ $0.4 = \dfrac{4}{10} = \dfrac{2}{5}$

Topic	Procedure	Examples
Adding and subtracting decimals, p. 31.	1. Carefully line up the decimal points as indicated for addition and subtraction. (Extra zeros may be added to the right-hand side of the decimals if desired.) 2. Add or subtract the appropriate digits.	Add. $\quad$ Subtract. $1.236 + 7.825 \quad\quad 2 - 1.32$ $\begin{array}{r} 1.236 \\ + 7.825 \\ \hline 9.061 \end{array} \quad\quad \begin{array}{r} 2.00 \\ - 1.32 \\ \hline 0.68 \end{array}$
Multiplying decimals, p. 32.	1. First multiply the digits. 2. Count the total number of decimal places in the numbers being multiplied. 3. Place the decimal point so that the number of decimal places appears in the answer.	$\begin{array}{r} 0.9 \text{ (one place)} \\ \times\ 0.7 \text{ (one place)} \\ \hline 0.63 \text{ (two places)} \end{array} \quad \begin{array}{r} 0.009 \text{ (three places)} \\ \times\ 0.07 \text{ (two places)} \\ \hline 0.00063 \text{ (five places)} \end{array}$
Dividing decimals, p. 33.	1. Count the number of decimal places in the divisor. 2. Move the decimal point to the right the same number of places in both the divisor and dividend. 3. Mark that position with a caret ($\wedge$). 4. Perform the division. Line up the decimal point in the quotient with the position indicated by the caret in the dividend.	Divide. $7.5 \div 0.6$. Move decimal point one place to the right. $\begin{array}{r} 12.5 \\ 0.6\,\overline{)\,7.5\,0} \\ \underline{6} \\ 15 \\ \underline{12} \\ 30 \\ \underline{30} \\ 0 \end{array}$ $\quad$ Therefore, $\quad 7.5 \div 0.6 = 12.5$
Multiplying by a multiple of 10, p. 35.	For every zero in the multiplier, move the decimal point one place to the right.	$3.86 \times 100 = 386$ $0.0072 \times 1000 = 7.2$
Dividing by a multiple of 10, p. 35.	For every zero in the divisor, move the decimal point one place to the left.	$12.36 \div 10 = 1.236$ $1.97 \div 1000 = 0.00197$
Changing a decimal to a percent, p. 39.	1. Move the decimal point two places to the right. 2. Add the % symbol.	$0.46 = 46\%$ $0.002 = 0.2\% \quad\quad 1.59 = 159\%$ $0.013 = 1.3\% \quad\quad 0.0007 = 0.07\%$
Changing a percent to a decimal, p. 40.	1. Remove the % symbol. 2. Move the decimal point two places to the left.	$49\% = 0.49 \quad\quad 180\% = 1.8$ $59.8\% = 0.598 \quad\quad 0.13\% = 0.0013$
Finding a percent of a number, p. 41.	1. Convert the percent to a decimal. 2. Multiply the decimal by the number.	Find 12% of 86. $12\% = 0.12$ $\begin{array}{r} 86 \\ \times\ 0.12 \\ \hline 1\,72 \\ 8\,6 \\ \hline 10.32 \end{array}$ $\quad$ Therefore, 12% of 86 = 10.32.
Finding what percent one number is of another number, p. 42.	1. Place the number after the word *of* in the denominator. 2. Place the other number in the numerator. 3. If possible, simplify the fraction. 4. Change the fraction to a decimal. 5. Express the decimal as a percent.	What percent of 8 is 7? $$\frac{7}{8} = 0.875 = 87.5\%$$ 42 is what percent of 12? $$\frac{42}{12} = \frac{7}{2} = 3.5 = 350\%$$

Topic	Procedure	Examples
Estimation, p. 47.	1. Round each number to an appropriate place. 2. Perform the calculation with the rounded numbers.	Estimate the number of square feet in a room that is 22 feet long and 13 feet wide. Assume that the room is rectangular. 1. We round 22 to 20. We round 13 to 10. 2. To find the area of a rectangle, we multiply length times width. We multiply $$20 \times 10 = 200.$$ We estimate that there are 200 square feet in the room.
Problem solving, p. 51.	In solving a real-life problem, you may find it helpful to complete the following steps. You will not use all of the steps all of the time. Choose the steps that best fit the conditions of the problem. 1. *Understand the problem.* (a) Read the problem carefully. (b) Draw a picture if this helps you. (c) Use the Mathematics Blueprint for Problem Solving. 2. *Solve and state the answer.* 3. *Check.* (a) Estimate to see if your answer is reasonable. (b) Repeat your calculation. (c) Work backward from your answer. Do you arrive at the original conditions of the problem?	Susan is installing wall-to-wall carpeting in her $10\frac{1}{2}$-ft-by-12-ft bedroom. How much will it cost at $20 a square yard? 1. *Understand the problem.* We need to find the area of the room in square yards. Then we can find the cost. 2. *Solve and state the answer.* Area: $10\frac{1}{2} \times 12 = \frac{21}{2} \times \frac{12}{1} = 126$ sq ft $126 \div 9 = 14$ sq yd Cost: $14 \times 20 = \$280$ The carpeting will cost \$280. 3. *Check.* Estimate: $10 \times 12 = 120$ sq ft $120 \div 9 = 13\frac{1}{3}$ sq yd $13 \times 20 = \$260$ Our answer is reasonable. ✓

Chapter 0 Review Problems

0.1 *In exercises 1–4, simplify.*

1. $\dfrac{36}{48}$ **2.** $\dfrac{15}{50}$ **3.** $\dfrac{36}{82}$ **4.** $\dfrac{18}{30}$

5. Write $2\dfrac{5}{7}$ as an improper fraction.   **6.** Write $\dfrac{34}{5}$ as a mixed number.   **7.** Write $\dfrac{27}{4}$ as a mixed number.

Change each fraction to an equivalent fraction with the specified denominator.

8. $\dfrac{5}{8} = \dfrac{?}{24}$ **9.** $\dfrac{1}{7} = \dfrac{?}{35}$ **10.** $\dfrac{4}{7} = \dfrac{?}{21}$ **11.** $\dfrac{2}{5} = \dfrac{?}{55}$

0.2 *Combine.*

12. $\dfrac{3}{5} + \dfrac{1}{4}$ **13.** $\dfrac{7}{12} + \dfrac{5}{8}$ **14.** $\dfrac{7}{20} - \dfrac{1}{12}$ **15.** $\dfrac{5}{12} - \dfrac{1}{16}$

16. $3\frac{1}{6} + 2\frac{3}{5}$

17. $1\frac{1}{4} + 2\frac{7}{10}$

18. $2\frac{1}{5} - 1\frac{2}{3}$

19. $3\frac{1}{15} - 1\frac{3}{20}$

0.3 *Multiply.*

20. $6 \times \frac{5}{11}$

21. $2\frac{1}{3} \times 4\frac{1}{2}$

22. $1\frac{1}{8} \times 2\frac{1}{9}$

23. $\frac{4}{7} \times 5$

Divide.

24. $\frac{2}{5} \div \frac{4}{3}$

25. $\frac{12}{17} \div \frac{18}{85}$

26. $\frac{15}{16} \div 6\frac{1}{4}$

27. $2\frac{6}{7} \div \frac{10}{21}$

0.4 *Combine.*

28. $1.634 + 3.007 + 2.560$

29. $24.831 - 17.094$

30. $26.762 - 5.19$

31. $1.9 + 2.53 + 0.006$

Multiply.

32. 0.007×5.35

33. 362.341×1000

34. $2.6 \times 0.03 \times 1.02$

35. $1.08 \times 0.06 \times 160$

Divide.

36. $0.186 \div 100$

37. $71.32 \div 1000$

38. $0.523 \div 0.4$

39. $1.35 \div 0.015$

40. $0.147 \div 2.1$

41. $0.19 \div 0.38$

42. Write as a percent. $\frac{3}{8}$

43. Write as a simplified fraction. 24%

0.5 *In exercises 44–47, write each percentage in decimal form.*

44. 1.4%

45. 36.1%

46. 0.02%

47. 125.3%

48. What is 85% of 600?

49. Find 250% of 36.

50. What is 1.8% of 1000?

51. What percent of 120 is 15?

52. What percent of 1250 is 750?

53. 36% of Americans have blood type B. If there are 273,000,000 Americans, how many of them have blood type B? (Source: U.S. Census Bureau.)

54. In a given university, 720 of the 960 freshmen had a math deficiency. What percentage of the class had a math deficiency?

0.6 *In exercises 55–60, estimate. Do not find an exact value.*

55. $234,897 \times 1,936,112$

56. $357 + 923 + 768 + 417$

57. $780,000 - 198,000$

58. $7\frac{1}{3} + 3\frac{5}{6} + 8\frac{3}{7}$

59. Find 18% of $56,297.

60. $12,482 \div 389$

61. Estimate the cost to have a chimney repaired by a stone mason who charges $18.00 per hour and says the job will take her $4\frac{1}{3}$ days. Assume that she works eight hours in one day.

62. Estimate the monthly cost for each of three roommates who want to share an apartment that costs $923.50 per month.

0.7 *Solve. You may use the Mathematics Blueprint for Problem Solving.*

▲ **63.** Mr. and Mrs. Carr are installing wall-to-wall carpeting in a room that measures $12\frac{1}{2}$ ft by $9\frac{2}{3}$ ft. How much will it cost if the carpet is $26.00 per square yard?

64. The population of Falmouth was 34,000 in 1980 and 36,720 in 2000. What was the percent of increase in population?

A six-passenger Piper Cub airplane has a gas tank that holds 240 gallons. Use this information to answer exercises 65 and 66.

65. When flying at cruising speed, the plane averages $7\frac{2}{3}$ miles per gallon. How far can the plane fly at cruising speed? If the pilot never plans to fly more than 80% of his maximum cruising distance, what is the longest trip he would plan to fly?

66. When flying at maximum speed, the plane averages $6\frac{1}{4}$ miles per gallon. How far can the plane fly at maximum speed? If the pilot never plans to fly more than 70% of his maximum flying distance when flying at full speed, what is the longest trip he would plan to fly at full speed?

67. When baking banana bread, Marcia always adds $\frac{1}{3}$ cup of sugar for every $\frac{3}{4}$ cup of flour. If she used 6 cups of flour, how much sugar did she use?

68. Melissa rode her mountain bike for $2\frac{2}{3}$ miles on Tuesday, $4\frac{1}{2}$ miles on Wednesday, and $3\frac{1}{6}$ miles on Thursday. She had wanted to bike 14 miles over the four-day period beginning on Tuesday. How many miles does she need to bike on the fourth day?

69. A tank of water holds $8\frac{1}{4}$ cubic feet. If it is filled with water, how much does it weigh? (Assume that one cubic foot of water weighs $62\frac{1}{2}$ pounds.)

70. Sam took to the post office several packages that weighed 0.75 pounds each. The total weight of all the packages was 34.5 pounds. How many packages did he bring to the post office?

71. Dick and Ann Wright purchased a new car. They took out a loan of $9214.50 to help pay for the car and paid the rest in cash. They paid off the loan with payments of $225 per month for four years. How much more did they pay back than the amount of the car loan? (This is the amount of interest they were charged for the car loan.)

72. Frank works as a clerk at Wal-Mart. He is paid $6.40 an hour for a 40-hour week. For any additional time he gets paid 1.5 times the normal rate. Last week he worked 52 hours. How much did he get paid last week?

Chapter 0 Test

1. _____

2. _____

3. _____

4. _____

5. _____

6. _____

7. _____

8. _____

9. _____

10. _____

11. _____

12. _____

13. _____

14. _____

15. _____

16. _____

17. _____

18. _____

19. _____

20. _____

21. _____

22. _____

23. _____

24. _____

25. _____

26. _____

27. _____

28. _____

29. _____

In exercises 1 and 2, simplify.

1. $\dfrac{16}{18}$

2. $\dfrac{35}{40}$

3. Write as an improper fraction. $6\dfrac{3}{7}$

4. Write as a mixed number. $\dfrac{108}{33}$

In exercises 5–12, perform the operations indicated. Simplify answers whenever possible.

5. $\dfrac{1}{3} + \dfrac{1}{4} + \dfrac{5}{6}$

6. $1\dfrac{1}{8} + 3\dfrac{3}{4}$

7. $3\dfrac{2}{3} - 2\dfrac{5}{6}$

8. $\dfrac{5}{7} \times \dfrac{28}{15}$

9. $\dfrac{5}{18} \times \dfrac{3}{4}$

10. $\dfrac{7}{4} \div \dfrac{1}{2}$

11. $2\dfrac{1}{2} \times 3\dfrac{1}{4}$

12. $8\dfrac{5}{6} \div 4\dfrac{1}{9}$

In exercises 13–18, perform the calculations indicated.

13. $1.6 + 3.24 + 9.8$

14. $2003.42 - 196.8$

15. 32.8×0.04

16. 0.1632×100

17. $12.88 \div 0.056$

18. $26{,}325.9 \div 100$

19. Write as a percent. 0.073

20. Write as a decimal. 196.5%

21. What is 18% of 350?

22. What is 2% of 16.8?

23. What percent of 360 is 18?

24. What percent of 460 is 138?

25. A 4-inch stack of computer chips is on the table. Each computer chip is $\dfrac{2}{9}$ of an inch thick. How many computer chips are in the stack?

In exercises 26–27, estimate. Round each number to one nonzero digit. Then calculate.

26. $52{,}344\overline{)4{,}678{,}987}$

27. $18\dfrac{1}{3} + 22\dfrac{6}{7} + 57\dfrac{1}{2}$

Solve. You may use the Mathematics Blueprint for Problem Solving.

28. Allison is paid $14,000 per year plus a sales commission of 3% of the value of her sales. Last year she sold $870,000 worth of products. What percent of her total income was her commission?

29. Fred and Melinda are laying wall tile in the kitchen. Each tile covers $3\dfrac{1}{2}$ square inches of space. They plan to cover 210 square inches of wall space. How many tiles will they need?

Real Numbers and Variables

D o you have a realistic sense of how much the cost of living has gone up in the last ten years? Do you know in what categories the increase has been the largest? If you moved to another city, do you know how you could determine the cost of living there? The federal government compiles a Consumer Price Index to keep a record of such information. Turn to the Putting Your Skills to Work problems on page 119 and see if you can answer these types of questions.

Pretest Chapter 1

1. _____

2. _____

3. _____

4. _____

5. _____

6. _____

7. _____

8. _____

9. _____

10. _____

11. _____

12. _____

13. _____

14. _____

15. _____

16. _____

17. _____

18. _____

19. _____

20. _____

21. _____

22. _____

23. _____

24. _____

25. _____

26. _____

27. _____

66

If you are familiar with the topics in this chapter, take this test now. Check your answers with those in the back of the book. If an answer is wrong or you can't answer a question, study the appropriate section of the chapter.

If you are not familiar with the topics in this chapter, don't take this test now. Instead, study the examples, work the practice problems, and then take the test.

This test will help you identify which concepts you have mastered and which you need to study further.

Sections 1.1–1.2

Combine.

1. $-3 - (-6)$ **2.** $12 - 16 + 3 - 14$ **3.** $-7 + (-11)$ **4.** $7 - (-13)$

Section 1.3

Multiply or divide.

5. $-7(-2)(+3)(-1)$ **6.** $\dfrac{-\dfrac{2}{3}}{-\dfrac{1}{4}}$ **7.** $-2.3(-4)$ **8.** $20 \div (-5)$

Section 1.4

Evaluate.

9. $(-2)^4$ **10.** 4^3 **11.** $\left(-\dfrac{2}{3}\right)^3$ **12.** -4^4

Section 1.5

Simplify.

13. $6 \cdot 2 + 8 \cdot 3 - 4 \cdot 2$ **14.** $3(4)^3 + 18 \div 3 - 1$

15. $-\dfrac{1}{6}\left(\dfrac{1}{2}\right) + \dfrac{3}{4} \div \dfrac{1}{12}$ **16.** $1.22 - 4.1(1.4) + (-3.3)^2$

Section 1.6

Simplify.

17. $-2x(3x - 2xy + z)$ **18.** $-(3x - 4y - 12)$

Section 1.7

Combine like terms.

19. $5x^2 - 3xy - 6x^2y - 8xy$ **20.** $11x - 3y + 7 - 4x + y$

21. $3(2x - 5y) - (x - 8y)$ **22.** $2y(x^2 + y) - 3(x^2y - 5y^2)$

Section 1.8

23. Evaluate $3x^2 - 5x - 4$ for $x = -2$.

24. Evaluate $ab + 3a^2 - 5b$ for $a = 3$ and $b = -1$.

25. Determine the Celsius temperature when the Fahrenheit temperature is 77°. Use the formula $C = \dfrac{5}{9}(F - 32)$.

Section 1.9

Simplify.

26. $3[(x - y) - (2x - y)] - 3(2x + y)$ **27.** $2x^2 - 3x[2x - (x + 2y)]$

1.1 Addition of Real Numbers

1 Different Types of Numbers

Student Learning Objectives

After studying this section, you will be able to:

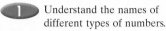

1 Understand the names of different types of numbers.

2 Recognize real-life situations for real numbers.

3 Add real numbers with the same sign.

4 Add real numbers with opposite signs.

Let's review some of the basic terms we use to talk about numbers.

> **Whole numbers** are numbers such as $0, 1, 2, 3, 4, \ldots$
>
> **Integers** are numbers such as $\ldots, -3, -2, -1, 0, 1, 2, 3, \ldots$.
>
> **Rational numbers** are numbers such as $\frac{3}{2}, \frac{5}{7}, -\frac{3}{8}, -\frac{4}{13}, \frac{6}{1}$, and $-\frac{8}{2}$.

Rational numbers can be written as one integer divided by another integer (as long as the denominator is not zero!). Integers can be written as fractions ($3 = \frac{3}{1}$, for example), so we can see that all integers are rational numbers. Rational numbers can be expressed in decimal form. For example, $\frac{3}{2} = 1.5$, $-\frac{3}{8} = -0.375$, and $\frac{1}{3} = 0.333\ldots$ or $0.\overline{3}$. It is important to note that rational numbers in decimal form are either terminating decimals or repeating decimals.

> **Irrational numbers** are numbers that cannot be expressed as one integer divided by another integer. The numbers π, $\sqrt{2}$, and $\sqrt[3]{7}$ are irrational numbers.

SSM   PH TUTOR CENTER   CD & VIDEO   MATH PRO   WEB

Irrational numbers can be expressed in decimal form. The decimal form of an irrational number is a nonterminating, nonrepeating decimal. For example, $\sqrt{2} = 1.414213\ldots$ can be carried out to an infinite number of decimal places with no repeating pattern of digits

Finally, **real numbers** are all the rational numbers and all the irrational numbers.

EXAMPLE 1 Classify as an integer, a rational number, an irrational number, and/or a real number.

(a) 5   **(b)** $-\dfrac{1}{3}$ **(c)** 2.85   **(d)** $\sqrt{2}$ **(e)** $0.777\ldots$

Make a table. Check off the description of the number that applies.

	Number	Integer	Rational Number	Irrational Number	Real Number
(a)	5	✓	✓		✓
(b)	$-\frac{1}{3}$		✓		✓
(c)	2.85		✓		✓
(d)	$\sqrt{2}$			✓	✓
(e)	$0.777\ldots$		✓		✓

Practice Problem 1 Classify.

(a) $-\dfrac{2}{5}$ **(b)** $1.515151\ldots$ **(c)** -8 **(d)** π

Any real number can be pictured on a **number line**.

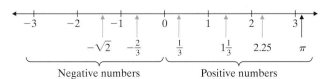

Negative numbers   Positive numbers

Positive numbers are to the right of 0 on the number line.

Negative numbers are to the left of 0 on the number line.

The **real numbers** include the positive numbers, the negative numbers, and zero.

2 Real-Life Situations for Real Numbers

We often encounter practical examples of number lines that include positive and negative rational numbers. For example, we can tell by reading the accompanying thermometer that the temperature is 20° below 0. From the stock market report, we see that the stock opened at 36 and closed at 34.5, and the net change for the day was −1.5.

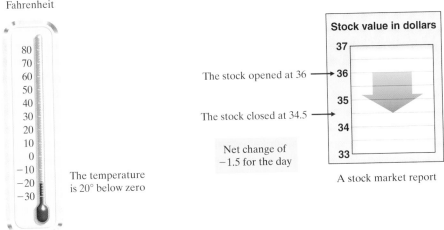

Temperature in degrees Fahrenheit

The temperature is 20° below zero

Stock value in dollars

The stock opened at 36

The stock closed at 34.5

Net change of −1.5 for the day

A stock market report

In the following example we use real numbers to represent real-life situations.

EXAMPLE 2   Use a real number to represent each situation.

(a) A temperature of 128.6°F below zero is recorded at Vostok, Antarctica.

(b) The Himalayan peak K2 rises 29,064 feet above sea level.

(c) The Dow gains 10.24 points.

(d) An oil drilling platform extends 328 feet below sea level.

A key word can help you to decide whether a number is positive or negative.

(a) 128.6°F *below* zero is −128.6.

(b) 29,064 feet *above* sea level is +29,064.

(c) A *gain* of 10.24 points is +10.24.

(d) 328 feet *below* sea level is −328.

Practice Problem 2   Use a real number to represent each situation.

(a) A population growth of 1,259

(b) A depreciation of $763

(c) A wind-chill factor of minus 10

In everyday life we consider positive numbers the opposite of negative numbers. For example, a gain of 3 yards in a football game is the opposite of a loss of 3 yards; a check written for $2.16 on a checking account is the opposite of a deposit of $2.16.

Each positive number has an opposite negative number. Similarly, each negative number has an opposite positive number. **Opposite numbers**, also called **additive inverses**, have the same magnitude but different signs and can be represented on the number line.

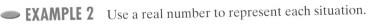

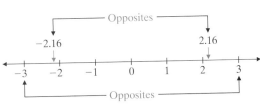

Opposites

−2.16   2.16

−3 −2 −1 0 1 2 3

Opposites

EXAMPLE 3 Find the additive inverse (that is, the opposite).   **(a)** -7 **(b)** $\frac{1}{4}$

(a) The opposite of -7 is $+7$.   **(b)** The opposite of $\frac{1}{4}$ is $-\frac{1}{4}$.

Practice Problem 3 Find the additive inverse (the opposite).

(a) $+\frac{2}{5}$ **(b)** -1.92 **(c)** a loss of 12 yards on a football play

3 *Adding Real Numbers with the Same Sign*

To use a real number, we need to be clear about its sign. When we write the number three as $+3$, the sign indicates that it is a positive number. The positive sign can be omitted. If someone writes three (3), it is understood that it is a positive three $(+3)$. To write a negative number such as negative three (-3), we must include the sign.

A concept that will help us add and subtract real numbers is the idea of ab-solute value. The **absolute value** of a number is the distance between that number and zero on the number line. The absolute value of 3 is written $|3|$.

Distance is always a positive number regardless of the direction we travel. This means that the absolute value of any number will be a positive value or zero. We place the symbols | and | around a number to mean the absolute value of the number.

The distance from 0 to 3 is 3, so $|3| = 3$. This is read "the absolute value of 3 is 3."

The distance from 0 to -3 is 3, so $|-3| = 3$. This is read "the absolute value of -3 is 3."

Some other examples are

$$|-22| = 22, \qquad |5.6| = 5.6, \qquad \text{and} \qquad |0| = 0.$$

Thus, the absolute value of a number can be thought of as the magnitude of the num-ber, without regard to its sign.

EXAMPLE 4 Find the absolute value.

(a) $|-4.62|$ **(b)** $\left|\frac{3}{7}\right|$ **(c)** $|0|$

(a) $|-4.62| = 4.62$ **(b)** $\left|\frac{3}{7}\right| = \frac{3}{7}$ **(c)** $|0| = 0$

Practice Problem 4 Find the absolute value.

(a) $|-7.34|$ **(b)** $\left|\frac{5}{8}\right|$ **(c)** $\left|\frac{0}{2}\right|$

Now let's look at addition of real numbers when the two numbers have the same sign. Suppose that you are keeping track of your checking account at a local bank. When you make a deposit of 5 dollars, you record it as $+5$. When you write a check for 4 dollars, you record it as -4, as a debit. Consider two situations.

Situation 1

You made a deposit of 20 dollars on one day and a deposit of 15 dollars the next day. You want to know the total value of your deposits.

Your record for situation 1.

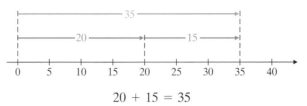

$$20 + 15 = 35$$

The amount of the deposit on the first day added to the amount of the deposit on the second day is the total of the deposits made over the two days.

Situation 2

You write a check for 25 dollars to pay one bill and two days later write a check for 5 dollars. You want to know the total value of debits to your account for the two checks.

Your record for situation 2.

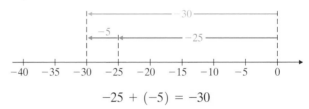

$$-25 + (-5) = -30$$

The value of the first check added to the value of the second check is the total debit to your account.

In each situation we found that we added the absolute value of each number. (That is, we added the numbers without regarding their sign.) The answer always contained the sign that was common to both numbers.

We will now state these results as a formal rule.

Addition Rule for Two Numbers with the Same Sign

To add two numbers with the same sign, add the absolute values of the numbers and use the common sign in the answer.

⬤ **EXAMPLE 5** Add.   **(a)** $14 + 16$ **(b)** $-8 + (-7)$

(a) $14 + 16$

$14 + 16 = 30$ Add the absolute values of the numbers.

$14 + 16 = +30$ Use the common sign in the answer. Here the common sign is the $+$ sign.

(b) $-8 + (-7)$

$8 + 7 = 15$ Add the absolute values of the numbers.

$-8 + (-7) = -15$ Use the common sign in the answer. Here the common sign is the $-$ sign.

Practice Problem 5 Add. $-23 + (-35)$ ⬤

⬤ **EXAMPLE 6** Add. $\dfrac{2}{3} + \dfrac{1}{7}$

$\dfrac{2}{3} + \dfrac{1}{7}$

$\dfrac{14}{21} + \dfrac{3}{21}$ Change each fraction to an equivalent fraction with a common denominator of 21.

$\dfrac{14}{21} + \dfrac{3}{21} = +\dfrac{17}{21}$ or $\dfrac{17}{21}$ Add the absolute values of the numbers. Use the common sign in the answer. Note that if no sign is written, the number is understood to be positive.

Practice Problem 6   Add. $-\dfrac{3}{5} + \left(-\dfrac{4}{7}\right)$

EXAMPLE 7   Add. $-4.2 + (-3.9)$

$$-4.2 + (-3.9)$$
$$4.2 + 3.9 = 8.1 \qquad \text{Add the absolute values of the numbers.}$$
$$-4.2 + (-3.9) = -8.1 \qquad \text{Use the common sign in the answer.}$$

Practice Problem 7   Add. $-12.7 + (-9.38)$

The rule for adding two numbers with the same signs can be extended to more than two numbers. If we add more than two numbers with the same sign, the answer will have the sign common to all.

EXAMPLE 8   Add. $-7 + (-2) + (-5)$

$$-7 + (-2) + (-5) \qquad \text{We are adding three real numbers all with the} \\ \text{same sign. We begin by adding the first two numbers.}$$
$$= -9 + (-5) \qquad \text{Add } -7 + (-2) = -9.$$
$$= -14 \qquad \text{Add } -9 + (-5) = -14.$$

Of course, this can be shortened by adding the three numbers without regard to sign and then using the common sign for the answer.

Practice Problem 8   Add. $-7 + (-11) + (-33)$

4 *Adding Real Numbers with Opposite Signs*

What if the signs of the numbers you are adding are different? Let's consider our checking account again to see how such a situation might occur.

Situation 3

You made a deposit of 30 dollars on one day. On the next day you write a check for 25 dollars. You want to know the result of your two transactions.

   Your record for situation 3.

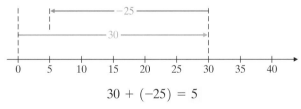

$$30 + (-25) = 5$$

A positive 30 for the deposit added to a negative 25 for the check, which is a debit, gives a net increase of 5 dollars in the account.

Situation 4

You made a deposit of 10 dollars on one day. The next day you write a check for 40 dollars. You want to know the result of your two transactions.

   Your record for situation 4.

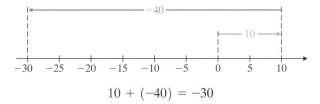

$$10 + (-40) = -30$$

A positive 10 for the deposit added to a negative 40 for the check, which is a debit, gives a net decrease of 30 dollars in the account.

The result is a negative thirty (-30), because the check was larger than the deposit. If you do not have at least 30 dollars in your account at the start of Situation 4, you have overdrawn your account.

What do we observe from situations 3 and 4? In each case, first we found the difference of the absolute values of the two numbers. Then the sign of the result was always the sign of the number with the greater absolute value. Thus, in situation 3, 30 is larger than 25. The sign of 30 is positive. The sign of the answer (5) is positive. In situation 4, 40 is larger than 10. The sign of 40 is negative. The sign of the answer (-30) is negative.

We will now state these results as a formal rule.

Addition Rule for Two Numbers with Different Signs

1. Find the difference between the larger absolute value and the smaller one.

2. Give the answer the sign of the number having the larger absolute value.

 EXAMPLE 9   Add. $8 + (-7)$

$8 + (-7)$	We are to add two numbers with opposite signs.
$8 - 7 = 1$	Find the difference between the two absolute values, which is 1.
$+8 + (-7) = +1$ or 1	The answer will have the sign of the number with the larger absolute value. That number is $+8$. Its sign is **positive**, so the answer will be $+1$.

Practice Problem 9   Add. $-9 + 15$

It is useful to know the following three properties of real numbers.

1. *Addition is commutative.*
  This property states that if two numbers are added, the result is the same no matter which number is written first. The order of the numbers does not affect the result.

$$3 + 6 = 6 + 3 = 9$$
$$-7 + (-8) = (-8) + (-7) = -15$$
$$-15 + 3 = 3 + (-15) = -12$$

2. *Addition of zero to any given number will result in that given number again.*

$$0 + 5 = 5$$
$$-8 + 0 = -8$$

3. *Addition is associative.*
  This property states that if three numbers are added, it does not matter which two numbers are grouped by parentheses and added first.

$3 + (5 + 7) = (3 + 5) + 7$	First combine numbers inside parentheses; then combine the remaining numbers. The results are the same no matter which numbers are grouped first.
$3 + (12) = (8) + 7$	
$15 = 15$	

We can use these properties along with the rules we have for adding real numbers to add three or more numbers. We go from left to right, adding two numbers at a time.

EXAMPLE 10 Add. $\dfrac{3}{17} + \left(-\dfrac{8}{17}\right) + \dfrac{4}{17}$

$$-\dfrac{5}{17} + \dfrac{4}{17} \qquad \text{Add } \tfrac{3}{17} + (-\tfrac{8}{17}) = -\tfrac{5}{17}.$$
The answer is negative since the larger of the two absolute values is negative.

$$= -\dfrac{1}{17} \qquad \text{Add} -\tfrac{5}{17} + \tfrac{4}{17} = -\tfrac{1}{17}.$$
The answer is negative since the larger of the two absolute values is negative.

Practice Problem 10 Add. $-\dfrac{5}{12} + \dfrac{7}{12} + \left(-\dfrac{11}{12}\right)$

Sometimes the numbers being added have the same signs; sometimes the signs are different. When adding three or more numbers, you may encounter both situations.

EXAMPLE 11 Add. $-1.8 + 1.4 + (-2.6)$

$$-0.4 + (-2.6) \qquad \text{We take the difference of 1.8 and 1.4 and use the sign of the number with the larger absolute value.}$$

$$= -3.0 \qquad \text{Add } -0.4 + (-2.6) = -3.0. \text{ The signs are the same; we add the absolute values of the numbers and use the common sign.}$$

Practice Problem 11 Add. $-6.3 + (-8.0) + 3.5$

If many real numbers are added, it is often easier to add numbers with like signs in a column format. Remember that addition is commutative; therefore, real numbers can be added *in any order*. You do *not* need to combine the first two numbers as your first step.

EXAMPLE 12 Add. $-8 + 3 + (-5) + (-2) + 6 + 5$

$$\begin{array}{l} -8 \\ -5 \\ \underline{-2} \\ -15 \end{array} \quad \begin{array}{l}\text{All the signs are the same.} \\ \text{Add the three negative} \\ \text{numbers to obtain } -15. \end{array} \qquad \begin{array}{l} +3 \\ +6 \\ \underline{+5} \\ +14 \end{array} \quad \begin{array}{l}\text{All the signs are the same.} \\ \text{Add the three positive} \\ \text{numbers to obtain } +14. \end{array}$$

Add the two results.

$$-15 + 14 = -1$$

The answer is negative because the number with the larger absolute value is negative.

Practice Problem 12 Add. $-6 + 5 + (-7) + (-2) + 5 + 3$

A word about notation: The only time we really need to show the sign of a number is when the number is negative—for example, -3. The only time we need to show parentheses when we add real numbers is when we have two different signs preceding a number. For example, $-5 + (-6)$.

EXAMPLE 13 Add.

(a) $2.8 + (-1.3)$ **(b)** $-\dfrac{2}{5} + \left(-\dfrac{3}{4}\right)$

(a) $2.8 + (-1.3) = 1.5$

(b) $-\dfrac{2}{5} + \left(-\dfrac{3}{4}\right) = -\dfrac{8}{20} + \left(-\dfrac{15}{20}\right) = -\dfrac{23}{20} \text{ or } -1\dfrac{3}{20}$

Practice Problem 13 Add.

(a) $-2.9 + (-5.7)$ **(b)** $\dfrac{2}{3} + \left(-\dfrac{1}{4}\right)$ **(c)** $-10 + (-3) + 15 + 4$

Verbal and Writing Skills

Check off any description of the number that applies.

	Number	Whole Number	Rational Number	Irrational Number	Real Number
1.	23				
2.	$-\frac{4}{5}$				
3.	π				
4.	2.34				
5.	$-6.666\ldots$				

	Number	Whole Number	Rational Number	Irrational Number	Real Number
6.	$-\frac{7}{9}$				
7.	$-2.3434\ldots$				
8.	14				
9.	$\sqrt{2}$				
10.	$3.232232223\ldots$				

Use a real number to represent each situation.

11. Jules Verne wrote a book with the title *20,000 Leagues under the Sea.*

12. The value of the dollar is up $0.07 with respect to the yen.

13. The Dow Jones average is down by $2\frac{3}{8}$.

14. Jon withdraws $102 from his account.

15. The temperature rises 7°F.

16. Maya won the game by 12 points.

Find the additive inverse (opposite).

17. $\frac{3}{4}$

18. -2

19. -2.73

20. 85.4

Find the absolute value.

21. $|-1.3|$

22. $|-5.9|$

23. $\left|\frac{5}{6}\right|$

24. $\left|\frac{7}{12}\right|$

Add.

25. $-6 + (-5)$

26. $-13 + (-3)$

27. $-\frac{1}{3} + \frac{2}{3}$

28. $-\frac{1}{5} + \left(-\frac{3}{5}\right)$

29. $-17 + (-14)$

30. $-12 + (-19)$

31. $-\frac{2}{13} + \left(-\frac{5}{13}\right)$

32. $-\frac{5}{14} + \frac{2}{14}$

33. $-1.5 + (-2.3)$

34. $-1.8 + (-1.4)$

35. $0.6 + (-0.2)$

36. $-0.8 + 0.5$

37. $-12 + (-13)$

38. $-17 + (-21)$

39. $-\frac{2}{5} + \frac{3}{7}$

40. $-\frac{2}{7} + \frac{3}{14}$

41. $-8 + 5 + (-3)$

42. $7 + (-8) + (-4)$

43. $-3 + 8 + 5 + (-7)$

Mixed Practice

Add.

44. $2 + (-17)$

45. $21 + (-4)$

46. $-83 + 42$

47. $-114 + 86$

48. $-\dfrac{4}{9} + \dfrac{5}{6}$

49. $-\dfrac{3}{5} + \dfrac{2}{3}$

50. $-\dfrac{1}{10} + \dfrac{1}{2}$

51. $-\dfrac{2}{3} + \left(-\dfrac{1}{4}\right)$

52. $4.3 + (-3.6)$

53. $5.7 + (-9.1)$

54. $4 + (-8) + 16$

55. $27 + (-11) + (-4)$

56. $-27 + 9 + (-54) + 30$

57. $18 + (-39) + 25 + (-3)$

58. $17.85 + (-2.06) + 0.15$

59. $23.17 + 5.03 + (-11.81)$

Applications

60. Hallie paid $47 for a vase at an estate auction. She resold it to an antiques dealer for $214. What was her profit or loss?

61. When we skied at Jackson Hole, Wyoming, yesterday, the temperature at the summit was $-12°$F. Today when we called the ski report, the temperature had risen $7°$F. What is the temperature at the summit today?

62. Donna is studying rain forest preservation in Central America. She stands on a ridge 126 feet below sea level. Then she and her team hike down to a gully 43 feet lower. Represent her distance below sea level as a real number.

63. Oceanographers studying effects of light on sea creatures are in a submarine 85 feet below sea level. Then they dive to a point 180 feet lower. Represent the submarine's distance below sea level as a real number.

64. On three successive football running plays, Jon gained 9 yards, lost 11 yards, and gained 5 yards. What was his total gain or loss?

65. Ureji's financial aid account at school held $643.85. She withdrew $185.50 to buy books for the semester. Does she have enough left in her account to pay the $475.00 registration fee for the next semester? If so, how much extra money does she have? If not, how much is she short?

66. The population of a particular butterfly species was 8000. Twenty years later there were 3000 fewer. Today, there are 1500 fewer. What is the new population?

67. Aaron owes $258 to a credit card company. He makes a purchase of $32 with the card and then makes a payment of $150 on the account. How much does he still owe?

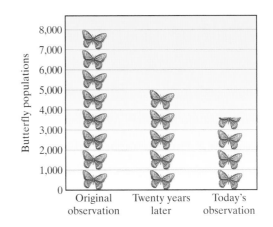

During the first five months of 2000, a regional Midwest airline posted profit and loss figures for each month of operation, as shown in the accompanying bar graph.

68. For the first three months of 2000, what were the total earnings of the airline?

69. For the first five months of 2000, what were the total earnings for the airline?

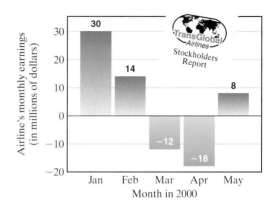

To Think About

70. What number must be added to −13 to get 5?

71. What number must be added to −18 to get 10?

72. Glen makes a deposit of $50 into his checking account and then writes out a check for $89. What is the least amount of money he needs in his checking account to cover the transactions?

73. The Roanoke Animal Shelter received 800 kittens from families whose cats had not been spayed. Only 328 kittens were placed with new families. Express as a percent how many kittens were able to find a home this year. If next year the same number of kittens are received, but 56% are placed in new homes, how many more kittens will find a home next year than did this year?

Cumulative Review Problems

Perform the indicated calcutions.

74. $\dfrac{3}{7} + \dfrac{5}{21}$

75. $\left(\dfrac{2}{5}\right)\left(\dfrac{20}{27}\right)$

76. $\dfrac{2}{15} - \dfrac{1}{20}$

77. $2\dfrac{1}{2} \div 3\dfrac{2}{5}$

1.2 Subtraction of Real Numbers

1 *Subtracting Real Numbers*

So far we have developed the rules for adding real numbers. We can use these rules to subtract real numbers. Let's look at a checkbook situation to see how.

Situation 5

You have a balance of 20 dollars in your checking account. The bank calls you and says that a deposit of 5 dollars that belongs to another account was erroneously added to your account. They say they will correct the account balance to 15 dollars. The bank tells you that since they cannot take away the erroneous credit, they will add a debit to your account. You want to keep track of what's happening to your account.

Your record for situation 5.

$$20 - (+5) = 15$$

From your present balance subtract the deposit to give the new balance. This equation shows what needs to be done to your account. The bank tells you that because the error happened in the past they cannot "take it away." However, they can add to your account a debit of 5 dollars. Here is the equivalent addition.

$$20 + (-5) = 15$$

To your present balance add a debit to give the new balance. Subtracting a positive 5 has the same effect as adding a negative 5.

Subtraction of Real Numbers

To subtract real numbers, add the opposite of the second number (that is, the number you are subtracting) to the first.

The rule tells us to do three things when we subtract real numbers. First, change subtraction to addition. Second, replace the second number by its opposite. Third, add the two numbers using the rules for addition of real numbers.

EXAMPLE 1 Subtract. $6 - (-2)$

$$6 \quad - \quad (-2)$$

Change subtraction to addition. Write the opposite of the second number.

$$= \quad 6 \quad + \quad (+2)$$

Add the two real numbers with the same sign.

$$= \quad 8$$

Practice Problem 1 Subtract. $9 - (-3)$

Student Learning Objectives

After studying this section, you will be able to:

1 Subtract real numbers.

SSM   PH TUTOR   CD & VIDEO   MATH PRO   WEB
CENTER

EXAMPLE 2 Subtract. $-8 - (-6)$

$$-8 \qquad - \qquad (-6)$$

Change subtraction to addition.   Write the opposite of the second number.

$$= \quad -8 \qquad + \qquad (+6)$$

Add the two real numbers with opposite signs.

$$= \qquad -2$$

Practice Problem 2 Subtract. $-12 - (-5)$

EXAMPLE 3 Subtract.   **(a)** $\dfrac{3}{7} - \dfrac{6}{7}$ **(b)** $-\dfrac{7}{18} - \left(-\dfrac{1}{9}\right)$

(a) $\dfrac{3}{7} - \dfrac{6}{7} = \dfrac{3}{7} + \left(-\dfrac{6}{7}\right)$ Change the subtraction problem to one of adding the opposite of the second number. We note that the problem has two fractions with the same denominator.

$\qquad = -\dfrac{3}{7}$ Add two numbers with different signs.

(b) $-\dfrac{7}{18} - \left(-\dfrac{1}{9}\right) = -\dfrac{7}{18} + \dfrac{1}{9}$ Change subtracting to adding the opposite.

$\qquad = -\dfrac{7}{18} + \dfrac{2}{18}$ Change $\frac{1}{9}$ to $\frac{2}{18}$ since LCD $= 18$.

$\qquad = -\dfrac{5}{18}$ Add two numbers with different signs.

Practice Problem 3 Subtract. $-\dfrac{1}{5} - \dfrac{1}{4}$

EXAMPLE 4 Subtract. $-5.2 - (-5.2)$

$$-5.2 - (-5.2) = -5.2 + 5.2$$ Change the subtraction problem to one of adding the opposite of the second number.

$$= 0$$ Add two numbers with different signs.

Example 4 illustrates what is sometimes called the **additive inverse property**. When you add two real numbers that are opposites of each other, you will obtain zero. Examples of this are the following:

$$5 + (-5) = 0 \qquad -186 + 186 = 0 \qquad -\dfrac{1}{8} + \dfrac{1}{8} = 0$$

Practice Problem 4 Subtract. $-17.3 - (-17.3)$

EXAMPLE 5   Calculate.

(a) $-8 - 2$ **(b)** $23 - 28$ **(c)** $5 - (-3)$ **(d)** $\dfrac{1}{4} - 8$

(a) $-8 - 2 = -8 + (-2)$ Notice that we are subtracting a positive 2.
                                            Change to addition.
            $= -10$ Add.
   In a similar fashion we have

(b) $23 - 28 = 23 + (-28) = -5$

(c) $5 - (-3) = 5 + 3 = 8$

(d) $\dfrac{1}{4} - 8 = \dfrac{1}{4} + (-8) = \dfrac{1}{4} + \left(-\dfrac{32}{4}\right) = -\dfrac{31}{4}$

Practice Problem 5   Calculate.   **(a)** $-21 - 9$ **(b)** $17 - 36$

EXAMPLE 6   A satellite is recording radioactive emissions from nuclear waste buried 3 miles below sea level. The satellite orbits the Earth at 98 miles above sea level. How far is the satellite from the nuclear waste?
     We want to find the difference between +98 miles and −3 miles.

$$98 - (-3) = 98 + 3$$
$$= 101$$

The satellite is 101 miles from the nuclear waste.

Practice Problem 6   A helicopter is directly over a sunken vessel. The helicopter is 350 feet above sea level. The vessel lies 186 feet below sea level. How far is the helicopter from the sunken vessel?

1.2 Exercises

Verbal and Writing Skills

1. Explain in your own words how you would perform the necessary steps to find $-8 - (-3)$.

2. Explain in your own words how you would perform the necessary steps to find $-10 - (-15)$.

Subtract by adding the opposite.

3. $20 - 46$

4. $15 - 28$

5. $-14 - (-3)$

6. $-24 - (-7)$

7. $-52 - (-60)$

8. $-48 - (-80)$

9. $0 - (-5)$

10. $0 - (-7)$

11. $15 - 20$

12. $18 - 24$

13. $-18 - (-18)$

14. $24 - (-24)$

15. $-11 - (-8)$

16. $-35 - (-10)$

17. $-0.6 - 0.3$

18. $-0.9 - 0.5$

19. $2.64 - (-1.83)$

20. $-0.03 - 0.06$

21. $2.3 - (-1.4)$

22. $\dfrac{1}{3} - \left(-\dfrac{2}{5}\right)$

23. $\dfrac{3}{4} - \left(-\dfrac{3}{5}\right)$

24. $-\dfrac{2}{3} - \dfrac{1}{4}$

25. $-\dfrac{3}{4} - \dfrac{5}{6}$

26. $-\dfrac{7}{10} - \dfrac{10}{15}$

Mixed Practice

Calculate.

27. $34 - 87$

28. $19 - 76$

29. $-67 - 32$

30. $-98 - 34$

31. $2.3 - (-4.8)$

32. $8.4 - (-2.7)$

33. $8 - \left(-\dfrac{3}{4}\right)$

34. $\dfrac{2}{3} - (-6)$

35. $\dfrac{3}{5} - (-8)$

36. $\dfrac{5}{6} - 4$

37. $-\dfrac{3}{10} - \dfrac{3}{4}$

38. $-\dfrac{11}{12} - \dfrac{5}{18}$

39. $-135 - (-126.5)$

40. $-97.6 - (-146)$

41. $0.0067 - (-0.0432)$

42. $0.0762 - (-0.0094)$

43. $\dfrac{1}{5} - 6$

44. $\dfrac{2}{7} - (-3)$

45. $-0.0023 - 6$

46. $-2 - 0.071$ | **47.** Subtract -9 from -2. | **48.** Subtract -12 from 20. | **49.** Subtract 13 from -35.

Change each subtraction operation to "adding the opposite." Then combine the numbers.

50. $9 + 6 - (-5)$ | **51.** $7 + (-6) - 3$ | **52.** $8 + (-4) - 10$ | **53.** $-10 + 6 - (-15)$

54. $18 - (-15) - 3$ | **55.** $7 + (-42) - 27$ | **56.** $-37 - (-18) + 5$ | **57.** $-21 - (-36) - 8$

58. $-3 - (-12) + 18 + 15 - (-6)$ | **59.** $42 - (-30) - 65 - (-11) + 20$

Applications

60. A rescue helicopter is 300 feet above sea level. The captain has located an ailing submarine directly below it that is 126 feet below sea level. How far is the helicopter from the submarine?

61. Yesterday Jackie had $112 in her checking account. Today her account reads "balance − $37." Find the difference in these two amounts.

62. On January 6, 1971, Hawley Lake, Arizona, had a record low temperature of $-23°$F. The next day the temperature at the same place was $-40°$F. What was the change in temperature from January 6 to January 7, 1971?

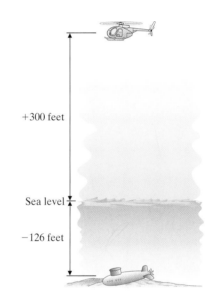

63. On January 19, 1937, the temperature at Boca, California, was $-29°$F. On January 20, the temperature at the same place was $-45°$F. What was the change in temperature from January 19 to January 20, 1937?

64. In 2000, Rachel had $1815 withheld from her paycheck in federal taxes. When she filed her tax return, she received a $265 refund. How much did she actually pay in taxes?

Cumulative Review Problems

In exercises 65–67, perform the indicated operations.

65. $-37 + 16$ | **66.** $-37 + (-14)$ | **67.** $-3 + (-6) + (-10)$

68. What is the temperature after a rise of $13°$C from a start of $-21°$C?

69. Sean and Khalid went hiking in the Blue Ridge Mountains. During their $8\frac{1}{3}$ mile hike, $\frac{4}{5}$ of the distance was covered with snow. How many miles were snow covered?

1.3 Multiplication and Division of Real Numbers

Student Learning Objectives

After studying this section, you will be able to:

 Multiply real numbers.

 Divide real numbers.

SSM
PH TUTOR
CENTER   CD & VIDEO   MATH PRO   WEB

1 *Multiplying Real Numbers*

We are familiar with the meaning of multiplication for positive numbers. For example, $5 \times 40 = 200$ might mean that you receive five weekly checks of 40 dollars each and you gain $200. Let's look at a situation that corresponds to $5 \times (-40)$. What might that mean?

Situation 6

You write a check for five weeks in a row to pay your weekly room rent of 40 dollars. You want to know the total impact on your checking account balance.

Your record for situation 6.

$(+5)$	$\times$	(-40)	$=$	-200
The number of checks you have written	times	negative 40, the value of each check that was a debit to your account,	gives	negative 200 dollars, a net debit to your account.

Note that a multiplication symbol is not needed between the $(+5)$ and the (-40) because the two sets of parentheses indicate multiplication. The multiplication $(5)(-40)$ is the same as repeated addition of five (-40)'s. Note that 5 multiplied by -40 can be written as $5(-40)$ or $(5)(-40)$.

$$\underbrace{(-40) + (-40) + (-40) + (-40) + (-40)}_{\text{repeated addition of five } (-40)\text{'s}} = -200$$

This example seems to show that a positive number multiplied by a negative number is negative.

What if the negative number is the one that is written first? If $(5)(-40) = -200$, then $(-40)(5) = -200$, by the commutative property of multiplication. This is an example showing that *when two numbers with opposite signs* (one positive, one negative) *are multiplied, the result is negative.*

But what if both numbers are negative? Consider the following situation.

Situation 7

Last year at college you rented a room at 40 dollars per week for 36 weeks, which included two semesters and summer school. This year you will not attend the summer session, so you will be renting the room for only 30 weeks. Thus the number of weekly rental checks will be six less than last year. You are making out your budget for this year. You want to know the financial impact of renting the room for six fewer weeks.

Your record for situation 7.

(-6)	$\times$	(-40)	$=$	240
The difference in the number of checks this year compared to last is -6, which is negative to show a decrease,	times	-40, the value of each check paid out,	gives	$+240$ dollars. The product is positive, because your financial situation will be 240 dollars better this year.

You could check that the answer is positive by calculating the total rental expenses.

	Dollars in rent last year	$(36)(40) =$	1440
(subtract)	Dollars in rent this year	$- (30)(40) =$	-1200
	Extra dollars available this year	$=$	$+240$

This agrees with our previous answer: $(-6)(-40) = +240$.

In this situation it seems reasonable that a negative number times a negative number yields a positive answer. We already know from arithmetic that a positive number times a positive number yields a positive answer. Thus we might see the general rule that *when two numbers with the same sign* (both positive or both negative) *are multiplied, the result is positive.*

We will now state our rule.

Multiplication of Real Numbers

To multiply two real numbers with **the same sign**, multiply the absolute values. The sign of the result is **positive**.

To multiply two real numbers with **opposite signs**, multiply the absolute values. The sign of the result is **negative**.

Note that negative 6 times −40 can be written as −6(−40) or (−6)(−40).

EXAMPLE 1 Multiply.

(a) $(3)(6)$ **(b)** $\left(-\dfrac{5}{7}\right)\left(-\dfrac{2}{9}\right)$ **(c)** $-4(8)$ **(d)** $\left(\dfrac{2}{7}\right)(-3)$

(a) $(3)(6) = 18$

When multiplying two numbers with the same sign, the result is a positive number.

(b) $\left(-\dfrac{5}{7}\right)\left(-\dfrac{2}{9}\right) = \dfrac{10}{63}$

(c) $-4(8) = -32$

When multiplying two numbers with opposite signs, the result is a negative number.

(d) $\left(\dfrac{2}{7}\right)(-3) = \left(\dfrac{2}{7}\right)\left(-\dfrac{3}{1}\right) = -\dfrac{6}{7}$

Practice Problem 1 Multiply.

(a) $(-6)(-2)$ **(b)** $(7)(9)$ **(c)** $\left(-\dfrac{3}{5}\right)\left(\dfrac{2}{7}\right)$ **(d)** $40(-20)$

To multiply more than two numbers, multiply two numbers at a time.

EXAMPLE 2 Multiply. $(-4)(-3)(-2)$

$(-4)(-3)(-2) = (+12)(-2)$ We begin by multiplying the first two numbers, (-4) and (-3). The signs are the same. The answer is positive 12.

$\qquad\qquad\qquad\ = -24$ Now we multiply, $(+12)$ and (-2). The signs are different. The answer is negative 24.

Practice Problem 2 Multiply. $(-5)(-2)(-6)$

EXAMPLE 3 Multiply.   **(a)** $-3(-8)$ **(b)** $\left(-\dfrac{1}{2}\right)(-1)(-4)$

(c) $-2(-2)(-2)(-2)$

Multiply two numbers at a time. See if you find a pattern.

(a) $-3(-8) = +24$ or 24

(b) $\left(-\dfrac{1}{2}\right)(-1)(-4) = +\dfrac{1}{2}(-4) = -2$

(c) $-2(-2)(-2)(-2) = +4(-2)(-2) = -8(-2) = +16$ or 16

What kind of answer would we obtain if we multiplied five negative numbers? If you guessed "negative," you probably see the pattern.

Practice Problem 3   Determine the sign of the product. Then multiply to check.

(a) $-2(-3)(-4)(-1)$ **(b)** $(-1)(-3)(-2)$ **(c)** $-4\left(-\dfrac{1}{4}\right)(-2)(-6)$

When you multiply two or more real numbers:
1. The result is always **positive** if there are an **even** number of negative signs.
2. The result is always **negative** if there are an **odd** number of negative signs.

For convenience, we will list the properties of multiplication.

1. *Multiplication is commutative.*
   This property states that if two real numbers are multiplied, the order of the numbers does not affect the result. The result is the same no matter which number is written first.

$$(5)(7) = (7)(5) = 35, \qquad \left(\frac{1}{3}\right)\left(\frac{2}{7}\right) = \left(\frac{2}{7}\right)\left(\frac{1}{3}\right) = \frac{2}{21}$$

2. *Multiplication of any real number by zero will result in zero.*

$$(5)(0) = 0, \qquad (-5)(0) = 0, \qquad (0)\left(\frac{3}{8}\right) = 0, \qquad (0)(0) = 0$$

3. *Multiplication of any real number by 1 will result in that same number.*

$$(5)(1) = 5, \qquad (1)(-7) = -7, \qquad (1)\left(-\frac{5}{3}\right) = -\frac{5}{3}$$

4. *Multiplication is associative.*
   This property states that if three real numbers are multiplied, it does not matter which two numbers are grouped by parentheses and multiplied first.

$2 \times (3 \times 4) = (2 \times 3) \times 4$ First multiply the numbers in parentheses. Then multiply the remaining numbers.

$2 \times (12) = (6) \times 4$ The results are the same no matter which numbers are grouped and multiplied first.

$24 = 24$

2   *Dividing Real Numbers*

What about division? Any division problem can be rewritten as a multiplication problem.

We know that $20 \div 4 = 5$ because $4(5) = 20$.
Similarly, $-20 \div (-4) = 5$ because $-4(5) = -20$.

In both division problems the answer is positive 5. Thus we see that *when you divide two numbers with the same sign* (both positive or both negative), *the answer is positive.* What if the signs are different?

We know that $-20 \div 4 = -5$ because $4(-5) = -20$.
Similarly, $20 \div (-4) = -5$ because $-4(-5) = 20$.

In these two problems the answer is negative 5. So we have reasonable evidence to see that *when you divide two numbers with different signs* (one positive and one negative), *the answer is negative.*

We will now state our rule for division.

Division of Real Numbers

To divide two real numbers with **the same sign**, divide the absolute values. The sign of the result is **positive**.

To divide two real numbers with **different signs**, divide the absolute values. The sign of the result is **negative**.

EXAMPLE 4 Divide.

(a) $12 \div 4$

(b) $(-25) \div (-5)$

(c) $\dfrac{-36}{18}$

(d) $\dfrac{42}{-7}$

(a) $12 \div 4 = 3$

When dividing two numbers with the same sign, the result is a positive number.

(b) $(-25) \div (-5) = 5$

(c) $\dfrac{-36}{18} = -2$

When dividing two numbers with different signs, the result is a negative number.

(d) $\dfrac{42}{-7} = -6$

Practice Problem 4 Divide.

(a) $-36 \div (-2)$

(b) $\dfrac{50}{-10}$

(c) $-49 \div 7$

EXAMPLE 5 Divide.   **(a)** $-36 \div 0.12$ **(b)** $-2.4 \div (-0.6)$

(a) $-36 \div 0.12$ Look at the problem to determine the sign. When dividing two numbers with different signs, the result will be a negative number.

We then divide the absolute values.

$$
\begin{array}{r}
3\,00. \\
0.12_{\wedge}\overline{)36.00_{\wedge}} \\
\underline{36} \\
00
\end{array}
$$

Thus $-36 \div 0.12 = -300$. The answer is a negative number.

(b) $-2.4 \div (-0.6)$ Look at the problem to determine the sign. When dividing two numbers with the same sign, the result will be positive.

We then divide the absolute values.

$$
\begin{array}{r}
4. \\
0.6_{\wedge}\overline{)2.4_{\wedge}} \\
\underline{2\,4}
\end{array}
$$

Thus $-2.4 \div (-0.6) = 4$. The answer is a positive number.

Practice Problem 5 Divide.

(a) $-1.242 \div (-1.8)$ **(b)** $0.235 \div (-0.0025)$

Note how similar the rules for multiplication and division are. When you **multiply** or **divide** two numbers with the **same** sign, you obtain **a positive** number. When you **multiply** or **divide** two numbers with **different** signs, you obtain a **negative** number.

EXAMPLE 6   Divide. $-\dfrac{12}{5} \div \dfrac{2}{3}$

$= \left(-\dfrac{12}{5}\right)\left(\dfrac{3}{2}\right)$ Divide two fractions. We invert the second fraction and multiply by the first fraction.

$= \left(-\dfrac{\overset{6}{\cancel{12}}}{5}\right)\left(\dfrac{3}{\underset{1}{\cancel{2}}}\right)$

$= -\dfrac{18}{5}$ or $-3\dfrac{3}{5}$ The answer is negative since the two numbers divided have different signs.

Practice Problem 6   Divide. $-\dfrac{5}{16} \div \left(-\dfrac{10}{13}\right)$

Note that division can be indicated by the symbol $\div$ or by the fraction bar $-$. $\frac{2}{3}$ means $2 \div 3$.

EXAMPLE 7   Divide.   **(a)** $\dfrac{\frac{7}{8}}{-21}$ **(b)** $\dfrac{-\frac{2}{3}}{-\frac{7}{13}}$

(a) $\dfrac{\frac{7}{8}}{-21}$

$= \dfrac{7}{8} \div \left(-\dfrac{21}{1}\right)$ Change -21 to a fraction. $-21 = -\frac{21}{1}$

$= \dfrac{\overset{1}{\cancel{7}}}{8}\left(-\dfrac{1}{\underset{3}{\cancel{21}}}\right)$ Change the division to multiplication. Cancel where possible.

$= -\dfrac{1}{24}$ Simplify.

(b) $\dfrac{-\frac{2}{3}}{-\frac{7}{13}} = -\dfrac{2}{3} \div \left(-\dfrac{7}{13}\right) = -\dfrac{2}{3}\left(-\dfrac{13}{7}\right) = \dfrac{26}{21}$ or $1\dfrac{5}{21}$

Practice Problem 7   Divide.   **(a)** $\dfrac{-12}{-\frac{4}{5}}$ **(b)** $\dfrac{-\frac{2}{9}}{\frac{8}{13}}$

1. *Division of 0 by any nonzero real number gives 0 as a result.*

$$0 \div 5 = 0, \qquad 0 \div \dfrac{2}{3} = 0, \qquad \dfrac{0}{5.6} = 0, \qquad \dfrac{0}{1000} = 0$$

You can divide zero by $5, \frac{2}{3}, 5.6, 1000$, or any number (except 0).

2. *Division of any real number by 0 is* **undefined**.

$$7 \div 0 \qquad\qquad \dfrac{64}{0} \qquad\qquad \dfrac{0}{0}$$
$$\uparrow \qquad\qquad\qquad \uparrow \qquad\qquad\qquad \uparrow$$

None of these operations is possible. **Division by zero is undefined.**

You may be wondering why division by zero is undefined. Let us think about it for a minute. We said that $7 \div 0$ is undefined. Suppose there were an answer. Let us call the answer a. So we assume for a minute that $7 \div 0 = a$. Then it would have to follow that $7 = 0(a)$. But this is impossible. Zero times any number is zero. So we see that if there were such a number, it would contradict known mathematical facts. Therefore there is no number a such that $7 \div 0 = a$. Thus we conclude that division by zero is undefined.

When combining two numbers, it is important to be sure you know which rule applies. Think about the concepts in the following chart. See if you agree with each example.

Operation	Two Real Numbers with the Same Sign	Two Real Numbers with Different Signs
Addition	Result may be positive or negative. $9 + 2 = 11$ $-5 + (-6) = -11$	Result may be positive or negative. $-3 + 7 = 4$ $4 + (-12) = -8$
Subtraction	Result may be positive or negative. $15 - 6 = 15 + (-6) = 9$ $-12 - (-3) = -12 + 3 = -9$	Result may be positive or negative. $-12 - 3 = -12 + (-3) = -15$ $5 - (-6) = 5 + 6 = 11$
Multiplication	Result is always positive. $9(3) = 27$ $-8(-5) = 40$	Result is always negative. $-6(12) = -72$ $8(-3) = -24$
Division	Result is always positive. $150 \div 6 = 25$ $-72 \div (-2) = 36$	Result is always negative. $-60 \div 10 = -6$ $30 \div (-6) = -5$

EXAMPLE 8   The Hamilton-Wenham Generals recently analyzed the 48 plays their team made while in the possession of the football during their last game. The following bar graph illustrates the number of plays made in each category. The team statistician prepared the following chart indicating the average number of yards gained or lost during each type of play.

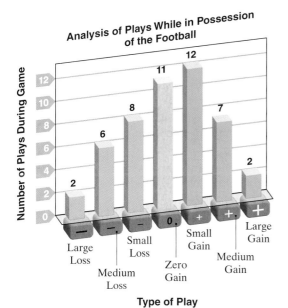

Analysis of Plays While in Possession of the Football

Type of Play	Average Yards Gained or Lost For Play
Large gain	+25
Medium gain	+15
Small gain	+5
Zero gain	0
Small loss	−5
Medium loss	−10
Large loss	−15

(a) How many yards were lost by the Generals in the plays that were considered small losses?

(b) How many yards were gained by the Generals in the plays that were considered small gains?

(c) If the total yards gained in small gains were combined with the total yards lost in small losses, what would be the result?

(a) We multiply the number of small losses by the average number of total yards lost on each small loss:

$$8(-5) = -40$$

The team lost approximately 40 yards with plays that were considered small losses.

(b) We multiply the number of small gains by the average number of yards gained on each small gain:

$$12(5) = 60$$

The team gained approximately 60 yards with plays that were considered small gains.

(c) We combine the results for (a) and (b):

$$-40 + 60 = 20$$

A total of 20 yards was gained during the plays that were small losses and small gains.

Practice Problem 8   Using the information provided in Example 8, answer the following:

(a) How many yards were lost by the Generals in the plays that were considered medium losses?

(b) How many yards were gained by the Generals in the plays that were considered medium gains?

(c) If the total yards gained in medium gains were combined with the total yards lost in medium losses, what would be the result?

Developing Your Study Skills

Reading the Textbook

Begin reading your textbook with a paper and pencil in hand. As you come across a new definition or concept, underline it in the text and/or write it down in your notebook. Whenever you encounter an unfamiliar term, look it up and make a note of it. When you come to an example, work through it step-by-step. Be sure to read each word and follow directions carefully.

Notice the helpful hints the author provides. They guide you to correct solutions and prevent you from making errors. Take advantage of these pieces of expert advice.

Be sure that you understand what you are reading. Make a note of any of those things that you do not understand and ask your instructor about them. Do not hurry through the material. Learning mathematics takes time.

1.3 Exercises

Verbal and Writing Skills

1. Explain in your own words the rule for determining the correct sign when multiplying two real numbers.

2. Explain in your own words the rule for determining the correct sign when multiplying three or more real numbers.

Multiply or divide. Be sure to write your answer in the simplest form.

3. $-8(3)$

4. $-5(11)$

5. $0(-12)$

6. $-3(150)$

7. $16(1.5)$

8. $24(2.5)$

9. $(-1.32)(-0.2)$

10. $(-2.3)(-0.11)$

11. $0.7(-2.5)$

12. $-6\left(\dfrac{3}{10}\right)$

13. $\left(\dfrac{12}{5}\right)(-10)$

14. $\left(\dfrac{3}{5}\right)\left(-\dfrac{15}{11}\right)$

15. $\left(-\dfrac{4}{9}\right)\left(-\dfrac{3}{5}\right)$

16. $\left(-\dfrac{5}{21}\right)\left(\dfrac{3}{10}\right)$

17. $12 \div (-4)$

18. $-36 \div (-9)$

19. $0 \div (-15)$

20. $-48 \div (-8)$

21. $-45 \div (9)$

22. $-220 \div (-11)$

23. $240 \div (-15)$

24. $156 \div (-13)$

25. $-0.6 \div 0.3$

26. $1.2 \div (-0.03)$

27. $-6.3 \div (0.7)$

28. $0.54 \div (-0.9)$

29. $-7.2 \div (8)$

30. $\dfrac{2}{7} \div \left(-\dfrac{3}{5}\right)$

31. $\left(-\dfrac{1}{5}\right) \div \left(\dfrac{2}{3}\right)$

32. $\left(-\dfrac{5}{6}\right) \div \left(-\dfrac{7}{18}\right)$

33. $\dfrac{5}{7} \div \left(-\dfrac{3}{28}\right)$

34. $\dfrac{4}{9} \div \left(-\dfrac{8}{15}\right)$

35. $\dfrac{12}{-\dfrac{2}{5}}$

36. $\dfrac{-\dfrac{3}{7}}{-6}$

37. $\dfrac{-\dfrac{3}{8}}{-\dfrac{2}{3}}$

38. $\dfrac{-\dfrac{1}{4}}{\dfrac{3}{8}}$

39. $\dfrac{\dfrac{7}{15}}{-\dfrac{3}{5}}$

Multiply. You may want to determine the sign of the product before you multiply.

40. $-6(2)(-3)(4)$

41. $-1(-2)(-3)(4)$

42. $-2(-1)(3)(-1)(-4)$

43. $-2(-2)(2)(-1)(-3)$

44. $-3(2)(-4)(0)(-2)$

45. $-3(-2)\left(\dfrac{1}{3}\right)(-4)(2)$

46. $60(-0.6)(-20)(0.5)$

47. $-3(-0.03)(100)(-2)$

48. $\left(\dfrac{3}{8}\right)\left(\dfrac{1}{2}\right)\left(-\dfrac{5}{6}\right)$

49. $\left(-\dfrac{4}{5}\right)\left(-\dfrac{6}{7}\right)\left(-\dfrac{1}{3}\right)$

50. $\left(-\dfrac{1}{2}\right)\left(\dfrac{4}{5}\right)\left(-\dfrac{7}{8}\right)\left(-\dfrac{2}{3}\right)$

51. $\left(-\dfrac{3}{4}\right)\left(-\dfrac{7}{15}\right)\left(-\dfrac{8}{21}\right)\left(-\dfrac{5}{9}\right)$

Mixed Practice

Take a minute to review the chart before Example 8. Be sure that you can remember the sign rules for each operation. Then do exercises 52–61. Perform the indicated calculations.

52. $-5 - (-2)$

53. $-20 \div 5$

54. $-3(-9)$

55. $5 + (-7)$

56. $-32 \div (-4)$

57. $8 - (-9)$

58. $-6 + (-3)$

59. $6(-12)$

60. $18 \div (-18)$

61. $-37 \div 37$

Applications

62. Debbie's monthly car loan payment is $250. If she pays by check, what is the net debit to her account after six months?

63. Phil pays a doctor bill for $100, but forgets to check the balance in his account. When he goes to the ATM, his balance reads −$70. How much was in his account before he wrote the check?

64. Ramon owes $6500 on his student loan. If $180 is automatically deducted from his bank account each month to pay the loan off, how much does he still owe after one year?

65. Keith's car loan of $15,768 is to be paid off in 48 equal monthly installments. What is his monthly bill?

The Beverly Panthers recently analyzed the 37 plays their team made while in the possession of the football during their last game. The team statistician prepared the following chart indicating the number of plays in each category and the average number of yards gained or lost during each type of play. Use this chart to answer exercises 66–73.

Type of Play	Number of Plays	Average Yards Gained or Lost per Play
Large gain	1	+25
Medium gain	6	+15
Small gain	4	+5
Zero gain	5	0
Small loss	10	−5
Medium loss	7	−10
Large loss	4	−15

66. How many yards were lost by the Panthers in the plays that were considered small losses?

67. How many yards were gained by the Panthers in the plays that were considered small gains?

68. If the total yards gained in small gains were combined with the total yards lost in small losses, what would be the result?

69. How many yards were lost by the Panthers in the plays that were considered medium losses?

70. How many yards were gained by the Panthers in the plays that were considered medium gains?

71. If the total yards gained in medium gains were combined with the total yards lost in medium losses, what would be the result?

72. The game being studied was lost by the Panthers. The coach said that if two of the large-loss plays had been avoided and the number of small-gain plays had been doubled, they would have won the game. If those actions listed by the coach had happened, how many additional yards would the Panthers have gained? Assume each play would reflect the average yards gained or lost per play.

73. The game being studied was lost by the Panthers. The coach of the opposing team said that they could have scored two more touchdowns against the Panthers if they had caused three more large-loss plays and avoided four medium-gain plays. If those actions had happened, what effect would this have had on total yards gained by the Panthers? Assume each play would reflect the average yards gained or lost per play.

Cumulative Review Problems

74. $-17.4 + 8.31 + 2.40$

75. $-13 + (-39) + (-20)$

76. $-3.7 - (-8.33)$

77. $-37 - 51$

78. A very famous sneaker company needs $104\frac{1}{2}$ square yards of white leather, $88\frac{2}{3}$ square yards of neon-yellow leather, and $72\frac{5}{6}$ square yards of neon-red leather to make today's batch of designer sneakers. What is today's required total square yardage of leather?

Developing Your Study Skills

Steps Toward Success In Mathematics

Mathematics is a building process, mastered one step at a time. The foundation of this process consists of a few basic requirements. Those who are successful in mathematics realize the absolute necessity for building a study of mathematics on the firm foundation of these six minimum requirements.

1. Attend class every day.

2. Read the textbook.

3. Take notes in class.

4. Do assigned homework every day.

5. Get help immediately when needed.

6. Review regularly.

1 *Writing Numbers in Exponent Form*

In mathematics, we use exponents as a way to abbreviate repeated multiplication.

Long Notation		**Exponent Form**
$2 \cdot 2 \cdot 2 \cdot 2 \cdot 2 \cdot 2$	$=$	2^6

There are two parts to exponent notation: (1) the **base** and (2) the **exponent**. The **base** tells you what number is being multiplied and the **exponent** tells you how many times this number is used as a factor. (A *factor*, you recall, is a number being multiplied.)

$$2 \cdot 2 \cdot 2 \cdot 2 \cdot 2 \cdot 2 = 2^6$$

The *base* is 2   the *exponent* is 6

(the number being multiplied)   (the number of times 2 is used as a factor)

If the base is a *positive* real number, the exponent appears to the right and slightly above the level of the number as in, for example, 5^6 and 8^3. If the base is a *negative* real number, then parentheses are used around the number and the exponent appears outside the parentheses. For example, $(-2)(-2)(-2) = (-2)^3$.

In algebra, if we do not know the value of a number, we use a letter to represent the unknown number. We call the letter a **variable**. This is quite useful in the case of exponents. Suppose we do not know the value of a number, but we know the number is multiplied by itself several times. We can represent this with a variable base and a whole-number exponent. For example, when we have an unknown number, represented by the variable x, and this number occurs as a factor four times, we have

$$(x)(x)(x)(x) = x^4.$$

Likewise if an unknown number, represented by the variable w, occurs as a factor five times, we have

$$(w)(w)(w)(w)(w) = w^5.$$

EXAMPLE 1   Write in exponent form.

(a) $9(9)(9)$ **(b)** $13(13)(13)(13)$ **(c)** $-7(-7)(-7)(-7)(-7)$

(d) $-4(-4)(-4)(-4)(-4)(-4)$ **(e)** $(x)(x)$ **(f)** $(y)(y)(y)$

(a) $9(9)(9) = 9^3$ **(b)** $13(13)(13)(13) = 13^4$

(c) The -7 is used as a factor five times. The answer must contain parentheses. Thus $-7(-7)(-7)(-7)(-7) = (-7)^5$.

(d) $-4(-4)(-4)(-4)(-4)(-4) = (-4)^6$ **(e)** $(x)(x) = x^2$ **(f)** $(y)(y)(y) = y^3$

Practice Problem 1   Write in exponent form.

(a) $6(6)(6)(6)$ **(b)** $-2(-2)(-2)(-2)(-2)$ **(c)** $108(108)(108)$

(d) $-11(-11)(-11)(-11)(-11)(-11)$ **(e)** $(w)(w)(w)$ **(f)** $(z)(z)(z)(z)$

If the base has an exponent of 2, we say the base is **squared**.

If the base has an exponent of 3, we say the base is **cubed**.

If the base has an exponent greater than 3, we say the base is raised **to the (exponent)-th power**.

x^2 is read "*x* squared."

y^3 is read "*y* cubed."

3^6 is read "three to the sixth power" or simply "three to the sixth."

2 *Evaluating Numerical Expressions That Contain Exponents*

EXAMPLE 2 Evaluate.   **(a)** 2^5 **(b)** $2^3 + 4^4$

(a) $2^5 = (2)(2)(2)(2)(2) = 32$ **(b)** First we evaluate each power.
$$2^3 = 8 \qquad 4^4 = 286$$
Then we add. $8 + 256 = 264$

Practice Problem 2 Evaluate.   **(a)** 3^5 **(b)** $2^2 + 3^3$

If the base is negative, be especially careful in determining the sign. Notice the following:
$$(-3)^2 = (-3)(-3) = +9 \qquad (-3)^3 = (-3)(-3)(-3) = -27$$
From Section 1.3 we know that when you multiply two or more real numbers, first you multiply their absolute values.

- The result is positive if there are an even number of negative signs.
- The result is negative if there are an odd number of negative signs.

Sign Rule for Exponents

Suppose that a number is written in exponent form and the base is negative. The result is **positive** if the exponent is **even**. The result is **negative** if the exponent is **odd**.

Be careful how you read expressions with exponents and negative signs.
$$(-3)^4 \text{ means } (-3)(-3)(-3)(-3) \text{ or } +81.$$
$$-3^4 \text{ means } -(3)(3)(3)(3) \text{ or } -81.$$

Calculator

Exponents

You can use a calculator to evaluate 3^5. Press the following keys:

$\boxed{3}$ $\boxed{y^x}$ $\boxed{5}$ $\boxed{=}$

The display should read
$\boxed{\qquad 243}$

Try the following.

(a) 4^6 **(b)** $(0.2)^5$
(c) 18^6 **(d)** 3^{12}

EXAMPLE 3 Evaluate.   **(a)** $(-2)^3$ **(b)** $(-4)^6$ **(c)** -3^6 **(d)** $-(5^4)$

(a) $(-2)^3 = -8$ The answer is negative since the base is negative and the exponent 3 is odd.

(b) $(-4)^6 = +4096$ The answer is positive since the exponent 6 is even.

(c) $-3^6 = -729$ The negative sign is not contained in parentheses. Thus we find 3 raised to the sixth power and then take the negative of that value.

(d) $-(5^4) = -625$ The negative sign is outside the parentheses.

Practice Problem 3 Evaluate. **(a)** $(-3)^3$ **(b)** $(-2)^6$ **(c)** -2^4 **(d)** $-(3^6)$

EXAMPLE 4 Evaluate.

(a) $\left(\dfrac{1}{2}\right)^4$ **(b)** $(0.2)^4$ **(c)** $\left(\dfrac{2}{5}\right)^3$ **(d)** $(3)^3(2)^5$ **(e)** $2^3 - 3^4$

(a) $\left(\dfrac{1}{2}\right)^4 = \left(\dfrac{1}{2}\right)\left(\dfrac{1}{2}\right)\left(\dfrac{1}{2}\right)\left(\dfrac{1}{2}\right) = \dfrac{1}{16}$ **(b)** $(0.2)^4 = (0.2)(0.2)(0.2)(0.2) = 0.0016$

(c) $\left(\dfrac{2}{5}\right)^3 = \left(\dfrac{2}{5}\right)\left(\dfrac{2}{5}\right)\left(\dfrac{2}{5}\right) = \dfrac{8}{125}$ **(d)** First we evalute each power.
$$3^3 = 27 \qquad 2^5 = 32$$
Then we multiply. $(27)(32) = 864$

(e) $2^3 - 3^4 = 8 - 81 = -73$

Practice Problem 4 Evaluate.

(a) $\left(\dfrac{1}{3}\right)^3$ **(b)** $(0.3)^4$ **(c)** $\left(\dfrac{3}{2}\right)^4$ **(d)** $(3)^4(4)^2$ **(e)** $4^2 - 2^4$

1.4 Exercises

Verbal and Writing Skills

1. Explain in your own words how to evaluate 4^4.

2. Explain in your own words how to evaluate 9^2.

3. Explain how you would determine whether $(-5)^3$ is negative or positive.

4. Explain how you would determine whether $(-2)^5$ is negative or positive.

5. Explain the difference between $(-2)^4$ and -2^4. What answers do you obtain when you evaluate the expressions?

6. Explain the difference between $(-3)^4$ and -3^4. What answers do you obtain when you evaluate the expressions?

Evaluate.

7. 3^3

8. 4^2

9. 3^4

10. 8^3

11. 7^3

12. 5^4

13. $(-3)^3$

14. $(-2)^3$

15. $(-2)^6$

16. $(-3)^4$

17. -5^3

18. -4^3

19. $\left(\dfrac{1}{4}\right)^2$

20. $\left(\dfrac{1}{2}\right)^3$

21. $\left(\dfrac{2}{5}\right)^3$

22. $\left(\dfrac{2}{3}\right)^4$

23. $(0.9)^2$

24. $(0.4)^2$

25. $(0.2)^4$

26. $(0.7)^3$

27. $(-16)^2$

28. $(-7)^4$

29. -16^2

30. -7^4

Write in exponent form.

31. $(6)(6)(6)(6)(6)$

32. $(8)(8)(8)(8)(8)(8)$

33. $(w)(w)$

34. $(w)(w)(w)$

35. $(x)(x)(x)(x)$

36. $(x)(x)(x)(x)(x)$

37. $(3q)(3q)(3q)$

38. $(6x)(6x)(6x)(6x)$

Evaluate.

39. $5^3 + 6^2$

40. $7^2 + 6^3$

41. $(-3)^3 - (8)^2$

42. $(-2)^3 - (-5)^4$

43. $5^3 - (-3)^3$

44. $9^2 - (-4)^2$

45. $(-4)^3(-3)^2$

46. $(-7)^3(-2)^4$

47. $8^2(-2)^3$

48. $9^2(-3)^3$

49. 4^{12}

50. 6^{11}

To Think About

51. What number to the seventh power equals 128?

52. What number to the fifth power equals -243?

Cumulative Review Problems

Evaluate.

53. $(-11) + (-13) + 6 + (-9) + 8$

54. $\dfrac{3}{4} \div \left(-\dfrac{9}{20}\right)$

55. $-17 - (-9)$

56. $(-2.1)(-1.2)$

57. Amanda decided to invest her summer job earnings of $1600 in a bank certificate of deposit, at an annual interest rate of 6% for nine months. How much money did Amanda have at the end of the nine months when her certificate of deposit matured?

1.5 Order of Arithmetic Operations

It is important to know *when* to do certain operations as well as how to do them. For example, to simplify the expression $2 - 4 \cdot 3$, should we subtract first or multiply first?

The following list will assist you. It tells which operations to do first: the correct **order of operations**. You might think of it as a *list of priorities*.

Order of Operations for Numbers

Follow this order of operations:

Do first   **1.** Combine numbers inside parentheses.

   2. Raise numbers to a power.

   3. Multiply and divide numbers from left to right.

Do last   **4.** Add and subtract numbers from left to right.

Let's return to the problem $2 - 4 \cdot 3$. There are no parentheses or numbers raised to a power, so multiplication comes next. We do that first. Then we subtract since this comes last on our list.

$$2 - 4 \cdot 3 = 2 - 12 \quad \text{Follow the order of operations by first multiplying } 4 \cdot 3 = 12.$$
$$= -10 \quad \text{Combine } 2 - 12 = -10.$$

EXAMPLE 1   Evaluate. $8 \div 2 \cdot 3 + 4^2$

$$8 \div 2 \cdot 3 + 4^2 = 8 \div 2 \cdot 3 + 16 \quad \text{Evaluate } 4^2 = 16 \text{ because the highest priority in this problem is raising to a power.}$$
$$= 4 \cdot 3 + 16 \quad \text{Next multiply and divide from left to right. So } 8 \div 2 = 4 \text{ and } 4 \cdot 3 = 12.$$
$$= 12 + 16$$
$$= 28 \quad \text{Finally, add.}$$

Practice Problem 1   Evaluate. $25 \div 5 \cdot 6 + 2^3$

Note: Multiplication and division have equal priority. We do not do multiplication first. Rather, we work from left to right, doing any multiplication or division that we encounter. Similarly, addition and subtraction have equal priority.

EXAMPLE 2   Evaluate. $(-3)^3 - 2^4$

The highest priority is to raise the expressions to the appropriate powers.

$$(-3)^3 - 2^4 = -27 - 16 \quad \text{In } (-3)^3 \text{ we are cubing the number } -3 \text{ to obtain } -27.$$
$$\text{Be careful; } -2^4 \text{ is not } (-2)^4!$$
$$\text{Raise 2 to the fourth power and take the negative of the result.}$$
$$= -43 \quad \text{The last step is to add and subtract from left to right.}$$

Practice Problem 2   Evaluate. $(-4)^3 - 2^6$

Student Learning Objectives

After studying this section, you will be able to:

 Use the order of operations to simplify numerical expressions involving addition, subtraction, multiplication, division, and exponents.

SSM   PH TUTOR   CD & VIDEO   MATH PRO   WEB
CENTER

EXAMPLE 3   Evaluate. $2 \cdot (2 - 3)^3 + 6 \div 3 + (8 - 5)^2$

$2 \cdot (2 - 3)^3 + 6 \div 3 + (8 - 5)^2$ Combine the numbers inside the parentheses.

$= 2 \cdot (-1)^3 + 6 \div 3 + 3^2$

$= 2 \cdot (-1) + 6 \div 3 + 9$ Next, raise to a power. Note that we need parentheses for -1 because of the negative sign, but they are not needed for 3.

$= -2 + 2 + 9$ Next, multiply and divide from left to right.

$= 9$ Finally, add and subtract from left to right.

Practice Problem 3   Evaluate. $6 - (8 - 12)^2 + 8 \div 2$

EXAMPLE 4   Evaluate. $\left(-\frac{1}{5}\right)\left(\frac{1}{2}\right) - \left(\frac{3}{2}\right)^2$

The highest priority is to raise $\frac{3}{2}$ to the second power.

$$\left(\frac{3}{2}\right)^2 = \left(\frac{3}{2}\right)\left(\frac{3}{2}\right) = \frac{9}{4} \quad \text{Next we multiply.}$$

$$\left(-\frac{1}{5}\right)\left(\frac{1}{2}\right) - \left(\frac{3}{2}\right)^2 = \left(-\frac{1}{5}\right)\left(\frac{1}{2}\right) - \frac{9}{4}$$

$$= -\frac{1}{10} - \frac{9}{4}$$

$$= -\frac{1 \cdot 2}{10 \cdot 2} - \frac{9 \cdot 5}{4 \cdot 5} \quad \text{We need to write each fraction as an equivalent fraction with the LCD of 20.}$$

$$= -\frac{2}{20} - \frac{45}{20}$$

$$= -\frac{47}{20} \quad \text{Add.}$$

Practice Problem 4   Evaluate. $\left(-\frac{1}{7}\right)\left(-\frac{14}{5}\right) + \left(-\frac{1}{2}\right) \div \left(\frac{3}{4}\right)$

Calculator

Order of Operations

Use your calculator to evaluate $3 + 4 \cdot 5$. Enter

3 $+$ 4 $\times$ 5 $=$

If the display is ⌷ 23, the correct order of operations is built in. If the display is not 23, you will need to modify the way you enter the problem. You should use

4 $\times$ 5 $+$ 3 $=$

Try $6 + 3 \cdot 4 - 8 \div 2$.

Developing Your Study Skills

Previewing New Material

Part of your study time each day should consist of looking over the sections in your text that are to be covered the following day. You do not necessarily need to study and learn the material on your own, but a survey of the concepts, terminology, diagrams, and examples will help the new ideas seem more familiar as the instructor presents them. You can look for concepts that appear confusing or difficult and be ready to listen carefully for your instructor's explanations. You can be prepared to ask the questions that will increase your understanding. Previewing new material enables you to see what is coming and prepares you to be ready to absorb it.

Verbal and Writing Skills

You have lost a game of UNO and are counting the points left in your hand. You announce that you have three fours and six fives.

1. Write this as a number expression.

2. How many points have you in your hand?

3. What answer would you get for the number expression if you simplified it by
(a) performing the operations from left to right?

(b) following the order of operations?

4. Which procedure in exercise 3 gives the correct number of total points?

Evaluate.

5. $(2 - 5)^2 \div 3 \times 4$

6. $2(3 - 5 + 6) + 5$

7. $8 - 2^3 \cdot 5 + 3$

8. $-14 \div (-7) - 8 \cdot 2 + 3^3$

9. $4 + 27 \div 3 \cdot 2 - 8$

10. $3 \cdot 5 + 7 \cdot 3 - 5 \cdot 3$

11. $8 - 5(2)^3 \div (-8)$

12. $11 - 3(4)^2 \div (-6)$

13. $3(5 - 7)^2 - 6(3)$

14. $-2(3 - 6)^2 - (-2)$

15. $5 \cdot 6 - (3 - 5)^2 + 8 \cdot 2$

16. $(-3)^2 \cdot 6 \div 9 + 4 \cdot 2$

17. $\dfrac{1}{2} \div \dfrac{2}{3} + 6 \cdot \dfrac{1}{4}$

18. $\dfrac{5}{6} \div \dfrac{2}{3} - 6 \cdot \left(\dfrac{1}{2}\right)^2$

19. $0.8 + 0.3(0.6 - 0.2)^2$

20. $0.05 + 1.4 - (0.5 - 0.7)^3$

21. $\dfrac{3}{4}\left(-\dfrac{2}{5}\right) - \left(-\dfrac{3}{5}\right)$

22. $-\dfrac{2}{3}\left(\dfrac{3}{5}\right) + \dfrac{5}{7} \div \dfrac{5}{3}$

23. $-6.3 - (-2.7)(1.1) + (3.3)^2$

24. $4.35 + 8.06 \div (-2.6) - (2.1)^2$

25. $\left(\dfrac{1}{2}\right)^3 + \dfrac{1}{4} - \left(\dfrac{1}{6} - \dfrac{1}{12}\right) - \dfrac{2}{3} \cdot \left(\dfrac{1}{4}\right)^2$

26. $(2.4 \cdot 1.2)^2 - 1.6 \cdot 2.2 \div 4.0 - 3.6$

Cumulative Review Problems   *Simplify.*

27. $(0.5)^3$

28. $-\dfrac{3}{4} - \dfrac{5}{6}$

29. -1^{20}

30. $3\dfrac{3}{5} \div 6\dfrac{1}{4}$

31. An Olympic weight lifter has been told by his trainer to increase his daily consumption of protein by 15 grams. The trainer suggested a health drink that provides 2 grams of protein for every 6 ounces consumed. How many ounces of this health drink would the weight lifter need to consume daily to reach his goal of 15 additional grams per day?

1.6 Using the Distributive Property to Simplify Expressions

Student Learning Objectives

After studying this section, you will be able to:

1. Use the distributive property to simplify expressions.

SSM
PH TUTOR CENTER | CD & VIDEO | MATH PRO | WEB

1 Using the Distributive Property to Simplify Expressions

As we learned previously, we use letters called *variables* to represent unknown numbers. If a number is multiplied by a variable we do not need any symbol between the number and variable. Thus, to indicate $(2)(x)$, we write $2x$. To indicate $3 \cdot y$, we write $3y$. If one variable is multiplied by another variable, we place the variables next to each other. Thus, $(a)(b)$ is written ab. We use exponent form if an unknown number (a variable) is used several times as a factor. Thus, $x \cdot x \cdot x = x^3$. Similarly, $(y)(y)(y)(y) = y^4$.

In algebra, we need to be familiar with several definitions. We will use them throughout the remainder of this book. Take some time to think through how each of these definitions is used.

An **algebraic expression** is a quantity that contains numbers and variables, such as $a + b$, $2x - 3$, and $5ab^2$. In this chapter we will be learning rules about adding and multiplying algebraic expressions. A **term** is a number, a variable, or a product of numbers and variables. 17, x, $5xy$, and $22xy^3$ are all examples of terms. We will refer to terms when we discuss the distributive property.

An important property of algebra is the **distributive property**. We can state it in an equation as follows:

Distributive Property

For all real numbers a, b, and c,

$$a(b + c) = ab + ac.$$

A numerical example shows that it does seem reasonable.

$$5(3 + 6) = 5(3) + 5(6)$$
$$5(9) = 15 + 30$$
$$45 = 45$$

We can use the distributive property to multiply any term by the sum of two or more terms. In Section 1.4, we defined the word *factor*. Two or more algebraic expressions that are multiplied are called **factors**. Consider the following examples of multiplying algebraic expressions.

EXAMPLE 1 Multiply. **(a)** $5(a + b)$ **(b)** $-1(3x + 2y)$

(a) $5(a + b) = 5a + 5b$ Multiply the factor $(a + b)$ by the factor 5.

(b) $-1(3x + 2y) = -1(3x) + (-1)(2y)$ Multiply the factor $(3x + 2y)$ by the factor -1.
$$= -3x - 2y$$

Practice Problem 1 Multiply.

(a) $-3(x + 2y)$ **(b)** $-a(a - 3b)$

If the parentheses are preceded by a negative sign, we consider this to be the product of (-1) and the expression inside the parentheses.

EXAMPLE 2 Multiply. $-(a - 2b)$

$$-(a - 2b) = (-1)(a - 2b) = (-1)(a) + (-1)(-2b) = -a + 2b$$

Practice Problem 2 Multiply. $-(-3x + y)$

In general, we see that in all these examples we have multiplied each term of the expression in the parentheses by the expression in front of the parentheses.

EXAMPLE 3   Multiply.   **(a)** $\frac{2}{3}(x^2 - 6x + 8)$ **(b)** $1.4(a^2 + 2.5a + 1.8)$

(a) $\frac{2}{3}(x^2 - 6x + 8) = \left(\frac{2}{3}\right)(1x^2) + \left(\frac{2}{3}\right)(-6x) + \left(\frac{2}{3}\right)(8)$

$$= \frac{2}{3}x^2 + (-4x) + \frac{16}{3}$$

$$= \frac{2}{3}x^2 - 4x + \frac{16}{3}$$

(b) $1.4(a^2 + 2.5a + 1.8) = 1.4(1a^2) + (1.4)(2.5a) + (1.4)(1.8)$

$$= 1.4a^2 + 3.5a + 2.52$$

Practice Problem 3   Multiply.

(a) $\frac{3}{5}(a^2 - 5a + 25)$ **(b)** $2.5(x^2 - 3.5x + 1.2)$

There are times we multiply a variable by itself and use exponent notation. For example, $(x)(x) = x^2$ and $(x)(x)(x) = x^3$. In other cases there will be numbers and variables multiplied at the same time.

We will see problems like $(2x)(x) = (2)(x)(x) = 2x^2$. Some expressions will involve multiplication of more than one variable. We will see problems like $(3x)(xy) = (3)(x)(x)(y) = 3x^2y$. There will be times when we use the distributive property and all of these methods will be used. For example,

$$2x(x - 3y + 2) = 2x(x) + (2x)(-3y) + (2x)(2)$$

$$= 2x^2 + (-6)(xy) + 4(x)$$

$$= 2x^2 - 6xy + 4x.$$

We will discuss this type of multiplication of variables with exponents in more detail in Section 4.1. At that point we will expand these examples and other similar examples to develop the general rule for multiplication $(x^a)(x^b) = x^{a+b}$.

EXAMPLE 4   Multiply. $-2x(3x + y - 4)$

$$-2x(3x + y - 4) = -2(x)(3)(x) + (-2)(x)(y) + (-2)(x)(-4)$$

$$= -2(3)(x)(x) + (-2)(xy) + (-2)(-4)(x)$$

$$= -6x^2 - 2xy + 8x$$

Practice Problem 4   Multiply. $-4x(x - 2y + 3)$

The distributive property can also be presented with the a on the right.

$$(b + c)a = ba + ca$$

The a is "distributed" over the b and c inside the parentheses.

EXAMPLE 5 Multiply. $(2x^2 - x)(-3)$

$$(2x^2 - x)(-3) = 2x^2(-3) + (-x)(-3)$$
$$= -6x^2 + 3x$$

Practice Problem 5 Multiply.

$$(3x^2 - 2x)(-4)$$

EXAMPLE 6 A farmer has a rectangular field that is 300 feet wide. One portion of the field is $2x$ feet long. The other portion of the field is $3y$ feet long. Use the distributive property to find an expression for the area of this field.

First we draw a picture of a field that is 300 feet wide and $2x + 3y$ feet long.

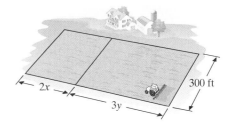

To find the area of the field, we multiply the width times the length.

$$300(2x + 3y) = 300(2x) + 300(3y) = 600x + 900y$$

Thus the area of the field in square feet is $600x + 900y$.

Practice Problem 6 A farmer has a rectangular field that is 400 feet wide. One portion of the field is $6x$ feet long. The other portion of the field is $9y$ feet long. Use the distributive property to find an expression for the area of this field.

Developing Your Study Skills

How to Do Homework

As you begin your homework assignments, read the directions carefully. You need to understand what is being asked for. Concentrate on each exercise, taking time to solve it accurately. Rushing through your work usually causes errors. Check your answers with those given in the back of the textbook. If your answer is incorrect, check to see that you are doing the right exercise. Redo the exercise, watching for little errors. If it is still wrong, check with a friend. Perhaps the two of you can figure it out.

Verbal and Writing Skills

In exercises 1 and 2, complete each sentence by filling in the blank.

1. A _____ is a symbol used to represent an unknown number.

2. When we write an expression with numbers and variables such as $7x$ it indicates that we are _____ 7 by x.

3. Explain in your own words how we multiply a problem like $(4x)(x)$.

4. Explain why you think the property $a(b + c) = ab + ac$ is called the distributive property. What does distribute mean?

5. Does the following distributive property work? $a(b - c) = ab - ac$ Why or why not? Give an example.

Multiply. Use the distributive property.

6. $-(a - 2b)$

7. $-(x + 4y)$

8. $-2(4a - 3b)$

9. $-3(2a - 5b)$

10. $3(3x - y + 5)$

11. $2(4x + y - 2)$

12. $-2b(a - 3b + c)$

13. $-3x(-2x + 3y - z)$

14. $3x(x - 3y - 7)$

15. $2x(4x - y - 6)$

16. $-9(9x - 5y + 8)$

17. $-5(3x + 9 - 7y)$

18. $\frac{1}{3}(3x^2 + 2x - 1)$

19. $\frac{1}{4}(x^2 + 2x - 8)$

20. $\frac{5}{6}(12x^2 - 24x + 18)$

21. $\frac{2}{3}(-27a^4 + 9a^2 - 21)$

22. $\frac{x}{5}(x + 10y - 4)$

23. $\frac{y}{3}(3y - 4x - 6)$

24. $5x(x + 2y + z - 1)$

25. $3a(2a + b - c - 4)$

26. $(2x - 3)(-2)$

27. $(5x + 1)(-4)$

28. $2x(3x + y - 4)$

29. $3x(4x - 5y - 6)$

30. $(3x + 2y - 1)(-xy)$

31. $(4a - 2b - 1)(-ab)$

32. $(2x + 3y - 2)3xy$

33. $(-2x + y - 3)4xy$

34. $1.5(2.8x^2 + 3.0x - 2.5)$

35. $2.5(1.5a^2 - 3.5a + 2.0)$

36. $-0.3x(-1.2x^2 - 0.3x + 0.5)$

37. $-0.9q(2.1q - 0.2r - 0.8s)$

38. $0.5x(0.6x + 0.8y - 5)$

Applications

▲ **39.** Blaine Johnson has a rectangular field that is 700 feet wide. One portion of the field is $12x$ feet long. The other portion of the field is $8y$ feet long. Use the distributive property to find an expression for the area of this field.

▲ **40.** Kathy DesMaris has a rectangular field that is 800 feet wide. One portion of the field is $5x$ feet long. The other portion of the field is $14y$ feet long. Use the distributive property to find an expression for the area of this field

To Think About

▲ **41.** The athletic field at Baxford College is $3x$ feet wide. It used to be 1500 feet long. However, due to the construction of a new dormitory, the field length was decreased by $4y$ feet. Use the distributive property to find an expression for the area of the new field after construction of the dormitory.

▲ **42.** The Beverly Airport runway is $4x$ feet wide. The airport was supposed to have a 3000-foot-long runway. However, some of the land was wetland, so a runway could not be built on all of it. Therefore, the length of the runway was decreased by $2y$ feet. Use the distributive property to find an expression for the area of the final runway.

Cumulative Review Problems

In exercises 43–45, evaluate.

43. $-18 + (-20) + 36 + (-14)$

44. $(-2)^6$

45. $-27 - (-41)$

46. The San Rafael Junior Soccer League would practice 365 days per year if it could. Last year because of the weather, the players could practice only 205 days. What percent of the days last year could players practice? (Round to the nearest whole percent.)

47. The coach of the team in exercise 46 predicts that this year the team will be able to practice 60% of the year. How many more days of practice will be possible this year than were possible last year?

1.7 Combining Like Terms

1 Identifying Like Terms

We can add or subtract quantities that are *like quantities*. This is called **combining** like quantities.

$$5 \text{ inches} + 6 \text{ inches} = 11 \text{ inches}$$
$$20 \text{ square inches} - 16 \text{ square inches} = 4 \text{ square inches}$$

However, we cannot combine things that are not the same.

$$16 \text{ square inches} - 4 \text{ inches} \text{ (Cannot be done!)}$$

Similarly, in algebra we can **combine like terms**. This means to add or subtract like terms. Remember, we cannot combine terms that are not the same. Recall that a *term* is a number, a variable, or a product of numbers and variables. **Like terms** are terms that have identical variables and exponents. In other words, like terms must have exactly the same letter parts.

EXAMPLE 1 List the like terms of each expression.

(a) $5x - 2y + 6x$ **(b)** $2x^2 - 3x - 5x^2 - 8x$

(a) $5x$ and $6x$ are like terms. These are the only like terms.

(b) $2x^2$ and $-5x^2$ are like terms.
$-3x$ and $-8x$ are like terms.
Note that x^2 and x are not like terms.

Practice Problem 1 List the like terms of each expression.

(a) $5a + 2b + 8a - 4b$ **(b)** $x^2 + y^2 + 3x - 7y^2$

Do you really understand what a term is? A term is a number, a variable, or a product of numbers and variables. Terms are the parts of an algebraic expression separated by plus signs. The sign in front of the term is considered part of the term.

2 Combining Like Terms

It is important to know how to combine like terms. Since

$$4 \text{ inches} + 5 \text{ inches} = 9 \text{ inches},$$

we would expect in algebra that

$$4x + 5x = 9x.$$

Why is this true? Let's take a look at the distributive property.

> Like terms may be added or subtracted by using the distributive property:
> $$ab + ac = a(b + c) \quad \text{and} \quad ba + ca = (b + c)a$$

For example,

$$-7x + 9x = (-7 + 9)x = 2x$$
$$5x^2 + 12x^2 = (5 + 12)x^2 = 17x^2.$$

EXAMPLE 2 Combine like terms. **(a)** $-4x^2 + 8x^2$ **(b)** $5x + 3x + 2x$

(a) Notice that each term contains the factor x^2. Using the distributive property, we have

$$-4x^2 + 8x^2 = (-4 + 8)x^2 = 4x^2.$$

(b) Note that each term contains the factor x. Using the distributive property, we have

$$5x + 3x + 2x = (5 + 3 + 2)x = 10x.$$

Practice Problem 2 Combine like terms.

(a) $5a + 7a + 4a$ **(b)** $16y^3 + 9y^3$

In this section, the direction *simplify* means to remove parentheses and/or combine like terms.

EXAMPLE 3 Simplify. $5a^2 - 2a^2 + 6a^2$

$$5a^2 - 2a^2 + 6a^2 = (5 - 2 + 6)a^2 = 9a^2$$

Practice Problem 3 Simplify. $-8y^2 - 9y^2 + 4y^2$

After doing a few problems, you will find that it is not necessary to write out the step of using the distributive property. We will omit this step for the remaining examples in this section.

EXAMPLE 4 Simplify.   **(a)** $5a + 2b + 7a - 6b$
(b) $3x^2y - 2xy^2 + 6x^2y$ **(c)** $2a^2b + 3ab^2 - 6a^2b^2 - 8ab$

(a) $5a + 2b + 7a - 6b = 12a - 4b$	We combine the a terms and the b terms separately.
(b) $3x^2y - 2xy^2 + 6x^2y = 9x^2y - 2xy^2$	**Note:** x^2y and xy^2 are not like terms because of different powers.
(c) $2a^2b + 3ab^2 - 6a^2b^2 - 8ab$	These terms cannot be combined; there are no like terms in this expression.

Practice Problem 4 Simplify.   **(a)** $-x + 3a - 9x + 2a$
(b) $5ab - 2ab^2 - 3a^2b + 6ab$ **(c)** $7x^2y - 2xy^2 - 3x^2y - 4xy^2 + 5x^2y$

The two skills in this section that a student must practice are identifying like terms and correctly adding and subtracting like terms. If a problem involves many terms, you may find it helpful to rearrange the terms so that like terms are together.

EXAMPLE 5 Simplify. $3a - 2b + 5a^2 + 6a - 8b - 12a^2$

There are three pairs of like terms.

$\underbrace{3a + 6a}_{a \text{ terms}} \;\; \underbrace{- 2b - 8b}_{b \text{ terms}} \;\; \underbrace{+ 5a^2 - 12a^2}_{a^2 \text{ terms}}$	You can rearrange the terms so that like terms are together, making it easier to combine them.
$= 9a - 10b - 7a^2$	Combine like terms.

The order of terms in an answer to this problem is not significant. These three terms can be rearranged in a different order. $-10b + 9a - 7a^2$ and $-7a^2 + 9a - 10b$ are also correct. However, we usually write polynomials in order of descending exponents. $-7a^2 + 9a - 10b$ would be the preferred way to write the answer.

Practice Problem 5 Simplify. $5xy - 2x^2y + 6xy^2 - xy - 3xy^2 - 7x^2y$

EXAMPLE 6 Simplify. $6(2x + 3xy) - 8x(3 - 4y)$

First remove the parentheses; then combine like terms.

$$6(2x + 3xy) - 8x(3 - 4y) = 12x + 18xy - 24x + 32xy \qquad \text{Use the distributive property.}$$

$$= -12x + 50xy \qquad \text{Combine like terms.}$$

Practice Problem 6 Simplify. $5a(2 - 3b) - 4(6a + 2ab)$

Use extra care with fractional values.

EXAMPLE 7 Simplify. $\dfrac{3}{4}x^2 - 5y - \dfrac{1}{8}x^2 + \dfrac{1}{3}y$

We need a least common denominator for the x^2 terms, which is 8.
Change $\dfrac{3}{4}$ to eighths by multiplying the numerator and denominator by 2.

$$\frac{3}{4}x^2 - \frac{1}{8}x^2 = \frac{3 \cdot 2}{4 \cdot 2}x^2 - \frac{1}{8}x^2 = \frac{6}{8}x^2 - \frac{1}{8}x^2 = \frac{5}{8}x^2$$

The least common denominator for the y terms is 3. Change 5 to thirds.

$$-\frac{5}{1}y + \frac{1}{3}y = \frac{-5 \cdot 3}{1 \cdot 3}y + \frac{1}{3}y = \frac{-15}{3}y + \frac{1}{3}y = -\frac{14}{3}y$$

Thus, our solution is $\dfrac{5}{8}x^2 - \dfrac{14}{3}y$.

Practice Problem 7 Simplify. $\dfrac{1}{7}a^2 - \dfrac{5}{12}b + 2a^2 - \dfrac{1}{3}b$

1.7 Exercises

Verbal and Writing Skills

1. Explain in your own words the mathematical meaning of the word *term*.

2. Explain in your own words the mathematical meaning of the phrase *like terms*.

3. Explain which terms are like terms in the expression $5x - 7y - 8x$.

4. Explain which terms are like terms in the expression $12a - 3b - 9a$.

5. Explain which terms are like terms in the expression $7xy - 9x^2y - 15xy^2 - 14xy$.

6. Explain which terms are like terms in the expression $-3a^2b - 12ab + 5ab^2 + 9ab$.

Simplify.

7. $-14b^2 - 11b^2$

8. $-17x^5 + 3x^5$

9. $10x^4 + 8x^4 + 7x^2$

10. $3a^3 - 6a^2 + 5a^3$

11. $2ab + 1 - 6ab - 8$

12. $2x^2 + 3x^2 - 7 - 5x^2$

13. $1.3x - 2.6y + 5.8x - 0.9y$

14. $3.1ab - 0.2b - 0.8ab + 5.3b$

15. $1.6x - 2.8y - 3.6x - 5.9y$

16. $1.9x - 2.4b - 3.8x - 8.2b$

17. $\dfrac{1}{2}x^2 - 3y - \dfrac{1}{3}y + \dfrac{1}{4}x^2$

18. $\dfrac{1}{5}a^2 - 2b - \dfrac{1}{2}a^2 - 3b$

19. $\dfrac{1}{3}x - \dfrac{2}{3}y - \dfrac{2}{5}x + \dfrac{4}{7}y$

20. $\dfrac{2}{5}s - \dfrac{3}{8}t - \dfrac{4}{15}s - \dfrac{5}{12}t$

21. $3p - 4q + 2p + 3 + 5q - 21$

22. $6x - 5y - 3y + 7 - 11x - 5$

23. $5x^2y - 10xy^6 + 6xy^2 - 7xy^2$

24. $5bcd - 8cd - 12bcd + cd$

25. $2ab + 5bc - 6ac - 2ab$

26. $5x^2y + 12xy^2 - 8x^2 - 12xy^2$

27. $2x^2 - 3x - 5 - 7x + 8 - x^2$

28. $5x + 7 - 6x^3 + 6 - 11x + 4x^3$

29. $2y^2 - 8y + 9 - 12y^2 - 8y + 3$

30. $5 - 2y^2 + 3y - 8y - 9y^2 - 12$

31. $ab + 3a - 4ab + 2a - 8b$

Simplify. Use the distributive property to remove parentheses; then combine like terms.

32. $5(a - 3b) + 2(-b - 4a)$

33. $3(x + y) - 5(-2y + 3x)$

34. $-3b(5a - 3b) + 4(-3ab - 5b^2)$

35. $2x(x - 3y) - 4(-3x^2 - 2xy)$

36. $-3(x^2 + 3y) + 5(-6y - x^2)$

37. $-3(7xy - 11y^2) - 2y(-2x + 3y)$

38. $4(2 - x) - 3(-5 - 12x)$

39. $7(3 - x) - 6(8 - 13x)$

To Think About

▲ **40.** A triangle has sides of length $2a$ centimeters, $7b$ centimeters, and $5a + 3$ centimeters. What is the perimeter of the triangle?

▲ **41.** A rectangle has sides of length $7x - 2$ meters and $3x + 4$ meters. What is the perimeter of the rectangle?

▲ **42.** A square has a side of length $9x - 2$ inches. Each side is shortened by 3 inches. What is the perimeter of the new smaller square?

▲ **43.** A triangle has sides of length $4a - 5$ feet, $3a + 8$ feet, and $9a + 2$ feet. Each side is doubled in length. What is the perimeter of the new enlarged triangle?

Cumulative Review Problems

Evaluate.

44. $-\dfrac{1}{3} - \left(-\dfrac{1}{5}\right)$

45. $\left(-\dfrac{5}{3}\right)\left(\dfrac{1}{2}\right)$

46. $\dfrac{4}{5} + \left(-\dfrac{1}{25}\right) + \left(-\dfrac{3}{10}\right)$

47. $\left(\dfrac{5}{7}\right) \div \left(-\dfrac{14}{3}\right)$

48. A 2-quart container of orange juice produces 9.5 average servings. The container holds approximately 1.9 liters of orange juice. How many liters are in one serving?

1.8 Using Substitution to Evaluate Expressions and Formulas

1 *Evaluating a Variable Expression for a Specified Value*

You will use the order of operations to **evaluate** variable expressions. Suppose we are asked to evaluate

$$6 + 3x \text{ for } x = -4.$$

In general, x represents some unknown number. Here we are told x has the value -4. We can replace x with -4. Use parentheses around -4. Note that we always put replacement values in parentheses.

$$6 + 3(-4) = 6 + (-12) = -6$$

When we replace a variable by a particular value, we say we have **substituted** the value for the variable. We then evaluate the expression (that is, find a value for it).

● EXAMPLE 1 Evaluate $\frac{2}{3}x - 5$ for $x = -6$.

$$\frac{2}{3}x - 5 = \frac{2}{3}(-6) - 5 \quad \text{Substitute } x = -6. \text{ Be sure to enclose the } -6 \text{ in parentheses.}$$

$$= -4 - 5 \quad \text{Multiply } \left(\frac{2}{3}\right)\left(-\frac{6}{1}\right) = -4.$$

$$= -9 \quad \text{Combine.}$$

Practice Problem 1 Evaluate $4 - \frac{1}{2}x$ for $x = -8$.

Compare parts (a) and (b) in the next example. The two parts illustrate that you must be careful what value you raise to a power. *Note:* In part (b) we will need parentheses within parentheses. To avoid confusion, we use brackets [ ] to represent the outside parentheses.

● EXAMPLE 2 **(a)** Evaluate $2x^2$ for $x = -3$.   **(b)** Evaluate $(2x)^2$ for $x = -3$.

(a) Here the value x is squared.

$$2x^2 = 2(-3)^2$$
$$= 2(9) \quad \text{First square } -3.$$
$$= 18 \quad \text{Then multiply.}$$

(b) Here the value $(2x)$ is squared.

$$(2x)^2 = [(2)(-3)]^2$$
$$= (-6)^2 \quad \text{First multiply the numbers inside the parentheses.}$$
$$= 36 \quad \text{Then square } -6.$$

Practice Problem 2 Evaluate for $x = -3$.   **(a)** $-x^4$ **(b)** $(-x)^4$

Carefully study the solutions to Example 2(a) and Example 2(b). You will find that taking the time to see *how* and *why* they are different is a good investment of study time.

● EXAMPLE 3 Evaluate $x^2 + x$ for $x = -4$.

$$x^2 + x = (-4)^2 + (-4) \quad \text{Replace } x \text{ by } -4 \text{ in the original expression.}$$
$$= 16 + (-4) \quad \text{Raise to a power.}$$
$$= 12 \quad \text{Finally, add.}$$

Practice Problem 3 Evaluate $(5x)^3 + 2x$ for $x = -2$.

2 Evaluating a Formula by Subtitution

We can *evaluate a formula* by substituting values for the variables. For example, the area of a triangle can be found using the formula $A = \frac{1}{2}ab$, where b is the length of the base of the triangle and a is the altitude of the triangle (see figure). If we know values for a and b, we can substitute those values into the formula to find the area. The units for area are *square units*.

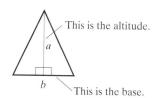
This is the altitude.

This is the base.

Because some of the examples and exercises in this section involve geometry, it may be helpful to review this topic.

However, before we proceed to do any examples, it may be helpful to review a little geometry. The following information is very important. If you have forgotten some of this material (or if you have never learned it), please take the time to learn it completely now. Throughout the entire book we will be using this information in solving applied problems.

Perimeter is the distance around a plane figure. Perimeter is measured in linear units (inches, feet, centimeters, miles). **Area** is a measure of the amount of surface in a region. Area is measured in square units (square inches, square feet, square centimeters).

In our sketches we will show angles of 90° by using a small square (⌐ ⌐). This indicates that the two lines are at right angles. All angles that measure 90° are called **right angles**. An **altitude** is perpendicular to the base of a figure. That is, the altitude forms right angles with the base. The small corner square in a sketch helps us identify the altitude of the figure.

The following box provides a handy guide to some facts and formulas you will need to know. Use it as a reference when solving word problems involving geometric figures.

Geometric Formulas: Two-Dimensional Figures

A **parallelogram** is a four-sided figure with opposite sides parallel. In a parallelogram, opposite sides are equal and opposite angles are equal.

Perimeter = the sum of all four sides

Area = ab

A **rectangle** is a parallelogram with all interior angles measuring 90°.

Perimeter = $2l + 2w$

Area = lw

A **square** is a rectangle with all four sides equal.

Perimeter = $4s$

Area = s^2

A **trapezoid** is a four-sided figure with two sides parallel. The parallel sides are called the *bases* of the trapezoid.

Perimeter = the sum of all four sides

Area = $\frac{1}{2}a(b_1 + b_2)$

A **triangle** is a closed plane figure with three sides.

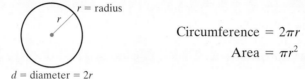

a = altitude

b = base

Perimeter = the sum of the three sides

$$\text{Area} = \frac{1}{2}\,ab$$

A **circle** is a plane curve consisting of all points at an equal distance from a given point called the center.
Circumference is the distance around a circle.

r = radius

d = diameter = $2r$

$$\text{Circumference} = 2\pi r$$
$$\text{Area} = \pi r^2$$

π (the number *pi*) is a constant associated with circles. It is an irrational number that is approximately 3.141592654. We usually use 3.14 as a sufficiently accurate approximation. Thus we write $\pi \approx 3.14$ for most of our calculations involving π.

▲ ◗ **EXAMPLE 4**   Find the area of a triangle with a base of 16 centimeters (cm) and a height of 12 centimeters (cm).
Use the formula

$$A = \frac{1}{2}\,ab.$$

Substitute 12 centimeters for a and 16 centimeters for b.

$$A = \frac{1}{2}\,(12\ \text{centimeters})\,(16\ \text{centimeters})$$

$$= \frac{1}{2}(12)(16)(\text{cm})(\text{cm}) \quad \text{If you take } \tfrac{1}{2} \text{ of 12 first, it will make your calculation easier.}$$

$$= (6)(16)(\text{cm})^2 = 96\ \text{square centimeters}$$

The area of the triangle is 96 square centimeters.

▲ **Practice Problem 4**   Find the area of a triangle with an altitude of 3 meters and a base of 7 meters. ◗

The area of a circle is given by

$$A = \pi r^2.$$

r

This is the radius.

We will use 3.14 as an approximation for the *irrational number* π.

▲ ◗ **EXAMPLE 5**   Find the area of a circle if the radius is 2 inches.

$$A = \pi r^2 = (3.14)(2\ \text{inches})^2 \quad \text{Write the formula and substitute the given values for the letters.}$$

$$= (3.14)(4)(\text{in.})^2 \quad \text{Raise to a power. Then multiply.}$$

$$= 12.56\ \text{square inches}$$

▲ **Practice Problem 5**   Find the area of a circle if the radius is 3 meters. ◗

The formula $C = \frac{5}{9}(F - 32)$ allows us to find the Celsius temperature if we know the Fahrenheit temperature. That is, we can substitute a value for F in degrees Fahrenheit into the formula to obtain a temperature C in degrees Celsius.

EXAMPLE 6 What is the Celsius temperature when the Fahrenheit temperature is $F = -22°$?
Use the formula.

$$C = \frac{5}{9}(F - 32)$$

$$= \frac{5}{9}((-22) - 32) \quad \text{Substitute } -22 \text{ for } F \text{ in the formula.}$$

$$= \frac{5}{9}(-54) \quad\quad \text{Combine the numbers inside the parentheses.}$$

$$= (5)(-6) \quad\quad\quad \text{Simplify.}$$
$$= -30 \quad\quad\quad\quad \text{Multiply.}$$

The temperature is $-30°$ Celsius.

Practice Problem 6 What is the Celsius temperature when the Fahrenheit temperature is $F = 68°$? Use the formula $C = \frac{5}{9}(F - 32)$.

When driving in Canada or Mexico, we must observe speed limits posted in kilometers per hour. A formula that converts r (miles per hour) to k (kilometers per hour) is $k \approx 1.61r$. Note that this is an approximation.

EXAMPLE 7 You are driving on a highway in Mexico. It has a posted maximum speed of 100 kilometers per hour. You are driving at 61 miles per hour. Are you exceeding the speed limit?
Use the formula.

$$k \approx 1.61r$$
$$= (1.61)(61) \quad \text{Replace } r \text{ by 61.}$$
$$= 98.21 \quad\quad \text{Multiply the numbers.}$$

You are driving at approximately 98 kilometers per hour. You are not exceeding the speed limit.

Practice Problem 7 You are driving behind a heavily loaded truck on a Canadian highway. The highway has a posted minimum speed of 65 kilometers per hour. When you travel at exactly the same speed as the truck ahead of you, you observe that the speedometer reads 35 miles per hour. Assuming that your speedometer is accurate, determine whether the truck is violating the minimum speed law.

1.8 Exercises

Evaluate.

1. $-2x + 1$ for $x = 3$

2. $-4x - 2$ for $x = 5$

3. $\frac{2}{3}x - 5$ for $x = -9$

4. $\frac{3}{4}x + 8$ for $x = -8$

5. $5x + 10$ for $x = \frac{1}{2}$

6. $7x + 20$ for $x = -\frac{1}{2}$

7. $2 - 4x$ for $x = 7$

8. $3 - 5x$ for $x = 8$

9. $x^2 - 3x$ for $x = -2$

10. $x^2 + 3x$ for $x = 4$

11. $3x^2$ for $x = -1$

12. $4x^2$ for $x = -1$

13. $-3x^3$ for $x = 2$

14. $-7x^2$ for $x = 5$

15. $9x + 13$ for $x = -\frac{3}{4}$

16. $5x + 7$ for $x = -\frac{2}{3}$

17. $2x^2 + 3x$ for $x = -3$

18. $18 - 5x$ for $x = -3$

19. $(2x)^2 + x$ for $x = 3$

20. $2 - x^2$ for $x = -2$

21. $2 - (-x)^2$ for $x = -2$

22. $2x - 3x^2$ for $x = -4$

23. $7x + (2x)^2$ for $x = -3$

24. $5x + (3x)^2$ for $x = -2$

25. $x^2 - 7x + 3$ for $x = 3$

26. $4x^2 - 3x + 9$ for $x = 2$

27. $\frac{1}{2}x^2 - 3x + 9$ for $x = -4$

28. $x^2 - 2y + 3y^2$ for $x = -3$ and $y = 4$

29. $2x^2 - 3xy + 2y$ for $x = 4$ and $y = -1$

30. $a^3 + 2abc - 3c^2$ for $a = 5, b = 9$, and $c = -1$

31. $a^2 - 2ab + 2c^2$ for $a = 3, b = 2$, and $c = -4$

Applications

▲ **32.** A sign is made in the shape of a parallelogram. The base measures 22 feet. The altitude measures 16 feet. What is the area of the sign?

▲ **33.** A field is shaped like a parallelogram. The base measures 92 feet. The altitude measures 54 feet. What is the area of the field?

▲ **34.** A square support unit in a television is made with a side measuring 3 centimeters. A new model being designed for next year will have a larger square with a side measuring 3.2 centimeters. By how much will the area of the square be increased?

▲ **35.** A square computer chip for last year's computer had a side measuring 23 millimeters. This year the computer chip has been reduced in size. The new square chip has a side of 20 millimeters. By how much has the area of the chip decreased?

▲ **36.** A carpenter cut out a small trapezoid as a wooden support for the front step. It has an altitude of 4 inches. One base of the trapezoid measures 9 inches and the other base measures 7 inches. What is the area of this support?

▲ **37.** The Media One signal tower has a small trapezoid frame on the top of the tower. The frame has an altitude of 9 inches. One base of the trapezoid is 20 inches and the other base measures 17 inches. What is the area of this small trapezoidal frame?

▲ **38.** Bradley Palmer State Park has a triangular piece of land on the border. The altitude of the triangle is 400 feet. The base of the triangle is 280 feet. What is the area of this piece of land?

▲ **39.** The ceiling in the Madisons' house has a leak. The roofer exposed a triangular region that needs to be sealed and then reroofed. The region has an altitude of 14 feet. The base of the region is 19 feet. What is the area of the region that needs to be reroofed?

▲ **40.** The radius of a circular opening of a chemistry flask is 4 cm. What is the area of the opening?

▲ **41.** An ancient outdoor sundial has a radius of 5 meters. What is its area?

For exercises 42 and 43, use the formula $F = \frac{9}{5}C + 32$ to find the Fahrenheit temperature.

42. Find the Fahrenheit temperature F when the Celsius temperature is 25°C.

43. It is not uncommon for parts of North Dakota to have a temperature of −10°C. Find the Fahrenheit temperature.

Solve.

▲ **44.** Find the total cost of making a triangular sail that has a base dimension of 12 feet and a height of 20 feet if the price for making the sail is $19.50 per square foot.

▲ **45.** A semicircular window of radius 15 inches is to be laminated with a sunblock coating that costs $0.85 per square inch to apply. What is the total cost of coating the window, to the nearest cent? (Use $\pi \approx 3.14$.)

46. Some new computers can be exposed to extreme temperatures (as high as 60°C and as low as −50°C). What is the temperature range in Fahrenheit that these computers can be exposed to? (Use the formula $F = \frac{9}{5}C + 32$.)

47. To deal with extreme temperatures while doing research at the South Pole, scientists have developed accommodations that can comfortably withstand an outside temperature of −60°C with no wind blowing, or −30°C with wind gusts of up to 50 miles per hour. What is the corresponding Fahrenheit temperature range? (Use the formula $F = \frac{9}{5}C + 32$.)

48. Bruce becomes exhausted while on a bicycle trip in Canada. He reads on the map that his present elevation is 2.3 kilometers above sea level. How many miles to the nearest tenth above sea level is he? Why is he so tired? Use the formula $r = 0.62k$ where r is the number of miles and k is the number of kilometers.

49. While biking down the Pacific coast of Mexico, you see on the map that it is 20 kilometers to the nearest town. Approximately how many miles is it to the nearest town? Use the formula $r = 0.62k$ where r is the number of miles and k is the number of kilometers.

Cumulative Review Problems

In exercises 50–51, simplify.

50. $(-2)^4 - 4 \div 2 - (-2)$

51. $3(x - 2y) - (x^2 - y) - (x - y)$

52. A 93-minute-long recordable compact disc is used to record 15 songs. Express in decimal form the average number of minutes available per song.

Putting Your Skills to Work

Measuring Your Level of Fitness

In recent years there has been a greater awareness of fitness, exercise, and weight control in our country. Often people would like a simple mathematical way to measure whether they are physically fit and of the proper weight.

A growing number of doctors and health fitness professionals refer to the government studies from the National Center for Health Statistics as supporting the BMI, or body mass index, as a helpful measure to determine whether a person is at the proper weight, overweight, or underweight.

To determine your BMI, multiply your weight in pounds by 0.45 to get your weight in kilograms. Next, convert your height to inches. Then multiply your height in inches by 0.0254 to get your height in meters. Now multiply that number by itself. Your weight in kilograms is divided by that number. (Your weight in kilograms is divided by the square of your height in meters.)

The result is your BMI. For most people this will be a number that ranges from the 20s to the low 30s. Federal guidelines suggest that you should keep your BMI under 25 in order not to be overweight. However, there is a limit to how low your BMI should be. Many doctors recommend that a person's BMI should not be lower than 18.

Problems for Individual Investigation and Study

Use the procedure for calculating BMI to answer the following questions. Round all measurements of BMI to the nearest tenth.

1. Find the BMI for a woman who weighs 125 pounds and is 5 feet, 4 inches tall.

2. Is a man who weighs 185 pounds and is 5 feet, 11 inches tall considered overweight by the government study? (Is his BMI 25 or higher?) What is his BMI?

Problems for Cooperative Group Activity

Together with other members of your class, see if you can complete the following.

3. Is a woman who weighs 95 pounds and is 1.6 meters tall considered underweight by the government study? (Is her BMI under 18?) What is her BMI?

4. Write a formula to find the BMI of a person if the height of the person is measured in meters (m) and the weight of the person is measured in pounds (p).

Internet Connections

Netsite: http://www.prenhall.com/tobey_beginning

Site: BMI Chart (Shape Up America!) or a related site

5. This site gives a shortcut method for calculating BMI. Write a formula based on this method to find the BMI of a person whose height is measured in inches (i) and weight is measured in pounds (p). Then use your formula to find the BMI of a person who is 5 feet, 10 inches tall and weighs 145 pounds.

1.9 Grouping Symbols

Simplifying Variable Expressions

 Many expressions in algebra use **grouping symbols** such as parentheses, brackets, and braces. Sometimes expressions are inside other expressions. Because it can be confusing to have more than one set of parentheses, brackets and braces are also used. How do we know what to do first when we see an expression like $2[5 - 4(a + b)]$?

To simplify the expression, we start with the innermost grouping symbols. Here it is a set of parentheses. We first use the distributive law to multiply.

$$2[5 - 4(a + b)] = 2[5 - 4a - 4b]$$

We use the distributive law again.

$$= 10 - 8a - 8b$$

There are no like terms, so this is our final answer.

Notice that we started with two sets of grouping symbols, but our final answer has none. So we can say we *removed* the grouping symbols. Of course, we didn't just take them away; we used the distributive law and the rules for real numbers to simplify as much as possible. Although simplifying expressions like this involves many steps, we sometimes say "remove parentheses" as a shorthand direction. Sometimes we say "simplify."

Remember to remove the innermost parentheses first. Keep working from the inside out.

 EXAMPLE 1 Simplify. $3[6 - 2(x + y)]$

We want to remove the innermost parentheses first. Therefore, we first use the distributive property to simplify $-2(x + y)$.

$$3[6 - 2(x + y)] = 3[6 - 2x - 2y] \quad \text{Use the distributive property.}$$

$$= 18 - 6x - 6y \quad \text{Use the distributive property again.}$$

Practice Problem 1 Simplify. $5[4x - 3(y - 2)]$

You recall that a negative sign in front of parentheses is equivalent to having a coefficient of negative 1. You can write the -1 and then multiply by -1 using the distributive property.

$$-(x + 2y) = -1(x + 2y) = -x - 2y$$

Notice that this has the effect of removing the parentheses. Each term in the result now has its sign changed.

Similarly, a positive sign in front of parentheses can be viewed as multiplication by $+1$.

$$+(5x - 6y) = +1(5x - 6y) = 5x - 6y$$

If a grouping symbol has a positive or negative sign in front, we mentally multiply by $+1$ or -1, respectively.

Fraction bars are also considered grouping symbols. Later in this book we will encounter problems where our first step will be to simplify expressions above and below fraction bars. This type of operation will have some similarities to the operation of removing parentheses.

 EXAMPLE 2 Simplify. $-2[3a - (b + 2c) + (d - 3e)]$

$$= -2[3a - b - 2c + d - 3e] \quad \text{Remove the two innermost sets of parentheses. Since one is not inside the other, we remove both sets at once.}$$

$$= -6a + 2b + 4c - 2d + 6e \quad \text{Now we remove the brackets by multiplying each term by } -2.$$

Practice Problem 2 Simplify. $3ab - [2ab - (2 - a)]$

Student Learning Objectives

After studying this section, you will be able to:

1 Simplify variable expressions with several grouping symbols.

SSM

PH TUTOR CENTER   CD & VIDEO   MATH PRO   WEB

115

EXAMPLE 3 Simplify. $2[3x - (y + w)] - 3[2x + 2(3y - 2w)]$

$= 2[3x - y - w] - 3[2x + 6y - 4w]$ In each set of brackets, remove the inner parentheses.

$= 6x - 2y - 2w - 6x - 18y + 12w$ Remove each set of brackets by multiplying by the appropriate number.

$= -20y + 10w$ or $10w - 20y$ Combine like terms.
(Note that $6x - 6x = 0x = 0$.)

Practice Problem 3   Simplify. $3[4x - 2(1 - x)] - [3x + (x - 2)]$

You can always simplify problems with many sets of grouping symbols by the method shown. Essentially, you just keep removing one level of grouping symbols at each step. Finally, at the end you add up the like terms if possible.

Sometimes it is possible to combine like terms at each step.

EXAMPLE 4   Simplify. $-3\{7x - 2[x - (2x - 1)]\}$

$= -3\{7x - 2[x - 2x + 1]\}$ Remove the inner parentheses by multiplying each term within the parentheses by -1.

$= -3\{7x - 2[-x + 1]\}$ Combine like terms by combining $+x - 2x$.

$= -3\{7x + 2x - 2\}$ Remove the brackets by multiplying each term within them by -2.

$= -3\{9x - 2\}$ Combine the x-terms.

$= -27x + 6$ Remove the braces by multiplying each term by -3.

Practice Problem 4   Simplify. $-2\{5x - 3x[2x - (x^2 - 4x)]\}$

Developing Your Study Skills

Exam Time: The Night Before

With adequate preparation, you can spend the night before an exam pulling together the final details.

1. Look over each section to be covered in the exam. Review the steps needed to solve each type of problem.

2. Review your list of terms, rules, and formulas that you are expected to know for the exam.

3. Take the Practice Test at the end of the chapter just as though you were taking the actual exam. Do not look in your text or get help in any way. Time yourself so that you know how long it takes you to complete the test.

4. Check the Practice Test. Redo the problems you missed.

5. Be sure you have ready the necessary supplies for taking your exam.

1.9 Exercises

Verbal and Writing Skills

1. Rewrite the expression $-3x - 2y$ using a negative sign and parentheses.

2. Rewrite the expression $-x + 5y$ using a negative sign and parentheses.

3. To simplify expressions with grouping symbols, we use the _____ property.

4. When an expression contains many grouping symbols, remove the _____ parentheses first.

Simplify. Remove grouping symbols and combine like terms.

5. $6x - 3(x - 2y)$

6. $-4x - 2(y - 3x)$

7. $2(a + 3b) - 3(b - a)$

8. $4(x - y) - 2(3x + y)$

9. $5[3 + 2(x - 26) + 3x]$

10. $-4[-(x + 3y) - 2(y - x)]$

11. $2x[4x^2 - 2(x - 3)]$

12. $4y[-3y^2 + 2(4 - y)]$

13. $2(x - 2y) - [3 - 2(x - y)]$

14. $3[x - y(3x + y) + y^2]$

15. $5[3a - 2a(3a + 6b) + 6a^2]$

16. $3(x + 2y) - [4 - 2(x + y)]$

17. $x(x^2 + 2x - 3) - 2(x^3 + 6)$

18. $5x^2(x + 6) - 2[x - 2(1 + 2x^2)]$

19. $3a^2 - 4[2b - 3b(b + 2)]$

20. $2a - \{6b - 4[a - (b - 3a)]\}$

21. $6b - \{5a - 2[a + (b - 2a)]\}$

22. $2b^2 - 3[5b + 2b(2 - b)]$

23. $-4\{3a^2 - 2[4a^2 - (b + a^2)]\}$

24. $-2\{x^2 - 3[x - (x - 2x^2)]\}$

To Think About

25. The dedicated job of a four-wheeled robot called Nomad, built by the Robotics Institute at Carnegie Mellon University, is to find samples of meteorites in Antarctica. If Nomad is successful in finding meteorites 2.5% of the time, out of four tries per day for six years, estimate the number of successful and unsuccessful meteorite search attempts.

26. Grandma and Grandpa Tobey had a tradition of eating out once a week. The average cost of the meal was $20. In Massachusetts there is a 5% state sales tax that is added to the cost of the meal. The Tobeys always left a 15% tip. They continued this pattern for ten years. Mrs. Tobey felt that they should calculate the tip on the total of the cost of the meal including the sales tax. Mr. Tobey felt they should calculate the tip on the cost of the meal alone. How much difference would this make over the ten-year period?

Cumulative Review Problems

27. Use $F = 1.8C + 32$ to find the Fahrenheit temperature equivalent to 36.4° Celsius.

▲ **28.** Use 3.14 as an approximation for π to compute the area covered by a circular irrigation system with radial arm of length 380 feet. $A = \pi r^2$.

▲ **29.** The base of an office building is in the shape of a trapezoid. The altitude of the trapezoid is 400 feet. The bases of the trapezoid are 700 feet and 800 feet, respectively. What is the area of the base of the office building? If the base has a marble floor that cost $55 per square foot, what was the cost of the marble floor?

▲ **30.** The Global Media Tower has a triangular signal tester at the top of the tower. The altitude of the triangle is 3.5 feet and the base is 6.5 feet. What is the area of the triangular signal tester? If the signal tester is coated with a special metallic surface paint that costs $122 per square foot, what was the cost of the amount of paint needed to coat one side of the triangle?

Putting Your Skills to Work

Consumer Price Index

The federal government produces a statistic that is helpful in determining how much the cost of living is increasing each year. It also produces different values for different regions of the country to track how the increase varies from city to city. This measure is called the Consumer Price Index (CPI). The CPI measures the average change in the price of goods and services purchased by all urban consumers. The following table shows the CPI for several years.

Type of Item	1975	1982–1984	1990	1995	2000*
Cost of all items	53.8	100	130.7	152.4	169.6
Food	59.8	100	132.4	148.4	188.4
Housing	50.7	100	128.5	148.5	173.2
Medical care	47.5	100	162.8	220.5	265.7
Entertainment	62.0	100	132.4	153.9	167.6

Source: U.S. Department of Labor
*estimated

The period of 1982–1984 is used as the focus of comparison, and the index for that period is assigned a value of 100. For example if the cost of food for a family was $100 in the period 1982–1984, then the cost of purchasing the same type of food in 1995 was $148.40. However, a similar purchase of the same type of food in 1975 would have amounted to only $59.80. Now use the table to answer the following questions.

Problems for Individual Investigation and Study

1. If a family spent $200 annually for entertainment during the 1982–1984 period, how much would the same entertainment have cost in 2000?

2. If an inpatient hospital operation cost $10,000 in the 1982–1984 period, how much would a similar operation have cost in 1995?

3. If a family found that three months' worth of food cost $1884.00 in 2000, how much would similar food have cost the family in 1975?

4. If a family found that renting an apartment for two months in 1995 cost $1485.00, how much would a similar apartment have cost for two months in 1975?

The CPI has increased more dramatically in certain parts of the country than in others. The following table records the CPI for five cities in 1998. For each city the index for each category is 100 for the 1982–1984 period.

CPI for Various Cities in 1998 (The CPI for each city in each category is 100 for 1982–1984.)

Type of Item	Anchorage	Boston	Houston	New York	Chicago
Cost of all items	146.9	171.7	146.8	173.6	165.0
Food	147.5	166.3	150.3	165.3	164.3
Housing	131.0	165.8	129.6	175.9	164.2
Medical care	255.7	313.9	235.4	255.0	244.3

Use the preceding table to answer the following questions.

Problems for Cooperative Group Activity and Investigation

5. What category for what city saw the biggest increase in the CPI since 1982–1984?

6. How many times greater was the increase in the cost of housing in Boston than that in Anchorage from the 1982–1984 period to 1998?

Internet Connections   *Site: Consumer Price Index provided by the Bureau of Labor Statistics*

 Netsite: `http://www.prenhall.com/tobey_beginning`

Use this Web site to obtain a CPI table for your city or region. Then answer the following questions.

7. Suppose you earned $22,000 during the 1982–1984 period. If your income increased at the same rate as the CPI for your city or region, how much would you have earned in 2000?

8. Suppose you spent $7000 on rent in 2000 in your city or region. How much would you have spent renting a similar location in your city or region during the 1982–1984 period?

Math in the Media

Readers Can Face Off with Pros and Darts

WSJ.com Staff Reporter

November 16, 2000. © 2000 Dow Jones & Company Inc. Reprinted by permission.

The *Wall Street Journal Interactive Edition* runs an investment contest, **Investment Dartboard**. Individual investors are invited to submit their best stock picks for the contest beginning in a given month and running for six months. Stock picks submitted by four readers will be selected in a drawing, and these picks will then be matched against the stocks selected by the pros and the darts.

There's no prize—just the glory, or the embarrassment, of publicly pitting your stock-picking skills against the investment professionals and the forces of chance.

The table shows Stock, Purchase Price, Latest Price, Day's Change, and % Gain/Loss.

Stock	Purchase Price	Latest Price	Day's Change	% Gain/Loss to Date
The Pros—professional investors				
A	38	$37\frac{3}{8}$	$-\frac{5}{8}$	−1.64%
B	$26\frac{9}{16}$	$26\frac{7}{8}$	$+\frac{5}{16}$	
C	$67\frac{3}{8}$	$65\frac{7}{8}$		−2.23%
D	$17\frac{3}{16}$	$17\frac{1}{8}$	$-\frac{1}{16}$	−0.36%
The Darts' Picks—Selected by Journal staffers flinging darts at the paper's stock tables.				
E	572	572	unch.	0
F	$5\frac{13}{16}$	$5\frac{1}{2}$	$-\frac{5}{16}$	−5.38%
G	$11\frac{1}{4}$	$10\frac{7}{8}$	$-\frac{3}{8}$	
H	$23\frac{1}{4}$	$22\frac{7}{8}$	$-\frac{3}{8}$	−1.61%
The Amateurs' Picks				
I	$14\frac{1}{2}$	$12\frac{11}{16}$	$-1\frac{13}{16}$	−12.50%
J	$29\frac{5}{16}$	$27\frac{29}{64}$	$-1\frac{55}{64}$	−6.34%
K	$41\frac{5}{16}$	40		−3.18%
L	$43\frac{5}{8}$	$43\frac{7}{8}$	$+\frac{1}{4}$	0.57%

EXERCISES

Investing can be risky business. Look at the table provided in the article. Test your skills. Try to answer questions 1–3 that follow.

1. How are the values in the Day's Change column achieved? Complete the tables for Stock C and Stock K.

2. How are the values in the % Gain/Loss column achieved? Complete the tables for Stock B and Stock G.

3. Compute the average % Gain/loss for the Pros, Darts' Pick, and Amateurs. Which was the best?

Chapter 1 Organizer

Topic	Procedure	Examples
Absolute value, p. 69.	The absolute value of a number is the distance between that number and zero on the number line. The absolute value of any number will be positive or zero.	$\|3\| = 3$ $\|-2\| = 2$ $\|0\| = 0$ $\left\|-\dfrac{5}{6}\right\| = \dfrac{5}{6}$ $\|-1.38\| = 1.38$
Adding real numbers with the same sign, p. 70.	If the signs are the same, add the absolute values of the numbers. Use the common sign in the answer.	$-3 + (-7) = -10$
Adding real numbers with opposite signs, p. 72.	If the signs are different: 1. Find the difference between the larger and the smaller absolute value. 2. Give the answer the sign of the number having the larger absolute value.	$(-7) + 13 = 6$ $7 + (-13) = -6$
Adding several real numbers, p. 73.	When adding several real numbers, separate them into two groups by sign. Find the sum of all the positive numbers and the sum of all the negative numbers. Combine these two subtotals by the method described above.	$-7 + 6 + 8 + (-11) + (-13) + 22$ $\begin{array}{rr} -7 & +6 \\ -11 & +8 \\ -13 & +22 \\ \hline -31 & +36 \end{array}$ $-31 + 36 = 5$ The answer is positive since 36 is positive.
Subtracting real numbers, p. 77.	Change the sign of the second number (the number you are subtracting) and then add.	$-3 - (-13) = -3 + (+13) = 10$
Multiplying and dividing real numbers, p. 83 and p. 85.	1. If the two numbers have the same sign, multiply (or divide) the absolute values. The result is positive. 2. If the two numbers have different signs, multiply (or divide) the absolute values. The result is negative.	$-5(-3) = +15$ $-36 \div (-4) = +9$ $28 \div (-7) = -4$ $-6(3) = -18$
Exponent form, p. 92.	The base tells you what number is being multiplied. The exponent tells you how many times this number is used as a factor.	$2^5 = 2 \cdot 2 \cdot 2 \cdot 2 \cdot 2 = 32$ $4^3 = 4 \cdot 4 \cdot 4 = 64$ $(-3)^4 = (-3)(-3)(-3)(-3) = 81$
Raising a negative number to a power, p. 93.	When the base is negative, the result is positive for even exponents, and negative for odd exponents.	$(-3)^3 = -27$ but $(-2)^4 = 16$
Order of operations, p. 95.	Remember the proper order of operations: 1. Perform operations inside parentheses. 2. Raise to powers. 3. Multiply and divide from left to right. 4. Add and subtract from left to right.	$3(5 + 4)^2 - 2^2 \cdot 3 \div (9 - 2^3)$ $= 3 \cdot 9^2 - 4 \cdot 3 \div (9 - 8)$ $= 3 \cdot 81 - 12 \div 1$ $= 243 - 12 = 231$
Removing parentheses, p. 98.	Use the distributive property to remove parentheses. $\qquad a(b + c) = ab + ac$	$3(5x + 2) = 15x + 6$

Topic	Procedure	Examples
Combining like terms, p. 103.	Combine terms that have identical letters and exponents.	$7x^2 - 3x + 4y + 2x^2 - 8x - 9y = 9x^2 - 11x - 5y$
Substituting into variable expressions, p. 108.	1. Replace each letter by the numerical value given for it. 2. Follow the order of operations in evaluating the expression.	Evaluate $2x^3 + 3xy + 4y^2$ for $x = -3$ and $y = 2$. $2(-3)^3 + 3(-3)(2) + 4(2)^2$ $\quad = 2(-27) + 3(-3)(2) + 4(4)$ $\quad = -54 - 18 + 16$ $\quad = -56$
Using formulas, p. 110.	1. Replace the variables in the formula by the given values. 2. Evaluate the expression. 3. Label units carefully.	Find the area of a circle with radius 4 feet. Use $A = \pi r^2$, with π as approximately 3.14. $A = (3.14)(4 \text{ feet})^2$ $\quad = (3.14)(16 \text{ feet}^2)$ $\quad = 50.24 \text{ feet}^2$ The area of the circle is approximately 50.24 square feet.
Removing grouping symbols, p. 115.	1. Remove innermost grouping symbols first. 2. Then remove remaining innermost grouping symbols. 3. Continue until all grouping symbols are removed. 4. Combine like terms.	$5\{3x - 2[4 + 3(x - 1)]\}$ $\quad = 5\{3x - 2[4 + 3x - 3]\}$ $\quad = 5\{3x - 8 - 6x + 6\}$ $\quad = 15x - 40 - 30x + 30$ $\quad = -15x - 10$

Developing Your Study Skills

Problems with Accuracy

Strive for accuracy. Mistakes are often made as a result of human error rather than from lack of understanding. Such mistakes are frustrating. A simple arithmetic or sign error can lead to an incorrect answer.

These five steps will help you cut down on errors.

1. Work carefully and take your time. Do not rush through a problem just to get it done.

2. Concentrate on one problem at a time. Sometimes problems become mechanical, and your mind begins to wander. You can become careless and make a mistake.

3. Check your problem. Be sure that you copied it correctly from the book.

4. Check your computations from step to step. Check the solution in the problem. Does it work? Does it make sense?

5. Keep practicing new skills. Remember the old saying, "Practice makes perfect." An increase in practice results in an increase in accuracy. Many errors are due simply to lack of practice.

There is no magic formula for eliminating all errors, but these five steps will be a tremendous help in reducing them.

Chapter 1 Review Problems

1.1 *Add.*

1. $-6 + (-2)$ **2.** $-12 + 7.8$ **3.** $5 + (-2) + (-12)$ **4.** $3.7 + (-1.8)$

5. $\dfrac{1}{2} + \left(-\dfrac{5}{6}\right)$ **6.** $-\dfrac{3}{11} + \left(-\dfrac{1}{22}\right)$ **7.** $\dfrac{3}{4} + \left(-\dfrac{1}{12}\right) + \left(-\dfrac{1}{2}\right)$ **8.** $-\dfrac{4}{15} + \dfrac{12}{5} + \left(-\dfrac{2}{3}\right)$

1.2 *Add or subtract.*

9. $5 - (-3)$ **10.** $-2 - (-15)$ **11.** $-30 - (+3)$ **12.** $8 - (-1.2)$

13. $-\dfrac{7}{8} + \left(-\dfrac{3}{4}\right)$ **14.** $-\dfrac{3}{14} + \dfrac{5}{7}$ **15.** $-20.8 - 1.9$ **16.** $-151 - (-63)$

Mixed Review

1.1–1.2 *Perform the operations indicated. Simplify all answers.*

17. $-5 + (-2) - (-3)$ **18.** $6 - (-4) + (-2) + 8$ **19.** $-16 + (-13)$ **20.** $-11 - (-12)$

1.3

21. $87 \div (-29)$ **22.** $-5(-6) + 4(-3)$ **23.** $\dfrac{-24}{-\dfrac{3}{4}}$ **24.** $-\dfrac{1}{2} \div \left(\dfrac{3}{4}\right)$

25. $\dfrac{5}{7} \div \left(-\dfrac{5}{25}\right)$ **26.** $-6(3)(4)$ **27.** $-1(-2)(-3)(-5)$ **28.** $(-5)\left(-\dfrac{1}{2}\right)(4)(-3)$

Mixed Review

1.1–1.3 *Perform the operations indicated. Simplify all answers.*

29. $-\dfrac{4}{3} + \dfrac{2}{3} + \dfrac{1}{6}$ **30.** $-\dfrac{6}{7} + \dfrac{1}{2} + \left(-\dfrac{3}{14}\right)$ **31.** $-3(-2)(-5)$ **32.** $-6 + (-2) - (-3)$

33. $3.5(-2.6)$ **34.** $-5.4 \div (-6)$ **35.** $5 - (-3.5) + 1.6$ **36.** $-8 + 2 - (-4.8)$

37. $17 + 3.4 + (-16) + (-2.5)$ **38.** $37 + (-44) + 12.5 + (-6.8)$

Solve.

39. The Dallas Cowboys football team had three plays in which they lost 8 yards each time. What was the total yardage lost?

40. The low temperature in Anchorage, Alaska, last night was $-34°F$. During the day the temperature rose $12°F$. What was the temperature during the day?

41. A mountain peak is 6895 feet above sea level. A location in Death Valley is 468 feet below sea level. What is the difference in height between these two locations?

42. During January 2000, IBM stock rose $1\frac{1}{2}$ points on Monday, dropped $3\frac{1}{4}$ points on Tuesday, rose 2 points on Wednesday, and dropped $2\frac{1}{2}$ points on Thursday. What was the total gain or loss on the value of the stock over this four-day period?

1.4 *Evaluate.*

43. $(-3)^5$

44. $(-2)^7$

45. $(-5)^4$

46. $\left(\dfrac{2}{3}\right)^3$

47. -9^2

48. $(0.6)^2$

49. $\left(\dfrac{5}{6}\right)^2$

50. $\left(\dfrac{3}{4}\right)^3$

1.5 *Simplify using the order of operations.*

51. $5(-4) + 3(-2)^3$

52. $20 - (-10) - (-6) + (-5) - 1$

53. $(7 - 9)^3 + -6(-2) + (-3)$

1.6 *Use the distributive property to multiply.*

54. $5(3x - 7y)$

55. $2x(3x - 7y + 4)$

56. $-(7x^2 - 3x + 11)$

57. $(2xy + x - y)(-3y)$

1.7 *Combine like terms.*

58. $3a^2b - 2bc + 6bc^2 - 8a^2b - 6bc^2 + 5bc$

59. $9x + 11y - 12x - 15y$

60. $4x^2 - 13x + 7 - 9x^2 - 22x - 16$

61. $-x + \dfrac{1}{2} + 14x^2 - 7x - 1 - 4x^2$

1.8 *Evaluate for the given value of the variable.*

62. $7x - 6$ for $x = -7$

63. $7 - \dfrac{3}{4}x$ for $x = 8$

64. $x^2 + 3x - 4$ for $x = -3$

65. $-3x^2 - 4x + 5$ for $x = 2$

66. $-3x^3 - 4x^2 + 2x + 6$ for $x = -2$

67. $vt - \dfrac{1}{2}at^2$ for $v = 24$, $t = 2$, and $a = 32$

68. $\dfrac{nRT}{V}$ for $n = 16$, $R = -2$, $T = 4$, and $V = -20$

Solve.

69. Find the simple interest on a loan of $6000 at an annual interest rate of 18% per year for $\frac{3}{4}$ of a year. Use $I = prt$, where $p =$ principal, $r =$ rate per year, and $t =$ time in years.

70. Find the Fahrenheit temperature if a radio announcer in Mexico City says that the high temperature today was 30°C. Use the formula $F = \dfrac{9C + 160}{5}$.

▲ **71.** How much will it cost to paint a circular sign with a radius of 15 meters if the painter charges $3 per square meter? Use $A = \pi r^2$, where π is approximately 3.14.

72. Find the daily profit P at a furniture factory if the initial cost of setting up the factory $C = \$1200$, rent $R = \$300$, and sales of furniture $S = \$56$. Use the profit formula $P = 180S - R - C$.

▲ **73.** A parking lot is in the shape of a trapezoid. The altitude of the trapezoid is 200 feet, and the bases of the trapezoid are 300 feet and 700 feet. What is the area of the parking lot? If the parking lot had a sealer applied that costs $2 per square foot, what was the cost of the amount of sealer needed for the entire parking lot?

▲ **74.** The Green Mountain Telephone Company has a triangular signal tester at the top of a communications tower. The altitude of the triangle is 3.8 feet and the base is 5.5 feet. What is the area of the triangular signal tester? If the signal tester is painted with a special metallic surface paint that costs $66 per square foot, what was the cost of the amount of paint needed to paint one side of the triangle?

1.9 *Simplify.*

75. $5x - 7(x - 6)$

76. $3(x - 2) - 4(5x + 3)$

77. $2[3 - (4 - 5x)]$

78. $-3x[x + 3(x - 7)]$

79. $2xy^3 - 6x^3y - 4x^2y^2 + 3(xy^3 - 2x^2y - 3x^2y^2)$

80. $-5(x + 2y - 7) + 3x(2 - 5y)$

81. $2\{x - 3(y - 2) + 4[x - 2(y + 3)]\}$

82. $-5\{2a - b[5a - b(3 + 2a)]\}$

83. $-3\{2x - [x - 3y(x - 2y)]\}$

84. $2\{3x + 2[x + 2y(x - 4)]\}$

1. _____

2. _____

3. _____

4. _____

5. _____

6. _____

7. _____

8. _____

9. _____

10. _____

11. _____

12. _____

13. _____

14. _____

15. _____

16. _____

17. _____

18. _____

19. _____

20. _____

21. _____

22. _____

23. _____

24. _____

25. _____

26. _____

Simplify.

1. $-3 + (-4) + 9 + 2$ **2.** $-0.6 - (-0.8)$ **3.** $-8(-12)$

4. $-5(-2)(7)(-1)$ **5.** $-12 \div (-3)$ **6.** $-1.8 \div (0.6)$

7. $(-4)^3$ **8.** $(1.3)^2$ **9.** $\left(\dfrac{2}{3}\right)^4$

10. $7 - (6 - 9)^2 + 5(2)$ **11.** $3(4 - 6)^3 + 12 \div (-4) + 2$

12. $-5x(x + 2y - 7)$ **13.** $-2ab^2(-3a - 2b + 7ab)$

14. $6ab - \dfrac{1}{2}a^2b + \dfrac{3}{2}ab + \dfrac{5}{2}a^2b$ **15.** $12a(a + b) - 4(a^2 - 2ab)$

16. $3(2 - a) - 4(-6 - 2a)$ **17.** $5(3x - 2y) - (x + 6y)$

In questions 18–20, evaluate for the value of the variable indicated.

18. $x^3 - 3x^2y + 2y - 5$ for $x = 3$ and $y = -4$

19. $3x^2 - 7x - 11$ for $x = -3$

20. $2a - 3b$ for $a = \dfrac{1}{3}$ and $b = -\dfrac{1}{2}$

21. If you are traveling 60 miles per hour on a highway in Canada, how fast are you traveling in kilometers per hour? (Use $k = 1.61r$, where r = rate in miles per hour and k = rate in kilometers per hour.)

▲ **22.** A field is in the shape of a trapezoid. The altitude of the trapezoid is 120 feet and the bases of the trapezoid are 180 feet and 200 feet. What is the area of the field?

▲ **23.** Jeff Slater's garage has a triangular roof support beam. The support beam is covered with a sheet of plywood. The altitude of the triangular region is 6.8 feet and the base is 8.5 feet. If the triangular piece of plywood was painted with paint that cost $0.80 per square foot, what was the cost of the amount of paint needed to coat one side of the triangle?

▲ **24.** The front lawn of Westwood High School is watered by a sprinkler that waters a circular region. The radius of the circle is 12 feet. What is the area of this region of the lawn? Use $\pi \approx 3.14$.

Simplify.

25. $-3\{a + b[3a - b(1 - a)]\}$ **26.** $3\{x - (5 - 2y) - 4[3 + (6x - 7y)]\}$

Equations and Inequalities

The Environmental Protection Agency monitors the level of emissions into the air of a number of pollutants. Progress has been made in reducing some pollutants, while others are still at significantly high levels. Could you analyze the level of emissions over the last 50 years and use your math skills to predict the level of emissions in the future? Turn to the Putting Your Skills to Work problems on page 161 to find out.

Pretest Chapter 2

1. _____

2. _____

3. _____

4. _____

5. _____

6. _____

7. _____

8. _____

9. _____

10. _____

11. _____

12. _____

13. (a) _____

(b) _____

14. (a) _____

(b) _____

15. _____

16. _____

17. _____

18. _____

19. _____

20. _____

If you are familiar with the topics in this chapter, take this test now. Check your answers with those in the back of the book. If an answer is wrong or you can't answer a question, study the appropriate section of the chapter.

If you are not familiar with the topics in this chapter, don't take this test now. Instead, study the examples, work the practice problems, and then take the test.

This test will help you identify which concepts you have mastered and which you need to study further.

Sections 2.1–2.3

Solve for x.

1. $x - 30 = -46$

2. $\dfrac{1}{4}x = 12$

3. $-5x = 20$

4. $7x + 2 = 30$

5. $3x - 8 = 7x + 6$

6. $5x + 3 - 6x = 4 - 8x - 2$

7. $5.2x + 0.9 = 2.8x + 3.3$

8. $2(x - 4) - 2 = 3(x - 2)$

Section 2.4

Solve for x.

9. $\dfrac{2}{5}x + \dfrac{1}{4} = \dfrac{1}{2}x$

10. $\dfrac{2}{3}(x - 2) + \dfrac{1}{2} = 5 - (3 + x)$

11. $x - \dfrac{3}{2} = -\dfrac{4}{3} - \dfrac{5}{6}$

12. $x - 2 = \dfrac{5}{7}(x + 4)$

Section 2.5

13. (a) Solve for F. $C = \dfrac{5}{9}(F - 32)$

(b) Find the temperature in degrees Fahrenheit if the Celsius temperature is $-15°$.

14. (a) Solve for r. $I = Prt$
(b) Find the simple interest rate r if $P = \$2000$, $t = 2$ years, and $I = \$240$.

Sections 2.6–2.7

Replace the question mark with $<$ or $>$.

15. $1 \, ? \, 10$

16. $2 \, ? \, -3$

17. $-12 \, ? \, -13$

18. $0 \, ? \, 0.9$

Solve and graph.

19. $-2x + 5 \le 4 - x + 3$

20. $\dfrac{1}{2}(x + 2) - 1 > \dfrac{x}{3} - 5 + x$

2.1 The Addition Principle

1 Using the Addition Principle

When we use an equal sign (=), we are indicating that two expressions are equal in value. Such a statement is called an **equation.** For example, $x + 5 = 23$ is an equation. A **solution** of an equation is a number that when substituted for the variable makes the equation true. Thus 18 is a solution of $x + 5 = 23$ because $18 + 5 = 23$. Equations that have exactly the same solutions are called **equivalent equations.** By following certain procedures, we can often transform an equation to a simpler equivalent one that has the form $x =$ some number. Then this number is a solution of the equation. The process of finding all solutions of an equation is called **solving the equation.**

One of the first procedures used in solving equations has an application in our everyday world. Suppose that we place a 10-kilogram box on one side of a seesaw and a 10-kilogram stone on the other side. If the center of the box is the same distance from the balance point as the center of the stone, we would expect the seesaw to balance. The box and the stone do not look the same, but their weights are equal. If we add a 2-kilogram lead weight to the center of weight of each object at the same time, the seesaw should still balance. The weights are still equal.

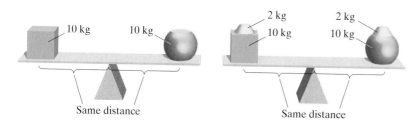

There is a similar principle in mathematics. We can state it in words as follows.

The Addition Principle

If the same number is added to both sides of an equation, the results on both sides are equal in value.

We can restate it in symbols this way.

For real numbers a, b, and c,   if $a = b$,   then $a + c = b + c$.

Here is an example.

$$\text{If } 3 = \frac{6}{2}, \quad \text{then } 3 + 5 = \frac{6}{2} + 5.$$

Since we added the same amount, 5, to both sides, the sides remain equal to each other.

$$3 + 5 = \frac{6}{2} + 5$$

$$8 = \frac{6}{2} + \frac{10}{2}$$

$$8 = \frac{16}{2}$$

$$8 = 8$$

129

We can use the addition principle to solve certain equations.

EXAMPLE 1 Solve for x. $x + 16 = 20$

$$x + 16 + (-16) = 20 + (-16) \qquad \text{Use the addition principle to add } -16 \text{ to both sides.}$$

$$x + 0 = 4 \qquad \text{Simplify.}$$

$$x = 4 \qquad \text{The value of } x \text{ is 4.}$$

We have just found a solution of the equation. A **solution** is a value for the variable that makes the equation true. We then say that the value 4 in our example **satisfies** the equation. We can easily verify that 4 is a solution by substituting this value into the original equation. This step is called **checking** the solution.

Check. $\qquad x + 16 = 20$

$$4 + 16 \overset{?}{=} 20$$

$$20 = 20 \quad ✓$$

When the same value appears on both sides of the equal sign, we call the equation an **identity.** Because the two sides of the equation in our check have the same value, we know that the original equation has been solved correctly. We have found a solution, and since no other number makes the equation true, it is the only solution.

Practice Problem 1 Solve for x and check your solution. $x + 0.3 = 1.2$

Notice that when you are trying to solve these types of equations, you must add a particular number to both sides of the equation. What is the number to choose? Look at the number that is on the same side of the equation with x, that is, the number added to x. Then think of the number that is **opposite in sign.** This is called the **additive inverse** of the number. The additive inverse of 16 is -16. The additive inverse of -3 is 3. The number to add to both sides of the equation is precisely this additive inverse.

It does not matter which side of the equation contains the variable. The x-term may be on the right or left. In the next example the x-term will be on the right.

EXAMPLE 2 Solve for x. $14 = x - 3$

$$14 + 3 = x - 3 + 3 \qquad \text{Notice that } -3 \text{ is being added to } x \text{ in the original equation. Add 3 to both sides, since 3 is the additive inverse of } -3. \text{ This will eliminate the } -3 \text{ on the right and isolate } x.$$

$$17 = x + 0 \qquad \text{Simplify.}$$

$$17 = x \qquad \text{The value of } x \text{ is 17.}$$

Check. $\qquad 14 = x - 3$

$$14 \overset{?}{=} 17 - 3 \qquad \text{Replace } x \text{ by 17.}$$

$$14 = 14 \quad ✓ \qquad \text{Simplify. It checks. The solution is 17.}$$

Practice Problem 2 Solve for x and check your solution. $17 = x - 5$

Before you add a number to both sides, you should always simplify the equation. The following example shows how combining numbers by addition—separately, on both sides of the equation—simplifies the equation.

EXAMPLE 3 Solve for x. $15 + 2 = 3 + x + 2$

$$17 = x + 5 \qquad \text{Simplify by adding.}$$

$$17 + (-5) = x + 5 + (-5) \qquad \text{Add the value } -5 \text{ to both sides, since } -5 \text{ is the additive inverse of 5.}$$

$$12 = x \qquad \text{Simplify. The value of } x \text{ is 12.}$$

Check.

$$15 + 2 = 3 + x + 2$$
$$15 + 2 \stackrel{?}{=} 3 + 12 + 2 \qquad \text{Replace } x \text{ by 12 in the original equation.}$$
$$17 = 17 \ \checkmark \qquad \text{It checks.}$$

Practice Problem 3 Solve for x and check your solution. $5 - 12 = x - 3$

In Example 3 we added -5 to each side. You could subtract 5 from each side and get the same result. In Chapter 1 we discussed how subtracting a 5 is the same as adding a negative 5. Do you see why?

Just as it is possible to add the same number to both sides of an equation, it is also possible to subtract the same number from both sides of an equation. This is so because any subtraction problem can be rewritten as an addition problem. For example, $17 - 5 = 17 + (-5)$. Thus the addition principle tells us that we can subtract the same number from both sides of the equation.

We can determine whether a value is the solution to an equation by following the same steps used to check an answer. Substitute the value to be tested for the variable in the original equation. We will obtain an identity if the value is the solution.

EXAMPLE 4 Is 10 the solution to the equation $-15 + 2 = x - 3$? If it is not, find the solution.

We substitute 10 for x in the equation and see if we obtain an identity.

$$-15 + 2 = x - 3$$
$$-15 + 2 \stackrel{?}{=} 10 - 3$$
$$-13 \neq 7 \qquad \text{The values are not equal. The statement is}$$
$$\text{not an identity.}$$

Thus, 10 is not the solution. Now we take the original equation and solve to find the solution.

$$-15 + 2 = x - 3$$
$$-13 = x - 3 \qquad \text{Simplify by adding.}$$
$$-13 + 3 = x - 3 + 3 \qquad \text{Add 3 to both sides. 3 is the additive}$$
$$\text{inverse of } -3.$$
$$-10 = x$$

Check to see if -10 is the solution. The value 10 was incorrect because of a sign error. We must be especially careful to write the correct sign for each number when solving equations.

Practice Problem 4 Is -2 the solution to the equation $x + 8 = -22 + 6$? If it is not, find the solution.

EXAMPLE 5 Find the value of x that satisfies the equation $\dfrac{1}{5} + x = -\dfrac{1}{10} + \dfrac{1}{2}$.

To be combined, the fractions must have common denominators. The least common denominator (LCD) of the fractions is 10.

$$\frac{1 \cdot 2}{5 \cdot 2} + x = -\frac{1}{10} + \frac{1 \cdot 5}{2 \cdot 5} \qquad \text{Change each fraction to an equivalent}$$
$$\text{fraction with a denominator of 10.}$$

$$\frac{2}{10} + x = -\frac{1}{10} + \frac{5}{10} \qquad \text{This is an equivalent equation.}$$

$$\frac{2}{10} + x = \frac{4}{10} \qquad \text{Simplify by adding.}$$

$$\frac{2}{10} + \left(-\frac{2}{10}\right) + x = \frac{4}{10} + \left(-\frac{2}{10}\right)$$

Add the additive inverse of $\frac{2}{10}$ to each side. You could also say that you are subtracting $\frac{2}{10}$ from each side.

$$x = \frac{2}{10}$$

Add the fractions.

$$x = \frac{1}{5}$$

Simplify the answer.

Check. We substitute $\frac{1}{5}$ for x in the original equation and see if we obtain an identity.

$$\frac{1}{5} + x = -\frac{1}{10} + \frac{1}{2}$$

$$\frac{1}{5} + \frac{1}{5} \overset{?}{=} -\frac{1}{10} + \frac{1}{2}$$

Substitute $\frac{1}{5}$ for x.

$$\frac{2}{5} \overset{?}{=} -\frac{1}{10} + \frac{5}{10}$$

$$\frac{2}{5} \overset{?}{=} \frac{4}{10}$$

$$\frac{2}{5} = \frac{2}{5} \quad \checkmark$$

It checks.

Practice Problem 5 Find the value of x that satisfies the equation $\frac{1}{20} - \frac{1}{2} = x + \frac{3}{5}$.

Developing Your Study Skills

Why Study Mathematics?

In our present-day, technological world, it is easy to see mathematics at work. Many vocational and professional areas—such as the fields of business, statistics, economics, psychology, finance, computer science, chemistry, physics, engineering, electronics, nuclear energy, banking, quality control, and teaching—require a certain level of expertise in mathematics. Those who want to work in these fields must be able to function at a given mathematical level. Those who cannot will not make it. So if your field of study requires you to take higher-level mathematics courses, be sure to master the basics of this course. Then you will be ready for the next one.

Verbal and Writing Skills

In exercises 1–3, fill in the blank with the appropriate word.

1. When we use the _____ sign, we indicate two expressions are _____ in value.

2. If the _____ _____ is added to both sides of an equation, the results on each side are equal in value.

3. The _____ of an equation is a value of the variable that makes the equation true.

4. What is the additive inverse of -20?

5. Why do we add the additive inverse of a to each side of $x + a = b$ to solve for x?

Find the value of x that satisfies each equation. Check your answers.

6. $x + 11 = 15$

7. $x + 12 = 18$

8. $13 = 4 + x$

9. $15 = x + 9$

10. $x - 3 = 14$

11. $x - 11 = 5$

12. $0 = x + 5$

13. $0 = x - 7$

14. $3 + 5 = x - 7$

15. $8 - 2 = x + 5$

16. $7 + 3 + x = 5 + 5$

17. $18 - 2 + 3 = x + 19$

18. $x - 6 = -19$

19. $x - 11 = -13$

20. $-12 + x = 50$

21. $-18 + x = 48$

22. $18 - 11 = x - 5$

23. $23 - 8 = x - 12$

24. $8 - 23 + 7 = 1 + x - 2$

25. $3 - 17 + 8 = 8 + x - 3$

In exercises 26–33, determine whether the given solution is correct. If it is not, find the solution.

26. Is 8 the solution to $-12 + x = 4$?

27. Is 12 the solution to $-19 + x = 7$?

28. Is -3 a solution to $-18 - 2 = x - 7$?

29. Is -5 a solution to $-16 + 5 = x - 6$?

30. Is -33 the solution to $x - 23 = -56$?

31. Is -8 the solution to $-39 = x - 47$?

32. Is 35 the solution to $15 - 3 + 20 = x - 3$?

33. Is -12 the solution to $x + 8 = 12 - 19 + 3$?

Find the value of x that satisfies each equation.

34. $-3 = x - 8$

35. $-11 + x = -7$

36. $-16 = x + 25$

37. $27 = 5 + x$

38. $1.3 + x + 1.8 = 0.2$

39. $3.6 + 1.2 = x + 1.3$

40. $2.5 + x = 0.7$

41. $4.2 + x = 1.3$

42. $x - \dfrac{1}{4} = \dfrac{3}{4}$

43. $x + \dfrac{1}{3} = \dfrac{2}{3}$

44. $\dfrac{2}{3} + x = \dfrac{1}{6} + \dfrac{1}{4}$

45. $\dfrac{2}{5} + x = \dfrac{1}{2} - \dfrac{3}{10}$

46. $\dfrac{1}{18} - \dfrac{5}{9} = x - \dfrac{1}{2}$

47. $\dfrac{7}{12} - \dfrac{2}{3} = x - \dfrac{5}{4}$

48. $5\dfrac{1}{6} + x = 8$

49. $7\dfrac{1}{8} = -20 + x$

50. $\dfrac{1}{2} + 3x = \dfrac{1}{4} + 2x - 3 - \dfrac{1}{2}$

51. $1.6 - 5x - 3.2 = -2x + 5.6 + 4x - 8x$

52. $x + 0.7513 = 2.2419$

53. $x - 0.2314 = -4.0144$

Cumulative Review Problems

Simplify by adding like terms.

54. $x + 3y - 5x - 7y + 2x$

55. $y^2 + y - 12 - 3y^2 - 5y + 16$

56. A telemarketer has been given the task of selling credit cards to people who subscribe to a certain medical journal. After making 85 calls, she has succeeded in selling 20 credit cards. What percent of the people did not choose to buy a credit card? (Round your answer to the nearest tenth of a percent.)

57. Trevor pays his monthly computer lease bill for $49.99, but forgets to look at his checking account balance before doing so. When he gets his checking account statement at the local ATM, his balance reads −$35.07. How much was in his account before he wrote the check?

58. A 90-meter-wide radar picture is taken of a swamp in northern Australia. The radar detects a rock outcrop that is 90 feet above sea level, and a vein of opal (a semiprecious stone) 27 feet below sea level. How far is the top of the rock from the location of the opal?

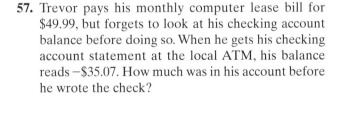

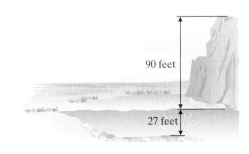

90 feet

27 feet

2.2 The Multiplication Principle

Solving Equations of the Form $\frac{1}{a}x = b$

The addition principle allows us to add the same number to both sides of an equation. What would happen if we multiplied each side of an equation by the same number? For example, what would happen if we multiplied each side of an equation by 3?

To answer this question, let's return to our simple example of the box and the stone on a balanced seesaw. If we triple the weight on each side (that is, multiply the weight on each side by 3), the seesaw should still balance. The weight values of both sides remain equal.

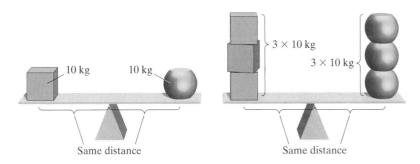

In words we can state this principle thus.

Multiplication Principle

If both sides of an equation are multiplied by the same nonzero number, the results on both sides are equal in value.

In symbols we can restate the multiplication principle this way.

For real numbers a, b, and c with $c \neq 0$, if $a = b$, then $ca = cb$.

It is important that we say $c \neq 0$. We will explore this idea in the To Think About exercises.

Let us look at an equation where it would be helpful to multiply each side by 3.

EXAMPLE 1 Solve for x. $\frac{1}{3}x = -15$

We know that $(3)(\frac{1}{3}) = 1$. We will multiply each side of the equation by 3 because we want to isolate the variable x.

$$3\left(\frac{1}{3}x\right) = 3(-15)$$ Multiply each side of the equation by 3 since $(3)(\frac{1}{3}) = 1$.

$$\left(\frac{3}{1}\right)\left(\frac{1}{3}\right)(x) = -45$$

$$1x = -45$$ Simplify.

$$x = -45$$ The solution is -45.

Check. $\quad \frac{1}{3}(-45) \stackrel{?}{=} -15$ Substitute -45 for x in the original equation.

$$-15 = -15 \ \checkmark \quad \text{It checks.}$$

Student Learning Objectives

After studying this section, you will be able to:

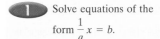 Solve equations of the form $\frac{1}{a}x = b$.

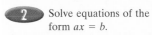 Solve equations of the form $ax = b$.

SSM
PH TUTOR CENTER   CD & VIDEO   MATH PRO   WEB

Practice Problem 1 Solve for x. $\dfrac{1}{8}x = -2$

Note that $\frac{1}{5}x$ can be written as $\frac{x}{5}$. To solve the equation $\frac{x}{5} = 3$, we could multiply each side of the equation by 5. Try it. Then check your solution.

2 Solving Equations of the Form $ax = b$

We can see that using the multiplication principle to multiply each side of an equation by $\frac{1}{2}$ is the same as dividing each side of the equation by 2. Thus, it would seem that the multiplication principle would allow us to divide each side of the equation by any nonzero real number. Is there a real-life example of this idea?

Let's return to our simple example of the box and the stone on a balanced seesaw. Suppose that we were to cut the two objects in half (so that the amount of weight of each was divided by 2). We then return the objects to the same places on the seesaw. The seesaw would still balance. The weight values of both sides remain equal.

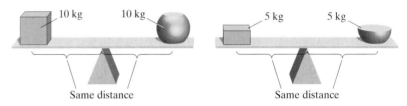

In words we can state this principle thus.

Division Principle

If both sides of an equation are divided by the same nonzero number, the results on both sides are equal in value.

Note: We put a restriction on the number by which we are dividing. We cannot divide by zero. We say that expressions like $\frac{2}{0}$ are not defined. Thus we restrict our divisor to *nonzero* numbers. We can restate the division principle this way.

For real numbers a, b, and c where $c \neq 0$, if $a = b$, then $\dfrac{a}{c} = \dfrac{b}{c}$.

EXAMPLE 2 Solve for x. $5x = 125$

$$\frac{5x}{5} = \frac{125}{5} \qquad \text{Divide both sides by 5.}$$
$$x = 25 \qquad \text{Simplify. The solution is 25.}$$

Check. $5x = 125$

$5(25) \overset{?}{=} 125$ Replace x by 25.

$125 = 125$ ✓   It checks.

Practice Problem 2 Solve for x. $9x = 72$

For equations of the form $ax = b$ (a number multiplied by x equals another number), we solve the equation by choosing to divide both sides by a particular number. What is the number to choose? We look at the side of the equation that contains x. We notice the number that is multiplied by x. We divide by that number. The division principle tells us that we can still have a true equation provided that we divide by that number *on both sides* of the equation.

The solution to an equation may be a proper fraction or an improper fraction.

◉ EXAMPLE 3   Solve for x. $4x = 38$

$$\frac{4x}{4} = \frac{38}{4} \qquad \text{Divide both sides by 4.}$$

$$x = \frac{19}{2} \qquad \text{Simplify. The solution is } \tfrac{19}{2}.$$

If you leave the solution as a fraction, it will be easier to check that solution in the original equation.

Check. $4x = 38$

$$\overset{2}{\cancel{4}}\left(\frac{19}{\cancel{2}}\right) \overset{?}{=} 38 \qquad \text{Replace } x \text{ by } \tfrac{19}{2}.$$

$$38 = 38 \quad ✓ \quad \text{It checks.}$$

Practice Problem 3   Solve for x. $6x = 50$ ◉

In Examples 2 and 3 we *divided by the number multiplied by x*. This procedure is followed regardless of whether the sign of that number is positive or negative. In equations of the form $ax = b$ the **coefficient** of x is a. A coefficient is a multiplier.

◉ EXAMPLE 4   Solve for x. $-3x = 48$

$$\frac{-3x}{-3} = \frac{48}{-3} \qquad \text{Divide both sides by } -3.$$

$$x = -16 \qquad \text{The solution is } -16.$$

Check.   Can you check this solution?

Practice Problem 4   Solve for x. $-27x = 54$ ◉

The coefficient of x may be 1 or -1. You may have to rewrite the equation so that the coefficient of 1 or -1 is obvious. With practice you may be able to recognize the coefficient without actually rewriting the equation.

◉ EXAMPLE 5   Solve for x. $-x = -24.$

$$-1x = -24 \qquad \begin{array}{l}\text{Rewrite the equation. } -1x \text{ is the same as } -x.\\ \text{Now the coefficient of } -1 \text{ is obvious.}\end{array}$$

$$\frac{-1x}{-1} = \frac{-24}{-1} \qquad \text{Divide both sides by } -1.$$

$$x = 24 \qquad \text{The solution is 24.}$$

Check.   Can you check this solution?

Practice Problem 5   Solve for x. $-x = 36$ ◉

The variable can be on either side of the equation. The equation $-78 = -3x$ can be solved in exactly the same way as $-3x = -78$.

EXAMPLE 6 Solve for x. $-78 = -3x$

$$\frac{-78}{-3} = \frac{-3x}{-3} \qquad \text{Divide both sides by } -3.$$

$$26 = x \qquad \text{The solution is 26.}$$

Check. $\qquad -78 = -3x$

$$-78 \stackrel{?}{=} -3(26) \qquad \text{Replace } x \text{ by 26.}$$

$$-78 = -78 \quad \checkmark \quad \text{It checks.}$$

Practice Problem 6 Solve for x. $-51 = 6x$

There is a mathematical concept that unites what we have learned in this section. The concept uses the idea of a multiplicative inverse. For any nonzero number a, the **multiplicative inverse** of a is $1/a$. Likewise, for any nonzero number a, the multiplicative inverse of $1/a$ is a. So to solve an equation of the form $ax = b$, we say that we need to multiply each side by the multiplicative inverse of a. Thus to solve $5x = 45$, we would multiply each side of the equation by the multiplicative inverse of 5, which is $1/5$. In similar fashion, if we wanted to solve the equation $(1/6)x = 4$, we would multiply each side of the equation by the multiplicative inverse of $1/6$, which is 6. In general, all the problems we have covered so far in this section can be solved by multiplying both sides of the equation by the multiplicative inverse of the coefficient of x.

EXAMPLE 7 Solve for x. $31.2 = 6.0x - 0.8x$

$$31.2 = 6.0x - 0.8x \qquad \text{There are like terms on the right side.}$$

$$31.2 = 5.2x \qquad \text{Collect like terms.}$$

$$\frac{31.2}{5.2} = \frac{5.2x}{5.2} \qquad \text{Divide both sides by 5.2 (which is the same as multiplying both sides by the multiplicative inverse of 5.2).}$$

$$6 = x \qquad \text{The solution is 6.}$$

Note: Be sure to place the decimal point in the quotient directly above the caret $(\wedge)$ when performing the division.

$$
\begin{array}{r}
6. \\
5.2_{\wedge}\overline{)31.2_{\wedge}} \\
\underline{31\ 2} \\
0
\end{array}
$$

Check. The check is up to you.

Practice Problem 7 Solve for x. $21 = 4.2x$

2.2 Exercises

Verbal and Writing Skills

1. To solve the equation $6x = -24$, divide each side of the equation by _____.

2. To solve the equation $-7x = 56$, divide each side of the equation by _____.

3. To solve the equation $\frac{1}{7}x = -2$, multiply each side of the equation by _____.

4. To solve the equation $\frac{1}{9}x = 5$, multiply each side of the equation by _____.

Solve for x. Be sure to reduce your answer. Check your solution.

5. $\frac{1}{7}x = 5$

6. $\frac{1}{5}x = 6$

7. $\frac{1}{3}x = -9$

8. $\frac{1}{4}x = -20$

9. $\frac{x}{5} = 16$

10. $\frac{x}{10} = 8$

11. $-3 = \frac{x}{5}$

12. $\frac{x}{3} = -12$

13. $12x = 48$

14. $8x = 72$

15. $-16 = 6x$

16. $-35 = 21x$

17. $1.5x = 75$

18. $2x = 0.36$

19. $-15 = -x$

20. $32 = -x$

21. $-84 = 12x$

22. $-72 = -9x$

23. $0.4x = 0.08$

24. $2.1x = 0.3$

Determine whether the given solution is correct. If it is not, find the correct solution.

25. Is 7 the solution for $-3x = 21$?

26. Is 8 the solution for $5x = -40$?

27. Is -6 the solution for $-11x = 66$?

28. Is -20 the solution for $-x = 20$?

Find the value of the variable that satisfies the equation.

29. $-3y = 2.4$

30. $5z = -1.8$

31. $-56 = -21t$

32. $34 = -51q$

33. $4.6y = -3.22$

34. $-2.8y = -3.08$

35. $4x + 3x = 21$

36. $5x + 4x = 36$

37. $2x - 7x = 20$

38. $3x - 9x = 18$

39. $-6x - 3x = -7$

40. $y - 11y = 7$

41. $-\frac{2}{3} = -\frac{4}{7}x$

42. $5.6 = -2.7x$

43. $3.6172x = -19.026472$

44. $-4.0518x = 14.505444$

139

To Think About

45. We have said that if $a = b$ and $c \neq 0$, then $ac = bc$. Why is it important that $c \neq 0$? What would happen if we tried to solve an equation by multiplying both sides by zero?

46. We have said that if $a = b$ and $c \neq 0$ then $\frac{a}{c} = \frac{b}{c}$. Why is it important that $c \neq 0$? What would happen if we tried to solve an equation by dividing both sides by zero?

Cumulative Review Problems

Evaluate using the correct order of operations. (Be careful to avoid sign errors.)

47. $(-6)(-8) + (-3)(2)$

48. $(-3)^3 + (-20) \div 2$

49. $5 + (2 - 6)^2$

50. An off-price clothing store chain specializes in last year's merchandise. Their contact in Hong Kong has purchased 12,000 famous designer men's sport coats for the stores. When the shipment is unloaded, 800 sport coats have no left sleeve. What percent of the shipment is acceptable?

51. In January, Keiko invested $600 in a certain stock. In February, the stock gained $82.00. In March, the stock lost $47.00. In April, the stock gained $103.00. In May, the stock lost $106.00. What was Keiko's stock holding worth after the May loss?

52. In 1995, the humpback whale calf population at Stellwagen Bank, near Gloucester, Massachusetts, was estimated at 12 calves. The population grew by 21 calves in 1996, 18 calves in 1997, and 51 calves in 1998. In 1999, the number of whale calves decreased by 4, and in 2000 it increased by 6. What was the whale calf population at the end of 2000?

Developing Your Study Skills

Getting Help

Getting the right kind of help at the right time can be a key ingredient in being successful in mathematics. When you have gone to class on a regular basis, taken careful notes, methodically read your textbook, and diligently done your homework—all of which means making every effort possible to learn the mathematics—you may find that you are still having difficulty. If this is the case, then you need to seek help. Make an appointment with your instructor to find out what help is available to you. The instructor, tutoring services, a mathematics lab, videotapes, and computer software may be among the resources you can draw on.

Once you discover the resources available in your school, you need to take advantage of them. Do not put it off, or you will find yourself getting behind. You cannot afford that. When studying mathematics, you must keep up with your work.

2.3 Using the Addition and Multiplication Principles Together

 Solving Equations of the Form $ax + b = c$

Jenny Crawford scored several goals in field hockey during April. Her teammates scored three more than five times the number she scored. Together the team scored 18 goals in April. How many did Jenny score? To solve this problem, we need to solve the equation $5x + 3 = 18$.

To solve an equation of the form $ax + b = c$, we must use both the addition principle and the multiplication principle.

EXAMPLE 1 Solve for x to determine how many goals Jenny scored and check your solution.

$$5x + 3 = 18$$

We first want to isolate the variable term.

$5x + 3 + (-3) = 18 + (-3)$ Use the addition principle to add -3 to both sides.

$$5x = 15$$ Simplify.

$$\frac{5x}{5} = \frac{15}{5}$$ Use the division principle to divide both sides by 5.

$$x = 3$$ The solution is 3. Thus Jenny scored 3 goals.

Check. $5(3) + 3 \overset{?}{=} 18$
$15 + 3 \overset{?}{=} 18$
$18 = 18$ ✓   It checks.

Practice Problem 1 Solve for x and check your solution. $9x + 2 = 38$

 Solving Equations with the Variable on Both Sides of the Equation

In some cases the variable appears on both sides of the equation. We would like to rewrite the equation so that all the terms containing the variable appear on one side. To do this, we apply the addition principle to the variable term.

EXAMPLE 2 Solve for x. $9x = 6x + 15$

$9x + (-6x) = 6x + (-6x) + 15$ Add $-6x$ to both sides. Notice $6x + (-6x)$ eliminates the variable on the right side.

$$3x = 15$$ Combine like terms.

$$\frac{3x}{3} = \frac{15}{3}$$ Divide both sides by 3.

$$x = 5$$ The solution is 5.

Check.   The check is left to the student.

Practice Problem 2 Solve for x. $13x = 2x - 66$

In many problems the variable terms and constant terms appear on both sides of the equations. You will want to get all the variable terms on one side and all the constant terms on the other side.

Student Learning Objectives

After studying this section, you will be able to:

1. Solve equations of the form $ax + b = c$.

2. Solve equations with the variable on both sides of the equation.

3. Solve equations with parentheses.

SSM   PH TUTOR   CD & VIDEO   MATH PRO   WEB
CENTER

EXAMPLE 3   Solve for x and check your solution. $9x + 3 = 7x - 2$.

$9x + (-7x) + 3 = 7x + (-7x) - 2$	Add $-7x$ to both sides of the equation.
$2x + 3 = -2$	Combine like terms.
$2x + 3 + (-3) = -2 + (-3)$	Add -3 to both sides.
$2x = -5$	Simplify.
$\dfrac{2x}{2} = \dfrac{-5}{2}$	Divide both sides by 2.
$x = -\dfrac{5}{2}$	The solution is $-\frac{5}{2}$.

Check.

$$9x + 3 = 7x - 2$$

$9\left(-\dfrac{5}{2}\right) + 3 \overset{?}{=} 7\left(-\dfrac{5}{2}\right) - 2$	Replace x by $-\dfrac{5}{2}$.
$-\dfrac{45}{2} + 3 \overset{?}{=} -\dfrac{35}{2} - 2$	Simplify.
$-\dfrac{45}{2} + \dfrac{6}{2} \overset{?}{=} -\dfrac{35}{2} - \dfrac{4}{2}$	Change to equivalent fractions with a common denominator.
$-\dfrac{39}{2} = -\dfrac{39}{2}$ ✓	It checks. The solution is $-\frac{5}{2}$.

Practice Problem 3   Solve for x and check your solution. $3x + 2 = 5x + 2$

In our next example we will study equations that need simplifying before any other steps are taken. Where it is possible, you should first collect like terms on one or both sides of the equation. The variable terms can be collected on the right side or the left side. In this example we will collect all the x-terms on the right side.

EXAMPLE 4   Solve for x. $5x + 26 - 6 = 9x + 12x$

$5x + 20 = 21x$	Combine like terms.
$5x + (-5x) + 20 = 21x + (-5x)$	Add $-5x$ to both sides.
$20 = 16x$	Combine like terms.
$\dfrac{20}{16} = \dfrac{16x}{16}$	Divide both sides by 16.
$\dfrac{5}{4} = x$	Don't forget to reduce the resulting fraction.

Check.   The check is left to the student.

Practice Problem 4   Solve for z. $-z + 8 - z = 3z + 10 - 3$

Do you really need all these steps? No. As you become more proficient you will be able to combine or eliminate some of these steps. However, it is best to write each step in its entirety until you are consistently obtaining the correct solution. It is much better to show every step than to take a lot of shortcuts and possibly obtain a wrong answer. This is a section of the algebra course where working neatly and accurately will help you—both now and as you progress through the course.

All the equations we have been studying so far are called **first-degree equations**. This means the variable terms are not squared (such as x^2 or y^2) or raised to some higher power. It is possible to solve equations with x^2 and y^2 terms by the same methods we have used so far. If x^2 or y^2 terms appear, try to collect them on one side of the equation. If the squared term drops out, you may solve the equation as a first-degree equation using the methods discussed in this section.

EXAMPLE 5 Solve for y. $5y^2 + 6y - 2 = -y + 5y^2 + 12$

$5y^2 - 5y^2 + 6y - 2 = -y + 5y^2 - 5y^2 + 12$	Subtract $5y^2$ from both sides.
$6y - 2 = -y + 12$	Combine, since $5y^2 - 5y^2 = 0$.
$6y + y - 2 = -y + y + 12$	Add y to each side.
$7y - 2 = 12$	Simplify.
$7y - 2 + 2 = 12 + 2$	Add 2 to each side.
$7y = 14$	Simplify.
$\dfrac{7y}{7} = \dfrac{14}{7}$	Divide each side by 7.
$y = 2$	Simplify. The solution is 2.

Check.　The check is left to the student.

Practice Problem 5　Solve for x. $2x^2 - 6x + 3 = -4x - 7 + 2x^2$

3　Solving Equations with Parentheses

The equations that you just solved are simpler versions of equations that we will now discuss. These equations contain parentheses. If the parentheses are first removed, the problems then become just like those encountered previously. We use the distributive property to remove the parentheses.

EXAMPLE 6　Solve for x and check your solution. $4(x + 1) - 3(x - 3) = 25$

$4(x + 1) - 3(x - 3) = 25$	
$4x + 4 - 3x + 9 = 25$	Multiply by 4 and -3 to remove parentheses. Be careful of the signs. Remember that $(-3)(-3) = 9$.

After removing the parentheses, it is important to collect like terms on each side of the equation. Do this before going on to isolate the variable.

$x + 13 = 25$	Collect like terms.
$x + 13 - 13 = 25 - 13$	Add -13 to both sides to isolate the variable.
$x = 12$	The solution is 12.
Check.　$4(12 + 1) - 3(12 - 3) \overset{?}{=} 25$	Replace x by 12.
$4(13) - 3(9) \overset{?}{=} 25$	Combine numbers inside parentheses.
$52 - 27 \overset{?}{=} 25$	Multiply.
$25 = 25$ ✓	Simplify. It checks.

Practice Problem 6　Solve for x and check your solution.

$$4x - (x + 3) = 12 - 3(x - 2)$$

EXAMPLE 7　Solve for x. $3(-x - 7) = -2(2x + 5)$

$-3x - 21 = -4x - 10$	Remove parentheses. Watch the signs carefully.
$-3x + 4x - 21 = -4x + 4x - 10$	Add $4x$ to both sides.
$x - 21 = -10$	Simplify.
$x - 21 + 21 = -10 + 21$	Add 21 to both sides.
$x = 11$	The solution is 11.

Check. The check is left to the student.

Practice Problem 7 Solve for x. $4(-2x - 3) = -5(x - 2) + 2$

In problems that involve decimals, great care should be taken. In some steps you will be multiplying decimal quantities, and in other steps you will be adding them.

EXAMPLE 8 Solve for x. $0.3(1.2x - 3.6) = 4.2x - 16.44$

$0.36x - 1.08 = 4.2x - 16.44$	Remove parentheses.
$0.36x - 0.36x - 1.08 = 4.2x - 0.36x - 16.44$	Subtract $0.36x$ from both sides.
$-1.08 = 3.84x - 16.44$	Collect like terms.
$-1.08 + 16.44 = 3.84x - 16.44 + 16.44$	Add 16.44 to both sides.
$15.36 = 3.84x$	Simplify.
$\dfrac{15.36}{3.84} = \dfrac{3.84x}{3.84}$	Divide both sides by 3.84.
$4 = x$	The solution is 4.

Check. The check is left to the student.

Practice Problem 8 Solve for x. $0.3x - 2(x + 0.1) = 0.4(x - 3) - 1.1$

EXAMPLE 9 Solve for z and check. $2(3z - 5) + 2 = 4z - 3(2z + 8)$

$6z - 10 + 2 = 4z - 6z - 24$	Remove parentheses.
$6z - 8 = -2z - 24$	Collect like terms.
$6z - 8 + 2z = -2z + 2z - 24$	Add $2z$ to each side.
$8z - 8 = -24$	Simplify.
$8z - 8 + 8 = -24 + 8$	Add 8 to each side.
$8z = -16$	Simplify.
$\dfrac{8z}{8} = \dfrac{-16}{8}$	Divide each side by 8.
$z = -2$	Simplify. The solution is -2.

Check.
$$2[3(-2) - 5] + 2 \stackrel{?}{=} 4(-2) - 3[2(-2) + 8] \quad \text{Replace } z \text{ by } -2.$$
$$2[-6 - 5] + 2 \stackrel{?}{=} -8 - 3[-4 + 8] \quad \text{Multiply.}$$
$$2[-11] + 2 \stackrel{?}{=} -8 - 3[4] \quad \text{Simplify.}$$
$$-22 + 2 \stackrel{?}{=} -8 - 12$$
$$-20 = -20 \quad \checkmark \quad \text{It checks.}$$

Practice Problem 9 Solve for z and check. $5(2z - 1) + 7 = 7z - 4(z + 3)$

Find the value of the variable that satisfies the equation. Check your solution. Answers that are not integers may be left in fractional form or decimal form.

1. $4x + 13 = 21$

2. $7x - 4 = 59$

3. $4x - 11 = 13$

4. $5x - 11 = 39$

5. $4x - 21 = -91$

6. $9x - 13 = -76$

7. $-4x + 17 = -35$

8. $-6x + 25 = -83$

9. $-15 + 2x = 15$

10. $-8 + 4x = 8$

11. $\frac{1}{5}x - 2 = 6$

12. $\frac{1}{3}x - 7 = 4$

13. $\frac{1}{6}x + 2 = -4$

14. $\frac{1}{5}x + 6 = -24$

15. $8x = 48 + 2x$

16. $5x = 22 + 3x$

17. $-6x = -27 + 3x$

18. $-7x = -26 + 6x$

19. $63 - x = 8x$

20. $56 - 3x = 5x$

21. $54 - 2x = -8x$

To Think About

22. Is 2 the solution for $2y + 3y = 12 - y$?

23. Is 4 the solution for $5y + 2 = 6y - 6 + y$?

24. Is 11 a solution for $7x + 6 - 3x = 2x - 5 + x$?

25. Is -12 a solution for $9x + 2 - 5x = -8 + 5x - 2$?

Solve for y by getting all the y-terms on the left. Then solve for y by getting all the y-terms on the right. Which approach is better?

26. $-3 + 10y + 6 = 15 + 12y - 18$

27. $7y + 21 - 5y = 5y - 7 + y$

Solve for the variable. You may move the variable terms to the right or to the left.

28. $14 - 2x = -5x + 11$

29. $8 - 3x = 7x + 8$

30. $x - 6 = 8 - x$

31. $2x + 5 = 4x - 5$

32. $6y - 5 = 8y - 7$

33. $11y - 8 = 9y - 16$

34. $5x - 9 + 2x = 3x + 23 - 4x$

35. $9x - 5 + 4x = 7x + 43 - 2x$

Remove the parentheses and solve for the variable. Check your solution. Answers that are not integers may be left in fractional form or decimal form.

36. $5(x + 3) = 35$

37. $6(x + 2) = 42$

38. $6(3x + 2) - 8 = -2$

39. $4(2x + 1) - 7 = 6 - 5$

40. $7x - 3(5 - x) = 10$

41. $6(3 - 4x) + 17 = 8x - 3(2 - 3x)$

42. $0.7x - 0.2(x + 1) = 0.16$

43. $3(x + 0.2) = 2(x - 0.3) + 4.3$

44. $5(x - 3) + 5 = 3(x + 2) - 4$

45. $3(x - 2) + 2 = 2(x - 4)$

46. $0.2(x + 3) - (x - 1.5) = 0.3(x + 2) - 2.9$

47. $3(x + 0.2) - (2x + 0.5) = 2(x + 0.3) - 0.5$

48. $-3(y - 3y) + 4 = -4(3y - y) + 6 + 13y$

49. $2(4x - x) + 6 = 2(2x + x) + 8 - x$

Mixed Practice

Solve for the variable.

50. $5.7x + 3 = 4.2x - 3$

51. $4x - 3.1 = 5.3 - 3x$

52. $5z + 7 - 2z = 32 - 2z$

53. $8 - 7z + 2z = 20 + 5z$

54. $-4w - 28 = -7 - w$

55. $-6w - 7 = -3 - 8w$

56. $6x + 8 - 3x = 11 - 12x - 13$

57. $4 - 7x - 13 = 8x - 3 - 5x$

58. $2x^2 - 3x - 8 = 2x^2 + 5x - 6$

59. $3x^2 + 4x - 7 = 3x^2 - 5x + 2$

60. $-3.5x + 1.3 = -2.7x + 1.5$

61. $2.8x - 0.9 = 5.2x - 3.3$

62. $5(4 + x) = 3(3x - 1) - 9$

63. $x - 0.8x + 4 = 2.6$

64. $17(y + 3) - 4(y - 10) = 13$

65. $3x + 2 - 1.7x = 0.6x + 31.4$

66. $3(x + 4) - 5(3x - 2) = 8$

67. $3(2z - 4) - 4(z + 5) = 5(z - 4)$

Solve for x. Round your answer to the nearest hundredth.

 68. $1.63x - 9.23 = 5.71x + 8.04$

69. $-2.21x + 8.65 = 3.69x - 7.78$

Cumulative Review Problems

Simplify.

70. $2x(3x - y) + 4(2x^2 - 3xy)$

71. $2\{x - 3[4 + 2(3 + x)]\}$

72. On March 30, 2000, William owned three different stocks: ABC, RTC, and JTJ. His portfolio contained the following:

> 4.0 shares of ABC stock valued at $\$42\frac{1}{8}$,
> 3.2 shares of RTC stock valued at $\$161\frac{1}{2}$, and
> 5.2 shares of JTJ stock valued at $\$102$.

Find the market value of William's stock holdings on March 30, 2000.

73. Bea works for a large department store and obtains a 10% discount on anything she buys from the store. One item Bea wishes to purchase costs $140 and is on sale at a 25% discount.
   (a) What is the price if Bea has a total discount of 35%? (Disregard sales tax.)
   (b) What is the price if Bea has a 10% discount on the 25% discount price? (Disregard sales tax.)

2.4 Equations with Fractions

1 Solving Equations with Fractions

Equations with fractions can be rather difficult to solve. This difficulty is simply due to the extra care we usually have to use when computing with fractions. The actual equation-solving procedures are the same, with fractions or without. To avoid unnecessary work, we transform the given equation with fractions to an equivalent equation that does not contain fractions. How do we do this? We multiply each side of the equation by the least common denominator of all the fractions contained in the equation. We then use the distributive property so that the LCD is multiplied by each term of the equation.

EXAMPLE 1 Solve for x. $\dfrac{1}{4}x - \dfrac{2}{3} = \dfrac{5}{12}x$

First we find that the LCD = 12.

$$12\left(\frac{1}{4}x - \frac{2}{3}\right) = 12\left(\frac{5}{12}x\right) \qquad \text{Multiply each side by 12.}$$

$$\left(\frac{12}{1}\right)\left(\frac{1}{4}\right)(x) - \left(\frac{12}{1}\right)\left(\frac{2}{3}\right) = \left(\frac{12}{1}\right)\left(\frac{5}{12}\right)(x) \qquad \text{Use the distributive property.}$$

$$3x - 8 = 5x \qquad \text{Simplify.}$$

$$3x + (-3x) - 8 = 5x + (-3x) \qquad \text{Add } -3x \text{ to each side.}$$

$$-8 = 2x \qquad \text{Simplify.}$$

$$\frac{-8}{2} = \frac{2x}{2} \qquad \text{Divide each side by 2.}$$

$$-4 = x \qquad \text{Simplify.}$$

Check.

$$\frac{1}{4}(-4) - \frac{2}{3} \overset{?}{=} \frac{5}{12}(-4)$$

$$-1 - \frac{2}{3} \overset{?}{=} -\frac{5}{3}$$

$$-\frac{3}{3} - \frac{2}{3} \overset{?}{=} -\frac{5}{3}$$

$$-\frac{5}{3} = -\frac{5}{3} \quad \checkmark \qquad \text{It checks.}$$

Practice Problem 1 Solve for x. $\dfrac{3}{8}x - \dfrac{3}{2} = \dfrac{1}{4}x$

In Example 1 we multiplied each side of the equation by the LCD. However, most students prefer to go immediately to the second step and multiply each term by the LCD. This avoids having to write out a separate step using the distributive property.

EXAMPLE 2 Solve for x and check your solution. $\dfrac{x}{3} + 3 = \dfrac{x}{5} - \dfrac{1}{3}$

$$15\left(\frac{x}{3}\right) + 15(3) = 15\left(\frac{x}{5}\right) - 15\left(\frac{1}{3}\right) \qquad \begin{array}{l}\text{The LCD is 15. Use the multiplication} \\ \text{principle to multiply each term by 15.}\end{array}$$

$$5x + 45 = 3x - 5 \qquad \text{Simplify.}$$

$$5x - 3x + 45 = 3x - 3x - 5 \qquad \text{Add } -3x \text{ to both sides.}$$

$$2x + 45 = -5 \qquad \text{Combine like terms.}$$

$$2x + 45 - 45 = -5 - 45 \qquad \text{Add } -45 \text{ to both sides.}$$

$$2x = -50 \qquad \text{Simplify.}$$

$$\frac{2x}{2} = \frac{-50}{2} \qquad \text{Divide both sides by 2.}$$

$$x = -25 \qquad \text{The solution is } -25.$$

Check.

$$\frac{-25}{3} + 3 \stackrel{?}{=} \frac{-25}{5} - \frac{1}{3}$$

$$-\frac{25}{3} + \frac{9}{3} \stackrel{?}{=} -\frac{5}{1} - \frac{1}{3}$$

$$-\frac{16}{3} \stackrel{?}{=} -\frac{15}{3} - \frac{1}{3}$$

$$-\frac{16}{3} = -\frac{16}{3} \quad \checkmark$$

Practice Problem 2 Solve for x and check your solution. $\dfrac{5x}{4} - 1 = \dfrac{3x}{4} + \dfrac{1}{2}$

EXAMPLE 3 Solve for x. $\dfrac{x + 5}{7} = \dfrac{x}{4} + \dfrac{1}{2}$

$$\frac{x}{7} + \frac{5}{7} = \frac{x}{4} + \frac{1}{2} \qquad \begin{array}{l}\text{First we rewrite the left side as two} \\ \text{fractions. This is actually multiplying} \\ \frac{1}{7}(x + 5) = \frac{x}{7} + \frac{5}{7}.\end{array}$$

$$28\left(\frac{x}{7}\right) + 28\left(\frac{5}{7}\right) = 28\left(\frac{x}{4}\right) + 28\left(\frac{1}{2}\right) \qquad \begin{array}{l}\text{We observe that the LCD is 28,} \\ \text{so we multiply each term by 28.}\end{array}$$

$$4x + 20 = 7x + 14 \qquad \text{Simplify.}$$

$$4x - 4x + 20 = 7x - 4x + 14 \qquad \text{Add } -4x \text{ to both sides.}$$

$$20 = 3x + 14 \qquad \text{Combine like terms.}$$

$$20 - 14 = 3x + 14 - 14 \qquad \text{Add } -14 \text{ to both sides.}$$

$$6 = 3x \qquad \text{Combine like terms.}$$

$$\frac{6}{3} = \frac{3x}{3} \qquad \text{Divide both sides by 3.}$$

$$2 = x \qquad \text{The solution is 2.}$$

Check. The check is left to the student.

Practice Problem 3 Solve for x. $\dfrac{5x}{6} - \dfrac{5}{8} = \dfrac{3x}{4} - \dfrac{1}{3}$

If a problem contains both parentheses and fractions, it is best to remove the parentheses first. Many students find it is helpful to have a written procedure to follow in solving these more involved equations.

Procedure to Solve Equations

1. Remove any parentheses.
2. If fractions exist, multiply all terms on both sides by the least common denominator of all the fractions.
3. Combine like terms if possible.
4. Add or subtract terms on both sides of the equation to get all terms with the variable on one side of the equation.
5. Add or subtract a constant value on both sides of the equation to get all terms not containing the variable on the other side of the equation.
6. Divide both sides of the equation by the coefficient of the variable.
7. Simplify the solution (if possible).
8. Check your solution.

Let's use each step in solving the next example.

● **EXAMPLE 4** Solve for x and check your solution. $\frac{1}{3}(x - 2) = \frac{1}{5}(x + 4) + 2$

Step 1 $\qquad\qquad \dfrac{x}{3} - \dfrac{2}{3} = \dfrac{x}{5} + \dfrac{4}{5} + 2$ $\qquad$ Remove parentheses.

Step 2 $\quad 15\left(\dfrac{x}{3}\right) - 15\left(\dfrac{2}{3}\right) = 15\left(\dfrac{x}{5}\right) + 15\left(\dfrac{4}{5}\right) + 15(2)$ $\quad$ Multiply by the LCD, 15.

$\qquad\qquad\qquad 5x - 10 = 3x + 12 + 30$ $\qquad$ Simplify.

Step 3 $\qquad\qquad 5x - 10 = 3x + 42$ $\qquad$ Combine like terms on each side.

Step 4 $\qquad 5x - 3x - 10 = 3x - 3x + 42$ $\qquad$ Add $-3x$ to both sides.

$\qquad\qquad\qquad\quad 2x - 10 = 42$ $\qquad$ Simplify.

Step 5 $\qquad 2x - 10 + 10 = 42 + 10$ $\qquad$ Add 10 to both sides.

$\qquad\qquad\qquad\qquad 2x = 52$ $\qquad$ Simplify.

Step 6 $\qquad\qquad\qquad \dfrac{2x}{2} = \dfrac{52}{2}$ $\qquad$ Divide both sides by 2.

Step 7 $\qquad\qquad\qquad\quad x = 26$ $\qquad$ Simplify the solution.

Step 8 Check. $\dfrac{1}{3}(26 - 2) \overset{?}{=} \dfrac{1}{5}(26 + 4) + 2$ $\qquad$ Replace x by 26.

$\qquad\qquad\quad \dfrac{1}{3}(24) \overset{?}{=} \dfrac{1}{5}(30) + 2$ $\qquad$ Combine values within parentheses.

$\qquad\qquad\qquad\quad 8 \overset{?}{=} 6 + 2$ $\qquad$ Simplify.

$\qquad\qquad\qquad\quad 8 = 8 \ \checkmark$ $\qquad$ The solution is 26.

Practice Problem 4 Solve for x and check your solution.

$$\frac{1}{3}(x - 2) = \frac{1}{4}(x + 5) - \frac{5}{3}$$

Remember that not every step will be needed in each problem. You can combine some steps as well, *as long as you are consistently obtaining the correct solution.* However, you are encouraged to write out every step as a way of helping you to avoid careless errors.

It is important to remember that when we write decimals these numbers are really fractions written in a special way. Thus, $0.3 = \frac{3}{10}$ and $0.07 = \frac{7}{100}$. It is possible to take an equation containing decimals and to multiply each term by the appropriate value to obtain integer coefficients.

● **EXAMPLE 5** Solve for x. $0.2(1 - 8x) + 1.1 = -5(0.4x - 0.3)$

$\qquad\qquad 0.2 - 1.6x + 1.1 = -2.0x + 1.5$ $\qquad$ Remove parentheses.

$10(0.2) - 10(1.6x) + 10(1.1) = 10(-2.0x) + 10(1.5)$ $\quad$ Multiply each term by 10.

$\qquad\qquad 2 - 16x + 11 = -20x + 15$ $\qquad$ Multiplying by 10 moves the decimal point one place to the right.

$\qquad\qquad\quad -16x + 13 = -20x + 15$ $\qquad$ Simplify.

$\qquad -16x + 20x + 13 = -20x + 20x + 15$ $\qquad$ Add $20x$ to each side.

$\qquad\qquad\qquad 4x + 13 = 15$ $\qquad$ Simplify.

$\qquad 4x + 13 + (-13) = 15 + (-13)$ $\qquad$ Add -13 to each side.

$\qquad\qquad\qquad\qquad 4x = 2$ $\qquad$ Simplify.

$\qquad\qquad\qquad\quad \dfrac{4x}{4} = \dfrac{2}{4}$ $\qquad$ Divide each side by 4.

$\qquad\qquad\qquad\quad x = \dfrac{1}{2} \ \text{ or } \ 0.5$ $\qquad$ Simplify.

Check. $0.2[1 - 8(0.5)] + 1.1 \overset{?}{=} -5[0.4(0.5) - 0.3]$

$0.2[1 - 4] + 1.1 \overset{?}{=} -5[0.2 - 0.3]$

$0.2[-3] + 1.1 \overset{?}{=} -5[-0.1]$

$-0.6 + 1.1 \overset{?}{=} 0.5$

$0.5 = 0.5$ ✓

Practice Problem 5   Solve for x. $2.8 = 0.3(x - 2) + 2(0.1x - 0.3)$

To Think About   Does every equation have one solution? Actually, no. There are some rare cases where an equation has no solution at all. Suppose we try to solve the equation

$$5(x + 3) = 2x - 8 + 3x.$$

If we remove the parentheses and collect like terms we have

$$5x + 15 = 5x - 8.$$

If we add $-5x$ to each side, we obtain

$$15 = -8.$$

Clearly this is impossible. There is no value of x for which these two numbers are equal. We would say this equation has **no solution.**

One additional surprise may happen. An equation may have an infinite number of solutions. Suppose we try to solve the equation

$$9x - 8x - 7 = 3 + x - 10.$$

If we combine like terms on each side, we have the equation

$$x - 7 = x - 7.$$

If we add $-x$ to each side, we obtain

$$-7 = -7.$$

Now this statement is always true, no matter what the value of x. We would say this equation has **an infinite number of solutions.**

In the To Think About exercises in this section, we will encounter some equations that have no solution or an infinite number of solutions.

Developing Your Study Skills

Taking Notes in Class

An important part of studying mathematics is taking notes. To take meaningful notes, you must be an active listener. Keep your mind on what the instructor is saying, and be ready with questions whenever you do not understand something.

If you have previewed the lesson material, you will be prepared to take good notes. The important concepts will seem somewhat familiar. If you frantically try to write all that the instructor says or copy all the examples done in class, you may find your notes nearly worthless when you are home alone. Write down *important* ideas and examples as the instructor lectures, making sure that you are listening and following the logic. Include any helpful hints or suggestions that your instructor gives you or refers to in your text.

In exercises 1–16, solve for the variable and check your answer. Noninteger answers may be left in fractional form or decimal form.

1. $\dfrac{1}{5}x + \dfrac{1}{10} = \dfrac{1}{2}$

2. $\dfrac{1}{3}x - \dfrac{1}{9} = \dfrac{8}{9}$

3. $\dfrac{3}{4}x = \dfrac{1}{2}x + \dfrac{5}{8}$

4. $\dfrac{1}{2} = \dfrac{3}{10} - \dfrac{2}{5}x$

5. $\dfrac{x}{2} + \dfrac{x}{5} = \dfrac{7}{10}$

6. $\dfrac{x}{5} - \dfrac{x}{3} = \dfrac{8}{15}$

7. $20 - \dfrac{1}{3}x = \dfrac{1}{2}x$

8. $\dfrac{1}{5}x - \dfrac{1}{2} = \dfrac{1}{6}x$

9. $2 + \dfrac{y}{2} = \dfrac{3y}{4} - 3$

10. $\dfrac{x}{3} - 1 = -\dfrac{1}{2} - x$

11. $\dfrac{y-1}{2} = 4 - \dfrac{y}{7}$

12. $\dfrac{x-7}{6} = -\dfrac{1}{2}$

13. $0.3x - 2.2 = 3.2$

14. $2.8 - 0.4x = 8$

15. $0.6x + 5.9 = 3.8$

16. $1.2x - 2.2 = 5.6$

17. Is 4 a solution to $\dfrac{1}{2}(y - 2) + 2 = \dfrac{3}{8}(3y - 4)$?

18. Is 2 a solution to $\dfrac{1}{5}(y + 2) = \dfrac{1}{10}y + \dfrac{3}{5}$?

19. Is $\dfrac{5}{8}$ a solution to $\dfrac{1}{2}\left(y - \dfrac{1}{5}\right) = \dfrac{1}{5}(y + 2)$?

20. Is $\dfrac{13}{3}$ a solution to $\dfrac{y}{2} - \dfrac{7}{9} = \dfrac{y}{6} + \dfrac{2}{3}$?

Solve for the variable. Noninteger answers may be left in fractional form or decimal form.

21. $\dfrac{3}{4}(3x + 1) = 2(3 - 2x) + 1$

22. $2(x - 2) = \dfrac{2}{5}(3x + 1) + 2$

23. $0.7x - 3.3 = 2.5 - 0.2x - 5.8$

24. $0.5 - 2.1x = 6.4 + 0.3x - 5.9$

25. $0.3x - 0.2(3 - 5x) = -0.5(x - 6)$

26. $0.3(x - 2) + 0.4x = -0.2(x - 6)$

27. $-5(0.2x + 0.1) - 0.6 = 1.9$

28. $0.3x + 1.7 = 0.2x - 0.4(5x + 1)$

Mixed Practice

Solve. Noninteger answers may be left in fractional form or decimal form.

29. $\dfrac{1}{3}(y + 2) = 3y - 5(y - 2)$

30. $\dfrac{2}{5}(y + 3) - \dfrac{1}{2} = \dfrac{1}{3}(y - 2) + \dfrac{1}{2}$

31. $\dfrac{1 + 2x}{5} + \dfrac{4 - x}{3} = \dfrac{1}{15}$

32. $\dfrac{x + 3}{4} = 4x - 2(x - 3)$

33. $\dfrac{1}{5}(x + 3) = 2x - 3(2 - x) - 3$

34. $\dfrac{2}{3}(x + 4) = 6 - \dfrac{1}{4}(3x - 2) - 1$

35. $\dfrac{1}{3}(x - 2) = 3x - 2(x - 1) + \dfrac{16}{3}$

36. $\dfrac{3}{4}(x - 2) + \dfrac{3}{5} = \dfrac{1}{5}(x + 1)$

37. $\dfrac{3}{2}x + \dfrac{1}{3} = \dfrac{2x - 3}{4}$

38. $\dfrac{5}{3} - \dfrac{1}{6}x = \dfrac{3x + 5}{4}$

39. $0.8(x - 3) = -5(2.1x + 0.4)$

40. $0.7(2x + 3) = -4(6.1x + 0.2)$

41. $\dfrac{1}{5}x + \dfrac{2}{3} + \dfrac{1}{15} = \dfrac{4}{3}x - \dfrac{7}{15}$

42. $\dfrac{1}{20}x - \dfrac{1}{4} + \dfrac{3}{5} = -\dfrac{1}{2}x + \dfrac{1}{4}x$

To Think About

Solve. Be careful to examine your work to see if the equation may have no solution or an infinite number of solutions.

43. $-1 + 5(x - 2) = 12x + 3 - 7x$

44. $x + 3x - 2 + 3x = -11 + 7(x + 2)$

45. $9(x + 3) - 6 = 24 - 2x - 3 + 11x$

46. $7(x + 4) - 10 = 3x + 20 + 4x - 2$

Cumulative Review Problems

47. Add. $\dfrac{3}{7} + 1\dfrac{5}{10}$

48. Subtract. $3\dfrac{1}{5} - 2\dfrac{1}{4}$

49. Multiply. $\left(2\dfrac{1}{5}\right)\left(6\dfrac{1}{8}\right)$

50. Divide. $5\dfrac{1}{2} \div 1\dfrac{1}{4}$

51. In 1975 there were 40 nesting pairs of the American peregrine falcon nesting in the United States and Canada. Between 1975 and 1985, the peregrine falcon population increased by 20 percent. Between 1985 and 2000, the population increased by 450 percent. How many pairs of peregrine falcons were thriving in 2000? (Source: U.S. Department of the Interior.)

▲ **52.** Tom Rourke needs to replace the sail on his sailboat. It is in the shape of a triangle with an altitude of 9 feet and a base of 8 feet. The material to make the sail costs $3 per square foot. How much will the material cost to make a new sail for the boat?

▲ **53.** The Newbury Elementary School needs a new air vent drilled into the wall of the maintenance room. The circular hole that is necessary will have a radius of 6 inches. The stainless steel grill that will cover the vent costs $2 per square inch. How much will the vent cost? Use $\pi \approx 3.14$.

▲ **54.** Sally constructed a new countertop for her kitchen. It is in the shape of a parallelogram. The altitude of the parallelogram is 19 inches. The base of the parallelogram is 24 inches. If the laminate used to make the countertop costs $0.60 per square inch, how much will the laminate cost?

2.5 Formulas

1 Solving a Formula for a Specified Variable

Student Learning
Objectives

After studying this section, you
will be able to:

 Solve a formula for a
specified variable.

Formulas are equations with one or more variables that are used to describe real-life situations. The formula describes the relationship that exists among the variables. For example, in the formula $d = rt$, distance (d) is equal to the rate of speed (r) multiplied by the time (t). We can use this formula to find distance if we know the rate and time. Sometimes, however, we are given the distance and the rate, and we are asked to find the time.

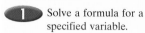

SSM
PH TUTOR   CD & VIDEO   MATH PRO   WEB
CENTER

EXAMPLE 1 Prior to a tragic accident in Paris in 2000, the Concorde was used extensively for business travel between New York and London. The British Airways Concorde had a range of approximately 3640 miles. The cruising speed for this plane was approximately 1300 miles per hour. How many hours did it take the Concorde to fly 3640 miles while traveling at cruising speed?

$$d = rt \qquad \text{Use the distance formula.}$$

$$3640 = 1300t \qquad \text{Substitute the known values for the variables.}$$

$$\frac{3640}{1300} = \frac{1300t}{1300} \qquad \text{Divide both sides of the equation by 1300 to solve for } t.$$

$$2.8 = t \qquad \text{Simplify.}$$

It took the Concorde 2.8 hours (2 hours and 48 minutes) to travel 3640 miles at cruising speed.

Practice Problem 1 On a flight in December 1999, the Concorde flew 3525 miles in 2.5 hours. Find the average rate of speed for the flight.

If we have many problems that ask us to find the time given the distance and rate, it may be worthwhile to rewrite the formula in terms of time.

EXAMPLE 2 Solve for t. $d = rt$

$$\frac{d}{r} = \frac{rt}{r} \qquad \text{We want to isolate } t. \text{ Therefore we divide both sides of the equation by the coefficient of } t, \text{ which is } r.$$

$$\frac{d}{r} = t \qquad \text{We have solved for the variable indicated.}$$

Practice Problem 2 Einstein's equation relating energy E to mass m and the speed of light c is $E = mc^2$. Solve it for m.

A straight line can be described by an equation of the form $Ax + By = C$ where A, B, and C are real numbers and A and B are not both zero. We will study this in later chapters. Often it is useful to solve such an equation for the variable y in order to make graphing the equation easier.

EXAMPLE 3 Solve for y. $3x - 2y = 6$

$$-2y = 6 - 3x \qquad \text{We want to isolate the term containing } y, \text{ so we subtract } 3x \text{ from both sides.}$$

$$\frac{-2y}{-2} = \frac{6 - 3x}{-2} \qquad \text{Divide both sides by the coefficient of } y.$$

$$y = \frac{6}{-2} + \frac{-3x}{-2} \qquad \text{Rewrite the fraction on the right side as two fractions.}$$

$$y = \frac{3}{2}x - 3 \qquad \text{Simplify and reorder the terms on the right.}$$

This is known as the slope–intercept form of the equation of a line.

Practice Problem 3 Solve for y. $8 - 2y + 3x = 0$

Our procedure for solving an equation can be rewritten to give us a procedure for solving a formula for a specified variable.

Procedure to Solve a Formula for a Specified Variable

1. Remove any parentheses.
2. If fractions exist, multiply all terms on both sides by the LCD of all the fractions.
3. Combine like terms on each side if possible.
4. Add or subtract terms on both sides of the equation to get all terms with the desired variable on one side of the equation.
5. Add or subtract the appropriate quantities to get all terms that do *not* have the desired variable on the other side of the equation.
6. Divide both sides of the equation by the coefficient of the desired variable.
7. Simplify if possible.

▲ ⬭ **EXAMPLE 4**   A trapezoid is a four-sided figure with two parallel sides. If the parallel sides are a and b and the altitude is h, the area is given by

$$A = \frac{h}{2}(a + b).$$

Solve this equation for a.

$$A = \frac{h}{2}(a + b)$$

$$A = \frac{ha}{2} + \frac{hb}{2} \qquad \text{Remove the parentheses.}$$

$$2(A) = 2\left(\frac{ha}{2}\right) + 2\left(\frac{hb}{2}\right) \qquad \text{Multiply all terms by the LCD of 2.}$$

$$2A = ha + hb \qquad \text{Simplify.}$$

$$2A - hb = ha \qquad \begin{array}{l}\text{We want to isolate the term containing } a. \\ \text{Therefore, we subtract } hb \text{ from both sides.}\end{array}$$

$$\frac{2A - hb}{h} = \frac{ha}{h} \qquad \text{Divide both sides by } h \text{ (the coefficient of } a\text{).}$$

$$\frac{2A - hb}{h} = a \qquad \text{The solution is obtained.}$$

Note: Although the solution is in simple form, it could be written in an alternative way. Since

$$\frac{2A - hb}{h} = \frac{2A}{h} - \frac{hb}{h} = \frac{2A}{h} - b,$$

we could also have written $\dfrac{2A}{h} - b = a$.

▲ **Practice Problem 4**   The relationship between the circumference C of a circle and the circle's diameter d is described by the equation $C = \pi d$. Solve it for d.   ⬭

Verbal and Writing Skills

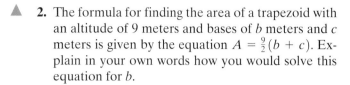

1. The formula for calculating the temperature in degrees Fahrenheit when you know the temperature in degrees Celsius is $F = \frac{9}{5}C + 32$. Explain in your own words how you would solve this equation for C.

2. The formula for finding the area of a trapezoid with an altitude of 9 meters and bases of b meters and c meters is given by the equation $A = \frac{9}{2}(b + c)$. Explain in your own words how you would solve this equation for b.

Applications

3. The formula for the area of a triangle is $A = \frac{1}{2}ab$, where b is the *base* of the triangle and a is the *altitude* of the triangle.
 (a) Use this formula to find the base of a triangle that has an area of 60 square meters and an altitude of 12 meters.

 (b) Use this formula to find the altitude of a triangle that has an area of 88 square meters and a base of 11 meters.

4. The formula for calculating simple interest is $I = Prt$, where P is the *principal* (amount of money invested), r is the *rate* at which the money is invested, and t is the *time*.
 (a) Use this formula to find how long it would take to earn $720 in interest on an investment of $3000 at the rate of 6%.

 (b) Use this formula to find the rate of interest if $5000 earns $400 interest in 2 years.

 (c) Use this formula to find the amount of money invested if the interest earned was $120 and the rate of interest was 5% over 3 years.

5. The equation $3x - 5y = 15$ describes a line and is written in standard form.
 (a) Solve for the variable y.

 (b) Use this result to find y with $x = -5$.

6. The equation $6x + 3y = -14$ describes a line and is written in standard form.
 (a) Solve for the variable x.

 (b) Use this result to find x with $y = -2$.

In each formula or equation, solve for the variable indicated.

Area of a triangle

▲ **7.** $A = \frac{1}{2}bh$ Solve for b.

▲ **8.** $A = \frac{1}{2}bh$ Solve for h.

Simple interest formula

9. $I = Prt$ Solve for P.

10. $I = Prt$ Solve for r.

Slope–intercept form of a line

11. $y = mx + b$ Solve for m.

12. $y = mx + b$ Solve for b.

Simple interest formula

13. $A = P(1 + rt)$ Solve for t.

Area of a trapezoid

▲ **14.** $A = \frac{1}{2}a(b_1 + b_2)$ Solve for b_1.

Standard form of a line

15. $5x - 6y = 6$ Solve for y.

16. $5x + 9y = -18$ Solve for y.

Slope–intercept form of a line

17. $y = -\frac{3}{4}x + 9$ Solve for x.

18. $y = \frac{6}{7}x - 12$ Solve for x.

Standard form of a line

19. $ax + by = c$ Solve for y.

20. $ax + by = c$ Solve for x.

Area of a circle

▲ **21.** $A = \pi r^2$ Solve for r^2.

Surface area of a sphere

▲ **22.** $s = 4\pi r^2$ Solve for r^2.

Distance of a falling object

23. $S = \frac{1}{2}gt^2$ Solve for g.

24. $S = \frac{1}{2}gt^2$ Solve for t^2.

Surface area of a right circular cylinder

▲ **25.** $S = 2\pi rh + 2\pi r^2$ Solve for h.

Perimeter of a square

▲ **26.** $P = 4s$ Solve for s.

Volume of a right circular cylinder

▲ **27.** $V = \pi r^2 h$ Solve for h.

▲ **28.** $V = \pi r^2 h$ Solve for r^2.

Volume of a rectangular prism

▲ **29.** $V = LWH$ Solve for L.

▲ **30.** $V = LWH$ Solve for H.

Volume of a cone

▲ **31.** $V = \frac{1}{3}\pi r^2 h$ Solve for r^2.

▲ **32.** $V = \frac{1}{3}\pi r^2 h$ Solve for h.

Perimeter of a rectangle

▲ **33.** $P = 2L + 2W$ Solve for W.

▲ **34.** $P = 2L + 2W$ Solve for L.

Pythagorean theorem

▲ **35.** $c^2 = a^2 + b^2$ Solve for a^2.

▲ **36.** $c^2 = a^2 + b^2$ Solve for b^2.

Temperature conversion formulas

37. $F = \frac{9}{5}C + 32$ Solve for C.

38. $C = \frac{5}{9}(F - 32)$ Solve for F.

Boyle's law for gases

39. $P = k\left(\dfrac{T}{V}\right)$ Solve for T.

40. $P = k\left(\dfrac{T}{V}\right)$ Solve for V.

Area of a sector of a circle

▲ **41.** $A = \dfrac{\pi r^2 S}{360}$ Solve for S.

▲ **42.** $A = \dfrac{\pi r^2 S}{360}$ Solve for r^2.

Applications

▲ **43.** Use the result you obtained in exercise 33 to solve the following problem. A farmer has a rectangular field with a perimeter of 5.8 miles and a length of 2.1 miles. Find the width of the field.

▲ **44.** Use the result you obtained in exercise 34 to solve the following problem. Smithfield High School has a rectangular athletic field with a perimeter of 1260 yards and a width of 280 yards. Find the length of the field.

▲ **45.** Use the result you obtained in exercise 29 to solve the following problem. The foundation of a house is in the shape of a rectangular solid. The volume held by the foundation is 5940 cubic feet. The height of the foundation is 9 feet and the width is 22 feet. What is the length of the foundation?

▲ **46.** Use the result you obtained in exercise 30 to solve the following problem. The fish tank at the Mandarin Danvers Restaurant is in the shape of a rectangular solid. The volume held by the tank is 3024 cubic inches. The length of the tank is 18 inches while the width of the tank is 14 inches. What is the height of the tank?

47. The number of foreign visitors measured in thousands (V) admitted to the United States for a pleasure trip for any given year can be predicted by the equation $V = 1100x + 7050$, where x is the number of years since 1985. For example, if $x = 3$ (this would be the year 1988), the predicted number of visitors in thousands would be $1100(3) + 7050 = 10,350$. Thus we would predict that in 1988, a total of 10,350,000 visitors came to the United States for a pleasure trip. (Source: U.S. Immigration and Naturalization Service.)

(a) Solve this equation for x.

(b) Use the result of your answer in (a) to find the year in which the number of visitors will be predicted to be 25,750,000. (*Hint:* Let $V = 25,750$ in your answer for (a).)

48. The number of foreign visitors from Europe measured in thousands (E) admitted to the United States from Europe for a pleasure trip for any given year can be predicted by the equation $E = 480x + 2400$, where x is the number of years since 1985. For example, if $x = 7$ (this would be the year 1992), the predicted number of visitors in thousands would be $480(7) + 2400 = 5760$. Thus we would predict that in 1992, a total of 5,760,000 visitors from Europe came to the United States for a pleasure trip. (Source: U.S. Immigration and Naturalization Service.)

(a) Solve this equation for x.

(b) Use the result of your answer in (a) to find the year in which the number of visitors from Europe will be predicted to be 11,520,000. (*Hint:* Let $E = 11,520$ in your answer for (a).)

To Think About

49. In $I = Prt$, if t doubles, what is the effect on I?

50. In $I = Prt$, if both r and t double, what is the effect on I?

▲ **51.** In $A = \pi r^2$, if r doubles, what is the effect on A?

▲ **52.** In $A = \pi r^2$, if r is halved, what is the effect on A?

Cumulative Review Problems

53. Find 12% of 260.

54. What is 0.2% of 48?

55. One day last year, 12 out of 30 people were able to climb Mt. Rainier in Washington. What percent were successful?

56. A major auto company received a shipment of car stereos. Four stereos out of 160 are defective. What percent of the shipment is defective?

57. A very popular handheld electronic game requires $3\frac{1}{4}$ square feet of a certain type of durable plastic in the manufacturing process. How many square feet of durable plastic does this company need to make 12,000 handheld games?

58. The Superstar Lighting Company rents out giant spotlights that shine up in the sky to mark the location of special events, such as the opening of a movie, a major sports play-off game, or a huge sales event at an auto dealership. A giant spotlight was used for $4\frac{1}{3}$ hours on Saturday, $2\frac{3}{4}$ hours on Tuesday, and $3\frac{1}{2}$ hours on Wednesday. What was the total number of hours that the spotlight was in use?

Putting Your Skills to Work

Controlling Emissions Levels

The U.S. Environmental Protection Agency monitors the emissions of a number of pollutants in the air in the United States. Each year, measurements are made to identify the number of tons of each type of pollutant. The following chart records in millions of tons the amount of emissions of nitrogen dioxide and volatile organic compounds.

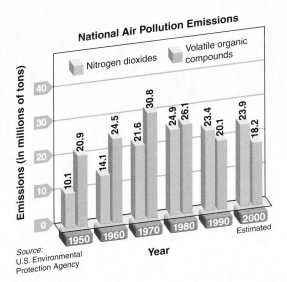

National Air Pollution Emissions

Emissions (in millions of tons)

Nitrogen dioxides / Volatile organic compounds

Source: U.S. Environmental Protection Agency

Year

Problems for Individual Study and Investigation

1. How many more tons of nitrogen dioxide were emitted in 1980 than in 1970?

2. How many fewer tons of volatile organic compounds were emitted in 2000 than in 1990?

Problems for Cooperative Study and Investigation

Scientists sometimes use the equation $n = 23.9 + 0.05x$, where x represents the number of years since the year 2000, to predict the number of millions of tons of nitrogen dioxide that will be emitted per year.

3. Use the formula to estimate how many tons of nitrogen dioxide will be emitted in 2005.

4. If the actual amount of nitrogen dioxide emitted in 2010 turns out to be 24.7 million tons, by how many tons will the formula be in error?

Scientists sometimes use the equation $v = 18.2 - 0.19x$, where x represents the number of years since the year 2000, to predict the number of millions of tons of volatile organic compounds that will be emitted per year.

5. Solve the formula for the variable x. Leave the answer in the form of a fractional expression. Do not divide out the decimals.

6. Use the result from question 5 to find the year when 16.68 million tons of volatile organic compounds will be emitted.

Internet Connections

 Netsite: http://www.prenhall.com/tobey_beginning

Site: U.S. Environmental Protection Agency

7. Examine the three separate tables for carbon monoxide, nitrogen oxides, and sulfur dioxide emissions estimates 1989–1998. For each of these, by what percent did total emissions increase or decrease in 1998 as compared with 1989?

8. In what year did the emissions of carbon monoxide from transportation sources have the greatest percent decrease from the previous year? What was that percent decrease?

2.6 Writing and Graphing Inequalities

1 Interpreting Inequality Statements

We frequently speak of one value being greater than or less than another value. We say that "5 is less than 7" or "9 is greater than 4." These relationships are called **inequalities.** We can write inequalities in mathematics by using symbols. We use the symbol **<** to represent the words "**is less than.**" We use the symbol **>** to represent the words "**is greater than.**"

Statement in Words	Statement in Algebra
5 is less than 7.	$5 < 7$
9 is greater than 4.	$9 > 4$

Note: "5 is less than 7" and "7 is greater than 5" have the same meaning. Similarly, $5 < 7$ and $7 > 5$ have the same meaning. They represent two equivalent ways of describing the same relationship between the two numbers 5 and 7.

We can better understand the concept of inequality if we examine a number line.

We say that one number is greater than another if it is to the right of the other on the number line. Thus $7 > 5$, since 7 is to the right of 5.

What about negative numbers? We can say "-1 is greater than -3" and write it in symbols as $-1 > -3$ because we know that -1 lies to the right of -3 on the number line.

EXAMPLE 1 In each statement, replace the question mark with the symbol $<$ or $>$.

(a) $3 \, ? \, -1$ **(b)** $-2 \, ? \, 1$ **(c)** $-3 \, ? \, -4$ **(d)** $0 \, ? \, 3$ **(e)** $-3 \, ? \, 0$

(a) $3 > -1$ Use $>$, since 3 is to the right of -1 on the number line.

(b) $-2 < 1$ Use $<$, since -2 is to the left of 1.
           (Or equivalently, we could say that 1 is to the right of -2.)

(c) $-3 > -4$ Note that -3 is to the right of -4.

(d) $0 < 3$

(e) $-3 < 0$

Practice Problem 1 In each statement, replace the question mark with the symbol $<$ or $>$.

(a) $7 \, ? \, 2$ **(b)** $-3 \, ? \, -4$ **(c)** $-1 \, ? \, 2$ **(d)** $-8 \, ? \, -5$ **(e)** $0 \, ? \, -2$ **(f)** $\dfrac{2}{5} \, ? \, \dfrac{3}{8}$

2 Graphing an Inequality on a Number Line

Sometimes we will use an inequality to express the relationship between a variable and a number. $x > 3$ means that x could have the value of *any number* greater than 3.

Any number that makes an inequality true is called a **solution** of the inequality. The set of all numbers that make the inequality true is called the **solution set.** A picture that represents all of the solutions of an inequality is called a **graph** of the inequality. The inequality $x > 3$ can be graphed on the number line as follows:

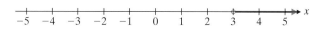

Note that all of the points to the right of 3 are shaded. The open circle at 3 indicates that we do not include the point for the number 3.

Similarly, we can graph $x < -2$ as follows:

Note that all of the points to the left of -2 are shaded.

Sometimes a variable will be either greater than or equal to a certain number. In the statement "x is greater than or equal to 3," we are implying that x could have the value of 3 or any number greater than 3. We write this as $x \geq 3$. We graph it as follows:

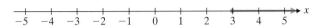

Note that the closed circle at 3 indicates that we *do* include the point for the number 3.

Similarly, we can graph $x \leq -2$ as follows:

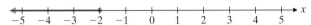

EXAMPLE 2   State each mathematical relationship in words and then graph it.

(a) $x < -2$ **(b)** $-3 < x$ **(c)** $x \geq -2$ **(d)** $x \leq -6$

(a) We state that "x is less than -2."

(b) We can state that "-3 is less than x" or, equivalently, that "x is greater than -3." Be sure you see that $-3 < x$ is equivalent to $x > -3$. Although both statements are correct, we *usually write the variable first* in a simple inequality containing a variable and a numerical value.

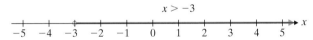

(c) We state that "x is greater than or equal to -2."

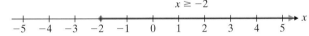

(d) We state that "x is less than or equal to -6."

Practice Problem 2   State each mathematical relationship in words and then graph it on a number line in the margin.

(a) $x > 5$

(b) $x \leq -2$

(c) $3 > x$

(d) $x \geq -\dfrac{3}{2}$

We can translate many everyday situations into algebraic statements with an unknown value and an inequality symbol. This is the first step in solving word problems using inequalities.

EXAMPLE 3 Translate each English statement into an algebraic statement.

(a) The police on the scene said that the car was traveling more than 80 miles per hour. (Use the variable s for speed.)

(b) The owner of the trucking company said that the payload of a truck must never exceed 4500 pounds. (Use the variable p for payload.)

(a) Since the speed must be greater than 80, we have $s > 80$.

(b) If the payload of the truck can never exceed 4500 pounds, then the payload must be always less than or equal to 4500 pounds. Thus we write $p \leq 4500$.

Practice Problem 3 Translate each English statement into an inequality.

(a) During the drying cycle, the temperature inside the clothes dryer must never exceed 180 degrees Fahrenheit. (Use the variable t for temperature.)

(b) The bank loan officer said that the total consumer debt incurred by Wally and Mary must be less than $15,000 if they want to qualify for a mortgage to buy their first home. (Use the variable d for debt.)

Developing Your Study Skills

Keep Trying

You may be one of those students who have had much difficulty with mathematics in the past and who are sure that you cannot do well in this course. Perhaps you are thinking, "I have never been any good at mathematics," or "I have always hated mathematics," or "Math always scares me," or "I have not had any math for so long that I have forgotten it all." You may have even picked up the label "math anxiety" and attached it to yourself. That is most unfortunate, and it is time for you to reprogram your thinking. Replace those negative thoughts with more positive ones. You need to say things like, "I will give this math class my best shot," or "I can learn mathematics if I work at it," or "I will try to do better than I have done in previous math classes." You will be pleasantly surprised at the difference a positive attitude makes!

We live in a highly technical world, and you cannot afford to give up on the study of mathematics. Dropping mathematics may prevent you from entering certain career fields that you may find interesting. You may not have to take math courses as high-level as calculus, but such courses as intermediate algebra, finite math, college algebra, and trigonometry may be necessary. Learning mathematics can open new doors for you.

Learning mathematics is a process that takes time and effort. You will find that regular study and daily practice are necessary to strengthen your skills and to help you to grow academically. This process will lead you toward success in mathematics. Then, as you become more successful, your confidence in your ability to do mathematics will grow.

Verbal and Writing Skills

1. Is the statement $5 > -6$ equivalent to the statement $-6 < 5$? Why?

2. Is the statement $-8 < -3$ equivalent to the statement $-3 > -8$? Why?

Replace the ? by < or >.

3. $9 \; ? \; -3$

4. $-2 \; ? \; 5$

5. $-4 \; ? \; -2$

6. $-3 \; ? \; -6$

7. $\dfrac{3}{5} \; ? \; \dfrac{4}{7}$

8. $\dfrac{4}{6} \; ? \; \dfrac{7}{9}$

9. $-1.2 \; ? \; 2.1$

10. $-3.6 \; ? \; 2.4$

11. $-\dfrac{13}{3} \; ? \; -4$

12. $-3 \; ? \; -\dfrac{15}{4}$

13. $-\dfrac{5}{8} \; ? \; -\dfrac{3}{5}$

14. $-\dfrac{2}{3} \; ? \; -\dfrac{3}{4}$

Which is greater?

 15. $\dfrac{123}{4986}$ or 0.0247?

16. $\dfrac{997}{6384}$ or 0.15613?

Graph each inequality on the number line.

17. $x \geq -6$

18. $x \leq -2$

19. $x > 7$

20. $x < 1$

21. $x > \dfrac{3}{4}$

22. $x \geq -\dfrac{5}{2}$

23. $x \leq -3.6$

24. $x < -2.2$

25. $25 < x$

26. $35 \geq x$

Translate each graph to an inequality using the variable x.

27.

28.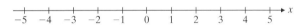

29.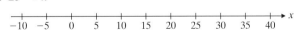

30.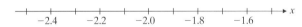

31.

32.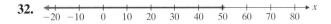

Translate each English statement into an inequality.

33. The speed of the rocket was greater than 580 kilometers per hour. (Use the variable V for speed.)

34. The cost of the hiking boots must be less than $56. (Use the variable c for cost.)

35. The number of hours for a full-time position at this company cannot be less than 37 in order to receive full-time benefits. (Use the variable h for hours.)

36. The number of nurses on duty on the floor can never exceed 6. (Use the variable n for the number of nurses.)

37. In order for you to be allowed to ride the roller coaster at the theme park, your height must be at least 48″. (Use h for height.)

38. In order for you to avoid paying extra tuition for a semester, the number of credits you are taking must not exceed 18. (Use C for credits.)

To Think About

39. Suppose that the variable x must satisfy *all* of these conditions.

$$x \le 2, \quad x > -3, \quad x < \frac{5}{2}, \quad x \ge -\frac{5}{2}$$

Graph on a number line the region that satisfies all of the conditions.

40. Suppose that the variable x must satisfy *all* of these conditions.

$$x < 4, \quad x > -4, \quad x \le \frac{7}{2}, \quad x \ge -\frac{9}{2}$$

Graph on a number line the region that satisfies all of the conditions.

Cumulative Review Problems

41. Find 16% of 38.

42. Find 1.3% of 1250.

43. For the most coveted graduate study positions, only 16 out of 800 students are accepted. What percent are accepted?

44. Write the fraction $\frac{3}{8}$ as a percent.

45. An Ecuadorian field scientist is attempting to breed hybrids of extremely rare wild orchid plants. The scientist estimates that 0.07% of all hybrid breeding attempts are successful. Last year, if 14,210,000 attempts were made worldwide by orchid lovers, how many attempts were successful and how many were unsuccessful?

46. A radar picture is taken of a portion of the Amazon rain forest. The radar detects that a cliff is 414 feet above sea level and that a giant tree trunk is directly below it in a body of water. The tree trunk is 81 feet below sea level. How far is the top of the cliff from the tree trunk?

Developing Your Study Skills

Exam Time: Getting Organized

Studying adequately for an exam requires careful preparation. Begin early so that you will be able to spread your review over several days. Even though you may still be learning new material at this time, you can be reviewing concepts previously learned in the chapter. Giving yourself plenty of time for review will take the pressure off. You need this time to process what you have learned and to tie concepts together.

Adequate preparation enables you to feel confident and to think clearly with less tension and anxiety.

2.7 Solving Inequalities

1 Solving an Inequality

As stated in Section 2.6, the possible values that make an inequality true are called its **solutions.** Thus, when we **solve an inequality,** we are finding *all* the values that make it true. To solve an inequality, we simplify it to the point where we can clearly see all possible values for the variable. We've solved equations by adding, subtracting, multiplying by, and dividing by a particular value on both sides of the equation. Here we perform similar operations with inequalities with one important exception. We'll show some examples so that you can see how these operations can be used with inequalities just as with equations.

We will first examine the pattern that occurs when we perform these operations *with a positive value* on both sides of an inequality.

● EXAMPLE 1

Original Inequality	Operations with a Positive Number	New Inequality
	Add 2 to both sides. ⟶	6 < 8
4 < 6	Subtract 2 from both sides. ⟶	2 < 4
	Multiply both sides by 2. ⟶	8 < 12
	Divide both sides by 2. ⟶	2 < 3

Notice that the inequality symbol remains the same when these operations are performed.

Practice Problem 1   Perform the given operation and write a new inequality.

(a) $9 > 6$ Add 4 to each side.

(b) $-2 < 5$ Subtract 3 from both sides.

(c) $1 > -3$ Multiply both sides by 2.

(d) $10 < 15$ Divide both sides by 5.

Now let us examine what happens when we perform these operations *with a negative value.*

Original Inequality	Operations with a Negative Number	New Inequality
	Add −2 to both sides. ⟶	2 < 4
4 < 6	Subtract −2 from both sides. ⟶	6 < 8
	Multiply both sides by −2. ⟶	−8 ? −12
	Divide both sides by −2. ⟶	−2 ? −3

What happens to the inequality sign when we multiply both sides by a negative number? Since -8 is to the right of -12 on the number line, we know that the new inequality should be $-8 > -12$ if we want the statement to remain true. Notice how we reverse the direction of the inequality from < (less than) to > (greater than). Thus we have the following.

$$4 < 6 \longrightarrow \text{Multiply both sides by } -2. \longrightarrow -8 > -12$$

The same thing happens when we divide by a negative number. The inequality is reversed from < to >. We know this since -2 is to the right of -3 on the number line.

$$4 < 6 \longrightarrow \text{Divide both sides by } -2. \longrightarrow -2 > -3$$

Similar reversals take place in the next example.

EXAMPLE 2

Original Inequality		*New Inequality*
(a) $-2 < -1$	$\longrightarrow$ Multiply both sides by -3. $\longrightarrow$	$6 > 3$
(b) $0 > -4$	$\longrightarrow$ Divide both sides by -2. $\longrightarrow$	$0 < 2$
(c) $8 \geq 4$	$\longrightarrow$ Divide both sides by -4. $\longrightarrow$	$-2 \leq -1$

Notice that we perform the arithmetic with signed numbers just as we always do. But the new inequality signs reversed (from those of the original inequalities). *Whenever both sides of an inequality are multiplied or divided by a negative quantity, the direction of the inequality is reversed.*

Practice Problem 2

(a) $7 > 2$ Multiply each side by -2.

(b) $-3 < -1$ Multiply each side by -1.

(c) $-10 \geq -20$ Divide each side by -10.

(d) $-15 \leq -5$ Divide each side by -5.

Procedure for Solving Inequalities

You may use the same procedures to solve inequalities that you did to solve equations *except* that the direction of an inequality is *reversed* if you *multiply* or *divide* both sides *by a negative number.*

It may be helpful to think over quickly what we have discussed here. The inequalities remain the same when we add a number to both sides or subtract a number from both sides of the equation. The inequalities remain the same when we multiply both sides by a positive number or divide both sides by a positive number.

Rules When Inequalities Remain the Same

For all real numbers a, b, and c:

1. If $a > b$, then $a + c > b + c$.

2. If $a > b$, then $a - c > b - c$.

3. If $a > b$ and c is a **positive number** $(c > 0)$, then $ac > bc$.

4. If $a > b$ and c is a **positive number** $(c > 0)$, then $a/c > b/c$.

However, if we multiply both sides of an inequality by a negative number or if we divide both sides of an inequality by a negative number, then the inequality is reversed.

Rules When Inequalities Are Reversed

For all real numbers a, b, and c:

1. If $a > b$, and c is a **negative number** $(c < 0)$, then $ac < bc$.

2. If $a > b$, and c is a **negative number** $(c < 0)$, then $a/c < b/c$.

The pattern is fairly simple. We could also make similar rule boxes for the cases where $a \geq b$, $a < b$, and $a \leq b$, but they are really not necessary. We simply remember that the inequality is reversed when we multiply or divide by a negative number. Otherwise the inequality remains unchanged.

EXAMPLE 3 Solve and graph $3x + 7 \geq 13$.

$3x + 7 - 7 \geq 13 - 7$ Subtract 7 from both sides.

$3x \geq 6$ Simplify.

$\dfrac{3x}{3} \geq \dfrac{6}{3}$ Divide both sides by 3.

$x \geq 2$ Simplify. Note that the direction of the inequality is not changed, since we have divided by a positive number.

The graph is as follows:

Practice Problem 3 Solve and graph $8x - 2 < 3$.

EXAMPLE 4 Solve and graph $5 - 3x > 7$.

$5 - 5 - 3x > 7 - 5$ Subtract 5 from both sides.

$-3x > 2$ Simplify.

$\dfrac{-3x}{-3} < \dfrac{2}{-3}$ Divide by -3 and **reverse the inequality** since you are dividing by a negative number.

$x < -\dfrac{2}{3}$ Note the direction of the inequality.

The graph is as follows:

Practice Problem 4 Solve and graph $4 - 5x > 7$.

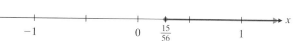

   Just like equations, some inequalities contain parentheses and fractions. The initial steps to solve these inequalities will be the same as those used to solve equations with parentheses and fractions. When the variable appears on both sides of the inequality, it is advisable to collect the x-terms on the left side of the inequality symbol.

EXAMPLE 5 Solve and graph $-\dfrac{13x}{2} \leq \dfrac{x}{2} - \dfrac{15}{8}$.

$8\left(\dfrac{-13x}{2}\right) \leq 8\left(\dfrac{x}{2}\right) - 8\left(\dfrac{15}{8}\right)$ Multiply all terms by LCD = 8. We do **not** reverse the direction of the inequality symbol since we are multiplying by a positive number.

$-52x \leq 4x - 15$ Simplify.

$-52x - 4x \leq 4x - 15 - 4x$ Subtract $4x$ from both sides.

$-56x \leq -15$ Combine like terms.

$\dfrac{-56x}{-56} \geq \dfrac{-15}{-56}$ Divide both sides by -56. We **reverse** the direction of the inequality when we divide both sides by a negative number.

$x \geq \dfrac{15}{56}$

The graph is as follows:

Practice Problem 5 Solve and graph $\dfrac{1}{2}x + 3 < \dfrac{2}{3}x$.

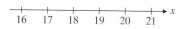

EXAMPLE 6   Solve and graph $\frac{1}{3}(3 - 2x) \le -4(x + 1)$.

$$1 - \frac{2x}{3} \le -4x - 4$$ Remove parentheses.

$$3(1) - 3\left(\frac{2x}{3}\right) \le 3(-4x) - 3(4)$$ Multiply all terms by LCD = 3.

$$3 - 2x \le -12x - 12$$ Simplify.

$$3 - 2x + 12x \le -12x + 12x - 12$$ Add $12x$ to both sides.

$$3 + 10x \le -12$$ Combine like terms.

$$3 - 3 + 10x \le -12 - 3$$ Subtract 3 from both sides.

$$10x \le -15$$ Simplify.

$$\frac{10x}{10} \le \frac{-15}{10}$$ Divide both sides by 10. Since we are dividing by a **positive** number, the inequality is **not** reversed.

$$x \le -\frac{3}{2}$$

The graph is as follows:

Practice Problem 6   Solve and graph $\frac{1}{2}(3 - x) \le 2x + 5$.

⊘ **CAUTION** The most common error students make in solving inequalities is forgetting to reverse the direction of the inequality symbol when multiplying or dividing both sides of the inequality by a negative number.

EXAMPLE 7   A hospital director has determined that the costs of operating one floor of the hospital for an eight-hour shift must never exceed $2370. An expression for the cost of operating one floor of the hospital is $130n + 1200$, where n is the number of nurses. This expression is based on an estimate of $1200 in fixed costs and a cost of $130 per nurse for an eight-hour shift. Solve the inequality $130n + 1200 \le 2370$ to determine the number of nurses that may be on duty on this floor during an eight-hour shift if the director's cost control measure is to be followed.

$$130n + 1200 \le 2370 \qquad \text{The inequality we must solve.}$$

$$130n + 1200 - 1200 \le 2370 - 1200 \qquad \text{Subtract 1200 from each side.}$$

$$130n \le 1170 \qquad \text{Simplify.}$$

$$\frac{130n}{130} \le \frac{1170}{130} \qquad \text{Divide each side by 130.}$$

$$n \le 9$$

The number of nurses on duty on this floor during an eight-hour shift must always be less than or equal to nine.

Practice Problem 7   The company president of Staywell, Inc., wants the monthly profits never to be less than $2,500,000. He has determined that an expression for monthly profit for the company is $2000n - 700,000$. In the expression, n is the number of exercise machines manufactured each month. The profit on each machine is $2000, and the −$700,000 in the expression represents the fixed costs of running the manufacturing division. Solve the inequality $2000n - 700,000 \ge 2,500,000$ to find how many machines must be made and sold each month to satisfy these financial goals.

Solve and graph the result.

1. $x + 7 \leq 4$

2. $x - 5 < -3$

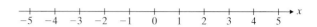

3. $-2x < 18$

4. $5x \leq 25$

5. $\dfrac{1}{2}x \geq 4$

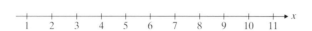

6. $-\dfrac{1}{5}x < 10$

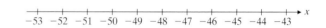

7. $2x - 3 < 4$

8. $3 - 3x > 12$

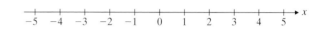

9. $-4 - 14x < 6 - 6x$

10. $7 - 8x \leq -6x - 5$

11. $\dfrac{5x}{6} - 5 > \dfrac{x}{6} - 9$

12. $\dfrac{x}{4} - 2 < \dfrac{3x}{4} + 5$

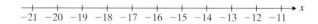

13. $2(3x + 4) > 3(x + 3)$

14. $5(x - 3) \leq 2(x - 3)$

Verbal and Writing Skills

15. Add -2 to both sides of the inequality $5 > 3$. What is the result? Why is the direction of the inequality not reversed?

16. Divide -3 into both sides of the inequality $-21 > -29$. What is the result? Why is the direction of the inequality reversed?

Mixed Practice

Solve. Collect the variable terms on the left side of the inequality.

17. $2x - 5 < 5x - 11$

18. $4x - 7 > 9x - 2$

19. $6x - 2 \geq 4x + 6$

20. $5x - 5 \leq 2x + 10$

21. $0.3(x - 1) < 0.1x - 0.5$

22. $0.2(3 - x) + 0.1 > 0.1(x - 2)$

23. $3 + 5(2 - x) \geq -3(x + 5)$

24. $7 - 2(x - 4) \leq 7(x - 3)$

25. $\dfrac{x + 6}{7} - \dfrac{6}{14} > \dfrac{x + 3}{2}$

26. $\dfrac{3x + 5}{4} + \dfrac{7}{12} > -\dfrac{x}{6}$

27. $\dfrac{1}{6} - \dfrac{1}{2}(3x + 2) < \dfrac{1}{3}\left(x - \dfrac{1}{2}\right)$

28. $\dfrac{2}{3}(2x - 5) + 3 \geq \dfrac{1}{4}(3x + 1) - 5$

Solve for x. Round your answer to the nearest hundredth.

 29. $1.96x - 2.58 < 9.36x + 8.21$

 30. $3.5(1.7x - 2.8) \leq 7.96x - 5.38$

Applications

31. To pass a course with a B grade, a student must have an average of 80 or greater. A student's grades on three tests are 75, 83, and 86. Solve the inequality $\dfrac{75 + 83 + 86 + x}{4} \geq 80$ to find what score the student must get on the next test to get a B average or better.

32. Sharon sells very expensive European sports cars. She may choose to receive $10,000.00 or 8% of her sales as payment for her work. Solve the inequality $0.08x > 10,000$ to find how much she needs to sell to make the 8% offer a better deal.

33. The average African elephant weighs 268 pounds at birth. During the first three weeks of life the baby elephant will usually gain about 4 pounds per day. Assuming that growth rate, solve the inequality $268 + 4x \geq 300$ to find how many days it will be until a baby elephant weighs at least 300 pounds.

34. Tess supervises a computer chip manufacturing facility. She has determined that her monthly profit factor is given by the expression $12.5n - 300,000$. Here n represents the number of chips manufactured each month. Each finished chip produces a profit of $12.50. The fixed costs (overhead) of the factory are $300,000 per month. Solve the inequality $12.5n - 300,000 \geq 650,000$ to find how many chips must be manufactured monthly to ensure a profit of $650,000.

Cumulative Review Problems

35. A rectangular tennis court measures 36 feet wide and 78 feet long. Robert is building a fence to surround the tennis court. He wants the fence to be 4 feet from each side of the court. How many feet of fence will he need?

36. Maria is looking at an enlargement of a rectangular photograph. The photograph has a perimeter of 29 inches. The length of the photograph is 10 inches. What is the width of the photograph?

37. Melinda is making a poster for her college basketball team. On the poster she is placing a life-sized picture of a basketball. When inflated properly, a basketball has a diameter of 9 inches. What is the area of the basketball on her poster? Use $\pi \approx 3.14$. Round your answer to the nearest tenth.

38. The seats of an outdoor amphitheater are arranged in the shape of a trapezoid. The altitude of the trapezoid is 120 feet. The bases of the trapezoid are 90 feet and 170 feet. What is the area of this seating area?

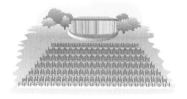

Math in the Media

Fastest Land Bird

Source: Guinness World Records.com

According to the *Guinness Book of World Records*®, the ostrich holds the world record as the fastest land bird. Its maximum running speed is 45 miles per hour. The ostrich lives in Africa. Despite its fast running speed, this bird cannot fly. Remarkably, the volume of just one ostrich egg is equal to the volume of approximately 25–40 chicken eggs.

EXERCISES

Can you use your skills from this chapter to perform some calculations in the following questions about this speedy bird?

1. An ostrich ran 20 miles per hour and traveled 10 miles. How long did it take?

2. If an ostrich ran for 45 minutes at a speed of 15 miles per hour, how far could it travel?

3. Could you arrive at a formula for calculating the maximum expected volume of one ostrich egg?

4. A man makes a huge order of scrambled eggs using 3 ostrich eggs. If he had to make this large of a quantity of scrambled eggs using chicken eggs, what is the smallest number of chicken eggs that he would need?

Chapter 2 Organizer

Topic	Procedure	Examples
Solving equations without parentheses or fractions, p. 142.	1. On each side of the equation, combine like terms if possible 2. Add or subtract terms on both sides of the equation in order to get all terms with the variable on one side of the equation. 3. Add or subtract a value on both sides of the equation to get all terms not containing the variable on the other side of the equation. 4. Divide both sides of the equation by the coefficient of the variable. 5. If possible, simplify the solution. 6. Check your solution by substituting the obtained value into the original equation.	Solve for x. $$5x + 2 + 2x = -10 + 4x + 3$$ $$7x + 2 = -7 + 4x$$ $$7x - 4x + 2 = -7 + 4x - 4x$$ $$3x + 2 = -7$$ $$3x + 2 - 2 = -7 - 2$$ $$3x = -9$$ $$\frac{3x}{3} = \frac{-9}{3}$$ $$x = -3$$ *Check:* Is -3 the solution of $$5x + 2 + 2x = -10 + 4x + 3?$$ $$5(-3) + 2 + 2(-3) \overset{?}{=} -10 + 4(-3) + 3$$ $$-15 + 2 - 6 \overset{?}{=} -10 + (-12) + 3$$ $$-13 - 6 \overset{?}{=} -22 + 3$$ $$-19 = -19 \quad \checkmark$$
Solving equations with parentheses and/or fractions, p. 143 and p. 148.	1. Remove any parentheses. 2. Simplify, if possible. 3. If fractions exist, multiply all terms on both sides by the least common denominator of all the fractions. 4. Now follow the remaining steps for solving an equation without parentheses or fractions.	Solve for y. $$5(3y - 4) = \frac{1}{4}(6y + 4) - 48$$ $$15y - 20 = \frac{3}{2}y + 1 - 48$$ $$15y - 20 = \frac{3}{2}y - 47$$ $$2(15y) - 2(20) = 2\left(\frac{3}{2}y\right) - 2(47)$$ $$30y - 40 = 3y - 94$$ $$30y - 3y - 40 = 3y - 3y - 94$$ $$27y - 40 = -94$$ $$27y - 40 + 40 = -94 + 40$$ $$27y = -54$$ $$\frac{27y}{27} = \frac{-54}{27}$$ $$y = -2$$

Topic	Procedure	Examples
Solving formulas, p. 156	1. Remove any parentheses. 2. If fractions exist, multiply all terms on both sides by the LCD, which may be a variable. 3. Add or subtract terms on both sides of the equation in order to get all terms containing the *desired variable* on one side of the equation. 4. Add or subtract terms on both sides of the equation in order to get all other terms on the opposite side of the equation. 5. Divide both side of the equation by the coefficient of the desired variable. This division may involve other variables. 6. Simplify, if possible. 7. (Optional) Check your solution by substituting the obtained expression into the original equation.	Solve for z. $B = \dfrac{1}{3}(hx + hz)$ First we remove parentheses. $$B = \dfrac{1}{3}hx + \dfrac{1}{3}hz$$ Now we multiply each term by 3. $$3(B) = 3\left(\dfrac{1}{3}hx\right) + 3\left(\dfrac{1}{3}hz\right)$$ $$3B = hx + hz$$ $$3B - hx = hx - hx + hz$$ $$3B - hx = hz$$ The coefficient of z is h, so we divide each side by h. $$\dfrac{3B - hx}{h} = z$$
Solving inequalities, p. 169	1. Follow the steps for solving an equation up until the division step. 2. If you divide both sides of the inequality by a *positive number*, the direction of the inequality is not reversed. 3. If you divide both sides of the inequality by a *negative number*, the direction of the inequality is reversed.	Solve for x and graph your solution. $$\dfrac{1}{2}(3x - 2) \le -5 + 5x - 3$$ First remove parentheses and simplify. $$\dfrac{3}{2}x - 1 \le -8 + 5x$$ Now multiply each term by 2. $$2\left(\dfrac{3}{2}x\right) - 2(1) \le 2(-8) + 2(5x)$$ $$3x - 2 \le -16 + 10x$$ $$3x - 10x - 2 \le -16 + 10x - 10x$$ $$-7x - 2 \le -16$$ $$-7x - 2 + 2 \le -16 + 2$$ $$-7x \le -14$$ When we divide both sides by a negative number, the inequality is reversed. $$\dfrac{-7x}{-7} \ge \dfrac{-14}{-7}$$ $$x \ge 2$$ Graphical solution:

Chapter 2 Review Problems

2.1–2.3 *Solve for the variable. Noninteger answers may be left in fractional form or decimal form.*

1. $5x + 20 = 3x$

2. $7x + 3 = 4x$

3. $6 - 18x = 4 - 17x$

4. $18 - 10x = 63 + 5x$

5. $6x - 2(x + 3) = 5$

6. $1 - 2(6 - x) = 3x + 2$

7. $x - (0.5x + 2.6) = 17.6$

8. $-0.2(x + 1) = 0.3(x + 11)$

9. $3(x - 2) = -4(5 + x)$

10. $\dfrac{1}{4}y = -16$

11. $y + 37 = 26$

12. $6(8x + 3) = 5(9x + 8)$

13. $3(x - 3) = 13x + 21$

14. $9x + 10 = 3x + 4$

15. $24 - 3x = 4(x - 1)$

16. $12 - x + 2 = 3x - 10 + 4x$

17. $36 = 9x - (3x - 18)$

18. $12 - 5x = -7x - 2$

19. $2(3 - x) = 1 - (x - 2)$

20. $4(x + 5) - 7 = 2(x + 3)$

21. $0.9y + 3 = 0.4y + 1.5$

22. $7y - 3.4 = 11.3$

23. $8(3x + 5) - 10 = 9(x - 2) + 13$

24. $8 - 3x + 5 = 13 + 4x + 2$

25. $3 = 2x + 5 - 3(x - 1)$

26. $-2(x - 3) = -4x + 3(3x + 2)$

27. $2(5x - 1) - 7 = 3(x - 1) + 5 - 4x$

2.4 *Solve for the variable. Noninteger answers may be left in fractional form or decimal form.*

28. $\dfrac{3}{4}x - 3 = \dfrac{1}{2}x + 2$

29. $1 = \dfrac{5x}{6} + \dfrac{2x}{3}$

30. $\dfrac{7x}{5} = 5 + \dfrac{2x}{5}$

31. $\dfrac{7x - 3}{2} - 4 = \dfrac{5x + 1}{3}$

32. $\dfrac{3x - 2}{2} + \dfrac{x}{4} = 2 + x$

33. $\dfrac{-3}{2}(x + 5) = 1 - x$

34. $\dfrac{-4}{3}(2x + 1) = -x - 2$

35. $\dfrac{1}{3}(x - 2) = \dfrac{x}{4} + 2$

36. $\dfrac{1}{5}(x - 3) = 2 - \dfrac{x}{2}$

37. $\dfrac{4}{5} + \dfrac{1}{2}x = \dfrac{1}{5}x + \dfrac{1}{2}$

38. $3x + \dfrac{6}{5} - x = \dfrac{6}{5}x - \dfrac{4}{5}$

39. $\dfrac{10}{3} - \dfrac{5}{3}x + x = \dfrac{2}{9} + \dfrac{1}{9}x$

40. $-\dfrac{8}{3}x - 8 + 2x - 5 = -\dfrac{5}{3}$

41. $\dfrac{1}{2} + \dfrac{5}{4}x = \dfrac{2}{5}x - \dfrac{1}{10} + 4$

42. $5 + 3x = \dfrac{7}{3}x + 5$

43. $\dfrac{1}{6}x - \dfrac{2}{3} = \dfrac{1}{3}(x - 4)$

44. $\dfrac{1}{2}(x - 3) = \dfrac{1}{4}(3x - 1)$

45. $\dfrac{7}{12}(x - 3) = \dfrac{1}{3}x + 4$

46. $\frac{1}{6} + \frac{1}{3}(x - 3) = \frac{1}{2}(x + 9)$ **47.** $\frac{1}{7}(x + 5) - \frac{6}{14} = \frac{1}{2}(x + 3)$ **48.** $\frac{1}{6}(8x + 3) = \frac{1}{2}(2x + 7)$

49. $-\frac{1}{3}(2x - 6) = \frac{2}{3}(3 + x)$ **50.** $\frac{7}{9}x + \frac{2}{3} = 5 + \frac{1}{3}x$

2.5 *Solve for the variable indicated.*

51. Solve for y. $3x - y = 10$ **52.** Solve for y. $5x + 2y + 7 = 0$

53. Solve for r. $A = P(1 + rt)$ **54.** Solve for h. $A = 4\pi r^2 + 2\pi rh$

55. Solve for p. $H = \frac{1}{3}(a + 2p + 3)$ **56.** Solve for y. $ax + by = c$

57. Solve for d. $H + 2d = 6c - 3d$ **58.** Solve for b. $H = \frac{3c + 2b}{4}$

59. (a) Solve for T. $C = \frac{WRT}{1000}$.
    (b) Use your result to find T if $C = 0.36$, $W = 30$, and $R = 0.002$.

60. (a) Solve for y. $5x - 3y = 12$
    (b) Use your result to find y if $x = 9$.

61. (a) Solve for R. $I = \frac{E}{R}$
    (b) Use your result to find R if $E = 100$ and $I = 20$.

2.6–2.7 *Solve each inequality and graph the result.*

62. $7 - 2x \geq 4x - 5$

63. $2 - 3x \leq -5 + 4x$

64. $2x - 3 + x > 5(x + 1)$

65. $-x + 4 < 3x + 16$

66. $4x \geq 2(12 - 2x)$

$-5 \quad -4 \quad -3 \quad -2 \quad -1 \quad 0 \quad 1 \quad 2 \quad 3 \quad 4 \quad 5$

67. $5 - \dfrac{1}{2}x > 4$

$-5 \quad -4 \quad -3 \quad -2 \quad -1 \quad 0 \quad 1 \quad 2 \quad 3 \quad 4 \quad 5$

68. $2(x - 1) \geq 3(2 + x)$

$-10 \quad -9 \quad -8 \quad -7 \quad -6 \quad -5 \quad -4 \quad -3 \quad -2 \quad -1 \quad 0$

69. $3x + 5 - 7x \leq -2x - 1$

$-5 \quad -4 \quad -3 \quad -2 \quad -1 \quad 0 \quad 1 \quad 2 \quad 3 \quad 4 \quad 5$

70. $-4x - 14 < 4 - 2(3x - 1)$

$7 \quad 8 \quad 9 \quad 10 \quad 11 \quad 12 \quad 13 \quad 14 \quad 15 \quad 16 \quad 17$

71. $3(x - 2) + 8 < 7x + 14$

$-5 \quad -4 \quad -3 \quad -2 \quad -1 \quad 0 \quad 1 \quad 2 \quad 3 \quad 4 \quad 5$

72. $\dfrac{1}{2}(2x + 3) > 10$

$5 \quad \frac{11}{2} \quad 6 \quad \frac{13}{2} \quad 7 \quad \frac{15}{2} \quad 8 \quad \frac{17}{2} \quad 9 \quad \frac{19}{2} \quad 10$

73. $\dfrac{1}{3}(x + 2) \leq \dfrac{1}{2}(3x - 5)$

$\frac{12}{7} \quad \frac{13}{7} \quad 2 \quad \frac{15}{7} \quad \frac{16}{7} \quad \frac{17}{7} \quad \frac{18}{7} \quad \frac{19}{7} \quad \frac{20}{7} \quad 3 \quad \frac{22}{7}$

74. $4(2 - x) - (-5x + 1) \geq -8$

$-18 \quad -17 \quad -16 \quad -15 \quad -14 \quad -13 \quad -12 \quad -11 \quad -10 \quad -9 \quad -8$

75. $5(1 - x) < 3(x - 1) - 2(3 - x)$

$1 \quad \frac{6}{5} \quad \frac{7}{5} \quad \frac{8}{5} \quad \frac{9}{5} \quad 2 \quad \frac{11}{5} \quad \frac{12}{5} \quad \frac{13}{5} \quad \frac{14}{5} \quad 3$

Use an inequality to solve.

76. The cost of a substitute teacher for one day at Central High School is $70. Let n = the number of substitute teachers. Set up an inequality to determine how many times a substitute teacher may be hired if the monthly budget for substitute teachers is $3220. What is the maximum number of substitute teachers that may be hired during the month? (*Hint:* Use $70n \leq 3220$.)

77. The cost of hiring a temporary secretary for a day is $85. Let n = the number of temporary secretaries. Set up an inequality to determine how many times a temporary secretary may be hired if the company budget for temporary secretaries is $1445 per month. What is the maximum number of days a temporary secretary may be hired during the month? (*Hint:* Use $85n \leq 1445$.)

Solve for the variable. Noninteger answers may be left in fractional form or decimal form.

1. $3x + 5.6 = 11.6$

2. $9x - 8 = -6x - 3$

3. $2(2y - 3) = 4(2y + 2)$

4. $\dfrac{1}{7}y + 3 = \dfrac{1}{2}y$

5. $5(8 - 2x) = -6x + 8$

6. $0.8x + 0.18 - 0.4x = 0.3(x + 0.2)$

7. $\dfrac{2y}{3} + \dfrac{1}{5} - \dfrac{3y}{5} + \dfrac{1}{3} = 1$

8. $3 - 2y = 2(3y - 2) - 5y$

9. $5(20 - x) + 10x = 165$

10. $2(x + 75) + 5x = 1025$

11. $-2(2 - 3x) = 76 - 2x$

12. $20 - (2x + 6) = 5(2 - x) + 2x$

In questions 13–17, solve for x.

13. $2x - 3 = 12 - 6x + 3(2x + 3)$

14. $\dfrac{1}{3}x - \dfrac{3}{4}x = \dfrac{1}{12}$

15. $\dfrac{3}{5}x + \dfrac{7}{10} = \dfrac{1}{3}x + \dfrac{3}{2}$

16. $\dfrac{15x - 2}{28} = \dfrac{5x - 3}{7}$

17. $\dfrac{1}{3}(7x - 1) + \dfrac{1}{4}(2 - 5x) = \dfrac{1}{3}(5 + 3x)$

18. Solve for w. $A = 3w + 2P$

19. Solve for w. $\dfrac{2w}{3} = 4 - \dfrac{1}{2}(x + 6)$

20. Solve for a. $A = \dfrac{1}{2}h(a + b)$

21. Solve for y. $5ax(2 - y) = 3axy + 5$

22. Solve for W. $P = 2L + 2W$

▲ **23.** Use your result from question 22 to find the width of a field if the perimeter is 120 feet and the length is 42 feet.

Solve and graph the inequality.

24. $4(x - 1) \geq 12x$

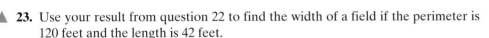

25. $2 - 7(x + 1) - 5(x + 2) < 0$

26. $5 + 8x - 4 < 2x + 13$

27. $\dfrac{1}{4}x + \dfrac{1}{16} \leq \dfrac{1}{8}(7x - 2)$

1.
2.
3.
4.
5.
6.
7.
8.
9.
10.
11.
12.
13.
14.
15.
16.
17.
18.
19.
20.
21.
22.
23.
24.
25.
26.
27.

181

1. _____

2. _____

3. _____

4. _____

5. _____

6. _____

7. _____

8. _____

9. _____

10. _____

11. _____

12. _____

13. _____

14. _____

15. _____

16. _____

17. _____

18. _____

19. _____

20. _____

21. _____

22. _____

23. _____

Approximately one-half of this test covers the content of Chapters 0 and 1. The remainder covers the content of Chapter 2. In questions 1–4, simplify.

1. $\dfrac{6}{7} - \dfrac{2}{3}$

2. $1\dfrac{3}{4} + 2\dfrac{1}{5}$

3. $3\dfrac{1}{5} \div 1\dfrac{1}{2}$

4. $(1.23)(0.56)$

5. Divide. $0.144 \div 1.2$

6. What is 26% of 450?

7. Multiply. $(-3)(-5)(-1)(2)(-1)$

8. Collect like terms. $5ab - 7ab^2 - 3ab - 12ab^2 + 10ab - 9ab^2$

9. Simplify. $(5x)^2$

10. Simplify. $2\{x + y[3 - 2x(1 - 4y)]\}$

11. Solve for x. $4(7 - 2x) = 3x - 12$

12. Solve for x. $\dfrac{1}{3}(x + 5) = 2x - 5$

13. Solve for y. $\dfrac{2y}{3} - \dfrac{1}{4} = \dfrac{1}{6} + \dfrac{y}{4}$

14. Solve for y. $3x - 7y + 2 = 0$

15. Solve for b. $H = \dfrac{2}{3}(b + 4a)$

16. Solve for t. $I = Prt$

17. Solve for a. $A = \dfrac{ha}{2} + \dfrac{hb}{2}$

In questions 18–22, solve and graph the inequality.

18. $-6x - 3 < 2x - 10x + 7$

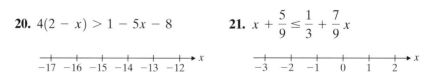

19. $\dfrac{1}{2}(x - 5) \geq x - 4$

20. $4(2 - x) > 1 - 5x - 8$

21. $x + \dfrac{5}{9} \leq \dfrac{1}{3} + \dfrac{7}{9}x$

22. $4 - 16x \leq 6 - 5(3x - 2)$

23. The football team will not let Chuck play unless he passes biology with a C (70 or better) average. There are five tests in the semester, and he has failed (0) the first one. However, he found a tutor and received an 82, an 89, and an 87 on the next three tests. Solve the inequality $\dfrac{0 + 82 + 89 + 87 + x}{5} \geq 70$ to find what his minimum score must be on the last test in order to pass the course and play football.

Solving
Applied
Problems

Has your state experienced greater precipitation than usual or are the weather conditions becoming more droughtlike? The National Oceanic and Atmospheric Administration maintains state and regional records that help analyze past weather patterns and help predict future ones. Can you use your mathematics to make some predictions based on past patterns? Turn to the Putting Your Skills to Work problems on page 228 to find out.

Pretest Chapter 3

If you are familiar with the topics in this chapter, take this test now. Check your answers with those in the back of the book. If an answer is wrong or you can't answer a question, study the appropriate section of the chapter.

If you are not familiar with the topics in this chapter, don't take this test now. Instead, study the examples, work the practice problems, and then take the test. This test will help you identify which concepts you have mastered and which you need to study further.

Section 3.1

Write an algebraic expression. Use the variable x to represent the unknown number.

1. Twice a number is then decreased by thirty.

2. Forty percent of a number is increased by 150.

Section 3.2

Translate the information into an equation, and solve.

3. A number is multiplied by seven and then is increased by nine. The result is forty-four. Find the original number.

4. Three numbers the following properties: The first number is double the second number. The third number is twenty more than the second number. The sum of the three numbers is fifty-six. Find the three numbers.

Section 3.3

▲ **5.** A triangular piece of insulation is placed in the peak of a roof of a house. One side of the triangle is 3 feet shorter than triple the second side. The third side is five feet shorter than the length of the first side. The perimeter of this piece of insulation is 38 feet. Find the length of each side of the piece of insulation.

6. Charlene took three packages to the post office to mail. The second package was $3\frac{1}{2}$ pounds less than the first package. The third package was 2 pounds less than the first package. The total weight of the three packages was 17 pounds. How much did each package weigh?

▲ **7.** A rectangular storage room in the school has a perimeter of 88 feet. The length of the room is 11 feet longer than twice the width. Find the dimensions of the room.

Section 3.4

8. Two investments totaling $1000 were made. The investments gained interest once each year. The first investment yielded 12% interest after one year. The second investment yielded 9% interest. The total interest for both investments was $102. How much was invested at each interest rate?

9. The population of Springville has grown 11% in the last five years. The town now has a population of 24,420 people. What was the population five years ago?

10. Enrique has $3.35 in change in his pocket. He has only nickels, dimes, and quarters. He has four more dimes than nickels. He has three more quarters than dimes. How many coins of each type does he have?

Section 3.5

▲ **11.** A ball is inflated so that it is a sphere. When fully inflated it has a radius of 6 inches. What is the volume of the ball? Use 3.14 for π.

▲ **12.** Find the volume of a cylinder whose height is 12 meters and whose radius is 3 meters. Use 3.14 for π.

▲ **13.** A square nut has a hole drilled in the center. Viewed from the top, the square measures 4 inches on a side. The hole has a radius of 1 inch. After the hole is drilled out, what is the area of the top of the nut? Use 3.14 for π.

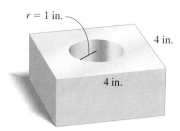

▲ **14.** A stainless steel plate in the shape of a trapezoid is to be manufactured according to the following diagram. The cost to manufacture the plate is $1.85 per square centimeter. What is the cost to manufacture this steel plate?

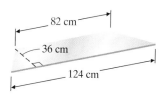

Section 3.6

Translate each statement into an inequality.

15. The width of the chair, w, must be no more than 28 inches.

16. The monthly parking fee, f, will be at least $250 a month.

17. The average class size, s, must be less than 23 students.

18. The cost of the new pickup truck, t, is more than $18,000.

Solve.

19. Alicia is making a commission of 4% of the amount of her sales. If she wants to earn more than $600 per week, what is the minimum amount of sales she should make during the week?

▲ **20.** Lauren is making a rectangular pen. The width is 11 feet. The perimeter cannot be more than 54 feet. What is the length?

11. _____

12. _____

13. _____

14. _____

15. _____

16. _____

17. _____

18. _____

19. _____

20. _____

Translating English Phrases into Algebra

One of the most useful applications of algebra is solving word problems. One of the first steps in solving word problems is translating the conditions of the problem into algebra. In this section we show you how to translate common English phrases into algebraic symbols. This process is similar to translating between languages like Spanish and French.

Several English phrases describe the operation of addition. If we represent an unknown number by the variable x, all of the following phrases can be translated into algebra as $x + 3$.

English Phrases Describing Addition	Algebraic Expression	Diagram
Three *more than* a number The *sum of* a number and three A number *increased by* three Three is *added to* a number. Three *greater than* a number A number *plus* three	$x + 3$	

In a similar way we can use algebra to express English phrases that describe the operations of subtraction, multiplication, and division.

English Phrases Describing Subtraction	Algebraic Expression	Diagram
A number *decreased by* four Four *less than* a number Four is *subtracted from* a number. Four *smaller than* a number Four *fewer than* a number A number *diminished by* four A number *minus* 4 The *difference between* a number and four	$x - 4$	

English Phrases Describing Multiplication	Algebraic Expression	Diagram
Double a number *Twice* a number The *product* of two and a number Two *of* a number Two *times* a number	$2x$	

English Phrases Describing Division	Algebraic Expression	Diagram
A number *divided by* five One-*fifth* of a number	$\dfrac{x}{5}$	

Often other words are used in English instead of the word *number*. We can use a variable, such as x, here also.

EXAMPLE 1

English Phrase	***Algebraic Expression***
(a) A *quantity* is increased by five.	$x + 5$
(b) Double the *value*	$2x$
(c) One-third of the *weight*	$\dfrac{x}{3}$ or $\dfrac{1}{3}x$
(d) Twelve *more than* a number	$x + 12$
(e) Seven *less than* a number	$x - 7$

Note that the algebraic expression for "seven less than a number" does not follow the order of the words in the English phrase. The variable or expression that follows the words *less than* always comes first.

$$\text{seven less than } x$$
$$x \; - \; 7$$

The variable or expression that follows the words *more than* technically comes before the plus sign. However, since addition is commutative, it also can be written after the plus sign.

Practice Problem 1 Write each English phrase as an algebraic expression.

(a) Four more than a number   **(b)** Triple a value

(c) Eight less than a number   **(d)** One-fourth of a height

More than one operation can be described in an English phrase. Sometimes parentheses must be used to make clear which operation is done first.

EXAMPLE 2

English Phrase	***Algebraic Expression***
(a) Seven more than double a number	$2x + 7$
	*Note that these are **not** the same.*
(b) The value of the number is increased by seven and then doubled.	$2(x + 7)$

Practice Problem 2 Write each English phrase as an algebraic expression.

(a) Eight more than triple a number

(b) A number is increased by eight and then it is tripled.

EXAMPLE 3 Write the following English sentence as an algebraic expression.
A number increased by five is subtracted from three times the same number.

We let x represent the number. We can visualize it this way:

$$\underbrace{\text{A number}}_{(x} \quad \underbrace{\text{increased by}}_{+} \quad \underbrace{5}_{5)}$$

Therefore, "a number increased by 5" is $(x + 5)$.
Now we must subtract the entire quantity $(x + 5)$ from three times the number.

A number increased by 5 ⎯⎯⎯⎯⎯⎯⎯⎯⎯⎯⎯⎯⎯⎯⎯

$3x \qquad - \qquad (x + 5)$

is subtracted from

⎯⎯⎯⎯⎯ three times the same number.

Compare the English sentence with the algebraic expression. Notice that the part that follows the word *from* in the English sentence comes first in the algebraic expression. It is also important to note the use of parentheses here. $3x - x + 5$ would be wrong.

Practice Problem 3   Write each English phrase as an algebraic expression.

(a) Four less than three times a number

(b) Two-thirds of the sum of a number and five (*Hint:* You will need parentheses.)

② Writing an Algebraic Expression to Compare Two or More Quantities

Often in a word problem two or more quantities are described in terms of another. We will want to use a variable to represent one quantity and then write an algebraic expression using *the same variable* to represent the other quantity. Which quantity should we let the variable represent? We usually let the variable represent the quantity that is the basis of comparison: the quantity that the others are being *compared to*.

EXAMPLE 4   Use a variable and an algebraic expression to describe the two quantities in the English sentence "Mike's salary is $2000 more than Fred's salary."

The two quantities that are being compared are Mike's salary and Fred's salary. Since Mike's salary is being *compared to* Fred's salary, we let the variable represent Fred's salary. The choice of the letter f helps us to remember that the variable represents Fred's salary.

Let f = Fred's salary.

Then $f + \$2000$ = Mike's salary.   *Since Mike's salary is $2000 more than Fred's.*

Practice Problem 4   Use a variable and an algebraic expression to describe the two quantities in the English sentence "Marie works 17 hours per week less than Ann."

EXAMPLE 5   The length of a rectangle is 3 meters shorter than twice the width. Use a variable and an algebraic expression to describe the length and the width. Draw a picture of the rectangle and label the length and width.

The length of the rectangle is being *compared to* the width. Use the letter w for width.

Let w = the width.

$$\underbrace{\text{Then } 2w - 3}_{\text{3 meters shorter than twice the width}} = \text{the length.}$$

A picture of the rectangle is shown.

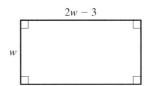

Practice Problem 5   The length of a rectangle is 5 meters longer than double the width. Use a variable and an algebraic expression to describe the length and the width. Draw a picture of the rectangle and label the length and width.

EXAMPLE 6   The first angle of a triangle is triple the second angle. The third angle of a triangle is 12° more than the second angle. Describe each angle algebraically. Draw a diagram of the triangle and label its parts.

Since the first and third angles are described in terms of the second angle, we let the variable represent the number of degrees in the second angle.

Let s = the number of degrees in the second angle.

Then $3s$ = the number of degrees in the first angle.

And $s + 12$ = the number of degrees in the third angle.

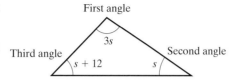

Practice Problem 6   The first angle of a triangle is 16° less than the second angle. The third angle is double the second angle. Describe each angle algebraically. Draw a diagram of the triangle and label its parts.

Some comparisons will involve fractions.

EXAMPLE 7   A theater manager was examining the records of attendance for last year. The number of people attending the theater in January was one-half of the number of people attending the theater in February. The number of people attending the theater in March was three-fifths of the number of people attending the theater in February. Use algebra to describe the attendance each month.

What are we looking for? The *number of people* who attended the theater *each month*. The basis of comparison is February. That is where we begin.

Let f = the number of people who attended in February.

Then $\frac{1}{2}f$ = the number of people who attended in January.

And $\frac{3}{5}f$ = the number of people who attended in March.

Practice Problem 7   The college dean noticed that in the spring the number of students on campus was two-thirds of the number of students on campus in the fall. She also noticed that in the summer the number of students on campus was one-fifth the number of students on campus in the fall. Use algebra to describe the number of students on campus in each of these three time periods.

Verbal and Writing skills

Write an algebraic expression for each quantity. Let x represent the unknown value.

1. a quantity increased by 5

2. nine greater than a number

3. six fewer than a quantity

4. a value decreased by seven

5. one-eighth of a quantity

6. twelve less than a number

7. seven less than a number

8. three more than half of a number

9. two divided by the sum of a number and eight

10. double a quantity increased by nine

11. six times the sum of a number and eight

12. one-third of the sum of a number and seven

13. four less than half a number

14. the sum of three and one-fifth of a number

15. four times a number added to one-half of the same number

16. triple a number added to one-half of the same number

17. seven times a number decreased by one-third of the same number

Write an algebraic expression for each of the quantities being compared.

18. The value of a share of IBM stock on that day was $74.50 more than the value of a share of AT&T stock.

19. The annual income from Dr. Smith's mutual fund was $833 less than the annual income from her retirement fund.

20. The length of the rectangle is 7 inches more than double the width.

21. The length of the rectangle is 3 meters more than triple the width.

22. The attendance on Monday was 1185 people more than on Tuesday. The attendance on Wednesday was 365 people fewer than on Tuesday.

23. The attendance on Friday was 1600 people more than on Saturday. The attendance on Thursday was 783 people fewer than on Saturday.

24. The first angle of a triangle is 16 degrees less than the second angle. The third angle of a triangle is double the second angle.

25. The first angle of a triangle is 19 degrees more than the third angle. The second angle is triple the third angle.

26. The value of the exports of Japan was twice the value of the exports of Canada.

27. Mark Spitz won two more Olympic medals than Carl Lewis did.

28. The first angle of a triangle is triple the second angle. The third angle of a triangle is 14 degrees less than the second angle.

29. The cost of Hiro's biology book was $13 more than the cost of his history book. The cost of his English book was $27 less than the cost of his history book.

Applications

Use algebra to describe the situation.

30. Kentucky has about half the land area of Minnesota. The land area of Maine is approximately two-fifths the land area of Minnesota. Describe the land area of each of these three states.

31. A census of El Cerrito Middle School found that the number of seventh graders was fifty more than the number of eighth graders. The number of sixth graders was three-fourths the number of eighth graders. Describe the population of each grade.

32. The mass of Jupiter is 32.5 more than triple the mass of Saturn. Write an expression for the mass of each planet.

33. The atomic weight of lead is 51.2 greater than triple the atomic weight of chromium. Write an expression for the atomic weight of each element.

To Think About

The following bar graph depicts income by age group in the United States in 1999. Use the graph to answer exercises 34 and 35.

34. Write an expression for the median weekly income for men in each age category. Start by using x for the median weekly income for men age 25–34.

35. Write an expression for the median weekly income for women in each age category. Start by using x for the median weekly income for women age 55–64.

Income by Age Group in 1999

1999 Median Weekly Income (in dollars) vs. Age of worker

Women / Men

16–24: 331, 377
25–34: 478, 593
35–44: 514, 713
45–54: 558, 780
55–64: 490, 728
65+: 391, 472

Source: U.S. Bureau of Labor Statistics

Cumulative Review Problems

Solve for the variable.

36. $x + 2(x + 2) = 55$

37. $2x + 2(3x - 4) = 72$

38. $5(x - 8) = 13 + x - 5$

39. $6(w - 1) - 3(2 + w) = 9$

3.2 Using Equations to Solve Word Problems

Student Learning Objectives

After studying this section, you will be able to:

1 Solve number problems.

2 Solve applied word problems.

3 Use formulas to solve word problems.

SSM
PH TUTOR CENTER   CD & VIDEO   MATH PRO   WEB

In Section 0.7 we introduced a simple three-step procedure to solve applied problems. You have had an opportunity to use that approach to solve word problems in Exercises 0.7 and in the Cumulative Review sections in Chapters 1 and 2. Now we are going to focus our attention on solving applied problems that require the use of variables, translating English phrases into algebraic expressions, and setting up equations. The process is a little more involved. Some students find the following outline a helpful way to keep organized while solving such problems.

1. *Understand the problem.*
   (a) Read the word problem carefully to get an overview.
   (b) Determine what information you will need to solve the problem.
   (c) Draw a sketch. Label it with the known information. Determine what needs to be found.
   (d) Choose a variable to represent one unknown quantity.
   (e) If necessary, represent other unknown quantities in terms of that very same variable.

2. *Write an equation.*
   (a) Look for key words to help you to translate the words into algebraic symbols and expressions.
   (b) Use a given relationship in the problem or an appropriate formula to write an equation.

3. *Solve and state the answer.*

4. *Check.*
   (a) Check the solution in the original equation. Is the answer reasonable?
   (b) Be sure the solution to the equation answers the question in the word problem. You may need to do some additional calculations if it does not.

1 Solving Number Problems

EXAMPLE 1 Two-thirds of a number is eighty-four. What is the number?

1. **Understand the problem.**
   Draw a sketch.
   Let x = the unknown number.

2. **Write an equation.**

$$\underbrace{\text{Two-thirds of a number}}_{\dfrac{2}{3}x} \quad \overset{\text{is}}{\underset{=}{\downarrow}} \quad \underbrace{\text{eighty-four.}}_{84}$$

3. **Solve and state the answer.**

$$\frac{2}{3}x = 84$$

$$3\left(\frac{2}{3}x\right) = 3(84) \qquad \text{Multiply both sides of the equation by 3.}$$

$$2x = 252$$

$$\frac{2x}{2} = \frac{252}{2} \qquad \text{Divide both sides by 2.}$$

$$x = 126$$

The number is 126.

4. **Check.**
   Is two-thirds of 126 eighty-four?

   $$\frac{2}{3}(126) \stackrel{?}{=} 84$$

   $$84 = 84 \quad \checkmark$$

Practice Problem 1 Three-fourths of a number is negative eighty-one. What is the number?

Learning to solve problems like Examples 1 and 2 is a very useful skill. You will find learning the material in Chapter 3 will be much easier if you can master the procedure used in these two examples.

EXAMPLE 2 Five more than six times a number is three hundred five. Find the number.

1. **Understand the problem.**
   Read the problem carefully. You may not need to draw a sketch.
   Let x = the unknown number.

2. **Write an equation.**

Five more than	six times a number	is	three hundred five.
5 +	6x	=	305

3. **Solve and state the answer.**
   You may want to rewrite the equation to make it easier to solve.

   $$6x + 5 = 305$$

   $$6x + 5 - 5 = 305 - 5 \quad \text{Subtract 5 from both sides.}$$

   $$6x = 300$$

   $$\frac{6x}{6} = \frac{300}{6} \quad \text{Divide both sides by 6.}$$

   $$x = 50$$

   The number is 50.

4. **Check.**
   Is five more than six times 50 three hundred five?

   $$6(50) + 5 \stackrel{?}{=} 305$$

   $$300 + 5 \stackrel{?}{=} 305$$

   $$305 = 305 \quad \checkmark$$

Practice Problem 2 Two less than triple a number is forty-nine. Find the number.

EXAMPLE 3 The larger of two numbers is three more than twice the smaller. The sum of the numbers is thirty-nine. Find each number.

1. *Understand the problem.*
   Read the problem carefully. The problem refers to *two* numbers. We must write an algebraic expression for *each number* before writing the equation. The larger number is being compared to the smaller number.

$$\text{Let } s = \text{the smaller number.}$$
$$\underbrace{\text{Then } 2s + 3 = \text{the larger number.}}_{\text{three more than twice the smaller number}}$$

2. *Write an equation.*

The sum of the numbers	is	thirty-nine.
$s + (2s + 3)$	$=$	39

3. *Solve.*

$$s + (2s + 3) = 39$$
$$3s + 3 = 39 \quad \text{\small Collect like terms.}$$
$$3s = 36 \quad \text{\small Subtract 3 from each side.}$$
$$s = 12 \quad \text{\small Divide both sides by 3.}$$

4. *Check.*

$$12 + \left[2(12) + 3\right] \overset{?}{=} 39$$
$$39 = 39 \quad \checkmark$$

The solution checks, but have we solved the word problem? We need to find *each* number. 12 is the smaller number. Substitute 12 into the expression $2s + 3$ to find the larger number.

$$2s + 3 = 2(12) + 3 = 27$$

The smaller number is 12. The larger number is 27.

Practice Problem 3 Consider two numbers. The second number is twelve less than triple the first number. The sum of the two numbers is twenty-four. Find each number.

2 Solving Applied Word Problems

To facilitate understanding more involved word problems we will use a Mathematics Blueprint similar to the one we used in Section 0.7. This format is a simple way to organize facts, determine what to set variables equal to, and select a method or approach that will assist you in finding the desired quantity. You will find using this form helpful, particularly in those cases when you read through a word problem and mentally say to yourself, "Now where do I begin?" You begin by responding to the headings of the blueprint. Soon a procedure for solving the problem will emerge.

Mathematics Blueprint For Problem Solving

Gather the Facts	Assign the Variable	Basic Formula or Equation	Key Points to Remember

EXAMPLE 4   The mean annual snowfall in Juneau, Alaska, is 105.8 inches. This is 20.2 inches less than three times the annual snowfall in Boston. What is the annual snowfall in Boston?

Understand the problem and write an equation.

Mathematics Blueprint For Problem Solving

Gather the Facts	Assign the Variable	Basic Formula or Equation	Key Points to Remember
Snowfall in Juneau is 105.8 inches. This is 20.2 inches less than three times the snowfall in Boston.	We do not know the snowfall in Boston. Let b = annual snowfall in Boston. Then $3b - 20.2$ = annual snowfall in Juneau.	Set $3b - 20.2$ equal to 105.8, which is the snowfall in Juneau.	All measurements of snowfall are recorded in inches.

Juneau's snowfall is 20.2 less than three times Boston's snowfall.

$$105.8 = 3b - 20.2$$

Solve and state the answer.
You may want to rewrite the equation to make it easier to solve.

$$3b - 20.2 = 105.8$$
$$3b = 126 \qquad \text{Add 20.2 to both sides.}$$
$$b = 42 \qquad \text{Divide both sides by 3.}$$

The annual snowfall in Boston is 42 inches.

Check.
Reread the word problem. Work backward.

Three times 42 is 126.

126 less 20.2 is 105.8.

Is this the annual snowfall in Juneau? Yes. ✓

Practice Problem 4   The maximum recorded rainfall for a 24-hour period in the United States occurred in Alvin, Texas, on July 25–26, 1979. This amount was 24 inches more than the maximum recorded rainfall for a 24-hour period in Canada, which occurred in Ucluelet Brynnor Mines, British Columbia, on October 6, 1977. The total rainfall from these two occurrences was 62 inches. How much rainfall was recorded for each location? (*Source:* National Oceanic and Atmospheric Administration)

Some word problems require a simple translation of the facts. Others require a little more detective work. You will not always need to use the Mathematics Blueprint to solve every word problem. As you gain confidence in problem solving, you will no doubt leave out some of the steps. We suggest that you use the procedure when you find yourself on unfamiliar ground. It is a powerful organizational tool.

3 *Using Formulas to Solve Word Problems*

Sometimes the relationship between two quantities is so well understood that we have developed a formula to describe that relationship. We have already done some work with formulas in Section 2.5. The following examples show how you can use a formula to solve a word problem.

EXAMPLE 5   Two people travel in separate cars. They each travel a distance of 330 miles on an interstate highway. To maximize fuel economy, Fred travels at exactly 50 mph. Sam travels at exactly 55 mph. How much time did the trip take each person?

Mathematics Blueprint For Problem Solving

Gather the Facts	Assign the Variable	Basic Formula or Equation	Key Points to Remember
Each person drives 330 miles. Fred drives at 50 mph. Sam drives at 55 mph.	Time is the unknown quantity for each driver. Use subscripts to denote different values of t. t_f = Fred's time t_s = Sam's time	distance = (rate)(time) or $d = rt$	The time is expressed in hours.

Substitute the known values into the formula and solve for t.

$$d = rt \qquad\qquad\qquad d = rt$$
$$330 = 50t_f \qquad\qquad 330 = 55t_s$$
$$6.6 = t_f \qquad\qquad\quad 6 = t_s$$

It took Fred 6.6 hours to drive 330 miles.   It took Sam 6 hours to drive 330 miles.

Check. Is this reasonable? Yes, you would expect Fred to take longer to drive the same distance because Fred is driving at a lower rate of speed.

Note: You may wish to express 6.6 hours in hours and minutes. To change 0.6 hours to minutes, proceed as follows:

$$0.6 \; \cancel{\text{hour}} \cdot \frac{60 \text{ minutes}}{1 \; \cancel{\text{hour}}} = (0.6)(60) \text{ minutes} = 36 \text{ minutes}$$

Thus, Fred drove for 6 hours and 36 minutes.

Practice Problem 5   Sarah left the city to visit her aunt and uncle, who live in a rural area north of the city. She traveled the 220-mile trip in 4 hours. On her way home she took a slightly longer route, which measured 225 miles on the car odometer. The return trip took 4.5 hours.

(a) What was her average speed on the trip leaving the city?

(b) What was her average speed on the return trip?

(c) On which trip did she travel faster and by how much?

EXAMPLE 6 A teacher told Melinda that she had a course average of 78 based on her six math tests. When she got home, Melinda found five of her tests. She had scores of 87, 63, 79, 71, and 96 on the five tests. She could not find her sixth test. What score did she obtain on that test?

Mathematics Blueprint For Problem Solving

Gather the Facts	Assign the Variable	Basic Formula or Equation	Key Points to Remember
Her five known test scores are 87, 63, 79, 71, and 96. Her course average is 78.	We do not know the score Melinda received on her sixth test. Let x = the score on the sixth test.	average = $\dfrac{\text{sum of scores}}{\text{number of scores}}$	Since there are six test scores, we will need to divide the sum by 6.

We now write the equation for the average and solve for x.

$$\frac{87 + 63 + 79 + 71 + 96 + x}{6} = 78$$

$$\frac{396 + x}{6} = 78 \qquad \text{Add the numbers in the numerator.}$$

$$6\left(\frac{396 + x}{6}\right) = 6(78) \qquad \begin{array}{l}\text{Multiply both sides of the equation by 6}\\\text{to remove the fraction.}\end{array}$$

$$396 + x = 468 \qquad \text{Simplify.}$$

$$x = 72 \qquad \text{Subtract 396 from both sides to find } x.$$

Melinda's score on the sixth test was 72.

Check. To verify that this is correct, we check that the average of the 6 tests is 78.

$$\frac{87 + 63 + 79 + 71 + 96 + 72}{6} \overset{?}{=} 78$$

$$\frac{468}{6} \overset{?}{=} 78$$

$$78 = 78 \quad \checkmark$$

The problem checks. We know that the score on the sixth test was 72.

Practice Problem 6 Barbara's math course has four tests and one final exam. The final exam counts as much as two tests. Barbara has test scores of 78, 80, 100, and 96 on her four tests. What grade does she need on the final exam if she wants to have a 90 average for the course?

Solve. Check your solution.

1. What number minus 543 gives 718?

2. What number added to 74 gives 265?

3. A number divided by eight is 296. What is the number?

4. Eighteen less than a number is 23. What is the number?

5. Seventeen greater than a number is 199. Find the number.

6. Three times a number is one. What is the number?

7. A number is doubled and then increased by seven. The result is ninety-three. What is the original number?

8. Four less than nine times a number is one hundred twenty-two. Find the original number.

9. When five is subtracted from one-third of a number, the result is 12. What is the original number?

10. A number is tripled and then increased by fourteen. The result is ninety-nine. What is the original number?

11. Five less than twice a number is the same as three times the number. Find the number.

12. Six less than five times a number is the same as seven times the number. Find the number.

13. A number, half of that number, and one-third of that number are added. The result is 22. What is the original number?

14. A number, twice that number, and one-third of that number are added. The result is 20. What is the original number?

Applications

Solve. Check to see if your answer is reasonable.

15. A Harley Davidson motorcycle shop maintains an inventory of four times as many new bikes as used bikes. If there are 60 new bikes, how many used bikes are now in stock?

16. In 1999, the population of Austria was one-third the population of Nepal. At that time the number of people living in Austria was 8,100,0000. How many people were living in Nepal at that time? (*Source: United Nations Statistics Division*)

17. At Center City Ford, the monthly rental on a Ford Focus is $310. The cost of a new car is $12,400. In how many months will the rental equal the cost of the car?

18. The cost of renting a subcompact car at Sudbay Chrysler is $20 a day plus 25 cents per mile. How far can Jeff drive in one day if he has only $75?

19. The sale price of a new Panasonic compact disc player is $218 at a local discount store. At the store where this sale is going on, each new CD is on sale for $11 each. If Kyle purchases a player and some CDs for $284, how many CDs did he purchase?

20. Suellen subscribes to an online computer service that charges $9.50 per month for 30 hours online and $1.50 for each hour online in excess of 30 hours. Last month her bill was $20. How many extra hours was she charged for?

21. Rob is on a diet and can have 500 calories for lunch. A 3-ounce hamburger on whole wheat bread has 315 calories, and 12 fluid ounces of soda has 145 calories. How many French fries can he eat if there are 10 calories in one French fry?

22. Lauren Lee makes $24 per hour for a 40-hour week and time and a half for every hour over 40 hours. If Lauren made $1140 last week, how many overtime hours did she work?

23. It has been shown that the force of gravity on a planet varies with the mass of the planet. The force of gravity on Jupiter, for example, is about two and a half times that of Earth. Using this information, approximately how much would a 220-lb astronaut weigh on Jupiter?

24. In 1958, the nuclear-powered submarine *Nautilus* took 6 days, 12 hours to travel submerged 5068 km across the Atlantic Ocean from Portsmouth, England, to New York City. What was its average speed, in kilometers per hour, for this trip? (Round to the nearest whole number.)

25. After closely following the advertisements, Dirk Pitt decided to change his phone service to the local cable company because they advertised a $14.00 monthly fee, plus $0.07 per minute for toll calls to anywhere. If Dirk's bill last month was $22.40, how many minutes of toll calls did he make?

26. Charles Schultz, the creator of *Peanuts*, once estimated that he had drawn close to 2600 Sunday comics over his career. At that time, about how many years had he been drawing *Peanuts*?

27. Two in-line skaters, Nell and Kristin, start from the same point and skate in the same direction. Nell skates at 12 miles per hour and Kristin skates at 14 miles per hour. If they can keep up that pace for 2.5 hours, how far apart will they be at the end of that time?

28. Two trains leave a train station at the same time. One train travels east at 50 miles per hour. The other train travels west at 55 miles per hour. In how many hours will the two trains be 315 miles apart?

29. Sammy drove from Albuquerque, New Mexico, to the Garden of the Gods rock formation in Colorado Springs. It took him six hours to travel 312 miles over the mountain road. He came home on the highway. On the highway he took five hours to travel 320 miles. How fast did he travel using the mountain route? How much faster (in miles per hour) did he travel using the highway route?

30. Allison drives 30 miles per hour through the city and 55 miles per hour on the New Jersey Turnpike. She drove 90 miles from Battery Park to the Jersey Shore. How much of the time was city driving if she spent 1.2 hours on the turnpike?

31. Juanita's chemistry class has taken seven of the eight quizzes planned for the term. Juanita's grades on the first seven quizzes are 75, 90, 85, 88, 92, 73, and 81. What score does she need to obtain on the last quiz to raise her quiz average to an 85?

32. By the end of Won Sun's math course, he will have taken four quizzes and two one-hour exams. Each exam will count as much as two quizzes. So far Won Sun has scored 91, 80, 86, and 83 on the four quizzes and a 93 on his first one-hour exam. What score does he need on his last one-hour exam to have a 90 average for the course?

33. The Ramirez family has five licensed drivers and three cars. The smallest family car is a subcompact and gets excellent fuel mileage of 38 miles per gallon (mpg) in city driving. The second car is a compact and gets 21 mpg in city driving. The third car is an ancient, heavy station wagon. Dad Ramirez does not like to admit the mileage rating of the "old beast." Rita calculated that if all three cars were driven the same number of miles in city driving each year, the *average* miles-per-gallon rating of the three Ramirez cars would be $22\frac{2}{3}$ mpg in city driving. What is the miles-per-gallon rating of the old station wagon?

34. A local hospital revealed that the five chief officers of the hospital had an average salary of $125,000 per year. Three of the annual salaries of these officers were revealed. They were $50,000, $60,000, and $65,000. The annual salaries for the executive vice president and the president were not revealed. It was disclosed, however, that the president makes double the salary of the executive vice president. Based on this information, find the salaries of the president and the executive vice president.

To Think About

35. In warmer climates, approximate temperature predictions can be made by counting the number of chirps of a cricket during a minute. The Fahrenheit temperature decreased by forty is equivalent to one-fourth of the number of cricket chirps.
(a) Write an equation for this relationship.
(b) Approximately how many chirps per minute should be recorded if the temperature is 90°F?
(c) If a person recorded 148 cricket chirps in a minute, what would be the Fahrenheit temperature according to this formula?

Cumulative Review Problems

Simplify.

36. $5x(2x^2 - 6x - 3)$

37. $-2a(ab - 3b + 5a)$

38. $7x - 3y - 12x - 8y + 5y$

39. $5x^2y - 7xy^2 - 8xy - 9x^2y$

40. The local Apple Factory produce market sells premium MacIntosh apples at the price of four apples for $3.60. They purchase the apples from the apple orchards for $5.40 per dozen. How many apples will they have to sell to make a profit of $1350?

41. Jennifer is hosting a party for her grandparents' 50th wedding anniversary, to be held in the church fellowship hall. She has a budget of $1425.00 to pay a caterer. If the caterer gives her a price of $23.75 per person, how many people can she invite?

3.3 Solving Word Problems: Comparisons

1 Solving Word Problems Involving Comparisons

Many real-life problems involve comparisons. We often compare quantities such as length, height, or income. Sometimes not all the information is known about the quantities that are being compared. You need to identify each quantity and write an algebraic expression that describes the situation in the word problem.

EXAMPLE 1 The Center City Animal Hospital treated a total of 18,360 dogs and cats last year. The hospital treated 1376 more dogs than cats. How many dogs were treated last year? How many cats were treated last year?

1. Understand the problem.

What information is given?   The combined number of dogs and cats is 18,360.

What is being compared?   There were 1376 more dogs than cats.

If you compare one quantity to another, usually the second quantity is represented by the variable. Since we are comparing the number of dogs to the number of cats, we start with the number of cats.

Let c = the number of cats treated at the hospital.

Then $c + 1376$ = the number of dogs treated at the hospital.

2. Write an equation.

The number of cats   plus   the number of dogs   is   18,360.

$$c \quad + \quad (c + 1376) \quad = \quad 18{,}360$$

3. Solve and state the answer.

$$c + c + 1376 = 18{,}360$$
$$2c + 1376 = 18{,}360 \qquad \text{Combine like terms.}$$
$$2c + 1376 - 1376 = 18{,}360 - 1376 \qquad \text{Subtract 1376 from both sides.}$$
$$2c = 16{,}984$$
$$\frac{2c}{2} = \frac{16{,}984}{2} \qquad \text{Divide both sides by 2.}$$
$$c = 8492 \qquad \text{The number of cats treated is 8492.}$$
$$c + 1376 = 8492 + 1376 = 9868 \qquad \text{The number of dogs treated is 9868.}$$

4. Check.

The number of dogs treated plus the number of cats treated should total 18,360.

$$8492 + 9868 \overset{?}{=} 18{,}360$$
$$18{,}360 = 18{,}360 \quad \checkmark$$

Practice Problem 1 A deck hand on a fishing boat is working with a rope that measures 89 feet. He needs to cut it into two pieces. The long piece must be 17 feet longer than the short piece. Find the length of each piece of rope.

If the word problem contains three unknown quantities, determine the basis of comparison for two of the quantities.

EXAMPLE 2 An airport filed a report showing the number of plane departures that took off from the airport during each month last year. The number of departures in March was 50 more than the number of departures in January. In July, the number of departures was 150 less than triple the number of departures in January. In those three months, the airport had 2250 departures. How many departures were recorded for each month?

1. **Understand the problem.**
   What is the basis of comparison?

   The number of departures in March is compared to the number in January.

   The number of departures in July is compared to the number in January.

   Express this algebraically. It may help to underline the key phrases.

$$\text{Let } j = \text{the departures in January.}$$

$$\underbrace{\text{March was 50 more than January}}$$

$$\text{Then } j + 50 = \text{the departures in March.}$$

$$\underbrace{\text{July was 150 less than triple January}}$$

$$\text{And } 3j - 150 = \text{the departures in July.}$$

2. **Write an equation.**

number of departures in January	+	number of departures in March	+	number of departures in July	=	three months' total departures
j	+	$(j + 50)$	+	$(3j - 150)$	=	2250

3. **Solve and state the answer.**

$$j + (j + 50) + (3j - 150) = 2250$$
$$5j - 100 = 2250 \quad \text{Collect like terms.}$$
$$5j = 2350 \quad \text{Add 100 to each side.}$$
$$j = 470 \quad \text{Divide both sides by 5.}$$

   Now, if $j = 470$, then

$$j + 50 = 470 + 50 = 520$$

   and

$$3j - 150 = 3(470) - 150 = 1410 - 150 = 1260.$$

   The number of departures in January was 470; the number of departures in March was 520; the number of departures in July was 1260.

4. **Check.**
   Do these answers seem reasonable? Yes. Do these answers agree with all the statements in the word problem?
   Is the number of departures in March 50 more than those in January?

$$520 \overset{?}{=} 50 + 470$$
$$520 = 520 \quad ✓$$

   Is the number of departures in July 150 less than triple those in January?

$$1260 \overset{?}{=} 3(470) - 150$$
$$1260 \overset{?}{=} 1410 - 150$$
$$1260 = 1260 \quad ✓$$

   Is the total number of departures in the three months equal to 2250?

$$470 + 520 + 1260 \overset{?}{=} 2250$$
$$2250 = 2250 \quad ✓$$

   Yes, all conditions are satisfied. The three answers are correct.

Practice Problem 2   A social services worker was comparing the cost incurred by three families in heating their homes for the year. The first family had an annual heating bill that was $360 more than that of the second family. The third family had a heating bill that was $200 less than double the heating bill of the second family. The total annual heating bill for the three families was $3960. What was the annual heating bill for each family?

▲ ● **EXAMPLE 3**   A small plot of land is in the shape of a rectangle. The length is 7 meters longer than the width. The perimeter of the rectangle is 86 meters. Find the dimensions of the rectangle.

1. **Understand the problem.**
   Read the problem: What information is given?

   The perimeter of a rectangle is 86 meters.

   What is being compared?

   The length is being compared to the width.

   Express this algebraically and draw a picture.

   Let w = the width.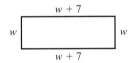
   Then $w + 7$ = the length.

   Reread the problem: What are you being asked to do?

   Find the dimensions of the rectangle. The dimensions of the rectangle are the length and the width of the rectangle.

2. **Write an equation.**
   The perimeter is the total distance around the rectangle.

   $$w + (w + 7) + w + (w + 7) = 86$$

3. **Solve and state the answer.**

   $$w + (w + 7) + w + (w + 7) = 86$$

   $$4w + 14 = 86 \quad \text{Combine like terms.}$$

   $$4w = 72 \quad \text{Subtract 14 from both sides.}$$

   $$w = 18 \quad \text{Divide both sides by 4.}$$

   The width of the rectangle is 18 meters. What is the length?

   $$w + 7 = \text{the length}$$

   $$18 + 7 = 25$$

   The length of the rectangle is 25 meters.

4. **Check.**
   Put the actual dimensions in your drawing and add the lengths of the sides. Is the sum 86 meters? ✓

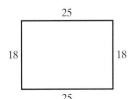

▲ **Practice Problem 3**   A farmer purchased 720 meters of wire fencing to enclose a pasture. The pasture is in the shape of a triangle. The first side of the triangle is 30 meters less than the second side. The third side is one-half as long as the second side. Find the dimensions of the triangle.

Applications

Solve. Check to see if your answer is reasonable. Have you answered the question that was asked?

1. For their homecoming parade, the students of Wheaton College have created a colorful banner, 47 meters in length, that is made of two pieces of parachute material. The short piece is 17 meters shorter than the long piece. Find the length of each piece.

2. A copper conducting wire measures 84 centimeters in length. It is cut into two pieces. The shorter piece is 5 centimeters shorter than the long piece. Find the length of each piece.

3. Three siblings—David, Kate, and Sarah—committed themselves to volunteer 100 hours for Habitat for Humanit to build new housing. Sarah worked 15 hours more than Dave. Kate worked 5 hours fewer than Dave. How many hours did each of these siblings volunteer?

4. Jerome counted 24 pieces of fruit in a fruit basket. There were twice as many apples as pears. There were four fewer peaches than pears. How many of each fruit were there?

5. Mt. McKinley in Alaska is 5826 feet taller than Mt. Whitney in California. Mt. Oxford in Colorado is 341 feet shorter than Mt. Whitney. The combined height of the three mountains is 48,967 feet. What is the height of each of the mountains? (*Source:* U.S. Department of the Interior)

6. The Sears Tower in Chicago is 200 feet taller than the Empire State Building in New York City. The Nations Bank Tower in Atlanta is 227 feet shorter than the Empire State Building. The total height of all three buildings is 3723 feet. What is the height of each of the three buildings? (*Source:* U.S. Department of the Interior)

▲ 7. The length of a polo field is 50 meters more than twice its width. If the perimeter is 700 meters, find the dimensions of the polo field.

8. The part of a diving board that is over the water is one meter longer than the part that is over the deck around the pool. The total length of the diving board is two meters less than three times the part that is over the deck. Find the length of the board.

▲ 9. A solid-gold jewelry box was found in the underwater palace of Cleopatra just off the shore of Alexandria. The length of the rectangular box is 35 centimeters less than triple the width. The perimeter of the box is 190 centimeters. Find the length and width of Cleopatra's solid-gold jewelry box.

▲ 10. A giant rectangular chocolate bar was made for a special promotion. The length was 6 meters more than half the width. The perimeter of the chocolate bar was 24 meters. Find the length and width of the giant chocolate bar.

11. The top running speed of a cheetah is double the top running speed of a jackal. The top running speed of an elk is 10 miles per hour faster than that of a jackal. If each of these three animals could run at top speed for an hour (which of course is not possible), they could run a combined distance of 150 miles. What is the top running speed of each of these three animals? (*Source:* American Museum of Natural History)

12. The Missouri River is 149 miles longer than double the length of the Snake River. The Potomac River is 796 miles shorter than the Snake River. The combined lengths of these three rivers is 3685 miles. What is the length of each of the three rivers? (*Source:* U.S. Department of the Interior)

▲ **13.** A small triangular piece of metal is welded onto the hull of a Gloucester whale watch boat. The perimeter of the metal piece is 46 inches. The shortest side of the triangle is four inches longer than one-half of the longest side. The second-longest side is 3 inches shorter than the longest side. Find the length of each side.

▲ **14.** Carmelina's uncle owns a triangular piece of land in Maryland. The perimeter fence that surrounds the land measures 378 yards. The shortest side is 30 yards longer than one-half of the longest side. The second-longest side is 2 yards shorter than the longest side. Find the length of each side.

15. A balloonist is trying to complete a nonstop trip around the world. During the course of his trip, he notes that he took 18 hours to travel 684 miles over land from Rockford, Illinois, to Washington, D.C. He took 21 hours to travel 1197 miles over water from Washington, D.C., to San Juan, Puerto Rico.
(a) How fast did he travel over land?
(b) How fast did he travel over water?
(c) How much faster did he travel over water than over land?

16. In a typical year in New York City, 600 waterline breaks are reported along the 6181 miles of water pipes that twist and turn underground. Water breaks occur five times more often during the period from October to March than they do during the period from April to September.
(a) On average, how many waterline breaks would be expected during the period from October to March?
(b) On average, how many waterline breaks would be expected during the period from April to September?
(c) A water department official once complained to city hall that he expects a break in the waterline every five miles during the course of a year. Is his comment correct? Why?

To Think About

▲ **17.** A small square is constructed. Then a new square is made by increasing each side by 2 meters. The perimeter of the new square is 3 meters shorter than five times the length of one side of the original square. Find the dimensions of the original square.

▲ **18.** A rectangle is constructed. The length of this rectangle is double the width. Then a new rectangle is made by increasing each side by 3 meters. The perimeter of the new rectangle is 2 meters greater than four times the length of the old rectangle. Find the dimensions of the original rectangle.

Cumulative Review Problems

Simplify.

19. $-4x(2x^2 - 3x + 8)$

20. $5a(ab + 6b - 2a)$

21. $-7x + 10y - 12x - 8y - 2$

22. $3x^2y - 6xy^2 + 7xy + 6x^2y$

23. Today there were 650 people shopping for hardware products in a home improvement store. Of the 650 shoppers, 117 were professional contractors and laborers. What percent of the shoppers were not professional contractors and laborers?

24. A dance instructor is doing her best to teach engaged couples how to dance well before their wedding days. After giving lessons to 60 couples, she notes that only 48 couples are able to dance well without coaching. What percent of couples are not dancing very well?

The problems we now present are frequently encountered in business. They deal with money: buying, selling, and renting items; earning and borrowing money; and the value of collections of stamps or coins. Many applications require an understanding of the use of percents and decimals. Review Sections 0.4 and 0.5 if you are weak in these skills.

Student Learning Objectives

After studying this section, you will be able to:

 Solve problems involving periodic rate charges.

Solve percent problems.

Solve investment problems involving simple interest.

Solve coin problems.

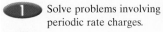

SSM PH TUTOR   CD & VIDEO   MATH PRO   WEB
       CENTER

Solving Problems Involving Periodic Rate Charges

EXAMPLE 1   A business executive rented a car. The Supreme Car Rental Agency charged $39 per day and $0.28 per mile. The executive rented the car for two days and the total rental cost was computed to be $176. How many miles did the executive drive the rented car?

1. **Understand the problem.**

   How do you calculate the cost of renting a car?

   total cost = per-day cost + mileage cost

   What is known?

   It cost $176 to rent the car for two days.

   What do you need to find?

   The number of miles the car was driven.

   Choose a variable:

   > Let m = the number of miles driven in the rented car.

2. **Write an equation.**
   Use the relationship for calculating the total cost.

$$\text{per-day cost} + \text{mileage cost} = \text{total cost}$$
$$(39)(2) + (0.28)m = 176$$

3. **Solve and state the answer.**

$$(39)(2) + (0.28)(m) = 176$$

$78 + 0.28m = 176$	Simplify the equation.
$0.28m = 98$	Subtract 78 from both sides.
$\dfrac{0.28m}{0.28} = \dfrac{98}{0.28}$	Divide both sides by 0.28.
$m = 350$	Simplify.

   The executive drove 350 miles.

4. **Check.**
   Does this seem reasonable? If he drove the car 350 miles in two days, would it cost $176?

$$(\text{Cost of \$39 per day for 2 days}) + (\text{cost of \$0.28 per mile for 350 miles})$$
$$\stackrel{?}{=} \text{total cost of \$176}$$
$$(\$39)(2) + (350)(\$0.28) \stackrel{?}{=} \$176$$
$$\$78 + \$98 \stackrel{?}{=} \$176$$
$$\$176 = \$176 \quad \checkmark$$

Practice Problem 1   Alfredo wants to rent a truck to move to Florida. He has determined that the cheapest rental rates for a truck of the correct size are from a local company that will charge him $25 per day and $0.20 per mile. He has not yet completed an estimate of the mileage of the trip, but he knows that he will need the truck for three days. He has allowed $350 in his moving budget for the truck. How far can he travel for a rental cost of exactly $350?

2 Solving Percent Problems

Many applied situations require finding a percent of an unknown number. If we want to find 23% of $400, we multiply 0.23 by 400: $0.23(400) = 92$. If we want to find 23% of an unknown number, we can express this using algebra by writing $0.23n$, where n represents the unknown number.

EXAMPLE 2   A sofa was marked with the following sign: "The price of this sofa has been reduced by 23%. You can save $138 if you buy now." What was the original price of the sofa?

1. Understand the problem.

Let s = the original price of the sofa.

Then $0.23s$ = the amount of the price reduction, which is $138.

2. Write an equation and solve.

$0.23s = 138$ Write the equation.

$\dfrac{0.23s}{0.23} = \dfrac{138}{0.23}$ Divide each side of the equation by 0.23.

$s = 600$ Simplify.

The original price of the sofa was $600.

3. Check:

Is $600 a reasonable answer?   ✓   Does 23% of $600 = $138?   ✓

Practice Problem 2   John earns a commission of 38% of the cost of every set of encyclopedias that he sells. Last year he earned $4560 in commissions. What was the cost of the encyclopedias that he sold last year?

EXAMPLE 3   Hector received a pay raise this year. The raise was 6% of last year's salary. This year he will earn $15,900. What was his salary last year before the raise?

1. Understand the problem.

What do we need to find?

Hector's salary last year.

What do we know?

Hector received a 6% pay raise and now earns $15,900.

What does this mean?

Reword the problem: *This year's salary of $15,900 is 6% more than last year's salary.*

Choose a variable:

Let x = Hector's salary last year.

Then $0.06x$ = the amount of the raise.

2. Write an equation and solve.

Last year's salary	+	the amount of his raise	=	this year's salary	
x	+	$0.06x$	=	$15,900$	Write the equation.
$1.00x$	+	$0.06x$	=	$15,900$	Rewrite x as $1.00x$.
		$1.06x$	=	$15,900$	Combine like terms.
x			=	$\dfrac{15,900}{1.06}$	Divide by 1.06.
x			=	$15,000$	Simplify.

Thus Hector's salary was $15,000 last year before the raise.

3. *Check.*

Does it seem reasonable that Hector's salary last year was $15,000? The check is up to you.

Practice Problem 3   The price of Betsy's new car is 7% more than the price of a similar model last year. She paid $13,910 for her car this year. What would a similar model have cost last year?

3   *Solving Investment Problems Involving Simple Interest*

Interest is a charge for borrowing money or an income from investing money. Interest rates affect our lives. They affect the national economy and they affect a consumer's ability to borrow money for big purchases. For these reasons, a student of mathematics should be able to solve problems involving interest.

There are two basic types of interest: simple and compound. **Simple interest** is computed by multiplying the amount of money borrowed or invested (which is called the *principal*) times the rate of interest times the period of time over which it is borrowed or invested (usually measured in years unless otherwise stated).

$$\text{Interest} = \text{principal} \times \text{rate} \times \text{time}$$
$$I = prt$$

You often hear of banks offering a certain interest rate *compounded* quarterly, monthly, weekly, or daily. In **compound interest** the amount of interest is added to the amount of the original principal at the end of each time period, so future interest is based on the sum of both principal and previous interest. Most financial institutions use compound interest in their transactions.

Problems involving compound interest may be solved by:

1. Repeated calculations using the simple interest formula.

2. Using a compound interest table.

3. Using exponential functions, a topic that is usually covered in a higher-level college algebra course.

*All examples and exercises in this chapter will involve **simple interest***.

EXAMPLE 4   Find the interest on $3000 borrowed at a simple interest rate of 18% for one year.

$I = prt$ The simple interest formula.

$I = (3000)(0.18)(1)$ Substitute the values of the variables: principal = 3000, the rate = 18% = 0.18, the time = one year.

$I = 540$

Thus the interest charge for borrowing $3000 for one year at a simple interest rate of 18% is $540.

Practice Problem 4   Find the interest on $7000 borrowed at a simple interest rate of 12% for one year.

Now we apply this concept to a word problem about investments.

EXAMPLE 5   A woman invested an amount of money in two accounts for one year. She invested some at 8% simple interest and the rest at 6% simple interest. Her total amount invested was $1250. At the end of the year she had earned $86 in interest. How much money had she invested in each account?

Calculator

Interest

You can use a calculator to find simple interest. Find the interest on $450 invested at an annual rate of 6.5% for 15 months. Notice that the time is in months. Since the interest formula $I = prt$ is in years, you need to change 15 months to years by dividing 15 by 12.

Enter: 15 [÷] 12 [=]

Display: [1.25]

Leave this on the display and multiply as follows:

1.25 [×] 450 [×]

6.5 [%] [=]

The display should read

[36.5625]

which would round to $36.56.

Try finding the simple interest in the following investments.

(a) $9516 invested at 12% for 30 months

(b) $593 borrowed at 8% for 5 months

Mathematics Blueprint For Problem Solving

Gather the Facts	Assign the Variable	Basic Formula or Equation	Key Points to Remember
$1250 is invested: part at 8% interest, part at 6% interest. The total interest for the year is $86.	Let x = the amount invested at 8%. $1250 - x$ = the amount invested at 6%. $0.08x$ = the amount of interest for x dollars at 8%. $0.06(1250 - x)$ = the amount of interest for $1250 - x$ dollars at 6%.	Interest earned at 8% + interest earned at 6% = total interest earned during the year, which is $86.	Be careful to write $1250 - x$ for the amount of money invested at 6%. The order is total $- x$. Do not use $x - 1250$.

$$\begin{array}{ccc} \text{interest} & & \text{interest} & & \text{total interest} \\ \text{earned} & + & \text{earned} & = & \text{earned during} \\ \text{at } 8\% & & \text{at } 6\% & & \text{the year} \\ 0.08x & + & 0.06(1250 - x) & = & 86 \end{array}$$

Note: Be sure you write $(1250 - x)$ for the amount of money invested at 6%. Students often write it backwards by mistake. It is *not* correct to use $(x - 1250)$ instead of $(1250 - x)$. The order of the terms is very important.

Solve and state the answer.

$0.08x + 75 - 0.06x = 86$	Remove parentheses.
$0.02x + 75 = 86$	Combine like terms.
$0.02x = 11$	Subtract 75 from both sides.
$\dfrac{0.02x}{0.02} = \dfrac{11}{0.02}$	Divide both sides by 0.02.
$x = 550$	The amount invested at 8% interest is $550.
$1250 - x = 1250 - 550 = 700$	The amount invested at 6% interest is $700.

Check.

Are these values reasonable? Yes. Do the amounts equal $1250?

$$\$550 + \$700 \overset{?}{=} \$1250$$
$$\$1250 = \$1250 \quad \checkmark$$

Would these amounts earn $86 interest in one year invested at the specified rates?

$$0.08(\$550) + 0.06(\$700) \overset{?}{=} \$86$$
$$\$44 + \$42 \overset{?}{=} \$86$$
$$\$86 = \$86 \quad \checkmark$$

Practice Problem 5 A woman invested her savings of $8000 in two accounts that each calculate interest only once per year. She placed one amount in a special notice account that yields 9% annual interest. The remainder she placed in a tax-free All-Savers account that yields 7% annual interest. At the end of the year, she had earned $630 in interest from the two accounts together. How much had she invested in each account?

4 Solving Coin Problems

Coin problems provide an unmatched opportunity to use the concept of *value*. We must make a distinction between how many coins there are and the *value* of the coins.

Consider the next example. Here we know *the value* of some coins, but do not know *how many* we have.

EXAMPLE 6 When Bob got out of math class, he had to make a long-distance call. He had exactly enough dimes and quarters to make a phone call that would cost $2.55. He had one less quarter than he had dimes. How many coins of each type did he have?

$$\text{Let } d = \text{the number of dimes.}$$
$$\text{Then } d - 1 = \text{the number of quarters.}$$

The total value of the coins was $2.55. How can we represent the value of the dimes and the value of the quarters? Think.

Each dime is worth $0.10.	Each quarter is worth $0.25.
5 dimes are worth $(5)(0.10) = 0.50.$	8 quarters are worth $(8)(0.25) = 2.00.$
d dimes are worth $(d)(0.10) = 0.10d.$	$(d - 1)$ quarters are worth $(d - 1)(0.25) = 0.25(d - 1).$

Now we can write an equation for the total value.

$$(\text{value of dimes}) + (\text{value of quarters}) = \$2.55$$
$$0.10d \quad + \quad 0.25(d - 1) \quad = \quad 2.55$$

$0.10d + 0.25d - 0.25 = 2.55$	Remove parentheses.
$0.35d - 0.25 = 2.55$	Combine like terms.
$0.35d = 2.80$	Add 0.25 to both sides.
$\dfrac{0.35d}{0.35} = \dfrac{2.80}{0.35}$	Divide both sides by 0.35.
$d = 8$	Simplify.
$d - 1 = 7$	

Thus Bob had eight dimes and seven quarters.
Check.
Is this answer reasonable? Yes. Does Bob have one less quarter than he has dimes?

$$8 - 7 \overset{?}{=} 1$$
$$1 = 1 \checkmark$$

Are eight dimes and seven quarters worth $2.55?

$$8(\$0.10) + 7(\$0.25) \overset{?}{=} \$2.55$$
$$\$0.80 + \$1.75 \overset{?}{=} \$2.55$$
$$\$2.55 = \$2.55 \checkmark$$

Practice Problem 6 Ginger has five more quarters than dimes. She has $5.10 in change. If she has only quarters and dimes, how many coins of each type does she have?

EXAMPLE 7 Michele and her two children returned from the grocery store with only $2.80 in change. She had twice as many quarters as nickels. She had two more dimes than nickels. How many nickels, dimes, and quarters did she have?

Mathematics Blueprint For Problem Solving

Gather the Facts	Assign the Variable	Basic Formula or Equation	Key Points to Remember
Michele had $2.80 in change. She had twice as many quarters as nickels. She had two more dimes than nickels.	Let x = the number of nickels. $2x$ = the number of quarters. $x + 2$ = the number of dimes. $0.05x$ = the value of the nickels. $0.25(2x)$ = the value of the quarters. $0.10(x + 2)$ = the value of the dimes.	The value of the nickels + the value of the dimes + the value of the quarters = $2.80.	Don't add the number of coins to get $2.80. You must add the value of the coins!

FOOD MARKET

(value of nickels) + (value of dimes) + (value of quarters) = $2.80
$$0.05x \quad + \quad 0.10(x + 2) \quad + \quad 0.25(2x) \quad = \quad 2.80$$

Solve.

$$0.05x + 0.10x + 0.20 + 0.50x = 2.80 \quad \text{Remove parentheses.}$$
$$0.65x + 0.20 = 2.80 \quad \text{Combine like terms.}$$
$$0.65x = 2.60 \quad \text{Subtract 0.20 from both sides.}$$
$$\frac{0.65x}{0.65} = \frac{2.60}{0.65} \quad \text{Divide both sides by 0.65.}$$
$$x = 4 \quad \text{Simplify. Michele had four nickels.}$$
$$2x = 8 \quad \text{She had eight quarters.}$$
$$x + 2 = 6 \quad \text{She had six dimes.}$$

When Michele left the grocery store she had four nickels, eight quarters, and six dimes.

Check.
Is the answer reasonable? Yes. Did Michele have twice as many quarters as nickels?

$$(4)(2) \overset{?}{=} 8 \qquad 8 = 8 \checkmark$$

Did she have two more dimes than nickels?

$$4 + 2 \overset{?}{=} 6 \qquad 6 = 6 \checkmark$$

Do four nickels, eight quarters, and six dimes have a value of $2.80?

$$4(\$0.05) + 8(\$0.25) + 6(\$0.10) \overset{?}{=} \$2.80$$
$$\$0.20 + \$2.00 + \$0.60 \overset{?}{=} \$2.80$$
$$\$2.80 = \$2.80 \checkmark$$

Practice Problem 7 A young boy told his friend that he had twice as many nickels as dimes in his pocket. He also said that he had four more quarters than dimes. He said that he had $2.35 in change in his pocket. Can you determine how many nickels, dimes, and quarters he had?

Developing Your Study Skills

Applications or Word Problems

Applications or word problems are the very life of mathematics! They are the reason for doing mathematics because they teach you how to put into use the mathematical skills you have developed. Learning mathematics without ever doing word problems is similar to learning all the skills of a sport without ever playing a game or learning all the notes on an instrument without ever playing a song.

The key to success is practice. Make yourself do as many problems as you can. If you need help organizing your facts, use the Mathematics Blueprint. You may not be able to do all problems correctly at first, but keep trying. Do not give up whenever you reach a difficult one. If you cannot solve it, just try another one. Then come back and try it again later.

A misconception among students when they begin studying word problems is that each problem is different. At first the problems may seem this way, but as you practice more and more, you will begin to see the similarities, the different "types." You will see patterns in solving problems, which will enable you to solve them more easily.

Applications

Solve. All problems involving interest refer to simple interest.

1. Paul has a job raking leaves for a neighbor. He makes $6.50 an hour, plus $0.75 for each bag he fills. Last Saturday he worked three hours and made a total of $27.75. How many bags of leaves did he fill?

2. Marybelle is contemplating a job as a waitress. She would be paid $4 per hour plus tips. The other waitresses have told her that an average tip at that restaurant is $3 per table served. If she works 20 hours per week, how many tables would she have to serve in order to make $191 during the week?

3. Ramon has a summer job as a lifeguard to earn his college tuition. He gets paid $6.00 per hour for the first 40 hours and $9.00 per hour for each hour in the week worked more than 40 hours. This summer his goal is to earn at least $303.00 per week. How many hours of overtime per week will he need to achieve his goal?

4. Maria has a summer job as a tennis instructor to earn her college tuition. She gets paid $6.50 per hour for the first 40 hours and $9.75 for each hour in the week worked more than 40 hours. This summer her goal is to earn at least $338.00 per week. How many hours of overtime per week will she need to achieve her goal?

5. The camera Melissa wanted for her birthday is on sale at 28% off the usual price. The amount of the discount is $100.80. What was the original price of the camera?

6. The number of women working full-time in Springfield has risen 7% this year. This means 126 more women have full-time jobs. What was the number of women working full-time last year?

7. In the bullish market at the end of 1999, two partners, Jim Brown and Walter Payton, split a profit of $9540 from the sale of stock. Jim had invested more, so he received 12% more of the profit than Walter did. How much money did each partner make?

8. A speculator bought stocks and later sold them for $5136, making a profit of 7%. How much did the stocks cost him?

9. Don Williams invested some money at 9% simple interest. At the end of the year, the total amount of his original principal and the interest was $6540. How much did he originally invest?

10. The cost of living last year went up 7%. Fortunately, Alice Swanson got a 7% raise in her salary from last year. This year she is earning $21,400. How much did she make last year?

11. Mr. and Mrs. Wright set up a trust fund for their children last year. Part of it is earning 7% simple interest per year while the rest of it is earning 5% simple interest per year. They placed $5000 in the trust fund. In one year the trust fund has earned $310. How much did they invest at each interest rate?

12. Anne and Michael invested $8000 last year in tax-free bonds. Some of the bonds earned 8% simple interest while the rest earned 6% simple interest. At the end of the year, they had earned $580 in interest. How much did they invest at each interest rate?

13. Plymouth Rock Bank invested $400,000 last year in mutual funds. The conservative fund earned 8% simple interest. The growth fund earned 12% simple interest. At the end of the year, the bank had earned $38,000 from these mutual funds. How much did it invest in each fund?

14. Millennium Securities last year invested $600,000 in mutual funds. The international fund earned 11% simple interest. The high-tech fund earned 7% simple interest. At the end of the year, the company had earned $50,000 in interest. How much did it invest in each fund?

15. William Tell invested half of his money at 5%, one-third of his money at 4%, and the rest of his money at 3.5%. If his total annual investment income was $530, how much had he invested?

16. Last year Pete Pfeffer decided to invest half of his money in a credit union paying 4.5% interest, one-third of his money in a mutual fund paying 5% interest, and the rest of his money in a bank CD paying 4% interest. If his annual investment income was $357.50 last year, how much money had he invested?

17. Little Melinda has nickels and quarters in her bank. She has four fewer nickels than quarters. She has $3.70 in the bank. How many coins of each type does she have?

18. Fred's younger brother had several coins when he returned from his paper route. He had a total of $5.35 in dimes and quarters. He had six more quarters than he had dimes. How many of each coin did he have?

19. A newspaper carrier has $3.75 in change. He has three more quarters than dimes but twice as many nickels as quarters. How many coins of each type does he have?

20. A baseball club brought 70 balls for $120. The balls used for practice cost $1.50 each and the balls used for games cost $2.25 each. How many of each kind were bought?

21. Charlie Saulnier cashed his paycheck and came home from the bank with $100 bills, $20 bills, and $10 bills. He has twice as many $20 bills as he has $10 bills. He has three more $100 bills than he has $10 bills. He is carrying $1500 in bills. How many of each denomination does he have?

22. Roberta Burgess came home with $325 in tips from two nights on her job as a waitress. She had $20 bills, $10 bills, and $5 bills. She discovered that she had three times as many $5 bills as she had $10 bills. She also found that she had 4 fewer $20 bills than she had $10 bills. How many of each denomination did she have?

▲ **23.** The sum of the measures of the three interior angles of a triangle is always 180°. In a certain triangle the measure of one angle is double the measure of the second angle but is 5 degrees less than the measure of the third angle. What is the measure of each angle?

24. In a citywide veterinary study of a group of cats over age 5 with no regular exercise and playtime, 136 of the cats, or 4%, had no health problems.
   (a) How many cats were in the original study?
   (b) 28% of the cats had minor health problems, and the rest had major health problems. How many cats had major health problems?

25. Walter invested $2969 for one year and earned $273.47 in interest. He had invested part of the $2969 at 7% and the remainder at 11%. How much had he invested at each rate?

26. Last year the town of Waterbury paid an interest payment of $52,396.08 for a one-year note. This represented 11% of the amount the town borrowed. How much had the town borrowed?

To Think About

27. The West Suburban Car Rental Agency will rent a compact car for $35 per day and an additional charge of $0.24 per mile. The Golden Gate Car Rental Agency charges only $0.16 per mile but charges $41 per day. If a salesperson wanted to rent a car for three days, how many miles would that person have to drive to make the Golden Gate Car Rental Agency car a better bargain?

28. A Peabody pumping station pumps 2000 gallons of water per hour into an empty reservoir tank for a town's drinking supply. The station pumps for three hours. Then a leak in the reservoir tank is created by a large crack. Some water flows out of the reservoir tank at a constant rate. The pumping station continues pumping for six more hours while the leak is undetected. At the end of nine hours the reservoir contains 17,640 gallons of water. During the last six hours, how many gallons per hour are leaking from the reservoir?

Cumulative Review Problems

Perform the operations in the proper order.

29. $5(3) + 6 \div (-2)$

30. $2 - (8 - 10)^3 + 5$

Evaluate for $x = -2$ and $y = 3$.

31. $2x^2 + 3xy - 2y^2$

32. $x^3 - 5x^2 + 3y - 6$

Brian sells high-tech parts to satellite communications companies. In his negotiations he originally offers to sell one company 200 parts for a total of $22,400. However, after negotiations, he offers to sell that company the same parts at a 15% discount if the company agrees to sign a purchasing contract for 200 additional parts at some future date.

33. What is the average cost per part if the parts are sold at the discounted price?

34. How much will the total bill be for the 200 parts at the discounted price?

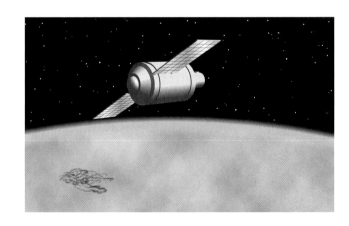

Putting Your Skills to Work

Automobile Loans and Installment Loans

The following table provides the amount of the monthly payment of a loan of $1000 at various interest rates for a period of 2 to 5 years. To find the monthly payment for a loan of a larger amount of money, merely multiply this payment by the number of thousands of the loan. For example, to borrow $14,000 at 8% interest for 3 years, you multiply $31.34 by 14 to obtain a payment of $438.76.

LOAN AMORTIZATION TABLE FOR REPRESENTATIVE INTEREST RATES
Monthly Payment per $1000 to Pay Principal and Interest

Period of Loan in Years	7%	8%	9%	10%	12%	16%
2	44.77	45.23	45.68	46.14	47.07	48.96
3	30.88	31.34	31.80	32.27	33.21	35.16
4	23.95	24.41	24.89	25.36	26.33	28.34
5	19.80	20.28	20.76	21.25	22.24	24.32

Problems for Individual Study and Investigation

Use the preceding table to solve the following problems.

1. Ricardo Sanchez wants to buy a new car. He will not make a down payment, but he will need to borrow $10,000 at a 9% interest rate for four years. What will his monthly payments be? How much interest will he pay over the four-year period? (That is, how much more than $10,000 will he pay over four years?)

2. If Alicia Wong wants to purchase a new stereo system for her apartment and borrow $3000 at 12% interest for two years, what will her monthly payments be? How much interest will she pay over the two-year period? (That is, how much more than $3000 will she pay over the two years?)

Problems for Cooperative Group Investigation

With some other members of your class, determine answers to the following.

3. If Walter Swenson plans to borrow $15,000 to purchase a new car at an interest rate of 7%, how much more in interest will he pay if he borrows the money for five years as opposed to a period of four years?

4. Ray and Shirley Peterson are planning to build a garage and a connecting family room for their house. It will cost $35,000. They have $17,000 in savings, and they plan to borrow the rest at 16% interest. They can afford a maximum monthly payment of $640. What time period for the loan will they have to select to meet that requirement? What will their monthly payment be?

Internet Connections

 Netsite: http://www.prenhall.com/tobey_beginning

Site: Mortgage Amortization Calculator or a related site

Alternate Site: Amortization Calculator

Kimberly Jones is buying a $135,000 house. She will make a down payment of 10%. The rest will be financed with a 30-year loan at an interest rate of 8.5%.

5. What will her monthly payment be? How much interest will she pay over the life of the loan?

6. Suppose she decides to pay an extra $200 every month. How much interest will she save by doing this? In how many years will the loan be paid off?

217

3.5 Solving Word Problems Using Geometric Formulas

1 Area, Perimeter, and Missing Angles

In Section 1.8 we reviewed a number of area and perimeter formulas. We repeat this list of formulas here for convenience.

Area and Perimeter Formulas

A parallelogram is a four-sided figure with opposite sides parallel. In a parallelogram, opposite sides are equal and opposite angles are equal.

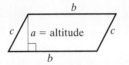

Perimeter = the sum of all four sides

Area = ab

A rectangle is a parallelogram with all interior angles measuring 90°.

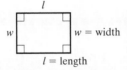

Perimeter = $2l + 2w$

Area = lw

A square is a rectangle with all four sides equal.

Perimeter = $4s$

Area = s^2

A trapezoid is a four-sided figure with two sides parallel. The parallel sides are called the bases of the trapezoid.

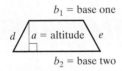

Perimeter = the sum of all four sides

Area = $\dfrac{1}{2}a(b_1 + b_2)$

A triangle is a closed plane figure with three sides.

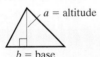

Perimeter = the sum of the three sides

Area = $\dfrac{1}{2}ab$

A circle is a plane curve consisting of all points at an equal distance from a given point called the center.

Circumference is the distance around a circle. The **radius** is a line segment from the center of the circle to a point on the circle. The **diameter** is a line segment across the circle that passes through the center.

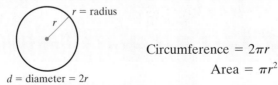

Circumference = $2\pi r$

Area = πr^2

π is a constant associated with circles. It is an irrational number that is approximately 3.141592654. We usually use 3.14 as a sufficiently accurate approximation. Thus we write $\pi \approx 3.14$ for most of our calculations involving π.

We frequently encounter triangles in word problems. There are four important facts about triangles, which we list for convenient reference.

Triangle Facts

1. The sum of the interior angles of any triangle is 180°. That is,

measure of ∠ A + measure of ∠ B + measure of ∠ C = 180°.

2. An **equilateral** triangle is a triangle with three sides equal in length and three angles that measure 60° each.

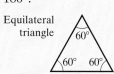

3. An **isosceles** triangle is a triangle with two equal sides. The two angles opposite the equal sides are also equal.

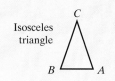

measure ∠ A = measure ∠ B

4. A **right** triangle is a triangle with one angle that measures 90°.

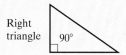

▲ 🔵 **EXAMPLE 1**   Find the area of a triangular window whose base is 16 inches and whose altitude is 19 inches.

We can use the formula for the area of a triangle. Substitute the known values in the formula and solve for the unknown.

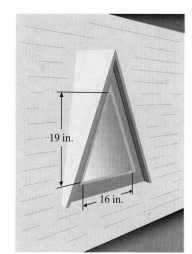

$A = \dfrac{1}{2} ab$ Write the formula for the area of a triangle.

$= \dfrac{1}{2} (16 \text{ in.})(19 \text{ in.})$ Substitute the known values in the formula.

$= \dfrac{1}{2} (16)(19)(\text{in.})(\text{in.})$ Simplify.

$A = 152 \text{ in.}^2$ or   152 square inches

The area of the triangle is 152 square inches.

▲ **Practice Problem 1**   Find the area of a triangle whose base is 14 inches and whose altitude is 20 inches.

▲ 🔵 **EXAMPLE 2**   The area of an NBA basketball court is 4700 square feet. Find the width of the court if the length is 94 feet.

Draw a diagram.

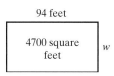

Write the formula for the area of a rectangle and solve for the unknown value.

$$A = lw$$
$$4700(\text{ft})^2 = (94 \text{ ft})(w) \quad \text{Substitute the known values into the formula.}$$
$$\frac{4700(\text{ft})(\text{ft})}{94 \text{ ft}} = \frac{94 \text{ ft}}{94 \text{ ft}} w \quad \text{Divide both sides by 94 feet.}$$
$$50 \text{ feet} = w$$

The width of the basketball court is 50 feet.

▲ **Practice Problem 2** The area of a rectangular field is 120 square yards. If the width of the field is 8 yards, what is the length?

▲ ⬭ **EXAMPLE 3** The area of a trapezoid is 400 square inches. The altitude is 20 inches and one of the bases is 15 inches. Find the length of the other base.

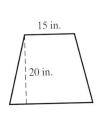

15 in.

20 in.

$$A = \frac{1}{2} a(b_1 + b_2) \qquad \text{Write the formula for the area of a trapezoid.}$$
$$400(\text{in.})^2 = \frac{1}{2}(20 \text{ in.})(15 \text{ in.} + b_2) \quad \text{Substitute the known values.}$$
$$400(\text{in.})(\text{in.}) = 10 \ (\text{in.})(15 \text{ in.} + b_2) \quad \text{Simplify.}$$
$$400(\text{in.})(\text{in.}) = 150(\text{in.})(\text{in.}) + (10 \text{ in.})b_2 \quad \text{Remove parentheses.}$$
$$250(\text{in.})(\text{in.}) = (10 \text{ in.})b_2 \qquad \qquad \text{Subtract 150 in.}^2 \text{ from both sides.}$$
$$25 \text{ in.} = b_2 \qquad \qquad \qquad \text{Divide both sides by 10 in.}$$

The other base is 25 inches long.

▲ **Practice Problem 3** The area of a trapezoid is 256 square feet. The bases are 12 feet and 20 feet. Find the altitude.

▲ ⬭ **EXAMPLE 4** Find the area of a circular sign whose diameter is 14 inches. (Use 3.14 for π.) Round to the nearest square inch.

$$A = \pi r^2 \quad \text{Write the formula for the area of a circle.}$$

Note that the length of the diameter is given in the word problem. We need to know the length of the radius to use the formula. Since $d = 2r$, then 14 in. $= 2r$ and $r = 7$ in.

$$A = \pi(7 \text{ in.})(7 \text{ in.}) \qquad \text{Substitute known values into the formula.}$$
$$= (3.14)(7 \text{ in.})(7 \text{ in.})$$
$$= (3.14)(49 \text{ in.}^2)$$
$$= 153.86 \text{ in.}^2$$

Rounded to the nearest square inch, the area of the circle is approximately 154 square inches.

In this example, you were asked to find the area of the circle. Do not confuse the formula for area with the formula for circumference. $A = \pi r^2$, while $C = 2\pi r$. Remember that area involves square units, so it is only natural that in the area formula you would square the radius.

▲ **Practice Problem 4** Find the circumference of a circle whose radius is 15 meters. (Use 3.14 for π.) Round your answer to the nearest meter.

▲ ◼ **EXAMPLE 5**   Find the perimeter of a parallelogram whose longer sides are 4 feet and whose shorter sides are 2.6 feet.

Draw a picture.

4 ft

2.6 ft

The perimeter is the distance around a figure. To find the perimeter, add the lengths of the sides. Since the opposite sides of a parallelogram are equal, we can write the following.

$$P = 2(4 \text{ feet}) + 2(2.6 \text{ feet})$$
$$= 8 \text{ feet} + 5.2 \text{ feet}$$
$$= 13.2 \text{ feet}$$

The perimeter of the parallelogram is 13.2 feet.

▲ **Practice Problem 5**   Find the perimeter of an equilateral triangle with a side that measures 15 centimeters.

▲ ◼ **EXAMPLE 6**   The smallest angle of an isosceles triangle measures 24°. The other two angles are larger. What are the measurements of the other two angles?
   We know that in an isosceles triangle the measures of two angles are equal. We know that the sum of the measures of all three angles is 180°. Both of the larger angles must be different from 24°, therefore these two larger angles must be equal.
   Let x = the measure in degrees of each of the larger angles.

Then we can write

$$24° + x + x = 180°$$
$$24° + 2x = 180° \quad \text{Add like terms.}$$
$$2x = 156° \quad \text{Subtract 24 from each side.}$$
$$x = 78° \quad \text{Divide each side by 2.}$$

Thus the measures of each of the other two angles of the triangle must be 78°.

▲ **Practice Problem 6**   The largest angle on an isosceles triangle measures 132°. The other two angles are smaller. What are the measurements of the other two angles?

② Volume and Surface Area

Now let's examine three-dimensional figures. **Surface area** is the total area of the faces of a figure. You can find surface area by calculating the area of each face and then finding the sum. **Volume** is the measure of the amount of space inside a figure. Some formulas for the surface area and the volume of regular figures can be found in the following table.

Geometric Formulas: Three-Dimensional Figures

Rectangular prism

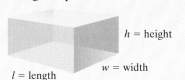

h = height

w = width

l = length

Note that all of the faces are rectangles.

Surface area = $2lw + 2wh + 2lh$

Volume = lwh

Geometric Formulas: Three-Dimensional Figures (continued)

Sphere

r = radius

Surface area $= 4\pi r^2$

Volume $= \dfrac{4}{3}\pi r^3$

Right circular cylinder

r = radius
h = height

Surface area $= 2\pi rh + 2\pi r^2$

Volume $= \pi r^2 h$

▲ ● **EXAMPLE 7**   Find the volume of a sphere with radius 4 centimeters. (Use 3.14 for π.) Round your answer to the nearest cubic centimeter.

$$V = \frac{4}{3}\pi r^3 \qquad \text{Write the formula for the volume of a sphere.}$$

$$= \frac{4}{3}(3.14)(4 \text{ cm})^3 \quad \text{Substitute the known values into the formula.}$$

$$= \frac{4}{3}(3.14)(64)\text{cm}^3 \approx 267.946667 \text{ cm}^3$$

Rounded to the nearest whole number, the volume of the sphere is 268 cubic centimeters.

▲ **Practice Problem 7**   Find the surface area of a sphere with radius 5 meters. Round your answer to the nearest square meter.   ●

▲ ● **EXAMPLE 8**   A can is made of aluminum. It has a flat top and a flat bottom. The height is 5 inches and the radius is 2 inches. How much aluminum is needed to make the can? How much aluminum is needed to make 10,000 cans?

Understand the problem.
What do we need to find? Reword the problem.

We need to find the total surface area of a cylinder.

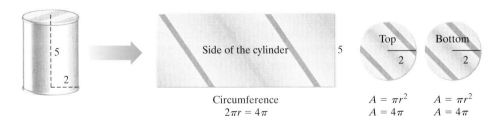

You may calculate the area of each piece and find the sum or you may use the formula. We will use 3.14 to approximate π.

$$\text{Surface area} = 2\pi rh + 2\pi r^2$$

$$= 2(3.14)(2 \text{ in.})(5 \text{ in.}) + 2(3.14)(2 \text{ in.})^2$$

$$= 62.8 \text{ in.}^2 + 25.12 \text{ in.}^2 = 87.92 \text{ in.}^2$$

87.92 square inches of aluminum is needed to make each can.

Have we answered all the questions in the word problem? Reread the problem. How much aluminum is needed to make 10,000 cans?

$$(\text{aluminum for 1 can})(10{,}000) = (87.92 \text{ in.}^2)(10{,}000) = 879{,}200 \text{ square inches}$$

It would take 879,200 square inches of aluminum to make 10,000 cans.

▲ **Practice Problem 8**   Sand is stored in a cylindrical drum that is 4 feet high and has a radius of 3 feet. How much sand can be stored in the drum?

3 More-Involved Geometric Problems

▲ ◖ **EXAMPLE 9**   A quarter-circle (of radius 1.5 yards) is connected to two rectangles with dimensions as labeled on the sketch in the margin. You need to lay a strip of carpet in your house according to this sketch. How many square yards of carpeting will be needed on the floor? (Use $\pi \approx 3.14$. Round your final answer to the nearest tenth.)

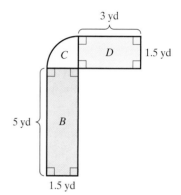

The desired area is the sum of three areas, which we will call B, C, and D. Area B and area D are rectangular in shape. It is relatively easy to find the areas of these shapes.

$$A_B = (5 \text{ yd})(1.5 \text{ yd}) = 7.5 \text{ yd}^2 \qquad A_D = (3 \text{ yd})(1.5 \text{ yd}) = 4.5 \text{ yd}^2$$

Area C is one-fourth of a circle. The radius of the circle is 1.5 yd.

$$A_C = \frac{\pi r^2}{4} \approx \frac{(3.14)(1.5 \text{ yd})^2}{4} = \frac{(3.14)(2.25)\text{yd}^2}{4} = 1.76625 \text{ yd}^2 \approx 1.8 \text{ yd}^2$$

The total area is $A_B + A_D + A_C \approx 7.5 \text{ yd}^2 + 4.5 \text{ yd}^2 + 1.8 \text{ yd}^2 \approx 13.8 \text{ yd}^2$.
13.8 square yards of carpeting will be needed to cover the floor.

▲ **Practice Problem 9**   John has a swimming pool that measures 8 feet by 12 feet. He plans to make a concrete walkway around the pool that is 3 feet wide. What will be the total cost of the walkway at $12 a square foot? ◖

▲ ◖ **EXAMPLE 10**   Find the weight of the water in a full cylinder containing water if the height of the cylinder is 5 feet and the radius is 2 feet. 1 cubic foot of water weighs 62.4 pounds. Round your answer to the nearest ten pounds. Use 3.14 for π.

Understand the problem.
Draw a diagram.

Write the formula for the volume of a cylinder.

$$V = \pi r^2 h$$
$$= (3.14)(2 \text{ ft})^2(5 \text{ ft})$$
$$= (3.14)(4 \text{ ft}^2)(5 \text{ ft})$$
$$= 62.8 \text{ ft}^3$$

The volume of the cylinder is 62.8 cubic feet.

Each cubic foot of water weighs 62.4 pounds. This can be written as 62.4 lb/ft³. Since the volume of the cylinder is 62.8 ft³, we multiply.

$$\text{Weight of the water} = (62.8 \; \cancel{\text{ft}^3})\left(\frac{62.4 \text{ lb}}{1 \; \cancel{\text{ft}^3}}\right)$$
$$= (62.8)(62.4) \text{ lb} = 3918.72 \text{ lb}$$

The weight of the water in the cylinder is 3920 rounded to the nearest ten pounds.

▲ **Practice Problem 10**   Find the weight of the water in a full rectangular container that is 5 feet wide, 6 feet long, and 8 feet high. Remember, 1 cubic foot of water weighs 62.4 pounds. Round your answer to the nearest 100 pounds. ◖

3.5 Exercises

Verbal and Writing Skills

Fill in the blank to complete each sentence.

▲ **1.** Perimeter is the _____ a plane figure.

▲ **2.** _____ is the distance around a circle.

▲ **3.** Area is a measure of the amount of _____ in a region.

▲ **4.** A _____ is a four-sided figure with exactly two sides parallel.

▲ **5.** The sum of the interior angles of any triangle is _____ .

Applications

▲ **6.** Find the area of a triangle whose altitude is 24 inches and whose base is 13 inches.

▲ **7.** Find the area of a parallelogram whose altitude is 14 inches and whose base is 7 inches.

▲ **8.** The area of a parallelogram is 345 square meters. The altitude is 15 meters. What is the base?

▲ **9.** The area of a triangle is 80 square feet. Find the altitude if the base is 16 feet.

▲ **10.** The area of a square is 25 square meters. What is its perimeter?

▲ **11.** Jaime found that a rope piece goes exactly once around a circle whose radius is 20 inches. How long is the rope? (Use 3.14 for π.)

▲ **12.** Find the area of a circular sign whose radius is 7.00 feet. (Use $\pi = 3.14$ as an approximate value.)

▲ **13.** Find the area of a circular flower bed whose diameter is 6 meters. (Use $\pi = 3.14$ as an approximate value.)

▲ **14.** Find the area of a trapezoid with an altitude of 12 meters and whose bases are 5 meters and 7 meters.

▲ **15.** The area of a trapezoid is 600 square inches and the bases are 20 inches and 30 inches. Find the altitude.

▲ **16.** In midtown Manhattan, the street blocks have a uniform size of 80 meters north-south by 280 meters east-west. If a typical New York City neighborhood is two blocks east-west by three blocks north-south, how much area does it cover?

▲ **17.** A computer manufacturing plant makes its computer cases from plastic. Two sides of a triangular scrap of plastic measure 7 cm and 25 cm.
 (a) If the perimeter of the plastic piece is 56 cm, how long is the third side?
 (b) If each side of the triangular piece is decreased by 0.1 cm, what would the new perimeter be?

▲ **18.** The circumference of a circle is 31.4 centimeters. Find the radius. (Use $\pi = 3.14$ as an approximate value.)

▲ **19.** The perimeter of an equilateral triangle is 27 inches. Find the length of each side of the triangle.

▲ **20.** An automobile tire has a diameter of 64 cm. What is the circumference of the tire? (Use $\pi = 3.14$ as an approximate value.)

▲ **21.** The ancient Greeks discovered that rectangles where the length-to-width ratio was 8 to 5 (known as the golden ratio) were most pleasing to the eye. Kayla wants to create a painting that will be framed by some antique molding that she found. If she calculates that the molding can be used to make a frame of 104 inches in perimeter, what dimensions should her painting be so that it is visually appealing?

▲ **22.** Two angles of a triangle measure 47 degrees and 59 degrees. What is the measure of the third angle?

▲ **23.** A right triangle has one angle that measures 77 degrees. What does the other acute angle measure?

▲ **24.** Each of the equal angles of an isosceles triangle is 4 times as large as the third angle. What is the measure of each angle?

▲ **25.** In a triangle, the measure of the first angle is twice the measure of the second angle. The measure of the third angle is 20 degrees less than the second angle. What is the measure of each angle?

▲ **26.** The smallest angle of an isosceles triangle used in the wood frame of a boat measures $38°$. The other two angles are larger. What are the measurements of the other two angles in this triangular part of the wood frame?

▲ **27.** The largest angle of an isosceles triangle used in holding a cable TV antenna measures $146°$. The other two angles are smaller. What are the measurements of the other two angles in the triangular piece of the cable TV antenna?

▲ **28.** A triangular region of the community college parking lot was measured. The measure of the first angle of this triangle is double the measure of the second angle. The measure of the third angle is $19°$ greater than double the measure of the first angle. What are the measurements of all three angles?

▲ **29.** A lobster pot used in Gloucester, Massachusetts, has a small triangular piece of wood at the entry point where the lobster enters. The measure of the first angle of this triangle is triple the measure of the second angle. The third angle of the triangle is $30°$ less than double the measure of the first angle. What are the measurements of all the three angles?

▲ **30.** The cost for paint to cover the exterior of the four sides of a shed was $40. A 5-gallon can of paint costs $20 and will cover 102 square feet. The shed consists of four sides, each shaped like a trapezoid. They each have one base of 8 feet and one base of 9 feet. What is the altitude of each trapezoid?

▲ **31.** The cost for paint to cover both sides of a large road-side sign (front side and back side) was $150. A 5-gallon can of paint costs $25 and will cover 250 square feet. The sign is constructed in the shape of a trapezoid. It has one base measuring 35 feet and one base measuring 40 feet. What is the altitude of this trapezoid?

▲ **32.** What is the volume of a cylinder whose height is 8 inches and whose radius is 10 inches? (Use $\pi = 3.14$ as an approximate value.)

▲ **33.** A cylinder holds propane gas. The volume of the cylinder is 235.5 cubic feet. Find the height if the radius is 5 feet. (Use $\pi = 3.14$ as an approximate value.)

▲ **34.** A beach ball is inflated to the shape of a sphere. Find the surface area of this sphere with radius 3 feet. (Use $\pi = 3.14$ as an approximate value.)

▲ **35.** A cabinet in the garage is shaped like a rectangular solid. The volume of the rectangular solid is 864 cubic inches. **(a)** Find the height of the solid if the length is 12 inches and the width is 9 inches. **(b)** What is the surface area?

▲ **36.** A plastic cylinder made to hold milk is constructed with a solid top and bottom. The radius is 6 centimeters and the height is 4 centimeters. **(a)** Find the volume of the cylinder. **(b)** Find the total surface area of the cylinder. (Use $\pi = 3.14$ as an approximate value.)

▲ **37.** A Pyrex glass sphere is made to hold liquids in a science lab. The radius of the sphere is 3 centimeters. **(a)** Find the volume of the sphere. **(b)** Find the total surface area of the sphere. (Use $\pi = 3.14$ as an approximate value.)

▲ **38.** Farmer Brown's goat is attached to the corner of the 8-foot-by-4-foot pump house by an 8-foot length of rope. The goat eats the grass around the pump house as shown in the diagram. This pattern consists of three-fourths of a circle of radius 8 feet and one-fourth of a circle of radius 4 feet. What area of grass can Farmer Brown's goat eat? (Use $\pi = 3.14$ as an approximate value.)

▲ **39.** For a party, Julian is making a seven-layer torte, which consists of seven 9-inch-diameter cake layers with a cream topping on each. He knows that one recipe of cream topping will cover 65 square inches of cake. How many batches of the topping must he make to complete his torte? (Use $\pi = 3.14$ as an approximate value.)

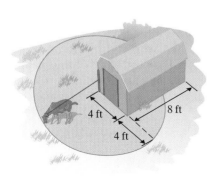

▲ **40.** A cement walkway is to be poured. It consists of the two rectangles with dimensions as shown in the diagram and a quarter of a circle with radius 1.5 yards. Use $\pi = 3.14$ as an approximate value. **(a)** How many square yards will the walkway be? **(b)** If a painter paints it for $2.50 per square yard, how much will the painting cost?

▲ **41.** An aluminum plate is made according to the following dimensions. The outer radius is 10 inches and the inner radius is 3 inches. Use $\pi = 3.14$ as an approximate value. **(a)** How many square inches in area is the aluminum plate (the shaded region in the sketch)? **(b)** If the cost to construct the plate is $7.50 per square inch, what is the construction cost?

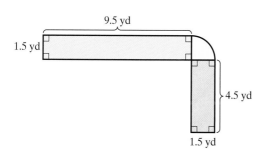

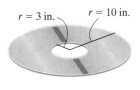

▲ **42.** It cost $120.93 to frame a painting with a flat aluminum strip. This was calculated at $1.39 per foot for materials and labor. Find the dimensions of the painting if its length is 3 feet less than four times its width.

▲ **43.** A concrete driveway measures 18 yards by 4 yards. The driveway will be 0.5 yards thick. **(a)** How many cubic yards of concrete will be needed? **(b)** How much will it cost if concrete is $3.50 per cubic yard?

To Think About

▲ **44.** An isosceles triangle has one side that measures 4 feet and another that measures 7.5 feet. The third side was not measured. Can you find one unique perimeter for this triangle? What are the two possible perimeters?

▲ **45.** Find the total surface area of **(a)** a sphere of radius 3 inches and **(b)** a right circular cylinder with radius 2 inches and height 0.5 inch. Which object has a greater surface area? Use $\pi = 3.14$ as an approximate value.

▲ **46.** Assume that a very long rope supported by poles that are 3 feet tall is stretched around the moon. (Neglect gravitational pull and assume that the rope takes the shape of a large circle.) The radius of the moon is approximately 1080 miles. How much longer would you need to make the rope if you wanted the rope to be supported by poles that are 4 feet tall? Use $\pi = 3.14$ as an approximate value.

Cumulative Review

Simplify.

47. $2(x - 6) - 3[4 - x(x + 2)]$

48. $-5(x + 3) - 2[4 + 3(x - 1)]$

49. $-\{3 - 2[x - 3(x + 1)]\}$

50. $-2\{x + 3[2 - 4(x - 3)]\}$

Putting Your Skills to Work

Examination of Weather Patterns

The National Oceanic and Atmospheric Administration maintains records of the climate for each state of the country for many years. Often these records are used to observe and predict patterns of heavy precipitation and severe drought. Consider the following graph, which is a record of the five-year average precipitation by month for the state of Rhode Island.

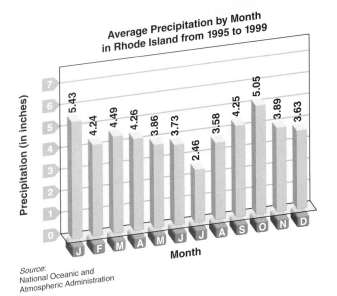

Average Precipitation by Month in Rhode Island from 1995 to 1999

5.43 4.24 4.49 4.26 3.86 3.73 2.46 3.58 4.25 5.05 3.89 3.63

Precipitation (in inches)

J F M A M J J A S O N D

Month

Source:
National Oceanic and
Atmospheric Administration

Problems for Individual Investigation and Analysis

1. What was the driest month recorded?

2. What was the wettest month recorded?

3. How many more inches of precipitation were recorded in January than in December?

4. How many more inches of precipitation were recorded in October than in July?

Problems for Group Investigation and Cooperative Activity

The annual precipitation for five consecutive years in Rhode Island is recorded in the following table.

Year	1995	1996	1997	1998	1999
Precipitation (in inches)	3.86	4.79	3.57	4.75	3.39

5. Some scientists have predicted increased precipitation for Rhode Island. They believe that when the precipitation value for 2000 is averaged in with those of the preceding five years, the average precipitation for the six-year period will be 4.50 inches per year. If this occurs, what will be the amount of precipitation in inches in 2000?

6. Some scientists have predicted a drier weather pattern for Rhode Island. They have predicted that when the data for 2000 and 2001 are averaged in with the data for the preceding five years, the average precipitation for the seven-year period will be 3.80 inches per year. If this occurs, what will be the average precipitation in inches per year for the two years 2000 and 2001?

Internet Connections

 www Netsite: http://www.prenhall.com/tobey_beginning

Site: National Climatic Data Center, sponsored and maintained by the National Oceanic and Atmospheric Administration

7. Find the wettest and driest month for your city or for a city close to yours. Make note of the time period covered by the data you use.

8. What was the average temperature for the warmest month in that period? For the coldest month?

For extremely interesting current weather information about your local weather, explore http://www.weathersite.com/

3.6 Using Inequalities to Solve Word Problems

① *Translating English Sentences into Inequalities*

Real-life situations often involve inequalities. For example, if a student wishes to maintain a certain grade point average (GPA), you may hear him or her say, "I have to get a grade of at least an 87 on the final exam."

What does "at least an 87" mean? What grade does the student need to achieve? 87 would fit the condition. What about 88? The student would be happy with an 88. In fact the student would be happy with any grade that *is greater than or equal to* an 87. We can write this as

$$\text{grade} \geq 87$$

Notice that the English phrase "at least an 87" translates mathematically as "greater than or equal to an 87."

Be careful when translating English phrases that involve inequalities into algebra. If you are not sure which symbol to use, try a few numbers to see if they fit the condition of the problem. Choose the mathematical symbol accordingly.

Student Learning Objectives

After studying this section, you will be able to:

① Translate English sentences into inequalities.

② Solve applied problems using inequalities.

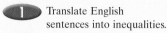
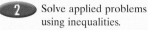

SSM PH TUTOR CD & VIDEO MATH PRO WEB
CENTER

EXAMPLE 1 Translate each sentence into an inequality.

(a) The perimeter of the rectangle must be no more than 70 inches.

(b) The average number of errors must be less than two mistakes per page.

(c) His income must be at least $950 per week.

(d) The new car will cost at most $12,070.

(a) Perimeter must be *no more than* 70.

$$\text{Perimeter} \quad \leq \quad 70$$

Try 80. Does 80 satisfy this requirement? That is, is 80 more than 70? No. Do 60 and 50 satisfy this requirement? Yes. Use the "less than or equal to" symbol.

(b) Average number must be *less than* 2.

$$\text{Average} \quad < \quad 2$$

(c) Income must be *at least* $950.

$$\text{Income} \quad \geq \quad 950$$

"At least" means, "not less than." The income must be $950 or more. Use ≥.

(d) The new car will cost *at most* $12,070.

$$\text{Car} \quad \leq \quad 12{,}070$$

"At most $12,070" means it is the top price for the car. The cost of the car cannot be more than $12,070. Use ≤.

Practice Problem 1 Translate each sentence into an inequality.

(a) The height of the person must be no greater than 6 feet.

(b) The speed of the car must have been greater than 65 miles per hour.

(c) The area of the room must be greater than or equal to 560 square feet.

(d) The profit margin cannot be less than $50 per television set.

2 *Solving Applied Problems Using Inequalities*

We can use the same problem-solving techniques we used previously to solve word problems involving inequalities.

EXAMPLE 2 A manufacturing company makes 60-watt light bulbs. For every 1000 bulbs manufactured, a random sample of five bulbs is selected and tested. To meet quality control standards, the average number of hours these five bulbs last must be at least 900 hours. The test engineer tested five bulbs. Four of them lasted 850 hours, 1050 hours, 1000 hours, and 950 hours, respectively. How many hours must the fifth bulb last in order for the batch to meet quality control standards?

1. **Understand the problem.**

   Let $n =$ the number of hours the fifth bulb must last

   The average that the five bulbs last *must be at least* 900 hours. We could write

$$\text{Average} \geq 900 \text{ hours.}$$

2. **Write an inequality.**

$$\text{Average of five bulbs} \geq 900 \text{ hours.}$$

$$\frac{850 + 1050 + 1000 + 950 + n}{5} \geq 900$$

3. **Solve and state the answer.**

$$\frac{850 + 1050 + 1000 + 950 + n}{5} \geq 900$$

$$\frac{3850 + n}{5} \geq 900 \qquad \text{Simplify the numerator.}$$

$$3850 + n \geq 4500 \qquad \text{Multiply both sides by 5.}$$

$$n \geq 650 \qquad \text{Subtract 3850 from each side.}$$

   If the fifth bulb burns for 650 hours or more, the quality control standards will be met.

4. **Check.**

   Any value of 650 or larger should satisfy the original inequality. Suppose we pick $n = 650$ and substitute it into the inequality.

$$\frac{850 + 1050 + 1000 + 950 + 650}{5} \overset{?}{\geq} 900$$

$$\frac{4500}{5} \overset{?}{\geq} 900$$

$$900 \geq 900 \quad \checkmark$$

   If we try $n = 700$, the left side of the inequality will be 910. $910 \geq 900$. In fact, if we try any number greater than 650 for n, the left side of the inequality will be greater than 900. The inequality will be valid for $n = 650$ or any larger value.

Practice Problem 2 A manufacturing company makes coils of $\frac{1}{2}$-inch maritime rope. For every 300 coils of rope made, a random sample of four coils of rope is selected and tested. To meet quality control standards, the average number of pounds that each coil will hold must be at least 1100 pounds. The test engineer tested four coils of rope. The first three held 1050 pounds, 1250 pounds, and 950 pounds, respectively. How many pounds must the last coil hold if the sample is to meet quality control standards?

EXAMPLE 3 Juan is selling supplies for restaurants in the Southwest. He gets paid $700 per month plus 8% of all the supplies he sells to restaurants. If he wants to earn more than $3100 per month, what value of products will he need to sell?

1. **Understand the problem.**
   Juan earns a fixed salary and a salary that depends on how much he sells (commission salary) each month.

   Let x = the value of the supplies he sells each month.
   Then $0.08x$ = the amount of the commission salary he earns each month.

   He wants his total income to be more than $3100 per month; thus

   $$\text{total income} > 3100.$$

2. **Write the inequality.**

   The fixed income added to the commission income must be greater than 3100.

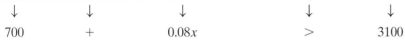

   | 700 | + | 0.08x | > | 3100 |

3. **Solve and state the answer.**

   $$700 + 0.08x > 3100$$
   $$0.08x > 2400 \quad \text{Subtract 700 from each side.}$$
   $$x > 30{,}000 \quad \text{Divide each side by 0.08.}$$

   Juan must sell more than $30,000 worth of supplies each month.

4. **Check.**
   Any value greater than 30,000 should satisfy the original inequality. Suppose we pick $x = 30{,}001$ and substitute it into the inequality.

   $$700 + 0.08(30{,}001) \overset{?}{>} 3100$$
   $$700 + 2400.08 \overset{?}{>} 3100$$
   $$3100.08 > 3100 \quad ✓$$

Practice Problem 3 Rita is selling commercial fire and theft alarm systems to local companies. She gets paid $1400 per month plus 2% of the cost of all the systems she sells in one month. She wants to earn more than $2200 per month. What value of products will she need to sell each month?

3.6 Exercises

Verbal and Writing Skills

Translate to an inequality.

1. The cost is greater than $67,000.

2. The clearance of the bridge is less than 12 feet.

3. The number of people is not more than 120.

4. The tax increase is more than $7890.

5. The perimeter is less than 34 miles.

6. The area is not greater than 550 square yards.

7. To earn an A in algebra, Ramon cannot get less than a 93 average.

8. The camp-ins at the Museum of Science in Boston cannot accommodate more than 600 people.

Applications *Solve using an inequality.*

▲ **9.** A triangular window on David Kirk's new Searay boat has two sides that measure 87 centimeters and 64 centimeters, respectively. The perimeter of the triangle must not exceed 291 centimeters. What are the possible values for the length of the third side of the window?

▲ **10.** The fencing around a farmer's triangular field needs to be installed. Two sides are 180 feet and 135 feet, respectively, and the farmer has 539 feet of fencing available. What are the possible values for the length of the third side of the field?

11. A video rental store needs to average 48 or more customers per day for six days each week in order to make a profit. On the first five days the following number of customers came into the store: 60, 36, 40, 52, and 59. What number of customers must come into the rental store on the last day in order for the store to make a profit?

12. Sophia has scored 85, 77, and 68 on three tests in her biology course. She has one final test to take and wants to obtain an average of 80 or more in the course. What possible values can she obtain on her final test and still achieve her goal?

▲ **13.** Joel plans to plant a flower garden in the front of his house, where he can fit a garden that is 8 feet wide. He wishes to use the can of wildflower mix that he got for his birthday. The can of seeds claims that it will luxuriously carpet an area of no more than 60 square feet. What are the possible dimensions for the depth of Joel's garden, so that it will be covered with wildflowers?

14. The Lee County Elementary School District noted that there was an increase of 1785 pupils during the fall semester. How many new teachers must they add to keep their average class size under 28? (Assume one teacher for each class.)

15. Suzie and Freddie decided that to save money, they would move themselves from their apartment to their new house. They had to rent a truck to do this. Because their budget was very strict, they could afford to spend no more than $150. The truck costs $37.50 per day and 19.5 cents per mile. If they decided they would need the truck for two days, what was the limit on the mileage they could put on the truck?

16. Robin just graduated from college and has taken her first job. She is working for a computer sales company, Compusell. While she hopes to eventually get into management, she is presently working as a salesperson. Her earnings consist of a monthly salary of $1300, plus 6% of all her sales. To meet her monthly obligations, Robin wants to earn at least $2800 per month. What is the amount of sales she must make monthly to meet her goal?

17. The coolant in Tom's car is effective for radiator temperatures of less than 110 degrees Celsius. What would be the corresponding temperature range in degrees Fahrenheit? (Use $C = \frac{5}{9}(F - 32)$ in your inequality.)

18. A person's IQ is found by multiplying mental age (m), as indicated by standard tests, by 100 and dividing this result by the chronological age (c). This gives the formula IQ $= 100\frac{m}{c}$. If a group of 14-year-old children has an IQ of not less than 80, what is the mental age range of this group?

19. The amount of federal aid in billions of dollars given to state and local governments can be approximated by the equation $A = 9.2x + 77.1$, where x is the number of years since 1980. Use this equation to determine those years for which the amount of federal aid to state and local governments will exceed $307.1 billion. (*Source:* U.S. Office of Management and Budget)

20. The number of people in millions who are receiving Social Security benefits can be approximated by the equation $N = 0.57x + 34.8$, where x is the number of years since 1980. Use this equation to determine those years for which the number of people receiving Social Security benefits will exceed 50.19 million. (*Source:* U.S. Social Security Administration)

21. Dr. Slater is purchasing computer tables and chairs for the tutoring center at Westmont College. The dean has told him that he cannot spend more than $850. Computer tables cost $55 and chairs cost $28. He has decided to purchase nine computer tables. How many chairs can he buy?

22. Franklin and Roberto are planning a pizza party for their college friends. They have decided they cannot spend more than $88 for food at their favorite pizza parlor. Each large pizza costs $11.50 and each six-pack of cold soda costs $3.50. They have decided to buy six large pizzas. How many six-packs of cold soda can they buy?

To Think About

23. A compact disc recording company's weekly costs can be described by the equation cost $= 5000 + 7n$, where n is the number of compact discs manufactured in one week. The equation for the amount of income produced from selling these n discs is income $= 18n$. How many discs need to be manufactured and sold in one week for the income to be greater than the cost?

24. A company that manufactures VCRs has weekly costs described by the equation cost $= 12{,}000 + 100n$, where n is the number of VCRs manufactured in one week. The equation for the amount of income produced from selling these n VCRs is income $= 210n$. How many VCRs need to be manufactured and sold in one week for the income to be greater than the cost?

Cumulative Review Problems

Solve.

25. $10 - 3x > 14 - 2x$ **26.** $2x - 3 \geq 7x - 18$ **27.** $30 - 2(x + 1) \leq 4x$ **28.** $2(x + 3) - 22 < 4(x - 2)$

29. Frank has a leaky faucet in his bathtub. He closed the drain and measured that the faucet drips 8 gallons of water in one 24-hour period.
   (a) If the faucet isn't fixed, how many gallons of water will be wasted in one year?
   (b) If the faucet isn't fixed and water costs $8.50 per 1000 gallons, how much extra money will Frank have to pay on his water bill for the year?

Math in the Media

Math Makes it to the Big Screen

Did you know that there are a surprising number of motion pictures that incorporate a math theme or a scene with real mathematics?

A popular Internet site that tracks and discusses such movies is the Math in the Movies Home Page. You can access it through a link on the Tobey/Slater Beginning Algebra book Web site.

One of the interesting uses of mathematics listed on the site involves the use of mathematics in the 1988 motion picture *Big* starring Tom Hanks. In that movie, Tom Hanks explains basic algebra to a child having trouble with his homework.

EXERCISES

Now that you have completed the chapter on solving applied problems, it's time for a break. Use your skills to research the following questions.

1. Can you find the algebra explanation that Tom Hanks provides in *Big*? If so, do you agree with his explanation? How would you answer the question?

2. Can you find an example of a movie incorporating a math theme related to the topics in beginning algebra? If so, what is it? What do you think of the portrayal of the math?

3. In the movie *Contact* starring Jodie Foster there is a detailed explanation of the number π. Is the explanation accurate? Is the use of the number π plausible if a person outside our solar system wished to communicate with us? Why?

Chapter 3 Organizer

Procedure for Solving Applied Problems

1. **Understand the problem.**
   (a) Read the word problem carefully to get an overview.
   (b) Determine what information you will need to solve the problem.
   (c) Draw a sketch. Label it with the known information. Determine what needs to be found.
   (d) Choose a variable to represent one unknown quantity.
   (e) If necessary, represent other unknown quantities in terms of that same variable.
2. **Write an equation.**
   (a) Look for key words to help you to translate the words into algebraic symbols.
   (b) Use a given relationship in the problem or an appropriate formula in order to write an equation.
3. **Solve and state the answer.**
4. **Check.**
   (a) Check the solution in the original equation. Is the answer reasonable?
   (b) Be sure the solution to the equation answers the question in the word problem. You may need to do some additional calculations if it does not.

Example A

The perimeter of a rectangle is 126 meters. The length of the rectangle is 6 meters less than double the width. Find the dimensions of the rectangle.

1. *Understand the problem.*
   We want to find the length and the width of a rectangle whose perimeter is 126 meters.

   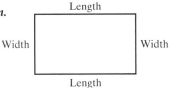

   The length is *compared to* the width, so we start with the width.

   Let w = width.

   The length is 6 meters less than double the width.

   Then $l = 2w - 6$.

2. *Write an equation.*
   The perimeter of a rectangle is $P = 2w + 2l$.

   $126 = 2w + 2(2w - 6)$

3. *Solve.*

   $126 = 2w + 4w - 12$
   $126 = 6w - 12$
   $138 = 6w$
   $23 = w$ The width is 23 meters.

   $2w - 6 = 2(23) - 6 =$
   $46 - 6 = 40$

   The length is 40 meters.

   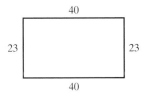

4. *Check:*
   Is this reasonable? Yes. A rectangle 23 meters wide and 40 meters long seems to be about right for the perimeter to be 126 meters.

   Is the perimeter exactly 126 meters?

   Is the length exactly 6 meters less than double the width?

   $2(23) + 2(40) \overset{?}{=} 126$
   $46 + 80 \overset{?}{=} 126$
   $126 = 126$ ✓

   $40 \overset{?}{=} 2(23) - 6$
   $40 \overset{?}{=} 46 - 6$
   $40 = 40$ ✓

Example B

Gina saved some money for college. She invested $2400 for one year and earned $225 in simple interest. She invested part of it at 12% and the rest of it at 9%. How much did she invest at each rate?

1. *Understand the problem.*
   We want to find each amount that was invested.

   interest earned at 12% + interest earned at 9% = total interest of $225

   We let x represent one quantity of money.

   Let x = amount of money invested at 12%.

   We started with $2400. If we invest x at 12%, we still have $(2400 - x)$ left.

   Then $2400 - x$ = the amount of money invested at 9%.

   Interest = prt
   Interest at 12% $I_1 = 0.12x$
   Interest at 9% $I_2 = 0.09(2400 - x)$

2. *Write an equation.*
   $I_1 + I_2 = 225$. $0.12x + 0.09(2400 - x) = 225$

3. *Solve.*

   $0.12x + 216 - 0.09x = 225$
   $0.03x + 216 = 225$
   $0.03x = 9$
   $x = \dfrac{9}{0.03}$
   $x = 300$

   $300 was invested at 12%.
   $2400 - x = 2400 - 300 = 2100$
   $2100 was invested at 9%.

4. *Check:*
   Is this reasonable? Yes.
   Are the conditions of the problem satisfied? Does the total amount invested equal $2400?

   $2100 + 300 \overset{?}{=} 2400$
   $2400 = 2400$ ✓

   Will $2100 at 9% and $300 at 12% yield $225 in interest?

   $0.09(2100) + 0.12(300) \overset{?}{=} 225$
   $189 + 36 \overset{?}{=} 225$
   $225 = 225$ ✓

235

Topic	Procedure	Examples
Geometric formulas, p. 218.	Know the definitions of and formulas related to basic geometric shapes such as the triangle, rectangle, parallelogram, trapezoid, and circle. Review these formulas from the beginning of Section 3.5 as needed. Formulas for the volume and total surface area of three solids are also given there.	The volume of a right circular cylinder found in a four-cylinder car engine is determined by the formula $V = \pi r^2 h$. Find V if $r = 6$ centimeters and $h = 5$ centimeters. Use $\pi = 3.14$. $$\begin{aligned} V &= 3.14(6 \text{ cm})^2(5 \text{ cm}) \\ &= 3.14(36)(5) \text{ cm}^3 \\ &= 565.2 \text{ cc (or cm}^3) \end{aligned}$$ Four such cylinders would make this a 2260.8-cc engine.
Using inequalities to solve word problems, p. 229.	Be sure that the inequality you use fits the conditions of the word problem. **1.** *At least 87* means ≥ 87. **2.** *No less than $150* means $\geq \$150$. **3.** *At most $500* means $\leq \$500$. **4.** *No more than 70* means ≤ 70. **5.** *Less than 2* means < 2. **6.** *More than 10* means > 10.	Julie earns 2% on sales over $1000. How much must Julie make in sales to earn at least $300? Let $x =$ amount of sales. $$\begin{aligned} (2\%)(x - \$1000) &\geq \$300 \\ 0.02(x - 1000) &\geq 300 \\ x - 1000 &\geq 15{,}000 \\ x &\geq 16{,}000 \end{aligned}$$ Julie must make sales of $16,000 or more (at least $16,000).

Chapter 3 Review Problems

3.1 *Write an algebraic expression. Use the variable x to represent the unknown value.*

1. 19 more than a number

2. two-thirds of a number

3. 56 less than a number

4. triple a number

5. double a number increased by seven

6. twice a number decreased by three

Write an expression for each of the quantities compared. Use the letter specified.

7. The number of working people is four times the number of retired people. The number of unemployed people is one-half the number of retired people. (Use the letter r.)

▲ **8.** The length of the rectangle is 5 meters more than triple the width. (Use the letter w.)

▲ **9.** A triangle has three angles A, B, and C. The number of degrees in angle A is double the number of degrees in angle B. The number of degrees in angle C is 17 degrees less than the number in angle B. (Use the letter b.)

10. There are 29 more students in biology class than in algebra. There are one-half as many students in geology class as in algebra. (Use the letter a.)

3.2 *Solve.*

11. Triple a number is decreased by seven. The result is 47. What is the original number?

12. Twice a number is increased by five. The result is 29. What is the original number?

13. Six chairs and a table cost $450. If the table costs $210, how much does one chair cost?

14. Jon is twice as old as his cousin David. If Jon is 32, how old is David?

15. Two cars left Center City and drove to the shore 330 miles away. One car was driven at 50 miles an hour. The other car was driven at 55 miles an hour. How long did it take each car to make the trip?

16. A student took five math tests. She obtained scores of 85, 78, 65, 92, and 70. She needs to complete one more test. If she wants to get an average of 80 on all six tests, what score does she need on her last test?

3.3

▲ **17.** The perimeter of a triangle is 40 yards. The second side is 1 yard shorter than double the length of the first side. The third side is 9 yards longer than the first side. Find the length of each side.

▲ **18.** The measure of the second angle of a triangle is three times the measure of the first angle. The measure of the third angle is 12 degrees less than twice the measure of the first. Find the measure of each of the three angles.

19. A piece of rope 50 yards long is cut into two pieces. One piece is three-fifths as long as the other. Find the length of each piece.

20. Jon and Lauren have a combined income of $48,000. Lauren earns $3000 more than half of what Jon earns. How much does each earn?

3.4

21. The electric bill at Jane's house this month was $71.50. The charge is based on a flat rate of $25 per month plus a charge of $0.15 per kilowatt-hour of electricity used. How many kilowatt-hours of electricity were used?

22. Abel rented a car from Sunshine Car Rentals. He was charged $39 per day and $0.25 per mile. He rented the car for three days and paid a rental cost of $187. How many miles did he drive the car?

23. Alfred has worked for his company for 10 years as a store manager. Last year he received a 9% raise in salary. He now earns $24,525 per year. What was his salary last year before the raise?

24. Jamie bought a new tape deck for her car. When she bought it, she found that the price had been decreased by 18%. She was able to buy the tape deck for $36 less than the original price. What was the original price?

25. Peter and Shelly invested $9000 for one year. Part of it was invested at 12% and the remainder was invested at 8%. At the end of one year the couple had earned exactly $1000 in simple interest. How much did they invest at each rate?

26. A man invests $5000 in a savings bank. He places part of it in a checking account that earns 4.5% and the rest in a regular savings account that earns 6%. His total annual income from this investment in simple interest is $270. How much was invested in each account?

27. Mary has $3.75 in nickels, dimes, and quarters. She has three more quarters than dimes. She has twice as many nickels as quarters. How many of each coin does she have?

28. Tim brought in $3.65 in coins. He had two more quarters than he had dimes. He had one fewer nickel than the number of dimes. How many coins of each type did he have?

3.5 *Solve.*

▲ **29.** Find the diameter of a circle whose circumference is 25.12 meters. (Use $\pi \approx 3.14$.)

▲ **30.** What is the perimeter of a rectangle whose length is 8 inches and whose width is $\frac{1}{4}$ inch?

▲ **31.** Find the third angle of a triangle if one angle measures 62 degrees and the second angle measures 47 degrees.

▲ **32.** Find the area of a triangle whose base is 8 miles and whose altitude is 10.5 miles.

▲ **33.** Find the volume of a rectangular solid whose width is 8 feet, whose length is 12 feet, and whose height is 20 feet.

▲ **34.** Find the volume of a sphere whose radius is 3 centimeters. (Use $\pi \approx 3.14$.)

▲ **35.** Find the area of a circle whose diameter is 18 centimeters. (Use $\pi \approx 3.14$.)

▲ **36.** The perimeter of an isosceles triangle is 46 inches. The two equal sides are each 17 inches long. How long is the third side?

▲ **37.** Jim wants to have the exterior siding on his house painted. Two sides of the house have a height of 18 feet and a length of 22 feet, and the other two sides of the house have a height of 18 feet and a length of 19 feet. The painter has quoted a price of $2.50 per square foot. The windows and doors of the house take up 100 square feet. How much will the painter charge to paint the siding?

▲ **38.** Find the cost to build a cylinder out of aluminum if the radius is 5 meters, the height is 14 meters, and the cylinder will have a flat top and flat bottom. The cost factor has been determined to be $40.00 per square meter. (*Hint:* First find the total surface area.) (Use $\pi \approx 3.14$.)

3.6 *Solve.*

39. Chelsea rents a car for $35 a day and 15 cents a mile. How far can she drive if she wants to rent the car for two days and does not want to spend more than $100?

40. Philip earns $12,000 per year in salary and 4% commission on his sales. How much were his sales if his annual income was more than $24,000?

41. Two cars start from the same point going in opposite directions. One car travels at 50 miles per hour. The other car travels at 55 miles per hour. How long must they travel to be at least 315 miles apart?

42. Michael plans to spend less than $70 on shirts and ties. He bought three shirts for $17.95 each. How much can he spend on ties?

Miscellaneous *Solve.*

43. Wally timed a popular "30-minute" TV show. The show lasted 4 minutes longer than four times the amount of time that was devoted to commercials. How many minutes long was the actual show?

44. Faye conducted an experiment. To verify the weight of four identical steel balls, she placed them on a balance. She placed three of the balls on one side of the balance. On the other side she placed an 8-ounce weight and one of the balls. She discovered that these two sides exactly balanced on the scale. How much does each steel ball weigh?

45. Fred and Nancy Sullivan are planning to purchase a new refrigerator. The most efficient model costs $870, but the electricity to operate it would cost only $41 per year. The less efficient model is the same size. It costs $744, but the electricity to operate it would cost $83 per year. How long will it be before the total cost is greater for the less efficient model?

46. Custom Computer Works of Gloucester rents its office building and warehouse for $1800 per month. The owner is thinking of buying the building. He could purchase it with a down payment of $42,000. He would then make monthly mortgage payments of $1100. How many months would it take for his total costs to be less to purchase the building rather than rent it?

▲ 47. Judy Carter is planning to build a bookcase for her home office. She has only 38 feet of lumber to make the unit. The bookcase is constructed of shelves as shown in the diagram. She wants the width of the bookcase to be 2.5 times the height. Find the width and the height of the bookcase she needs to construct.

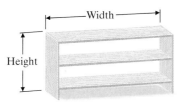

48. Rick Ponticelli is investing some money in a retirement plan. He plans to invest $100 a month the first year, $200 a month the second year, $300 a month the third year, and continue with that pattern. How many years will it be before he has invested $66,000 in his retirement plan?

49. In a recent basketball game the Boston Celtics scored 99 points. They scored three times as many field goals as free throws. They also made 12 three-point baskets. How many field goals did they make? How many free throws did they make? (A field goal is worth two points and a free throw is worth one point.)

▲ 50. The perimeter of a rectangular room in Sally's house is 76 feet. The length of the room is 10 feet less than double the width of the room. Find the dimensions of the room.

51. Michael and Scotty decided to split the cost of a large pizza from Papa John's based on how many slices they ate. Michael ate three slices and Scotty ate five slices. They had no pizza left over. The pizza cost $16. How much should each person pay for his share of the pizza?

52. Mr. and Mrs. Traicoff invested $12,000 in mutual funds. Part of it was invested in high-tech funds that earned 7% simple interest over the course of a year. The rest was invested in bond funds that earned 5% simple interest during the year. If the total amount of interest earned in a year was $780, how much did they invest in each mutual fund?

▲ 53. The first angle of a triangle measures 20° larger than the second angle of a triangle. The third angle measures twice as large as the second angle. Find the measure of these three angles.

▲ 54. A sign is made in the shape of a trapezoid. The altitude of the trapezoid is 12 feet. The two bases of the trapezoid are 15 feet and 19 feet. The sign is constructed of prime cedar that costs $6 per square foot. How much did the cedar that was used to make the sign cost?

55. Thelma has a math final exam that counts twice as much as a test. She had four tests with grades of 88, 77, 95, and 92. What score does she need to obtain on the final exam to get an average of 90 for the course?

56. Barbara drove her car 27,000 miles last year. She drove twice as much for pleasure as she did for business. She can deduct $0.33 per mile for all business miles. Last year she claimed a business mileage deduction of $2000. Was her deduction correct? Please explain your answer. If her deduction was not correct, please indicate the correct deduction.

57. A jet plane travels 32 miles in three minutes. How far can the plane travel in one hour if it continues at that speed? How many minutes would it take the jet to travel from Boston to Denver (approximately 1800 miles) at that same speed?

58. A running back in a Wheaton College football game catches a pass and runs for the goal line 80 yards away. He can run a 50-yard dash in six seconds. If he runs at 80% of his speed in the 50-yard dash, how long will it take him teach to reach the goal line?

59. A Ford Explorer can hold a driver and four passengers. A Dodge Caravan can hold a driver and six passengers. The biology department is taking a total of 95 people on a field trip. A total of 20 of the 95 people are available to drive. Not all potential drivers will be needed. The college wants to use twice as many Caravans as Explorers because the Caravans get better gas mileage. How many of each kind of vehicle are needed?

Putting Your Skills to Work

Using Mathematics for Search and Rescue

The Coast Guard employs modern computer methods to determine the time required to reach the area where help is needed. These search-and-rescue questions are determined by mathematical relationships. Answering important questions regarding rescue operations requires that a person solve equations by hand or by using a computer.

Problems for Individual Study and Investigation

A sinking boat reports that it is 100 miles from the Coast Guard rescue station. The boat is sinking and will stay afloat for only one more hour. It is slowly heading toward the rescue station at 5 miles per hour. A rescue helicopter flying at 120 miles per hour leaves the station five minutes after the call comes in.

1. Let t = the time elapsed from the time of the call to the time of the helicopter reaching the ship. Write two equations using $d = rt$ to describe the distance traveled by the sinking boat and the distance traveled by the helicopter.

2. Write an equation using the two distance expressions in question 1 and the total distance of 100 miles. Solve this equation for t. Will the helicopter reach the boat before it sinks?

Problems for Cooperative Group Investigation

Together with other members of your class, determine an answer to the following.

3. A sinking boat is 76 miles from a Coast Guard station. The boat is traveling toward the Coast Guard station battling heavy seas. It is able to move toward the station at 2 miles per hour. At the same time a rescue helicopter leaves the Coast Guard station and flies directly toward the boat at 150 miles per hour. If the boat does not sink, how long will it take the helicopter to reach the boat?

4. When the helicopter arrives, the boat has already sunk. The Coast Guard orders a search by eight rescue helicopters. The search area is defined by the trapezoid shown. How many square miles are in the search area? If a helicopter can search 15 square miles in an hour, how long will it take the eight helicopters to search the entire area?

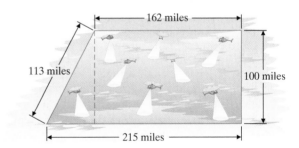

Internet Connections

  Netsite: http://www.prenhall.com/tobey_beginning

Site: U.S. Coast Guard Air Station, San Francisco, or a related site

This site gives general information about the Dolphin helicopters used by the U.S. Coast Guard.

5. Suppose the crew of the Dolphin needs to rescue someone who is 150 miles away from the air station. Assume that they need to stay at the rescue location for 30 minutes. How long will the complete rescue operation take, including round-trip travel?

6. Explain why it might not be practical for the Dolphin crew to rescue someone who is 200 nautical miles away, even though the range of the helicopter is 400 nautical miles.

Solve.

1. A number is doubled and then decreased by 11. The result is 59. What is the original number?

2. The sum of one-half of a number, one-ninth of the number, and one-twelfth of the number is twenty-five. Find the original number.

3. Triple a number is increased by six. The result is the same as when the original number was diminished by three and then doubled. Find the original number.

▲ 4. A triangular region has a perimeter of 66 meters. The first side is two-thirds of the second side. The third side is 14 meters shorter than the second side. What are the lengths of the three sides of the triangular region?

▲ 5. A rectangle has a length 7 meters longer than double the width. The perimeter is 134 meters. Find the dimensions of the rectangle.

6. Three harmful pollutants were measured by a consumer group in the city. The total sample contained 15 parts per million of three harmful pollutants. The amount of the first pollutant was double the second. The amount of the third pollutant was 75% of the second. How many parts per million of each pollutant were found?

7. Raymond has a budget of $1940 to rent a computer for his company office. The computer company he wants to rent from charges $200 for installation and service as a one-time fee. Then they charge $116 per month rental for the computer. How many months will Raymond be able to rent a computer with this budget?

8. Last year the yearly tuition at Westmont College went up 8%. This year's charge for tuition for the year is $9180. What was it last year before the increase went into effect?

9. Franco invested $4000 in money market funds. Part was invested at 14% interest, the rest at 11% interest. At the end of each year the fund company pays interest. After one year he earned $482 in simple interest. How much was invested at each interest rate?

10. Mary has $3.50 in change. She has twice as many nickels as quarters. She has one less dime than she has quarters. How many of each coin does she have?

1. _____

2. _____

3. _____

4. _____

5. _____

6. _____

7. _____

8. _____

9. _____

10. _____

11. Find the circumference of a circle with radius 34 inches. Use $\pi = 3.14$ as an approximation.

12. Find the area of a trapezoid if the two bases are 10 inches and 14 inches and the altitude is 16 inches.

13. Find the volume of a sphere with radius 10 inches. Use $\pi = 3.14$ as an approximation and round your answer to the nearest cubic inch.

14. Find the area of a parallelogram with a base of 12 centimeters and an altitude of 8 centimeters.

15. How much would it cost to carpet the area shown in the figure if carpeting costs $12 per square yard?

16. Cathy scored 76, 84, and 78 on three tests. How much must she score on the next test to have at least an 80 average?

17. Carol earns $15,000 plus 5% commission on sales over $10,000. How much must she make in sales to earn more than $20,000?

18. Andrew is putting down tile for a new gymnasium. He can lay 25 square feet of tile in an hour. How long will it take him to do a gymnasium 150 feet long and 100 feet wide?

Approximately one-half of this test covers the content of Chapters 0–2. The remainder covers the content of Chapter 3.

1. Divide.

$2.4\overline{)8.856}$

2. Add.

$$\frac{3}{8} + \frac{5}{12} + \frac{1}{2}$$

3. Simplify.
   $3(2x - 3y + 6) + 2(-3x - 7y)$

4. Evaluate.
   $12 - 3(2 - 4) + 12 \div 4$

5. Solve for b.
   $$H = \frac{1}{2}(3a + 5b)$$

6. Solve for x and graph your solution.
   $$5x - 3 \le 2(4x + 1) + 4$$

7. A number is first tripled and then decreased by 17. The result is 103. Find the original number.

8. The number of students in the psychology class is 34 fewer than the number of students in world history. The total enrollment for the two classes is 134 students. How many students are in each class?

▲ **9.** A rectangle has a perimeter of 78 centimeters. The length of the rectangle is 11 centimeters longer than triple the width. Find the dimensions of the rectangle.

10. This year the sales for Homegrown Video are 35% higher than last year. This year the sales are $182,250. What were the sales last year?

11. Hassan invested $7000 for one year. He invested part of it in a high-risk fund that pays 15% interest. He placed the rest in a safe, low-risk fund that pays 7% interest. At the end of one year he earned $730 in simple interest. How much did he invest in each of the two funds?

12. Linda has some dimes, nickels, and quarters in her purse. The total value of these coins is $2.55. She has three times as many dimes as quarters. She has three more nickels than quarters. How many coins of each type does she have?

▲ **13.** Find the area of a triangle with an altitude of 13 meters and a base of 25 meters. What is the cost to construct such a triangle out of sheet metal that costs $4.50 per square meter?

▲ **14.** Find the volume of a sphere that has a radius of 3.00 inches. How much will the contents of the sphere weigh if it is filled with a liquid that weighs 1.50 pounds per cubic inch? Use $\pi = 3.14$ as an approximation.

15. Melinda Tobey saw a glacier in Alaska that moves 65 feet per day. How long will it take the glacier to reach a cabin that is 1625 feet away?

1. _____

2. _____

3. _____

4. _____

5. _____

6. _____

7. _____

8. _____

9. _____

10. _____

11. _____

12. _____

13. _____

14. _____

15. _____

Exponents and Polynomials

Polynomials are an important tool of mathematics. They are used to create formulas that model many real-world phenomena. For example, using records of past student enrollments in the United States, one can create a polynomial that predicts future student enrollments. Do you think you could use your mathematical skills to predict the number of students who will be in school in the year 2009? Turn to the Putting Your Skills to Work problems on page 291 to find out.

Pretest Chapter 4

1. _____

2. _____

3. _____

4. _____

5. _____

6. _____

7. _____

8. _____

9. _____

10. _____

11. _____

12. _____

13. _____

14. _____

15. _____

16. _____

Section 4.1

Multiply. Leave your answer in exponent form.

1. $(3^6)(3^{10})$

2. $(-5x^3)(2x^4)$

3. $(-4ab^2)(2a^3b)(3ab)$

Simplify. Assume that all variables are nonzero.

4. $\dfrac{x^{26}}{x^3}$

5. $\dfrac{12xy^2}{-6x^3y^4}$

6. $\dfrac{-25a^0b^2c^3}{-15abc^2}$

7. $\left(x^5\right)^{10}$

8. $\left(-2x^2y\right)^3$

9. $\left(\dfrac{4a^3b}{c^2}\right)^3$

Section 4.2

In questions 10–12, express with positive exponents.

10. $3x^2y^{-3}z^{-4}$

11. $\dfrac{-2a^{-3}}{b^{-2}c^4}$

12. Evaluate. $(-3)^{-2}$

13. Write in scientific notation. 0.000638

14. Write in decimal notation. 1.894×10^{12}

Section 4.3

Combine.

15. $(3x^2 - 6x - 9) + (-5x^2 + 13x - 20)$

16. $(2x^3 - 5x^2 + 7x) - (x^3 - 4x + 12)$

Section 4.4

Multiply.

17. $-3x(2 - 7x)$

18. $2x^2(x^2 + 4x - 1)$

19. $(x^3 - 3xy + y^2)(5xy^2)$

20. $(x + 5)(x + 9)$

21. $(3x - 2y)(4x + 3y)$

22. $(2x^2 - y^2)(x^2 - 3y^2)$

Section 4.5

Multiply.

23. $(8x - 11y)(8x + 11y)$

24. $(5x - 3y)^2$

25. $(5ab - 6)^2$

26. $(x^2 - 3x + 2)(4x - 1)$

27. $(2x - 3)(2x + 3)(4x^2 + 9)$

Section 4.6

Divide.

28. $(42x^4 - 36x^3 + 66x^2) \div (6x)$

29. $(15x^3 - 28x^2 + 18x - 7) \div (3x - 2)$

17.

18.

19.

20.

21.

22.

23.

24.

25.

26.

27.

28.

29.

Student Learning Objectives

After studying this section, you will be able to:

1 Multiply exponential expressions with like bases.

2 Divide exponential expressions with like bases.

3 Raise exponential expressions to a power.

SSM
PH TUTOR CENTER
CD & VIDEO
MATH PRO
WEB

1 *Multiplying Exponential Expressions with Like Bases*

Recall that x^2 means $x \cdot x$. That is, x appears as a factor two times. The 2 is called the **exponent**. The **base** is the variable x. The expression x^2 is called an **exponential expression**. What happens when we multiply $x^2 \cdot x^2$? Is there a pattern that will help us form a general rule?

Notice that

$$\overbrace{(2^2)(2^3) = (2 \cdot 2)(2 \cdot 2 \cdot 2)}^{5 \text{ twos}} = 2^5$$

The exponent means 2 occurs 5 times as a factor.

$$\overbrace{(3^3)(3^4) = (3 \cdot 3 \cdot 3)(3 \cdot 3 \cdot 3 \cdot 3)}^{7 \text{ threes}} = 3^7$$

Notice that $3 + 4 = 7$.

$$\overbrace{(x^3)(x^5) = (x \cdot x \cdot x)(x \cdot x \cdot x \cdot x \cdot x)}^{8 \; x\text{'s}} = x^8$$

The sum of the exponents is $3 + 5 = 8$.

$$\overbrace{(y^4)(y^2) = (y \cdot y \cdot y \cdot y)(y \cdot y)}^{6 \; y\text{'s}} = y^6$$

The sum of the exponents is $4 + 2 = 6$.

We can state the pattern in words and then use variables.

The Product Rule

To multiply two exponential expressions that have the same base, keep the base and *add the exponents*.

$$x^a \cdot x^b = x^{a+b}$$

Be sure to notice that this rule applies only to expressions that have the *same base*. Here x represents the base, while the letters a and b represent the exponents that are added.

It is important that you apply this rule even when an exponent is 1. Every variable that does not have a written exponent is understood to have an exponent of 1. Thus $x^1 = x$, $y^1 = y$, and so on.

EXAMPLE 1 Multiply. **(a)** $x^3 \cdot x^6$ **(b)** $x \cdot x^5$

(a) $x^3 \cdot x^6 = x^{3+6} = x^9$

(b) $x \cdot x^5 = x^{1+5} = x^6$ Note that the exponent of the first x is 1.

Practice Problem 1 Multiply. **(a)** $a^7 \cdot a^5$ **(b)** $w^{10} \cdot w$

EXAMPLE 2 Simplify. **(a)** $y^5 \cdot y^{11}$ **(b)** $2^3 \cdot 2^5$ **(c)** $x^6 \cdot y^8$

(a) $y^5 \cdot y^{11} = y^{5+11} = y^{16}$

(b) $2^3 \cdot 2^5 = 2^{3+5} = 2^8$ Note that the base does not change! Only the exponent changes.

(c) $x^6 \cdot y^8$ The rule for multiplying exponential expressions does not apply since the bases are not the same. This cannot be simplified.

Practice Problem 2 Simplify, if possible.

(a) $x^3 \cdot x^9$ **(b)** $3^7 \cdot 3^4$ **(c)** $a^3 \cdot b^2$

We can now look at multiplying expressions such as

$$(2x^5)(3x^6).$$

The number 2 in $2x^5$ is called the **numerical coefficient**. Recall that a numerical coefficient is a number that is multiplied by a variable. When we multiply two expressions such as $2x^5$ and $3x^6$, we first multiply the numerical coefficients; we multiply the variables with exponents separately.

EXAMPLE 3   Multiply.   **(a)** $(2x^5)(3x^6)$ **(b)** $(5x^3)(x^6)$ **(c)** $(-6x)(-4x^5)$

(a) $(2x^5)(3x^6) = (2 \cdot 3)(x^5 \cdot x^6)$ Multiply the numerical coefficients.

$\qquad\qquad\quad = 6(x^5 \cdot x^6)$ Use the rule for multiplying expressions with exponents. Add the exponents.

$\qquad\qquad\quad = 6x^{11}$

(b) Every variable that does not have a visible numerical coefficient is understood to have a numerical coefficient of 1. Thus x^6 has a numerical coefficient of 1.

$$(5x^3)(x^6) = (5 \cdot 1)(x^3 \cdot x^6) = 5x^9$$

(c) $(-6x)(-4x^5) = (-6)(-4)(x^1 \cdot x^5) = 24x^6$ Remember that x has an exponent of 1.

Practice Problem 3   Multiply.

(a) $(-a^8)(a^4)$ **(b)** $(3y^2)(-2y^3)$ **(c)** $(-4x^3)(-5x^2)$

Problems of this type may involve more than one variable or more than two factors.

EXAMPLE 4   Multiply. $(5ab)(-\frac{1}{3}a)(9b^2)$

$$(5ab)(-\tfrac{1}{3}a)(9b^2) = (5)(-\tfrac{1}{3})(9)(a \cdot a)(b \cdot b^2)$$
$$= -15a^2b^3$$

As you do the following problems, keep in mind the rule for multiplying numbers with exponents and the rules for multiplying signed numbers.

Practice Problem 4   Multiply.

$(2xy)(-\frac{1}{4}x^2y)(6xy^3)$

2   *Dividing Exponential Expressions with Like Bases*

Frequently, we must divide exponential expressions. Since division by zero is undefined, in all problems in this chapter we assume that the denominator of any variable expression is not zero. We'll look at division in three separate parts.

Suppose that we want to simplify $x^5 \div x^2$. We could do the division the long way.

$$\frac{x^5}{x^2} = \frac{(x)(x)(x)\cancel{(x)}\cancel{(x)}}{\cancel{(x)}\cancel{(x)}} = x^3$$

Here we are using the arithmetical property of reducing fractions (see Section 0.1). When the same factor appears in both numerator and denominator, that factor can be removed.

A simpler way is to *subtract the exponents*. Notice that the base remains the same.

The Quotient Rule (Part 1)

$$\frac{x^a}{x^b} = x^{a-b}$$ Use this form if the larger exponent is in the numerator and $x \neq 0$.

EXAMPLE 5 Divide. **(a)** $\dfrac{2^{16}}{2^{11}}$ **(b)** $\dfrac{x^5}{x^3}$ **(c)** $\dfrac{y^{16}}{y^7}$

(a) $\dfrac{2^{16}}{2^{11}} = 2^{16-11} = 2^5$ Note that the base does *not* change.

(b) $\dfrac{x^5}{x^3} = x^{5-3} = x^2$ **(c)** $\dfrac{y^{16}}{y^7} = y^{16-7} = y^9$

Practice Problem 5 Divide. **(a)** $\dfrac{10^{13}}{10^7}$ **(b)** $\dfrac{x^{11}}{x}$

Now we consider the situation where the larger exponent is in the denominator. Suppose that we want to simplify $x^2 \div x^5$.

$$\frac{x^2}{x^5} = \frac{\cancel{(x)}\cancel{(x)}}{\cancel{(x)}\cancel{(x)}(x)(x)(x)} = \frac{1}{x^3}$$

The Quotient Rule (Part 2)

$\dfrac{x^a}{x^b} = \dfrac{1}{x^{b-a}}$ Use this form if the larger exponent is in the denominator and $x \neq 0$.

EXAMPLE 6 Divide. **(a)** $\dfrac{12^{17}}{12^{20}}$ **(b)** $\dfrac{b^7}{b^9}$ **(c)** $\dfrac{x^{20}}{x^{24}}$

(a) $\dfrac{12^{17}}{12^{20}} = \dfrac{1}{12^{20-17}} = \dfrac{1}{12^3}$ Note that the base does *not* change.

(b) $\dfrac{b^7}{b^9} = \dfrac{1}{b^{9-7}} = \dfrac{1}{b^2}$ **(c)** $\dfrac{x^{20}}{x^{24}} = \dfrac{1}{x^{24-20}} = \dfrac{1}{x^4}$

Practice Problem 6 Divide. **(a)** $\dfrac{c^3}{c^4}$ **(b)** $\dfrac{10^{31}}{10^{56}}$

When there are numerical coefficients, use the rules for dividing signed numbers to reduce fractions to lowest terms.

EXAMPLE 7 Divide. **(a)** $\dfrac{5x^5}{25x^7}$ **(b)** $\dfrac{-12x^8}{4x^3}$ **(c)** $\dfrac{-16x^7}{-24x^8}$

(a) $\dfrac{5x^5}{25x^7} = \dfrac{1}{5x^{7-5}} = \dfrac{1}{5x^2}$ **(b)** $\dfrac{-12x^8}{4x^3} = -3x^{8-3} = -3x^5$ **(c)** $\dfrac{-16x^7}{-24x^8} = \dfrac{2}{3x^{8-7}} = \dfrac{2}{3x}$

Practice Problem 7 Divide. **(a)** $\dfrac{-7x^7}{-21x^9}$ **(b)** $\dfrac{15x^{11}}{-3x^4}$

You have to work very carefully if two or more variables are involved. Treat the coefficients and each variable separately.

EXAMPLE 8 Divide. **(a)** $\dfrac{x^3y^2}{5xy^6}$ **(b)** $\dfrac{-3x^2y^5}{12x^6y^8}$

(a) $\dfrac{x^3y^2}{5xy^6} = \dfrac{x^2}{5y^4}$ **(b)** $\dfrac{-3x^2y^5}{12x^6y^8} = -\dfrac{1}{4x^4y^3}$

Practice Problem 8 Divide. **(a)** $\dfrac{x^7 y^9}{y^{10}}$ **(b)** $\dfrac{12x^5 y^6}{-24x^3 y^8}$

Suppose that a given base appears with the same exponent in the numerator and denominator of a fraction. In this case we can use the fact that *any nonzero number divided by itself is* 1.

EXAMPLE 9 Divide. **(a)** $\dfrac{x^6}{x^6}$ **(b)** $\dfrac{3x^5}{x^5}$

(a) $\dfrac{x^6}{x^6} = 1$ **(b)** $\dfrac{3x^5}{x^5} = 3\left(\dfrac{x^5}{x^5}\right) = 3(1) = 3$

Practice Problem 9 Divide. **(a)** $\dfrac{10^7}{10^7}$ **(b)** $\dfrac{12a^4}{15a^4}$

Do you see that if we had subtracted exponents when simplifying $\dfrac{x^6}{x^6}$ we would have obtained x^0 in Example 9? So we can surmise that any number (except 0) to the 0 power equals 1. We can write this fact as a separate rule.

The Quotient Rule (Part 3)

$$\frac{x^a}{x^a} = x^0 = 1 \quad \text{if } x \neq 0 \quad (0^0 \text{ remains undefined}).$$

To Think About What about 0^0? Why is it undefined? $0^0 = 0^{1-1}$. If we use the quotient rule, $0^{1-1} = \dfrac{0}{0}$. Since division by zero is undefined, we must agree that 0^0 is undefined.

EXAMPLE 10 Divide. **(a)** $\dfrac{4x^0 y^2}{8^0 y^5 z^3}$ **(b)** $\dfrac{5x^2 y}{10x^2 y^3}$

(a) $\dfrac{4x^0 y^2}{8^0 y^5 z^3} = \dfrac{4(1)y^2}{(1)y^5 z^3} = \dfrac{4y^2}{y^5 z^3} = \dfrac{4}{y^3 z^3}$ **(b)** $\dfrac{5x^2 y}{10x^2 y^3} = \dfrac{1x^0}{2y^2} = \dfrac{(1)(1)}{2y^2} = \dfrac{1}{2y^2}$

Practice Problem 10 Divide. $\dfrac{-20a^3 b^8 c^4}{28a^3 b^7 c^5}$

We can combine all three parts of the quotient rule we have developed.

The Quotient Rule

$\dfrac{x^a}{x^b} = x^{a-b}$ Use this form if the larger exponent is in the numerator and $x \neq 0$.

$\dfrac{x^a}{x^b} = \dfrac{1}{x^{b-a}}$ Use this form if the larger exponent is in the denominator and $x \neq 0$.

$\dfrac{x^a}{x^a} = x^0 = 1$ if $x \neq 0$.

We can combine the product rule and the quotient rule to simplify algebraic expressions that involve both multiplication and division.

EXAMPLE 11   Simplify. $\dfrac{(8x^2y)(-3x^3y^2)}{-6x^4y^3}$

$$\frac{(8x^2y)(-3x^3y^2)}{-6x^4y^3} = \frac{-24x^5y^3}{-6x^4y^3} = 4x$$

Practice Problem 11   Simplify. $\dfrac{(-6ab^5)(3a^2b^4)}{16a^5b^7}$

3 **Raising Exponential Expressions to a Power**

How do we simplify an expression such as $(x^4)^3$? $(x^4)^3$ is x^4 raised to the third power. For this type of problem we say that we are raising a power to a power. A problem such as $(x^4)^3$ could be done by writing the following.

$$(x^4)^3 = x^4 \cdot x^4 \cdot x^4 \quad \text{By definition}$$
$$= x^{12} \quad \text{By adding exponents}$$

Notice that when we add the exponents we get $4 + 4 + 4 = 12$. This is the same as multiplying 4 by 3. That is, $4 \cdot 3 = 12$. This process can be summarized by the following rule.

Raising a Power to a Power

To raise a power to a power, keep the same base and multiply the exponents.

$$(x^a)^b = x^{ab}$$

Recall what happens when you raise a negative number to a power. $(-1)^2 = 1$. $(-1)^3 = -1$. In general,

$$(-1)^n = \begin{cases} +1 & \text{if } n \text{ is even} \\ -1 & \text{if } n \text{ is odd.} \end{cases}$$

EXAMPLE 12   Simplify.   **(a)** $(x^3)^5$ **(b)** $(2^7)^3$ **(c)** $(-1)^8$

(a) $(x^3)^5 = x^{3 \cdot 5} = x^{15}$ **(b)** $(2^7)^3 = 2^{7 \cdot 3} = 2^{21}$ **(c)** $(-1)^8 = +1$

Note that in both parts (a) and (b) the base does not change.

Practice Problem 12   Simplify.

(a) $(a^4)^3$ **(b)** $(10^5)^2$ **(c)** $(-1)^{15}$

Here are two rules involving products and quotients that are very useful. We'll illustrate each with an example.

If a product in parentheses is raised to a power, the parentheses indicate that *each factor* must be raised to that power.

$$(xy)^2 = x^2y^2 \qquad (xy)^3 = x^3y^3$$

$$(xy)^a = x^ay^a$$

EXAMPLE 13   Simplify.   **(a)** $(ab)^8$ **(b)** $(3x)^4$ **(c)** $(-2x^2)^3$

(a) $(ab)^8 = a^8b^8$ **(b)** $(3x)^4 = (3)^4x^4 = 81x^4$ **(c)** $(-2x^2)^3 = (-2)^3 \cdot (x^2)^3 = -8x^6$

Practice Problem 13   Simplify.

(a) $(3xy)^3$ **(b)** $(yz)^{37}$

If a fractional expression within parentheses is raised to a power, the parentheses indicate that both numerator and denominator must be raised to that power.

$$\left(\frac{x}{y}\right)^5 = \frac{x^5}{y^5} \qquad \left(\frac{x}{y}\right)^2 = \frac{x^2}{y^2} \qquad \text{if } y \neq 0$$

$$\left(\frac{x}{y}\right)^a = \frac{x^a}{y^a} \qquad \text{if } y \neq 0.$$

● EXAMPLE 14 Simplify. **(a)** $\left(\dfrac{x}{y}\right)^5$ **(b)** $\left(\dfrac{7}{w}\right)^4$

(a) $\left(\dfrac{x}{y}\right)^5 = \dfrac{x^5}{y^5}$ **(b)** $\left(\dfrac{7}{w}\right)^4 = \dfrac{7^4}{w^4} = \dfrac{2401}{w^4}$

Practice Problem 14 Simplify. $\left(\dfrac{4a}{b}\right)^6$

Many expressions can be simplified by using the previous rules involving exponents. Be sure to take particular care to determine the correct sign, especially if there is a negative numerical coefficient.

● EXAMPLE 15 Simplify. $\left(\dfrac{-3x^2z^0}{y^3}\right)^4$

$\left(\dfrac{-3x^2z^0}{y^3}\right)^4 = \left(\dfrac{-3x^2}{y^3}\right)^4$ Simplify inside the parentheses first. Note that $z^0 = 1$.

$\qquad = \dfrac{(-3)^4 x^8}{y^{12}}$ Apply the rules for raising a power to a power. Notice that we wrote $(-3)^4$ and not -3^4. We are raising -3 to the fourth power.

$\qquad = \dfrac{81x^8}{y^{12}}$ Simplify the coefficient: $(-3)^4 = +81$.

Practice Problem 15 Simplify. $\left(\dfrac{-2x^3y^0z}{4xz^2}\right)^5$

Developing Your Study Skills

Why Is Review Necessary?

You master a course in mathematics by learning the concepts one step at a time. Thus the study of mathematics is built step-by-step, with each step a supporting foundation for the next. The process is a carefully designed procedure, so no steps can be skipped. A student of mathematics needs to realize the importance of this building process to succeed.

Because new concepts depend on those previously learned, students often need to take time to review. The reviewing process will strengthen understanding and skills, which may be weak due to a lack of mastery or the passage of time. Review at the right time on the right concepts can strengthen previously learned skills and make progress possible.

Timely, periodic review of previously learned mathematical concepts is absolutely necessary for mastery of new concepts. You may have forgotten a concept or grown a bit rusty in applying it. Reviewing is the answer. Make use of the cumulative review problems in your textbook, whether they are assigned or not. Look back to previous chapters whenever you have forgotten how to do something. Review the chapter organizers from previous chapters. Study the examples and practice some exercises to refresh your understanding.

Be sure that you understand and can perform the computations of each new concept. This will enable you to move successfully on to the next one.

Remember, mathematics is a step-by-step building process. Learn each concept and reinforce and strengthen with review whenever necessary.

Verbal and Writing Skills

1. Write in your own words the product rule for exponents.

2. To be able to use the rules of exponents, what must be true of the bases?

3. If the larger exponent is in the denominator, the quotient rule states that $\dfrac{x^a}{x^b} = \dfrac{1}{x^{b-a}}$. Provide an example to show why this is true.

In exercises 4 and 5, identify the numerical coefficient, the base(s), and the exponent(s).

4. $-8x^5y^2$

5. $6x^{11}y$

6. Evaluate **(a)** $3x^0$ and **(b)** $(3x)^0$. **(c)** Why are the results different?

Write in simplest exponent form.

7. $2 \cdot 2 \cdot a \cdot a \cdot a \cdot b$

8. $5 \cdot x \cdot x \cdot x \cdot y \cdot y$

9. $(-3)(a)(a)(b)(c)(b)(c)(c)$

10. $(-7)(x)(y)(z)(y)(x)$

Multiply. Leave your answer in exponent form.

11. $(3^8)(3^7)$

12. $(2^5)(2^8)$

13. $(5^{10})(5^{16})$

14. $(5^3)(2^6)$

15. $(3^5)(8^2)$

Multiply.

16. $-5x^4(4x)$

17. $6x^2(-9x^3)$

18. $(5x)(10x^2)$

19. $(-4x^2)(-3x^3)$

20. $(-2a^5)(-5a^3)$

21. $(4x^8)(-3x^3)$

22. $\left(\dfrac{2}{5}xy^3\right)\left(\dfrac{1}{3}x^2y^2\right)$

23. $\left(\dfrac{4}{5}x^5y\right)\left(\dfrac{15}{16}x^2y^4\right)$

24. $(1.1x^2z)(-2.5xy)$

25. $(2.3x^4w)(-3.5xy^4)$

26. $(-3x)(2y)(5x^3y)$

27. $(8a)(2a^3b)(0)$

28. $(-5ab)(2a^2)(0)$

29. $(-16x^2y^4)(-5xy^3)$

30. $(-12x^4y)(-7x^5y^3)$

31. $(14a^5)(-2b^6)$

32. $(5a^3b^2)(-ab)$

33. $(-2x^3y^2)(0)(-3x^4y)$

34. $(-4x^8y^2)(13y^3)(0)$

35. $(4x^3y)(0)(-12x^4y^2)$

36. $(5x^3y)(-2w^4z)$

37. $(6w^5z^6)(-4xy)$

38. $(3ab)(5a^2c)(-2b^2c^3)$

Divide. Leave your answer in exponent form. Assume that all variables in any denominator are nonzero.

39. $\dfrac{y^5}{y^8}$ **40.** $\dfrac{x^{13}}{x^3}$ **41.** $\dfrac{y^{12}}{y^5}$ **42.** $\dfrac{b^{20}}{b^{23}}$ **43.** $\dfrac{13^{20}}{13^{30}}$ **44.** $\dfrac{5^{19}}{5^{11}}$

45. $\dfrac{-3^{18}}{-3^{14}}$ **46.** $\dfrac{-5^{16}}{-5^{12}}$ **47.** $\dfrac{a^{13}}{4a^5}$ **48.** $\dfrac{4b^{16}}{b^{13}}$ **49.** $\dfrac{x^7}{y^9}$ **50.** $\dfrac{x^{20}}{y^3}$

51. $\dfrac{-12x^5y^3}{-24xy^3}$ **52.** $\dfrac{-45a^4b^3}{-15a^4b^2}$ **53.** $\dfrac{a^5b^6}{a^5b^6}$ **54.** $\dfrac{-36x^3y^7}{72x^5y}$ **55.** $\dfrac{48x^7y}{-24x^3y^6}$ **56.** $\dfrac{84x^5y^2}{4.2xy^3}$

57. $\dfrac{3.1s^5t^3}{62s^8t}$ **58.** $\dfrac{-51x^6y^8z^{12}}{17x^3y^8z^7}$ **59.** $\dfrac{30x^5y^4}{5x^3y^4}$ **60.** $\dfrac{27x^3y^2}{3xy^2}$ **61.** $\dfrac{8^0x^2y^3}{16x^5y}$ **62.** $\dfrac{3^2x^3y^7}{3^0x^5y^2}$

63. $\dfrac{18a^6b^3c^0}{24a^5b^3}$ **64.** $\dfrac{12a^7b^8}{16a^3b^8c^0}$ **65.** $\dfrac{25x^6}{35y^8}$ **66.** $\dfrac{24y^5}{16x^3}$

Simplify. Multiply the numerator first. Then divide the two expressions.

67. $\dfrac{(4x)(9x^4)}{12x^3}$ **68.** $\dfrac{(6x^7)(3x^2)}{9x^4}$ **69.** $\dfrac{(9a^2b)(2a^3b^6)}{-27a^8b^7}$

To Think About

70. What expression can be multiplied by $(-3x^3yz)$ to obtain $81x^8y^2z^4$?

71. $63a^5b^6$ is divided by an expression and the result is $-9a^4b$. What is this expression?

Find a value for the variable to show that the two expressions are not equivalent.

72. $(x^2)(x^3);\ x^6$ **73.** $\dfrac{y^8}{y^4};\ y^2$ **74.** $\dfrac{z^3}{z^6};\ z^3$

Simplify.

75. $\dfrac{x^{5a}}{x^{2a}}$ **76.** $y^{7b}\cdot y^{2b}$ **77.** $\dfrac{(c^{3y})(c^4)}{c^{2y+2}}$

Simplify.

78. $(x^2)^6$ **79.** $(w^5)^8$ **80.** $(xy^2)^7$ **81.** $(a^3b)^4$

82. $\left(a^5b^2\right)^6$

83. $\left(m^3n^2p\right)^5$

84. $\left(3a^3b^2c\right)^3$

85. $\left(3^2xy^2\right)^4$

86. $\left(-3a^4\right)^2$

87. $\left(-2a^5\right)^4$

88. $\left(\dfrac{7}{w^2}\right)^8$

89. $\left(\dfrac{12x}{y^2}\right)^5$

90. $\left(\dfrac{5x}{7y^2}\right)^2$

91. $\left(\dfrac{2a^4}{3b^3}\right)^4$

92. $\left(-3a^2b^3c^0\right)^4$

93. $\left(-2a^5b^2c\right)^5$

94. $\left(-2x^3yz\right)^3$

95. $\left(-4xy^0z^4\right)^3$

96. $\dfrac{(2x)^4}{(2x)^5}$

97. $\dfrac{\left(4a^2b\right)^2}{\left(4ab^2\right)^3}$

98. $\left(3ab^2\right)^3(ab)$

99. $\left(-2a^2b^3\right)^3\left(ab^2\right)$

100. $\left(\dfrac{8}{y^5}\right)^2$

101. $\left(\dfrac{4}{x^6}\right)^3$

102. $\left(\dfrac{ab^2}{c^3d^4}\right)^4$

103. $\left(\dfrac{a^3b}{c^5d}\right)^5$

104. $\left(5xy^2\right)^3\left(xy^2\right)$

105. $\left(4x^3y\right)^2\left(x^3y\right)$

To Think About

106. What expression raised to the third power is $-27x^9y^{12}z^{21}$?

107. What expression raised to the fourth power is $16x^{20}y^{16}z^{28}$?

Cumulative Review Problems   *Simplify.*

108. $-3 - 8$

109. $-17 + (-32) + (-24) + 27$

110. $\left(\dfrac{2}{3}\right)\left(-\dfrac{21}{8}\right)$

111. $\dfrac{-3}{4} \div \dfrac{-12}{80}$

A recent medical study documenting the link between smoking marijuana and heart attacks has been published. The study shows that a middle-aged person's risk of having a heart attack rises 480% during the first hour after smoking marijuana. The risk of this same person having a heart attack rises 170% during the second hour after smoking marijuana. Suppose in a healthy population of 50,000 middle-aged people, the risk of having a heart attack during a given year is 3%. (Source: Health Resources and Services Administration)

112. How many people would be expected to have a heart attack if the entire population did not smoke marijuana?

113. If the entire population smoked marijuana, how many people would be expected to have a heart attack in the first hour after smoking marijuana?

114. If the entire population smoked marijuana, how many people would be expected to have a heart attack in the second hour after smoking marijuana?

4.2 Negative Exponents and Scientific Notation

1 Using Negative Exponents

If n is an integer, and $x \neq 0$, then x^{-n} is defined as follows:

Definition of a Negative Exponent

$$x^{-n} = \frac{1}{x^n}, \quad x \neq 0$$

EXAMPLE 1 Write with positive exponents.

(a) y^{-3} **(b)** z^{-6} **(c)** w^{-1}

(a) $y^{-3} = \dfrac{1}{y^3}$ **(b)** $z^{-6} = \dfrac{1}{z^6}$ **(c)** $w^{-1} = \dfrac{1}{w^1} = \dfrac{1}{w}$

Practice Problem 1 Write with positive exponents.

(a) x^{-12} **(b)** w^{-5} **(c)** z^{-2}

To evaluate a numerical expression with a negative exponent, first write the expression with a positive exponent. Then simplify.

EXAMPLE 2 Evaluate. 2^{-5}

$$2^{-5} = \frac{1}{2^5} = \frac{1}{32}$$

Practice Problem 2 Evaluate. 4^{-3}

All the previously studied laws of exponents are true for any integer exponent. These laws are summarized in the following box. Assume that $x, y \neq 0$.

Laws of Exponents

The Product Rule

$$x^a \cdot x^b = x^{a+b}$$

The Quotient Rule

$$\frac{x^a}{x^b} = x^{a-b} \quad \text{Use if } a > b, \qquad \frac{x^a}{x^b} = \frac{1}{x^{b-a}} \quad \text{Use if } a < b.$$

Power Rules

$$(xy)^a = x^a y^a, \qquad \left(x^a\right)^b = x^{ab}, \qquad \left(\frac{x}{y}\right)^a = \frac{x^a}{y^a}$$

By using the definition of a negative exponent and the properties of fractions, we can derive two more helpful properties of exponents. Assume that $x, y \neq 0$.

Properties of Negative Exponents

$$\frac{1}{x^{-n}} = x^n \qquad \frac{x^{-m}}{y^{-n}} = \frac{y^n}{x^m}$$

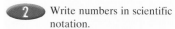
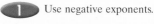

SSM   PH TUTOR   CD & VIDEO   MATH PRO   WEB
CENTER

EXAMPLE 3 Simplify. Write the expression with no negative exponents.

(a) $\dfrac{1}{x^{-6}}$ **(b)** $\dfrac{x^{-3}y^{-2}}{z^{-4}}$ **(c)** $x^{-2}y^3$

(a) $\dfrac{1}{x^{-6}} = x^6$ **(b)** $\dfrac{x^{-3}y^{-2}}{z^{-4}} = \dfrac{z^4}{x^3y^2}$ **(c)** $x^{-2}y^3 = \dfrac{y^3}{x^2}$

Practice Problem 3 Simplify. Write the expression with no negative exponents.

(a) $\dfrac{3}{w^{-4}}$ **(b)** $\dfrac{x^{-6}y^4}{z^{-2}}$ **(c)** $x^{-6}y^{-5}$

EXAMPLE 4 Simplify. Write the expression with no negative exponents.

(a) $\left(3x^{-4}y^2\right)^{-3}$ **(b)** $\dfrac{x^2y^{-4}}{x^{-5}y^3}$

(a) $\left(3x^{-4}y^2\right)^{-3} = 3^{-3}x^{12}y^{-6} = \dfrac{x^{12}}{3^3y^6} = \dfrac{x^{12}}{27y^6}$

(b) $\dfrac{x^2y^{-4}}{x^{-5}y^3} = \dfrac{x^2x^5}{y^4y^3} = \dfrac{x^7}{y^7}$ First rewrite the expression so that only positive exponents appear. Then simplify using the product rule.

Practice Problem 4 Simplify. Write the expression with no negative exponents.

(a) $\left(2x^4y^{-5}\right)^{-2}$ **(b)** $\dfrac{y^{-3}z^{-4}}{y^2z^{-6}}$

2 Writing Numbers in Scientific Notation

One common use of negative exponents is in writing numbers in scientific notation. Scientific notation is most useful in expressing very large and very small numbers.

Scientific Notation

A positive number is written in **scientific notation** if it is in the form $a \times 10^n$, where $1 \le a < 10$ and n is an integer.

EXAMPLE 5 Write in scientific notation. **(a)** 4567 **(b)** 157,000,000

(a) $4567 = 4.567 \times 1000$ To change 4567 to a number that is greater than 1 but less than 10, we move the decimal point three places to the left. We must then multiply the number by a power of 10 so that we do not change the value of the number. Use 1000.

$= 4.567 \times 10^3$

(b) $157,000,000 = \underset{\text{8 places}}{1.57000000} \times \underset{\text{8 zeros}}{100000000}$

$= 1.57 \times 10^8$

Practice Problem 5 Write in scientific notation.

(a) 78,200 **(b)** 4,786,000

Numbers that are smaller than 1 will have a negative power of 10 if they are written in scientific notation.

EXAMPLE 6   Write in scientific notation.

(a) 0.061   **(b)** 0.000052

(a) We need to write 0.061 as a number that is greater than 1 but less than 10. In which direction do we move the decimal point?

$0.061 = 6.1 \times 10^{-2}$ Move the decimal point 2 places to the right. Then multiply by 10^{-2}.

(b) $0.000052 = 5.2 \times 10^{-5}$ Why?

Practice Problem 6   Write in scientific notation.

(a) 0.98   **(b)** 0.000092

The reverse procedure transforms scientific notation into ordinary decimal notation.

EXAMPLE 7   Write in decimal notation.   **(a)** 1.568×10^2 **(b)** 7.432×10^{-3}

(a) $1.568 \times 10^2 = 1.568 \times 100$
$$= 156.8$$

Alternative Method

$1.568 \times 10^2 = 156.8$ The exponent 2 tells us to move the decimal point 2 places to the right.

(b) $7.432 \times 10^{-3} = 7.432 \times \dfrac{1}{1000}$
$$= 0.007432$$

Alternative Method

$7.432 \times 10^{-3} = 0.007432$ The exponent -3 tells us to move the decimal point 3 places to the left.

Practice Problem 7   Write in decimal notation.

(a) 2.96×10^3 **(b)** 1.93×10^6 **(c)** 5.43×10^{-2} **(d)** 8.562×10^{-5}

The distance light travels in one year is called a *light-year*. A light-year is a convenient unit of measure to use when investigating the distances between stars.

EXAMPLE 8   A light-year is a distance of 9,460,000,000,000,000 meters. Write this in scientific notation.

$$9,460,000,000,000,000 = 9.46 \times 10^{15} \text{ meters}$$

Practice Problem 8   Astronomers measure distances to faraway galaxies in parsecs. A parsec is a distance of 30,900,000,000,000,000 meters. Write this in scientific notation.

To perform a calculation involving very large or very small numbers, it is usually helpful to write the numbers in scientific notation and then use the laws of exponents to do the calculation.

Calculator

Scientific Notation

Most calculators can display only eight digits at one time. Numbers with more than eight digits are usually shown in scientific notation. 1.12 E 08 or 1.12 8 means 1.12×10^8. You can use the calculator to compute with large numbers by entering the numbers using scientific notation. For example,

$$(7.48 \times 10^{24}) \times (3.5 \times 10^8)$$

is entered as follows.

7.48 [EXP]

24 [×] 3.5

[EXP] 8 [=]

Display: [2.618 E 33]

or [2.618   33]

Note: Some calculators have an [EE] key instead of [EXP].

Compute on a calculator.

1. 35,000,000,000 + 77,000,000,000
2. $(6.23 \times 10^{12}) \times (4.9 \times 10^5)$
3. $(2.5 \times 10^7)^5$
4. $3.3284 \times 10^{32} \div (6.28 \times 10^{24})$
5. How many seconds are there in 1000 years?

EXAMPLE 9 Use scientific notation and the laws of exponents to find the following. Leave your answer in scientific notation.

(a) $(32{,}000{,}000)(1{,}500{,}000{,}000{,}000)$

(b) $\dfrac{0.00063}{0.021}$

(a) $(32{,}000{,}000)(1{,}500{,}000{,}000{,}000)$

$\qquad = (3.2 \times 10^7)(1.5 \times 10^{12})$ Write each number in scientific notation.

$\qquad = 3.2 \times 1.5 \times 10^7 \times 10^{12}$ Rearrange the order. Remember that multiplication is commutative.

$\qquad = 4.8 \times 10^{19}$ Multiply 3.2×1.5. Multiply $10^7 \times 10^{12}$.

(b) $\dfrac{0.00063}{0.021} = \dfrac{6.3 \times 10^{-4}}{2.1 \times 10^{-2}}$ Write each number in scientific notation.

$\qquad = \dfrac{6.3}{2.1} \times \dfrac{10^{-4}}{10^{-2}}$ Rearrange the order. We are actually using the definition of multiplication of fractions.

$\qquad = \dfrac{6.3}{2.1} \times \dfrac{10^2}{10^4}$ Rewrite with positive exponents.

$\qquad = 3.0 \times 10^{-2}$

Practice Problem 9 Use scientific notation and the laws of exponents to find the following. Leave your answer in scientific notation.

(a) $(56{,}000)(1{,}400{,}000{,}000)$

(b) $\dfrac{0.000111}{0.00000037}$

When we use scientific notation, we are writing approximate numbers. We must include some zeros so that the decimal point can be properly located. However, all other digits except for these zeros are considered **significant digits**. The number 34.56 has four significant digits. The number 0.0049 has two significant digits. The zeros are considered placeholders. The number 634,000 has three significant digits (unless we have specific knowledge to the contrary). The zeros are considered placeholders. We sometimes round numbers to a specific number of significant digits. For example, 0.08746 rounded to two significant digits is 0.087. When we round 1,348,593 to three significant digits, we obtain 1,350,000.

EXAMPLE 10 The approximate distance from Earth to the star Polaris is 208 parsecs. A parsec is a distance of approximately 3.09×10^{13} kilometers. How long would it take a space probe traveling at 40,000 kilometers per hour to reach the star? Round to three significant digits.

1. *Understand the problem.*
   Recall that the distance formula is

$$\text{distance} = \text{rate} \times \text{time}.$$

We are given the distance and the rate. We need to find the time.

Let's take a look at the distance. The distance is given in parsecs, but the rate is given in kilometers per hour. We need to change the distance to kilometers. We are told that a parsec is approximately 3.09×10^{13} kilometers. That is, there are 3.09×10^{13} kilometers per parsec. We use this information to change 208 parsecs to kilometers.

$$208 \text{ parsecs} = \frac{(208 \text{ parsecs})(3.09 \times 10^{13} \text{ kilometers})}{1 \text{ parsec}} = 642.72 \times 10^{13} \text{ kilometers}$$

2. *Write an equation.*

Use the distance formula.

$$d = r \times t$$

3. *Solve the equation and state the answer.*

Substitute the known values into the formula and solve for the unknown, time.

$$642.72 \times 10^{13} \text{ km} = \frac{40{,}000 \text{ km}}{1 \text{ hr}} \times t$$

$$6.4272 \times 10^{15} \text{ km} = \frac{4 \times 10^4 \text{ km}}{1 \text{ hr}} \times t \qquad \text{Change the numbers to}$$
$$\text{scientific notation.}$$

$$\frac{6.4272 \times 10^{15} \text{ km}}{\dfrac{4 \times 10^4 \text{ km}}{1 \text{ hr}}} = t \qquad \text{Divide both sides by } \frac{4 \times 10^4 \text{ km}}{1 \text{ hr}}.$$

$$\frac{(6.4272 \times 10^{15} \text{ km})(1 \text{ hr})}{4 \times 10^4 \text{ km}} = t$$

$$1.6068 \times 10^{11} \text{ hr} = t$$

1.6068×10^{11} is 160.68×10^9 or 160.68 billion hours. The space probe will take approximately 160.68 billion hours to reach the star.

Reread the problem. Are we finished? What is left to do? We need to round the answer to three significant digits. Rounding to three significant digits, we have

$$160.68 \times 10^9 \approx 161 \times 10^9.$$

This is approximately 161 billion hours or a little more than 18 million years.

4. *Check.*

Unless you have had a great deal of experience working in astronomy, it would be difficult to determine whether this is a reasonable answer. You may wish to reread your analysis and redo your calculations as a check.

Practice Problem 10 The average distance from Earth to the star Betelgeuse is 159 parsecs. How many hours would it take a space probe to travel from Earth to Betelgeuse at a speed of 50,000 kilometers per hour? Round to three significant digits.

Simplify. Express your answer with positive exponents. Assume that all variables are nonzero.

1. $3x^{-2}$

2. $4xy^{-4}$

3. $(4x^2y)^{-2}$

4. $(2x^3y^5)^{-3}$

5. $\dfrac{3xy^{-2}}{z^{-3}}$

6. $\dfrac{4x^{-2}y^{-3}z^0}{y^4}$

7. $\dfrac{(3x)^{-2}}{(3x)^{-3}}$

8. $\dfrac{(2ab^2)^{-3}}{(2ab^2)^{-4}}$

9. $wx^{-5}y^3z^{-2}$

10. $a^5b^{-3}c^{-4}d^0$

11. $(8^{-2})(2^3)$

12. $(9^2)(3^{-3})$

13. $\left(\dfrac{3xy^2}{z^4}\right)^{-2}$

14. $\left(\dfrac{2a^3b^0}{c^2}\right)^{-3}$

15. $\dfrac{x^{-2}y^{-3}}{x^4y^{-2}}$

16. $\dfrac{a^{-6}b^3}{a^{-2}b^{-5}}$

Write in scientific notation.

17. 123,780

18. 0.063

19. 0.000742

20. 889,610,000,000

21. 7,652,000,000

22. 0.00000001963

In exercises 23–28, write in decimal notation.

23. 3.02×10^5

24. 8.137×10^7

25. 3.3×10^{-5}

26. 1.99×10^{-1}

27. 9.83×10^5

28. 3.5×10^{-8}

29. The mass of a proton is 0.000000000000000000000000000167 kilogram. Write this in scientific notation.

30. The speed of light is 3.00×10^8 meters per second. Write this in decimal notation.

31. A single human red blood cell is about 7×10^{-6} meters in diameter. Write this in decimal notation.

32. The average volume of an atom of gold is 0.0000000000000000000001695 cubic centimeters. Write this in scientific notation.

Evaluate by using scientific notation and the laws of exponents. Leave your answer in scientific notation.

33. $\dfrac{(5,000,000)(16,000)}{8,000,000,000}$

34. $(0.0075)(0.0000002)(0.001)$

35. $(0.0002)^5$

36. $(40,000,000)^3$

37. $(150,000,000)(0.00005)(0.002)(30,000)$

38. $\dfrac{(1,600,000)(0.00003)}{2400}$

Applications

For fiscal year 1999, the national debt was determined to be approximately 5.614×10^{12} *dollars.*

39. The census bureau estimates that in 1999, the entire population of the United States was 2.76×10^8 people. If the national debt were evenly divided among every person in the country, how much debt would be assigned to each individual? Round to three significant digits.

40. The census bureau estimates that in 1999, the number of people in the United States who were over age 18 was approximately 2.05×10^8 people. If the national debt were evenly divided among every person over age 18 in the country, how much debt would be assigned to each individual? Round to three significant digits.

A parsec is a distance of approximately 3.09×10^{13} *kilometers.*

41. How long would it take a space probe to travel from Earth to the star Rigel, which is 276 parsecs from Earth? Assume that the space probe travels at 45,000 kilometers per hour. Round to three significant digits.

42. How long would it take a space probe to travel from Earth to the star Hadar, which is 150 parsecs from Earth? Assume that the space probe travels at 55,000 kilometers per hour. Round to three significant digits.

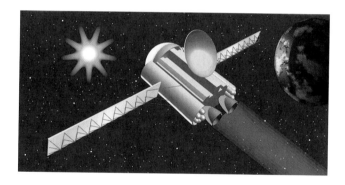

43. Antares, a supergiant star, is one of the brightest stars seen by the naked eye. It is 130 parsecs from Earth. How long would it take an experimental robotic probe to reach Antares if we assume that the probe travels at 35,000 kilometers per hour? Round to three significant digits.

44. The mass of a neutron is approximately 1.675×10^{-27} kilogram. Find the mass of 180,000 neutrons.

45. The sun radiates energy into space at the rate of 3.9×10^{26} joules per second. How many joules are emitted in two weeks?

46. Avogadro's number says that there are approximately 6.02×10^{23} molecules/mole. How many molecules can one expect in 0.00483 mole?

47. In 1990 the cost for construction of new private buildings was estimated at $\$3.61 \times 10^{11}$. By 2000 the estimated cost for construction of new private buildings had risen to $\$5.28 \times 10^{11}$. What was the percent of increase from 1990 to 2000? Round to the nearest tenth of a percent. (*Source:* U.S. Census Bureau)

48. In 1990 the cost for construction of new public buildings was estimated at $\$1.07 \times 10^{11}$. By 2000 the estimated cost for construction of new public buildings had risen to $\$1.53 \times 10^{11}$. What was the percent of increase from 1990 to 2000? Round to the nearest tenth of a percent. (*Source:* U.S. Census Bureau)

Cumulative Review Problems

Simplify.

49. $-2.7 - (-1.9)$

50. $(-1)^{33}$

51. $\dfrac{-3}{4} + \dfrac{5}{7}$

52. A recent debate between two political candidates attracted 3540 people to a local gymnasium. At the conclusion of the debate, each person went to one side of the gymnasium to shake the hand of candidate #1 or to the other side of the room to shake the hand of candidate #2. Candidate #1 shook 524 less than triple the number of hands shaken by candidate #2. How many hands did each candidate shake?

53. Gina has a bachelor's degree. She earns $\$12,460$ more a year than Mario, who holds an associate's degree. Alfonso, who has not yet been able to attend college, earns $\$8742$ a year less than Mario. The combined annual salaries of the three people is $\$112,000$. What is the annual salary of each person?

Putting Your Skills to Work

The Mathematics of Forests

One-third of the world's area that was originally covered by forests and woodlands no longer has either. The welfare of people and animals in future generations on Earth may well depend on the ability of people to stop the destruction of forests worldwide. Some countries have a very significant amount of the present total of 9.258×10^9 acres of forests in the world.

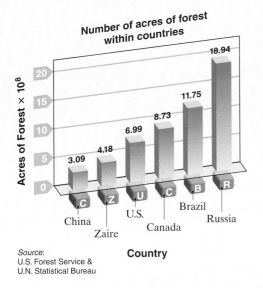

Number of acres of forest within countries

Acres of Forest $\times 10^8$

- China (C): 3.09
- Zaire (Z): 4.18
- U.S. (U): 6.99
- Canada (C): 8.73
- Brazil (B): 11.75
- Russia (R): 18.94

Country

Source:
U.S. Forest Service &
U.N. Statistical Bureau

Problems for Individual Investigation and Analysis

1. How many more acres of forests are in Brazil than are in Canada?

2. What is the combined acreage of forests in Russia and China?

Problems for Group Investigation and Cooperative Group Activity

Together with some members of your class, see if you can determine the answers to the following.

3. If the current population of China is 1.256×10^9 people, how many people are there for each acre of forest in China? Round to the nearest whole number.

4. For quite some time rainforests have been destroyed at the rate of 40 million acres per year. If we assume that this rate has been the same for the last 25 years, approximately how many acres of forests were in the world 25 years ago?

Internet Connections

 Netsite: http://www.prenhall.com/tobey_beginning

Site: Rates of Rainforest Loss (Rainforest Action Network) or a related site

This site presents information about rates of rainforest destruction. Use the information to answer the following questions. (There is some difference of opinion among scientists as to the rate of destruction.)

5. About how many acres of rainforest are destroyed in one hour?

6. About how many acres of Brazilian rainforest were destroyed during the last 12 years?

7. The current indigenous population of Brazil's forests is about what percent of the indigenous population in 1500?

4.3 Fundamental Polynomial Operations

Student Learning Objectives

After studying this section, you will be able to:

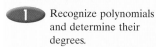

1 Recognize polynomials and determine their degrees.

2 Add polynomials.

3 Subtract polynomials.

4 Evaluate polynomials.

SSM

PH TUTOR
CENTER   CD & VIDEO   MATH PRO   WEB

1 Recognizing Polynomials and Determining Their Degrees

A **polynomial** in x is the sum of a finite number of terms of the form ax^n, where a is any real number and n is a whole number. Usually these polynomials are written in descending powers of the variable, as in

$$5x^3 + 3x^2 - 2x - 5 \quad \text{and} \quad 3.2x^2 - 1.4x + 5.6.$$

A **multivariable polynomial** is a polynomial with more than one variable. The following are multivariable polynomials:

$$5xy + 8, \quad 2x^2 - 7xy + 9y^2, \quad 17x^3y^9$$

The **degree of a term** is the sum of the exponents of all of the variables in the term. For example, the degree of $7x^3$ is three. The degree of $4xy$ is two. The degree of $10x^4y^2$ is six.

The **degree of a polynomial** is the highest degree of all of the terms in the polynomial. For example, the degree of $5x^3 + 8x^2 - 20x - 2$ is three. The degree of $6xy - 4x^2y + 2xy^3$ is four.

The polynomial 0 is said to have **no degree**.

There are special names for polynomials with one, two, or three terms.

A **monomial** has *one* term:

$$5a, \quad 3x^3yz^4, \quad 12xy$$

A **binomial** has *two* terms:

$$7x + 9y, \quad -6x - 4, \quad 5x^4 + 2xy^2$$

A **trinomial** has *three* terms:

$$8x^2 - 7x + 4, \quad 2ab^3 - 6ab^2 - 15ab, \quad 2 + 5y + y^4$$

EXAMPLE 1 State the degree of the polynomial, and whether it is a monomial, a binomial, or a trinomial.

(a) $5xy + 3x^3$ **(b)** $-7a^5b^2$ **(c)** $8x^4 - 9x - 15$

(a) This polynomial is of degree 3. It has two terms, so it is a binomial.

(b) The sum of the exponents is $5 + 2 = 7$. Therefore this polynomial is of degree 7. It has one term, so it is a monomial.

(c) This polynomial is of degree 4. It has three terms, so it is a trinomial.

Practice Problem 1 State the degree of the polynomial, and whether it is a monomial, a binomial, or a trinomial.

(a) $-7x^5 - 3xy$ **(b)** $22a^3b^4$ **(c)** $-3x^3 + 5x^2 - 6x$

2 Adding Polynomials

We usually write a polynomial in x so that the exponents on x decrease from left to right. You can add, subtract, multiply, and divide polynomials. Let us take a look at addition. To add two polynomials, we add their like terms.

EXAMPLE 2 Add. $(5x^2 - 6x - 12) + (-3x^2 - 9x + 5)$

$$
\begin{aligned}
(5x^2 - 6x - 12) + (-3x^2 - 9x + 5) &= [5x^2 + (-3x^2)] + [-6x + (-9x)] + [-12 + 5] \\
&= [(5 - 3)x^2] + [(-6 - 9)x] + [-12 + 5] \\
&= 2x^2 + (-15x) + (-7) \\
&= 2x^2 - 15x - 7
\end{aligned}
$$

Practice Problem 2 Add. $(-8x^3 + 3x^2 + 6) + (2x^3 - 7x^2 - 3)$

The numerical coefficients of the polynomials may be any real number. Thus the polynomials may have numerical coefficients that are decimals or fractions.

EXAMPLE 3 Add. $(\frac{1}{2}x^2 - 6x + \frac{1}{3}) + (\frac{1}{5}x^2 - 2x - \frac{1}{2})$

$$
\begin{aligned}
(\tfrac{1}{2}x^2 - 6x + \tfrac{1}{3}) + (\tfrac{1}{5}x^2 - 2x - \tfrac{1}{2}) &= [\tfrac{1}{2}x^2 + \tfrac{1}{5}x^2] + [-6x + (-2x)] + [\tfrac{1}{3} + (-\tfrac{1}{2})] \\
&= [(\tfrac{1}{2} + \tfrac{1}{5})x^2] + [(-6 - 2)x] + [\tfrac{1}{3} + (-\tfrac{1}{2})] \\
&= [(\tfrac{5}{10} + \tfrac{2}{10})x^2] + [-8x] + [\tfrac{2}{6} - \tfrac{3}{6}] \\
&= \tfrac{7}{10}x^2 - 8x - \tfrac{1}{6}
\end{aligned}
$$

Practice Problem 3 Add. $(-\frac{1}{3}x^2 - 6x - \frac{1}{12}) + (\frac{1}{4}x^2 + 5x - \frac{1}{3})$

EXAMPLE 4 Add. $(1.2x^3 - 5.6x^2 + 5) + (-3.4x^3 - 1.2x^2 + 4.5x - 7)$
Group like terms.

$$
\begin{aligned}
(1.2x^3 - 5.6x^2 + 5) + (-3.4x^3 - 1.2x^2 + 4.5x - 7) &= (1.2 - 3.4)x^3 + (-5.6 - 1.2)x^2 + 4.5x + (5 - 7) \\
&= -2.2x^3 - 6.8x^2 + 4.5x - 2
\end{aligned}
$$

Practice Problem 4 Add.

$(3.5x^3 - 0.02x^2 + 1.56x - 3.5) + (-0.08x^2 - 1.98x + 4)$

As mentioned previously, polynomials may involve more than one variable.

3 Subtracting Polynomials

Recall that subtraction of real numbers can be defined as adding the opposite of the second number. Thus $a - b = a + (-b)$. That is, $3 - 5 = 3 + (-5)$. A similar method is used to subtract two polynomials.

To subtract two polynomials, change the sign of each term in the second polynomial and then add.

EXAMPLE 5 Subtract. $(7x^2 - 6x + 3) - (5x^2 - 8x - 12)$

We change the sign of each term in the second polynomial and then add.

$$
\begin{aligned}
(7x^2 - 6x + 3) - (5x^2 - 8x - 12) &= (7x^2 - 6x + 3) + (-5x^2 + 8x + 12) \\
&= (7 - 5)x^2 + (-6 + 8)x + (3 + 12) \\
&= 2x^2 + 2x + 15
\end{aligned}
$$

Practice Problem 5 Subtract.
$(5x^3 - 15x^2 + 6x - 3) - (-4x^3 - 10x^2 + 5x + 13)$

EXAMPLE 6 Subtract. $(-6x^2y - 3xy + 7xy^2) - (5x^2y - 8xy - 15x^2y^2)$

Change the sign of each term in the second polynomial and add. Look for like terms.

$$(-6x^2y - 3xy + 7xy^2) + (-5x^2y + 8xy + 15x^2y^2) = (-6 - 5)x^2y + (-3 + 8)xy + 7xy^2 + 15x^2y^2$$

$$= -11x^2y + 5xy + 7xy^2 + 15x^2y^2$$

Nothing further can be done to combine these four terms.

Practice Problem 6 Subtract.
$(x^3 - 7x^2y + 3xy^2 - 2y^3) - (2x^3 + 4xy - 6y^3)$

4 Evaluating Polynomials

Sometimes polynomials are used to predict a value. In such cases we need to **evaluate** the polynomial. We do this by substituting a known value for the variable and determining the value of the polynomial.

EXAMPLE 7 Automobiles sold in the United States have become more fuel efficient over the years due to regulations from Congress. The number of miles per gallon obtained by the average automobile in the United States can be described by the polynomial

$$0.3x + 12.9,$$

where x is the number of years since 1970. (*Source:* U.S. Federal Highway Administration.) Use this polynomial to estimate the number of miles per gallon obtained by the average automobile in **(a)** 1972 **(b)** 2004.

(a) The year 1972 is two years later than 1970, so $x = 2$.
Thus the number of miles per gallon obtained by the average automobile in 1972 can be estimated by evaluating $0.3x + 12.9$ when $x = 2$.

$$0.3(2) + 12.9 = 0.6 + 12.9$$
$$= 13.5$$

We estimate that the average car in 1972 obtained 13.5 miles per gallon.

(b) The year 2004 will be 34 years after 1970, so $x = 34$.
Thus the estimated number of miles per gallon obtained by the average automobile in 2004 can be predicted by evaluating $0.3x + 12.9$ when $x = 34$.

$$0.3(34) + 12.9 = 10.2 + 12.9$$
$$= 23.1$$

We therefore predict that the average car in 2004 will obtain 23.1 miles per gallon.

Practice Problem 7 The number of miles per gallon obtained by the average truck in the United States can be described by the polynomial $0.03x + 5.4$, where x is the number of years since 1970. (*Source:* U.S. Federal Highway Administration.) Use this polynomial to estimate the number of miles per gallon obtained by the average truck in

(a) 1974 **(b)** 2006.

4.3 Exercises

Verbal and Writing Skills

1. State in your own words a definition for a polynomial in x and give an example.

2. State in your own words a definition for a multivariable polynomial and give an example.

3. State in your own words how to determine the degree of a polynomial in x.

4. State in your own words how to determine the degree of a multivariable polynomial.

State the degree of the polynomial and whether it is a monomial, a binomial, or a trinomial.

5. $6x^3y$

6. $5xy^6$

7. $20x^5 + 6x^3 - 7x$

8. $13x^4 - 12x + 20$

9. $5xy^2 - 3x^2y^3$

10. $7x^3y + 5x^4y^4$

Add.

11. $(-3x + 15) + (8x - 43)$

12. $(5x - 11) + (-7x + 34)$

13. $(2x^2 - 8x + 7) + (-5x^2 - 2x + 3)$

14. $(x^2 - 6x - 3) + (-4x^2 + x - 4)$

15. $\left(\frac{1}{2}x^2 + \frac{1}{3}x - 4\right) + \left(\frac{1}{3}x^2 + \frac{1}{6}x - 5\right)$

16. $\left(\frac{1}{4}x^2 - \frac{2}{3}x - 10\right) + \left(-\frac{1}{3}x^2 + \frac{1}{9}x + 2\right)$

17. $(3.4x^3 - 5.6x^2 - 7.1x + 3.4) + (-1.7x^3 + 2.2x^2 - 6.1x - 8.8)$

18. $(-4.6x^3 - 2.9x^2 + 5.6x - 0.3) + (-2.6x^3 + 9.8x^2 + 4.5x - 1.7)$

Subtract.

19. $(2x - 19) - (-3x + 5)$

20. $(5x - 5) - (6x - 3)$

21. $\left(\frac{1}{2}x^2 + 3x - \frac{1}{5}\right) - \left(\frac{2}{3}x^2 - 4x + \frac{1}{10}\right)$

22. $\left(\frac{1}{8}x^2 - 5x + \frac{2}{7}\right) - \left(\frac{1}{4}x^2 - 6x + \frac{1}{14}\right)$

23. $(-3x^2 + 5x) - (2x^3 - 3x^2 + 10)$

24. $(7x^3 - 3x^2 - 5x + 4) - (x^3 - 7x + 3)$

25. $(0.5x^4 - 0.7x^2 + 8.3) - (5.2x^4 + 1.6x + 7.9)$

26. $(1.3x^4 - 3.1x^3 + 6.3x) - (x^4 - 5.2x^2 + 6.5x)$

Perform the indicated operations.

27. $(7x - 8) - (5x - 6) + (8x + 12)$

28. $(-8x + 3) + (-6x - 20) - (-3x + 9)$

29. $(5x^2y - 6xy^2 + 2) + (-8x^2y + 12xy^2 - 6)$

30. $(7x^2y^2 - 6xy + 5) + (-15x^2y^2 - 6xy + 18)$

31. $\left(3x^4 - 4x^2 - 18\right) - \left(2x^4 + 3x^3 + 6\right)$

32. $\left(2b^3 + 3b - 5\right) - \left(-3b^3 + 5b^2 + 7b\right)$

To Think About

Buses operated in this country have become more fuel efficient in recent years. The number of miles per gallon achieved by the average bus operated in the United States can be described by the polynomial $0.04x + 5.2$, where x is the number of years since 1970. (Source: U.S. Federal Highway Administration)

33. Estimate the number of miles per gallon obtained by the average bus in 1975.

34. Estimate the number of miles per gallon obtained by the average bus in 1995.

35. Estimate in what year the rating of the average bus will be 7.2 miles per gallon.

36. Estimate in what year the rating of the average bus will be 6.8 miles per gallon.

The average number of prisoners held in federal and state prisons increases each year. The number of prisoners measured in thousands can be described by the polynomial $1.8x^2 + 22.2x + 325$, where x is the number of years since 1980. (Source: U.S. Bureau of Justice Statistics)

37. Estimate the number of prisoners in 1990.

38. Estimate the number of prisoners in 1995.

39. According to the polynomial, by how much will the prison population increase from 2002 to 2006?

40. According to the polynomial, by how much will the prison population increase from 2004 to 2010?

Applications

▲ **41.** The lengths and the widths of the following three rectangles are labeled. Create a polynomial that describes the sum of the *area* of these three rectangles.

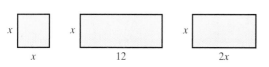

▲ **42.** The dimensions of the sides of the following figure are labeled. Create a polynomial that describes the *perimeter* of this figure.

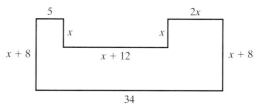

Cumulative Review Problems

43. Solve for x. $3y - 8x = 2$

44. Solve for B. $3A = 2BCD$

45. Solve for b. $A = \dfrac{1}{2}h(b + c)$

46. Solve for d. $B = \dfrac{5xy}{d}$

47. The total approximate expenditure for health care in the United States increased by 90% over the years 1990 to 2000. In 2000 a total of 1.324 trillion dollars was spent. How much was spent in 1990? (*Source:* U.S. Department of Health and Human Services)

4.4 Multiplication of Polynomials

1 Multiplying a Monomial by a Polynomial

We use the distributive property to multiply a monomial by a polynomial. Remember, the distributive property states that for real numbers a, b, and c,

$$a(b + c) = ab + ac.$$

Student Learning Objectives

After studying this section, you will be able to:

 Multiply a monomial by a polynomial.

 Multiply two binomials.

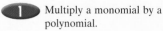

SSM   PH TUTOR   CD & VIDEO   MATH PRO   WEB
CENTER

 EXAMPLE 1   Multiply. $3x^2(5x - 2)$

$$3x^2(5x - 2) = 3x^2(5x) + 3x^2(-2) \qquad \text{Use the distributive property.}$$
$$= (3 \cdot 5)(x^2 \cdot x) + (3)(-2)x^2$$
$$= 15x^3 - 6x^2$$

Practice Problem 1   Multiply. $4x^3(-2x^2 + 3x)$

Try to do as much of the multiplication as you can mentally.

 EXAMPLE 2   Multiply. **(a)** $2x(x^2 + 3x - 1)$ **(b)** $-2xy^2(x^2 - 2xy - 3y^2)$

(a) $2x(x^2 + 3x - 1) = 2x^3 + 6x^2 - 2x$

(b) $-2xy^2(x^2 - 2xy - 3y^2) = -2x^3y^2 + 4x^2y^3 + 6xy^4$

Notice in part (b) that you are multiplying each term by the negative number $-2xy^2$. This will affect the sign of each term in the product.

Practice Problem 2   Multiply. **(a)** $-3x(x^2 + 2x - 4)$ **(b)** $6xy(x^3 + 2x^2y - y^2)$

When we multiply by a monomial, the monomial may be on the right side.

 EXAMPLE 3   Multiply. $(x^2 - 2x + 6)(-2xy)$

$$(x^2 - 2x + 6)(-2xy) = -2x^3y + 4x^2y - 12xy$$

Practice Problem 3   Multiply.

$$(-6x^3 + 4x^2 - 2x)(-3xy)$$

2 Multiplying Two Binomials

We can build on our knowledge of the distributive property and our experience with multiplying monomials to learn how to multiply two binomials. Let's suppose that we want to multiply $(x + 2)(3x + 1)$. We can use the distributive property. Since a can represent any number, let $a = x + 2$. Then let $b = 3x$ and $c = 1$. We now have the following.

$$a(b + c) \quad = \quad ab \quad + \quad ac$$
$$(x + 2)(3x + 1) = (x + 2)(3x) + (x + 2)(1)$$
$$= 3x^2 + 6x + x + 2$$
$$= 3x^2 + 7x + 2$$

271

Let's take another look at the original problem, $(x + 2)(3x + 1)$. This time we will assign a letter to each term in the binomials. That is, let $a = x$, $b = 2$, $c = 3x$, and $d = 1$. Using substitution, we have the following.

$$(x + 2)(3x + 1) = (a + b)(c + d)$$
$$= (a + b)c + (a + b)d$$
$$= ac + bc + ad + bd$$
$$= (x)(3x) + (2)(3x) + (x)(1) + (2)(1) \quad \text{By substitution}$$
$$= 3x^2 + 6x + x + 2$$
$$= 3x^2 + 7x + 2$$

How does this compare with the preceding result?

The distributive property shows us *how* the problem can be done and *why* it can be done. In actual practice there is a somewhat easier approach to obtain the answer. It is often referred to as the FOIL method. The letters FOIL stand for the following.

F   multiply the *First* terms

O   multiply the *Outer* terms

I   multiply the *Inner* terms

L   multiply the *Last* terms

The FOIL letters are simply a way to remember the four terms in the final product and how they are obtained. Let's return to our original problem.

$(x + 2)(3x + 1)$ F   Multiply the *first* terms to obtain $3x^2$.

$(x + 2)(3x + 1)$ O   Multiply the *outer* terms to obtain x.

$(x + 2)(3x + 1)$ I   Multiply the *inner* terms to obtain $6x$.

$(x + 2)(3x + 1)$ L   Multiply the *last* terms to obtain 2.

The result so far is $3x^2 + x + 6x + 2$. These four terms are the same four terms that we obtained when we multiplied using the distributive property. We can combine the like terms to obtain the final answer: $3x^2 + 7x + 2$. Now let's study the FOIL method in a few examples.

EXAMPLE 4   Multiply. $(2x - 1)(3x + 2)$

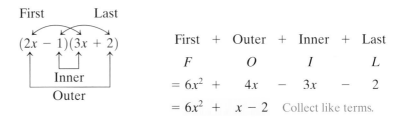

First   Last

$(2x - 1)(3x + 2)$

Inner

Outer

First   +   Outer   +   Inner   +   Last

    F   O   I   L

$= 6x^2 + \quad 4x \quad - \quad 3x \quad - \quad 2$

$= 6x^2 + \quad x - 2$ Collect like terms.

Practice Problem 4   Multiply. $(5x - 1)(x - 2)$

● **EXAMPLE 5**   Multiply. $(3a + 2b)(4a - b)$

First   Last

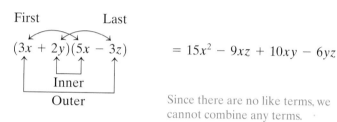

$\qquad\qquad = 12a^2 - 3ab + 8ab - 2b^2$

$\qquad\qquad = 12a^2 + 5ab + 2b^2$

Practice Problem 5   Multiply. $(8a - 5b)(3a - b)$ ●

After you have done several problems, you may be able to combine the outer and inner products mentally.

In some problems the inner and outer products cannot be combined.

● **EXAMPLE 6**   Multiply. $(3x + 2y)(5x - 3z)$

First   Last

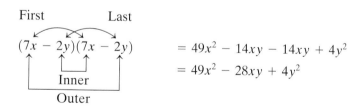

$\qquad\qquad = 15x^2 - 9xz + 10xy - 6yz$

Since there are no like terms, we cannot combine any terms.

Practice Problem 6   Multiply. $(3a + 2b)(2a - 3c)$ ●

● **EXAMPLE 7**   Multiply. $(7x - 2y)^2$

$(7x - 2y)(7x - 2y)$ When we square a binomial, it is the same as multiplying the binomial by itself.

First   Last

$(7x - 2y)(7x - 2y)$
Inner
Outer

$\qquad\qquad = 49x^2 - 14xy - 14xy + 4y^2$

$\qquad\qquad = 49x^2 - 28xy + 4y^2$

Practice Problem 7   Multiply. $(3x - 2y)^2$ ●

We can multiply binomials containing exponents that are greater than 1. That is, we can multiply binomials containing x^2 or y^3, and so on.

● **EXAMPLE 8**   Multiply. $(3x^2 + 4y^3)(2x^2 + 5y^3)$

$(3x^2 + 4y^3)(2x^2 + 5y^3) = 6x^4 + 15x^2y^3 + 8x^2y^3 + 20y^6$

$\qquad\qquad\qquad = 6x^4 + 23x^2y^3 + 20y^6$

Practice Problem 8   Multiply. $(2x^2 + 3y^2)(5x^2 + 6y^2)$ ●

▲ ● **EXAMPLE 9** The width of a living room is $(x + 4)$ feet. The length of the room is $(3x + 5)$ feet. What is the area of the room in square feet?

$3x + 5$

$x + 4$

$$A = (\text{length})(\text{width}) = (3x + 5)(x + 4)$$
$$= 3x^2 + 12x + 5x + 20$$
$$= 3x^2 + 17x + 20$$

There are $(3x^2 + 17x + 20)$ square feet in the room.

Practice Problem 9 What is the area in square feet of a room that is $(2x - 1)$ feet wide and $(7x + 3)$ feet long?

Developing Your Study Skills

Exam Time: How to Review

Reviewing adequately for an exam enables you to bring together the concepts you have learned over several sections. For your review, you will need to:

1. Reread your textbook. Make a list of any terms, rules, or formulas you need to know for the exam. Be sure you understand them all.

2. Reread your notes. Go over returned homework and quizzes. Redo the problems you missed.

3. Practice some of each type of problem covered in the chapter(s) you are to be tested on.

4. Use the end-of-chapter materials provided in your textbook. Read carefully through the chapter organizer. Do the review problems. Take the chapter test. When you are finished, check your answers. Redo any problems you missed.

5. Get help if any concepts give you difficulty.

Multiply.

1. $-2x(6x^3 - x)$

2. $5x(-3x^4 + 4x)$

3. $-5x(3x^2 - 2x + 5)$

4. $3x(-2x^2 + 7x - 11)$

5. $2x^3(3x^4 - 2x^3 + 5x - 1)$

6. $5x^2(8 - 4x + 7x^5 - 8x^7)$

7. $\frac{1}{2}(2x + 3x^2 + 5x^3)$

8. $\frac{2}{3}(4x + 6x^2 - 2x^3)$

9. $(5x^3 - 2x^2 + 6x)(-3xy^2)$

10. $(3b^2 - 6b + 8ab)(5b^3)$

11. $(2b^2 + 3b - 4)(2b^2)$

12. $(2x^3 + x^2 - 6x)(2xy)$

13. $(x^3 - 3x^2 + 5x - 2)(3x)$

14. $(-2x^3 + 4x^2 - 7x + 3)(2x)$

15. $(x^2y^2 - 6xy + 8)(-2xy)$

16. $(x^2y^2 + 5xy - 9)(-3xy)$

17. $(-7x^3 + 3x^2 + 2x - 1)(4x^2y)$

18. $(-5x^3 - 6x^2 + x - 1)(5xy^2)$

19. $(3d^4 - 4d^2 + 6)(-2c^2d)$

20. $(-4x^3 + 6x^2 - 5x)(-7xy^2)$

21. $6x^3(2x^4 - x^2 + 3x + 9)$

22. $8x^3(-2x^4 + 3x^2 - 5x - 14)$

23. $-4x^3(6x^4 - 3x^2 + 2)$

24. $-3x^5(-5x^4 + 2x^2 - 3)$

Multiply. Try to do most of the exercises mentally without writing down intermediate steps.

25. $(x + 10)(x + 3)$

26. $(x + 3)(x + 4)$

27. $(x + 6)(x + 2)$

28. $(x - 8)(x + 2)$

29. $(x + 3)(x - 6)$

30. $(x - 5)(x - 4)$

31. $(x - 6)(x - 5)$

32. $(5x - 2)(-x - 3)$

33. $(7x + 1)(-2x - 3)$

34. $(7x - 4)(x + 2y)$

35. $(x - 9)(3x + 4y)$

36. $(2y + 3)(5y - 2)$

37. $(3y + 2)(4y - 3)$

38. $(7y - 2)(3y - 1)$

39. $(5y - 3)(4y - 2)$

To Think About

40. What is wrong with this multiplication?
$(x - 2)(-3) = 3x - 6$

41. What is wrong with this answer?
$-(3x - 7) = -3x - 7$

42. What is the missing term?

$(5x + 2)(5x + 2) = 25x^2 + $ _____ $ + 4$

43. Multiply the binomials and write a brief description of what is special about the result.

$(5x - 1)(5x + 1)$

Multiply.

44. $(4x^2 - 3y)(5x^2 - 2y)$

45. $(3b^2 - 5c)(2b^2 - 7c)$

46. $(8x^2 - 3y^2)(8x^2 - 3y^2)$

47. $(8x - 2)^2$

48. $(5x - 3)^2$

49. $(5a^2 - 3b^2)^2$

50. $(6x^2 + 5y^3)^2$

51. $(0.2x + 3)(4x - 0.3)$

52. $(0.5x - 2)(6x - 0.2)$

53. $(5x + 8y)(6x - y)$

54. $\left(\dfrac{1}{2}x + \dfrac{1}{3}\right)\left(\dfrac{1}{2}x - \dfrac{1}{4}\right)$

55. $\left(\dfrac{1}{3}x + \dfrac{1}{5}\right)\left(\dfrac{1}{3}x - \dfrac{1}{2}\right)$

56. $(2a - 3b)(x - 5b)$

57. $(5b - 7c)(x - 2b)$

58. $(x - 3y)(z - 8y)$

Find the area of the rectangle.

▲ **59.**

5x + 2

2x − 3

▲ **60.**

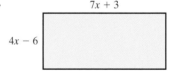

7x + 3

4x − 6

Cumulative Review Problems

61. Solve for x. $3(x - 6) = -2(x + 4) + 6x$

62. Solve for w. $3(w - 7) - (4 - w) = 11w$

63. La Tanya has three more quarters than dimes, giving her $3.55. How many of each coin does she have? (Use d for the number of dimes.)

64. A number decreased by three is seven more than three times the number. Find the number.

65. Heather returned from the bank with $375. She had one more $20 bill than she had $10 bills. The number of $5 bills she had was one less than triple the number of $10 bills. How many of each denomination did she have?

The polynomial $1.5x + 16$ can be used to describe the number of hours per week that young people age 12–18 spend playing video games, where x is the number of years since 1992. (Source: U.S. Department of Health and Human Services)

66. What was the average number of hours per week young people spent playing video games in 1994?

67. What was the average number of hours per week young people spent playing video games in 1996?

68. Predict the number of hours per week young people will spend playing video games in 2002.

69. Predict the number of hours per week young people will spend playing video games in 2003.

4.5 Multiplication: Special Cases

1 ⬤ *Multiplying Binomials of the Type* $(a + b)(a - b)$

The case when you multiply $(x + y)(x - y)$ is interesting and deserves special consideration. Using the FOIL method, we find

$$(x + y)(x - y) = x^2 - xy + xy - y^2 = x^2 - y^2.$$

Notice that the sum of the inner product and the outer product is zero. We see that

$$(x + y)(x - y) = x^2 - y^2.$$

This works in all cases when the binomials are the sum and difference of the same two terms. That is, in one factor the terms are added, while in the other factor the same two terms are subtracted.

$$(5a + 2b)(5a - 2b) = 25a^2 - 10ab + 10ab - 4b^2$$
$$= 25a^2 - 4b^2$$

The product is the difference of the squares of the terms. That is, $(5a)^2 - (2b)^2$ or $25a^2 - 4b^2$.

Many students find it helpful to memorize this equation.

Multiplying Binomials: A Sum and a Difference

$$(a + b)(a - b) = a^2 - b^2$$

You may use this relationship to find the product quickly in cases where it applies. The terms must be the same and there must be a sum and a difference.

⬤ **EXAMPLE 1** Multiply. $(7x + 2)(7x - 2)$

$$(7x + 2)(7x - 2) = (7x)^2 - (2)^2 = 49x^2 - 4$$

Practice Problem 1 Multiply. $(6x + 7)(6x - 7)$ ⬤

⬤ **EXAMPLE 2** Multiply. $(5x - 8y)(5x + 8y)$

$$(5x - 8y)(5x + 8y) = (5x)^2 - (8y)^2 = 25x^2 - 64y^2$$

Practice Problem 2 Multiply. $(3x + 5y)(3x - 5y)$ ⬤

2 ⬤ *Multiplying Binomials of the Type* $(a + b)^2$ *and* $(a - b)^2$

A second case that is worth special consideration is a binomial that is squared. Consider the following problem.

$$(3x + 2)^2 = (3x + 2)(3x + 2)$$
$$= 9x^2 + 6x + 6x + 4$$
$$= 9x^2 + 12x + 4$$

If you complete enough problems of this type, you will notice a pattern. The answer always contains the square of the first term added to double the product of the first and last terms added to the square of the last term.

3x is the first term	2 is the last term	Square the first term: $(3x)^2$	Double the product of the first and last terms: $2(3x)(2)$	Square the last term: $(2)^2$
↓	↓	↓	↓	↓

$$(3x \quad + \quad 2)^2 \quad = \quad 9x^2 \quad + \quad 12x \quad + \quad 4$$

We can show the same steps using variables instead of words.

$$(a + b)^2 = a^2 + 2ab + b^2$$

There is a similar formula for the square of a difference:

$$(a - b)^2 = a^2 - 2ab + b^2$$

We can use this formula to simplify $(2x - 3)^2$.

$$(2x - 3)^2 = (2x)^2 - 2(2x)(3) + (3)^2$$
$$= 4x^2 - 12x + 9$$

You may wish to multiply this product using FOIL to verify.

These two types of products, the square of a sum and the square of a difference, can be summarized as follows.

A Binomial Squared

$$(a + b)^2 = a^2 + 2ab + b^2$$
$$(a - b)^2 = a^2 - 2ab + b^2$$

EXAMPLE 3 Multiply. **(a)** $(5y - 2)^2$ **(b)** $(8x + 9y)^2$

(a) $(5y - 2)^2 = (5y)^2 - (2)(5y)(2) + (2)^2$
$$= 25y^2 - 20y + 4$$

(b) $(8x + 9y)^2 = (8x)^2 + (2)(8x)(9y) + (9y)^2$
$$= 64x^2 + 144xy + 81y^2$$

Practice Problem 3 Multiply. **(a)** $(5x + 4)^2$ **(b)** $(4a - 9b)^2$

Warning

$(a + b)^2 \neq a^2 + b^2$! The two sides are not equal! Squaring the sum $(a + b)$ does not give $a^2 + b^2$! Beginning algebra students often make this error. Make sure you remember that when you square a binomial there is always a *middle term*.

$$(a + b)^2 = a^2 + 2ab + b^2$$

Sometimes a numerical example helps you to see this.

$$(3 + 4)^2 \neq 3^2 + 4^2$$
$$7^2 \neq 9 + 16$$
$$49 \neq 25$$

Notice that what is missing on the right is $2ab = 2 \cdot 3 \cdot 4 = 24$.

3   *Multiplying Polynomials with More Than Two Terms*

We used the distributive property to multiply two binomials $(a + b)(c + d)$, and we obtained $ac + ad + bc + bd$. We could also use the distributive property to multiply the polynomials $(a + b)$ and $(c + d + e)$, and we would obtain $ac + ad + ae + bc + bd + be$. Let us see if we can find a direct way to multiply products such as $(3x - 2)(x^2 - 2x + 3)$. It can be done quickly using an approach similar to that used in arithmetic for multiplying whole numbers. Consider the following arithmetic problem.

$$
\begin{array}{r}
128 \\
\times \quad 43 \\
\hline
384 \\
512 \\
\hline
5504
\end{array}
$$

$384 \leftarrow$ The product of 128 and 3

$512 \leftarrow$ The product of 128 and 4 moved one space to the left

$5504 \leftarrow$ The sum of the two partial products

Let us follow a similar format to multiply the two polynomials. For example, multiply $(x^2 - 2x + 3)$ and $(3x - 2)$.

$$
\begin{array}{r}
x^2 - 2x + 3 \\
3x - 2 \\
\hline
-\,2x^2 + 4x - 6 \\
3x^3 - 6x^2 + 9x \\
\hline
3x^3 - 8x^2 + 13x - 6
\end{array}
$$

$-2x^2 + 4x - 6 \leftarrow$ The product $(x^2 - 2x + 3)(-2)$ This is often called **vertical multiplication**.

$3x^3 - 6x^2 + 9x \leftarrow$ The product $(x^2 - 2x + 3)(3x)$ moved one space to the left so that like terms are underneath each other

$3x^3 - 8x^2 + 13x - 6 \leftarrow$ The sum of the two partial products

EXAMPLE 4   Multiply vertically. $(3x^3 + 2x^2 + x)(x^2 - 2x - 4)$

$$
\begin{array}{r}
3x^3 + 2x^2 + x \\
x^2 - 2x - 4 \\
\hline
-12x^3 - 8x^2 - 4x \\
-6x^4 - 4x^3 - 2x^2 \\
3x^5 + 2x^4 + x^3 \\
\hline
3x^5 - 4x^4 - 15x^3 - 10x^2 - 4x
\end{array}
$$

We place one polynomial over the other.

$-12x^3 - 8x^2 - 4x \leftarrow$ The product $(3x^3 + 2x^2 + x)(-4)$

$-6x^4 - 4x^3 - 2x^2 \leftarrow$ The product $(3x^3 + 2x^2 + x)(-2x)$

$3x^5 + 2x^4 + x^3 \leftarrow$ The product $(3x^3 + 2x^2 + x)(x^2)$

$3x^5 - 4x^4 - 15x^3 - 10x^2 - 4x \leftarrow$ The sum of the three partial products

Note that the answers for each partial product are placed so that like terms are underneath each other.

Practice Problem 4   Multiply vertically. $(3x^2 - 2xy + 4y^2)(x - 2y)$

Alternative Method

Some students prefer to do this type of multiplication using a horizontal format similar to the FOIL method. The following example illustrates this approach.

EXAMPLE 5   Multiply horizontally. $(x^2 + 3x + 5)(x^2 - 2x - 6)$

We will use the distributive property repeatedly.

$$
\begin{aligned}
(x^2 + 3x + 5)(x^2 - 2x - 6) &= x^2(x^2 - 2x - 6) + 3x(x^2 - 2x - 6) + 5(x^2 - 2x - 6) \\
&= x^4 - 2x^3 - 6x^2 + 3x^3 - 6x^2 - 18x + 5x^2 - 10x - 30 \\
&= x^4 + x^3 - 7x^2 - 28x - 30
\end{aligned}
$$

Practice Problem 5   Multiply horizontally. $(2x^2 - 5x + 3)(x^2 + 3x - 4)$

Some problems may need to be done in two or more separate steps.

EXAMPLE 6 Multiply. $(2x - 3y)(x + 2y)(x + y)$

We first need to multiply any two of the binomials. Let us select the first pair.

$$\underbrace{(2x - 3y)(x + 2y)}_{\text{Find this product first.}}(x + y)$$

$$(2x - 3y)(x + 2y) = 2x^2 + 4xy - 3xy - 6y^2$$
$$= 2x^2 + xy - 6y^2$$

Now we replace the first two factors with their resulting product.

$$\underbrace{(2x^2 + xy - 6y^2)}_{\text{Result of first product}}(x + y)$$

We then multiply again.

$$(2x^2 + xy - 6y^2)(x + y) = (2x^2 + xy - 6y^2)x + (2x^2 + xy - 6y^2)y$$
$$= 2x^3 + x^2y - 6xy^2 + 2x^2y + xy^2 - 6y^3$$
$$= 2x^3 + 3x^2y - 5xy^2 - 6y^3$$

The vertical format of Example 4 is an alternative method for this type of problem.

$$\begin{array}{r} 2x^2 + xy - 6y^2 \\ x + y \\ \hline 2x^2y + xy^2 - 6y^3 \\ 2x^3 + x^2y - 6xy^2 \\ \hline 2x^3 + 3x^2y - 5xy^2 - 6y^3 \end{array}$$

$\left.\begin{array}{c} \\ \\ \end{array}\right\}$ Be sure to use special care in writing the exponents correctly in problems with more than one variable.

Thus we have

$$(2x - 3y)(x + 2y)(x + y) = 2x^3 + 3x^2y - 5xy^2 - 6y^3.$$

Note that it does not matter which two binomials are multiplied first. For example, you could first multiply $(2x - 3y)(x + y)$ to obtain $2x^2 - xy - 3y^2$ and then multiply that product by $(x + 2y)$ to obtain the same result.

Practice Problem 6 Multiply. $(3x - 2)(2x + 3)(3x + 2)$
(*Hint:* Rearrange the factors.)

4.5 Exercises

Verbal and Writing Skills

1. In the special case of $(a + b)(a - b)$, a binomial times a binomial is a _____.

2. Identify which of the following could be the answer to a problem using the formula for $(a + b)(a - b)$. Why?
   (a) $9x^2 - 16$ (b) $4x^2 + 25$
   (c) $9x^2 + 12x + 4$ (d) $x^4 - 1$

3. A student evaluated $(4x - 7)^2$ as $16x^2 + 49$. What is missing? State the correct answer.

4. The square of a binomial, $(a - b)^2$, always produces which of the following?
   (a) binomial   (b) trinomial
   (c) four-term polynomial

Use the formula $(a + b)(a - b) = a^2 - b^2$ to multiply.

5. $(y - 7)(y + 7)$

6. $(x + 5)(x - 5)$

7. $(x - 9)(x + 9)$

8. $(x + 6)(x - 6)$

9. $(8x + 3)(8x - 3)$

10. $(6x + 5)(6x - 5)$

11. $(2x - 7)(2x + 7)$

12. $(3x - 10)(3x + 10)$

13. $(5x - 3y)(5x + 3y)$

14. $(8a - 3b)(8a + 3b)$

15. $(0.6x + 3)(0.6x - 3)$

16. $(5x - 0.2)(5x + 0.2)$

Use the formula for a binomial squared to multiply.

17. $(3y + 1)^2$

18. $(4x - 1)^2$

19. $(5x - 4)^2$

20. $(6x + 5)^2$

21. $(9x + 5)^2$

22. $(5x - 7)^2$

23. $(3x - 7)^2$

24. $(2x + 3y)^2$

25. $\left(\dfrac{2}{3}x + \dfrac{1}{4}\right)^2$

26. $\left(\dfrac{3}{4}x + \dfrac{1}{2}\right)^2$

27. $(6w + 5z)^2$

28. $(5xy - 6z)^2$

Multiply. Use the special formula that applies.

29. $(7x + 3y)(7x - 3y)$

30. $(12a - 5b)(12a + 5b)$

31. $(7c^3 - 6d)^2$

32. $(4f^2 - 5g^2)^2$

Multiply.

33. $(x^2 + 3x - 2)(x - 3)$

34. $(x^2 + 3x - 1)(x + 3)$

35. $(4x + 1)(x^3 - 2x^2 + x - 1)$

36. $(3x - 1)(x^3 + x^2 - 4x - 2)$

37. $(x + 3)(x - 1)(3x - 8)$

38. $(x - 7)(x + 4)(2x - 5)$

39. $(3x + 5)(x - 2)(x - 4)$

40. $(2x - 7)(x + 1)(x - 2)$

41. $(x + 4)(2x - 7)(x - 4)$

42. $(x - 2)(5x + 3)(x - 2)$

43. $(a^2 - 3a + 2)(a^2 + 4a - 3)$

44. $(x^2 + 4x - 5)(x^2 - 3x + 4)$

To Think About

Find the volume.

▲ **45.**

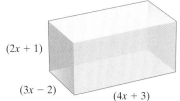

$(2x + 1)$
$(3x - 2)$
$(4x + 3)$

▲ **46.**

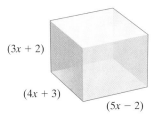

$(3x + 2)$
$(4x + 3)$
$(5x - 2)$

Cumulative Review Problems

47. An executive invested $18,000 in two accounts, one a cash-on-reserve account that yielded 7% simple interest, and the other a long-term certificate paying 11% simple interest. At the end of one year, the two accounts had earned her a total of $1540. How much had she invested in each account?

▲ **48.** The perimeter of a rectangular room measures 34 meters. The width is 2 meters more than half the length. Find the dimensions of the room.

49. Ammonia has a boiling point of $-33.4°C$. What is the boiling point of ammonia measured in degrees Fahrenheit? (Use $F = 1.8C + 32$.)

50. A man with 5.5 liters of blood has approximately 2.75×10^{10} red blood cells. Each red blood cell is a small disk that measures 7.5×10^{-6} meters in diameter. If all of the red blood cells were arranged in a long straight line, how long would that line be?

Putting Your Skills to Work

The Mathematics of DNA

The genetic material that organisms inherit from their parents consists of DNA. Each time a cell reproduces itself by dividing, its DNA is copied and passed on to the next generation. The DNA molecule consists of two strands twisted in the shape of a double helix. Each strand is composed of fundamental units called nucleotides. Each nucleotide on one strand is bonded (base paired) to a complementary nucleotide on the opposing strand. (*Source:* Dr. Russ Camp, Biology Department, Gordon College, Wenham, MA.)

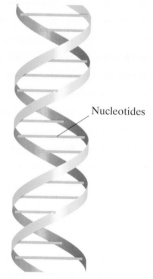

Nucleotides

Problems for Individual Investigation and Analysis

1. *E. coli*, the most popular bacterial cell used in molecular biology research, has a single chromosome made up of one double-stranded DNA molecule. The linear distance from one nucleotide pair to the next is 0.34 nanometer (one nanometer = 1×10^{-9} meter). The entire *E. coli* chromosome contains 4.5×10^6 nucleotide pairs (or base pairs). How long is the chromosome in nanometers?

2. How long is the chromosome in millimeters?

Problems for Cooperative Group Investigation

Together with other members of your class, see if you can determine the following.

3. If a human cell has 46 chromosomes containing a total of 8.0×10^9 nucleotide pairs, and if all the DNA in these chromosomes were added together, how long would the DNA be in meters?

4. How many times larger is the DNA in a human cell than the DNA in an *E. coli* cell?

Internet Connections

 Netsite: http://www.prenhall.com/tobey_beginning

Introduction to Primer on Molecular Genetics (Department of Energy) or a related site

This site provides some basic information about genes and DNA. Use the information to answer the following questions.

▲ 5. Estimate the volume of DNA in a human genome by assuming that, when the strands of DNA are end to end unwound and tied together, the resulting singular strand has the shape of a right circular cylinder.

6. Tell what percent of a human genome is made up of intron sequences and other noncoding regions.

4.6 Division of Polynomials

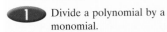
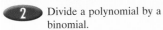

1 Dividing a Polynomial by a Monomial

To divide a polynomial by a monomial, divide each term of the numerator by the denominator; then write the sum of the results. We are using the property of fractions that states that

$$\frac{a + b}{c} = \frac{a}{c} + \frac{b}{c}.$$

Dividing a Polynomial by a Monomial

1. Divide each term of the polynomial by the monomial.

2. When dividing variables, use the property $\dfrac{x^a}{x^b} = x^{a-b}$.

EXAMPLE 1 Divide. $\dfrac{8y^6 - 8y^4 + 24y^2}{8y^2}$

$$\frac{8y^6 - 8y^4 + 24y^2}{8y^2} = \frac{8y^6}{8y^2} - \frac{8y^4}{8y^2} + \frac{24y^2}{8y^2} = y^4 - y^2 + 3$$

Practice Problem 1 Divide. $\dfrac{15y^4 - 27y^3 - 21y^2}{3y^2}$

2 Dividing a Polynomial by a Binomial

Division of a polynomial by a binomial is similar to long division in arithmetic. Notice the similarity in the following division problems.

***Division of a three-digit number
by a two-digit number***

$$
\begin{array}{r}
32 \\
21\overline{)672} \\
63 \\
\hline
42 \\
42 \\
\hline
0
\end{array}
$$

***Division of a polynomial
by a binomial***

$$
\begin{array}{r}
3x + 2 \\
2x + 1\overline{)6x^2 + 7x + 2} \\
6x^2 + 3x \\
\hline
4x + 2 \\
4x + 2 \\
\hline
0
\end{array}
$$

Dividing a Polynomial by a Binomial

1. Place the terms of the polynomial and binomial in descending order. Insert a 0 for any missing term.
2. Divide the first term of the polynomial by the first term of the binomial. The result is the first term of the answer.
3. Multiply the first term of the answer by the binomial and subtract the result from the first two terms of the polynomial. Bring down the next term to obtain a new polynomial.
4. Divide the new polynomial by the binomial using the process described in step 2.
5. Continue dividing, multiplying, and subtracting until the degree of the remainder is less than the degree of the binomial divisor.
6. Write the remainder as the numerator of a fraction that has the binomial divisor as its denominator.

EXAMPLE 2 Divide. $(x^3 + 5x^2 + 11x + 14) \div (x + 2)$

Step 1 The terms are arranged in descending order. No terms are missing.

Step 2 Divide the first term of the polynomial by the first term of the binomial. In this case, divide x^3 by x to get x^2.

$$
\begin{array}{r}
x^2 \\
x + 2 \overline{)x^3 + 5x^2 + 11x + 14}
\end{array}
$$

Step 3 Multiply x^2 by $x + 2$ and subtract the result from the first two terms of the polynomial, $x^3 + 5x^2$ in this case.

$$
\begin{array}{r}
x^2 \\
x + 2 \overline{)x^3 + 5x^2 + 11x + 14} \\
\underline{x^3 + 2x^2} \downarrow \\
3x^2 + 11x
\end{array}
$$
Bring down the next term.

Step 4 Continue to use the step 2 process. Divide $3x^2$ by x. Write the resulting $3x$ as the next term of the answer.

$$
\begin{array}{r}
x^2 + 3x \\
x + 2 \overline{)x^3 + 5x^2 + 11x + 14} \\
\underline{x^3 + 2x^2} \\
3x^2 + 11x
\end{array}
$$

Step 5 Continue multiplying, dividing, and subtracting until the degree of the remainder is less than the degree of the divisor. In this case, we stop when the remainder does not have an x.

$$
\begin{array}{r}
x^2 + 3x + 5 \\
x + 2 \overline{)x^3 + 5x^2 + 11x + 14} \\
\underline{x^3 + 2x^2} \\
3x^2 + 11x \\
\underline{3x^2 + 6x} \\
5x + 14 \\
\underline{5x + 10} \\
4
\end{array}
$$
← The remainder is 4.

Step 6 The answer is $x^2 + 3x + 5 + \dfrac{4}{x + 2}$.

To check the answer, we multiply $(x + 2)(x^2 + 3x + 5)$ and add the remainder 4.

$$(x + 2)(x^2 + 3x + 5) + 4 = x^3 + 5x^2 + 11x + 10 + 4 = x^3 + 5x^2 + 11x + 14$$

This is the original polynomial. It checks.

Practice Problem 2 Divide. $(x^3 + 10x^2 + 31x + 35) \div (x + 4)$

Take great care with the subtraction step when negative numbers are involved.

→ ● **EXAMPLE 3** Divide. $(5x^3 - 24x^2 + 9) \div (5x + 1)$

We must first insert $0x$ to represent the missing x-term. Then we divide $5x^3$ by $5x$.

$$
\begin{array}{r}
x^2 \\
5x + 1 \overline{)\,5x^3 - 24x^2 + 0x + 9} \\
\underline{5x^3 + x^2} \\
-25x^2
\end{array}
$$

Note that we are subtracting:
$-24x^2 - (+1x^2) = -24x^2 - 1x^2$
$= -25x^2$

Next we divide $-25x^2$ by $5x$.

$$
\begin{array}{r}
x^2 - 5x \\
5x + 1 \overline{)\,5x^3 - 24x^2 + 0x + 9} \\
\underline{5x^3 + x^2} \\
-25x^2 + 0x \\
\underline{-25x^2 - 5x} \\
5x
\end{array}
$$

Note that we are subtracting:
$0x - (-5x) = 0x + 5x = 5x$

Finally, we divide $5x$ by $5x$.

$$
\begin{array}{r}
x^2 - 5x + 1 \\
5x + 1 \overline{)\,5x^3 - 24x^2 + 0x + 9} \\
\underline{5x^3 + x^2} \\
-25x^2 + 0x \\
\underline{-25x^2 - 5x} \\
5x + 9 \\
\underline{5x + 1} \\
8
\end{array}
$$

←—— The remainder is 8.

The answer is $x^2 - 5x + 1 + \dfrac{8}{5x + 1}$.

To check, multiply $(5x + 1)(x^2 - 5x + 1)$ and add the remainder 8.

$$(5x + 1)(x^2 - 5x + 1) + 8 = 5x^3 - 24x^2 + 1 + 8 = 5x^3 - 24x^2 + 9$$

This is the original polynomial. Our answer is correct.

Practice Problem 3 Divide. $(2x^3 - x^2 + 1) \div (x - 1)$

Now we will perform the division by writing a minimum of steps. See if you can follow each step.

EXAMPLE 4 Divide and check. $(12x^3 - 11x^2 + 8x - 4) \div (3x - 2)$

$$
\begin{array}{r}
4x^2 - x + 2 \\
3x - 2 \overline{\smash{)}12x^3 - 11x^2 + 8x - 4} \\
\underline{12x^3 - 8x^2} \\
-3x^2 + 8x \\
\underline{-3x^2 + 2x} \\
6x - 4 \\
\underline{6x - 4} \\
0
\end{array}
$$

Check. $(3x - 2)(4x^2 - x + 2) = 12x^3 - 3x^2 + 6x - 8x^2 + 2x - 4$

$\qquad\qquad\qquad\qquad\qquad = 12x^3 - 11x^2 + 8x - 4$ Our answer is correct.

Practice Problem 4 Divide and check. $(20x^3 - 11x^2 - 11x + 6) \div (4x - 3)$

Developing Your Study Skills

Exam Time: Taking the Exam

Allow yourself plenty of time to get to your exam. You may even find it helpful to arrive a little early in order to collect your thoughts and ready yourself. This will help you feel more relaxed.

After you get your exam, you will find it helpful to do the following.

1. Take two or three moderately deep breaths. Inhale, then exhale slowly. You will feel your entire body begin to relax.

2. Write down on the back of the exam any formulas or ideas that you need to remember.

3. Look over the entire test quickly in order to pace yourself and use your time wisely. Notice how many points each question is worth. Spend more time on items of greater worth.

4. Read directions carefully and be sure to answer all questions clearly. Keep your work neat and easy to read.

5. Ask your instructor about anything that is not clear to you.

6. Answer the questions that are easiest for you first. Then come back to the more difficult ones.

7. Do not get bogged down on one question for too long because it may jeopardize your chances of finishing other problems. Leave the tough question and come back to it when you have time later.

8. Check your work. This will help you to catch minor errors.

9. Stay calm if others leave before you do. You are entitled to use the full amount of allotted time. You will do better on the exam if you take your time and work carefully.

Divide.

1. $\dfrac{25x^4 - 15x^2 + 20x}{5x}$

2. $\dfrac{18b^5 - 12b^3 + 6b^2}{3b^2}$

3. $\dfrac{8y^4 - 12y^3 - 4y^2}{4y^2}$

4. $\dfrac{10y^4 - 35y^3 + 5y^2}{5y^2}$

5. $\dfrac{49x^6 - 21x^4 + 56x^2}{7x^2}$

6. $\dfrac{36x^6 + 54x^4 - 6x^2}{6x^2}$

7. $\left(48x^7 - 54x^4 + 36x^3\right) \div 6x^3$

8. $\left(72x^8 - 56x^5 - 40x^3\right) \div 8x^3$

Divide. Check your answers for exercises 9–16 by multiplication.

9. $\dfrac{6x^2 + 13x + 5}{2x + 1}$

10. $\dfrac{12x^2 + 19x + 5}{3x + 1}$

11. $\dfrac{x^2 - 9x - 6}{x - 7}$

12. $\dfrac{x^2 - 8x - 4}{x - 6}$

13. $\dfrac{3x^3 - x^2 + 4x - 2}{x + 1}$

14. $\dfrac{2x^3 - 3x^2 - 3x + 6}{x - 1}$

15. $\dfrac{4x^3 + 4x^2 - 19x - 15}{2x + 5}$

16. $\dfrac{6x^3 + 11x^2 - 8x + 5}{2x + 5}$

17. $\dfrac{6x^3 + x^2 - 7x + 2}{3x - 1}$

18. $\dfrac{3x^3 + 8x^2 - 6x + 2}{3x - 1}$

19. $\dfrac{12y^3 - 12y^2 - 25y + 31}{2y - 3}$

20. $\dfrac{9y^3 - 30y^2 + 31y - 4}{3y - 5}$

21. $\left(y^3 - y^2 - 13y - 12\right) \div (y + 3)$

22. $\left(4y^3 - 17y^2 + 7y + 10\right) \div (4y - 5)$

23. $\left(y^4 - 9y^2 - 5\right) \div (y - 2)$

24. $\left(2y^4 + 3y^2 - 5\right) \div (y - 2)$

To Think About

25. $(8y^3 + 3y - 7) \div (4y - 1)$

(*Hint:* The answer contains fractions.)

26. $(6y^3 - 3y^2 + 4) \div (2y + 1)$

(*Hint:* The answer contains fractions.)

Cumulative Review Problems

27. The average home in the United States uses 110,000 gallons of water per year. Most environmental groups feel that with improved efficiency of water-using devices (washers, toilets, faucets, shower heads), this could be reduced by 30%. (*Source:* U.S. Energy Information Administration.) How many gallons of water per year would be used in the average home if this reduction were accomplished?

28. From 1900 to 2000, the population of New York City increased by 123% to a new total of 7.6 million. (*Source:* U.S. Census Bureau.) What was the population of the city in 1900?

29. In a veterinary study of a group of cats over age 5 who experienced regular exercise and playtime, 184 of the cats, or 92%, had no health problems. How many cats were in the study?

30. Marlena's job requires doing research on the Internet for a marketing company. Today, she has downloaded a huge file. She has been reading all morning and has just completed two consecutive pages whose page numbers add to 1039. What are the page numbers?

Putting Your Skills to Work

The Enrollment Polynomial

The U.S. National Center for Education Statistics maintains records of past enrollments of students in school as well as projected enrollments for the future. The following bar graph below indicates the number of students in school for several selected years and includes some projected figures for the future.

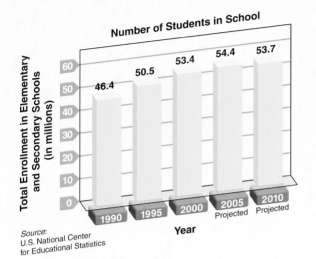

Number of Students in School

Source:
U.S. National Center
for Educational Statistics

Problems for Individual Investigation and Study

1. How many more students were enrolled in school in 1995 than in 1990?

2. How many fewer students are projected to be enrolled in school in 2010 than in 2005?

Problems for Group Investigation and Cooperative Learning

The approximate total number of students enrolled in the United States during a given year can be obtained by using the Enrollment Polynomial. It is written as

$$E = -0.00031x^3 - 0.024268x^2 + 0.975171x + 46.3761,$$

where x is the number of years since 1990 and E is the total number of students in millions. Use a scientific calculator, a graphing calculator, or a computer to evaluate the Enrollment Polynomial and answer the following questions. Round to the nearest tenth of a million.

3. How many students were enrolled in 1998?

4. How many students were enrolled in 1994?

5. What is the projected number of students who will be enrolled in 2006?

6. What is the projected number of students who will be enrolled in 2009?

Internet Connections

 Netsite: http://www.prenhall.com/tobey_beginning

Site: The U.S. Department of Education

7. Based on the current projections, what is the number of classroom teachers that will be needed in 2006? Give the figures for the low, medium, and high estimates.

8. Comparing elementary teachers with secondary teachers, which group will have experienced the higher percentage growth in the period 1994–2006? The higher absolute growth? Consider the low, medium, and high estimates.

Math in the Media

Tallest Hotel in the World

Source: Guinness World Book®.com

According to the *Guinness World Book*®, the Burj Al Arab Hotel (Arabian Tower) is the world's tallest hotel. It is located South of Dubai, United Arab Emirates. The hotel stands 1052 feet tall and is shaped like a sail. This structure also features the world's fastest elevators which travel at 15 miles per hour, 22 feet per second.

EXERCISES

If an object is dropped from the top of the hotel, could you calculate the height of the object at specified times? Try the following questions.

1. Neglecting air resistance, the height of the object at a time t seconds after the object is dropped is given by the polynomial $-16t^2 + 1052$. Find the height of the object when $t = 1$ second and when $t = 5$ seconds.

2. Using the polynomial from exercise 1, how long before the object hits the ground? Use a scientific calculator or a graphing calculator to assist you. Find the answer to the nearest tenth of a second.

3. Use your results from exercise 1. Suppose an elevator is traveling at maximum speed from the height at $t = 1$ second to the height when $t = 5$ seconds. How much longer does it take the elevator to travel this distance than it took the object to fall the same distance? Round your answer to the nearest tenth of a second.

Chapter 4 Organizer

Topic	Procedure	Examples
Multiplying monomials, p. 248.	$x^a \cdot x^b = x^{a+b}$ 1. Multiply the numerical coefficients. 2. Add the exponents of a given base.	$3^{12} \cdot 3^{15} = 3^{27}$ $x^3 \cdot x^4 = x^7$ $(-3x^2)(6x^3) = -18x^5$ $(2ab)(4a^2b^3) = 8a^3b^4$
Dividing monomials, p. 250.	$\dfrac{x^a}{x^b} = \begin{cases} x^{a-b} & \text{Use if } a \text{ is greater than } b. \\ \dfrac{1}{x^{b-a}} & \text{Use if } b \text{ is greater than } a. \end{cases}$ 1. Divide or reduce the fraction created by the quotient of the numerical coefficients. 2. Subtract the exponents of a given base.	$\dfrac{16x^7}{8x^3} = 2x^4$ $\dfrac{5x^3}{25x^5} = \dfrac{1}{5x^2}$ $\dfrac{-12x^5y^7}{18x^3y^{10}} = -\dfrac{2x^2}{3y^3}$
Exponent of zero, p. 251.	$x^0 = 1 \quad$ if $x \neq 0$	$5^0 = 1 \qquad \dfrac{x^6}{x^6} = 1$ $w^0 = 1 \qquad 3x^0y = 3y$
Raising a power to a power, p. 252.	$(x^a)^b = x^{ab}$ $(xy)^a = x^a y^a$ $\left(\dfrac{x}{y}\right)^a = \dfrac{x^a}{y^a} \quad (y \neq 0)$ 1. Raise the numerical coefficient to the power outside the parentheses. 2. Multiply the exponent outside the parentheses times the exponent inside the parentheses.	$(x^9)^3 = x^{27}$ $(3x^2)^3 = 27x^6$ $\left(\dfrac{2x^2}{y^3}\right)^3 = \dfrac{8x^6}{y^9}$ $(-3x^4y^5)^4 = 81x^{16}y^{20}$ $(-5ab)^3 = -125a^3b^3$
Negative exponents, p. 257.	If $x \neq 0$ and $y \neq 0$, then $x^{-n} = \dfrac{1}{x^n}$ $\dfrac{1}{x^{-n}} = x^n$ $\dfrac{x^{-m}}{y^{-n}} = \dfrac{y^n}{x^m}$	Write with positive exponents. $3^{-4} = \dfrac{1}{3^4} = \dfrac{1}{81}$ $x^{-6} = \dfrac{1}{x^6}$ $\dfrac{1}{w^{-3}} = w^3$ $\dfrac{w^{-12}}{z^{-5}} = \dfrac{z^5}{w^{12}}$ $(2x^2)^{-3} = 2^{-3}x^{-6} = \dfrac{1}{2^3 x^6} = \dfrac{1}{8x^6}$
Scientific notation, p. 258.	A positive number is written in scientific notation if it is in the form $a \times 10^n$, where $1 \leq a < 10$ and n is an integer.	$128 = 1.28 \times 10^2$ $2{,}568{,}000 = 2.568 \times 10^6$ $13{,}200{,}000{,}000 = 1.32 \times 10^{10}$ $0.16 = 1.6 \times 10^{-1}$ $0.00079 = 7.9 \times 10^{-4}$ $0.0000034 = 3.4 \times 10^{-6}$
Add polynomials, p. 266.	To add two polynomials, we add their like terms.	$(-7x^3 + 2x^2 + 5) + (x^3 + 3x^2 + x)$ $\qquad = -6x^3 + 5x^2 + x + 5$

Topic	Procedure	Examples
Subtracting polynomials, p. 267.	To subtract the polynomials, change all signs of the second polynomial and add the result to the first polynomial. $(a) - (b) = (a) + (-b)$	$(5x^2 - 6) - (-3x^2 + 2) = (5x^2 - 6) + (+3x^2 - 2)$ $= 8x^2 - 8$
Multiplying a monomial by a polynomial, p. 271.	Use the distributive property. $a(b + c) = ab + ac$ $(b + c)a = ba + ca$	Multiply. $-5x(2x^2 + 3x - 4) = -10x^3 - 15x^2 + 20x$ $(6x^3 - 5xy - 2y^2)(3xy) = 18x^4y - 15x^2y^2 - 6xy^3$
Multiplying two binomials, p. 272.	1. The product of the sum and difference of the same two terms yields the difference of their squares. $(a + b)(a - b) = a^2 - b^2$ 2. The square of a binomial yields a trinomial: the square of the first term plus twice the product of the first and second terms, plus the square of the second term. $(a + b)^2 = a^2 + 2ab + b^2$ $(a - b)^2 = a^2 - 2ab + b^2$ 3. Use FOIL for other binomial multiplication. The middle terms can often be combined, giving a trinomial answer.	Multiply. $(3x + 7y)(3x - 7y) = 9x^2 - 49y^2$ $(3x + 7y)^2 = 9x^2 + 42xy + 49y^2$ $(3x - 7y)^2 = 9x^2 - 42xy + 49y^2$ $(3x - 5)(2x + 7) = 6x^2 + 21x - 10x - 35$ $= 6x^2 + 11x - 35$
Multiplying two polynomials, p. 279.	To multiply two polynomials, multiply each term of one by each term of the other. This method is similar to the multiplication of many-digit numbers.	Vertical method: $3x^2 - 7x + 4$ $\times \quad 3x - 1$ $\overline{-3x^2 + 7x - 4}$ $9x^3 - 21x^2 + 12x$ $\overline{9x^3 - 24x^2 + 19x - 4}$ Horizontal method: $(5x + 2)(2x^2 - x + 3)$ $= 10x^3 - 5x^2 + 15x + 4x^2 - 2x + 6$ $= 10x^3 - x^2 + 13x + 6$
Multiplying three or more polynomials, p. 280.	1. Multiply any two polynomials. 2. Multiply the result by any remaining polynomials.	$(2x + 1)(x - 3)(x + 4) = (2x^2 - 5x - 3)(x + 4)$ $2x^2 - 5x - 3$ $x + 4$ $\overline{8x^2 - 20x - 12}$ $2x^3 - 5x^2 - 3x$ $\overline{2x^3 + 3x^2 - 23x - 12}$
Dividing a polynomial by a monomial, p. 284.	1. Divide each term of the polynomial by the monomial. 2. When dividing variables, use the property $\dfrac{x^a}{x^b} = x^{a-b}.$	Divide. $(15x^3 + 20x^2 - 30x) \div (5x)$ $= \dfrac{15x^3}{5x} + \dfrac{20x^2}{5x} + \dfrac{-30x}{5x}$ $= 3x^2 + 4x - 6$

Topic	Procedure	Examples
Dividing a polynomial by a binomial, p. 285.	1. Place the terms of the polynomial and binomial in descending order. Insert a 0 for any missing term. 2. Divide the first term of the polynomial by the first term of the binomial. 3. Multiply the partial answer by the binomial, and subtract the result from the first two terms of the polynomial. Bring down the next term to obtain a new polynomial. 4. Divide the new polynomial by the binomial using the process described in step 2. 5. Continue dividing, multiplying, and subtracting until the degree of the remainder is less than the degree of the binomial divisor. 6. Write the remainder as the numerator of a fraction that has the binomial divisor as its denominator.	Divide. $$(8x^3 - 13x + 2x^2 + 7) \div (4x - 1)$$ We rearrange the terms. $$\begin{array}{r} 2x^2 + x - 3 \\ 4x - 1 \overline{\smash{)}8x^3 + 2x^2 - 13x + 7} \\ \underline{8x^3 - 2x^2} \\ 4x^2 - 13x \\ \underline{4x^2 - x} \\ -12x + 7 \\ \underline{-12x + 3} \\ 4 \end{array}$$ The answer is $$2x^2 + x - 3 + \frac{4}{4x - 1}.$$

Chapter 4 Review Problems

4.1 *Simplify. Leave your answer in exponent form.*

1. $(-6a^2)(3a^5)$

2. $(5^{10})(5^{13})$

3. $(3xy^2)(2x^3y^4)$

4. $\dfrac{8^{20}}{8^3}$

5. $\dfrac{7^{15}}{7^{27}}$

6. $\dfrac{x^{12}}{x^{17}}$

7. $\dfrac{y^{30}}{y^{16}}$

8. $\dfrac{3x^8y^0}{9x^4}$

9. $\dfrac{-15xy^2}{25x^6y^6}$

10. $\dfrac{-12a^3b^6}{18a^2b^{12}}$

11. $(x^3)^8$

12. $(5xy^2)^3$

13. $(-3a^3b^2)^2$

14. $\dfrac{2x^4}{3y^2}$

15. $\left(\dfrac{5ab^2}{c^3}\right)^2$

16. $\left(\dfrac{x^0y^3}{4w^5z^2}\right)^3$

4.2   *Simplify. Write with positive exponents.*

17. x^{-3}

18. $x^{-5}y^{-11}$

19. $\dfrac{2x^{-6}}{y^{-3}}$

20. $2^{-1}x^5y^{-6}$

21. $\left(2x^3\right)^{-2}$

22. $\dfrac{3x^{-3}}{y^{-2}}$

23. $\dfrac{4x^{-5}y^{-6}}{w^{-2}z^8}$

24. $\dfrac{3^{-3}a^{-2}b^5}{c^{-3}d^{-4}}$

Write in scientific notation.

25. 156,340,200,000

26. 179,632

27. 0.0078

28. 0.00006173

Write in decimal notation.

29. 1.2×10^5

30. 8.367×10^{10}

31. 3×10^6

32. 2.5×10^{-1}

33. 5.708×10^{-8}

34. 6×10^{-9}

Perform the indicated calculation. Leave your answer in scientific notation.

35. $\dfrac{(28,000,000)(5,000,000,000)}{7000}$

36. $\left(3.12 \times 10^5\right)\left(2.0 \times 10^6\right)\left(1.5 \times 10^8\right)$

37. $\left(1.6 \times 10^{-3}\right)\left(3.0 \times 10^{-5}\right)\left(2.0 \times 10^{-2}\right)$

38. $\dfrac{(0.00078)(0.000005)(0.00004)}{0.002}$

39. If a space probe travels at 40,000 kilometers per hour for 1 year, how far will it travel? (Assume that 1 year = 365 days.)

40. An atomic clock is based on the fact that cesium emits 9,192,631,770 cycles of radiation in one second. How many of these cycles occur in one day? Round to three significant digits.

41. Today's fastest modern computers can perform one operation in 1×10^{-8} second. How many operations can such a computer perform in 1 minute?

4.3   *Combine.*

42. $\left(2x^2 - 3x + 5\right) + \left(-7x^2 - 8x - 23\right)$

43. $\left(1.2x^2 - 3.4x + 6\right) + \left(5.5x^2 - 7.6x - 3\right)$

44. $\left(x^3 + x^2 - 6x + 2\right) - \left(2x^3 - x^2 - 5x - 6\right)$

45. $\left(4x^3 - x^2 - x + 3\right) - \left(-3x^3 + 2x^2 + 5x - 1\right)$

46. $\left(9x^3y^3 + 3xy - 4\right) - \left(4x^3y^3 + 2x^2y^2 - 7xy\right)$

47. $\dfrac{1}{2}x^2 - \dfrac{3}{4}x + \dfrac{1}{5} - \left(\dfrac{1}{4}x^2 - \dfrac{1}{2}x + \dfrac{1}{10}\right)$

48. $\left(5x^2 + 3x\right) + \left(-6x^2 + 2\right) - (5x - 8)$

49. $\left(2x^2 - 7\right) - \left(3x^2 - 4\right) + \left(-5x^2 - 6x\right)$

4.4 *Multiply.*

50. $(3x + 1)(5x - 1)$

51. $(7x - 2)(4x - 3)$

52. $(2x + 3)(10x + 9)$

53. $5x(2x^2 - 6x + 3)$

54. $(3x^2y^2 - 5xy + 6)(-2xy)$

55. $(x^3 - 3x^2 + 5x - 2)(4x)$

56. $(5a + 7b)(a - 3b)$

57. $(2x^2 - 3)(4x^2 - 5y)$

58. $-3x^2y(5x^4y + 3x^2 - 2)$

4.5 *Multiply.*

59. $(3x - 2)^2$

60. $(5x + 3)(5x - 3)$

61. $(7x + 6y)(7x - 6y)$

62. $(5a - 2b)^2$

63. $(8x + 9y)^2$

64. $(x^2 + 7x + 3)(4x - 1)$

65. $(x - 6)(2x - 3)(x + 4)$

4.6 *Divide.*

66. $(12y^3 + 18y^2 + 24y) \div (6y)$

67. $(30x^5 + 35x^4 - 90x^3) \div (5x^2)$

68. $(16x^3y^2 - 24x^2y + 32xy^2) \div (4xy)$

69. $(106x^6 - 24x^5 + 38x^4 + 26x^3) \div (2x^3)$

70. $(16x^2 - 8x - 3) \div (4x - 3)$

71. $(15x^2 + 41x + 14) \div (5x + 2)$

72. $(6x^3 + x^2 + 6x + 5) \div (2x - 1)$

73. $(2x^3 - x^2 + 3x - 1) \div (x + 2)$

74. $(12x^2 + 11x + 2) \div (3x + 2)$

75. $(8x^2 - 6x + 6) \div (2x + 1)$

76. $(x^3 - x - 24) \div (x - 3)$

77. $(2x^3 - 3x + 1) \div (x - 2)$

Applications

Solve. Express your answer in scientific notation.

78. In 2000, the estimated population of China was 1.256×10^9 people, while the estimated population of Brazil was 1.74×10^8 people. (*Source:* United Nations Statistical Bureau.) What was the total population in those two countries?

79. In 2000, the estimated population of India was 1.018×10^9 people, while the estimated population of Bangladesh was 1.29×10^8 people. (*Source:* United Nations Statistical Bureau.) What was the total population in those two countries?

80. The mass of an electron is approximately 9.11×10^{-28} gram. Find the mass of 30,000 electrons.

81. The sun radiates energy into space at the rate of 3.9×10^{26} joules per second. How many joules are emitted in a day?

To Think About

Find a polynomial that describes the shaded area.

▲ **82.**

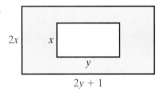

▲ **83.**

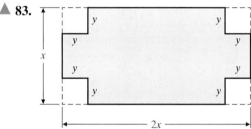

Chapter 4 Test

Simplify. Leave your answer in exponent form.

1. $(3^{10})(3^{24})$

2. $\dfrac{3^{35}}{3^7}$

3. $(8^4)^6$

In questions 4–8, simplify.

4. $(-3xy^4)(-4x^3y^6)$

5. $\dfrac{-35x^8y^{10}}{25x^5y^{10}}$

6. $(-5xy^6)^3$

7. $\left(\dfrac{7a^7b^2}{3c^0}\right)^2$

8. $\dfrac{4a^5b^6}{16a^{10}b^{12}}$

9. Evaluate. 4^{-3}

In questions 10 and 11, write with only positive exponents.

10. $6a^{-4}b^{-3}c^5$

11. $\dfrac{2x^{-3}y^{-4}}{w^{-6}z^8}$

12. Write in scientific notation. 0.0005482

13. Write in decimal notation. 5.82×10^8

14. Multiply. Leave your answer in scientific notation.
$(4.0 \times 10^{-3})(3.0 \times 10^{-8})(2.0 \times 10^4)$

Combine.

15. $(2x^2 - 3x - 6) + (-4x^2 + 8x + 6)$

16. $(5x^2 + 6xy - 7y^2) - (2x^2 + 3xy + 6y)$

Multiply.

17. $-7x^2(3x^3 - 4x^2 + 6x - 2)$

18. $(5x^2y^2 - 6xy + 2)(3x^2y)$

19. $(5a - 4b)(2a + 3b)$

20. $(3x + 2)(2x + 1)(x - 3)$

21. $(7x^2 + 2y^2)^2$

22. $(9x - 2y)(9x + 2y)$

23. $(3x - 2)(4x^3 - 2x^2 + 7x - 5)$

24. $(3x^2 - 5xy)(x^2 + 3xy)$

Divide.

25. $15x^6 - 5x^4 + 25x^3 \div 5x^3$

26. $(8x^3 - 22x^2 - 5x + 12) \div (4x + 3)$

27. $(2x^3 - 6x - 36) \div (x - 3)$

Solve. Express your answer in scientific notation. Round to the nearest hundredth.

28. The estimated population of the United States in 2000 was 2.749×10^8 people. The area of the United States is approximately 3.50×10^6 square miles. (*Source:* U.S. Census Bureau.) How many people per square mile were there in the United States in 2000?

29. A space probe is traveling from Earth to the planet Pluto at a speed of 2.49×10^4 miles per hour. How far would this space probe travel in one week?

1. _____

2. _____

3. _____

4. _____

5. _____

6. _____

7. _____

8. _____

9. _____

10. _____

11. _____

12. _____

13. _____

14. _____

15. _____

16. _____

17. _____

18. _____

19. _____

20. _____

21. _____

22. _____

23. _____

24. _____

25. _____

26. _____

27. _____

28. _____

29. _____

299

1. _____

2. _____

3. _____

4. _____

5. _____

6. _____

7. _____

8. _____

9. _____

10. _____

11. _____

12. _____

13. _____

14. _____

15. _____

16. _____

17. _____

18. _____

19. _____

20. _____

21. _____

22. _____

23. _____

24. _____

25. _____

26. _____

Approximately one-half of this test covers the content of Chapters 0 through 3. The remainder covers the content of Chapter 4.

In questions 1–5, simplify.

1. $\dfrac{5}{12} - \dfrac{7}{8}$

2. $(-3.7) \times (0.2)$

3. $\left(-4\dfrac{1}{2}\right) \div \left(5\dfrac{1}{4}\right)$

4. 7% of $128.00

5. $7x(3x - 4) - 5x(2x - 3) - (3x)^2$

6. Evaluate $2x^2 - 3xy + y^2$ when $x = -2$ and $y = 3$.

In questions 7–9, solve.

7. $7x - 3(4 - 2x) = 14x - (3 - x)$

8. $\dfrac{x - 5}{3} = \dfrac{1}{4}x + 2$

9. $4 - 7x < 11$

10. Solve for f. $B = \frac{1}{2}a(c + 3f)$

11. A national walkout of 11,904 employees of the VBM Corp. occurred last month. This was 96% of the total number of employees. How many employees does VBM have?

12. How much interest would $3320 earn in one year if invested at 6% simple interest?

13. Multiply. $(3x - 7)(5x - 4)$

14. Multiply. $(3x - 5)^2$

15. Multiply. $(3x + 2)(2x + 1)(x - 4)$

In questions 16–18, simplify.

16. $(-4x^4y^5)(5xy^3)$

17. $\dfrac{14x^8y^3}{-21x^5y^{12}}$

18. $(-3xy^4z^2)^3$

19. Write with only positive exponents. $\dfrac{9x^{-3}y^{-4}}{w^2z^{-8}}$

20. Write in scientific notation. 1,360,000,000,000,000

21. Write in scientific notation. 0.00056

22. Calculate. Leave your answer in scientific notation. $\dfrac{(2.0 \times 10^{-12})(8.0 \times 10^{-20})}{4.0 \times 10^3}$

23. Subtract. $(x^3 - 3x^2 - 5x + 20) - (-4x^3 - 10x^2 + x - 30)$

Multiply.

24. $-6xy^2(6x^2 - 3xy + 8y^2)$

25. $(x^2 - 6x + 3)(2x^2 - 3x + 4)$

26. $(x^2 + 2x - 12) \div (x - 3)$

Factoring

I f you buy new office equipment or a new vehicle and you use it for your business, you will need to keep track of how much these items depreciate in value each year for tax purposes. The IRS has three methods for calculating the depreciation. One of them is called the sum-of-the-year's-digits method. This common math calculation often gives people some difficulty. Do you think you could perform this calculation? Turn to the Putting Your Skills to Work problems on page 348 to find out.

Pretest Chapter 5

1. _____

2. _____

3. _____

4. _____

5. _____

6. _____

7. _____

8. _____

9. _____

10. _____

11. _____

12. _____

13. _____

14. _____

15. _____

16. _____

17. _____

18. _____

19. _____

20. _____

21. _____

22. _____

23. _____

24. _____

If you are familiar with the topics in this chapter, take this test now. Check your answers with those in the back of the book. If an answer is wrong or you can't answer a question, study the appropriate section of the chapter.

If you are not familiar with the topics in this chapter, don't take this test now. Instead, study the examples, work the practice problems, and then take the test.

This test will help you to identify which concepts you have mastered and which you need to study further.

Section 5.1

Factor out the greatest common factor.

1. $2x^2 - 6xy + 12xy^2$

2. $3x(a - 2b) + 4y(a - 2b)$

3. $36ab^2 - 18ab$

Section 5.2

Factor.

4. $5a - 10b - 3ax + 6xb$

5. $3x^2 - 4y + 3xy - 4x$

6. $21x^2 - 14x - 9x + 6$

Section 5.3

Factor.

7. $x^2 - 22x - 48$

8. $x^2 - 8x + 15$

9. $x^2 + 9x + 8$

10. $2x^2 + 8x - 24$

11. $3x^2 - 6x - 189$

Section 5.4

Factor.

12. $15x^2 - 16x + 4$

13. $6y^2 + 5yz - 6z^2$

14. $12x^2 + 44x + 40$

Section 5.5

Factor completely.

15. $81x^4 - 16$

16. $49x^2 - 28xy + 4y^2$

17. $25x^2 + 80x + 64$

Section 5.6

Factor completely. If not possible, so state.

18. $6x^3 + 15x^2 - 9x$

19. $32x^2y^2 - 48xy^2 + 18y^2$

20. $25x^2 + 81$

Section 5.7

Solve for the roots of each quadratic equation.

21. $2x^2 + x - 3 = 0$

22. $x^2 + 30 = 11x$

23. $\dfrac{3x^2 - 7x}{2} = 3$

▲ **24.** The walking trail to the top of Mount Washington is marked by triangular signs. The area of each sign is 60 square centimeters. The base of each sign is 1 centimeter less than double the altitude. Find the altitude and the base of the triangular signs.

5.1 Introduction to Factoring

1 *Factoring Polynomials Whose Terms Contain a Common Factor*

Recall that when two or more numbers, variables, or algebraic expressions are multiplied, each is called a **factor**.

$$\underbrace{3 \cdot 2}_{\text{factor } \text{factor}} \qquad \underbrace{3x^2 \cdot 5x^3}_{\text{factor } \text{factor}} \qquad \underbrace{(2x - 3)(x + 4)}_{\text{factor } \text{factor}}$$

When you are asked **to factor** a number or an algebraic expression, you are being asked, "What factors, when multiplied, will give that number or expression?"

For example, you can factor 6 as $3 \cdot 2$ since $3 \cdot 2 = 6$. You can factor $15x^5$ as $3x^2 \cdot 5x^3$ since $3x^2 \cdot 5x^3 = 15x^5$. Factoring is simply the reverse of multiplying. 6 and $15x^5$ are simple expressions to factor and can be factored in different ways.

The factors of the polynomial $2x^2 + x - 12$ are not so easy to recognize. In this chapter we will be learning techniques for finding the factors of a polynomial. We will begin with **common factors**.

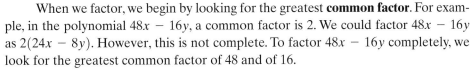

⬤ **EXAMPLE 1** Factor. **(a)** $3x - 6y$ **(b)** $9x + 2xy$

Begin by looking for a common factor, a factor that both terms have in common. Then rewrite the expression as a product.

(a) $3x - 6y = 3(x - 2y)$ This is true because $3(x - 2y) = 3x - 6y$.
(b) $9x + 2xy = x(9 + 2y)$ This is true because $x(9 + 2y) = 9x + 2xy$.

Some people find it helpful to think of factoring as the distributive property in reverse. When we write $3x - 6y = 3(x - 2y)$, we are doing the reverse of distributing the 3.

Practice Problem 1 Factor. **(a)** $21a - 7b$ **(b)** $p + prt$

When we factor, we begin by looking for the greatest **common factor**. For example, in the polynomial $48x - 16y$, a common factor is 2. We could factor $48x - 16y$ as $2(24x - 8y)$. However, this is not complete. To factor $48x - 16y$ completely, we look for the greatest common factor of 48 and of 16.

$$48x - 16y = 16(3x - y)$$

⬤ **EXAMPLE 2** Factor. $24xy + 12x^2 + 36x^3$. Be careful to remove the greatest common factor.

Find the greatest common factor of 24, 12, and 36. You may want to factor each number, or you may notice that 12 is a common factor. 12 is the greatest numerical common factor.

Notice also that x is a factor of each term. Thus, $12x$ is the greatest common factor.

$$24xy + 12x^2 + 36x^3 = 12x(2y + x + 3x^2)$$

Practice Problem 2 Factor. $12a^2 + 16ab^2 - 12a^2b$. Be careful to remove the greatest common factor.

Student Learning Objectives

After studying this section, you will be able to:

1 Factor polynomials whose terms contain a common factor.

SSM   PH TUTOR   CD & VIDEO   MATH PRO   WEB
CENTER

303

Common Factors of a Polynomial

1. You can determine the greatest common numerical factor by asking. "What is the largest integer that will divide into the coefficient of all the terms?"

2. You can determine the greatest common variable factor by asking, "What variables are common to all the terms, and what is the smallest exponent on each of those variables?"

EXAMPLE 3 Factor. **(a)** $12x^2 + 18y^2$ **(b)** $x^2y^2 + 3xy^2 + y^3$

(a) Note that the largest integer that is common to both terms is 6 (not 3 or 2).

$$12x^2 + 18y^2 = 6(2x^2 + 3y^2)$$

(b) Although y is common to all of the terms, we factor out y^2 since 2 is the largest exponent of y that is common to all terms. We do not factor out x, since x is not common to all of the terms.

$$x^2y^2 + 3xy^2 + y^3 = y^2(x^2 + 3x + y)$$

Practice Problem 3 Factor. **(a)** $16a^3 - 24b^3$ **(b)** $r^3s^2 - 4r^4s + 7r^5$

Checking

You can check any factoring problem by multiplying the factors you obtain. The result should be the same as the original polynomial.

EXAMPLE 4 Factor. $8x^3y + 16x^2y^2 + 24x^3y^3$

We see that 8 is the largest integer that will divide evenly into the three numerical coefficients. We can factor an x^2 out of each term. We can also factor y out of each term.

$$8x^3y + 16x^2y^2 + 24x^3y^3 = 8x^2y(x + 2y + 3xy^2)$$

Check.

$$8x^2y(x + 2y + 3xy^2) = 8x^3y + 16x^2y^2 + 24x^3y^3 \ \checkmark$$

Practice Problem 4 Factor. $18a^3b^2c - 27ab^3c^2 - 45a^2b^2c^2$

EXAMPLE 5 Factor. $9a^3b^2 + 9a^2b^2$

We observe that both terms contain a common factor of 9. We can factor a^2 and b^2 out from each term.

$$9a^3b^2 + 9a^2b^2 = 9a^2b^2(a + 1)$$

⊘ **WARNING** Don't forget to include the 1 inside the parentheses in Example 5. The solution is wrong without it. You will see why if you try to check a result written without the 1.

Practice Problem 5 Factor and check. $30x^3y^2 - 24x^2y^2 + 6xy^2$

EXAMPLE 6   Factor. $3x(x - 4y) + 2(x - 4y)$

Be sure you understand what are terms and what are factors of the polynomial in this example. There are two terms. The expression $3x(x - 4y)$ is one term. The expression $2(x - 4y)$ is the second term. Each term is made up of two factors. Observe that the binomial $(x - 4y)$ is a common factor of the terms. A common factor may be any type of polynomial. Thus we can factor out the common factor $(x - 4y)$.

$$3x(x - 4y) + 2(x - 4y) = (x - 4y)(3x + 2)$$

Practice Problem 6   Factor. $3(a + b) + x(a + b)$

EXAMPLE 7   Factor. $7x^2(2x - 3y) - (2x - 3y)$

The common factor of the terms is $(2x - 3y)$. What happens when we factor out $(2x - 3y)$? What are we left with in the second term?

Recall that $(2x - 3y) = 1(2x - 3y)$. Thus

$$7x^2(2x - 3y) - (2x - 3y) = 7x^2(2x - 3y) - 1(2x - 3y)$$ Rewrite the original expression.

$$= (2x - 3y)(7x^2 - 1)$$ Factor out $(2x - 3y)$.

Practice Problem 7   Factor. $8y(9y^2 - 2) - (9y^2 - 2)$

▲ **EXAMPLE 8**   A computer programmer is writing a program to find the area of 4 circles. She uses the formula $A = \pi r^2$. The radii of the circles are a, b, c, and d, respectively. She wants the final answer to be in factored form with the value of π occurring only once, in order to minimize round-off error. Write the total area with a formula that has π occurring only once.

For each circle, $A = \pi r^2$, where $r = a, b, c,$ or d.

The total area is $\pi a^2 + \pi b^2 + \pi c^2 + \pi d^2$.

In factored form the total area $= \pi(a^2 + b^2 + c^2 + d^2)$.

▲ **Practice Problem 8**   Use $A = \pi r^2$ to find the shaded area. The radius of the larger circle is b. The radius of the smaller circle is a. Write the total area formula in factored form so that π appears only once.

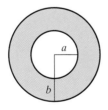

5.1 Exercises

Verbal and Writing Skills

In exercises 1 and 2, write a word or words to complete each sentence.

1. In the expression $3x^2 \cdot 5x^3$, $3x^2$ and $5x^3$ are called _____

2. In the expression $3x^2 + 5x^3$, $3x^2$ and $5x^3$ are called _____

3. We can factor $30a^4 + 15a^3 - 45a^2$ as $5a(6a^3 + 3a^2 - 9a)$. Is the factoring complete? Why or why not?

Remove the largest possible common factor. Check your answers for exercises 4–15 by multiplication.

4. $3a^2 + 3a$

5. $2c^2 + 2c$

6. $21abc - 14ab^2$

7. $18wz - 27w^2z$

8. $5x^3 + 25x^2 - 15x$

9. $8x^3 - 10x^2 - 14x$

10. $12ab - 28bc + 20ac$

11. $12xy - 18yz - 36xz$

12. $3xy^2 - 2ay + 5xy - 2y$

13. $2ab^3 + 3xb^2 - 5b^4 + 2b^2$

14. $60x^3 - 50x^2 + 25x$

15. $6x^9 - 8x^7 + 4x^5$

16. $2\pi rh + 2\pi r^2$

17. $9a^2b^2 - 36ab$

18. $14x^2y - 35xy - 63x$

19. $40a^2 - 16ab - 24a$

20. $54x^2 - 45xy + 18x$

21. $48xy - 24y^2 + 40y$

Hint: In exercises 22–35, refer to Examples 6 and 7.

22. $7a(x + 2y) - b(x + 2y)$

23. $6(3a + b) - z(3a + b)$

24. $3x(x - 4) - 2(x - 4)$

25. $5x(x - 7) + 3(x - 7)$

26. $6b(2a - 3c) - 5d(2a - 3c)$

27. $7x(3y + 5z) - 6t(3y + 5z)$

28. $3(x^2 + 1) + 2y(x^2 + 1) + w(x^2 + 1)$

29. $5a(bc - 1) + b(bc - 1) + c(bc - 1)$

30. $2a(xy - 3) - 4(xy - 3) - z(xy - 3)$

31. $3c(bc - 3a) - 2(bc - 3a) - 6b(bc - 3a)$

32. $4y(x + 2y) + (x + 2y)$

33. $3x^2(x - 2y) - (x - 2y)$

34. $(2a + 3) - 7x(2a + 3)$

35. $d(5x - 3) - (5x - 3)$

To Think About

▲ **36.** Find a formula for the area of four rectangles of width 2.786 inches. The lengths of the rectangles are *a, b, c*, and *d* inches. Write the formula in factored form.

37. Find a formula for the total cost of all purchases by four people. Each person went to the local whole-sale warehouse and spent $29.95 per item. Harry bought *a* items, Richard bought *b* items, Lyle bought *c* items, and Selena bought *d* items. Write the formula in factored form.

Cumulative Review Problems

38. The sum of three consecutive odd integers is 40 more than the smallest of these integers. Find each of these three consecutive odd integers. (*Hint*: Consecutive odd integers are numbers included in the following pattern: $-3, -1, 1, 3, 5, 7, 9, \ldots, x, x + 2, \ldots$).

39. Tuition, room, and board at a college in Colorado cost Lisa $27,040. This was a 4% increase over the cost of these items last year. What did Lisa pay last year for tuition, room, and board?

40. A videotape plays 6.72 feet per minute for 2 hours at standard speed. It plays for 6 hours at extended long play speed. What is the rate of the tape at the slower speed?

In a recent poll, we asked average people what they would do if they won the lottery. Specifically, we asked what would be the first thing that they would do with the money.

28% said that they would buy a new car.

20% said that they would buy themselves a house.

18% said that they would travel.

13% said that they would pay off their debts.

12% said that they would buy a house for their parents.

7% said that they would give the money to a church or to a charity.

2% said that they would invest the money.

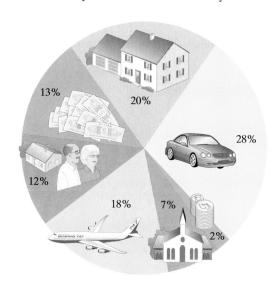

41. It has been estimated that 650 people won a state lottery with a prize of one million dollars or more last year. If that is true, how many people first thought of buying a house for themselves or for their parents?

42. Lottery officials estimate that next year 750 people will win a state lottery with a prize of one million dollars or more. If this happens, how many people will first think of buying a car or traveling?

1 *Factoring Expressions with Four Terms by Grouping*

A common factor of a polynomial can be a number, a variable, or an algebraic expression. Sometimes the polynomial is written so that it is easy to recognize the common factor. This is especially true when the common factor is enclosed by parentheses.

EXAMPLE 1   Factor. $x(x - 3) + 2(x - 3)$

Observe each term:

$$\underbrace{x(x - 3)}_{\substack{\text{first} \\ \text{term}}} + \underbrace{2(x - 3)}_{\substack{\text{second} \\ \text{term}}}$$

The common factor of the first and second terms is the quantity $(x - 3)$, so we have

$$x(x - 3) + 2(x - 3) = (x - 3)(x + 2).$$

Practice Problem 1   Factor. $3y(2x - 7) - 8(2x - 7)$

Suppose the polynomial in Example 1, $x(x - 3) + 2(x - 3)$, were written in the form $x^2 - 3x + 2x - 6$. (Note that this form is obtained by multiplying the factors of the first and second terms.) How would we factor a four-term polynomial like this?

In such cases we remove a common factor from the first two terms and a different common factor from the second two terms. That is, we would factor x from $x^2 - 3x$ and 2 from $2x - 6$.

$$x^2 - 3x + 2x - 6 = x(x - 3) + 2(x - 3)$$

Because the resulting terms have a common factor (the binomial enclosed by the parentheses), we would then proceed as we did in Example 1. This procedure for factoring is often called **factoring by grouping**.

EXAMPLE 2   Factor. $2x^2 + 3x + 6x + 9$

$$\begin{array}{cc} 2x^2 + 3x & + \qquad 6x + 9 \\ \text{Factor out a common} & \text{Factor out a common} \\ \text{factor of } x \text{ from} & \text{factor of 3 from} \\ \text{the first two terms.} & \text{the second two terms.} \\ \underbrace{x(2x + 3)} & \underbrace{3(2x + 3)} \end{array}$$

Note that the sets of parentheses in the two terms contain the same expression at this step.

The expression in parentheses is now a common factor of the terms. Now we finish the factoring.

$$2x^2 + 3x + 6x + 9 = x(2x + 3) + 3(2x + 3) = (2x + 3)(x + 3)$$

Practice Problem 2   Factor. $6x^2 - 15x + 4x - 10$

EXAMPLE 3 Factor. $4x + 8y + ax + 2ay$

Factor out a common
factor of 4 from
the first two terms.

$$4x + 8y + ax + 2ay = 4(x + 2y) + a(x + 2y)$$

Factor out a common
factor of a from
the second two terms.

$4(x + 2y) + a(x + 2y) = (x + 2y)(4 + a)$ The common factor of the terms
is the expression in parentheses,
$x + 2y$.

Practice Problem 3 Factor by grouping. $ax + 2a + 4bx + 8b$

In practice, these problems are done in just two steps.
In some problems the terms are out of order. We have to rearrange the order
of the terms first so that the first two terms have a common factor.

EXAMPLE 4 Factor. $bx + 4y + 4b + xy$

$bx + 4y + 4b + xy = bx + 4b + xy + 4y$ Rearrange the terms so that the first
terms have a common factor.

$\qquad\qquad\qquad = b(x + 4) + y(x + 4)$ Factor out the common factor of b
from the first two terms.
Factor out the common factor of y
from the second two terms.

$\qquad\qquad\qquad = (x + 4)(b + y)$

Practice Problem 4 Factor. $6a^2 + 5bc + 10ab + 3ac$

Sometimes you will need to factor out a negative common factor from the second two terms to obtain two terms that contain the same parenthetical expression.

EXAMPLE 5 Factor. $2x^2 + 5x - 4x - 10$

$2x^2 + 5x - 4x - 10 = x(2x + 5) - 2(2x + 5)$ Factor out the common factor of
x from the first two terms and
the common factor of -2 from
the second two terms.

$\qquad\qquad\qquad\qquad = (2x + 5)(x - 2)$

Notice that if you factored out a common factor of $+2$ in the first step, the two
resulting terms would not contain the same parenthetical expression. If the expressions inside the two sets of parentheses are not exactly the same, you cannot express
the polynomial as a product of two factors!

Practice Problem 5 Factor. $6xy + 14x - 15y - 35$

EXAMPLE 6 Factor. $2ax - a - 2bx + b$

$2ax - a - 2bx + b = a(2x - 1) - b(2x - 1)$ Factor out the common factor of a from the first two terms. Factor out the common factor of $-b$ from the second two terms.

$\qquad\qquad\qquad\qquad = (2x - 1)(a - b)$ Since the two resulting terms contain the same parenthetical expression, we can complete the factoring.

Practice Problem 6 Factor. $3x - 10ay + 6y - 5ax$

⊘ **WARNING** Many students find that they make a factoring error in the first step of problems like Example 6. Multiplying the results of the first step (even if you do it in your head) will usually help you to detect any error you may have made.

EXAMPLE 7 Factor and check your answer. $8ad + 21bc - 6bd - 28ac$

We observe that the first two terms do not have a common factor.

$8ad + 21bc - 6bd - 28ac = 8ad - 6bd - 28ac + 21bc$ Rearrange the order using the commutative property of addition.

$\qquad\qquad\qquad\qquad = 2d(4a - 3b) - 7c(4a - 3b)$ Factor out the common factor of $2d$ from the first two terms and the common factor of $-7c$ from the last two terms.

$\qquad\qquad\qquad\qquad = (4a - 3b)(2d - 7c)$ Factor out the common factor of $(4a - 3b)$.

To check, we multiply the two binomials using the FOIL procedure.

$(4a - 3b)(2d - 7c) = 8ad - 28ac - 6bd + 21bc$

$\qquad\qquad\qquad\quad = 8ad + 21bc - 6bd - 28ac$ ✓   Rearrange the order of the terms. This is the original problem. Thus it checks.

Practice Problem 7 Factor and check your answer.

$10ad + 27bc - 6bd - 45ac$

Factor by grouping. Check your answers for exercises 1–12.

1. $ab - 3a + 4b - 12$

2. $xy - x + 4y - 4$

3. $2ax + 6bx - ay - 3by$

4. $4x + 8y - 3wx - 6wy$

5. $x^3 - 4x^2 + 3x - 12$

6. $x^3 - 6x^2 + 2x - 12$

7. $3ax + bx - 6a - 2b$

8. $4ax + bx - 28a - 7b$

9. $5a + 12bc + 10b + 6ac$

10. $2x + 15yz + 6y + 5xz$

11. $5a - 5b - 2ax + 2xb$

12. $xy - 4x - 3y + 12$

13. $y^2 - 2y - 3y + 6$

14. $12 + 3x - 4x - x^2$

15. $14 - 7y + 2y - y^2$

16. $xa + 2bx - a - 2b$

17. $6ax - y + 2ay - 3x$

18. $6tx - 3t - 2rx + r$

19. $2x^2 + 8x - 3x - 12$

20. $3y^2 - y + 9y - 3$

21. $28x^2 + 8xy^2 + 21xw + 6y^2w$

22. $8xw + 10x^2 + 35xy^2 + 28y^2w$

Verbal and Writing Skills

23. Although $6a^2 - 12bd - 8ad + 9ab = 6(a^2 - 2bd) - a(8d - 9b)$ is true, it is not the correct solution to the problem "Factor $6a^2 - 12bd - 8ad + 9ab$." Explain. Can this expression be factored?

Cumulative Review Problems

24. A train was traveling at 73 miles per hour when the engineer spotted a stalled truck on the railroad crossing ahead. He jammed on the brakes but was unable to stop the train before it collided with the stalled truck. For every second the engineer applied the brakes, the train slowed down by 4 miles per hour. The accident reconstruction team found that the train was still traveling at 41 miles per hour at the time of impact. For how many seconds were the brakes applied?

25. Tim Brown made an average profit of $320 per car for each car he sold at Donahue Motors over the first 5 months of the year. In January and February, he averaged only $200 per car. In March, he averaged $300 per car, but in April he averaged $480 per car. Assuming that he sells around the same number of cars each month, what was his average profit per car for the month of May?

26. In 1998, the Recording Industry Association of America reported a 6.8% increase from the previous year in U.S. unit sales of all musical recordings. This translates to an 11.9% increase in dollar value, or a 15.1 million increase. (*Source*: Bureau of Economic Analysis.) If the dollar value increase is equal to $15.1 million dollars, what was the total dollar value of musical recordings in the United States in 1997? Round to the nearest tenth of a million.

27. Using the data from exercise 26, if there was a 20% increase in the dollar value of sales from 1998 to 2000, what was the dollar value of sales for the year 2000?

1 *Factoring Polynomials of the Form $x^2 + bx + c$*

Suppose that you wanted to factor $x^2 + 5x + 6$. After some trial and error you *might* obtain $(x + 2)(x + 3)$, or you might get discouraged and not get an answer. If you did get these factors, you could check this answer by the FOIL method.

$$(x + 2)(x + 3) = x^2 + 3x + 2x + 6$$
$$= x^2 + 5x + 6$$

But trial and error can be a long process. There is another way. Let's look at the preceding equation again.

$$\overset{\text{F}\quad\text{O}\quad\text{I}\quad\text{L}}{(x + 2)(x + 3) = x^2 + \underbrace{3x + 2x}_{} + 6}$$
$$= x^2 \quad + 5x \quad + 6$$

The first thing to notice is that the product of the first terms in the factors gives the first term of the polynomial. That is, $x \cdot x = x^2$.

The first term is the product of these terms.

$$x^2 + 5x + 6 \quad = \quad (x + 2)(x + 3)$$

The next thing to notice is that the sum of the products of the outer and inner terms in the factors produces the middle term of the polynomial. That is, $(x \cdot 3) + (2 \cdot x) = 3x + 2x = 5x$. Thus we see that the sum of the second terms in the factors, $2 + 3$, gives the coefficient of the middle term, 5.

Finally, note that the product of the last terms of the factors gives the last term of the polynomial. That is, $2 \cdot 3 = 6$.

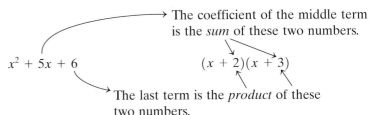

The coefficient of the middle term is the *sum* of these two numbers.

$$x^2 + 5x + 6 \qquad (x + 2)(x + 3)$$

The last term is the *product* of these two numbers.

Let's summarize our observations in general terms and then try a few examples.

Factoring Trinomials of the Form $x^2 + bx + c$

1. The answer will be of the form $(x + m)(x + n)$.

2. m and n are numbers such that:
   (a) When you multiply them, you get the last term, which is c.
   (b) When you add them, you get the coefficient of the middle term, which is b.

EXAMPLE 1 Factor. $x^2 + 7x + 12$

The answer is of the form $(x + m)(x + n)$. We want to find the two numbers, m and n, that you can multiply to get 12 and add to get 7. The numbers are 3 and 4.

$$x^2 + 7x + 12 = (x + 3)(x + 4)$$

Practice Problem 1 Factor. $x^2 + 8x + 12$

EXAMPLE 2 Factor. $x^2 + 12x + 20$

We want two numbers that have a product of 20 and a sum of 12. The numbers are 10 and 2.

$$x^2 + 12x + 20 = (x + \underline{10})(x + \underline{2})$$

Note: If you cannot think of the numbers in your head, write down the possible factors whose product is 20.

Product	*Sum*
$1 \cdot 20 = 20$	$1 + 20 = 21$
$2 \cdot 10 = 20$	$2 + 10 = 12$
$4 \cdot 5 = 20$	$4 + 5 = 9$

Then select the pair whose sum is 12. Select this pair.

Practice Problem 2 Factor. $x^2 + 17x + 30$

You may find that it is helpful to list all the factors whose product is 30 first.

So far we have factored only trinomials of the form $x^2 + bx + c$, where b and c are positive numbers. The same procedure applies if b is a negative number and c is positive. Because m and n have a positive product and a negative sum, they must both be negative.

EXAMPLE 3 Factor. $x^2 - 8x + 15$

We want two numbers that have a product of +15 and a sum of −8. They must be negative numbers since the sign of the middle term is negative and the sign of the last term is positive.

the sum − 5 + (−3)
$$x^2 - 8x + 15 = (x - 5)(x - 3)$$
the product (−5)(−3)

Think: $(-5)(-3) = +15$ and $-5 + (-3) = -8$.

Multiply using FOIL to check.

Practice Problem 3 Factor. $x^2 - 11x + 18$

EXAMPLE 4 Factor. $x^2 - 9x + 14$

We want two numbers whose product is 14 and whose sum is −9. The numbers are −7 and −2. So

$$x^2 - 9x + 14 = (x - 7)(x - 2) \text{ or } (x - 2)(x - 7).$$

Practice Problem 4 Factor. $y^2 - 11y + 24$

All the examples so far have had a positive last term. What happens when the last term is negative? If the last term is negative, one of the numbers m or n must be a positive and the other must be a negative. Why? The product of a positive number and a negative number is negative.

EXAMPLE 5 Factor. $x^2 - 3x - 10$

We want two numbers whose product is -10 and whose sum is -3. The two numbers are -5 and $+2$.

$$x^2 - 3x - 10 = (x - 5)(x + 2)$$

Practice Problem 5 Factor. $a^2 - 5a - 24$

What if we made a sign error and *incorrectly* factored the trinomial $x^2 - 3x - 10$ as $(x + 5)(x - 2)$? We could detect the error immediately since the sum of $+5$ and -2 is 3. We need a sum of -3!

EXAMPLE 6 Factor. $x^2 + 10x - 24$ Check your answer.

The two numbers whose product is -24 and whose sum is $+10$ are the numbers $+12$ and -2.

$$x^2 + 10x - 24 = (x + 12)(x - 2)$$

🚫 **WARNING** It is very easy to make a sign error in these problems. Make sure that you mentally multiply your answer back to obtain the original expression. Check each sign carefully.

Check. $(x + 12)(x - 2) = x^2 - 2x + 12x - 24 = x^2 + 10x - 24$ ✓

Practice Problem 6 Factor. $x^2 + 17x - 60$ Multiply your answer to check.

EXAMPLE 7 Factor. $x^2 - 16x - 36$

We want two numbers whose product is -36 and whose sum is -16.

List all the possible factors of 36 (without regard to sign). Find the pair that has a difference of 16. We are looking for a difference because the signs of the factors are different.

Factors of 36	*The Difference between the Factors*
36 and 1	35
18 and 2	16 ← This is the value we want.
12 and 3	9
9 and 4	5
6 and 6	0

Once we have picked the pair of numbers (18 and 2), it is easy to find the signs. For the coefficient of the middle term to be -16, we will have to add the numbers -18 and $+2$.

$$x^2 - 16x - 36 = (x - 18)(x + 2)$$

Practice Problem 7 Factor. $x^2 - 7x - 60$ You may find it helpful to list the pairs of numbers whose product is 60.

At this point you should work several problems to develop your factoring skill. This is one section where you really need to drill by doing many problems.

Feel a little confused about the signs? If you do, you may find these facts helpful.

Facts about Factoring Trinomials of the Form $x^2 + bx + c$

The *two numbers m* and *n* will have the *same sign* if the last term of the polynomial is *positive*.

$$x^2 + bx + c = (x \quad m)(x \quad n)$$

1. They will both be *positive* if the *coefficient* of the *middle* term is *positive*.

$$x^2 + 5x + 6 = (x + 2)(x + 3)$$

2. They will both be *negative* if the *coefficient* of the *middle* term is *negative*.

$$x^2 - 5x + 6 = (x - 2)(x - 3)$$

The two numbers *m* and *n* will have *opposite signs* if the last term is *negative*.

1. The *larger* of the absolute values of the two numbers will be given a plus sign if the coefficient of the *middle term* is *positive*.

$$x^2 + 6x - 7 = (x + 7)(x - 1)$$

2. The larger of the absolute values of the two numbers will be given a negative sign if the coefficient of the *middle term* is *negative*.

$$x^2 - 6x - 7 = (x - 7)(x + 1)$$

Do not memorize these facts; rather, try to understand the pattern.

Sometimes the exponent of the first term of the polynomial will be greater than 2. If the exponent is an even power, it is a square. For example, $x^4 = (x^2)(x^2)$. Likewise, $x^6 = (x^3)(x^3)$.

EXAMPLE 8   Factor. $y^4 - 2y^2 - 35$

Think: $y^4 = (y^2)(y^2)$ This will be the first term of each parentheses.

$(y^2 \quad)(y^2 \quad)$

$(y^2 + \ )(y^2 - \ )$ The last term of the polynomial is negative.

$(y^2 + 5)(y^2 - 7)$ Thus the signs of *m* and *n* will be different.
                       Now think of factors of 35 whose difference is 2.

Practice Problem 8   Factor. $a^4 + a^2 - 42$

---

2   *Factoring Polynomials That Have a Common Factor and a Factor of the Form $x^2 + bx + c$*

Some factoring problems require two steps. Often we must first factor out a common factor from each term of the polynomial. Once this is done, we may find that the other factor is a trinomial that can be factored using the methods previously discussed in this section.

EXAMPLE 9 Factor. $2x^2 + 36x + 160$

$2x^2 + 36x + 160 = 2(x^2 + 18x + 80)$ First factor out the common factor of 2 from each term of the polynomial.

$= 2(x + 8)(x + 10)$ Then factor the remaining polynomial.

The final answer is $2(x + 8)(x + 10)$. *Be sure to list all parts of the answer.*

Check: $2(x + 8)(x + 10) = 2(x^2 + 18x + 80) = 2x^2 + 36x + 160$ ✓

Thus we are sure that the answer is $2(x + 8)(x + 10)$.

Practice Problem 9   Factor. $3x^2 + 45x + 150$

EXAMPLE 10   Factor. $3x^2 + 9x - 162$

$3x^2 + 9x - 162 = 3(x^2 + 3x - 54)$ First factor out the common factor of 3 from each term of the polynomial.

$= 3(x - 6)(x + 9)$ Then factor the remaining polynomial.

The final answer is $3(x - 6)(x + 9)$.

Check: $3(x - 6)(x + 9) = 3(x^2 + 3x - 54) = 3x^2 + 9x - 162$ ✓

Thus we are sure that the answer is $3(x - 6)(x + 9)$.

Practice Problem 10   Factor. $4x^2 - 8x - 140$

It is quite easy to forget to look for a greatest common factor as the first step of factoring a trinomial. Therefore, it is a good idea to examine your final answer in any factoring problem and ask yourself, "Can I factor out a common factor from any binomial contained inside a set of parentheses?" Often you will be able to see a common factor at that point if you missed it in the first step of the problem.

▲ **EXAMPLE 11**   Find a polynomial in factored form for the shaded area in the figure.

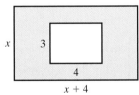

To obtain the shaded area, we find the area of the larger rectangle and subtract from it the area of the smaller rectangle. Thus we have the following:

$$\text{shaded area} = x(x + 4) - (4)(3)$$
$$= x^2 + 4x - 12$$

Now we factor this polynomial to obtain the shaded area $= (x + 6)(x - 2)$.

▲ **Practice Problem 11**   Find a polynomial in factored form for the shaded area in the figure.

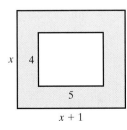

Verbal and Writing Skills

Fill in the blanks.

1. To factor $x^2 + 5x + 6$, find two numbers whose _____ is 6 and whose _____ is 5.

2. To factor $x^2 + 5x - 6$, find two numbers whose _____ is -6 and whose _____ is 5.

Factor.

3. $x^2 + 2x + 1$

4. $x^2 + 11x + 30$

5. $x^2 + 12x + 35$

6. $x^2 + 11x + 24$

7. $x^2 - 4x + 3$

8. $x^2 - 6x + 8$

9. $x^2 - 11x + 28$

10. $x^2 - 13x + 12$

11. $x^2 + x - 12$

12. $x^2 + 4x - 5$

13. $x^2 - 13x - 14$

14. $x^2 - 6x - 16$

15. $x^2 + 2x - 35$

16. $x^2 - 4x - 12$

17. $x^2 - 2x - 24$

18. $x^2 - 11x - 26$

Look over your answers to exercises 3–18 carefully. Be sure that you are clear on your sign rules. Exercises 19–42 contain a mixture of all the types of problems in this section. Make sure you can do them all. Check your answers by multiplication.

19. $x^2 + 5x - 14$

20. $x^2 - 2x - 15$

21. $x^2 - 10x + 24$

22. $x^2 - 13x + 42$

23. $x^2 + 13x + 30$

24. $x^2 - 3x - 28$

25. $y^2 - 4y - 5$

26. $y^2 - 8y + 7$

27. $a^2 + 6a - 16$

28. $a^2 - 13a + 30$

29. $x^2 - 12x + 32$

30. $x^2 - 6x - 27$

31. $x^2 + 4x - 21$

32. $x^2 - 9x + 18$

33. $x^2 + 13x + 40$

34. $x^2 + 15x + 50$

35. $x^2 - 21x - 46$

36. $x^2 + 12x - 45$

37. $x^2 + 9x - 36$

38. $x^2 - 13x + 36$

39. $x^2 - 2xy - 15y^2$

40. $x^2 - 2xy - 35y^2$

41. $x^2 - 16xy + 63y^2$

42. $x^2 + 19xy + 48y^2$

In exercises 43–54, first factor out the greatest common factor from each term. Then factor the remaining polynomial. Refer to Examples 9 and 10.

43. $2x^2 - 12x + 16$

44. $2x^2 - 14x + 24$

45. $3x^2 - 6x - 72$

46. $3x^2 - 12x - 63$

47. $4x^2 + 24x + 20$

48. $4x^2 + 28x + 40$

49. $7x^2 + 21x - 70$

50. $7x^2 + 7x - 84$

51. $6x^2 + 18x + 12$

52. $6x^2 + 24x + 18$

53. $3x^2 - 18x + 15$

54. $3x^2 - 33x + 54$

Find a polynomial in factored form for the shaded area.

▲ **55.** The circle has radius $2x$. The square has diagonals of $4x$.

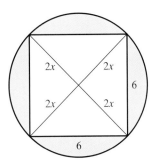

▲ **56.** Both figures are rectangles with dimensions as labeled.

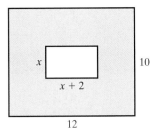

Cumulative Review Problems

57. Solve for t. $A = P + Prt$

58. Solve $2 - 3x \leq 7$ for x. Graph the solution on the number line.

59. A new car that maintains a constant speed travels from Watch Hill, Rhode Island, to Greenwich, Connecticut, in 2 hours. A train, traveling 20 mph faster, makes the trip in $1\frac{1}{2}$ hours. How far is it from Watch Hill to Greenwich? (*Hint*: Let $c =$ the car's speed. First find the speed of the car and the speed of the train. Then be sure to answer the question.)

60. Kerri works as a radio advertising sales rep. She earns a guaranteed minimum salary of $500 per month plus 3% commission on her sales. She wants to earn $4400 or more this month. At least how much must she generate in sales?

The Golden Sounds DJ Service charges $65 per hour plus $85 for each 30-minute period after midnight.

61. If the bill for Marcia's wedding reception was $515 and the reception started at 8 P.M., when did the reception end?

62. If the bill for Melissa's wedding reception was $535 and the reception ended at 2 A.M., when did the reception start?

The equation $T = 19 + 2M$ has been used by some meteorologists to predict the monthly average temperature for the small island of Menorca off the coast of Spain during the first 6 months of the year. The variable T represents the average monthly temperature measured in degrees Celsius. The variable M represents the number of months since January.

63. What is the average temperature of Menorca during the month of April?

64. During what month will the average temperature be $29°C$?

① Using the Trial-and-Error Method

When the coefficient of the x^2-term in a trinomial of the form $ax^2 + bx + c$ is not 1, the trinomial is more difficult to factor. Several possibilities must be considered.

EXAMPLE 1 Factor. $2x^2 + 5x + 3$

In order for the coefficient of the x^2-term of the polynomial to be 2, the coefficients of the x-terms in the factors must be 2 and 1. Thus $2x^2 + 5x + 3 = (2x \quad)(x \quad)$.

In order for the last term of the polynomial to be 3, the constants in the factors must be 3 and 1.

Since all signs in the polynomial are positive, we know that each factor in parentheses will contain only positive signs. However, we still have two possibilities. They are as follows.

$$(2x + 3)(x + 1)$$
$$(2x + 1)(x + 3)$$

We check them by multiplying by the FOIL method.

$(2x + 1)(x + 3) = 2x^2 + 7x + 3$ Wrong middle term
$(2x + 3)(x + 1) = 2x^2 + 5x + 3$ Correct middle term

Thus the correct answer is

$$(2x + 3)(x + 1) \text{ or } (x + 1)(2x + 3).$$

Practice Problem 1 Factor. $2x^2 - 7x + 5$

Some problems have many more possibilities.

EXAMPLE 2 Factor. $4x^2 - 13x + 3$

The Different Factors of 4 Are:	**The Factors of 3 Are:**
2 and 2	1 and 3
1 and 4	

Let us list the possible factoring combinations and compute the middle term by the FOIL method. Note that the signs of the constants in both factors will be negative. Why?

Possible Factors	**Middle Term**	**Correct?**
$(2x - 3)(2x - 1)$	$-8x$	No
$(4x - 3)(x - 1)$	$-7x$	No
$(4x - 1)(x - 3)$	$-13x$	Yes

The correct answer is $(4x - 1)(x - 3)$ or $(x - 3)(4x - 1)$.
This method is called the **trial-and-error method**.

Practice Problem 2 Factor. $9x^2 - 64x + 7$

Student Learning Objectives

After studying this section, you will be able to:

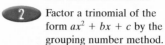

①   Factor a trinomial of the form $ax^2 + bx + c$ by the trial-and-error method.

②   Factor a trinomial of the form $ax^2 + bx + c$ by the grouping number method.

③   Factor a trinomial of the form $ax^2 + bx + c$ after a common factor has been factored out of each term.

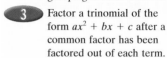

SSM   PH TUTOR CENTER   CD & VIDEO   MATH PRO   WEB

319

EXAMPLE 3   Factor. $3x^2 - 2x - 8$

Factors of 3	***Factors of 8***
3 and 1	8 and 1
	4 and 2

Let us list only one-half of the possibilities. We'll let the constant in the first factor of each product be positive.

Possible Factors	***Middle Term***	***Correct Factors?***
$(x + 8)(3x - 1)$	$+23x$	No
$(x + 1)(3x - 8)$	$-5x$	No
$(x + 4)(3x - 2)$	$+10x$	No
$(x + 2)(3x - 4)$	$+2x$	No (but only because the sign is wrong)

So we just *reverse* the signs of the constants in the factors.

	Middle Term	***Correct Factor?***
$(x - 2)(3x + 4)$	$-2x$	Yes

The correct answer is

$$(x - 2)(3x + 4) \text{ or } (3x + 4)(x - 2).$$

Practice Problem 3   Factor. $3x^2 - x - 14$

It takes a good deal of practice to readily factor problems of this type. The more problems you do, the more proficient you will become. The following method will help you factor more quickly.

2   *Using the Grouping Number Method*

One way to factor a trinomial of the form $ax^2 + bx + c$ is to write it with four terms and factor by grouping, as we did in Section 5.2. For example, the trinomial $2x^2 + 13x + 20$ can be written as $2x^2 + 5x + 8x + 20$. Using the methods of Section 5.2, we factor it as follows.

$$2x^2 + 5x + 8x + 20 = x(2x + 5) + 4(2x + 5)$$
$$= (2x + 5)(x + 4)$$

We can factor all factorable trinomials of the form $ax^2 + bx + c$ in this way. We will use the following procedure.

Grouping Number Method for Factoring Trinomials of the Form $ax^2 + bx + c$

1. Obtain the grouping number ac.
2. Find the two numbers whose product is the grouping number and whose sum is b.
3. Use those numbers to write bx as the sum of two terms.
4. Factor by grouping.
5. Multiply to check.

Let's try the problem from Example 1.

EXAMPLE 4 Factor by grouping. $2x^2 + 5x + 3$

1. The grouping number is $(2)(3) = 6$.
2. The factors of 6 are $6 \cdot 1$ and $3 \cdot 2$. We choose the numbers 3 and 2 because their product is 6 and their sum is 5.
3. We write $5x$ as the sum $3x + 2x$.
4. Factor by grouping.

$$
\begin{aligned}
2x^2 + 5x + 3 &= 2x^2 + 2x + 3x + 3 \\
&= 2x(x + 1) + 3(x + 1) \\
&= (x + 1)(2x + 3)
\end{aligned}
$$

5. Multiply to check.

$$
\begin{aligned}
(x + 1)(2x + 3) &= 2x^2 + 3x + 2x + 3 \\
&= 2x^2 + 5x + 3 \quad \checkmark
\end{aligned}
$$

Practice Problem 4 Factor by grouping. $2x^2 - 7x + 5$

EXAMPLE 5 Factor by grouping. $4x^2 - 13x + 3$

1. The grouping number is $(4)(3) = 12$.
2. The factors of 12 are $(12)(1)$ or $(4)(3)$ or $(6)(2)$. Note that the middle term of the polynomial is negative. Thus we choose the numbers -12 and -1 because their product is still 12 and their sum is -13.
3. We write $-13x$ as the sum $-12x + (-1x)$.
4. Factor by grouping.

$$
\begin{aligned}
4x^2 - 13x + 3 &= 4x^2 - 12x - 1x + 3 \\
&= 4x(x - 3) - 1(x - 3) \\
&= (x - 3)(4x - 1)
\end{aligned}
$$

Remember to factor out a -1 from the last two terms so that both sets of parentheses contain the same expression.

Practice Problem 5 Factor by grouping. $9x^2 - 64x + 7$

EXAMPLE 6 Factor by grouping. $3x^2 - 2x - 8$

1. The grouping number is $(3)(-8) = -24$.
2. We want two numbers whose product is -24 and whose sum is -2. They are -6 and 4.
3. We write $-2x$ as a sum $-6x + 4x$.
4. Factor by grouping.

$$
\begin{aligned}
3x^2 - 6x + 4x - 8 &= 3x(x - 2) + 4(x - 2) \\
&= (x - 2)(3x + 4)
\end{aligned}
$$

Practice Problem 6 Factor by grouping. $3x^2 + 4x - 4$

To factor polynomials of the form $ax^2 + bx + c$, use the method, either trial-and-error or grouping, that works best for you.

3 ▸ Using the Common Factor Method

Some problems require first factoring out a common factor and then factoring the trinomial by one of the two methods of this section.

▸ **EXAMPLE 7** Factor. $9x^2 + 3x - 30$

$9x^2 + 3x - 30 = 3(3x^2 + 1x - 10)$ We first factor out the common factor of 3 from each term of the trinomial.

$\qquad\qquad\qquad\quad = 3(3x - 5)(x + 2)$ We then factor the trinomial by the grouping method or by the trial-and-error method.

Practice Problem 7 Factor. $8x^2 - 8x - 6$

▸ **EXAMPLE 8** Factor. $32x^2 - 40x + 12$

$32x^2 - 40x + 12 = 4(8x^2 - 10x + 3)$ We first factor out the greatest common factor of 4 from each term of the trinomial.

$\qquad\qquad\qquad\qquad = 4(2x - 1)(4x - 3)$ We then factor the trinomial by the grouping method or by the trial-and-error method.

Practice Problem 8 Factor. $24x^2 - 38x + 10$

Factor. Check your answers for exercises 1–20 using the FOIL method.

1. $4x^2 + 13x + 3$

2. $3x^2 + 11x + 10$

3. $2x^2 - 5x + 2$

4. $3x^2 - 8x + 4$

5. $3x^2 - 4x - 7$

6. $5x^2 + 7x - 6$

7. $2x^2 - 5x - 3$

8. $2x^2 - x - 6$

9. $5x^2 + 3x - 2$

10. $6x^2 + x - 2$

11. $15x^2 - 34x + 15$

12. $10x^2 - 29x + 10$

13. $2x^2 + 3x - 20$

14. $6x^2 + 11x - 10$

15. $9x^2 + 9x + 2$

16. $4x^2 + 11x + 6$

17. $6x^2 - 5x - 6$

18. $3x^2 - 13x - 10$

19. $6x^2 - 19x + 10$

20. $10x^2 - 19x + 6$

21. $7x^2 - 5x - 18$

22. $9x^2 - 22x - 15$

23. $9y^2 - 13y + 4$

24. $5y^2 - 11y + 2$

25. $5a^2 - 13a - 6$

26. $3a^2 - 10a - 8$

27. $12x^2 - 20x + 3$

28. $9x^2 + 5x - 4$

29. $15x^2 + 4x - 4$

30. $8x^2 - 11x + 3$

31. $12x^2 + 28x + 15$

32. $24x^2 + 17x + 3$

33. $12x^2 - 16x - 3$

34. $12x^2 + x - 6$

35. $2x^4 + 15x^2 - 8$

36. $4x^4 - 11x^2 - 3$

37. $4x^2 + 8xy - 5y^2$

38. $3x^2 + 8xy + 4y^2$

39. $5x^2 + 16xy - 16y^2$

40. $12x^2 + 11xy - 5y^2$

Factor by first factoring out the greatest common factor. See Examples 7 and 8.

41. $10x^2 + 22x + 12$ **42.** $4x^2 + 34x + 42$ **43.** $12x^2 - 24x + 9$ **44.** $8x^2 - 26x + 6$

45. $10x^2 - 25x - 15$ **46.** $20x^2 - 25x - 30$ **47.** $6x^3 - 16x^2 - 6x$ **48.** $6x^3 + 9x^2 - 60x$

Factor.

49. $12x^2 + 16x - 35$ **50.** $20x^2 - 53x + 18$ **51.** $20x^2 - 27x + 9$ **52.** $12x^2 - 29x + 15$

Cumulative Review Problems

53. Solve. $7x - 3(4 - 2x) = 2(x - 3) - (5 - x)$

54. In the 2000–2001 school year, 33.8 million children were enrolled in kindergarten through grade 8 in the United States. This represented an increase of 80% in the number of enrolled children since the school year 1939–1940. (*Source*: U.S. National Center for Education Statistics.) How many children were enrolled in grades K–8 during 1939–1940? Round to the nearest tenth of a million.

55. In the 2002–2003 school year, it is projected that 18% more students will be enrolled in kindergarten through grade 8 in the United States than there were in the year 1992–1993. In the 1992–1993 school year, 31.1 million children were enrolled in kindergarten through grade 8. (*Source*: U.S. National Center for Education Statistics.) How many children will be enrolled in grades K–8 during 2002–2003? Round to the nearest tenth of a million.

5.5 Special Cases of Factoring

As we proceed in this section you will be able to reduce the time it takes you to factor a polynomial by quickly recognizing and factoring two special types of polynomials: the difference of two squares and perfect-square trinomials.

Student Learning Objectives

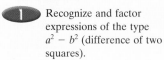

After studying this section, you will be able to:

1 Recognize and factor expressions of the type $a^2 - b^2$ (difference of two squares).

2 Recognize and factor expressions of the type $a^2 + 2ab + b^2$ (perfect-square trinomial).

3 Recognize and factor expressions that require factoring out a common factor and then using a special-case formula.

SSM   PH TUTOR   CD & VIDEO   MATH PRO   WEB
CENTER

1 Factoring the Difference of Two Squares

Recall the formula from Section 4.5:

$$(a + b)(a - b) = a^2 - b^2.$$

In reverse form we can use it for factoring.

Difference of Two Squares

$$a^2 - b^2 = (a + b)(a - b)$$

We can state it in words in this way: "The difference of two squares can be factored into the sum and difference of those values that were squared."

EXAMPLE 1 Factor. $9x^2 - 1$

We see that the problem is in the form of the difference of two squares. $9x^2$ is a square and 1 is a square. So using the formula we can write the following.

$$9x^2 - 1 = (3x + 1)(3x - 1) \quad \text{Because } 9x^2 = (3x)^2 \text{ and } 1 = (1)^2$$

Practice Problem 1 Factor. $1 - 64x^2$

EXAMPLE 2 Factor. $25x^2 - 16$

Again we use the formula for the difference of squares.

$$25x^2 - 16 = (5x + 4)(5x - 4) \quad \text{Because } 25x^2 = (5x)^2 \text{ and } 16 = (4)^2$$

Practice Problem 2 Factor. $36x^2 - 49$

Sometimes the polynomial contains two variables.

EXAMPLE 3 Factor. $4x^2 - 49y^2$

We see that

$$4x^2 - 49y^2 = (2x + 7y)(2x - 7y).$$

Practice Problem 3 Factor. $100x^2 - 81y^2$

325

Some problems may involve more than one step.

EXAMPLE 4 Factor. $81x^4 - 1$

We see that

$$81x^4 - 1 = (9x^2 + 1)(9x^2 - 1) \quad \text{Because } 81x^4 = (9x^2)^2 \text{ and } 1 = (1)^2$$

Is the factoring complete? We can factor $9x^2 - 1$.

$$81x^4 - 1 = (9x^2 + 1)(3x - 1)(3x + 1) \quad \text{Because } (9x^2 - 1) = (3x - 1)(3x + 1)$$

Practice Problem 4 Factor. $x^8 - 1$

2 Factoring Perfect-Square Trinomials

There is a formula that will help us to factor very quickly certain trinomials, called **perfect-square trinomials**. Recall from Section 4.5 the formulas for a binomial squared.

$$(a + b)^2 = a^2 + 2ab + b^2$$
$$(a - b)^2 = a^2 - 2ab + b^2$$

We can use these two equations in reverse form for factoring.

Perfect-Square Trinomials

$$a^2 + 2ab + b^2 = (a + b)^2$$
$$a^2 - 2ab + b^2 = (a - b)^2$$

A perfect-square trinomial is a trinomial that is the result of squaring a binomial. How can we recognize a perfect-square trinomial?

1. The first and last terms are *perfect squares*.
2. The middle term is twice the product of the values whose squares are the first and last terms.

EXAMPLE 5 Factor. $x^2 + 6x + 9$

This is a perfect-square trinomial.

1. The first and last terms are perfect squares because $x^2 = (x)^2$ and $9 = (3)^2$.
2. The middle term, $6x$, is twice the product of x and 3.

Since $x^2 + 6x + 9$ is a perfect-square trinomial, we can use the formula

$$a^2 + 2ab + b^2 = (a + b)^2$$

with $a = x$ and $b = 3$. So we have

$$x^2 + 6x + 9 = (x + 3)^2.$$

Practice Problem 5 Factor. $16x^2 + 8x + 1$

EXAMPLE 6 Factor. $4x^2 - 20x + 25$

This is a perfect-square trinomial. Note that $20x = 2(2x \cdot 5)$. Also note the negative sign. Thus we have the following.

$$4x^2 - 20x + 25 = (2x - 5)^2 \quad \text{Since } a^2 - 2ab + b^2 = (a - b)^2$$

Practice Problem 6 Factor. $25x^2 - 30x + 9$

A polynomial may have more than one variable and its exponents may be higher than 2. The same principles apply.

● EXAMPLE 7 Factor. **(a)** $49x^2 + 42xy + 9y^2$ **(b)** $36x^4 - 12x^2 + 1$

(a) This is a perfect-square trinomial. Why?

$$49x^2 + 42xy + 9y^2 = (7x + 3y)^2 \qquad \text{Because } 49x^2 = (7x)^2, 9y^2 = (3y)^2,$$
$$\text{and } 42xy = 2(7x \cdot 3y)$$

(b) This is a perfect-square trinomial. Why?

$$36x^4 - 12x^2 + 1 = (6x^2 - 1)^2 \qquad \text{Because } 36x^4 = (6x^2)^2, 1 = (1)^2,$$
$$\text{and } 12x^2 = 2(6x^2 \cdot 1)$$

Practice Problem 7 Factor. **(a)** $25x^2 - 60xy + 36y^2$ **(b)** $64x^6 - 48x^3 + 9$ ●

Some polynomials appear to be perfect-square trinomials but are not. They were factored in other ways in Section 5.4.

● EXAMPLE 8 Factor. $49x^2 + 35x + 4$

This is *not* a perfect-square trinomial! Although the first and last terms are perfect squares since $(7x)^2 = 49x^2$ and $(2)^2 = 4$, the middle term, $35x$, is not double the product of 2 and $7x$! $35x \neq 28x$! So we must factor by trial and error or by grouping to obtain

$$49x^2 + 35x + 4 = (7x + 4)(7x + 1).$$

Practice Problem 8 Factor. $9x^2 - 15x + 4$ ●

3 *Factoring Out a Common Factor and Then Using a Special-Case Formula*

For some polynomials, we will first factor out a common factor. Then we will find an opportunity to use the difference-of-two-squares formula or one of the perfect-square trinomial formulas.

● EXAMPLE 9 Factor. $12x^2 - 48$

$$12x^2 - 48 = 12(x^2 - 4) \qquad \text{First we factor out the greatest}$$
$$\text{common factor, 12.}$$

$$= 12(x + 2)(x - 2) \qquad \text{Then we use the difference-of-two-squares}$$
$$\text{formula, } a^2 - b^2 = (a + b)(a - b).$$

Practice Problem 9 Factor. $20x^2 - 45$ ●

● EXAMPLE 10 Factor. $24x^2 + 72x + 54$

$$24x^2 + 72x + 54 = 6(4x^2 + 12x + 9) \qquad \text{First we factor out the greatest common}$$
$$\text{factor, 6.}$$

$$= 6(2x + 3)^2 \qquad \text{Then we use the perfect-square trinomial}$$
$$\text{formula, } a^2 + 2ab + b^2 = (a + b)^2.$$

Practice Problem 10 Factor. $75x^2 - 60x + 12$ ●

Factor.

1. $81x^2 - 16$

2. $100x^2 - 49$

3. $16 - 9x^2$

4. $49 - 25x^2$

5. $9x^2 - 25$

6. $81x^2 - 1$

7. $4x^2 - 25$

8. $16x^2 - 25$

9. $36x^2 - 25$

10. $1 - 25x^2$

11. $1 - 49x^2$

12. $1 - 36x^2$

13. $16x^2 - 49y^2$

14. $25x^4 - 81y^4$

15. $25 - 121x^2$

16. $9x^2 - 49$

17. $81x^2 - 100y^2$

18. $25a^2 - 1$

19. $25a^2 - 49$

20. $9x^2 - 49y^2$

21. $9x^2 + 6x + 1$

22. $25x^2 + 10x + 1$

23. $y^2 - 6y + 9$

24. $y^2 - 8y + 16$

25. $9x^2 - 24x + 16$

26. $4x^2 + 20x + 25$

27. $49x^2 + 28x + 4$

28. $25x^2 + 30x + 9$

29. $x^2 + 14x + 49$

30. $x^2 + 8x + 16$

31. $25x^2 - 40x + 16$

32. $49x^2 - 42x + 9$

33. $81x^2 + 36xy + 4y^2$

34. $36x^2 + 60xy + 25y^2$

35. $25x^2 - 30xy + 9y^2$

36. $4x^2 - 28xy + 49y^2$

37. $16a^2 + 72ab + 81b^2$

38. $169a^2 + 26ab + b^2$

39. $9x^4 - 6x^2y + y^2$

40. $y^4 - 22y^2 + 121$

41. $49x^2 + 70x + 9$

42. $25x^2 - 50x + 16$

43. $16x^4 - 1$

44. $81x^4 - 1$

45. $x^{10} - 36y^{10}$

46. $x^4 - 49y^6$

47. $9x^{10} - 12x^5 + 4$

48. $36x^8 - 36x^4 + 9$

To Think About

49. In Example 4, first we factored $81x^4 - 1$ as $(9x^2 + 1)(9x^2 - 1)$, and then we factored $9x^2 - 1$ as $(3x + 1)(3x - 1)$. Show why you cannot factor $9x^2 + 1$.

50. What two numbers could replace the b in $25x^2 + bx + 16$ so that the resulting trinomial would be a perfect square? (*Hint*: One number is negative.)

51. What value could you give to c so that $16y^2 - 56y + c$ would become a perfect-square trinomial? Is there only one answer or more than one?

52. Jerome says that he can find two values of b so that $100x^2 + bx - 9$ will be a perfect square. Kesha says there is only one that fits, and Larry says there are none. Who is correct and why?

Factor by first looking for a greatest common factor. See Examples 9 and 10.

53. $16x^2 - 36$

54. $27x^2 - 75$

55. $147x^2 - 3y^2$

56. $16y^2 - 100x^2$

57. $12x^2 - 36x + 27$

58. $125x^2 - 100x + 20$

59. $98x^2 + 84x + 18$

60. $128x^2 + 96x + 18$

Mixed Practice

Factor. Be sure to look for common factors first.

61. $x^2 - 9x + 14$

62. $x^2 - 9x - 36$

63. $2x^2 + 5x - 3$

64. $15x^2 - 11x + 2$

65. $16x^2 - 121$

66. $9x^2 - 100y^2$

67. $9x^2 + 42x + 49$

68. $9x^2 + 30x + 25$

69. $3x^2 + 6x - 45$

70. $4x^2 + 24x + 32$

71. $5x^2 - 80$

72. $13x^2 - 13$

73. $5x^2 + 20x + 20$

74. $8x^2 + 48x + 72$

75. $2x^2 - 32x + 126$

76. $2x^2 - 32x + 110$

Cumulative Review Problems

77. Divide. $(x^3 + x^2 - 2x - 11) \div (x - 2)$

78. Divide. $(6x^3 + 11x^2 - 11x - 20) \div (3x + 4)$

The green iguana can reach a length of 6 feet and weigh up to 18 pounds. Of the basic diet of the iguana, 40% should consist of greens such as lettuce, spinach, and parsley; 35% should consist of bulk vegetables such as broccoli, zucchini, and carrots; and 25% should consist of fruit.

79. If a certain iguana weighing 150 ounces has a daily diet equal to 2% of its body weight, compose a diet for it in ounces that will meet the iguana's one-day requirement for nutrition.

80. If another iguana weighing 120 ounces has a daily diet equal to 3% of its body weight, compose a diet for it in ounces that will meet the iguana's one-day requirement for nutrition.

The peak of Mount Washington is at an altitude of 6288 feet above sea level. The altitude A in feet of a car driving down the mountain road from the peak a distance M measured in miles is given by $A = 6288 - 700M$.

81. What is the altitude of a car that has driven from the mountain peak down a distance of 3.5 miles?

82. A car drives from the peak down the mountain road to a point where the altitude is 2788 feet above sea level. How many miles down the road has the car driven?

1 Identifying and Factoring Polynomials

Often the various types of factoring problems are all mixed together. We need to be able to identify each type of polynomial quickly. The following table summarizes the information we have learned about factoring.

Many polynomials require more than one factoring method. When you are asked to factor a polynomial, it is expected that you will factor it completely. Usually, the first step is factoring out a common factor; then the next step will become apparent.

Carefully go through each example in the following Factoring Organizer. Be sure you understand each step that is involved.

Factoring Organizer

Number of Terms in the Polynomial	Identifying Name and/or Formula	Example
A. Any number of terms	**Common factor** The terms have a common factor consisting of a number, a variable, or both.	$2x^2 - 16x = 2x(x - 8)$ $3x^2 + 9y - 12 = 3(x^2 + 3y - 4)$ $4x^2y + 2xy^2 - wxy + xyz = xy(4x + 2y - w + z)$
B. Two terms	**Difference of two squares** First and last terms are perfect squares. $a^2 - b^2 = (a + b)(a - b)$	$16x^2 - 1 = (4x + 1)(4x - 1)$ $25y^2 - 9x^2 = (5y + 3x)(5y - 3x)$
C. Three terms	**Perfect-square trinomial** First and last terms are perfect squares. $a^2 + 2ab + b^2 = (a + b)^2$ $a^2 - 2ab + b^2 = (a - b)^2$	$25x^2 - 10x + 1 = (5x - 1)^2$ $16x^2 + 24x + 9 = (4x + 3)^2$
D. Three terms	**Trinomial of the form $x^2 + bx + c$** It starts with x^2. The constants of the two factors are numbers whose product is c and whose sum is b.	$x^2 - 7x + 12 = (x - 3)(x - 4)$ $x^2 + 11x - 26 = (x + 13)(x - 2)$ $x^2 - 8x - 20 = (x - 10)(x + 2)$
E. Three terms	**Trinomial of the form $ax^2 + bx + c$** It starts with ax^2, where a is any number but 1.	Use trial-and-error or the grouping number method to factor $12x^2 - 5x - 2$. 1. The grouping number is -24. 2. The two numbers whose product is -24 and whose sum is -5 are -8 and 3. 3. $12x^2 - 5x - 2 = 12x^2 + 3x - 8x - 2$ $= 3x(4x + 1) - 2(4x + 1)$ $= (4x + 1)(3x - 2)$
F. Four terms	**Factor by grouping** Rearrange the order if the first two terms do not have a common factor.	$wx - 6yz + 2wy - 3xz = wx + 2wy - 3xz - 6yz$ $= w(x + 2y) - 3z(x + 2y)$ $= (x + 2y)(w - 3z)$

 EXAMPLE 1 Factor.

(a) $25x^3 - 10x^2 + x$ (b) $20x^2y^2 - 45y^2$

(c) $2ax + 4ay + 4x + 8y$ (d) $15x^2 - 3x^3 + 18x$

(a) $25x^3 - 10x^2 + x = x(25x^2 - 10x + 1)$ Factor out the common factor of x. The other factor is a perfect-square trinomial.

$$= x(5x - 1)^2$$

(b) $20x^2y^2 - 45y^2 = 5y^2(4x^2 - 9)$ Factor out the common factor of $5y^2$. The other factor is a difference of squares.

$$= 5y^2(2x - 3)(2x + 3)$$

(c) $2ax + 4ay + 4x + 8y = 2[ax + 2ay + 2x + 4y]$ Factor out the common factor of 2.

$$= 2[a(x + 2y) + 2(x + 2y)]$$ Factor the terms inside the bracket by the grouping method.

$$= 2[(x + 2y)(a + 2)]$$ Factor out the common factor of $(x + 2y)$.

(d) $15x^2 - 3x^3 + 18x = -3x^3 + 15x^2 + 18x$ Rearrange the terms in descending order of powers of x.

$$= -3x(x^2 - 5x - 6)$$ Factor out the common factor of $-3x$.

$$= -3x(x - 6)(x + 1)$$ Factor the trinomial.

Practice Problem 1   Factor. Be careful. These practice problems are mixed.

(a) $9x^4y^2 - 9y^2$ **(b)** $12x - 9 - 4x^2$

(c) $3x^2 - 36x + 108$ **(d)** $5x^3 - 15x^2y + 10x^2 - 30xy$

2 Determining Whether a Polynomial Is Prime

Not all polynomials can be factored using the methods in this chapter. If we cannot factor a polynomial by elementary methods, we will identify it as a **prime** polynomial. If, after you have mastered the factoring techniques in this chapter, you encounter a polynomial that you cannot factor with these methods, you should feel comfortable enough to say, "The polynomial cannot be factored with the methods in this chapter, so it is prime," rather than "I can't do it—I give up!"

EXAMPLE 2   Factor, if possible. $x^2 + 6x + 12$

The factors of 12 are

$$(1)(12) \text{ or } (2)(6) \text{ or } (3)(4).$$

None of these pairs add up to 6, the coefficient of the middle term. Thus the problem cannot be factored by the methods of this chapter. It is prime.

Practice Problem 2   Factor. $x^2 - 9x - 8$

EXAMPLE 3   Factor, if possible. $25x^2 + 4$

We have a formula to factor the difference of two squares. There is no way to factor the sum of two squares. That is, $a^2 + b^2$ cannot be factored. Thus

$$25x^2 + 4 \text{ is prime.}$$

Practice Problem 3   Factor, if possible. $25x^2 + 82x + 4$

Review the six basic types of factoring in the Factoring Organizer on page 330. Each of the six types is included in exercises 1–12. Be sure you can find two of each type.

Factor. Check your answer by multiplying.

1. $6a^2 + 2ab - 3a$

2. $6x^2 - 3xy + 5x$

3. $36x^2 - 9y^2$

4. $100x^2 - 1$

5. $9x^2 - 12xy + 4y^2$

6. $16x^2 + 24xy + 9y^2$

7. $x^2 + 8x + 15$

8. $x^2 + 15x + 54$

9. $15x^2 + 7x - 2$

10. $6x^2 + 13x - 5$

11. $ax - 3ay - 6by + 2bx$

12. $ax - 20y - 5x + 4ay$

Factor, if possible. Be sure to factor completely. Always factor out the greatest common factor first, if one exists.

13. $3x^4 - 12$

14. $y^2 + 16y + 64$

15. $4x^2 - 12x + 9$

16. $108x^2 - 3$

17. $2x^2 - 11x + 12$

18. $2xy^2 - 50x$

19. $x^2 - 3xy - 70y^2$

20. $2x^3 - 7x^2 + 4x - 14$

21. $ax - 5a + 3x - 15$

22. $by + 7b - 6y - 42$

23. $45x - 5x^3$

24. $18y^2 + 3y - 6$

25. $5x^3y^3 - 10x^2y^3 + 5xy^3$

26. $12x^2 - 36x + 27$

27. $27xyz^2 - 12xy$

28. $12x^2 - 2x - 18x^3$

29. $3x^2 + 6x - 105$

30. $4x^2 - 28x - 72$

31. $5x^2 - 30x + 40$

32. $7x^2 + 3x - 2$

33. $7x^2 - 2x^4 + 4$

34. $2x^4 - 9x^2 - 5$

35. $6x^2 - 3x + 2$

36. $4x^3 + 8x^2 - 60x$

Remove the greatest common factor first. Then continue to factor.

37. $5x^2 + 10xy - 30y$

38. $7a^2 + 21b - 42$

39. $30x^3 + 3x^2y - 6xy^2$

40. $56x^2 - 14xy - 7y^2$

41. $8x^2 + 28x - 16$

42. $12x^2 - 30x + 12$

To Think About

43. A polynomial that cannot be factored by the methods of this chapter is called _____.

44. A binomial of the form $x^2 - d$ can be quickly factored or identified as prime. If it can be factored, what is true of the number d?

Cumulative Review Problems

45. When Dave Barry decided to leave the company and work as an independent contractor, he took a pay cut of 14%. He earned $24,080 this year. What did he earn in his previous job?

46. A major pharmaceutical company is testing a new, powerful antiviral drug. It kills 13 strains of virus every hour. If there are presently 294 live strains of virus in the test container, how many live strains were there 6 hours ago?

47. Gary loves to read. In his living room he has hardcover books, softcover books, and magazines. He has 37 fewer hardcover books than softcover books. He has twice as many softcover books as magazines. If there are 198 total books and magazines in his bookcase, how many of each type did he have?

48. Gary took some of the items listed in exercise 47 from his bookcase. It now contains a total of 193 books and magazines. The statements comparing the numbers of hardcover and softcover books and the numbers of softcover books and magazines given in exercise 47 still apply. How many of each type does he have now?

1 Solving a Quadratic Equation by Factoring

In Chapter 2 we learned how to solve linear equations such as $3x + 5 = 0$ by finding the root (or value of x) that satisfied the equation. Now we turn to the question of how to solve equations like $3x^2 + 5x + 2 = 0$. Such equations are called **quadratic equations**. A quadratic equation is a polynomial equation in one variable that contains a variable term of degree 2 and no terms of higher degree.

> The *standard form* of a quadratic equation is $ax^2 + bx + c = 0$, where a, b, and c are real numbers and $a \neq 0$.

In this section, we will study quadratic equations in standard form, where a, b, and c are integers.

Many quadratic equations have two real number solutions (also called real **roots**). But how can we find them? The most direct approach is the factoring method. This method depends on a very powerful property.

Zero Factor Property

If $a \cdot b = 0$, then $a = 0$ or $b = 0$.

Notice the word "or" in the zero factor property. When we make a statement in mathematics using this word, we intend it to mean *one or the other or both*. Therefore, the zero factor property states that if the product $a \cdot b$ is zero, then a can equal zero or b can equal zero or *both a and b can equal zero*. We can use this principle to solve a quadratic equation. Before you start, make sure that the equation is in standard form.

1. Factor, if possible, the quadratic expression that equals 0.
2. Set each factor equal to 0.
3. Solve the resulting equations to find each root.
4. Check each root.

EXAMPLE 1 Solve the equation to find the two roots. $3x^2 + 5x + 2 = 0$

$$3x^2 + 5x + 2 = 0 \qquad \text{The equation is in standard form.}$$
$$(3x + 2)(x + 1) = 0 \qquad \text{Factor the quadratic expression.}$$
$$3x + 2 = 0 \qquad x + 1 = 0 \qquad \text{Set each factor equal to 0.}$$
$$3x = -2 \qquad x = -1 \qquad \text{Solve the equations to find the two roots.}$$
$$x = -\frac{2}{3}$$

The two roots (that is, solutions) are $-\frac{2}{3}$ and -1.

Check. We can determine if the two numbers $-\frac{2}{3}$ and -1 are solutions to the equation. Substitute $-\frac{2}{3}$ for x in the *original equation*. If an identity results, $-\frac{2}{3}$ is a solution. Do the same for -1.

$$3x^2 + 5x + 2 = 0 \qquad\qquad 3x^2 + 5x + 2 = 0$$
$$3\left(-\frac{2}{3}\right)^2 + 5\left(-\frac{2}{3}\right) + 2 \overset{?}{=} 0 \qquad\qquad 3(-1)^2 + 5(-1) + 2 \overset{?}{=} 0$$
$$3\left(\frac{4}{9}\right) + 5\left(-\frac{2}{3}\right) + 2 \overset{?}{=} 0 \qquad\qquad 3(1) + 5(-1) + 2 \overset{?}{=} 0$$
$$\frac{4}{3} - \frac{10}{3} + 2 \overset{?}{=} 0 \qquad\qquad 3 - 5 + 2 \overset{?}{=} 0$$
$$\frac{4}{3} - \frac{10}{3} + \frac{6}{3} \overset{?}{=} 0 \qquad\qquad -2 + 2 \overset{?}{=} 0$$
$$0 = 0 \ \checkmark \qquad\qquad\qquad 0 = 0 \ \checkmark$$

Thus $-\frac{2}{3}$ and -1 are both roots of the equation $3x^2 + 5x + 2 = 0$.

Practice Problem 1 Solve the equation by factoring to find the two roots and check.
$10x^2 - x - 2 = 0$

EXAMPLE 2 Solve the equation to find the two roots. $2x^2 + 13x - 7 = 0$

$2x^2 + 13x - 7 = 0$ — The equation is in standard form.

$(2x - 1)(x + 7) = 0$ — Factor.

$2x - 1 = 0 \qquad x + 7 = 0$ — Set each factor equal to 0.

$2x = 1 \qquad\qquad x = -7$ — Solve the equations to find the two roots.

$x = \dfrac{1}{2}$

The two roots are $\frac{1}{2}$ and -7.

Check. If $x = \frac{1}{2}$, then we have the following.

$$2\left(\frac{1}{2}\right)^2 + 13\left(\frac{1}{2}\right) - 7 = 2\left(\frac{1}{4}\right) + 13\left(\frac{1}{2}\right) - 7$$

$$= \frac{1}{2} + \frac{13}{2} - \frac{14}{2} = 0 \checkmark$$

If $x = -7$, then we have the following.

$$2(-7)^2 + 13(-7) - 7 = 2(49) + 13(-7) - 7$$

$$= 98 - 91 - 7 = 0 \checkmark$$

Thus $\frac{1}{2}$ and -7 are both roots of the equation $2x^2 + 13x - 7 = 0$.

Practice Problem 2 Solve the equation to find the two roots. $3x^2 - 5x + 2 = 0$

If the quadratic equation $ax^2 + bx + c = 0$ has no visible constant term, then $c = 0$. All such quadratic equations can be solved by factoring out a common factor and then using the zero factor property to obtain two solutions that are real numbers.

EXAMPLE 3 Solve the equation to find the two roots. $7x^2 - 3x = 0$

$7x^2 - 3x = 0$ — The equation is in standard form. Here $c = 0$.

$x(7x - 3) = 0$ — Factor out the common factor.

$x = 0 \qquad 7x - 3 = 0$ — Set each factor equal to 0 by the zero factor property.

$7x = 3$ — Solve the equations to find the two roots.

$x = \dfrac{3}{7}$

The two roots are 0 and $\frac{3}{7}$.

Check. Verify that 0 and $\frac{3}{7}$ are the roots of $7x^2 - 3x = 0$.

Practice Problem 3 Solve the equation to find the two roots. $7x^2 + 11x = 0$

If the quadratic equation is not in standard form, we use the same basic algebraic methods we studied in Sections 2.1–2.3 to place the terms on one side and zero on the other so that we can use the zero factor property.

EXAMPLE 4 Solve. $x^2 = 12 - x$

$$x^2 = 12 - x \qquad \text{The equation is not in standard form.}$$

$$x^2 + x - 12 = 0 \qquad \text{Add } x \text{ and } -12 \text{ to both sides of the equation so that the left side is equal to zero; we can now factor.}$$

$$(x - 3)(x + 4) = 0 \qquad \text{Factor.}$$

$$x - 3 = 0 \qquad x + 4 = 0 \qquad \text{Set each factor equal to 0 by the zero factor property.}$$

$$x = 3 \qquad x = -4 \qquad \text{Solve the equations for } x.$$

Check. If $x = 3$: $\quad (3)^2 \overset{?}{=} 12 - 3 \qquad\qquad$ If $x = -4$: $\quad (-4)^2 \overset{?}{=} 12 - (-4)$

$$9 \overset{?}{=} 12 - 3 \qquad\qquad\qquad\qquad 16 \overset{?}{=} 12 + 4$$

$$9 = 9 \;\checkmark \qquad\qquad\qquad\qquad\qquad 16 = 16 \;\checkmark$$

Both roots check.

Practice Problem 4 Solve. $x^2 - 6x + 4 = -8 + x$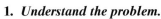

EXAMPLE 5 Solve. $\dfrac{x^2 - x}{2} = 6$

We must first clear the fractions from the equation.

$$2\left(\frac{x^2 - x}{2}\right) = 2(6) \qquad \text{Multiply each side by 2.}$$

$$x^2 - x = 12 \qquad \text{Simplify.}$$

$$x^2 - x - 12 = 0 \qquad \text{Place in standard form.}$$

$$(x - 4)(x + 3) = 0 \qquad \text{Factor.}$$

$$x - 4 = 0 \qquad x + 3 = 0 \qquad \text{Set each factor equal to zero.}$$

$$x = 4 \qquad x = -3 \qquad \text{Solve the equations for } x.$$

The check is left to the student.

Practice Problem 5 Solve. $\dfrac{2x^2 - 7x}{3} = 5$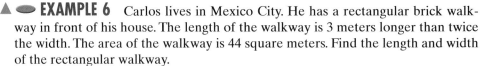

2 *Using Quadratic Equations to Solve Applied Problems*

Certain types of word problems—for example, some geometry applications—lead to quadratic equations. We'll show how to solve such word problems in this section.

It is particularly important to check the apparent solutions to the quadratic equation with conditions stated in the word problem. Often a particular solution to the quadratic equation will be eliminated by the conditions of the word problem.

▲ **EXAMPLE 6** Carlos lives in Mexico City. He has a rectangular brick walkway in front of his house. The length of the walkway is 3 meters longer than twice the width. The area of the walkway is 44 square meters. Find the length and width of the rectangular walkway.

1. *Understand the problem.*
   Draw a picture.

   Let $w =$ the width in meters

   Then $2w + 3 =$ the length in meters

2. *Write an equation.*

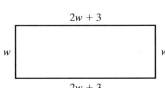

$$\text{area} = (\text{width})(\text{length})$$

$$44 = w(2w + 3)$$

3. *Solve and state the answer.*

$44 = w(2w + 3)$

$44 = 2w^2 + 3w$ Remove parentheses.

$0 = 2w^2 + 3w - 44$ Subtract 44 from both sides.

$0 = (2w + 11)(w - 4)$ Factor.

$2w + 11 = 0$ $w - 4 = 0$ Set each factor equal to 0.

$2w = -11$ $w = 4$ Simplify and solve.

$w = -5\dfrac{1}{2}$ Although $-5\dfrac{1}{2}$ is a solution to the quadratic equation, it is not a valid solution to the word problem. It would not make sense to have a rectangle with a negative number as a width.

Since $w = 4$, the width of the walkway is 4 meters. The length is $2w + 3$, so we have $2(4) + 3 = 8 + 3 = 11$. Thus the length of the walkway is 11 meters.

4. *Check.*   Is the length 3 meters more than twice the width?

$$11 \overset{?}{=} 3 + 2(4) \qquad 11 = 3 + 8 \quad ✓$$

Is the area of the rectangle 44 square meters?

$$4 \times 11 \overset{?}{=} 44 \qquad 44 = 44 \quad ✓$$

▲ **Practice Problem 6**   The length of a rectangle is 2 meters longer than triple the width. The area of the rectangle is 85 square meters. Find the length and width of the rectangle.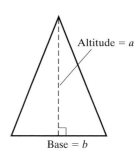

 EXAMPLE 7   The top of a local cable television tower has several small triangular reflectors. The area of each triangle is 49 square centimeters. The altitude of each triangle is 7 centimeters longer than the base. Find the altitude and the base of one of the triangles.

Let b = the length of the base in centimeters

$b + 7$ = the length of the altitude in centimeters

To find the area of a triangle, we use

$$\text{area} = \frac{1}{2}(\text{altitude})(\text{base}) = \frac{1}{2}ab = \frac{ab}{2}.$$

$\dfrac{ab}{2} = 49$ Write an equation.

$\dfrac{(b + 7)(b)}{2} = 49$ Substitute the expressions for altitude and base.

$\dfrac{b^2 + 7b}{2} = 49$ Simplify.

$b^2 + 7b = 98$ Multiply each side of the equation by 2.

$b^2 + 7b - 98 = 0$ Place the quadratic equation in standard form.

$(b - 7)(b + 14) = 0$ Factor.

$b - 7 = 0$ $b + 14 = 0$ Set each factor equal to zero.

$b = 7$ $b = -14$ Solve the equations for b.

We cannot have a base of -14 centimeters, so we reject the negative answer. The only possible solution is 7. So the base is 7 centimeters. The altitude is $b + 7 = 7 + 7 = 14$. The altitude is 14 centimeters. The triangular reflector has a base of 7 centimeters and an altitude of 14 centimeters.

Altitude = a

Base = b

Check. When you do the check, answer the following two questions.

1. Is the altitude 7 centimeters longer than the base?

2. Is the area of a triangle with a base of 7 centimeters and an altitude of 14 centimeters actually 49 square centimeters?

▲ **Practice Problem 7** A triangle has an area of 35 square centimeters. The altitude of the triangle is 3 centimeters shorter than the base. Find the altitude and the base of the triangle.

Many problems in the sciences require the use of quadratic equations. You will study these in more detail if you take a course in physics or calculus in college. Often a quadratic equation is given as part of the problem.

When an object is thrown upward, its height (S) in meters is given, approximately, by the quadratic equation

$$S = -5t^2 + vt + h.$$

The letter h represents the initial height in meters. The letter v represents the initial velocity of the object thrown. The letter t represents the time in seconds starting from the time the object is thrown.

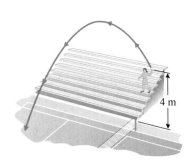

4 m

EXAMPLE 8 A tennis ball is thrown upward with an initial velocity of 8 meters/second. Suppose that the initial height above the ground is 4 meters. At what time t will the ball hit the ground?

In this case $S = 0$ since the ball will hit the ground. The initial upward velocity is $v = 8$ meters/second. The initial height is 4 meters, so $h = 4$.

$S = -5t^2 + vt + h$	Write an equation.
$0 = -5t^2 + 8t + 4$	Substitute all values into the equation.
$5t^2 - 8t - 4 = 0$	Isolate the terms on the left side. (Most students can factor more readily if the squared variable is positive.)
$(5t + 2)(t - 2) = 0$	Factor.
$5t + 2 = 0 \qquad t - 2 = 0$	Set each factor $= 0$.
$5t = -2 \qquad\qquad t = 2$	Solve the equations for t.
$t = -\dfrac{2}{5}$	

We want a positive time t in seconds; thus we do not use $t = -\frac{2}{5}$. Therefore the ball will strike the ground 2 seconds after it is thrown.

Check. Verify the solution.

Practice Problem 8 A Mexican cliff diver does a dive from a cliff 45 meters above the ocean. This constitutes free fall, so the initial velocity is $v = 0$, and if there is no upward spring, then $h = 45$ meters. How long will it be until he breaks the water's surface?

Exercises

Using the factoring method, solve for the roots of each quadratic equation. Be sure to place the equation in standard form before factoring. Check your answers.

1. $x^2 - 4x - 21 = 0$ **2.** $x^2 - x - 20 = 0$ **3.** $2x^2 - 5x - 3 = 0$ **4.** $3x^2 - 5x - 2 = 0$

5. $2x^2 - 7x + 6 = 0$ **6.** $2x^2 - 11x + 12 = 0$ **7.** $6x^2 - 13x = -6$ **8.** $10x^2 + 19x = 15$

9. $x^2 + 13x = 0$ **10.** $8x^2 - x = 0$ **11.** $8x^2 = 72$ **12.** $9x^2 = 81$

13. $5x^2 + 3x = 8x$ **14.** $6x^2 - 4x = 3x$

15. $(x - 5)(x + 2) = -4(x + 1)$ **16.** $(x - 5)(x + 4) = 2(x - 5)$

17. $4x^2 - 3x + 1 = -7x$ **18.** $9x^2 - 2x + 4 = 10x$ **19.** $\dfrac{x^2}{2} - 8 + x = -8$ **20.** $4 + \dfrac{x^2}{3} = 2x + 4$

21. $\dfrac{x^2 + 7x}{4} = -3$ **22.** $\dfrac{x^2 + 5x}{6} = 4$ **23.** $\dfrac{9x^2 - 12x}{3} = 15$ **24.** $\dfrac{4x^2 - 10x}{2} = 12$

To Think About

25. Why can an equation in standard form with $c = 0$ (that is, an equation of the form $ax^2 + bx = 0$) always be solved?

26. Martha solved $(x + 3)(x - 2) = 14$ as follows:

$$x + 3 = 14 \quad \text{or} \quad x - 2 = 14$$
$$x = 11 \quad \text{or} \quad x = 16$$

Josette said this had to be wrong because these values do not check. Explain what is wrong with Martha's method.

Applications

▲ **27.** The area of a rectangular garden is 140 square meters. The width is 3 meters longer than one-half of the length. Find the length and the width of the garden.

▲ **28.** The area of a triangular sign is 33 square meters. The base of the triangle is 1 meter less than double the altitude. Find the altitude and the base of the sign.

Suppose the number of teams competing in a sports league is x. In this league each team plays each other team twice. The total number G of games to be played is given by the equation $G = x^2 - x$. Use this information in solving exercises 29–32.

29. A women's basketball league has a total of 14 teams. How many games will be played during the season?

30. A men's baseball league has a total of 12 teams. How many games will be played during the season?

31. A city tennis league has a total of 210 games. How many teams are in the league?

32. A midget football league has a total of 156 games. How many teams are in the league?

Use the following information for exercises 33 and 34. When an object is thrown upward, its height (S), in meters, is given (approximately) by the quadratic equation

$$S = -5t^2 + vt + h,$$

where v = the upward initial velocity in meters/second,

t = the time of flight in seconds, and

h = the height above level ground from which the object is thrown.

33. Johnny is standing on a platform 6 meters high and throws a ball straight up as high as he can at a velocity of 13 meters/second. At what time t will the ball hit the ground? How far from the ground is the ball after 2 seconds have elapsed from the time of the throw? (Assume that the ball is 6 meters from the ground when it leaves Johnny's hand.)

34. You are standing on the edge of a cliff near Acapulco, overlooking the ocean. The place where you stand is 180 meters from the ocean. You drop a pebble into the water. ("Dropping" the pebble implies that there is no initial velocity, so $v = 0$.) How many seconds will it take to hit the water? How far has the pebble dropped after 3 seconds?

35. Paying overtime to employees can be very expensive, but if profits are realized, it is worth putting a shift on overtime. In a high-tech helicopter company, the extra hourly cost in dollars for producing x additional helicopters is given by the cost equation $C = 2x^2 - 7x$. If the extra hourly cost is \$15, how many additional helicopters are produced?

36. A boat generator on a Gloucester fishing boat is required to produce 64 watts of power. The amount of current I measured in amperes needed to produce the power for this generator is given by the equation $P = 40I - 4I^2$. What is the *minimum* number of amperes required to produce the necessary power?

The technology and communication office of a local company has set up a new telephone system so that each employee has a separate telephone and extension number. They are studying the possible number of telephone calls that can be made from people in the office to other people in the office. They have discovered that the total number of possible telephone calls T is described by the equation $T = 0.5(x^2 - x)$, where x is the number of people in the office. Use this information to answer exercises 37–40.

37. If 70 people are presently employed at the office, how many possible telephone calls can be made between these 70 people?

38. If the company hires 10 new employees next year, how many possible telephone calls can be made between the 80 people that will be employed next year?

39. One Saturday, only a small number of employees were working at the office. It has been determined that on that day, a total of 153 different phone calls could have been made from people working in the office to other people working in the office. How many people worked on that Saturday?

40. On the day after Thanksgiving, only a small number of employees were working at the office. It has been determined that on that day, a total of 105 different phone calls could have been made from people working in the office to other people working in the office. How many people worked on the day after Thanksgiving?

Cumulative Review Problems

Simplify.

41. $(2x^2y^3)(-6xy^4)$

42. $(3a^4b^5)(4a^6b^8)$

43. $\dfrac{21a^5b^{10}}{-14ab^{12}}$

44. $\dfrac{18x^3y^6}{54x^8y^{10}}$

Putting Your Skills to Work

Predicting Total Earnings by a Mathematical Series

When assessing the financial health and future of a company or person, it is often helpful to look at the total income of that company or person over a long period of time. If the income tends to increase or decrease at a fairly constant rate, then the sum of this income can be determined by a formula for a mathematical series.

Angela Sanchez has been employed by West Coast Cable Vision for 9 years. She started at an annual salary of $18,000. Her income has increased about $1200 per year for the last 9 years. She is now earning $27,600. To find her total earnings for 9 years we can use the formula in factored form

$$T = 0.5n(s + e),$$

where n = the number of years she has worked,

s = her starting salary, and

e = her salary at the end of the time period.

Thus we have the following.

$$T = 0.5(9)(18{,}000 + 27{,}600)$$
$$T = 0.5(9)(45{,}600)$$
$$T = \$205{,}200$$

Thus Angela has earned $205,200 in the last 9 years.

Use the preceding formula to determine the following.

Problems for Individual Study and Investigation

1. Monique Bussell has worked for the last 20 years at a salary that has increased about $1500 per year. Her starting salary was $24,000. Approximately how much has she earned in this 20-year period?

2. The Software Service Center has been in operation for 8 years. During the first year they made a profit of $150,000. Each year the profit has increased by $70,000. What is their total profit for 8 years?

Problems for Cooperative Group Investigation

3. James Kerr opened his own travel business. Last year he earned $40,000. This year he earned $56,000. If his business earnings continue to increase each year by the same amount, how many years will it be until he has earned a total income of $1 million?

4. A national chain of department stores made a profit of $780,000,000 ten years ago. Over the past ten years, the profit has been decreasing at the rate of $15,000,000 per year. If this rate of decrease continues, what will be the total earnings of the store for the next 15 years (starting with this present year)?

Internet Connections

  Netsite: http://www.prenhall.com/tobey_beginning

Site: DataMasters Salary Survey or a related site

This site gives salary information for executives and professionals in the computer industry.

5. Karl is an experienced software engineer whose salary has kept up with the regional median for the West Coast since 1990. Use the DataMasters site to determine Karl's salary for 1990 and for the current year. Then use the formula $T = 0.5n(s + e)$ to find his total salary for all years from 1990 to the current year, assuming that his income has increased by a constant amount each year.

Math in the Media

Media Literacy Tips

We receive information daily from a number of media sources—radio, newspapers, television, and on-line content providers. It's important to be able to use math skills to understand the data presented. But using your math skills alone is not enough. You also need to be able to interpret the message in context of the article.

The media can be very persuasive when promoting a viewpoint. There are many factors to consider when evaluating the article. Following are a few tips that may help you to filter out the "hype" and focus on the meaning.

1. Consider the source. Who is the author? Does he or she have a particular platform that may influence the message?

2. Look for the key points. Has anything been done to either attract attention to certain issues and facts or distract the reader from them?

3. Be alert for key omissions. Sometimes what is left unsaid is the most compelling statement.

4. Consider the supporting information. Does the author support his or her argument with facts, opinions, both? How is math used to support the argument?

EXERCISES

1. Locate an article or news feature that interests you. Read the article and write down your responses to each of the 4 tips above.

2. What response do you think the author is trying to elicit?

3. Suppose you read an article with the following information: "Data released yesterday by the FBI indicates that the number of murders committed in the United States has decreased by 33.7% in the last five years. The murder rate for 1998 was 6.3% per 100,000 inhabitants. This rate is the lowest the rate has been since 1979. What a wonderful improvement. Truly the United States is becoming a safe place to live." What concerns or objections would you raise concerning this article?

Chapter 5 Organizer

Topic	Procedure	Examples
A. Common factor, p. 303.	Factor out the largest common factor from each term.	$2x^2 - 2x = 2x(x - 1)$ $3a^2 + 3ab - 12a = 3a(a + b - 4)$ $8x^4y - 24x^3 = 8x^3(xy - 3)$
Special cases **B. Difference of squares, p. 325.** **C. Perfect-square trinomials, p. 326.**	If you recognize the special cases, you will be able to factor quickly. $$a^2 - b^2 = (a + b)(a - b)$$ $$a^2 + 2ab + b^2 = (a + b)^2$$ $$a^2 - 2ab + b^2 = (a - b)^2$$	$25x^2 - 36y^2 = (5x + 6y)(5x - 6y)$ $16x^4 - 1 = (4x^2 + 1)(2x + 1)(2x - 1)$ $25x^2 + 10x + 1 = (5x + 1)^2$ $49x^2 - 42xy + 9y^2 = (7x - 3y)^2$
D. Trinomials of the form $x^2 + bx + c$, p. 312.	Factor trinomials of the form $x^2 + bx + c$ by asking what two numbers have a product of c and a sum of b. If each term of the trinomial has a common factor, factor it out as the first step.	$x^2 - 18x + 77 = (x - 7)(x - 11)$ $x^2 + 7x - 18 = (x + 9)(x - 2)$ $5x^2 - 10x - 40 = 5(x^2 - 2x - 8)$ $\qquad\qquad = 5(x - 4)(x + 2)$
E. Trinomials of the form $ax^2 + bx + c$, where $a \neq 1$, p. 319.	Factor trinomials of the form $ax^2 + bx + c$ by the grouping number method or by the trial-and-error method.	$6x^2 + 11x - 10$ Grouping number $= -60$ Two numbers whose product is -60 and whose sum is $+11$ are $+15$ and -4. $6x^2 + 15x - 4x - 10 = 3x(2x + 5) - 2(2x + 5)$ $\qquad\qquad\qquad = (2x + 5)(3x - 2)$
F. Four terms. **Factor by grouping, p. 308.**	Rearrange the terms if necessary so that the first two terms have a common factor. Then factor out the common factors. $ax + ay - bx - by = a(x + y) - b(x + y)$ $\qquad\qquad\qquad = (a - b)(x + y)$	$2ax^2 + 21 + 3a + 14x^2$ $= 2ax^2 + 14x^2 + 3a + 21$ $= 2x^2(a + 7) + 3(a + 7)$ $= (a + 7)(2x^2 + 3)$
Prime polynomials, p. 331.	A polynomial that is not factorable is called prime.	$x^2 + y^2$ is prime. $x^2 + 5x + 7$ is prime.
Multistep factoring, p. 331.	Many problems require two or three steps of factoring. Always try to factor out a common factor as the first step.	$3x^2 - 21x + 36 = 3(x^2 - 7x + 12)$ $\qquad\qquad\qquad = 3(x - 4)(x - 3)$ $2x^3 - x^2 - 6x = x(2x^2 - x - 6)$ $\qquad\qquad\qquad = x(2x + 3)(x - 2)$ $25x^3 - 49x = x(25x^2 - 49)$ $\qquad\qquad\qquad = x(5x + 7)(5x - 7)$ $8x^2 - 24x + 18 = 2(4x^2 - 12x + 9)$ $\qquad\qquad\qquad = 2(2x - 3)^2$
Solving quadratic equations by factoring, p. 334.	1. Write as $ax^2 + bx + c = 0$. 2. Factor. 3. Set each factor equal to 0. 4. Solve the resulting equations.	Solve: $3x^2 + 5x = 2$ $\qquad 3x^2 + 5x - 2 = 0$ $\qquad (3x - 1)(x + 2) = 0$ $\qquad 3x - 1 = 0 \quad\text{or}\quad x + 2 = 0$ $\qquad x = \dfrac{1}{3} \quad\text{or}\qquad x = -2$

Topic	Procedure	Examples
Using quadratic equations to solve applied problems, p. 336.	Some word problems, like those involving the product of two numbers, area, and formulas with a squared variable, can be solved using the factoring methods we have shown.	The length of a rectangle is 4 less than three times the width. Find the length and width if the area is 55 square inches. Let w = width. Then $3w - 4$ = length. $$55 = w(3w - 4)$$ $$55 = 3w^2 - 4w$$ $$0 = 3w^2 - 4w - 55$$ $$0 = (3w + 11)(w - 5)$$ $$w = -\tfrac{11}{3} \quad \text{or} \quad w = 5$$ $-\tfrac{11}{3}$ is not a valid solution. Thus width = 5 inches and length = 11 inches.

Chapter 5 Review Problems

5.1 *Factor out the greatest common factor.*

1. $15x^3y - 9x^2y^2$

2. $40x^2 - 32x$

3. $7x^2y - 14xy^2 - 21x^3y^3$

4. $50a^4b^5 - 25a^4b^4 + 75a^5b^5$

5. $27x^3 - 9x^2$

6. $2x - 4y + 6z + 12$

7. $2a(a + 3b) - 5(a + 3b)$

8. $15x^3y + 6xy^2 + 3xy$

5.2 *Factor by grouping.*

9. $3ax - 7a - 6x + 14$

10. $a^2 + 5ab - 4a - 20b$

11. $x^2y + 3y - 2x^2 - 6$

12. $4ad - 25c - 5d + 20ac$

13. $15x^2 - 3x + 10x - 2$

14. $30w^2 - 18w + 5wz - 3z$

5.3 *Factor completely. Be sure to factor out any common factors as your first step.*

15. $x^2 - 2x - 35$

16. $x^2 - 10x + 24$

17. $x^2 + 14x + 48$

18. $x^2 + 8xy + 15y^2$

19. $x^4 + 13x^2 + 42$

20. $x^2 - 14x - 51$

21. $5x^2 + 20x + 15$

22. $3x^2 + 39x + 36$

23. $2x^2 - 28x + 96$

24. $4x^2 - 44x + 120$

5.4 *Factor completely. Be sure to factor out any common factors as your first step.*

25. $4x^2 + 7x - 15$

26. $12x^2 + 11x - 5$

27. $15x^2 + 7x - 4$

28. $6x^2 - 13x + 6$

29. $2x^2 - x - 3$

30. $3x^2 + 2x - 8$

31. $20x^2 + 48x - 5$

32. $20x^2 + 21x - 5$

33. $4a^2 - 11a - 3$

34. $4a^2 - a - 3$

35. $6x^2 + 4x - 10$

36. $6x^2 - 4x - 10$

37. $4x^2 - 26x + 30$

38. $4x^2 - 20x - 144$

39. $12x^2 + 5x - 3$

40. $16x^2 + 14x - 30$

41. $6x^2 - 19xy + 10y^2$

42. $6x^2 - 32xy + 10y^2$

5.5 *Factor these special cases. Be sure to factor out any common factors.*

43. $49x^2 - y^2$

44. $16x^2 - 36y^2$

45. $9x^2 - 12x + 4$

46. $64x^2 - 1$

47. $25x^2 - 36$

48. $100x^2 - 9$

49. $1 - 49x^2$

50. $4 - 49x^2$

51. $36x^2 + 12x + 1$

52. $25x^2 - 20x + 4$

53. $16x^2 - 24xy + 9y^2$

54. $49x^2 - 28xy + 4y^2$

55. $2x^2 - 18$

56. $3x^2 - 75$

57. $8x^2 + 40x + 50$

58. $50x^2 - 120x + 72$

5.6 *If possible, factor each polynomial completely. If a polynomial cannot be factored, state that it is prime.*

59. $4x^2 - 9y^2$

60. $x^2 + 6x + 9$

61. $x^2 - 9x + 18$

62. $x^2 + 13x - 30$

63. $6x^2 + x - 7$

64. $10x^2 + x - 2$

65. $12x + 16$

66. $8x^2y^2 - 4xy$

67. $50x^3y^2 + 30x^2y^2 - 10x^2y^2$

68. $26a^3b - 13ab^3 + 52a^2b^4$

69. $x^3 - 16x^2 + 64x$

70. $2x^2 + 40x + 200$

71. $3x^2 - 18x + 27$

72. $25x^3 - 60x^2 + 36x$

73. $7x^2 - 9x - 10$

74. $4x^2 - 13x - 12$

75. $9x^3y - 4xy^3$

76. $3x^3a^3 - 11x^4a^2 - 20x^5a$

77. $12a^2 + 14ab - 10b^2$

78. $16a^2 - 40ab + 25b^2$

79. $7a - 7 - ab + b$

80. $3d - 4 - 3cd + 4c$

81. $2x - 1 + 2bx - b$

82. $5xb - 35x + 4by - 28y$

83. $2a^2x - 15ax + 7x$

84. $x^5 - 17x^3 + 16x$

85. $x^4 - 81y^{12}$

86. $6x^4 - x^2 - 15$

87. $28yz - 16xyz + x^2yz$

88. $12x^3 + 17x^2 + 6x$

89. $16w^2 - 2w - 5$

90. $12w^2 - 12w + 3$

91. $4y^3 + 10y^2 - 6y$

92. $10y^2 + 33y - 7$

93. $8y^{10} - 16y^8$

94. $49x^4 - 49$

95. $x^2 + 13x + 54$

96. $8x^2 - 19x - 6$

97. $8y^5 + 4y^3 - 60y$

98. $9xy^2 + 3xy - 42x$

99. $16x^4y^2 - 56x^2y + 49$

100. $128x^3y - 2xy$

101. $2ax + 5a - 10b - 4bx$

102. $2x^3 - 9 + x^2 - 18x$

5.7 *Solve the following equations by factoring.*

103. $x^2 - 3x - 18 = 0$

104. $x^2 + 6x - 27 = 0$

105. $5x^2 = 2x - 7x^2$

106. $8x^2 + 5x = 2x^2 - 6x$

107. $2x^2 + 9x - 5 = 0$

108. $x^2 + 11x + 24 = 0$

109. $x^2 + 14x + 45 = 0$

110. $5x^2 = 7x + 6$

111. $3x^2 + 6x = 2x^2 - 9$

112. $4x^2 + 9x - 9 = 0$

113. $5x^2 - 11x + 2 = 0$

Solve.

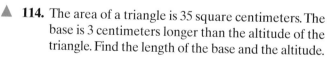

114. The area of a triangle is 35 square centimeters. The base is 3 centimeters longer than the altitude of the triangle. Find the length of the base and the altitude.

115. The area of a rectangle is 105 square feet. The length of the rectangle is 1 foot longer than double the width. Find the length and width of the rectangle.

116. The height in feet that a model rocket attains is given by $h = -16t^2 + 80t + 96$, where t is the time measured in seconds. How many seconds will it take until the rocket finally reaches the ground? (*Hint:* At ground level $h = 0$.)

117. An electronic technician is working with a 100-volt electric generator. The output power of the generator is given by the equation $p = -5x^2 + 100x$, where x is the amount of current measured in amperes and p is measured in watts. The technician wants to find the value for x when the power is 480 watts. Can you find the two answers?

Putting Your Skills to Work

The Sum-of-the-Digits Depreciation Method

All of us find that cars and trucks depreciate in value too quickly. The amount of time from when a car is new to when it ends up in the scrap heap is all too brief. However, in mathematics, we need to understand precisely how much decrease in value each year occurs for a given car or truck. The following method is one possible way to measure this.

Many people have a car or a truck that they use for business. When figuring out their annual income tax, it is important that they determine how much these vehicles have decreased in value. One such method used by the IRS is called the sum-of-the-digits depreciation method. Suppose that someone bought a new Ford Taurus for $20,500 and wanted to use it in a business for six years and then sell it for $5,500. The usable value of the vehicle for six years would be $20,500 − $5,500, which is $15,000.

Now how do we spread this usable value over six years?

The sum-of-the-digits depreciation method states that we can spread a usable value over n years by multiplying this value by a series of fractions. We form these fractions in the following way.

Find the sum of the integers from 1 to n. The result will become the denominators of the fractions. The numerators are the integers from 1 to n in descending order. Thus we have the fractions

$$\frac{n}{1 + 2 + \ldots + n}, \quad \frac{n - 1}{1 + 2 + \ldots + n}, \quad \ldots,$$

$$\frac{1}{1 + 2 + \ldots + n}.$$

To find the depreciation for:

the first year, multiply the usable value by the first fraction.

the second year, multiply the usable value by the second fraction.

(continue each year following the same pattern)

the last year, multiply the usable value by the last fraction.

So, for the example given, we can spread the usable value of $15,000 over six years by first finding the sum of the integers from 1 to 6.

$$1 + 2 + 3 + 4 + 5 + 6 = 21$$

Next we multiply the $15,000 by the fractions $\frac{6}{21}$, $\frac{5}{21}$, $\frac{4}{21}$, $\frac{3}{21}$, $\frac{2}{21}$, and $\frac{1}{21}$.

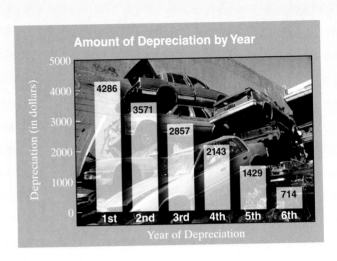

Amount of Depreciation by Year

We will round each value to the nearest dollar, a method suggested by the IRS.

First year's depreciation: $\frac{6}{21} \times \$15,000 = \$4286$

Second year's depreciation: $\frac{5}{21} \times \$15,000 = \$3571$

Third year's depreciation: $\frac{4}{21} \times \$15,000 = \$2857$

Fourth year's depreciation: $\frac{3}{21} \times \$15,000 = \$2143$

Fifth year's depreciation: $\frac{2}{21} \times \$15,000 = \$1429$

Sixth year's depreciation: $\frac{1}{21} \times \$15,000 = \$714$

The dollar amounts by which the car will depreciate each year are shown on the bar graph above.

Problems for Individual Investigation and Analysis

1. How many times greater is the first year's depreciation than the fifth year's depreciation? Round to the nearest tenth.

2. How many times greater is the second year's depreciation than the sixth year's depreciation? Round to the nearest tenth.

Problems for Group Investigation and Cooperative Study

3. If a new Honda Civic is purchased for Sal's Delivery Service for $16,750, used for five years, and then sold for $5,750, what will be the depreciation each year for these five years? Round to the nearest dollar.

4. If a new Ford F-250 truck is purchased for Camp Property Care for $26,250, used for seven years, and then sold for $12,250, what will be the depreciation each year for these seven years? Round to the nearest dollar.

Some of the students at North Shore Community College wrote an equation that described the approximate trade-in value of a three-year-old Ford Taurus. The equation was $V = 5700 - 30x$, where V represents the value of the car in dollars and x represents the number of miles in thousands in excess of 20,000 miles. For example, if the car had 50,000 miles on the odometer, it would have 30,000 miles in excess of the 20,000 figure. Measured in thousands, then, the value of x would be 30. To find the trade-in value of the car, substitute $x = 30$ and obtain

$$V = 5700 - 30(30) = 5700 - 900 = 4800.$$

A used Ford Taurus with 50,000 miles would have an approximate trade-in value of $4800.

5. To simplify calculations write the equation $V = 5700 - 30x$ in factored form. Use the factored form to find the trade-in value for a used three-year-old Ford Taurus with 80,000 miles on the odometer.

6. Use your result from problem 5 to find the number of miles that would be registered on the odometer of a used three-year-old Taurus that had a trade-in value of $4500.

Internet Connections

 Netsite: http://www.prenhall.com/tobey_beginning

Site: Kelley Blue Book of Car and Truck Values

7. Use this site to determine the amount of depreciation for a one-year-old Honda Civic. (Subtract the list price from the wholesale value after one year.) What percent of the value of the Honda Civic is lost in one year?

8. Use this site to determine the amount of depreciation for a one-year-old Ford F-250 truck. (Subtract the list price from the wholesale value after one year.) What percent of the value of the truck is lost in one year?

Chapter 5 Test

1. _____

2. _____

3. _____

4. _____

5. _____

6. _____

7. _____

8. _____

9. _____

10. _____

11. _____

12. _____

13. _____

14. _____

15. _____

16. _____

17. _____

18. _____

19. _____

20. _____

21. _____

22. _____

23. _____

24. _____

If possible, factor each polynomial completely. If a polynomial cannot be factored, state that it is prime.

1. $x^2 + 12x - 28$

2. $25x^2 - 49y^2$

3. $10x^2 + 27x + 5$

4. $9a^2 - 30ab + 25b^2$

5. $7x - 9x^2 + 14xy$

6. $3x^2 - 4wy - 2wx + 6xy$

7. $6x^3 - 20x^2 + 16x$

8. $5a^2c - 11abc + 2b^2c$

9. $100x^4 - 16y^4$

10. $9x^2 - 15xy + 4y^2$

11. $7x^2 - 42x$

12. $36x^2 + 1$

13. $3x^2 + 5x + 1$

14. $60xy^2 - 20x^2y - 45y^3$

15. $81x^2 - 1$

16. $x^{16} - 1$

17. $2ax + 6a - 5x - 15$

18. $aw^2 - 8b + 2bw^2 - 4a$

19. $3x^2 - 3x - 90$

20. $2x^3 - x^2 - 15x$

Solve.

21. $x^2 + 14x + 45 = 0$

22. $14 + 3x(x + 2) = -7x$

23. $2x^2 + x - 10 = 0$

Solve using a quadratic equation.

▲ **24.** The park service is studying a rectangular piece of land that has an area of 91 square miles. The length of this piece of land is 1 mile shorter than double the width. Find the length and width of this rectangular piece of land.

Approximately one-half of this test covers the content of Chapters 0–4. The remainder covers the content of Chapter 5.

1. What % of 480 is 72?

2. What is 13% of 3.8?

Simplify.

3. $-3.2 - 6.4 + 0.24 - 1.8 + 0.8$

4. $(-2x^3y^4)(-4xy^6)$

5. $(-3)^4$

6. $(9x - 4)(3x + 2)$

7. $(2x^2 - 6x + 1)(x - 3)$

Solve.

8. $3x - 4 \geq 6x + 5$

9. $3x - (7 - 5x) = 3(4x - 5)$

10. $\dfrac{1}{2}x - 3 = \dfrac{1}{4}(3x + 3)$

11. Solve for t: $s = \dfrac{1}{2}(2a + 3t)$.

Factor each polynomial completely. If a polynomial cannot be factored, state that it is prime.

12. $6x^2 - 5x + 1$

13. $6x^2 + 5x - 4$

14. $9x^2 + 3x - 2$

15. $121x^2 - 64y^2$

16. $4x + 120 - 80x^2$

17. $x^2 + 5x + 9$

18. $16x^3 + 40x^2 + 25x$

19. $81x^4 - 16b^4$

20. $2ax - 4bx + 3a - 6b$

21. $x^4 + 8x^2 + 15$

Solve.

22. $x^2 + 5x - 24 = 0$

23. $3x^2 - 11x + 10 = 0$

Solve using a quadratic equation.

▲ **24.** The park service is studying a triangular piece of land that contains 57 square miles. The altitude of this triangle is 7 miles longer than double the base. Find the altitude and the base of this triangular piece of land.

1. _____

2. _____

3. _____

4. _____

5. _____

6. _____

7. _____

8. _____

9. _____

10. _____

11. _____

12. _____

13. _____

14. _____

15. _____

16. _____

17. _____

18. _____

19. _____

20. _____

21. _____

22. _____

23. _____

24. _____

Rational
Expressions
and Equations

Do you realize that the United States might have a
surplus of $1.3 trillion by 2010? Furthermore, if
economic growth drops just a little to 2.1% per year,
we might have instead a deficit of $286 billion by 2010?
Do you think you could use mathematical equations to
predict the future of the United States' budget surplus or
deficit? Turn to the Putting Your Skills to Work problems
on page 374 to find out.

Pretest Chapter 6

1. _____

2. _____

3. _____

4. _____

5. _____

6. _____

7. _____

8. _____

9. _____

10. _____

11. _____

12. _____

13. _____

14. _____

15. _____

16. _____

17. _____

18. _____

If you are familiar with the topics in this chapter, take this test now. Check your answers with those in the back of the book. If an answer is wrong or you can't answer a question, study the appropriate section of the chapter.

If you are not familiar with the topics in this chapter, don't take this test now. Instead, study the examples, work the practice problems, and then take the test.

This test will help you identify which concepts you have mastered and which you need to study further.

Simplify the algebraic expressions in questions 1–14.

Section 6.1

1. $\dfrac{6a - 4b}{2b - 3a}$

2. $\dfrac{x^2 - x - 6}{2x^2 + 7x + 6}$

3. $\dfrac{a^2b + 2ab^2}{2a^3 + 3a^2b - 2ab^2}$

Section 6.2

4. $\dfrac{4a^2 - b^2}{6a - 6b} \cdot \dfrac{3a - 3b}{6a + 3b}$

5. $\dfrac{x^2 - 6x + 9}{x^2 - x - 6} \div \dfrac{x^2 + 2x - 15}{x^2 + 2x}$

6. $\dfrac{x^2 + 5x + 6}{3x^2 + 8x + 4} \cdot \dfrac{6x^2 - 11x - 10}{2x^2 + x - 15}$

7. $\dfrac{xy + 3y}{x^2 - x} \div \dfrac{x + 3}{x}$

Section 6.3

8. $\dfrac{3y + 1}{y + 2} + \dfrac{5}{y + 2}$

9. $\dfrac{2y - 1}{2y^2 + y - 3} - \dfrac{2}{y - 1}$

10. $\dfrac{4x + 10}{x^2 - 25} - \dfrac{3}{x - 5}$

11. $\dfrac{x + 3}{x^2 - 6x + 9} + \dfrac{2x + 3}{3x^2 - 9x}$

Section 6.4

12. $\dfrac{\dfrac{2}{a} - \dfrac{3}{a^2}}{5 + \dfrac{1}{a}}$

13. $\dfrac{\dfrac{a}{a + 1} - \dfrac{2}{a}}{3a}$

14. $\dfrac{\dfrac{x - y}{x} - \dfrac{x + y}{y}}{\dfrac{x - y}{x} + \dfrac{x + y}{y}}$

Section 6.5

Solve for x. If the equation has no solution, so state.

15. $\dfrac{5}{2x} = 2 - \dfrac{2x}{x + 1}$

16. $\dfrac{24}{x^2 - 3x - 10} = \dfrac{2}{x - 5} + \dfrac{3}{x + 2}$

Section 6.6

17. Solve for x and round to the nearest tenth. $\dfrac{5}{x} = \dfrac{7}{13}$

18. When Marcia went to England, she found that $3.10 in U.S. money was needed at the current exchange rate to receive 2 British pounds. How much U.S. money was needed to receive 170 British pounds?

6.1 Simplifying Rational Expressions

Recall that a rational number is a number that can be written as one integer divided by another integer, such as $3 \div 4$ or $\frac{3}{4}$. We usually use the word *fraction* to mean $\frac{3}{4}$. We can extend this idea to algebraic expressions. A **rational expression** is a polynomial divided by another polynomial, such as

$$(3x + 2) \div (x + 4) \quad \text{or} \quad \frac{3x + 2}{x + 4}.$$

The last fraction is sometimes also called a **fractional algebraic expression**. There is a special restriction for all fractions, including fractional algebraic expressions. The denominator of the fraction cannot be 0. For example, in the expression

$$\frac{3x + 2}{x + 4},$$

the denominator cannot be 0. Therefore, the value of x cannot be -4. The following important restriction will apply throughout this chapter. We state it here to avoid having to mention it repeatedly throughout this chapter.

Student Learning Objectives

After studying this section, you will be able to:

1 Simplify rational expressions by factoring.

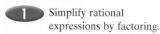

SSM   PH TUTOR   CD & VIDEO   MATH PRO   WEB
CENTER

Restriction

The denominator of a rational expression cannot be zero. Any value of the variable that would make the denominator zero is not allowed.

We have discovered that fractions can be simplified (or reduced) in the following way.

$$\frac{15}{25} = \frac{3 \cdot \cancel{5}}{5 \cdot \cancel{5}} = \frac{3}{5}$$

This is sometimes referred to as the **basic rule of fractions** and can be stated as follows.

Basic Rule of Fractions

For any rational expression $\frac{a}{b}$ and any polynomial a, b, and c (where $b \neq 0$ and $c \neq 0$),

$$\frac{ac}{bc} = \frac{a}{b}.$$

We will examine several examples where a, b, and c are real numbers, as well as more involved examples where a, b, and c are polynomials. In either case we shall make extensive use of our factoring skills in this section.

One essential property is revealed by the basic rule of fractions: If the numerator and denominator of a given fraction are multiplied by the same nonzero quantity, an equivalent fraction is obtained. The rule can be used two ways. You can start with $\frac{ac}{bc}$ and end with the equivalent fraction $\frac{a}{b}$. Or, you can start with $\frac{a}{b}$ and end with the equivalent fraction $\frac{ac}{bc}$.

⬤ EXAMPLE 1

(a) Write a fraction equivalent to $\frac{3}{5}$ with a denominator of 10.

(b) Reduce $\frac{21}{39}$.

(a) $\dfrac{3}{5} = \dfrac{3 \cdot 2}{5 \cdot 2} = \dfrac{6}{10}$ Use the rule $\frac{a}{b} = \frac{ac}{bc}$. Let $c = 2$ since $5 \cdot 2 = 10$.

(b) $\dfrac{21}{39} = \dfrac{7 \cdot 3}{13 \cdot 3} = \dfrac{7}{13}$ Use the rule $\frac{ac}{bc} = \frac{a}{b}$. Let $c = 3$ because 3 is the greatest common factor of 21 and 39.

Practice Problem 1

(a) Find a fraction equivalent to $\dfrac{7}{3}$ with a denominator of 18.

(b) Reduce $\dfrac{28}{63}$.

❶ Simplifying Rational Expressions by Factoring

The process of reducing the fraction shown previously is sometimes called *dividing out* common factors. Remember, only factors of both the numerator and the denominator can be divided out. To apply the basic rule of fractions, it is usually necessary that the numerator and denominator of the fraction be completely factored. You will need to use your factoring skills from Chapter 5 to accomplish this step. When you apply this rule, you are **simplifying the fraction**.

EXAMPLE 2 Simplify. $\dfrac{4x + 12}{5x + 15}$

$\dfrac{4x + 12}{5x + 15} = \dfrac{4(x + 3)}{5(x + 3)}$ Factor 4 from the numerator.
                            Factor 5 from the denominator.

$\phantom{\dfrac{4x + 12}{5x + 15}} = \dfrac{4\cancel{(x + 3)}}{5\cancel{(x + 3)}}$ Apply the basic rule of fractions.

$\phantom{\dfrac{4x + 12}{5x + 15}} = \dfrac{4}{5}$

Practice Problem 2 Simplify. $\dfrac{12x - 6}{14x - 7}$

EXAMPLE 3 Simplify. $\dfrac{x^2 + 9x + 14}{x^2 - 4}$

$= \dfrac{(x + 7)(x + 2)}{(x - 2)(x + 2)}$ Factor the numerator.
                        Factor the denominator.

$= \dfrac{(x + 7)\cancel{(x + 2)}}{(x - 2)\cancel{(x + 2)}}$ Apply the basic rule of fractions.

$= \dfrac{x + 7}{x - 2}$

Practice Problem 3 Simplify. $\dfrac{4x - 6}{2x^2 - x - 3}$

Some problems may involve more than one step of factoring. Always remember to factor out any common factors as the first step, if it is possible to do so.

EXAMPLE 4   Simplify. $\dfrac{9x - x^3}{x^3 + x^2 - 6x}$

$= \dfrac{x\left(9 - x^2\right)}{x\left(x^2 + x - 6\right)}$ Factor out a common factor from the polynomials in the numerator and the denominator.

$= \dfrac{\cancel{x}\cancel{(3 + x)}(3 - x)}{\cancel{x}\cancel{(3 + x)}(x - 2)}$ Factor each polynomial and apply the basic rule of fractions. Note that $(3 + x)$ is equivalent to $(x + 3)$ since addition is commutative.

$= \dfrac{3 - x}{x - 2}$

Practice Problem 4   Simplify. $\dfrac{x^3 + 11x^2 + 30x}{3x^3 + 17x^2 - 6x}$

When you are simplifying, be on the lookout for the special situation where *a factor in the denominator is the opposite of a factor in the numerator*. In such a case you should factor a negative number from one of the factors so that it becomes equivalent to the other factor and it can be divided out. Look carefully at the following two examples.

EXAMPLE 5   Simplify. $\dfrac{5x - 15}{6 - 2x}$

Notice that the variable term in the numerator, $5x$, and the variable term in the denominator, $-2x$, *are opposite in sign*. Likewise, the numerical terms -15 and 6 *are opposite in sign*. Factor out a negative number from the denominator.

$\dfrac{5x - 15}{6 - 2x} = \dfrac{5\left(x - 3\right)}{-2\left(-3 + x\right)}$ Factor 5 from the numerator.

   Factor -2 from the denominator.

   Note that $(x - 3)$ and $(-3 + x)$ are equivalent since $(+x - 3) = (-3 + x)$.

$= \dfrac{5(x - 3)}{-2(-3 + x)}$ Apply the basic rule of fractions.

$= -\dfrac{5}{2}$

Note that $\dfrac{5}{-2}$ is not considered to be in simple form. We usually avoid leaving a negative number in the denominator. Therefore, to simplify, give the result as $-\dfrac{5}{2}$ or $\dfrac{-5}{2}$.

Practice Problem 5   Simplify. $\dfrac{2x - 5}{5 - 2x}$

EXAMPLE 6   Simplify. $\dfrac{2x^2 - 11x + 12}{16 - x^2}$

$= \dfrac{(x - 4)(2x - 3)}{(4 - x)(4 + x)}$ Factor the numerator and the denominator. Observe that $(x - 4)$ and $(4 - x)$ are opposites.

$= \dfrac{(x - 4)(2x - 3)}{-1\left(-4 + x\right)(4 + x)}$ Factor -1 out of $(+4 - x)$ to obtain $-1(-4 + x)$.

$= \dfrac{\cancel{(x - 4)}(2x - 3)}{-1\cancel{(-4 + x)}(4 + x)}$ Apply the basic rule of fractions since $(x - 4)$ and $(-4 + x)$ are equivalent.

$= \dfrac{2x - 3}{-1(4 + x)}$

$= -\dfrac{2x - 3}{4 + x}$

Practice Problem 6 Simplify. $\dfrac{4x^2 + 3x - 10}{25 - 16x^2}$

After doing Examples 5 and 6, you will notice a pattern. Whenever the factor in the numerator and the factor in the denominator are opposites, the value -1 results. We could actually make this a definition of that property.

For all monomials A and B where $A \neq B$, it is true that

$$\dfrac{A - B}{B - A} = -1.$$

You may use this definition in reducing fractions if it is helpful to you. Otherwise, you may use the factoring method discussed in Examples 5 and 6.

Some problems will involve two or more variables. In such cases, you will need to factor carefully and make sure that each set of parentheses contains the correct letters.

EXAMPLE 7 Simplify. $\dfrac{x^2 - 7xy + 12y^2}{2x^2 - 7xy - 4y^2}$

$$= \dfrac{(x - 4y)(x - 3y)}{(2x + y)(x - 4y)} \qquad \text{Factor the numerator.}$$
$$\text{Factor the denominator.}$$

$$= \dfrac{\cancel{(x - 4y)}(x - 3y)}{(2x + y)\cancel{(x - 4y)}} \qquad \text{Apply the basic rule of fractions.}$$

$$= \dfrac{x - 3y}{2x + y}$$

Practice Problem 7 Simplify. $\dfrac{4x^2 - 9y^2}{4x^2 + 12xy + 9y^2}$

EXAMPLE 8 Simplify. $\dfrac{6a^2 + ab - 7b^2}{36a^2 - 49b^2}$

$$= \dfrac{(6a + 7b)(a - b)}{(6a + 7b)(6a - 7b)} \qquad \text{Factor the numerator.}$$
$$\text{Factor the denominator.}$$

$$= \dfrac{\cancel{(6a + 7b)}(a - b)}{\cancel{(6a + 7b)}(6a - 7b)} \qquad \text{Apply the basic rule of fractions.}$$

$$= \dfrac{a - b}{6a - 7b}$$

Practice Problem 8 Simplify. $\dfrac{12x^3 - 48x}{6x - 3x^2}$

Simplify.

1. $\dfrac{3a - 9b}{a - 3b}$

2. $\dfrac{5x + 2y}{35x + 14y}$

3. $\dfrac{6x + 18}{x^2 + 3x}$

4. $\dfrac{8x - 12}{2x^3 - 3x^2}$

5. $\dfrac{2x - 8}{x^2 - 8x + 16}$

6. $\dfrac{4x^2 + 4x + 1}{1 - 4x^2}$

7. $\dfrac{xy(x + y^2)}{x^2 y^2}$

8. $\dfrac{6x^2}{2x(x - 3y)}$

9. $\dfrac{x^2 + x - 2}{x^2 - x}$

10. $\dfrac{x^2 + x - 12}{2x^2 - 3x - 9}$

11. $\dfrac{x^2 - 3x - 10}{3x^2 + 5x - 2}$

12. $\dfrac{4x^2 - 10x + 6}{2x^2 + x - 3}$

13. $\dfrac{x^3 - 8x^2 + 16x}{x^2 + 2x - 24}$

14. $\dfrac{x^2 - 10x - 24}{x^3 + 9x^2 + 14x}$

15. $\dfrac{3x^2 + 7x - 6}{x^2 + 7x + 12}$

16. $\dfrac{x^2 - 5x - 14}{2x^2 - x - 10}$

17. $\dfrac{3x^2 - 8x + 5}{4x^2 - 5x + 1}$

18. $\dfrac{3y^2 + 10y + 3}{3y^2 - 14y - 5}$

19. $\dfrac{5x^2 - 27x + 10}{5x^2 + 3x - 2}$

20. $\dfrac{2x^2 + 2x - 12}{x^2 + 3x - 4}$

21. $\dfrac{6 - 3x}{2x - 4}$

22. $\dfrac{5 - ay}{ax^2 y - 5x^2}$

23. $\dfrac{2x^2 - 7x - 15}{25 - x^2}$

24. $\dfrac{49 - x^2}{2x^2 - 9x - 35}$

25. $\dfrac{(4x + 5)^2}{8x^2 + 6x - 5}$

26. $\dfrac{6x^2 - 13x - 8}{(3x - 8)^2}$

27. $\dfrac{2x^2 + 9x - 18}{30 - x - x^2}$

28. $\dfrac{4y^2 - y - 3}{8 - 7y - y^2}$

29. $\dfrac{a^2 + 2ab - 3b^2}{2a^2 + 5ab - 3b^2}$

30. $\dfrac{a^2 + 3ab - 10b^2}{3a^2 - 7ab + 2b^2}$

31. $\dfrac{9x^2 - 4y^2}{9x^2 + 12xy + 4y^2}$

32. $\dfrac{16x^2 - 24xy + 9y^2}{16x^2 - 9y^2}$

33. $\dfrac{6x^4 - 9x^3 - 6x^2}{12x^3 + 42x^2 + 18x}$

34. $\dfrac{xa - yb - ya + xb}{xa - ya + 2bx - 2by}$

Cumulative Review Problems

Multiply.

35. Multiply. $(3x - 7)^2$

36. $(10x + 9y)(10x - 9y)$

37. $(2x + 3)(x - 4)(x - 2)$

38. $(x^2 - 3x - 5)(2x^2 + x + 3)$

39. Walter and Ann Perkins wish to divide $4\frac{7}{8}$ acres of farmland into three equal-sized house lots. What will be the acreage of each lot?

40. Ron and Mary Larson are planning to cook a $17\frac{1}{2}$-pound turkey. The directions suggest a cooking time of 22 minutes per pound for turkeys that weigh between 16 and 20 pounds. How many hours and minutes should they allow for an approximate cooking time?

The number of people on our planet who speak the Chinese Mandarin language is 221,000,000 more than twice the number of people who speak Spanish. The number of people on Earth who speak English is 10,000,000 less than the number of those who speak Spanish. There are 322,000,000 people who speak English on our planet.

41. How many people speak Chinese Mandarin?

42. How many people speak Spanish?

1   *Multiplying Rational Expressions*

To multiply two rational expressions, we multiply the numerators and we multiply the denominators. As before, the denominators cannot equal zero.

Student Learning Objectives

After studying this section, you will be able to:

1   Multiply rational expressions.

2   Divide rational expressions.

SSM   PH TUTOR   CD & VIDEO   MATH PRO   WEB
CENTER

For any two rational expressions $\dfrac{a}{b}$ and $\dfrac{c}{d}$ where $b \neq 0$ and $d \neq 0$,

$$\frac{a}{b} \cdot \frac{c}{d} = \frac{ac}{bd}.$$

Simplifying or reducing fractions *prior to multiplying them* usually makes the computations easier to do. Leaving the reducing step until the end makes the simplifying process longer and increases the chance for error. This long approach should be avoided.

As an example, let's do the same problem two ways to see which one is easier. Let's simplify the following problem by multiplying first and then reducing the result.

$$\frac{5}{7} \times \frac{49}{125}$$

$$\frac{5}{7} \times \frac{49}{125} = \frac{245}{875} \qquad \text{Multiply the numerators and multiply the denominators.}$$

$$= \frac{7}{25} \qquad \text{Reduce the fraction. (\textit{Note:} It takes a bit of trial and error to discover how to reduce it.)}$$

Compare this with the following method, where we reduce the fractions prior to multiplying them.

$$\frac{5}{7} \times \frac{49}{125}$$

$$\frac{5}{7} \times \frac{7 \cdot 7}{5 \cdot 5 \cdot 5} \qquad \textbf{Step 1.} \text{ It is easier to factor first. We factor the numerator and denominator of the second fraction.}$$

$$\frac{5 \cdot 7 \cdot 7}{7 \cdot 5 \cdot 5 \cdot 5} \qquad \textbf{Step 2.} \text{ We express the product as one fraction (by the definition of multiplication of fractions).}$$

$$\frac{\cancel{5} \cdot \cancel{7} \cdot 7}{\cancel{7} \cdot \cancel{5} \cdot 5 \cdot 5} = \frac{7}{25} \qquad \textbf{Step 3.} \text{ Then we apply the basic rule of fractions to divide the common factors of 5 and 7 that appear in the numerator and in the denominator.}$$

A similar approach can be used with the multiplication of rational expressions. We first factor the numerator and denominator of each fraction. Then we divide out any factor that is common to a numerator and a denominator. Finally, we multiply the remaining numerators and the remaining denominators.

EXAMPLE 1   Multiply. $\dfrac{x^2 - x - 12}{x^2 - 16} \cdot \dfrac{2x^2 + 7x - 4}{x^2 - 4x - 21}$

$$\frac{(x - 4)(x + 3)}{(x - 4)(x + 4)} \cdot \frac{(x + 4)(2x - 1)}{(x + 3)(x - 7)} \qquad \text{Factoring is always the first step.}$$

$$= \frac{\cancel{(x - 4)}\cancel{(x + 3)}}{\cancel{(x - 4)}\cancel{(x + 4)}} \cdot \frac{\cancel{(x + 4)}(2x - 1)}{\cancel{(x + 3)}(x - 7)} \qquad \text{Apply the basic rule of fractions. (Three pairs of factors divide out.)}$$

$$= \frac{2x - 1}{x - 7} \qquad \text{The final answer.}$$

361

Practice Problem 1 Multiply. $\dfrac{10x}{x^2 - 7x + 10} \cdot \dfrac{x^2 + 3x - 10}{25x}$

In some cases, a given numerator can be factored more than once. You should always check for a *common factor* as your first step.

EXAMPLE 2 Multiply. $\dfrac{x^4 - 16}{x^3 + 4x} \cdot \dfrac{2x^2 - 8x}{4x^2 + 2x - 12}$

$= \dfrac{(x^2 + 4)(x^2 - 4)}{x(x^2 + 4)} \cdot \dfrac{2x(x - 4)}{2(2x^2 + x - 6)}$ Factor each numerator and denominator. Factoring out the common factor first is very important.

$= \dfrac{(x^2 + 4)(x + 2)(x - 2)}{x(x^2 + 4)} \cdot \dfrac{2x(x - 4)}{2(x + 2)(2x - 3)}$ Factor again where possible.

$= \dfrac{\cancel{(x^2 + 4)}\,\cancel{(x + 2)}\,(x - 2)}{\cancel{x}\,\cancel{(x^2 + 4)}} \cdot \dfrac{\cancel{2x}\,(x - 4)}{\cancel{2}\,\cancel{(x + 2)}\,(2x - 3)}$ Divide out factors that appear in both the numerator and the denominator. (There are four such pairs of factors.)

$= \dfrac{(x - 2)(x - 4)}{(2x - 3)}$ or $\dfrac{x^2 - 6x + 8}{2x - 3}$ Write the answer as one fraction. (Usually, if there is more than one factor in a numerator, the answer is left in factored form.)

Practice Problem 2 Multiply. $\dfrac{2y^2 - 6y - 8}{y^2 - y - 2} \cdot \dfrac{y^2 - 5y + 6}{2y^2 - 4y - 6}$.

2 Dividing Rational Expressions

For any two fractions $\frac{a}{b}$ and $\frac{c}{d}$, the operation of division can be performed by inverting the second fraction and multiplying it by the first fraction. When we invert a fraction, we are finding its *reciprocal*. Two numbers are **reciprocals** of each other if their product is 1. The reciprocal of $\frac{3}{5}$ is $\frac{5}{3}$. The reciprocal of 7 is $\frac{1}{7}$. The reciprocal of $\frac{a}{b}$ is $\frac{b}{a}$. Sometimes people state the rule for dividing fractions this way: "To divide two fractions, keep the first fraction unchanged and multiply by the reciprocal of the second fraction."

The definition for division of fractions is

$$\frac{a}{b} \div \frac{c}{d} = \frac{a}{b} \cdot \frac{d}{c}.$$

This property holds whether a, b, c, and d are polynomials or numerical values. (It is assumed, of course, that no denominator is zero.)

In the first step for dividing two rational expressions, invert the second fraction and rewrite the quotient as a product. Then follow the procedure for multiplying rational expressions.

EXAMPLE 3 Divide. $\dfrac{6x + 12y}{2x - 6y} \div \dfrac{9x^2 - 36y^2}{4x^2 - 36y^2}$

$= \dfrac{6x + 12y}{2x - 6y} \cdot \dfrac{4x^2 - 36y^2}{9x^2 - 36y^2}$ Invert the second fraction and write the problem as the product of two fractions.

$$= \frac{6(x + 2y)}{2(x - 3y)} \cdot \frac{4(x^2 - 9y^2)}{9(x^2 - 4y^2)}$$

Factor each numerator and denominator.

$$= \frac{(3)(2)(x + 2y)}{2(x - 3y)} \cdot \frac{(2)(2)(x + 3y)(x - 3y)}{(3)(3)(x + 2y)(x - 2y)}$$

Factor again where possible.

$$= \frac{\cancel{(3)}\cancel{(2)}\cancel{(x + 2y)}}{2\cancel{(x - 3y)}} \cdot \frac{(2)(2)(x + 3y)\cancel{(x - 3y)}}{\cancel{(3)}(3)\cancel{(x + 2y)}(x - 2y)}$$

Divide out factors that appear in both numerator and denominator.

$$= \frac{(2)(2)(x + 3y)}{3(x - 2y)}$$

Write the result as one fraction.

$$= \frac{4(x + 3y)}{3(x - 2y)}$$

Simplify. (Usually, answers are left in this form.)

Although it is correct to write this answer as $\dfrac{4x + 12y}{3x - 6y}$, it is customary to leave the answer in factored form to ensure that the final answer is simplified.

Practice Problem 3 Divide. $\dfrac{x^2 + 5x + 6}{x^2 + 8x} \div \dfrac{2x^2 + 5x + 2}{2x^2 + x}$

A polynomial that is not in fraction form can be written as a fraction if you give it a denominator of 1.

EXAMPLE 4 Divide. $\dfrac{15 - 3x}{x + 6} \div (x^2 - 9x + 20)$

Note that $x^2 - 9x + 20$ can be written as $\dfrac{x^2 - 9x + 20}{1}$.

$$= \frac{15 - 3x}{x + 6} \cdot \frac{1}{x^2 - 9x + 20}$$

Invert and multiply.

$$= \frac{-3(-5 + x)}{x + 6} \cdot \frac{1}{(x - 5)(x - 4)}$$

Factor where possible. Note that we had to factor -3 from the first numerator so that it would contain a factor in common with the second denominator.

$$= \frac{-3\cancel{(-5 + x)}}{x + 6} \cdot \frac{1}{\cancel{(x - 5)}(x - 4)}$$

Divide out the common factor. $(-5 + x)$ is equivalent to $(x - 5)$.

$$= \frac{-3}{(x + 6)(x - 4)}$$

The final answer. Note that the answer can be written in several equivalent forms.

or $-\dfrac{3}{(x + 6)(x - 4)}$ or $\dfrac{3}{(x + 6)(4 - x)}$

Practice Problem 4 Divide. $\dfrac{x + 3}{x - 3} \div (9 - x^2)$

A word of caution:

It is logical to assume that the problems included in Section 6.2 have at least one common factor that can be divided out. Therefore if, after factoring, you do not observe any common factors, you should be somewhat suspicious. In such cases, it would be wise to double check your factoring steps to see if an error has been made.

Verbal and Writing Skills

1. Before multiplying rational expressions, we should always first try to

   _____

   _____ .

2. Division of two rational expressions is done by keeping the first fraction unchanged and then

   _____ .

Perform the operation indicated.

3. $\dfrac{2x - 10}{x - 4} \cdot \dfrac{x^2 + 5x + 4}{x^2 - 4x - 5}$

4. $\dfrac{7x + 7}{x + 4} \cdot \dfrac{x^2 - x - 20}{7x^2 - 42x - 49}$

5. $\dfrac{x^2 + 2x}{6x} \cdot \dfrac{3x^2}{x^2 - 4}$

6. $\dfrac{3x + 12}{8x^3} \cdot \dfrac{16x^2}{9x + 36}$

7. $\dfrac{x^2 + 3x - 10}{x^2 + x - 20} \cdot \dfrac{x^2 - 3x - 4}{x^2 + 4x + 3}$

8. $\dfrac{x^2 - x - 20}{x^2 - 3x - 10} \cdot \dfrac{x^2 + 7x + 10}{x^2 + 4x - 5}$

9. $(6x - 5) \div \dfrac{36x^2 - 25}{6x^2 + 17x + 10}$

10. $\dfrac{4x^2 - 9}{4x^2 + 12x + 9} \div (6x - 9)$

11. $\dfrac{3x^2 + 12xy + 12y^2}{x^2 + 4xy + 3y^2} \div \dfrac{4x + 8y}{x + y}$

12. $\dfrac{5x^2 + 10xy + 5y^2}{x^2 + 5xy + 6y^2} \div \dfrac{3x + 3y}{x + 2y}$

13. $\dfrac{x^2 + 5x - 14}{x - 5} \div \dfrac{x^2 + 12x + 35}{15 - 3x}$

14. $\dfrac{3x^2 + 13x + 4}{16 - x^2} \div \dfrac{3x^2 - 5x - 2}{3x - 12}$

15. $\dfrac{(x + 5)^2}{3x^2 - 7x + 2} \cdot \dfrac{x^2 - 4x + 4}{x + 5}$

16. $\dfrac{3x^2 - 10x - 8}{(4x + 5)^2} \cdot \dfrac{4x + 5}{(x - 4)^2}$

17. $\dfrac{y^2 + 4y - 12}{y^2 + 2y - 24} \cdot \dfrac{y^2 - 16}{y^2 + 2y - 8}$

18. $\dfrac{5y^2 + 17y + 6}{10y^2 + 9y + 2} \cdot \dfrac{4y^2 - 1}{2y^2 + 5y - 3}$

To Think About

19. Consider the problem $\dfrac{x+5}{x-2} \div \dfrac{x+7}{x-6}$. Explain why 2, −7, and 6 are not allowable replacements for the variable x.

20. Consider the problem $\dfrac{x-8}{x+5} \div \dfrac{x-9}{x+4}$. Explain why −5, 9, and −4 are not allowable replacements for the variable x.

Cumulative Review Problems

21. Solve. $5x^2 - 7x + 11 = 5x^2 - x + 2$

22. Multiply. $(7x^2 - x - 1)(x - 3)$

23. Suzanne Hartling is a nurse at Meadowbrook Rehabilitation Center. She must give an 80-kilogram patient a medication based on his weight. The dosage is $\frac{1}{16}$ milligram of medication for each kilogram of patient weight. Find the amount of medication she should give this patient.

▲ 24. The Golden Gate Bridge has a total length (including approaches) of 8981 feet and a road width of 90 feet. The width of the sidewalk is 10.5 feet. (The sidewalk spans the entire length of the bridge.) Assume it would cost $x per square foot to resurface the road or the sidewalk. Write an expression for how much more it would cost to resurface the road than the sidewalk.

25. Denise Abrahamson purchased 4 gallons of milk. The price of 4 gallons of milk is the same as the price of 2 gallons of milk plus some other groceries that cost $4.76. What is the price of a gallon of milk?

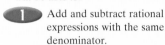

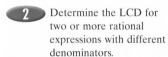
Adding and Subtracting Rational Expressions with the Same Denominator

If rational expressions have the same denominator, they can be combined in a fashion similar to that used for fractions in arithmetic. The numerators are added or subtracted and the denominator remains the same.

Adding Rational Expressions

For any rational expressions $\dfrac{a}{b}$ and $\dfrac{c}{b}$,

$$\frac{a}{b} + \frac{c}{b} = \frac{a+c}{b} \qquad \text{where } b \neq 0.$$

EXAMPLE 1   Add. $\dfrac{5a}{a+2b} + \dfrac{6a}{a+2b}$

$$\frac{5a}{a+2b} + \frac{6a}{a+2b} = \frac{5a+6a}{a+2b} = \frac{11a}{a+2b} \qquad$$ Note that the denominators are the same. Add the numerators.

Practice Problem 1   Add. $\dfrac{2s+t}{2s-t} + \dfrac{s-t}{2s-t}$

Subtracting Rational Expressions

For any rational expressions $\dfrac{a}{b}$ and $\dfrac{c}{b}$,

$$\frac{a}{b} - \frac{c}{b} = \frac{a-c}{b} \qquad \text{where } b \neq 0.$$

EXAMPLE 2   Subtract. $\dfrac{3x}{(x+y)(x-2y)} - \dfrac{8x}{(x+y)(x-2y)}$

$$\frac{3x}{(x+y)(x-2y)} - \frac{8x}{(x+y)(x-2y)} = \frac{3x-8x}{(x+y)(x-2y)} \qquad \text{Write as one fraction.}$$

$$= \frac{-5x}{(x+y)(x-2y)} \qquad \text{Simplify.}$$

Practice Problem 2   Subtract. $\dfrac{b}{(a-2b)(a+b)} - \dfrac{2b}{(a-2b)(a+b)}$

 Determining the LCD for Two or More Rational Expressions with Different Denominators

How do we add or subtract rational expressions when the denominators are not the same? First we must find the **least common denominator** (LCD). You need to be clear on how to find a least common denominator and how to add and subtract fractions from arithmetic before you attempt this section. Review Sections 0.1 and 0.2 if you have any questions about this topic.

How to Find the LCD of Two or More Rational Expressions

1. Factor each denominator completely.
2. The LCD is a product containing each *different factor*.
3. If a factor occurs more than once in any one denominator, the LCD will contain that factor repeated the greatest number of times that it occurs in any one denominator.

EXAMPLE 3   Find the LCD. $\dfrac{5}{2x-4}, \dfrac{6}{3x-6}$

Factor each denominator.

$$2x - 4 = 2(x-2) \qquad 3x - 6 = 3(x-2)$$

The different factors are 2, 3, and $(x-2)$. Since no factor appears more than once in any one denominator, the LCD is the product of these three factors.

$$LCD = (2)(3)(x-2) = 6(x-2)$$

Practice Problem 3   Find the LCD. $\dfrac{7}{6x+21}, \dfrac{13}{10x+35}$

EXAMPLE 4   Find the LCD.   **(a)** $\dfrac{5}{12ab^2c}, \dfrac{13}{18a^3bc^4}$

(b) $\dfrac{8}{x^2-5x+4}, \dfrac{12}{x^2+2x-3}$

If a factor occurs more than once in any one denominator, the LCD will contain that factor repeated the greatest number of times that it occurs in any one denominator.

(a) $12ab^2c = 2 \cdot 2 \cdot 3 \cdot \quad a \cdot \qquad b \cdot b \cdot c$

$18a^3bc^4 = \;\begin{vmatrix} 2 \cdot 3 \cdot 3 \cdot a \cdot a \cdot a \cdot b \cdot & c \cdot c \cdot c \cdot c \end{vmatrix}$

$\qquad\qquad\;\; 2 \cdot 2 \cdot 3 \cdot 3 \cdot a \cdot a \cdot a \cdot b \cdot b \cdot c \cdot c \cdot c \cdot c$

$$LCD = 2^2 \cdot 3^2 \cdot a^3 \cdot b^2 \cdot c^4 = 36a^3b^2c^4$$

(b) $x^2 - 5x + 4 = (x-4)(x-1)$

$x^2 + 2x - 3 = \qquad (x-1)(x+3)$

$$LCD = (x-4)(x-1)(x+3)$$

Practice Problem 4   Find the LCD.   **(a)** $\dfrac{3}{50xy^2z}, \dfrac{19}{40x^3yz}$

(b) $\dfrac{2}{x^2+5x+6}, \dfrac{6}{3x^2+5x-2}$

3 Adding and Subtracting Rational Expressions with Different Denominators

If two rational expressions have different denominators, we first change them to equivalent rational expressions with the least common denominator. Then we add or subtract the numerators and keep the common denominator.

EXAMPLE 5 Add. $\dfrac{5}{xy} + \dfrac{2}{y}$

The denominators are different. We must find the LCD. The two factors are x and y. We observe that the LCD is xy.

$$\frac{5}{xy} + \frac{2}{y} = \frac{5}{xy} + \frac{2}{y} \cdot \frac{x}{x} \qquad \text{Multiply the second fraction by } \frac{x}{x}.$$

$$= \frac{5}{xy} + \frac{2x}{xy} \qquad \text{Now each fraction has a common denominator of of } xy.$$

$$= \frac{5 + 2x}{xy} \qquad \text{Write the sum as one fraction.}$$

Practice Problem 5 Add. $\dfrac{7}{a} + \dfrac{3}{abc}$

EXAMPLE 6 Add. $\dfrac{3x}{(x + y)(x - y)} + \dfrac{5}{x + y}$

The factors of the denominators are $(x + y)$ and $(x - y)$. We observe that the LCD $= (x + y)(x - y)$.

$$\frac{3x}{(x + y)(x - y)} + \frac{5}{(x + y)} \cdot \frac{x - y}{x - y} \qquad \text{Multiply the second fraction by } \frac{x - y}{x - y}.$$

$$= \frac{3x}{(x + y)(x - y)} + \frac{5x - 5y}{(x + y)(x - y)} \qquad \begin{array}{l}\text{Now each fraction has a common} \\ \text{denominator of } (x + y)(x - y).\end{array}$$

$$= \frac{3x + 5x - 5y}{(x + y)(x - y)} \qquad \begin{array}{l}\text{Write the sum of the numerators over one} \\ \text{common denominator.}\end{array}$$

$$= \frac{8x - 5y}{(x + y)(x - y)} \qquad \text{Collect like terms.}$$

Practice Problem 6 Add. $\dfrac{2a - b}{(a + 2b)(a - 2b)} + \dfrac{2}{(a + 2b)}$

It is important to remember that the LCD is the smallest algebraic expression into which each denominator can be divided. For rational expressions the LCD must contain *each factor* that appears in any denominator. If the factor is repeated, the LCD must contain that factor the greatest number of times that it appears in any one denominator.

In many cases, the denominators in an addition or subtraction problem are not in factored form. You must factor each denominator to determine the LCD. Collect like terms in the numerator; then determine whether that final numerator can be factored. If so, you may be able to simplify the fraction.

⬤ **EXAMPLE 7**   Add. $\dfrac{5}{x^2 - y^2} + \dfrac{3x}{x^3 + x^2y}$

$\dfrac{5}{x^2 - y^2} + \dfrac{3x}{x^3 + x^2y}$

Factor the two denominators. Observe that the LCD is $x^2(x + y)(x - y)$.

$= \dfrac{5}{(x + y)(x - y)} + \dfrac{3x}{x^2(x + y)}$

$= \dfrac{5}{(x + y)(x - y)} \cdot \dfrac{x^2}{x^2} + \dfrac{3x}{x^2(x + y)} \cdot \dfrac{x - y}{x - y}$

Multiply each fraction by the appropriate value to obtain a common denominator of $x^2(x + y)(x - y)$.

$= \dfrac{5x^2}{x^2(x + y)(x - y)} + \dfrac{3x^2 - 3xy}{x^2(x + y)(x - y)}$

$= \dfrac{5x^2 + 3x^2 - 3xy}{x^2(x + y)(x - y)}$

Write the sum of the numerators over one common denominator.

$= \dfrac{8x^2 - 3xy}{x^2(x + y)(x - y)}$

Collect like terms.

$= \dfrac{x(8x - 3y)}{x^2(x + y)(x - y)}$

Divide out the common factor x in the numerator and denominator and simplify.

$= \dfrac{8x - 3y}{x(x + y)(x - y)}$

Practice Problem 7   Add. $\dfrac{7a}{a^2 + 2ab + b^2} + \dfrac{4}{a^2 + ab}$ ⬤

It is very easy to make a sign mistake when subtracting two fractions. You will find it helpful to place parentheses around the numerator of the second fraction so that you will not forget to subtract the entire numerator.

⬤ **EXAMPLE 8**   Subtract. $\dfrac{3x + 4}{x - 2} - \dfrac{x - 3}{2x - 4}$

Factor the second denominator.

$= \dfrac{3x + 4}{x - 2} - \dfrac{x - 3}{2(x - 2)}$

Observe that the LCD is $2(x - 2)$.

$= \dfrac{2}{2} \cdot \dfrac{(3x + 4)}{x - 2} - \dfrac{x - 3}{2(x - 2)}$

Multiply the first fraction by $\frac{2}{2}$ so that the resulting fraction will have the common denominator.

$= \dfrac{2(3x + 4) - (x - 3)}{2(x - 2)}$

Write the indicated subtraction as one fraction. Note the parentheses around $x - 3$.

$= \dfrac{6x + 8 - x + 3}{2(x - 2)}$

Remove the parentheses in the numerator.

$= \dfrac{5x + 11}{2(x - 2)}$

Collect like terms.

Practice Problem 8   Subtract. $\dfrac{x + 7}{3x - 9} - \dfrac{x - 6}{x - 3}$

To avoid making errors when subtracting two fractions, some students find it helpful to change subtraction to addition of the opposite of the second fraction.

EXAMPLE 9   Subtract. $\dfrac{8x}{x^2 - 16} - \dfrac{4}{x - 4}$

$$\dfrac{8x}{(x + 4)(x - 4)} + \dfrac{-4}{x - 4}$$

Factor the first denominator. Use the property that $\dfrac{a}{b} - \dfrac{c}{b} = \dfrac{a}{b} + \dfrac{-c}{b}$.

$$= \dfrac{8x}{(x + 4)(x - 4)} + \dfrac{-4}{x - 4} \cdot \dfrac{x + 4}{x + 4}$$

Multiply the second fraction by $\dfrac{x + 4}{x + 4}$.

$$= \dfrac{8x + (-4)(x + 4)}{(x + 4)(x - 4)}$$

Write the sum of the numerators over one common denominator.

$$= \dfrac{8x - 4x - 16}{(x + 4)(x - 4)}$$

Remove parentheses.

$$= \dfrac{4x - 16}{(x + 4)(x - 4)}$$

Collect like terms. Note that the numerator can be factored.

$$= \dfrac{4\cancel{(x - 4)}}{(x + 4)\cancel{(x - 4)}}$$

Since $(x - 4)$ is a *factor* of the numerator *and* the denominator, we may divide out the common factor.

$$= \dfrac{4}{x + 4}$$

Practice Problem 9   Subtract and simplify. $\dfrac{x - 2}{x^2 - 4} - \dfrac{x + 1}{2x^2 + 4x}$

Verbal and Writing Skills

1. Suppose two rational expressions have denominators of $(x + 3)(x + 5)$ and $(x + 3)^2$. Explain how you would determine the LCD.

2. Suppose two rational expressions have denominators of $(x - 4)^2(x + 7)$ and $(x - 4)^3$. Explain how you would determine the LCD.

Perform the operation indicated. Be sure to simplify.

3. $\dfrac{x}{x + 5} + \dfrac{2x + 1}{5 + x}$

4. $\dfrac{8}{7 + 2x} + \dfrac{x + 3}{2x + 7}$

5. $\dfrac{x}{x - 4} - \dfrac{x + 1}{x - 4}$

6. $\dfrac{x + 3}{x + 5} - \dfrac{x - 8}{x + 5}$

7. $\dfrac{8x + 3}{5x + 7} - \dfrac{6x + 10}{5x + 7}$

Find the LCD. Do not combine fractions.

8. $\dfrac{13}{3ab}, \dfrac{7}{a^2b^2}$

9. $\dfrac{12}{5a^2}, \dfrac{9}{a^3}$

10. $\dfrac{5}{18x^2y^5}, \dfrac{7}{30x^3y^3}$

11. $\dfrac{11}{16x^2y^3}, \dfrac{17}{56xy^4}$

12. $\dfrac{8}{x + 3}, \dfrac{15}{x^2 - 9}$

13. $\dfrac{13}{x^2 - 16}, \dfrac{7}{x - 4}$

14. $\dfrac{6}{2x^2 + x - 3}, \dfrac{5}{4x^2 + 12x + 9}$

15. $\dfrac{3}{16x^2 - 8x + 1}, \dfrac{6}{4x^2 + 7x - 2}$

Add or subtract.

16. $\dfrac{3}{x + 7} + \dfrac{8}{x^2 - 49}$

17. $\dfrac{5}{x^2 - 2x + 1} + \dfrac{3}{x - 1}$

18. $\dfrac{3y}{y + 2} + \dfrac{y}{y - 2}$

19. $\dfrac{2}{y - 1} + \dfrac{2}{y + 1}$

20. $\dfrac{4}{a + 3} + \dfrac{2}{3a}$

21. $\dfrac{5}{2ab} + \dfrac{1}{2a + b}$

22. $\dfrac{2}{3xy} + \dfrac{1}{6yz}$

23. $\dfrac{x - 3}{4x} + \dfrac{6}{x^2}$

24. $\dfrac{2}{x^2 + 5x + 6} + \dfrac{3}{x^2 + 7x + 10}$

25. $\dfrac{3}{x^2 - x - 12} + \dfrac{2}{x^2 + x - 20}$

26. $\dfrac{3x - 8}{x^2 - 5x + 6} + \dfrac{x + 2}{x^2 - 6x + 8}$

27. $\dfrac{3x + 5}{x^2 + 4x + 3} + \dfrac{-x + 5}{x^2 + 2x - 3}$

28. $\dfrac{a + b}{3} - \dfrac{a - 2b}{4}$

29. $\dfrac{a + 1}{2} - \dfrac{a - 1}{3}$

30. $\dfrac{8}{2x - 3} - \dfrac{6}{x + 2}$

31. $\dfrac{6}{3x - 4} - \dfrac{5}{4x - 3}$

32. $\dfrac{x}{x^2 + 2x - 3} - \dfrac{x}{x^2 - 5x + 4}$

33. $\dfrac{1}{x^2 - 2x} - \dfrac{5}{x^2 - 4x + 4}$

34. $\dfrac{3y}{8y^2 + 2y - 1} - \dfrac{5y}{2y^2 - 9y - 5}$

35. $\dfrac{2x}{x^2 + 5x + 6} - \dfrac{x + 1}{x^2 + 2x - 3}$

36. $\dfrac{6a}{a - 7} - \dfrac{3a}{7 - a}$

37. $\dfrac{7y}{y - 3} - \dfrac{2y}{3 - y}$

38. $\dfrac{4y}{y^2 + 4y + 3} + \dfrac{2}{y + 1}$

39. $\dfrac{y - 23}{y^2 - y - 20} + \dfrac{2}{y - 5}$

40. $\dfrac{2x}{x - 3} + \dfrac{3x}{x + 2} + \dfrac{7}{x^2 - x - 6}$

41. $\dfrac{x}{x - 4} + \dfrac{5}{x + 4} + \dfrac{10}{x^2 - 16}$

42. $\dfrac{3x}{x^2 + 3x - 10} + \dfrac{5}{4 - 2x}$

43. $\dfrac{2y}{3y^2 - 8y - 3} + \dfrac{1}{6y - 2y^2}$

44. $\dfrac{6}{7x + 14} - \dfrac{2}{5x + 10}$

45. $\dfrac{8}{15x + 10} - \dfrac{3}{6x + 4}$

Cumulative Review Problems

46. Solve. $\dfrac{1}{3}(x - 2) + \dfrac{1}{2}(x + 3) = \dfrac{1}{4}(3x + 1)$

47. Solve for y. $5ax = 2(ay - 3bc)$

48. Solve for x. $\dfrac{1}{2}x < \dfrac{1}{3}x + \dfrac{1}{4}$

49. Simplify. $\left(3x^3 y^4\right)^4$

50. A subway token costs \$1.50. A monthly unlimited ride subway pass costs \$50. How many days per month would you have to use the subway to go to work (assume one subway token to get to work and one subway token to get back home) in order for it to be cheaper to buy a monthly subway pass?

51. In Finland the unemployment rate went from 3.5% in 1990 to 14.6% in 1999. (*Source:* United Nations Statistics Division.) If the working age population of Finland was 5,100,000 during this period, how many more people were unemployed in 1999 than in 1990?

52. During the 1972–1973 El Niño weather pattern, the waters around the country of Peru became very warm. The cool-water-loving anchovy fish population dropped from 20 million to 2 million. This particular El Niño lasted 20 months. (*Source:* National Oceanic and Atmospheric Administration.) **(a)** If during this time the population level dropped at a constant rate, how much did it drop per month? **(b)** Fishermen worried that the entire anchovy population would be destroyed. If the weather pattern had continued beyond the 20 months and if the anchovy population had continued to decline at the same rate, how many more months would it have been before the entire anchovy population was destroyed?

53. The government of Finland is hoping to lower the number of people unemployed in Finland to 400,000 or less by the year 2003. (*Source:* United Nations Statistics Division.) Assuming the population of Finland remains constant during the entire time period from 1999 to 2003, what will be the unemployment rate in Finland in 2003 if this goal is met? Refer to exercise 51.

Putting Your Skills to Work

Estimating the Surplus or Deficit

In March 1999, the Congressional Budget Office issued the following surprising projection: assuming a moderate level of economic growth, we can expect a surplus of $489 billion in the U.S. budget in 2010. Then the office further amazed people with the statement that the surplus could grow to $1.3 trillion if the U.S. experiences stronger economic growth. Finally, the office projected a deficit of $286 billion by 2010 if the United States experiences slow economic growth. A three-dimensional bar graph visualizing these three possibilities is shown for 2000, 2005, and 2010.

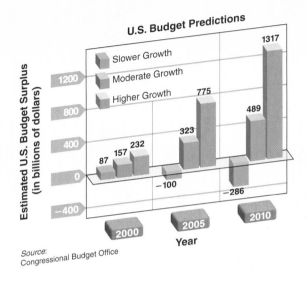

Source:
Congressional Budget Office

The three possibilities are defined more precisely as follows: The higher surplus figures would result from strong economic growth (3.2% per year) and limited congressional spending (that is, holding to agreed-upon budgeted spending limits). The moderate or lower surplus figures would result from moderate economic growth (2.65% per year) and increased congressional spending (whose annual rate of increase would equal that of inflation). The deficit figures would result from slow economic growth (2.1% per year) and increased congressional spending (whose annual rate of increase would equal that of inflation).

Each of these projections can be represented by an equation that predicts the budget surplus (or deficit) in billions of dollars in terms of the number of years x since 1999.

For strong economic growth, the equation is

$$y = 124 + 108.45x.$$

For moderate economic growth, the equation is

$$y = 124 + 33.18x.$$

For slow economic growth, the equation is

$$y = 124 - 37.27x.$$

Problems for Individual Study and Investigation

1. Use one of the preceding equations to find the budget surplus for 2006 if there is strong economic growth. Round to the nearest billion.

2. Use one of the preceding equations to find the budget surplus for 2008 if there is moderate economic growth. Round to the nearest billion.

Problems for Cooperative Group Investigation and Analysis

3. Use one of the preceding equations to find, for 2004, the difference between the budget surplus that would result from moderate economic growth and the budget deficit that would result from slow economic growth. Round to the nearest billion.

4. Use one of the preceding equations to find, for 2007, the difference between the budget surplus that would result from strong economic growth and the budget deficit that would result from slow economic growth. Round to the nearest billion.

One of the major factors in predicting a government deficit or surplus is Social Security. With larger numbers of older people retiring and drawing Social Security funds, accurately predicting the income to the Social Security trust fund in the coming years becomes an extremely critical governmental function. Under one model studied by the Social Security Administration, the net income to the Social Security trust fund is described by the equation

$$I = \frac{5.6x + 10.8}{x + 3},$$

where I is measured in hundreds of billions of dollars and x is the number of years since 2000. For example, in 2000 x would be 0. Thus

$$I = \frac{5.6(0) + 10.8}{0 + 3} = \frac{10.8}{3} = 3.6.$$

Thus in 2000 the projected income was \$3.6 hundred billion or \$360,000,000,000. (*Source:* U.S. Social Security Administration)

5. Use the preceding equation to find how much larger the income to the Social Security trust fund will be in 2005 than in 2001.

6. Use the preceding equation to find how much larger the income to the Social Security trust fund will be in 2012 than in 2009.

Internet Connections

 Netsite: http://www.prenhall.com/tobey_beginning

Site: U.S. Congressional Budget Office

7. Use the site to find projections about the size of the national debt. The national debt is "Debt Held by the Public." According to current projections, how much will the national debt decrease from 2005 to 2010?

8. Use the above site to find data about the revenues and spending for social insurance (Social Security and Medicare together). How much less are the projected revenues associated with social insurance in 2008 than the projected costs?

6.4 Simplifying Complex Rational Expressions

 Simplifying Complex Rational Expressions by Adding or Subtracting in the Numerator and Denominator

A **complex rational expression** (also called a **complex fraction**) has a fraction in the numerator or in the denominator, or both.

$$\frac{3 + \dfrac{2}{x}}{\dfrac{x}{7} + 2} \qquad \frac{\dfrac{x}{y} + 1}{2} \qquad \frac{\dfrac{a+b}{3}}{\dfrac{x-2y}{4}}$$

The bar in a complex rational expression is both a grouping symbol and a symbol for division.

$$\frac{\dfrac{a+b}{3}}{\dfrac{x-2y}{4}} \quad \text{is equivalent to} \quad \left(\frac{a+b}{3}\right) \div \left(\frac{x-2y}{4}\right)$$

We need a procedure for simplifying complex rational expressions.

> **Procedure to Simplify a Complex Rational Expression Adding and Subtracting**
>
> 1. Add or subtract so that you have a single fraction in the numerator and in the denominator.
> 2. Divide the fraction in the numerator by the fraction in the denominator. This is done by inverting the fraction in the denominator and multiplying it by the numerator.

 **EXAMPLE 1** Simplify. $\dfrac{\dfrac{1}{x}}{\dfrac{2}{y^2} + \dfrac{1}{y}}$

Step 1 Add the two fractions in the denominator.

$$\frac{\dfrac{1}{x}}{\dfrac{2}{y^2} + \dfrac{1}{y} \cdot \dfrac{y}{y}} = \frac{\dfrac{1}{x}}{\dfrac{2+y}{y^2}}$$

Step 2 Divide the fraction in the numerator by the fraction in the denominator.

$$\frac{1}{x} \div \frac{2+y}{y^2} = \frac{1}{x} \cdot \frac{y^2}{2+y} = \frac{y^2}{x(2+y)}$$

Practice Problem 1 Simplify. $\dfrac{\dfrac{1}{a} + \dfrac{1}{b}}{\dfrac{2}{ab^2}}$

A complex rational expression may contain two or more fractions in the numerator and the denominator.

376

EXAMPLE 2 Simplify. $\dfrac{\dfrac{1}{x} + \dfrac{1}{y}}{\dfrac{3}{a} - \dfrac{2}{b}}$

We observe that the LCD of the fractions in the numerator is xy. The LCD of the fractions in the denominator is ab.

$= \dfrac{\dfrac{1}{x} \cdot \dfrac{y}{y} + \dfrac{1}{y} \cdot \dfrac{x}{x}}{\dfrac{3}{a} \cdot \dfrac{b}{b} - \dfrac{2}{b} \cdot \dfrac{a}{a}}$ Multiply each fraction by the appropriate value to obtain common denominators.

$= \dfrac{\dfrac{y + x}{xy}}{\dfrac{3b - 2a}{ab}}$ Add the two fractions in the numerator.

    Subtract the two fractions in the denominator.

$= \dfrac{y + x}{xy} \cdot \dfrac{ab}{3b - 2a}$ Invert the fraction in the denominator and multiply it by the numerator.

$= \dfrac{ab(y + x)}{xy(3b - 2a)}$ Write the answer as one fraction.

Practice Problem 2 Simplify. $\dfrac{\dfrac{1}{a} + \dfrac{1}{b}}{\dfrac{1}{a} - \dfrac{1}{b}}$

For some complex rational expressions, factoring may be necessary to determine the LCD and to combine fractions.

EXAMPLE 3 Simplify. $\dfrac{\dfrac{1}{x^2 - 1} + \dfrac{2}{x + 1}}{x}$

We need to factor $x^2 - 1$.

$= \dfrac{\dfrac{1}{(x + 1)(x - 1)} + \dfrac{2}{(x + 1)} \cdot \dfrac{x - 1}{x - 1}}{x}$ The LCD for the fractions in the numerator is $(x + 1)(x - 1)$.

$= \dfrac{\dfrac{1 + 2x - 2}{(x + 1)(x - 1)}}{x}$ Add the two fractions in the numerator.

$= \dfrac{2x - 1}{(x + 1)(x - 1)} \cdot \dfrac{1}{x}$ Simplify the numerator. Invert the fraction in the denominator and multiply.

$= \dfrac{2x - 1}{x(x + 1)(x - 1)}$ Write the answer as one fraction.

Practice Problem 3 Simplify. $\dfrac{\dfrac{x}{x^2 + 4x + 3} + \dfrac{2}{x + 1}}{x + 1}$

When simplifying complex rational expressions, always check to see if the final fraction can be reduced or simplified.

EXAMPLE 4 Simplify. $\dfrac{\dfrac{3}{a+b} - \dfrac{3}{a-b}}{\dfrac{5}{a^2-b^2}}$

The LCD of the two fractions in the numerator is $(a+b)(a-b)$.

$$\dfrac{\dfrac{3}{a+b} \cdot \dfrac{a-b}{a-b} - \dfrac{3}{a-b} \cdot \dfrac{a+b}{a+b}}{\dfrac{5}{a^2-b^2}}$$

$$= \dfrac{\dfrac{3a-3b}{(a+b)(a-b)} - \dfrac{3a+3b}{(a+b)(a-b)}}{\dfrac{5}{a^2-b^2}}$$

Study carefully how we combine the two fractions in the numerator. Do you see how we obtain $-6b$?

$$= \dfrac{\dfrac{-6b}{(a+b)(a-b)}}{\dfrac{5}{(a+b)(a-b)}} \qquad \text{Factor } a^2 - b^2 \text{ as } (a+b)(a-b).$$

$$= \dfrac{-6b}{(a+b)(a-b)} \cdot \dfrac{(a+b)(a-b)}{5} \qquad \begin{array}{l}\text{Since } (a+b)(a-b) \text{ are} \\ \text{factors in both numerator} \\ \text{and denominator, they may be} \\ \text{divided out.}\end{array}$$

$$= \dfrac{-6b}{5} \quad \text{or} \quad -\dfrac{6b}{5}$$

Practice Problem 4 Simplify. $\dfrac{\dfrac{6}{x^2-y^2}}{\dfrac{1}{x-y} + \dfrac{3}{x+y}}$

② Simplifying Complex Rational Expressions by Multiplying by the LCD of All the Denominators

There is another way to simplify complex rational expressions: Multiply the numerator and denominator of the complex fraction by the least common denominator of all the denominators appearing in the complex fraction.

Procedure to Simplify a Complex Rational Expression: Multiplying by the LCD

1. Determine the LCD of all individual denominators occurring in the numerator and denominator of the complex rational expression.
2. Multiply both the numerator and the denominator of the complex rational expression by the LCD.
3. Simplify, if possible.

● **EXAMPLE 5**   Simplify by multiplying by the LCD. $\dfrac{\dfrac{5}{ab^2} - \dfrac{2}{ab}}{3 - \dfrac{5}{2a^2b}}$

The LCD of all the denominators in the complex rational expression is $2a^2b^2$.

$$\dfrac{2a^2b^2\left(\dfrac{5}{ab^2} - \dfrac{2}{ab}\right)}{2a^2b^2\left(3 - \dfrac{5}{2a^2b}\right)}$$

$$= \dfrac{2a^2b^2\left(\dfrac{5}{ab^2}\right) - 2a^2b^2\left(\dfrac{2}{ab}\right)}{2a^2b^2(3) - 2a^2b^2\left(\dfrac{5}{2a^2b}\right)} \qquad \text{Multiply each term by } 2a^2b^2.$$

$$= \dfrac{10a - 4ab}{6a^2b^2 - 5b} \qquad \text{Simplify.}$$

Practice Problem 5   Simplify by multiplying by the LCD. $\dfrac{\dfrac{2}{3x^2} - \dfrac{3}{y}}{\dfrac{5}{xy} - 4}$ ●

So that you can compare the two methods, we will redo Example 4 by multiplying by the LCD.

● **EXAMPLE 6**   Simplify by multiplying by the LCD. $\dfrac{\dfrac{3}{a+b} - \dfrac{3}{a-b}}{\dfrac{5}{a^2-b^2}}$

The LCD of all individual fractions contained in the complex fraction is $(a+b)(a-b)$.

$$\dfrac{(a+b)(a-b)\left(\dfrac{3}{a+b}\right) - (a+b)(a-b)\left(\dfrac{3}{a-b}\right)}{(a+b)(a-b)\left(\dfrac{5}{(a+b)(a-b)}\right)} \qquad \text{Multiply each term by the LCD.}$$

$$= \dfrac{3(a-b) - 3(a+b)}{5} \qquad \text{Simplify.}$$

$$= \dfrac{3a - 3b - 3a - 3b}{5} \qquad \text{Remove parentheses.}$$

$$= -\dfrac{6b}{5} \qquad \text{Simplify.}$$

Practice Problem 6   Simplify by multiplying by the LCD. $\dfrac{\dfrac{6}{x^2-y^2}}{\dfrac{7}{x-y} + \dfrac{3}{x+y}}$ ●

Simplify.

1. $\dfrac{\dfrac{5}{xy}}{\dfrac{7}{y}}$

2. $\dfrac{\dfrac{4a}{b}}{\dfrac{6}{ab}}$

3. $\dfrac{\dfrac{1}{x} + \dfrac{1}{y}}{\dfrac{1}{xy}}$

4. $\dfrac{\dfrac{1}{x} + 1}{\dfrac{1}{x}}$

5. $\dfrac{\dfrac{1}{x} + y}{\dfrac{1}{y} + x}$

6. $\dfrac{\dfrac{1}{x} + \dfrac{1}{y}}{x + y}$

7. $\dfrac{1 - \dfrac{9}{x^2}}{\dfrac{3}{x} + 1}$

8. $\dfrac{\dfrac{4}{x} + 1}{1 - \dfrac{16}{x^2}}$

9. $\dfrac{\dfrac{2}{x + 1} - 2}{3}$

10. $\dfrac{1 - \dfrac{3}{xy}}{x + 2y}$

11. $\dfrac{a + \dfrac{3}{a}}{\dfrac{a^2 + 2}{3a}}$

12. $\dfrac{a + \dfrac{1}{a}}{\dfrac{3}{a} - a}$

13. $\dfrac{y - \dfrac{4}{y}}{1 + \dfrac{3}{2y + 1}}$

14. $\dfrac{a + 1 - \dfrac{12}{a - 2}}{\dfrac{-2}{a - 2} + a - 1}$

15. $\dfrac{\dfrac{x}{6} - \dfrac{1}{3}}{\dfrac{2}{3x} + \dfrac{5}{6}}$

16. $\dfrac{\dfrac{7}{5x} - \dfrac{1}{x}}{\dfrac{3}{5} + \dfrac{2}{x}}$

17. $\dfrac{\dfrac{1}{x^2 - 9} + \dfrac{2}{x + 3}}{\dfrac{3}{x - 3}}$

18. $\dfrac{\dfrac{5}{x + 4}}{\dfrac{1}{x - 4} - \dfrac{2}{x^2 - 16}}$

19. $\dfrac{\dfrac{2}{y - 1} + 2}{\dfrac{2}{y + 1} - 2}$

20. $\dfrac{\dfrac{y}{y + 1} + 1}{\dfrac{2y + 1}{y - 1}}$

To Think About

21. Consider the complex fraction $\dfrac{\dfrac{4}{x + 3}}{\dfrac{5}{x} - 1}$. What values are not allowable replacements for the variable x?

22. Consider the complex fraction $\dfrac{\dfrac{5}{x - 2}}{\dfrac{6}{x} + 1}$. What values are not allowable replacements for the variable x?

Cumulative Review Problems

23. Solve for w. $P = 2(l + w)$

24. Solve and graph. $7 + x < 11 + 5x$

25. Manuela is paying back a $4000 loan with no interest from her favorite sister. If Manuela has already paid $125 per month for 17 months, how much of the loan does she have yet to pay?

26. A neighborhood in Hopewell, New Jersey, is upset because a big office complex is being built in their neighborhood. Many of the citizens of Hopewell moved to their town to get away from urban sprawl. An investment firm will be building a complex planned to cover 3.5 million square feet. This initial estimate could grow to 5.5 million square feet. What percent of increase would this be?

6.5 Equations Involving Rational Expressions

① Solving Equations Involving Rational Expressions That Have Solutions

In Section 2.4 we developed procedures to solve linear equations containing fractions whose denominators were numerical values. In this section we use a similar approach to solve equations containing fractions whose denominators are polynomials. It would be wise for you to review Section 2.4 briefly *before you begin this section*. It will be especially helpful to carefully study Examples 1 and 2.

To Solve an Equation Containing Rational Expressions

1. Determine the LCD of all the denominators.
2. Multiply each term of the equation by the LCD.
3. Solve the resulting equation.
4. Check your solution. Exclude from your solution any value that would make the LCD equal to zero. If such a value is obtained, there is *no solution*.

EXAMPLE 1 Solve for x and check your solution. $\dfrac{5}{x} + \dfrac{2}{3} = 2 - \dfrac{2}{x} - \dfrac{1}{6}$

$6x\left(\dfrac{5}{x}\right) + 6x\left(\dfrac{2}{3}\right) = 6x(2) - 6x\left(\dfrac{2}{x}\right) - 6x\left(\dfrac{1}{6}\right)$ Observe that the LCD is $6x$. Multiply each term by $6x$.

$30 + 4x = 12x - 12 - x$ Simplify. Do you see how each term is obtained?

$30 + 4x = 11x - 12$ Collect like terms.

$30 = 7x - 12$ Subtract $4x$ from both sides.

$42 = 7x$ Add 12 to both sides.

$6 = x$ Divide both sides by 7.

Check. $\dfrac{5}{6} + \dfrac{2}{3} \stackrel{?}{=} 2 - \dfrac{2}{6} - \dfrac{1}{6}$ Replace each x by 6.

$\dfrac{5}{6} + \dfrac{4}{6} \stackrel{?}{=} \dfrac{12}{6} - \dfrac{2}{6} - \dfrac{1}{6}$

$\dfrac{9}{6} = \dfrac{9}{6}$ ✓ It checks.

Practice Problem 1 Solve for x and check your solution. $\dfrac{2}{x} + \dfrac{4}{x} = 3 - \dfrac{1}{x} - \dfrac{17}{8}$

EXAMPLE 2 Solve and check. $\dfrac{6}{x + 3} = \dfrac{3}{x}$

Observe that the LCD $= x(x + 3)$.

$x(x + 3)\left(\dfrac{6}{x + 3}\right) = x(x + 3)\left(\dfrac{3}{x}\right)$ Multiply both sides by $x(x + 3)$.

$6x = 3(x + 3)$ Simplify. Do you see how this is done?

$6x = 3x + 9$ Remove parentheses.

$3x = 9$ Subtract $3x$ from both sides.

$x = 3$ Divide both sides by 3.

Student Learning Objectives

After studying this section, you will be able to:

① Solve equations involving rational expressions that have solutions.

② Determine whether an equation involving rational expressions has no solution.

SSM

PH TUTOR CD & VIDEO MATH PRO WEB
CENTER

381

Check. $\dfrac{6}{3+3} \overset{?}{=} \dfrac{3}{3}$ Replace each x by 3.

$\dfrac{6}{6} = \dfrac{3}{3}$ ✓ It checks.

Practice Problem 2   Solve and check. $\dfrac{4}{2x+1} = \dfrac{6}{2x-1}$

It is sometimes necessary to factor denominators before the correct LCD can be determined.

EXAMPLE 3   Solve and check. $\dfrac{3}{x+5} - 1 = \dfrac{4-x}{2x+10}$

$\dfrac{3}{x+5} - 1 = \dfrac{4-x}{2(x+5)}$ Factor $2x+10$. We determine that the LCD is $2(x+5)$.

$2(x+5)\left(\dfrac{3}{x+5}\right) - 2(x+5)(1) = 2(x+5)\left[\dfrac{4-x}{2(x+5)}\right]$ Multiply each term by the LCD.

$2(3) - 2(x+5) = 4 - x$ Simplify.

$6 - 2x - 10 = 4 - x$ Remove parentheses.

$-2x - 4 = 4 - x$ Collect like terms.

$-4 = 4 + x$ Add $2x$ to both sides.

$-8 = x$ Subtract 4 from both sides.

Check. $\dfrac{3}{-8+5} - 1 \overset{?}{=} \dfrac{4-(-8)}{2(-8)+10}$ Replace each x in the original equation by -8.

$\dfrac{3}{-3} - 1 \overset{?}{=} \dfrac{4+8}{-16+10}$

$-1 - 1 \overset{?}{=} \dfrac{12}{-6}$

$-2 = -2$ ✓   It checks. The solution is -8.

Practice Problem 3   Solve and check. $\dfrac{x-1}{x^2-4} = \dfrac{2}{x+2} + \dfrac{4}{x-2}$

2 *Determining Whether an Equation Involving Rational Expressions Has No Solution*

Equations containing rational expressions sometimes appear to have solutions when in fact they do not. By this we mean that the "solutions" we get by using completely correct methods are, in actuality, not solutions.

In the case where a value makes a denominator in the equation equal to zero, we say it is not a solution to the equation. Such a value is called an **extraneous solution**. An extraneous solution is an apparent solution that does *not* satisfy the original equation. If all of the apparent solutions of an equation are extraneous solutions, we say that the equation has **no solution**. It is important that you check all apparent solutions in the original equation.

EXAMPLE 4 Solve and check. $\dfrac{y}{y-2} - 4 = \dfrac{2}{y-2}$

Observe that the LCD is $y - 2$.

$$(y-2)\left(\dfrac{y}{y-2}\right) - (y-2)(4) = (y-2)\left(\dfrac{2}{y-2}\right) \qquad \text{Multiply each term by } (y-2).$$

$$y - 4(y-2) = 2 \qquad \text{Simplify. Do you see how this is done?}$$

$$y - 4y + 8 = 2 \qquad \text{Remove parentheses.}$$

$$-3y + 8 = 2 \qquad \text{Collect like terms.}$$

$$-3y = -6 \qquad \text{Subtract 8 from both sides.}$$

$$\dfrac{-3y}{-3} = \dfrac{-6}{-3} \qquad \text{Divide both sides by } -3.$$

$$y = 2 \qquad \text{2 is only an apparent solution.}$$

This equation has no solution.
Why? We can see immediately that $y = 2$ is not a solution for the original equation. When we substitute 2 for y in a denominator, the denominator is equal to zero and the expression is undefined.

Check. $\dfrac{y}{y-2} - 4 = \dfrac{2}{y-2}$ Suppose that you try to check the apparent solution by substituting 2 for y.

$$\dfrac{2}{2-2} - 4 \overset{?}{=} \dfrac{2}{2-2}$$

$$\dfrac{2}{0} - 4 = \dfrac{2}{0} \qquad \text{This does not check since you do not obtain a real number when you divide by zero.}$$

$$\uparrow \qquad\quad \uparrow$$

These expressions are not defined.

There is no such number as $2 \div 0$. We see that 2 does *not* check. This equation has **no solution**.

Practice Problem 4 Solve and check. $\dfrac{2x}{x+1} = \dfrac{-2}{x+1} + 1$

Solve and check.

1. $\dfrac{x+1}{2x} = \dfrac{2}{3}$

2. $\dfrac{6}{3x-5} = \dfrac{3}{2x}$

3. $\dfrac{1}{x+6} = \dfrac{4}{x}$

4. $\dfrac{x-2}{4x} = \dfrac{1}{6}$

5. $\dfrac{5}{3x-4} = \dfrac{3}{x-3}$

6. $\dfrac{5}{x-1} = \dfrac{3}{x+1}$

7. $\dfrac{2}{x} + \dfrac{x}{x+1} = 1$

8. $\dfrac{5}{2} = 3 + \dfrac{2x+7}{x+6}$

9. $\dfrac{x+1}{x} = 1 + \dfrac{x-2}{2x}$

10. $\dfrac{7x-4}{5x} = \dfrac{9}{5} - \dfrac{4}{x}$

11. $\dfrac{x+1}{x} = \dfrac{1}{2} - \dfrac{4}{3x}$

12. $\dfrac{2x-1}{3x} + \dfrac{1}{9} = \dfrac{x+2}{x} + \dfrac{1}{9x}$

13. $\dfrac{2}{x-6} - 5 = \dfrac{2(x-5)}{x-6}$

14. $7 - \dfrac{x}{x+5} = \dfrac{5}{5+x}$

15. $\dfrac{2}{x+1} - \dfrac{1}{x-1} = \dfrac{2x}{x^2-1}$

16. $\dfrac{8x}{4x^2-1} = \dfrac{3}{2x+1} + \dfrac{3}{2x-1}$

17. $\dfrac{y+1}{y^2+2y-3} = \dfrac{1}{y+3} - \dfrac{1}{y-1}$

18. $\dfrac{2y}{y+1} + \dfrac{1}{3y-2} = 2$

19. $\dfrac{79-x}{x} = 5 + \dfrac{7}{x}$

20. $\dfrac{6}{x-3} = \dfrac{-5}{x-2} + \dfrac{-5}{x^2-5x+6}$

21. $\dfrac{x+11}{x^2-5x+4} + \dfrac{3}{x-1} = \dfrac{5}{x-4}$

22. $\dfrac{52-x}{x} = 9 + \dfrac{2}{x}$

23. $\dfrac{2x}{x+4} - \dfrac{8}{x-4} = \dfrac{2x^2+32}{x^2-16}$

24. $\dfrac{4x}{x+3} - \dfrac{12}{x-3} = \dfrac{4x^2+36}{x^2-9}$

25. $\dfrac{6}{x-5} + \dfrac{3x+1}{x^2-2x-15} = \dfrac{5}{x+3}$

26. $\dfrac{4}{x^2-1} + \dfrac{7}{x+1} = \dfrac{5}{x-1}$

To Think About

In each of the following equations, what values are not allowable replacements for the variable x? Do not solve the equation.

27. $\dfrac{3x}{x-2} - \dfrac{4x}{x-4} = \dfrac{3}{x^2 - 6x + 8}$

28. $\dfrac{2x}{x+2} - \dfrac{x}{x+3} = \dfrac{-3}{x^2 + 5x + 6}$

Cumulative Review Problems

29. Factor. $6x^2 - x - 12$

30. Solve. $4 - (3 - x) = 7(x + 3)$

▲ **31.** The perimeter of a rectangular sign is 54 meters. The length is 1 meter less than triple the width. Find the dimensions of the sign.

32. Wally and Adele Panzas plan to purchase a new home and borrow $115,000. They plan to take out a 25-year mortgage at an annual interest rate of 8.75%. The bank will charge them $8.23 per month for each $1000 of mortgage. What will their monthly payments be?

The accompanying bar graph depicts the number of immigrants to the United States from Europe and Latin America for each decade from 1960 to 2000. Use the graph to answer exercises 33–36.

33. What was the percent of decrease in immigrants from Europe from the decade 1960-1969 to the decade 1970-1979? Round to the nearest tenth.

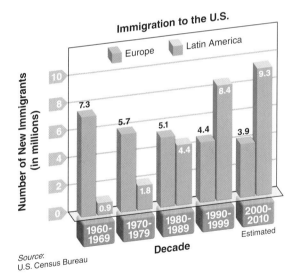

34. What was the percent of increase in immigrants from Latin America from the decade 1980-1989 to the decade 1990-1999? Round to the nearest tenth.

35. If the rate of increase in immigrants from Latin America from the decade starting in 1990 to the decade starting in 2000 continues for another two decades, what will be the expected number of immigrants from Latin America to the United States in 2010? Round to the nearest tenth.

36. If the rate of decrease in immigrants from Europe from the decade starting in 1990 to the decade starting in 2000 continues for another two decades, what will be the expected number of immigrants from Europe to the United States in the decade starting in 2010?

6.6 Ratio, Proportion, and Other Applied Problems

Student Learning Objectives

After studying this section, you will be able to:

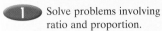
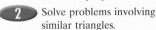

1. Solve problems involving ratio and proportion.

2. Solve problems involving similar triangles.

3. Solve distance problems involving rational expressions.

4. Solve work problems.

SSM

PH TUTOR   CD & VIDEO   MATH PRO   WEB
CENTER

1 Solving Problems Involving Ratio and Proportion

A **ratio** is a comparison of two quantities. You may be familiar with ratios that compare miles to hours or miles to gallons. A ratio is often written as a quotient in the form of a fraction. For example, the ratio of 7 to 9 can be written as $\frac{7}{9}$.

A **proportion** is an equation that states that two ratios are equal. For example,

$$\frac{7}{9} = \frac{21}{27}, \quad \frac{2}{3} = \frac{10}{15}, \quad \text{and} \quad \frac{a}{b} = \frac{c}{d} \quad \text{are proportions.}$$

Let's take a closer look at the last proportion. We can see that the LCD of the fractional equation is bd.

$$(bd)\frac{a}{b} = (bd)\frac{c}{d} \qquad \text{Multiply each side by the LCD.}$$

$$da = bc$$

$$ad = bc \qquad \text{Since multiplication is commutative, } da = ad.$$

Thus we have proved the following.

The Proportion Equation

If

$$\frac{a}{b} = \frac{c}{d}, \quad \text{then} \quad ad = bc$$

for all real numbers a, b, c, and d, where $b \neq 0$ and $d \neq 0$.

This is sometimes called **cross multiplying**. It can be applied only if you have *one* fraction and nothing else on each side of the equation.

EXAMPLE 1 Michael took 5 hours to drive 245 miles on the turnpike. At the same rate, how many hours will it take him to drive a distance of 392 miles?

1. *Understand the problem.*
   Let $x =$ the number of hours it will take to drive 392 miles. If 5 hours are needed to drive 245 miles, then x hours are needed to drive 392 miles.

2. *Write an equation.*
   We can write this as a proportion. Compare time to distance in each ratio.

   $$\text{Time} \longrightarrow \frac{5 \text{ hours}}{245 \text{ miles}} = \frac{x \text{ hours}}{392 \text{ miles}} \longleftarrow \text{Time}$$
   $$\text{Distance} \longrightarrow \qquad\qquad\qquad \longleftarrow \text{Distance}$$

3. *Solve and state the answer.*

   $$5(392) = 245x \qquad \text{Cross multiply.}$$

   $$\frac{1960}{245} = x \qquad \text{Divide both sides by 245.}$$

   $$8 = x$$

   It will take Michael 8 hours to drive 392 miles.

4. *Check.* Is $\frac{5}{245} = \frac{8}{392}$? Do the computation and see.

Practice Problem 1 It took Brenda 8 hours to drive 420 miles. At the same rate, how long would it take her to drive 315 miles?

EXAMPLE 2 If $\frac{3}{4}$ inch on a map represents an actual distance of 20 miles, how long is the distance represented by $4\frac{1}{8}$ inches on the same map?

Let x = the distance represented by $4\frac{1}{8}$ inches.

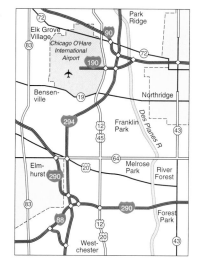

Initial measurement on map $\longrightarrow \dfrac{\dfrac{3}{4}}{20} = \dfrac{4\dfrac{1}{8}}{x} \longleftarrow$ Second measurement on the map

Initial distance $\longrightarrow$ $\phantom{\dfrac{3}{4}}$ $\longleftarrow$ Second distance

$$\left(\frac{3}{4}\right)(x) = (20)\left(4\frac{1}{8}\right) \qquad \text{Cross multiply.}$$

$$\left(\frac{3}{4}\right)(x) = (\overset{5}{\cancel{20}})\left(\frac{33}{\underset{2}{\cancel{8}}}\right) \qquad \text{Write } 4\frac{1}{8} \text{ as } \frac{33}{8} \text{ and simplify.}$$

$$\frac{3x}{4} = \frac{165}{2} \qquad \text{Multiplication of fractions.}$$

$$4\left(\frac{3x}{4}\right) = \overset{2}{\cancel{4}}\left(\frac{165}{\cancel{2}}\right) \qquad \text{Multiply each side by 4.}$$

$$3x = 330 \qquad \text{Simplify.}$$

$$x = 110 \qquad \text{Divide both sides by 3.}$$

$4\frac{1}{8}$ inches on the map represents an actual distance of 110 miles.

Practice Problem 2 If $\frac{5}{8}$ inch on a map represents an actual distance of 30 miles, how long is the distance represented by $2\frac{1}{2}$ inches on the same map?

2 *Solving Problems Involving Similar Triangles*

Similar triangles are triangles that have the same shape, but may be different sizes. For example, if you draw a triangle on a sheet of paper, place the paper in a photocopy machine, and make a copy that is reduced by 25%, you would create a triangle that is similar to the original triangle. The two triangles will have the same shape. The corresponding sides of the triangles will be proportional. The corresponding angles of the triangles will also be equal.

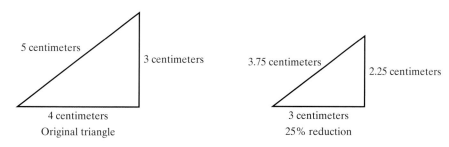

You can use the proportion equation to show that the corresponding sides of the preceding triangles are proportional. In fact, you can use the proportion equation to find an unknown length of a side of one of the two similar triangles.

▲ **EXAMPLE 3** A ramp is 32 meters long and rises up 15 meters. A ramp at the same angle is 9 meters long. How high is the second ramp? To answer this question, we find the length of side x in the following two similar triangles.

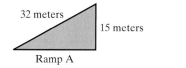

Ramp A, longest side $\longrightarrow$ $\dfrac{32}{9} = \dfrac{15}{x}$ $\longleftarrow$ Shortest side, ramp A

Ramp B, longest side $\longrightarrow$ $\longleftarrow$ Shortest side, ramp B

$$32x = (9)(15) \qquad \text{Cross multiply.}$$
$$32x = 135 \qquad \text{Multiply.}$$
$$x = \frac{135}{32} \qquad \text{Divide both sides by 32.}$$
$$\text{or} \quad x = 4\frac{7}{32} \text{ meters}$$

▲ **Practice Problem 3** Triangle C is similar to triangle D. Find the length of side x. Leave your answer as a fraction.

13 centimeters   16 centimeters

Triangle C

x 18 centimeters

Triangle D

We can also use similar triangles for indirect measurement—for instance, to find the measure of an object that is too tall to measure using standard measuring devices. When the sun shines on two vertical objects at the same time, the shadows and the objects form similar triangles.

▲ ◗ **EXAMPLE 4** A woman who is 5 feet tall casts a shadow that is 8 feet long. At the same time of day, a building casts a shadow that is 72 feet long. How tall is the building?

1. Understand the problem.
First we draw a sketch. We do not know the height of the building, so we call it x.

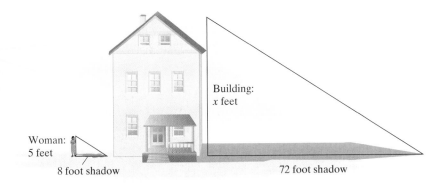

Building: x feet

Woman: 5 feet

8 foot shadow

72 foot shadow

2. Write an equation and solve.

Height of woman $\longrightarrow$ $\dfrac{5}{8} = \dfrac{x}{72}$ $\longleftarrow$ Height of building

Length of woman's shadow $\longrightarrow$ $\longleftarrow$ Length of building's shadow

$$(5)(72) = 8x \quad \text{Cross multiply.}$$
$$360 = 8x$$
$$45 = x$$

The height of the building is 45 feet.

▲ **Practice Problem 4** A man who is 6 feet tall casts a shadow that is 7 feet long. At the same time of day, a large flagpole casts a shadow that is 38.5 feet long. How tall is the flagpole?

In problems such as Example 4, we are assuming that the building and the person are standing exactly perpendicular to the ground. In other words, each triangle is assumed to be a right triangle. In other similar triangle problems, if the triangles are not right triangles you must be careful that the angles between the objects and the ground are the same.

3 Solving Distance Problems Involving Rational Expressions

Some distance problems are solved using equations with rational expressions. We will need the formula Distance = Rate × Time, $D = RT$, which we can write in the form $T = \dfrac{D}{R}$.

EXAMPLE 5 Plane A flies at a speed that is 50 kilometers per hour faster than plane B. Plane A flies 500 kilometers in the amount of time that plane B flies 400 kilometers. Find the speed of each plane.

1. **Understand the problem.**
   Let s = the speed of plane B in kilometers per hour.
   Let $s + 50$ = the speed of plane A in kilometers per hour.
   Make a simple table for D, R, and T.

	D	R	$T = \dfrac{D}{R}$
Plane A	500	$s + 50$	?
Plane B	400	s	?

   Since $T = \dfrac{D}{R}$, for each plane we divide the expression for D by the expression for R and write it in the table in the column for time.

	D	R	$T = \dfrac{D}{R}$
Plane A	500	$s + 50$	$\dfrac{500}{s + 50}$
Plane B	400	s	$\dfrac{400}{s}$

2. **Write an equation and solve.**
   Each plane flies for the same amount of time. That is, the time for plane A equals the time for plane B.

   $$\frac{500}{s + 50} = \frac{400}{s}$$

   You can solve this equation using the methods in Section 6.5 or you may cross multiply. Here we will cross multiply.

$500s = (s + 50)(400)$	Cross multiply.
$500s = 400s + 20{,}000$	Remove parentheses.
$100s = 20{,}000$	Subtract $400s$ from each side.
$s = 200$	Divide each side by 100.

   Plane B travels 200 kilometers per hour. Since

   $$s + 50 = 200 + 50 = 250,$$

   plane A travels 250 kilometers per hour.

Practice Problem 5 Two European freight trains traveled toward Paris for the same amount of time. Train A traveled 180 kilometers, while train B traveled 150 kilometers. Train A traveled 10 kilometers per hour faster than train B. What was the speed of each train?

4 Solving Work Problems

Some applied problems involve the length of time needed to do a job. These problems are often referred to as work problems.

EXAMPLE 6 Reynaldo can sort a huge stack of mail on an old sorting machine in 9 hours. His brother Carlos can sort the same amount of mail using a newer sorting machine in 8 hours. How long would it take them to do the job working together? Express your answer in hours and minutes. Round to the nearest minute.

1. *Understand the problem.*
    Let's do a little reasoning.
    If Reynaldo can do the job in 9 hours, then in *1 hour* he could do $\frac{1}{9}$ of the job.
    If Carlos can do the job in 8 hours, then in *1 hour* he could do $\frac{1}{8}$ of the job.
    Let x = the number of hours it takes Reynaldo and Carlos to do the job together. In *1 hour* together they could do $\frac{1}{x}$ of the job.

2. *Write an equation and solve.*
    The amount of work Reynaldo can do in 1 hour plus the amount of work Carlos can do in 1 hour must be equal to the amount of work they could do together in 1 hour.

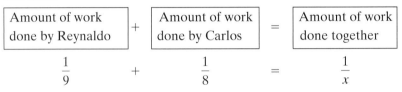

Amount of work done by Reynaldo		Amount of work done by Carlos		Amount of work done together
$\frac{1}{9}$	$+$	$\frac{1}{8}$	$=$	$\frac{1}{x}$

Let us solve for x. We observe that the LCD is $72x$.

$$72x\left(\frac{1}{9}\right) + 72x\left(\frac{1}{8}\right) = 72x\left(\frac{1}{x}\right) \qquad \text{Multiply each term by the LCD.}$$
$$8x + 9x = 72 \qquad \text{Simplify.}$$
$$17x = 72 \qquad \text{Collect like terms.}$$
$$x = \frac{72}{17} \qquad \text{Divide each side by 17.}$$
$$x = 4\frac{4}{17}$$

To change $\frac{4}{17}$ of an hour to minutes we multiply.
$$\frac{4}{17} \; \cancel{\text{hour}} \times \frac{60 \text{ minutes}}{1 \; \cancel{\text{hour}}} = \frac{240}{17} \text{ minutes, which is approximately } 14.118 \text{ minutes}$$
To the nearest minute this is 14 minutes. Thus doing the job together will take 4 hours and 14 minutes.

Practice Problem 6 John Tobey and Dave Wells obtained night custodian jobs at a local factory while going to college part time. Using the buffer machine, John can buff all the floors in the building in 6 hours. Dave takes a little longer and can do all the floors in the building in 7 hours. Their supervisor bought another buffer machine. How long will it take John and Dave to do all the floors in the building working together, each with his own machine? Express your answer in hours and minutes. Round to the nearest minute.

Calculator

Reciprocals
You can find $\frac{1}{x}$ for any value of x on a scientific calculator by using the key labeled $\boxed{x^{-1}}$ or the key labeled $\boxed{1/x}$. For example, to find $\frac{1}{9}$, we use $9 \boxed{x^{-1}}$ or $9 \boxed{1/x}$. The display will read 0.11111111. Therefore we can solve Example 6 as follows:

$9 \boxed{x^{-1}} \boxed{+} 8 \boxed{x^{-1}} \boxed{=}$

The display will read 0.2361111.
Thus we have obtained the equation
$0.2361111 = \frac{1}{x}$.
Now this is equivalent to
$x = \frac{1}{0.2361111}$.
(Do you see why?)
Thus we enter 0.2361111 $\boxed{x^{-1}}$, and the display reads 4.235294118.
If we round to the nearest hundredth, we have
$x \approx 4.24$ hours, which is approximately equal to our answer of $4\frac{4}{17}$ hours.

Solve.

1. $\dfrac{4}{9} = \dfrac{8}{x}$

2. $\dfrac{5}{12} = \dfrac{x}{8}$

3. $\dfrac{x}{17} = \dfrac{12}{5}$

4. $\dfrac{16}{x} = \dfrac{3}{4}$

5. $\dfrac{5}{3} = \dfrac{x}{8}$

6. $\dfrac{150}{70} = \dfrac{9}{x}$

7. $\dfrac{7}{x} = \dfrac{40}{130}$

8. $\dfrac{x}{18} = \dfrac{13}{2}$

Applications

Use a proportion to answer exercises 9–16.

9. The scale on the AAA map of Colorado is approximately $\frac{3}{4}$ inch to 15 miles. If the distance from Denver to Pueblo measures 5.5 inches on the map, how far apart are the two cities?

10. Nella Coastes' recipe for Shoofly Pie contains $\frac{3}{4}$ cup of unsulfured molasses. This recipe makes a small pie that serves 8 people. If she makes the larger pie that serves 12 people and the ratio of molasses to people remains the same, how much molasses will she need for the larger pie?

11. James LeBlanc spent a summer semester in England. When he arrived in England he converted $800 to pounds. The posted exchange rate the day he arrived that summer was that 1 British pound was worth $1.53 in American currency.
  (a) How many pounds did James receive for his $800?
  (b) At the end of the summer he had 140 pounds left and he changed it back into dollars. The exchange was the same that day as it was the day he arrived. How many dollars did he receive?

12. Christine Maney spent a summer studying in Paris as an exchange student. The exchange rate the day she arrived that summer was 5.56 francs per American dollar. When she arrived she brought $900.
  (a) How many French francs was her $900 worth?
  (b) At the end of the summer she had 350 francs left and exchanged them for American dollars. The exchange rate was the same that day as it was the day she arrived. How many dollars did she receive?

13. Alfonse and Melinda are taking a drive in Mexico. They know that a speed of 100 kilometers per hour is approximately equal to 62 miles per hour. They are now driving on a Mexican road that has a speed limit of 90 kilometers per hour. How many miles per hour is the speed limit? Round to the nearest mile per hour.

14. Dick and Anne took a trip to France. Their suitcases were weighed at the airport and the weight recorded was 39 kilograms. If 50 kilograms is equivalent to 110 pounds, how many pounds did their suitcases weigh? Round to the nearest pound.

15. On a map the distance between two mountains is $3\frac{1}{2}$ inches. The actual distance between the mountains is 136 miles. Russ is camped at a location that on the map is $\frac{3}{4}$ inch from the base of the mountain. How many miles is he from the base of the mountain? Round to the nearest mile.

16. Maria is adding a porch to her house that is 18 feet long. On the drawing done by the carpenter the length is shown as 11 inches. The drawing shows that the width of the porch is 8 inches. How many feet wide will the porch be? Round to the nearest foot.

Triangles A and B are similar. Use them to answer exercises 17 and 18. Leave your answers as fractions.

Triangle A

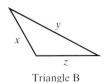

Triangle B

▲ **17.** If $x = 20$ in., $y = 29$ in., and $m = 13$ in., find the length of side n.

▲ **18.** If $p = 14$ in., $m = 17$ in., and $z = 23$ in., find the length of side x.

Just as we have discussed similar triangles, other geometric shapes can be similar. Similar geometric shapes will have sides that are proportional. Quadrilaterals abcd and ghjk are similar. Use them to answer exercises 19–22. Leave your answers in fractions.

▲ **19.** If $a = 7$ in., $g = 9$ in., and $k = 12$ in., find the length of side d.

▲ **20.** If $b = 8$ ft, $c = 7$ ft, and $j = 11$ ft, find the length of side h.

▲ **21.** If $b = 20$ m, $h = 24$ m, and $d = 32$ m, find the length of side k.

▲ **22.** If $a = 16$ cm, $d = 19$ cm, and $k = 23$ cm, find the length of side g.

Use a proportion to solve.

▲ **23.** A rectangle whose width-to-length ratio is approximately 5 to 8 is called a **golden rectangle** and is said to be pleasing to the eye. Using this ratio, what should the length of a rectangular picture be if its width is to be 30 inches?

▲ **24.** A 5-foot-tall woman casts a shadow of 4 feet. At the same time, a tree casts a shadow of 31 feet. How tall is the tree? Round to the nearest foot.

▲ **25.** A kite is held out on a line that is almost perfectly straight. When the kite is held out on 7 meters of line, it is 5 meters off the ground. How high would the kite be if it is held out on 120 meters of line at the same angle from the ground? Round to the nearest meter.

▲ **26.** A wire line helps to secure a radio transmission tower. The wire measures 23 meters from the tower to the ground anchor pin. The wire is secured 14 meters up on the tower. If a second wire is secured 130 meters up on the tower and is extended from the tower at the same angle as the first wire, how long would the second wire need to be to reach an anchor pin on the ground? Round to the nearest meter.

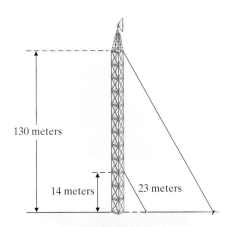

130 meters

14 meters

23 meters

27. Ben Hale is driving his new Toyota Camry on Interstate 90 at 45 miles per hour. He accelerates at the rate of 3 miles per hour every 2 seconds. How fast will he be traveling after accelerating for 11 seconds?

28. Tim Newitt is driving a U-Haul truck to Chicago. He is driving at 55 miles per hour and has to hit the brakes because of heavy traffic. His truck slows at the rate of 2 miles per hour for every 3 seconds. How fast will he be traveling 10 seconds after he hits the brakes?

29. A Montreal commuter airliner travels 40 kilometers per hour faster than the television news helicopter over the city. The commuter airliner travels 1250 kilometers during the same time that the television news helicopter travels only 1050 kilometers. How fast does the commuter airliner fly? How fast does the television news helicopter fly?

30. Melissa drove to Dallas while Marcia drove to Houston in the same amount of time. Melissa drove 360 kilometers, while Marcia drove 280 kilometers. Melissa traveled 20 kilometers per hour faster than Marcia on her trip. What was the average speed in kilometers per hour for each woman?

31. A famous brand of men's cologne is priced at $75 for a 3.4-ounce size. A 1.7-ounce bottle is priced at $42.
   (a) How much does the cologne in the larger bottle cost per ounce? Round to the nearest cent.

   (b) How much does the cologne in the smaller bottle cost per ounce? Round to the nearest cent.

   (c) How much would you save if you purchased 10.2 ounces of the cologne in the larger bottles rather than the smaller bottles? Find the exact answer; do not round off.

32. Won Ling is a Chinese tea importer in Boston's Chinatown. He charges $12.25 for four sample packs of his famous green tea. The packs are in the following sizes: 25 grams, 40 grams, 50 grams, and 60 grams.
   (a) How much is Won Ling charging per gram for his green tea?
   (b) How much would you pay for a 60-gram pack if he were willing to sell that one by itself?
   (c) How much would you pay for an 800-gram package of green tea if it cost the same amount per gram?

33. It takes a person using a large rotary mower 4 hours to mow all the lawns at the town park. It takes a person using a small rotary mower 5 hours to mow these same lawns. How long should it take two people using these mowers to mow these lawns together? Round to the nearest minute.

34. It takes a secretary with a typewriter 7 hours to address envelopes for a business mailing list. It takes 3 hours for the computer to print out the addresses and the secretary to put the computer labels on the envelopes. The company boss is in a big rush and wants one secretary to do part of the job with the computer and another secretary to do the rest of the job with a typewriter. How long should it take the two secretaries working together? Round to the nearest minute.

35. When all the leaves have fallen at Fred and Suzie's house in Concord, New Hampshire, Suzie can rake the entire yard in 6 hours. When Fred does it alone, it takes him 8 hours. How long would it take them to rake the yard together? Round to the nearest minute.

36. Professor Matthews can type his course syllabus in 12 hours. The departmental secretary can type it in 7 hours. How long would it take the two people to type it together? Round to the nearest minute.

Cumulative Review Problems

37. Write in scientific notation. 0.0000006316

38. Write in decimal notation. 5.82×10^8

39. Write with positive exponents. $\dfrac{x^{-3}y^{-2}}{z^4w^{-8}}$

40. Evaluate. $\left(\dfrac{2}{3}\right)^{-3}$

Putting Your Skills to Work

Mathematical Measurement of Planet Orbit Time

The closer a planet is to the sun, the fewer the number of days it takes to complete one orbit around the sun. The Earth takes approximately 365 days to complete one orbit. Only two of the nine planets have a shorter orbit time. The orbit times of the four planets closest to the sun are shown in the bar graph below. Use this bar graph to answer the following questions. Round all answers to the nearest whole number.

Number of Days for a Planet to Orbit the Sun

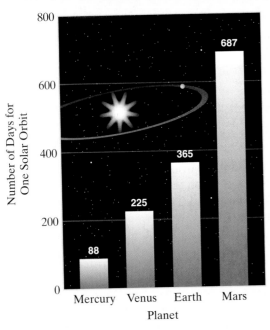

Problems for Individual Investigation

1. How many more days does it take Mars than Venus to complete one rotation around the sun?

2. In 5 years, the Earth will rotate around the sun five times. In that time period, how many orbits will Mercury make around the sun?

Problems for Group Investigation and Cooperative Learning

Together with some other members of your class, see if you can answer the following.

The **synodic period** of a planet is the number of days it takes the planet to gain one orbit on a planet farther from the Sun. The synodic period (S) can be found by the formula

$$\frac{1}{S} = \frac{1}{a} - \frac{1}{b}$$

where $a =$ the number of days it takes the planet closer to the sun to complete one orbit and

$b =$ the number of days it takes the planet farther from the sun to complete one orbit.

3. What is the synodic period for Mercury to gain one orbit on Venus?

4. What is the synodic period for the Earth to gain one orbit on Jupiter if the number of days it takes Jupiter to orbit the sun is approximately 4333 days?

Internet Connections

  Netsite: http://www.prenhall.com/tobey_beginning

Site: The Nine Planets or a related site

This site provides a wealth of information about our solar system. Note that many of the numbers here are given in an abbreviated form of scientific notation; for example, 1.35e23 means 1.35×10^{23}.

5. Tell how many days it takes each of the planets Saturn, Uranus, Neptune, and Pluto to complete one rotation about the sun. This information is listed as "O-Period" in the Solar System Data appendix of the Web site.

6. How many more days does it take Neptune than Saturn to complete one orbit about the sun?

7. Use the formula to find the synodic period for Uranus to gain one orbit on Pluto.

Math in the Media

Math Behind the Scenes

At close to 200 million dollars, *Titanic* is one of the most expensive movies ever made. For the movie, a replica of the original ship was constructed. *Entertainment Weekly Online* reported the replica to be 770 feet long—close to 90% to scale of the real ship.

Details of the original ship were followed closely to create the replica. A special studio was built for filming. The ship "sailed" in an enormous tank that held 17 million gallons of water.

Although movies frequently use models when filming, replicas tend to be on a much smaller scale than that used for the *Titanic*. In the questions that follow, you can use your own calculations to compare model and actual scales.

EXERCISES

1. The original Titanic was 880 feet long and 92 feet high. Use the exact length of the original Titanic and the exact length of the model to determine the percentage to scale of the real ship. Using this information, determine the height of the replica.

2. Does your calculation in Exercise 1 verify that the replica is 90% to scale of the actual ship? How much does your answer differ from the original estimate of 90%?

3. In the movie *Thu Hunt for Red October*, the story centers on the search for a Russian ballistic missile submarine that is reported to be 610 feet long and has a beam (width) of 46 feet. In filming some of the underwater sequences, a model of the hull of a submarine was used that was reported to be 14 feet wide. If that report is true what percent to scale is the model to the actual submarine in the story? If the entire model was built to that scale, what would be the expected length of the model submarine used in the filming?

Chapter 6 Organizer

Topic	Procedure	Examples
Simplifying rational expressions, p. 355.	1. Factor the numerator and denominator. 2. Divide out any factor common to both the numerator and denomiator.	$$\frac{36x^2 - 16y^2}{18x^2 + 24xy + 8y^2} = \frac{4(3x - 2y)(3x + 2y)}{2(3x + 2y)(3x + 2y)}$$ $$= \frac{2(3x - 2y)}{3x - 2y}$$
Multiplying rational expressions, p. 361.	1. Factor all numerators and denominators and rewrite the product as one fraction. 2. Simplify the resulting rational expression as described above.	$$\frac{x^2 - y^2}{x^2 + 2xy + y^2} \cdot \frac{x^2 + 4xy + 3y^2}{x^2 - 4xy + 3y^2}$$ $$= \frac{(x + y)(x - y)(x + y)(x + 3y)}{(x + y)(x + y)(x - y)(x - 3y)}$$ $$= \frac{x + 3y}{x - 3y}$$
Dividing rational expressions, p. 362.	1. Invert the second fraction and rewrite the problem as a product. 2. Multiply the rational expressions.	$$\frac{14x^2 + 17x - 6}{x^2 - 25} \div \frac{4x^2 - 8x - 21}{x^2 + 10x + 25}$$ $$= \frac{(2x + 3)(7x - 2)}{(x + 5)(x - 5)} \cdot \frac{(x + 5)(x + 5)}{(2x - 7)(2x + 3)}$$ $$= \frac{(7x - 2)(x + 5)}{(x - 5)(2x - 7)}$$
Adding rational expressions, p. 366.	1. If the denominators differ, factor them and determine the least common denominator (LCD). 2. Change each fraction by multiplication into an equivalent one with the LCD. 3. Add numerators; put the answer over the LCD. 4. Simplify as needed.	$$\frac{x - 1}{x^2 - 4} + \frac{x - 1}{3x + 6} = \frac{x - 1}{(x + 2)(x - 2)} + \frac{x - 1}{3(x + 2)}$$ $$\text{LCD} = 3(x + 2)(x - 2)$$ $$\frac{x - 1}{(x + 2)(x - 2)} = \frac{?}{3(x + 2)(x - 2)}$$ Need to multiply by $\frac{3}{3}$. $$\frac{(x - 1)}{3(x + 2)} = \frac{?}{3(x + 2)(x - 2)}$$ Need to multiply by $\frac{x - 2}{x - 2}$. $$\frac{x - 1}{(x + 2)(x - 2)} + \frac{x - 1}{3(x + 2)}$$ $$= \frac{(x - 1) \cdot 3}{3(x + 2)(x - 2)} + \frac{(x - 1)(x - 2)}{3(x + 2)(x - 2)}$$ $$= \frac{3x - 3 + x^2 - 3x + 2}{3(x + 2)(x - 2)}$$ $$= \frac{x^2 - 1}{3(x + 2)(x - 2)}$$
Subtracting rational expressions, p. 369.	Move a subtraction sign to the numerator of the second fraction. Add. Simplify if possible. $$\frac{a}{b} - \frac{c}{b} = \frac{a}{b} + \frac{-c}{b}$$	$$\frac{5x}{x - 2} - \frac{3x + 4}{x - 2} = \frac{5x}{x - 2} + \frac{-(3x + 4)}{x - 2}$$ $$= \frac{5x - 3x - 4}{x - 2}$$ $$= \frac{2x - 4}{x - 2}$$ $$= \frac{2(x - 2)}{x - 2} = 2$$

Topic	Procedure	Examples
Simplifying complex rational expressions, p. 376.	1. Add the two fractions in the numerator. 2. Add the two fractions in the denominator. 3. Divide the fraction in the numerator by the fraction in the denominator. This is done by inverting the fraction in the denominator and multiplying by the numerator. 4. Simplify.	$$\dfrac{\dfrac{x}{x^2-4}+\dfrac{1}{x+2}}{\dfrac{3}{x+2}-\dfrac{4}{x-2}}$$ $$=\dfrac{\dfrac{x}{(x+2)(x-2)}+\dfrac{1(x-2)}{(x+2)(x-2)}}{\dfrac{3(x-2)}{(x+2)(x-2)}+\dfrac{-4(x+2)}{(x+2)(x-2)}}$$ $$=\dfrac{\dfrac{x+x-2}{(x+2)(x-2)}}{\dfrac{3x-6-4x-8}{(x+2)(x-2)}}$$ $$=\dfrac{2x-2}{(x+2)(x-2)}\div\dfrac{-x-14}{(x+2)(x-2)}$$ $$=\dfrac{2(x-1)}{(x+2)(x-2)}\cdot\dfrac{(x+2)(x-2)}{-x-14}$$ $$=\dfrac{2x-2}{-x-14}\ \text{or}\ \dfrac{-2x+2}{x+14}$$
Solving equations involving rational expressions, p. 381.	1. Determine the LCD of all denominators. 2. Note what values will make the LCD equal to 0. These are excluded from your solutions. 3. Multiply each side by the LCD, distributing as needed. 4. Solve the resulting polynomial equation. 5. Check. Be sure to exclude those values found in step 2.	$$\dfrac{3}{x-2}=\dfrac{4}{x+2}$$ LCD $=(x-2)(x+2)$. Since LCD $\neq 0$, then $x\neq 2,-2$. $$(x-2)(x+2)\,\dfrac{3}{x-2}=\dfrac{4}{x+2}\,(x-2)(x+2)$$ $$3(x+2)=4(x-2)$$ $$3x+6=4x-8$$ $$-x=-14$$ $$x=14$$ (Since $x\neq 2,-2$, this solution should check unless an error has been made.) $$Check:\quad \dfrac{3}{14-2}\overset{?}{=}\dfrac{4}{14+2}$$ $$\dfrac{3}{12}\overset{?}{=}\dfrac{4}{16}$$ $$\dfrac{1}{4}=\dfrac{1}{4}\ \checkmark$$
Solving applied problems with proportions, p. 386.	1. Organize the data. 2. Write a proportion equating the respective parts. Let x represent the value that is not known. 3. Solve the proportion.	Renee can make five cherry pies with 3 cups of flour. How many cups of flour does she need to make eight cherry pies? $$\dfrac{5\text{ cherry pies}}{3\text{ cups flour}}=\dfrac{8\text{ cherry pies}}{x\text{ cups flour}}$$ $$\dfrac{5}{3}=\dfrac{8}{x}$$ $$5x=24$$ $$x=\dfrac{24}{5}$$ $$x=4\dfrac{4}{5}$$ $4\dfrac{4}{5}$ cups of flour are needed for eight cherry pies.

Chapter 6 Review Problems

6.1 *Simplify.*

1. $\dfrac{4x - 4y}{5y - 5x}$

2. $\dfrac{bx}{bx - by}$

3. $\dfrac{2x^2 + 5x - 3}{2x^2 - 9x + 4}$

4. $\dfrac{3x^2 + 7x + 2}{3x^2 + 13x + 4}$

5. $\dfrac{x^2 - 9}{x^2 - 10x + 21}$

6. $\dfrac{2x^2 + 18x + 40}{3x + 15}$

7. $\dfrac{4x^2 + 4x - 3}{4x^2 - 2x}$

8. $\dfrac{x^3 + 3x^2}{x^3 - 2x^2 - 15x}$

9. $\dfrac{2x^2 - 2xy - 24y^2}{2x^2 + 5xy - 3y^2}$

10. $\dfrac{4 - y^2}{3y^2 + 5y - 2}$

11. $\dfrac{5x^3 - 10x^2}{25x^4 + 5x^3 - 30x^2}$

12. $\dfrac{16x^2 - 4y^2}{4x - 2y}$

6.2 *Multiply or divide.*

13. $\dfrac{3x^2 - 13x - 10}{3x^2 + 2x} \cdot \dfrac{x^2 - 25x}{x^2 - 25}$

14. $\dfrac{2y^2 - 18}{3y^2 + 3y} \div \dfrac{y^2 + 6y + 9}{y^2 + 4y + 3}$

15. $\dfrac{2y^2 + 3y - 2}{2y^2 + y - 1} \div \dfrac{2y^2 + y - 1}{2y^2 - 3y - 2}$

16. $\dfrac{6y^2 + 13y - 5}{9y^2 + 3y} \div \dfrac{4y^2 + 20y + 25}{12y^2}$

17. $\dfrac{3xy^2 + 12y^2}{2x^2 - 11x + 5} \div \dfrac{2xy + 8y}{8x^2 + 2x - 3}$

18. $\dfrac{11}{x - 2} \cdot \dfrac{2x^2 - 8}{44}$

19. $\dfrac{2x^2 + 10x + 2}{8x - 8} \cdot \dfrac{3x - 3}{4x^2 + 20x + 4}$

20. $\dfrac{x^2 - 5xy - 24y^2}{2x^2 - 2xy - 24y^2} \cdot \dfrac{4x^2 + 4xy - 24y^2}{x^2 - 10xy + 16y^2}$

6.3 *Add or subtract.*

21. $\dfrac{7}{x + 1} + \dfrac{4}{2x}$

22. $5 + \dfrac{1}{x} + \dfrac{1}{x + 1}$

23. $\dfrac{2}{x^2 - 9} + \dfrac{x}{x + 3}$

24. $\dfrac{7}{x + 2} + \dfrac{3}{x - 4}$

25. $\dfrac{x}{y} + \dfrac{3}{2y} + \dfrac{1}{y + 2}$

26. $\dfrac{4}{a} + \dfrac{2}{b} + \dfrac{3}{a + b}$

27. $\dfrac{3x + 1}{3x} - \dfrac{1}{x}$

28. $\dfrac{x + 4}{x + 2} - \dfrac{1}{2x}$

29. $\dfrac{1}{x^2 + 7x + 10} - \dfrac{x}{x + 5}$

30. $\dfrac{27}{x^2 - 81} + \dfrac{3}{2(x + 9)}$

6.4 *Simplify.*

31. $\dfrac{\dfrac{3}{2y} - \dfrac{1}{y}}{\dfrac{4}{y} + \dfrac{3}{2y}}$

32. $\dfrac{\dfrac{2}{x} + \dfrac{1}{2x}}{x + \dfrac{x}{2}}$

33. $\dfrac{w - \dfrac{4}{w}}{1 + \dfrac{2}{w}}$

34. $\dfrac{1 - \dfrac{w}{w - 1}}{1 + \dfrac{w}{1 - w}}$

35. $\dfrac{1 + \dfrac{1}{y^2 - 1}}{\dfrac{1}{y + 1} - \dfrac{1}{y - 1}}$

36. $\dfrac{\dfrac{1}{y} + \dfrac{1}{x + y}}{1 + \dfrac{2}{x + y}}$

37. $\dfrac{\dfrac{1}{a + b} - \dfrac{1}{a}}{b}$

38. $\dfrac{\dfrac{2}{a + b} - \dfrac{3}{b}}{\dfrac{1}{a + b}}$

39. $\left(\dfrac{1}{x + 2y} - \dfrac{1}{x - y} \right) \div \dfrac{2x - 4y}{x^2 - 3xy + 2y^2}$

40. $\dfrac{x + 5y}{x - 6y} \div \left(\dfrac{1}{5y} - \dfrac{1}{x + 5y} \right)$

6.5 *Solve for the variable. If there is no solution, say so.*

41. $\dfrac{8}{a - 3} = \dfrac{12}{a + 3}$

42. $\dfrac{8a - 1}{6a + 8} = \dfrac{3}{4}$

43. $\dfrac{2x - 1}{x} - \dfrac{1}{2} = -2$

44. $\dfrac{5 - x}{x} - \dfrac{7}{x} = -\dfrac{3}{4}$

45. $\dfrac{5}{2} - \dfrac{2y + 7}{y + 6} = 3$

46. $\dfrac{5}{4} - \dfrac{1}{2x} = \dfrac{1}{x} + 2$

47. $\dfrac{7}{8x} - \dfrac{3}{4} = \dfrac{1}{4x} + \dfrac{1}{2}$

48. $\dfrac{1}{3x} + 2 = \dfrac{5}{6x} - \dfrac{1}{2}$

49. $\dfrac{3}{y - 3} = \dfrac{3}{2} + \dfrac{y}{y - 3}$

50. $\dfrac{x - 8}{x - 2} = \dfrac{2x}{x + 2} - 2$

51. $\dfrac{3y - 1}{3y} - \dfrac{6}{5y} = \dfrac{1}{y} - \dfrac{4}{15}$

52. $\dfrac{9}{2} - \dfrac{7y - 4}{y + 2} = -\dfrac{1}{4}$

53. $\dfrac{y + 18}{y^2 - 16} = \dfrac{y}{y + 4} - \dfrac{y}{y - 4}$

54. $\dfrac{4}{x^2 - 1} = \dfrac{2}{x - 1} + \dfrac{2}{x + 1}$

55. $\dfrac{9y - 3}{y^2 + 2y} - \dfrac{5}{y + 2} = \dfrac{3}{y}$

56. $\dfrac{2}{3 - 3y} + \dfrac{2}{2y - 1} = \dfrac{4}{3y - 3}$

6.6 *Solve. Round to the nearest tenth.*

57. $\dfrac{8}{5} = \dfrac{2}{x}$

58. $\dfrac{x}{4} = \dfrac{12}{17}$

59. $\dfrac{33}{10} = \dfrac{x}{8}$

60. $\dfrac{5}{x} = \dfrac{22}{9}$

61. $\dfrac{13.5}{0.6} = \dfrac{360}{x}$

62. $\dfrac{2\frac{1}{2}}{3\frac{1}{4}} = \dfrac{7}{x}$

Use a proportion to answer each question.

63. A 5-gallon can of paint will cover 240 square feet. How many gallons of paint will be needed to cover 400 square feet? Round to the nearest tenth of a gallon.

64. Aunt Lexie uses 3 pounds of sugar to make 100 cookies. How many cookies can she make with 5 pounds of sugar? Round to the nearest whole cookie.

65. Ron found that his car used 7 gallons of gas to travel 200 miles. He plans to drive 1300 miles from his home to Denver, Colorado. How many gallons of gas will he use if his car continues to consume gas at the same rate? Round to the nearest gallon.

66. The distance on a map between two cities is 4 inches. The actual distance between these cities is 122 miles. Two lakes are 3 inches apart on the same map. How many miles apart are these two lakes?

67. A train travels 180 miles in the same time that a car travels 120 miles. The speed of the train is 20 miles per hour faster than the speed of the car. Find the speed of the train and the speed of the car.

68. A professional painter can paint the interior of the Jacksons' house in 5 hours. John Jackson can do the same job in 8 hours. How long would it take these two people working together on the painting job? Round to the nearest minute.

▲ 69. A flagpole that is 8 feet tall casts a shadow of 3 feet. At the same time of day, a tall office building in the city casts a shadow of 450 feet. How tall is the office building?

▲ 70. Mary takes a walk across a canyon in New Mexico. She stands 5.75 feet tall and her shadow is 3 feet long. At the same time, the shadow from the peak of the canyon wall casts a shadow that is 95 feet long. How tall is the peak of the canyon? Round to the nearest foot.

71. Fred is an experienced painter. He can paint the sides of an average house in 5 hours. His new assistant is still being trained. It takes the assistant 10 hours to paint the sides of an average house. How long would it take Fred and his assistant to paint the sides of an average house if they worked together?

72. Sally runs the family farm in Boone, Iowa. She can plow the fields of the farm in 20 hours. Her daughter Brenda can plow the fields of the farm in 30 hours. If they have two identical tractors, how long would it take Brenda and Sally to plow the fields of the farm if they worked together?

Perform the operation indicated. Simplify.

1. $\dfrac{2ac + 2ad}{3a^2c + 3a^2d}$

2. $\dfrac{8x^2 - 2x^2y^2}{y^2 + 4y + 4}$

3. $\dfrac{x^2 + 2x}{2x - 1} \cdot \dfrac{10x^2 - 5x}{12x^3 + 24x^2}$

4. $\dfrac{x + 2y}{12y^2} \cdot \dfrac{4y}{x^2 + xy - 2y^2}$

5. $\dfrac{2a^2 - 3a - 2}{4a^2 + a - 14} \div \dfrac{2a^2 + 5a + 2}{16a^2 - 49}$

6. $\dfrac{3}{x^2 + x - 6} + \dfrac{1}{x^2 + 3x - 10}$

7. $\dfrac{x - y}{xy} - \dfrac{a - y}{ay}$

8. $\dfrac{3x}{x^2 - 3x - 18} - \dfrac{x - 4}{x - 6}$

9. $\dfrac{\dfrac{x}{3y} - \dfrac{1}{2}}{\dfrac{4}{3y} - \dfrac{2}{x}}$

10. $\dfrac{\dfrac{2}{x + 3} + \dfrac{1}{x}}{3x + 9}$

11. $\dfrac{2x^2 + 3xy - 9y^2}{4x^2 + 13xy + 3y^2}$

12. $\dfrac{1}{x + 4} - \dfrac{2}{x^2 + 6x + 8}$

In questions 13–18, solve for x. Check your answers. If there is no solution, say so.

13. $\dfrac{15}{x} + \dfrac{9x - 7}{x + 2} = 9$

14. $\dfrac{x - 3}{x - 2} = \dfrac{2x^2 - 15}{x^2 + x - 6} - \dfrac{x + 1}{x + 3}$

15. $3 - \dfrac{7}{x + 3} = \dfrac{x - 4}{x + 3}$

16. $\dfrac{3}{3x - 5} = \dfrac{7}{5x + 4}$

17. $\dfrac{9}{x} = \dfrac{13}{5}$

18. $\dfrac{9.3}{2.5} = \dfrac{x}{10}$

19. Katie is typing a term paper for her English class. She typed 3 pages of the paper in 55 minutes. If she continues at the same rate, how long will it take her to type the entire 21-page paper? Express your answer in hours and minutes.

20. In northern Michigan the Gunderson family heats their home with firewood. They used $100 worth of wood in 25 days. Mr. Gunderson estimates that he needs to burn wood at that rate for about 92 days during the winter. If that is so, how much will the 92-day supply of wood cost?

21. A hiking club is trying to construct a rope bridge across a canyon. A 6-foot construction pole held upright casts a 7-foot shadow. At the same time of day, a tree at the edge of the canyon casts a shadow that exactly covers the distance that is needed for the rope bridge. The tree is exactly 87 feet tall. How long should the rope bridge be? Round to the nearest foot.

1. _____

2. _____

3. _____

4. _____

5. _____

6. _____

7. _____

8. _____

9. _____

10. _____

11. _____

12. _____

13. _____

14. _____

15. _____

16. _____

17. _____

18. _____

19. _____

20. _____

21. _____

1. _____

2. _____

3. _____

4. _____

5. _____

6. _____

7. _____

8. _____

9. _____

10. _____

11. _____

12. _____

13. _____

14. _____

15. _____

16. _____

17. _____

18. _____

19. _____

20. _____

21. _____

22. _____

Approximately one-half of this test covers the content of Chapters 0–5. The remainder covers the content of Chapter 6.

1. To the nearest thousandth, how much error can be tolerated in the length of a wire that is supposed to be 2.57 centimeters long if specifications allow an error of no more than 0.25%?

2. Fifteen percent of what number is $56.25?

3. Carla is thinking of buying a new car for $18,500. Her state presently has a sales tax of 5.5%. In a few weeks the state will raise the sales tax to 6%. How much will she save in sales tax if she purchases the car before the sales tax rate is raised?

4. Solve. $5(x - 3) - 2(4 - 2x) = 7(x - 1) - (x - 2)$

5. Solve for h. $A = \pi r^2 h$

6. Solve and graph on a number line. $4(2 - x) < 3$

7. Solve for x. $\dfrac{1}{4}x + \dfrac{3}{4} < \dfrac{2}{3}x - \dfrac{4}{3}$

8. Factor. $3ax + 3bx - 2ay - 2by$

9. Factor. $8a^3 - 38a^2b - 10ab^2$

10. Simplify. $(-3x^2y)(4x^3y^6)$

11. Simplify. $\dfrac{4x^2 - 25}{2x^2 + 9x - 35}$

Perform the indicated operations.

12. $\dfrac{x^2 - 4}{x^2 - 25} \cdot \dfrac{3x^2 - 14x - 5}{3x^2 + 6x}$

13. $\dfrac{2x^2 - 9x + 9}{8x - 12} \div \dfrac{x^2 - 3x}{2x}$

14. $\dfrac{5}{2x + 4} + \dfrac{3}{x - 3}$

15. $\dfrac{2x + 1}{x^2 + x - 12} - \dfrac{x - 1}{x^2 - x - 6}$

Solve for x.

16. $\dfrac{3x - 2}{3x + 2} = 2$

17. $\dfrac{x - 3}{x} = \dfrac{x + 2}{x + 3}$

In questions 18 and 19, simplify.

18. $\dfrac{\dfrac{1}{x - 3} + \dfrac{5}{x^2 - 9}}{\dfrac{6x}{x + 3}}$

19. $\dfrac{\dfrac{3}{a} + \dfrac{2}{b}}{\dfrac{5}{a^2} - \dfrac{2}{b^2}}$

20. Solve for x. $\dfrac{5}{7} = \dfrac{14}{x}$

21. Jane is looking at a road map. The distance between two cities is 130 miles. This distance is represented by $2\frac{1}{2}$ inches on the map. She sees that the distance she has to drive today is a totally straight interstate highway that is 4 inches long on the map. How many miles does she have to drive?

22. Roberto is working as a telemarketing salesperson for a large corporation. In the last 22 telephone calls he has made, he was able to make a sale five times. His goal is to make 110 sales this month. If his rate for making a sale continues as in the past, how many phone calls must he make?

Graphing and Functions

A rainforest is one of the most amazing ecosystems in the entire world. It is a huge region with its own climate, an amazing variety of animals and plant life, and home to many species that do not exist anywhere else in the world. However, all is not well. The amount of land covered by rainforests is decreasing. Can you predict how rapidly this decrease is taking place? Turn to the Putting Your Skills to Work problems on page 461 to find out.

1. _____

2. _____

3. (a) _____

(b) _____

4. _____

5. _____

6. _____

7. _____

8. (a) _____

(b) _____

If you are familiar with the topics in this chapter, take this test now. Check your answers with those in the back of the book. If an answer is wrong or you can't answer a question, study the appropriate section of the chapter.

If you are not familiar with the topics in this chapter, don't take this test now. Instead, study the examples, work the Practice Problems, and then take the test.

This test will help you to identify which concepts you have mastered and which you need to study further.

Section 7.1

1. Graph the points.
$(-3, 1), (4, -4), (-2, -6)$

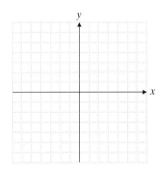

2. Write the coordinates of each point.

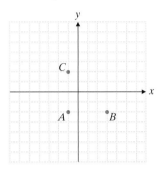

3. Complete each ordered pair so that it is a solution to the equation $y = -3x + 7$.
(a) $(-1, \quad)$ **(b)** $(\quad, 1)$

Section 7.2

4. Graph $5x - 2y = -10$.

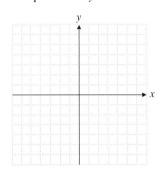

5. Graph $y = 4$.

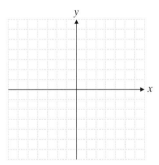

Sections 7.3 and 7.4

6. What is the slope of the line passing through $(4, -2)$ and $(-4, 6)$?

7. Find the slope of the line $4x - 3y - 7 = 0$.

8. Find the equation of the line with slope 4 and y-intercept $(0, -5)$.
(a) Write the equation in slope–intercept form.
(b) Write the equation in the form $Ax + By = C$.

9. Graph the equation $y = 2x + 1$.

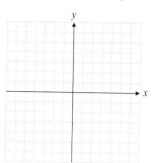

10. Write the equation of the graph.

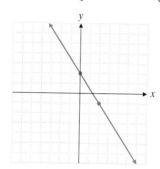

11. Line p has a slope of $\frac{3}{5}$.
   (a) What is the slope of a line parallel to line p?
   (b) What is the slope of a line perpendicular to line p?

12. Find the equation of a line with slope $\frac{2}{3}$ that passes through the point $(-3, -3)$.

13. Find the equation of a line that passes through $(-1, 6)$ and $(2, 3)$.

Section 7.5

14. Graph $y \geq 3x - 2$.

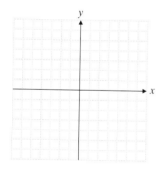

15. Graph $4x + 2y < -12$.

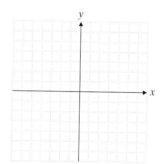

Section 7.6

Determine whether each of the following is a function.

16. Circle with radius r and circumference C

r	1	5	10	15
C	6.28	31.4	62.8	94.2

17.

Find the indicated values for the function.

18. $f(x) = 3x - 7$
   (a) $f(-2)$ **(b)** $f(7)$

19. $g(x) = 2x^2$
   (a) $g(-6)$ **(b)** $g\left(\frac{1}{2}\right)$

9. _____

10. _____

11. (a) _____

(b) _____

12. _____

13. _____

14. _____

15. _____

16. _____

17. _____

18. (a) _____

(b) _____

19. (a) _____

(b) _____

1 Plotting a Point, Given the Coordinates

Oftentimes we can better understand an idea if we see a picture. This is the case with many mathematical concepts, including those relating to algebra. We can illustrate algebraic relationships with drawings called **graphs**. Before we can draw a graph, however, we need a frame of reference.

In Chapter 1 we showed that any real number can be represented on a number line. Look at the following number line. The arrow indicates the positive direction.

To form a **rectangular coordinate system**, we draw a second number line vertically. We construct it so that the 0 point on each number line is exactly at the same place. We refer to this location as the **origin**. The horizontal number line is often called the *x*-**axis**. The vertical number line is often called the *y*-**axis**. Arrows show the positive direction for each axis.

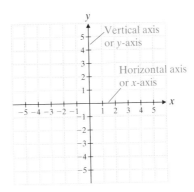

We can represent a point in this rectangular coordinate system by using an **ordered pair** of numbers. For example, $(5, 2)$ is an ordered pair that represents a point in the rectangular coordinate system. The numbers in an ordered pair are often referred to as the **coordinates** of the point. The first number is called the *x*-**coordinate** and it represents the distance from the origin measured along the horizontal or *x*-axis. If the *x*-coordinate is positive, we count the proper number of squares to the right (that is, in the positive direction). If the *x*-coordinate is negative, we count to the left. The second number in the pair is called the *y*-**coordinate** and it represents the distance from the origin measured along the *y*-axis. If the *y*-coordinate is positive, we count the proper number of squares upward (that is, in the positive direction). If the *y*-coordinate is negative, we count downward.

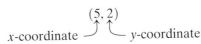

x-coordinate ⎯⎯⎯ y-coordinate

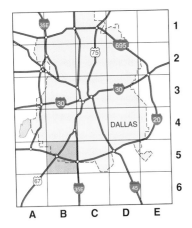

Suppose the directory for the map on the left indicated that you would find a certain street in the region B5. To find the street you would first scan across the horizontal scale until you found section B; from there you would scan up the map until you hit section 5 along the vertical scale. As we will see in the next example, plotting a point in the rectangular coordinate system is much like finding a street on a map with grids.

EXAMPLE 1 Plot the point $(5, 2)$ on a rectangular coordinate system. Label this as point A.

Since the *x*-coordinate is 5, we first count 5 units to the right on the *x*-axis. Then, because the *y*-coordinate is 2, we count 2 units up from the point where we stopped on the *x*-axis. This locates the point corresponding to $(5, 2)$. We mark this point with a dot and label it A.

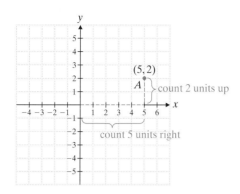

Practice Problem 1   Plot the point $(3, 4)$ on the preceding coordinate system. Label this as point B.

It is important to remember that the first number in an ordered pair is the *x*-coordinate and the second number is the *y*-coordinate. The ordered pairs $(5, 2)$ and $(2, 5)$ represent different points.

EXAMPLE 2   Plot each point on the following coordinate system. Label the points F, G, and H, respectively.

(a) $(-5, 3)$ **(b)** $(2, -6)$ **(c)** $(-4, -5)$

(a) Notice that the *x*-coordinate, -5, is negative. On the coordinate grid, negative *x*-values appear to the left of the origin. Thus, we will begin by counting 5 squares to the left starting at the origin. Since the *y*-coordinate, 3, is positive, we will count 3 units up from the point where we stopped on the *x*-axis.

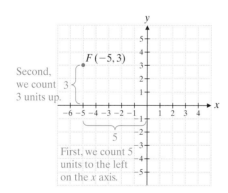

(b) The *x*-coordinate is positive. Begin by counting 2 squares to the right of the origin. Then count down because the *y*-coordinate is negative.

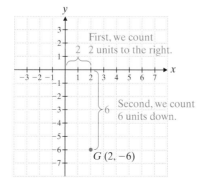

(c) The *x*-coordinate is negative. Begin by counting 4 squares to the left of the origin. Then count down because the *y*-coordinate is negative.

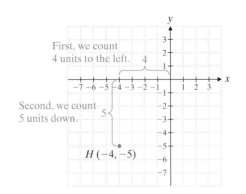

PRACTICE PROBLEM 2

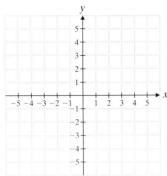

Practice Problem 2 Use the coordinate system in the margin to plot each point. Label the points *I*, *J*, and *K*, respectively.

(a) $(-2, -4)$ (b) $(-4, 5)$ (c) $(4, -2)$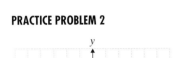

EXAMPLE 3 Plot the following points.

$F: (0, 5)$ $G: \left(3, \frac{3}{2}\right)$ $H: (-6, 4)$ $I: (-3, -4)$
$J: (-4, 0)$ $K: (2, -3)$ $L: (6.5, -7.2)$

These points are plotted in the figure.

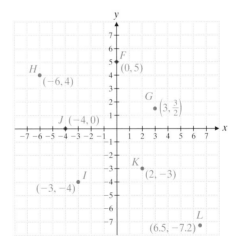

PRACTICE PROBLEM 3

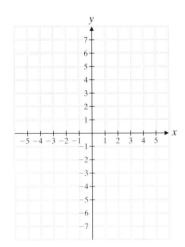

Note: When you are plotting decimal values like $(6.5, -7.2)$, plot the point halfway between 6 and 7 in the *x*-direction (for the 6.5) and at your best approximation of -7.2 in the *y*-direction.

Practice Problem 3 Plot the following points. Label each point with both the letter and the ordered pair. Use the coordinate system provided in the margin.
$A: (3, 7)$ $B: (0, -6)$ $C: (3, -4.5)$ $D: \left(-\frac{7}{2}, 2\right)$

2 *Determining the Coordinates of a Plotted Point*

Sometimes we need to find the coordinates of a point that has been plotted. First, we count the units we need on the *x*-axis to get as close as possible to the point. Next we count the units up or down we need to go from the *x*-axis to reach the point.

EXAMPLE 4   What ordered pair of numbers represents point A in the graph below?

If we move along the x-axis until we get as close as possible to A, we end up at the number 5. Thus we obtain 5 as the first number of the ordered pair. Then we count 4 units upward on a line parallel to the y-axis to reach A. So we obtain 4 as the second number of the ordered pair. Thus point A is represented by the ordered pair $(5, 4)$.

Practice Problem 4   What ordered pair of numbers represents point B in the graph below?

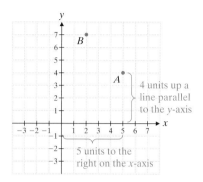

EXAMPLE 5   Write the coordinates of each point plotted in the following graph.

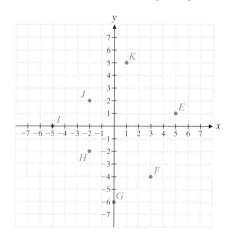

The coordinates of each point are as follows.

$E = (5, 1)$ $I = (-5, 0)$
$F = (3, -4)$ $J = (-2, 2)$
$G = (0, -6)$ $K = (1, 5)$
$H = (-2, -2)$

Be very careful that you put the x-coordinate first and the y-coordinate second. Be careful that each sign is correct.

PRACTICE PROBLEM 5

Practice Problem 5   Give the coordinates of each point plotted in the graph in the margin.

3   *Finding Ordered Pairs for a Given Linear Equation*

Equations such as $3x + 2y = 5$ and $6x + y = 3$ are called linear equations in two variables.

A **linear equation in two variables** is an equation that can be written in the form $Ax + By = C$ where A, B, and C are real numbers but A and B are not *both* zero.

Replacement values for x and y that make *true mathematical statements* of the equation are called *truth values*, and an ordered pair of these truth values is called a **solution**.

Consider the equation $3x + 2y = 5$. The ordered pair $(1, 1)$ is a solution to the equation because when we replace x by 1 and y by 1 in the equation, we obtain a true statement.

$$3(1) + 2(1) = 5 \quad \text{or} \quad 3 + 2 = 5$$

Likewise $(-1, 4)$, $(3, -2)$, $(5, -5)$, and $(7, -8)$ are also solutions to the equation. In fact, there are an infinite number of solutions for any given linear equation in two variables.

If one value of an ordered-pair solution to a linear equation is known, the other can be quickly obtained. To do so, we replace the proper variable in the equation by the known value. Then using the methods learned in Chapter 2, we solve the resulting equation for the other variable.

EXAMPLE 6 Find the missing coordinate to complete the following ordered-pair solutions for the equation $2x + 3y = 15$.

(a) $(0, ?)$ **(b)** $(?, 1)$

(a) For the ordered pair $(0, ?)$ we know that $x = 0$. Replace x by 0 in the equation.

$$2x + 3y = 15$$
$$2(0) + 3y = 15$$
$$0 + 3y = 15$$
$$y = 5$$

Thus we have the ordered pair $(0, 5)$.

(b) For the ordered pair $(?, 1)$, we *do not know* the value of x. However, we do know that $y = 1$. So we start by replacing the variable y by 1. We will end up with an equation with one variable, x. We can then solve for x.

$$2x + 3y = 15$$
$$2x + 3(1) = 15$$
$$2x + 3 = 15$$
$$2x = 12$$
$$x = 6$$

Thus we have the ordered pair $(6, 1)$.

Practice Problem 6 Find the missing coordinate to complete the following ordered-pair solutions for the equation $3x - 4y = 12$.

(a) $(0, ?)$ **(b)** $(?, 3)$ **(c)** $(?, -6)$

The linear equations that we work with are not always written in the form $Ax + By = C$, but are sometimes solved for y, as in $y = -6x + 3$. Consider the equation $y = -6x + 3$. The ordered pair $(2, -9)$ is a solution to the equation. When we replace x by 2 and y by -9 we obtain a true mathematical statement:

$$(-9) = -6(2) + 3 \quad \text{or} \quad -9 = -12 + 3.$$

In examining data from real-world situations, we often find that plotting data points shows useful trends. In such cases, it is often necessary to use a different scale, one that displays only positive values.

EXAMPLE 7   The number of motor vehicle accidents in millions is recorded in the following table for the years 1980 to 2000.

(a) Plot points that represent this data on the given coordinate system.

(b) What trends are apparent from the plotted data?

Number of Years Since 1980	Number of Motor Vehicle Accidents (in Millions)
0	18
5	19
10	12
15	11
20	15

Source: U.S. National Highway Traffic Safety Administration

(a)

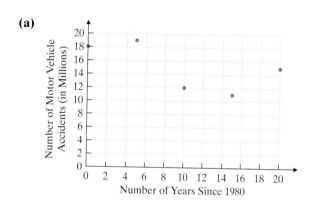

(b) From 1980 to 1985, there was a slight increase in the number of accidents. From 1985 to 1995, there was a significant decrease in the number of accidents. From 1995 to 2000, there was a moderate increase in the number of accidents.

Practice Problem 7   The number of motor vehicle deaths in thousands is recorded in the following table for the years 1980 to 2000.

(a) Plot points that represent this data on the given coordinate system.

(b) What trends are apparent from the plotted data?

Number of Years Since 1980	Number of Motor Vehicle Deaths (in Thousands)
0	51
5	44
10	45
15	42
20	43

Source: U.S. National Highway Traffic Safety Administration

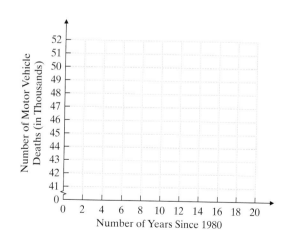

1. Plot the following points.

$J: (-4, 3.5)$ $K: (6, 0)$

$L: (5, -6)$ $M: (0, -4)$

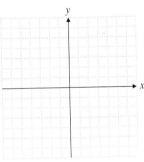

2. Plot the following points.

$R: (-3, 0)$ $S: (3.5, 4)$

$T: (-2, -2.5)$ $V: (0, 5)$

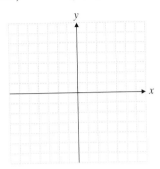

Consider the points plotted in the graph at right.

3. Give the coordinates for points R, S, X, and Y.

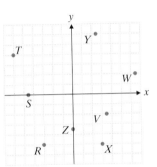

4. Give the coordinates for points T, V, W, and Z.

Verbal and Writing Skills

5. What is the x-coordinate of the origin?

6. What is the y-coordinate of the origin?

7. Explain why $(5, 1)$ is referred to as an *ordered* pair of numbers.

In exercises 8 and 9, 10 points are plotted in the figure. List all the ordered pairs needed to represent the points.

8.

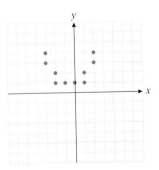

9.

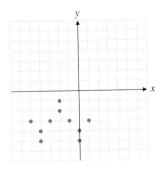

Complete the ordered pairs so that each is a solution to the given linear equation.

10. $y = 2x + 5$
    (a) $(0, \quad)$ **(b)** $(3, \quad)$

11. $y = 3x + 8$
    (a) $(0, \quad)$ **(b)** $(4, \quad)$

12. $y = -4x + 2$
    (a) $(-5, \quad)$ **(b)** $(4, \quad)$

13. $y = -2x + 3$
    (a) $(-6, \quad)$ **(b)** $(3, \quad)$

14. $3x - 4y = 11$
    (a) $(-3, \quad)$ **(b)** $(\quad, 1)$

15. $5x - 2y = 9$
    (a) $(7, \quad)$ **(b)** $(\quad, -7)$

16. $2y + 3x = -6$
    (a) $(-2, \quad)$ **(b)** $(\quad, 3)$

17. $-4x + 5y = -20$
    (a) $(10, \quad)$ **(b)** $(\quad, -8)$

18. $3x + \dfrac{1}{2}y = 2$
    (a) $(\quad, 16)$ **(b)** $\left(\dfrac{5}{2}, \quad\right)$

19. $4x + \dfrac{1}{3}y = 8$ **(a)** $(\quad, -12)$ **(b)** $\left(\dfrac{3}{2}, \quad\right)$

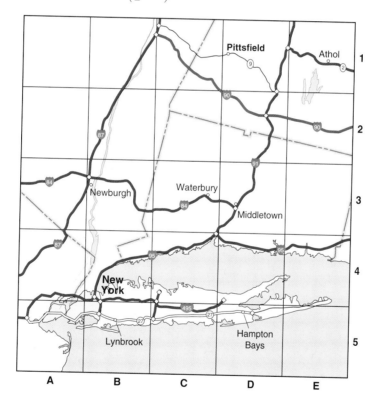

The preceding map shows a portion of New York, Connecticut, and Massachusetts. Like many maps used in driving or flying, it has horizontal and vertical grid markers for ease of use. For example, Newburgh, New York, is located in grid B3. Use the grid labels to indicate the locations of the following cities.

20. Lynbrook, New York

21. Hampton Bays, New York

22. Athol, Massachusetts

23. Pittsfield, Massachusetts

24. Middletown, Connecticut

25. Waterbury, Connecticut

26. The number of farms in the United States for selected years starting in 1940 is recorded in the following table. The number of farms is measured in hundred thousands. For example, 61 in the second column means 61 hundred thousand or 6,100,000.

(a) Plot points that represent this data on the given rectangular coordinate system.

(b) What trends are apparent from the plotted data?

Number of Years Since 1940	Number of Farms (in Hundred Thousands)
0	61
10	54
20	40
30	30
40	24
50	21
60	22

Source: U.S. Department of Agriculture

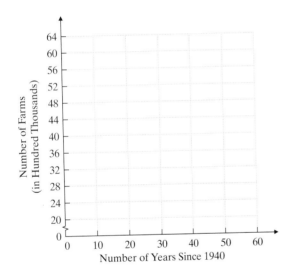

27. The number of barrels of oil produced in the U.S. for selected years starting in 1940 is recorded in the following table. The number of barrels of oil is measured in hundred thousands. For example, 14 in the second column means 14 hundred thousand, or 1,400,000.

(a) Plot points on the given rectangular coordinate system that represent this data.

(b) What trends are apparent from the plotted data?

Number of Years Since 1940	Number of Barrels of Crude Oil (in Hundred Thousands)
0	14
10	20
20	26
30	35
40	31
50	27
60	21

Source: U.S. Department of Energy

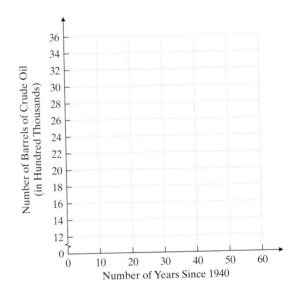

Cumulative Review Problems

28. Simplify. $\dfrac{8x}{x + y} + \dfrac{8y}{x + y}$

29. Simplify. $\dfrac{x^2 - y^2}{x^2 + 4xy + 3y^2} \cdot \dfrac{x^2 - 9y^2}{x^2 - 6xy + 9y^2}$

▲ **30.** The circular pool at the hotel where Bob and Linda stayed in Orlando, Florida, has a radius of 19 yards. What is the area of the pool? (Use $\pi \approx 3.14$).

31. The membership at David's gym has increased by 16% over the last 2 years. It used to have 795 members. How many members does it have now?

32. A major Russian newspaper called *Izvestia* had a circulation of 6,109,005 before the dissipation of the Soviet Union. Currently the circulation is 262,045. What is the percent of decrease of readership of *Izvestia*?

▲ **33.** An expensive Persian rug that measures 30 feet by 22 feet is priced at \$44,020.00. A customer negotiates with the owner of the rug, and they agree upon a price of \$36,300.
   (a) What is the cost per square foot of the rug at the discounted price?
   (b) The rug dealer has a smaller matching Persian rug measuring 14 feet by 8 feet. He says he will sell it to the customer at the same cost per square foot as the larger one. How much will the small rug cost?

Student Learning Objectives

After studying this section, you will be able to:

 Graph a linear equation by plotting three ordered pairs.

 Graph a straight line by plotting its intercepts.

 Graph horizontal and vertical lines.

SSM

PH TUTOR   CD & VIDEO   MATH PRO   WEB
CENTER

1 Graphing a Linear Equation by Plotting Three Ordered Pairs

We have seen that a solution to a linear equation in two variables is an ordered pair. The graph of an ordered pair is a point. Thus we can graph an equation by graphing the points corresponding to its ordered-pair solutions.

A linear equation in two variables has an infinite number of ordered-pair solutions. We can see that this is true by noting that we can substitute any number for x in the equation and solve it to obtain a y-value. For example, if we substitute $x = 0$, 1, 2, 3, ... into the equation $y = -x + 3$ and solve for y, we obtain the ordered-pair solutions $(0, 3), (1, 2), (2, 1), (3, 0), \ldots$. (Substitute these values into the equation to convince yourself.) If we plot these points on a rectangular coordinate system, we notice that they form a straight line.

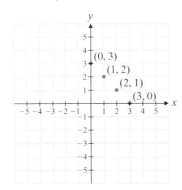   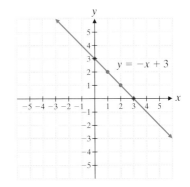

It turns out that all of the points corresponding to the ordered-pair solutions of $y = -x + 3$ lie on this line, and the line extends forever in both directions. A similar statement can be made about any linear equation in two variables.

The graph of any linear equation in two variables is a straight line.

From geometry we know that two points determine a line. Thus to graph a linear equation in two variables, we need only graph two ordered-pair solutions of the equation and then draw the line that passes through them. Having said this, we recommend that you use three points to graph a line. Two points will determine where the line is. The third point verifies that you have drawn the line correctly. For ease in plotting, it is better if the ordered pairs contain integers.

To Graph a Linear Equation

1. Look for three ordered pairs that are solutions to the equation.
2. Plot the points.
3. Draw a line through the points.

EXAMPLE 1 Find three ordered pairs that satisfy $2x + y = 4$. Then graph the resulting straight line.

Since we can choose any value for x, choose numbers that are convenient. To organize the results, we will make a table of values. We will let $x = 0$, $x = 1$, and $x = 3$, respectively. We write these numbers under x in our table of values. For each of these x-values, we find the corresponding y-value in the equation $2x + y = 4$.

Table of Values	
x	y
0	4
1	2
3	−2

$$2x + y = 4 \qquad\qquad 2x + y = 4 \qquad\qquad 2x + y = 4$$
$$2(0) + y = 4 \qquad\qquad 2(1) + y = 4 \qquad\qquad 2(3) + y = 4$$

$$0 + y = 4 \qquad\qquad 2 + y = 4 \qquad\qquad 6 + y = 4$$
$$y = 4 \qquad\qquad\quad y = 2 \qquad\qquad\quad y = -2$$

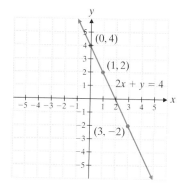

We record these results by placing each y-value in the table next to its corresponding x-value. Keep in mind that these values represent ordered pairs, each of which is a solution to the equation. To make calculating and graphing easier, we choose integer values whenever possible. If we plot these ordered pairs and connect the three points, we get a straight line that is the graph of the equation $2x + y = 4$. The graph of the equation is shown in the figure at the right.

Practice Problem 1   Graph $x + y = 10$ on the given coordinate system.

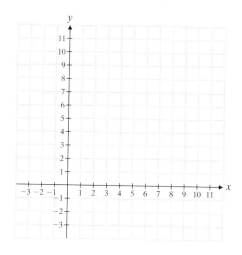

⬭ **EXAMPLE 2**   Graph $5x - 4y + 2 = 2$.

First, we simplify the equation $5x - 4y + 2 = 2$ by adding -2 to each side.

$$5x - 4y + 2 - 2 = 2 - 2$$
$$5x - 4y = 0$$

Since we are free to choose any value of x, $x = 0$ is a natural choice. Calculate the value of y when $x = 0$.

$$5(0) - 4y = 0$$
$$-4y = 0$$
$$y = 0 \quad \text{Remember: Any number times 0 is 0.}$$
$$\qquad\quad \text{Since } -4y = 0, y \text{ must equal 0.}$$

Now let's see what happens when $x = 1$.

$$5(1) - 4y = 0$$
$$5 - 4y = 0$$
$$-4y = -5$$

$$y = \frac{-5}{-4} \quad \text{or} \quad \frac{5}{4} \quad \text{This is not an easy number to graph.}$$

A better choice for a replacement of x is a number that is divisible by 4. Let's see why. Let $x = 4$ and let $x = -4$.

$$5(4) - 4y = 0 \qquad\qquad\qquad 5(-4) - 4y = 0$$
$$20 - 4y = 0 \qquad\qquad\qquad\quad -20 - 4y = 0$$
$$-4y = -20 \qquad\qquad\qquad\quad -4y = 20$$
$$y = \frac{-20}{-4} \quad \text{or} \quad 5 \qquad\qquad y = \frac{20}{-4} \quad \text{or} \quad -5$$

Now we can put these numbers into our table of values and graph the line.

PRACTICE PROBLEM 2

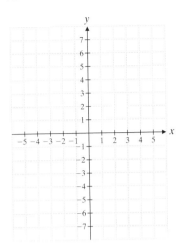

Table of Values	
x	y
0	0
4	5
-4	-5

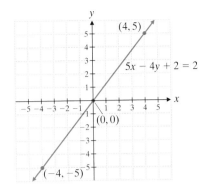

Practice Problem 2 Graph $7x + 3 = -2y + 3$ on the coordinate system in the margin.

To Think About In Example 2, we picked values of x and found the corresponding values for y. An alternative approach is to solve the equation for the variable y *first*.

$$5x - 4y = 0$$
$$-4y = -5x \quad \text{Add } -5x \text{ to each side.}$$
$$\frac{-4y}{-4} = \frac{-5x}{-4} \quad \text{Divide each side by } -4.$$
$$y = \frac{5}{4}x$$

Now let $x = -4$, $x = 0$, and $x = 4$, and find the corresponding values of y. Explain why you would choose multiples of 4 as replacements of x in this equation. Graph the equation and compare it to the graph in Example 2.

In the previous two examples we began by picking values for x. We could just as easily have chosen values for y.

② Graphing a Straight Line by Plotting Its Intercepts

What values should we pick for x and y? Which points should we use for plotting? For many straight lines it is easiest to pick the two *intercepts*. Some lines have only one intercept. We will discuss these separately.

The ***x*-intercept** of a line is the point where the line crosses the *x*-axis; it has the form $(a, 0)$. The ***y*-intercept** of a line is the point where the line crosses the *y*-axis; it has the form $(0, b)$.

Intercept Method of Graphing

To graph an equation using intercepts, we:

1. Find the *x*-intercept by letting $y = 0$ and solving for *x*.

2. Find the *y*-intercept by letting $x = 0$ and solving for *y*.

3. Find one additional ordered pair so that we have three points with which to plot the line.

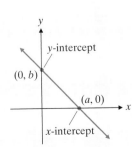

EXAMPLE 3   Use the intercept method to graph $5y - 3x = 15$.

Let $y = 0$. $5(0) - 3x = 15$ Replace y by 0.

$-3x = 15$ Divide both sides by -3.

$x = -5$

The ordered pair $(-5, 0)$ is the x-intercept.

Let $x = 0$. $5y - 3x = 15$

$5y - 3(0) = 15$ Replace x by 0.

$5y = 15$ Divide both sides by 5.

$y = 3$

The ordered pair $(0, 3)$ is the y-intercept.
      We find another ordered pair to have a third point.

Let $y = 6$. $5(6) - 3x = 15$ Replace y by 6.

$30 - 3x = 15$ Simplify.

$-3x = -15$ Subtract 30 from both sides.

$x = \dfrac{-15}{-3}$ or   5

The ordered pair is $(5, 6)$.
      Our table of values is

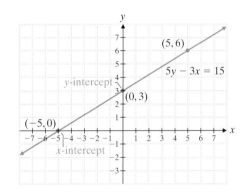

x	y
0	3
-5	0
5	6

Practice Problem 3   Use the intercept method to graph $2y - x = 6$. Use the given coordinate system.

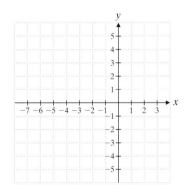

Sidelight

Can you draw all straight lines by the intercept method? Not really. Some straight lines may go through the origin and have only one intercept. If a line goes through the origin, it will have an equation of the form $Ax + By = 0$, where $A \neq 0$ or $B \neq 0$ or both. Examples are $3x + 4y = 0$ and $5x - 2y = 0$. In such cases you should plot two additional points besides the origin. Be sure to simplify each equation before attempting to graph it.

3 *Graphing Horizontal and Vertical Lines*

You will notice that the x-axis is a horizontal line. It is the line $y = 0$, since for any value of x, the value of y is 0. Try a few points. The points $(1, 0)$, $(3, 0)$, and $(-2, 0)$ all lie on the x-axis. Any horizontal line will be parallel to the x-axis. Lines such as $y = 5$ and $y = -2$ are horizontal lines. What does $y = 5$ mean? It means that for any value of x, y is 5. Likewise $y = -2$ means that for any value of x, $y = -2$.

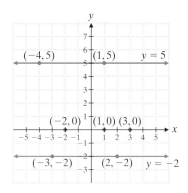

How can we recognize the equation of a line that is horizontal, that is, parallel to the x-axis?

If the graph of an equation is a straight line that is parallel to the x-axis (that is, a horizontal line), the equation will be of the form $y = b$, where b is some real number.

◗ EXAMPLE 4   Graph $y = -3$.

You could write the equation as $0x + y = -3$. Then it is clear that for any value of x that you substitute, you will always obtain $y = -3$. Thus, as shown in the figure, $(4, -3)$, $(0, -3)$, and $(-3, -3)$ are all ordered pairs that satisfy the equation $y = -3$. Since the y-coordinate of every point on this line is -3, it is easy to see that the horizontal line will be 3 units below the x-axis.

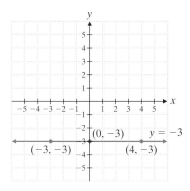

Practice Problem 4   Graph $2y - 3 = 0$ on the given coordinate system.

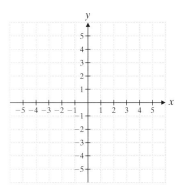

Notice that the y-axis is a vertical line. This is the line $x = 0$, since for any y, x is 0. Try a few points. The points $(0, 2)$, $(0, -3)$, and $\left(0, \frac{1}{2}\right)$ all lie on the y-axis. Any vertical line will be parallel to the y-axis. Lines such as $x = 2$ and $x = -3$ are vertical lines. Think of what $x = 2$ means. It means that for any value of y, x is 2. The graph of $x = 2$ is a vertical line two units to the right of the y-axis.

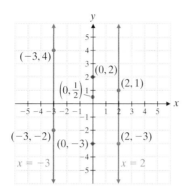

How can we recognize the equation of a line that is vertical, that is, parallel to the y-axis?

If the graph of an equation is a straight line that is parallel to the y-axis (that is, a vertical line), the equation will be of the form $x = a$, where a is some real number.

EXAMPLE 5   Graph $x = 5$.

This can be done immediately by drawing a vertical line 5 units to the right of the origin. The x-coordinate of every point on this line is 5.

The equation $x - 5 = 0$ can be rewritten as $x = 5$ and graphed as shown.

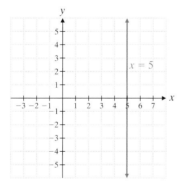

Practice Problem 5   Graph $x + 3 = 0$ on the following coordinate system.

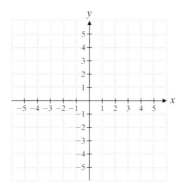

Verbal and Writing Skills

1. Is the point $(-2, 5)$ a solution to the equation $2x + 5y = 0$? Why or why not?

2. The graph of a linear equation in two variables is a _____ _____ .

3. The x-intercept of a line is the point where the line crosses the _____ .

4. The graph of the equation $y = b$ is a _____ line.

Complete the ordered pairs so that each is a solution of the given linear equation. Then plot each solution and graph the equation by connecting the points by a straight line.

5. $y = -2x + 1$
$(0, \ )$
$(-2, \ )$
$(1, \ )$

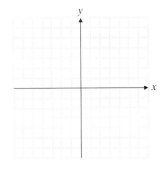

6. $y = -3x - 4$
$(-2, \ )$
$(-1, \ )$
$(0, \ )$

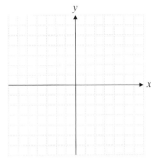

7. $y = x - 4$
$(0, \ )$
$(2, \ )$
$(4, \ )$

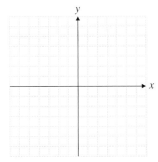

Graph each equation by plotting three points and connecting them.

8. $y = -2x + 3$

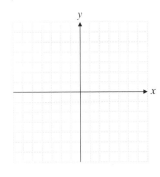

9. $y = 2x - 5$

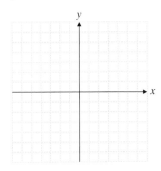

10. $y = 3x + 2$

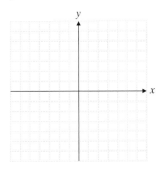

11. $y = 2x + 1$

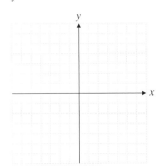

12. $3x - 2y = 0$

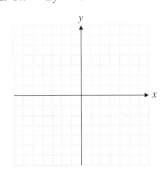

13. $2y - 5x = 0$

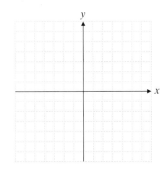

Graph each equation by plotting the intercepts and one other point.

14. $y = -\frac{1}{2}x + 3$

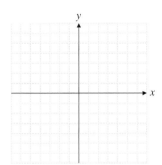

15. $y = -\frac{2}{3}x + 2$

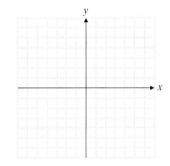

16. $4x + 3y = 12$

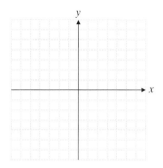

17. $y = 6 - 2x$

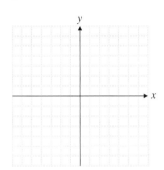

18. $y = 4 - 2x$

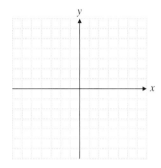

19. $x + 3 = 6y$

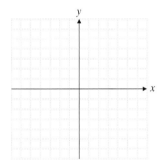

20. $x - 6 = 2y$

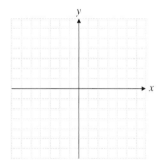

Graph the equation. Be sure to simplify the equation before graphing it.

21. $3x + 2y = 6$

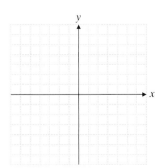

22. $y - 2 = 3y$

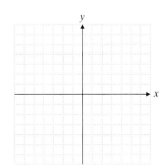

23. $2x + 5y - 2 = -12$

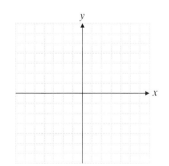

24. $3x - 4y - 5 = -17$ **25.** $3y + 1 = 7$ **26.** $3x - 4 = -13$

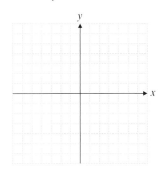

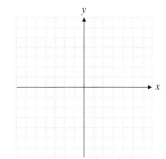

Applications

27. The number of calories burned by an average person while cross-country skiing is given by the equation $C = 8m$, where m is the number of minutes. (*Source:* National Center for Health Statistics.) Find the values of C for $m = 0, 15, 30, 45, 60,$ and 75. Graph your result.

Calories Burned While Cross Country Skiing

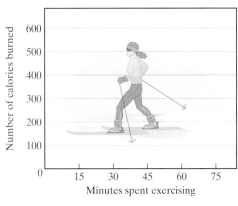

28. The number of calories burned by an average person while jogging is given by the equation $C = \dfrac{28}{3}m$, where m is the number of minutes. (*Source:* National Center for Health Statistics.) Find the values of C for $m = 0, 15, 30, 45, 60,$ and 75. Graph your result.

Calories Burned While Jogging

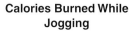
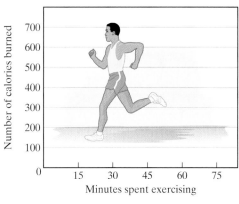

29. The maximum amount of carbon dioxide in the air in Mauna Loa, Hawaii, is approximated by the equation $C = 1.6t + 343$, where C is the concentration of CO_2 measured in parts per million and t is the number of years since 1982. (*Source: Environmental Science: The Way the World Works*, Seventh Edition, B. Nebel and R. Wright, Upper Saddle River, NJ, Prentice-Hall.) Graph the equation for $t = 0, 4, 8$ and 18.

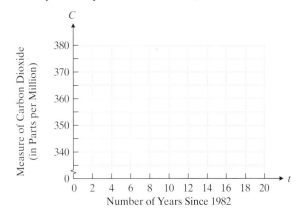

30. The approximate population of the United States is given by the equation $P = 2.3t + 179$ where P is the population in millions and t is the number of years since 1960. (*Source:* Bureau of the Census, U.S. Department of Commerce.) Graph the equation for $t = 0, 10, 30, 40.$

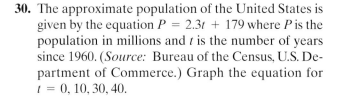
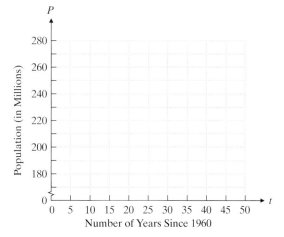

Cumulative Review Problems

31. Solve. $\dfrac{x}{x+4} + \dfrac{x-1}{x-4} = 2$

32. Solve and graph on a number line. $4 - 3x \leq 18$

▲ **33.** A rectangle's width is 1 meter more than half of its length. Find the dimensions if the perimeter measures 53 meters.

34. Last semester there were 29 students for every 2 faculty members at Skyline University. At that time the school had 3074 students. How many faculty members were there last semester?

35. Hank is working for six months in Japan while his wife is back in California with their two children. Each of them sends the other a letter each week. The cost of mailing a letter from Japan and the cost of mailing a letter from the United States totals $1.84. The cost of mailing a letter from the United States is two cents less than half the cost of mailing a letter from Japan. What is the cost of mailing a letter from Japan?

36. Each day Melinda has a cup of instant coffee at home in the morning and a cup of percolated coffee at the office in the afternoon. She consumes 194 milligrams of caffeine each day with these two cups. If the instant coffee has seventeen more milligrams than half the number of milligrams of caffeine in the percolated coffee, how many milligrams of caffeine does a cup of the instant coffee contain?

7.3 Slope of a Line

Student Learning Objectives

After studying this section, you will be able to:

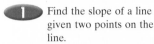 Find the slope of a line given two points on the line.

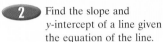 Find the slope and *y*-intercept of a line given the equation of the line.

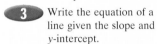 Write the equation of a line given the slope and *y*-intercept.

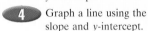 Graph a line using the slope and *y*-intercept.

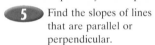 Find the slopes of lines that are parallel or perpendicular.

SSM
PH TUTOR CENTER   CD & VIDEO   MATH PRO   WEB

Finding the Slope of a Line Given Two Points on the Line

We often use the word *slope* to describe the incline (the steepness) of a hill. A carpenter or a builder will refer to the *pitch* or *slope* of a roof. The slope is the change in the vertical distance compared to the change in the horizontal distance as you go from one point to another point along the roof. If the change in the vertical distance is greater than the change in the horizontal distance, the slope will be steep. If the change in the horizontal distance is greater than the change in the vertical distance, the slope will be gentle.

In a coordinate plane, the **slope** of a straight line is defined by the change in *y* divided by the change in *x*.

$$\text{Slope} = \frac{\text{change in } y}{\text{change in } x}$$

Consider the line drawn through points *A* and *B* in the figure. If we measure the change from point *A* to point *B* in the *x*-direction and the *y*-direction, we will have an idea of the steepness (or the slope) of the line. From point *A* to point *B* the change in *y* values is from 2 to 4, a *change of* 2. From point *A* to point *B* the change in *x* values is from 1 to 5, a *change of* 4. Thus

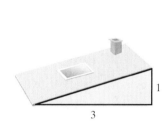

$$\text{slope} = \frac{\text{change in } y}{\text{change in } x} = \frac{2}{4} = \frac{1}{2}$$

Informally, we can describe this move as the rise over the run: $\text{slope} = \frac{\text{rise}}{\text{run}}$.

We now state a more formal (and more frequently used) definition.

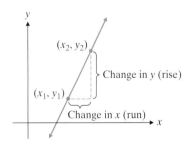

Definition of Slope of a Line

The **slope** of any *nonvertical* straight line that contains the points with coordinates (x_1, y_1) and (x_2, y_2) has a slope defined by the difference ratio

$$\text{slope} = m = \frac{y_2 - y_1}{x_2 - x_1} \qquad \text{where } x_2 \neq x_1.$$

The use of subscripted terms such as x_1, x_2, and so on, is just a way of indicating that the first *x*-value is x_1 and the second *x*-value is x_2. Thus (x_1, y_1) are the coordinates of the first point and (x_2, y_2) are the coordinates of the second point. The letter *m* is commonly used for the slope.

426

EXAMPLE 1   Find the slope of the line that passes through $(2, 0)$ and $(4, 2)$. Let $(2, 0)$ be the first point (x_1, y_1) and $(4, 2)$ be the second point (x_2, y_2).

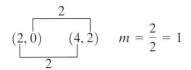

By the formula,   slope $= m = \dfrac{y_2 - y_1}{x_2 - x_1} = \dfrac{2 - 0}{4 - 2} = \dfrac{2}{2} = 1.$

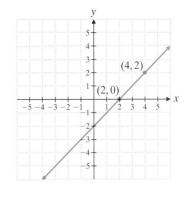

The sketch of the line is shown in the figure at the right.

Note that the slope of the line will be the same if we let $(4, 2)$ be the first point (x_1, y_1) and $(2, 0)$ be the second point (x_2, y_2).

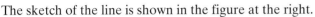

$$m = \frac{y_2 - y_1}{x_2 - x_1} = \frac{0 - 2}{2 - 4} = \frac{-2}{-2} = 1$$

Thus, given two points, it does not matter which you call (x_1, y_1) and which you call (x_2, y_2).

🚫 **WARNING** Be careful, however, not to put the x's in one order and the y's in another order when finding the slope from two points on a line.

Practice Problem 1   Find the slope of the line that passes through $(6, 1)$ and $(-4, -1)$.

It is a good idea to have some concept of the values of slopes. In downhill skiing, a very gentle slope used for teaching beginning skiers might drop one foot vertically for each 10 feet horizontally. The slope would be $\frac{1}{10}$. The speed of a skier on a hill with such a gentle slope would be only about 6 miles per hour.

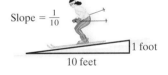

A triple diamond slope for experts might drop 11 feet vertically for each 10 feet horizontally. The slope would be $\frac{11}{10}$. The speed of a skier on such an expert trail would be in the range of 60 miles per hour.

It is important to see how positive and negative slopes affect the graphs of lines.

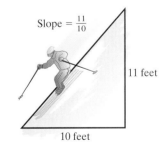

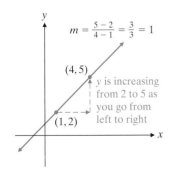

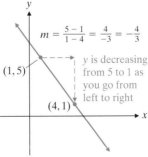

1. If the y-values increase as you go from left to right, the slope of the line is positive.

2. If the y-values decrease as you go from left to right, the slope of the line is negative.

EXAMPLE 2 Find the slope of the line that passes through $(-3, 2)$ and $(2, -4)$. Let $(-3, 2)$ be (x_1, y_1) and $(2, -4)$ be (x_2, y_2).

$$m = \frac{y_2 - y_1}{x_2 - x_1} = \frac{-4 - 2}{2 - (-3)} = \frac{-4 - 2}{2 + 3} = \frac{-6}{5} = -\frac{6}{5}$$

The slope of this line is negative. We would expect this, since the y-value decreased from 2 to -4 as the x-value increased. What does the graph of this line look like? Plot the points and draw the line to verify.

Practice Problem 2 Find the slope of the line that passes through $(2, 0)$ and $(-1, 1)$.

To Think About Describe the line in Practice Problem 2 by looking at its slope. Then verify by drawing the graph.

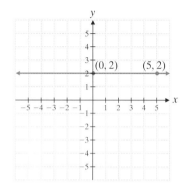

EXAMPLE 3 Find the slope of the line that passes through the given points.

(a) $(0, 2)$ and $(5, 2)$ **(b)** $(-4, 0)$ and $(-4, -4)$

(a) Take a moment to look at the y-values. What do you notice? What does this tell you about the line? Now calculate the slope.

$$m = \frac{2 - 2}{5 - 0} = \frac{0}{5} = 0$$

Since any two points on a horizontal line will have the same y-value, the slope of a horizontal line is 0.

(b) Take a moment to look at the x-values. What do you notice? What does this tell you about the line? Now calculate the slope.

$$m = \frac{-4 - 0}{-4 - (-4)} = \frac{-4}{0}$$

Recall that division by 0 is undefined.

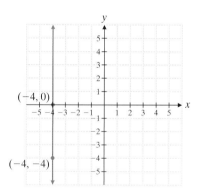

The slope of a vertical line is undefined. We say that a vertical line has **no slope**. Notice in our definition of slope that $x_2 \neq x_1$. Thus it is not appropriate to use the formula for slope for the points in (b). We did so to illustrate what would happen if $x_2 = x_1$. We get an impossible situation, $\frac{y_2 - y_1}{0}$. Now you can see why we include the restriction $x_2 \neq x_1$ in our definition.

Practice Problem 3 Find the slope of the line that passes through the given points.

(a) $(-5, 6)$ and $(-5, 3)$ **(b)** $(-7, -11)$ and $(3, -11)$

Slope of a Straight Line

Positive slope
Line goes upward to the right

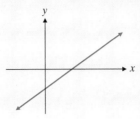

1. *Lines with positive slopes go upward as x increases* (that is, as you go from left to right).

Negative slope
Line goes downward to the right

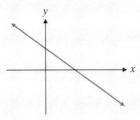

2. *Lines with negative slopes go downward as x increases* (that is, as you go from left to right).

Zero slope
Horizontal line

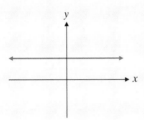

3. Horizontal lines have a slope of 0.

Undefined slope
Vertical line

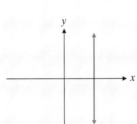

4. A vertical line is said to have undefined slope. The slope of a vertical line is not defined. In other words, a vertical line has no slope.

2 *Finding the Slope and y-Intercept of a Line Given the Equation of the Line*

Recall that the equation of a line is a linear equation in two variables. This equation can be written in several different ways. A very useful form of the equation of a straight line is the slope–intercept form. This form can be derived in the following way. Suppose that a straight line with slope m crosses the y-axis at a point $(0, b)$. Consider any other point on the line and label the point (x, y). Then we have the following.

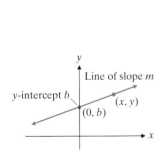

$$\frac{y_2 - y_1}{x_2 - x_1} = m \qquad \text{Definition of slope.}$$

$$\frac{y - b}{x - 0} = m \qquad \text{Substitute } (0, b) \text{ for } (x_1, y_1) \text{ and } (x, y) \text{ for } (x_2, y_2).$$

$$\frac{y - b}{x} = m \qquad \text{Simplify.}$$

$$y - b = mx \qquad \text{Multiply both sides by } x.$$

$$y = mx + b \qquad \text{Add } b \text{ to both sides.}$$

This form of a linear equation immediately reveals the slope of the line, m, and the y-coordinate of the point where the line intercepts (crosses) the y-axis, b.

Slope–Intercept Form of a Line

The slope–intercept form of the equation of the line that has slope m and the y-intercept $(0, b)$ is given by

$$y = mx + b.$$

By using algebraic operations, we can write any linear equation in slope–intercept form and use this form to identify the slope and the y-intercept of the line.

EXAMPLE 4 What is the slope and the y-intercept of the line $5x + 3y = 2$?
We want to solve for y and get the equation in the form $y = mx + b$. We need to isolate the y-variable.

$$5x + 3y = 2$$
$$3y = -5x + 2 \qquad \text{Subtract } 5x \text{ from both sides.}$$
$$y = \frac{-5x + 2}{3} \qquad \text{Divide both sides by 3.}$$
$$y = -\frac{5}{3}x + \frac{2}{3} \qquad \text{Using the property } \frac{a + b}{c} = \frac{a}{c} + \frac{b}{c},$$
$$\text{write the right-hand side as two fractions.}$$

The *slope* is $-\dfrac{5}{3}$. The y-intercept is $\left(0, \dfrac{2}{3}\right)$.

Practice Problem 4 What is the slope and the y-intercept of the line $4x - 2y = -5$?

3 Writing the Equation of a Line Given the Slope and y-Intercept

If we know the slope of a line and the y-intercept, we can write the equation of the line, $y = mx + b$.

EXAMPLE 5 Find the equation of the line with slope $\frac{2}{5}$ and y-intercept $(0, -3)$.

(a) Write the equation in slope–intercept form, $y = mx + b$.

(b) Write the equation in the form $Ax + By = C$.

(a) We are given that $m = \frac{2}{5}$ and $b = -3$. Thus we have the following.

$$y = mx + b$$
$$y = \frac{2}{5}x + (-3)$$
$$y = \frac{2}{5}x - 3$$

(b) Recall, for the form $Ax + By = C$, that A, B, and C are integers. We first clear the equation of fractions. Then we move the x-term to the left side.

$$5y = 5\left(\frac{2x}{5}\right) - 5(3) \qquad \text{Multiply each term by 5.}$$
$$5y = 2x - 15 \qquad \text{Simplify.}$$
$$-2x + 5y = -15 \qquad \text{Subtract } 2x \text{ from each side.}$$
$$2x - 5y = 15 \qquad \text{Multiply each term by } -1. \text{ The form } Ax + By = C$$
$$\text{is usually written with } A \text{ as a positive integer.}$$

Practice Problem 5   Find the equation of the line with slope $-\frac{3}{7}$ and y-intercept $\left(0, \frac{2}{7}\right)$.

(a) Write the equation in slope–intercept form.

(b) Write the equation in the form $Ax + By = C$.

4 *Graphing a Line Using the Slope and y-Intercept*

If we know the slope of a line and the y-intercept, we can draw the graph of the line.

EXAMPLE 6   Graph the line with slope $m = \frac{2}{3}$ and y-intercept $(0, -3)$. Use the given coordinate system.

Recall that the y-intercept is the point where the line crosses the y-axis. We plot the point $(0, -3)$ on the y-axis.

Recall that slope $= \dfrac{\text{rise}}{\text{run}}$. Since the slope for this line is $\frac{2}{3}$, we will go up (rise) 2 units and go over (run) to the right 3 units from the point $(0, -3)$. Look at the figure below. This is the point $(3, -1)$. Plot the point. Draw a line that connects the two points $(0, -3)$ and $(3, -1)$.

This is the graph of the line with slope $\frac{2}{3}$ and y-intercept $(0, -3)$.

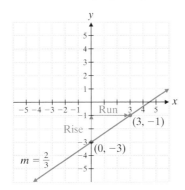

PRACTICE PROBLEM 6

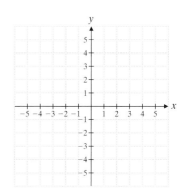

Practice Problem 6   Graph the line with slope $= \frac{3}{4}$ and y-intercept $(0, -1)$. Use the coordinate system in the margin.

EXAMPLE 7   Graph the equation $y = -\frac{1}{2}x + 4$. Use the following coordinate system.

Begin with the y-intercept. Since $b = 4$, plot the point $(0, 4)$. Now look at the slope $-\frac{1}{2}$. This can be written as $\frac{-1}{2}$. Begin at $(0, 4)$ and go *down* 1 unit and to the right 2 units. This is the point $(2, 3)$. Plot the point. Draw the line that connects the points $(0, 4)$ and $(2, 3)$.

This is the graph of the equation $y = -\frac{1}{2}x + 4$.

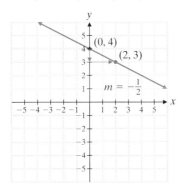

PRACTICE PROBLEM 7

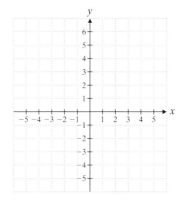

Practice Problem 7   Graph the equation $y = -\frac{2}{3}x + 5$. Use the coordinate system in the margin.

To Think About Explain why you count down 1 unit and move to the right 2 units to represent the slope $-\frac{1}{2}$. Could you have done this in another way? Try it. Verify that this is the same line.

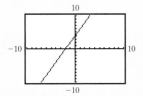
5 ▸ *Finding the Slopes of Lines That Are Parallel or Perpendicular*

Parallel lines are two straight lines that never touch. Look at the parallel lines in the figure below. Notice that the slope of line *a* is -3 and the slope of line *b* is also -3. Why do you think the slopes must be equal? What would happen if the slope of line *b* were -1? Graph it and see.

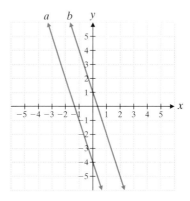

Parallel Lines

Parallel lines are two straight lines that never touch.
Parallel lines have the same slope.

$$m_1 = m_2$$

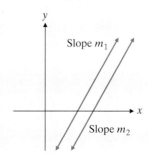

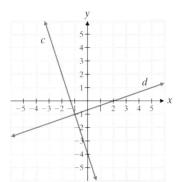

Perpendicular lines are two lines that meet at a $90°$ angle. Look at the perpendicular lines in the figure at left. The slope of line *c* is -3. The slope of line *d* is $\frac{1}{3}$. Notice that

$$(-3)\left(\frac{1}{3}\right) = \left(-\frac{3}{1}\right)\left(\frac{1}{3}\right) = -1.$$

You may wish to draw several pairs of perpendicular lines to determine whether the product of their slopes is always -1.

Perpendicular Lines

Perpendicular lines are two lines that meet at a 90° angle.
Perpendicular lines have slopes whose product is –1. If m_1 and m_2 are slopes of perpendicular lines, then

$$m_1 m_2 = -1 \quad \text{or} \quad m_1 = -\frac{1}{m_2}$$

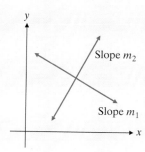

EXAMPLE 8 Line e has a slope of $-\frac{2}{3}$.

(a) If line f is parallel to line e, what is its slope?

(b) If line g is perpendicular to line e, what is its slope?

(a) Parallel lines have the same slope. Line f has a slope of $-\frac{2}{3}$.

(b) Perpendicular lines have slopes whose product is -1.

$$m_1 m_2 = -1$$

$$-\frac{2}{3} m_2 = -1 \qquad \text{Substitute } -\frac{2}{3} \text{ for } m_1.$$

$$\left(-\frac{3}{2}\right)\left(-\frac{2}{3}\right)m_2 = -1\left(-\frac{3}{2}\right) \qquad \text{Multiply both sides by } -\frac{3}{2}.$$

$$m_2 = \frac{3}{2}$$

Thus line g has a slope of $\frac{3}{2}$.

Practice Problem 8 Line h has a slope of $\frac{1}{4}$.

(a) If line j is parallel to line h, what is its slope?

(b) If line k is perpendicular to line h, what is its slope?

EXAMPLE 9 The equation of line l is $y = -2x + 3$.

(a) What is the slope of a line that is parallel to line l?

(b) What is the slope of a line that is perpendicular to line l?

(a) Looking at the equation, we can see that the slope of line l is -2. The slope of a line that is parallel to line l is -2.

(b) Perpendicular lines have slopes whose product is -1.

$$m_1 m_2 = -1$$

$$(-2)m_2 = -1 \qquad \text{Substitute } -2 \text{ for } m_1.$$

$$m_2 = \frac{1}{2} \qquad \text{Because } (-2)\left(\frac{1}{2}\right) = -1.$$

The slope of a line that is perpendicular to line l is $\frac{1}{2}$.

Practice Problem 9 The equation of line n is $y = \frac{1}{4}x - 1$.

(a) What is the slope of a line that is parallel to line n?

(b) What is the slope of a line that is perpendicular to line n?

Graphing Calculator

Graphing Parallel Lines

If two equations are in the form $y = mx + b$, then it will be obvious that they are parallel because the slope will be the same. On a graphing calculator graph both of these equations:

$$y = -2x + 6$$

$$y = -2x - 4$$

Use the window of -10 to 10 for both x and y. Display:

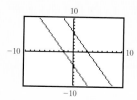

7.3 Exercises

Find the slope of a straight line that passes through the given pair of points.

1. $(6, 6)$ and $(9, 3)$

2. $(4, 1)$ and $(6, 7)$

3. $(-2, 1)$ and $(3, 4)$

4. $(5, 6)$ and $(-3, 1)$

5. $(-7, -4)$ and $(3, -8)$

6. $(-4, -6)$ and $(5, -9)$

7. $(-3, 0)$ and $(0, -4)$

8. $(0, 5)$ and $(5, 3)$

9. $\left(\frac{3}{4}, -4\right)$ and $(2, -8)$

10. $\left(\frac{5}{3}, -2\right)$ and $(3, 6)$

Verbal and Writing Skills

11. Can you find the slope of the line passing through $(5, -12)$ and $(5, -6)$? Why or why not?

12. Can you find the slope of the line passing through $(6, -2)$ and $(-8, -2)$? Why or why not?

Find the slope and the y-intercept.

13. $y = 8x + 9$

14. $y = 2x + 10$

15. $y = -3x + 4$

16. $y = -8x - 7$

17. $y = \frac{5}{6}x - \frac{2}{9}$

18. $y = -\frac{3}{4}x + \frac{5}{6}$

19. $y = -6x$

20. $y = -2$

21. $6x + y = \frac{4}{5}$

22. $2x + y = -\frac{3}{4}$

23. $5x + 2y = 3$

24. $7x + 3y = 4$

25. $7x - 3y = 4$

26. $9x - 4y = 18$

Write the equation of the line (a) in slope–intercept form and (b) in the form $Ax + By = C$.

27. $m = \frac{3}{4}, b = 2$

28. $m = \frac{4}{5}, b = 3$

29. $m = 6, b = -3$

30. $m = 5, b = -6$

31. $m = -\dfrac{5}{4}, b = -\dfrac{3}{4}$

32. $m = -4, b = \dfrac{1}{2}$

Graph the line $y = mx + b$ for the given values.

33. $m = \dfrac{1}{2}, b = -3$

34. $m = \dfrac{2}{3}, b = -4$

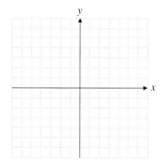

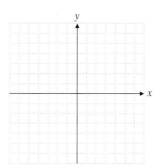

35. $m = -\dfrac{5}{3}, b = 2$

36. $m = -\dfrac{3}{2}, b = 4$

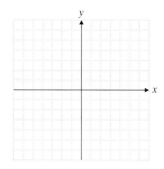

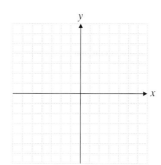

In exercises 37–41, graph the line.

37. $y = \dfrac{3}{4}x - 5$

38. $y = \dfrac{2}{3}x - 6$

39. $y + 2x = 3$

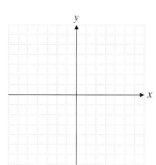

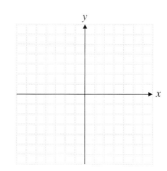

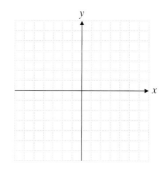

40. $y + 4x = 5$

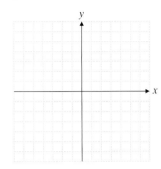

41. $y = 2x$

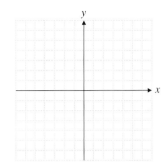

42. A line has a slope of $\dfrac{2}{3}$.

(a) What is the slope of a line parallel to it?

(b) What is the slope of a line perpendicular to it?

43. A line has a slope of $\dfrac{13}{5}$.

(a) What is the slope of a line parallel to it?

(b) What is the slope of a line perpendicular to it?

44. A line has a slope of 6.

(a) What is the slope of the line parallel to it?

(b) What is the slope of the line perpendicular to it?

45. A line has a slope of $-\dfrac{5}{8}$.

(a) What is the slope of the line parallel to it?

(b) What is the slope of the line perpendicular to it?

46. The equation of a line is $y = \dfrac{1}{3}x + 2$.

(a) What is the slope of a line parallel to it?

(b) What is the slope of a line perpendicular to it?

47. Do the points $(3, -4)$, $(18, 6)$, and $(9, 0)$ all lie on the same line? If so, what is the equation of the line?

To Think About

48. During the years from 1980 to 2000, the total income for the U.S. federal budget can be approximated by the equation $y = 14(4x + 35)$, where x is the number of years since 1980 and y is the amount of money in billions of dollars. (*Source:* U.S. Office of Management and Budget)

(a) Write the equation in slope–intercept form.

(b) Find the slope and the y-intercept.

(c) In this specific equation, what is the meaning of the slope? What does it indicate?

49. During the years from 1970 to 1990, the approximate number of civilians employed in the United States could be predicted by the equation $y = \frac{1}{10}(22x + 830)$, where x is the number of years since 1970 and y is the number of civilians employed, measured in millions. (*Source:* U.S. Bureau of Labor Statistics)

(a) Write the equation in slope–intercept form.

(b) Find the slope and the y-intercept.

(c) In this specific equation, what is the meaning of the slope? What does it indicate?

Cumulative Review Problems

Solve for x and graph the solution.

50. $3x + 8 > 2x + 12$

51. $\frac{1}{4}x + 3 > \frac{2}{3}x + 2$

52. $\frac{1}{2}(x + 2) \leq \frac{1}{3}x + 5$

53. $7x - 2(x + 3) \leq 4(x - 7)$

54. Captain Charles Allen has a 35-foot charter offshore fishing boat in Falmouth, Massachusetts. His boat has a cruising range of 390 miles when traveling at 25 miles per hour. However, the cruising range is decreased by 62% at the boat's maximum speed of 45 miles per hour. What is the cruising range at maximum speed? (Round to the nearest mile.)

55. New England Camp Cherith tries to maintain a ratio of 3 counselors for every 25 campers. One week last summer there were 189 campers. What is the minimum number of counselors who should have been there that week?

56. New York City subway ridership rose from 206 million a year in 1990 to 240 million in 1999. (*Source:* Bureau of Transportation Statistics.) The mayor predicts it will rise to 288 million by 2008. How much larger is the projected percent of increase from 1999 to 2008 than the observed percent of increase from 1990 to 1999?

Putting Your Skills to Work

Underwater Pressure

The pressure on a submarine or other submersible object deep under the ocean's surface is very significant. In 1960 a record was set for a manned submersible by a special research vessel named the *Trieste*, which descended to a depth of more than 35,800 feet. The pressure on the hull of the *Trieste* at that depth was more than 8 tons per square inch!

The underwater pressure in pounds per square inch on an object submerged in the ocean is given by the equation $p = \frac{5}{11}d + 15$, where d is the depth in feet below the surface.

Problems for Individual Study and Investigation

1. Find the underwater pressure at a depth of 22 feet.

2. At a certain location, the underwater pressure was found to be 35 pounds per square inch. What was the depth at which this measurement was taken?

3. Using the coordinates of the two points obtained in questions 1 and 2, calculate the slope of the line.

4. From the linear equation, determine the coordinates of a third point. Plot this point and the points obtained in questions 1 and 2. Draw a line.

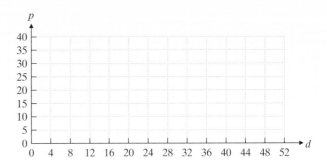

5. From the graph, determine the slope of the line. Does this agree with the slope in question 3? Does this agree with the equation? Why? Why not?

6. Complete the following table of values.

d	0	11	22	33	44
p					

(a) What is the difference between successive p values?

(b) What is the difference between successive d values?

(c) How is this related to the slope? Why?

Problems For Cooperative Group Activity and Analysis

Together with members of your class, see if you can determine the following.

Scientists are working to find a pressure equation similar to $p = \frac{5}{11}d + 15$ for a liquid that has different properties than water. They have compiled the following table of values in which p is the pressure in pounds per square inch and d is the depth in feet below the surface of this liquid.

d	0	5	12
p	18	22	27.6

7. Using the values $(0, 18)$ and $(5, 22)$, find the slope of the line.

8. Using the values $(5, 22)$ and $(12, 27.6)$, find the slope of the line. How does this compare with your answer from question 7? What does this imply?

9. Write the equation of the pressure for this new liquid in the form $p = md + b$ by determining the slope m and the p-intercept $(0, b)$.

10. What is the pressure on an object if it is submerged in this liquid at a depth of 44 feet? How does this compare with the pressure on an object submerged in water at a depth of 44 feet?

11. Is this new liquid more or less dense than water? What evidence do you have to support your conclusions?

Internet Connections

  Netsite: http://www.prenhall.com/tobey_beginning

Site: Boat Dives (Kona Coast Divers) or a related site

This site provides information about diving locations in Hawaii. Use the information provided to find the underwater pressure that you would experience in a dive to the maximum possible depth at each of the following locations.

12. Robert's Reef

13. Garden Eel Cove

14. Fish Rock (Ka'iwi Point)

15. Long Lava Tube

16. Amphitheatre

7.4 Obtaining the Equation of a Line

Student Learning Objectives

After studying this section, you will be able to:

 1 Write the equation of a line given a point and a slope.

 2 Write the equation of a line given two points.

3 Write the equation of a line given a graph of the line.

SSM
PH TUTOR   CD & VIDEO   MATH PRO   WEB
CENTER

---

1 ## Writing the Equation of a Line Given a Point and a Slope

If we know the slope of a line and the y-intercept, we can write the equation of the line in slope–intercept form. Sometimes we are given the slope and a point on the line. We use the information to find the y-intercept. Then we can write the equation of the line.

It may be helpful to summarize our approach.

To Find the Equation of a Line Given a Point and a Slope

1. Substitute the given values of x, y, and m into the equation $y = mx + b$.
2. Solve for b.
3. Use the values of b and m to write the equation in the form $y = mx + b$.

EXAMPLE 1 Find an equation of the line that passes through $(-3, 6)$ with slope $-\frac{2}{3}$.

We are given the values $m = -\frac{2}{3}$, $x = -3$, and $y = 6$.

$$y = mx + b$$
$$6 = \left(-\frac{2}{3}\right)(-3) + b \quad \text{Substitute known values.}$$
$$6 = 2 + b$$
$$4 = b$$

The equation of the line is $y = -\frac{2}{3}x + 4$.

Practice Problem 1 Find an equation of the line that passes through $(-8, 12)$ with slope $-\frac{3}{4}$.

2 ## Writing the Equation of a Line Given Two Points

Our procedure can be extended to the case for which two points are given.

EXAMPLE 2 Find an equation of the line that passes through $(2, 5)$ and $(6, 3)$.

We first find the slope of the line. Then we proceed as in Example 1.

$$m = \frac{y_2 - y_1}{x_2 - x_1}$$
$$m = \frac{3 - 5}{6 - 2} \quad \text{Substitute } (x_1, y_1) = (2, 5) \text{ and } (x_2, y_2) = (6, 3) \text{ into the formula.}$$
$$= \frac{-2}{4} = -\frac{1}{2}$$

440

Choose either point, say $(2, 5)$, to substitute into $y = mx + b$ as in Example 1.

$$5 = -\frac{1}{2}(2) + b$$

$$5 = -1 + b$$

$$6 = b$$

The equation of the line is $y = -\frac{1}{2}x + 6$.

Note: We could have substituted the slope and the other point, $(6, 3)$, into the slope–intercept form and arrived at the same answer. Try it.

Practice Problem 2 Find an equation of the line that passes through $(3, 5)$ and $(-1, 1)$.

3 Writing the Equation of a Line Given a Graph of the Line

EXAMPLE 3 What is the equation of the line in the figure at right?

First, look for the y-intercept. The line crosses the y-axis at $(0, 4)$. Thus $b = 4$.

Second, find the slope.

$$m = \frac{\text{change in } y}{\text{change in } x}$$

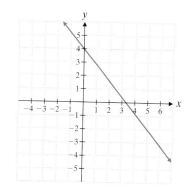

Look for another point on the line. We choose $(5, -2)$. Count the number of vertical units from 4 to -2 (rise). Count the number of horizontal units from 0 to 5 (run).

$$m = \frac{-6}{5}$$

Now using $m = -\frac{6}{5}$ and $b = 4$, we can write the equation of the line.

$$y = mx + b$$

$$y = -\frac{6}{5}x + 4$$

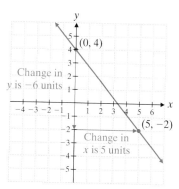

PRACTICE PROBLEM 3

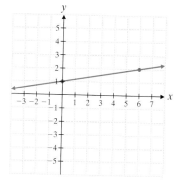

Practice Problem 3 What is the equation of the line in the figure at the right?

Find an equation of the line that has the given slope and passes through the given point.

1. $m = 4, (-3, 0)$　　　　**2.** $m = 3, (2, -2)$　　　　**3.** $m = -2, (4, 3)$　　　　**4.** $m = -4, (5, 7)$

5. $m = -3, \left(\dfrac{1}{2}, 2\right)$　　　**6.** $m = -2, \left(3, \dfrac{1}{3}\right)$　　　**7.** $m = -\dfrac{2}{5}, (5, -3)$　　　**8.** $m = \dfrac{2}{3}, (3, -2)$

Write an equation of the line passing through the given points.

9. $(3, -12)$ and $(-4, 2)$　　**10.** $(-3, 9)$ and $(2, -11)$　　**11.** $(2, -3)$ and $(-1, 6)$　　**12.** $(1, -8)$ and $(2, -14)$

13. $(3, 5)$ and $(-1, -15)$　　**14.** $(-1, -19)$ and $(2, 2)$　　**15.** $\left(1, \dfrac{5}{6}\right)$ and $\left(3, \dfrac{3}{2}\right)$　　**16.** $(2, 0)$ and $\left(\dfrac{3}{2}, \dfrac{1}{2}\right)$

Write an equation of each line.

17.

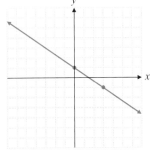

18.

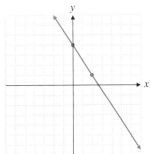

19.

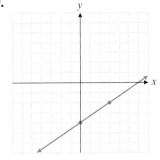

20.

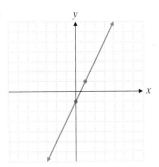

21.

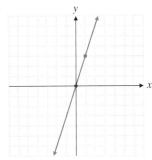

22.

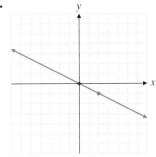

23.

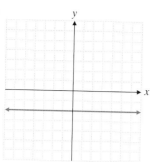

24.

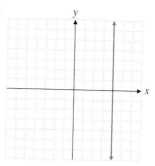

To Think About

Find an equation of the line that fits each description.

25. Passes through $(7, -2)$ and has zero slope

26. Passes through $(4, 6)$ and has undefined slope

27. Passes through $(0, 2)$ and is perpendicular to the y-axis

28. Passes through $(-1, 4)$ and is perpendicular to the x-axis

29. Passes through $(0, -4)$ and is parallel to $y = \dfrac{3}{4}x + 2$

30. Passes through $(0, -4)$ and is perpendicular to $y = \dfrac{3}{4}x + 2$

31. Passes through $(2, 3)$ and is perpendicular to $y = 2x - 9$

32. Passes through $(2, 9)$ and is parallel to $y = 5x - 3$

33. The growth of the population of the United States during the period from 1980 to 2000 can be approximated by an equation of the form $y = mx + b$, where x is the number of years since 1980 and y is the population measured in millions. (*Source:* U.S. Census Bureau.) Find the equation if two ordered pairs that satisfy it are $(0, 227)$ and $(10, 251)$.

34. The amount of debt outstanding on home equity loans in the United States during the period from 1993 to 2000 can be approximated by an equation of the form $y = mx + b$, where x is the number of years since 1993 and y is the debt measured in billions of dollars. (*Source:* Board of Governors of the Federal Reserve System.) Find the equation if two ordered pairs that satisfy it are $(1, 280)$ and $(6, 500)$.

Cumulative Review Problems

35. Solve. $\dfrac{1}{8} + \dfrac{1}{t} = \dfrac{1}{3}$

36. Simplify. $\dfrac{4x^2 - 25}{4x^2 - 20x + 25} \cdot \dfrac{x^2 - 9}{2x^2 + x - 15}$

37. A pair of basketball sneakers sells for $80. The next week the sneakers go on sale for 15% off. The third week there is a coupon in the newspaper offering a 10% discount off the second week's price. How much would you have paid for the sneakers during the third week if you had used the coupon?

38. Dave and Jane Wells have a cell phone. The plan they subscribe to costs $50 per month. It includes 200 free minutes of calling time. Each minute after the 200 is charged at the rate of $0.21 per minute. Last month their cell phone bill was $68.90. How many total minutes did they use their cell phone during the month?

39. A Scandinavian furniture wholesaler had a shipment of oak entertainment centers shipped by boat across the Atlantic. There were 35 cargo containers, each containing 60 entertainment centers. The entire shipment was found to contain 104 defective entertainment centers. What percent of the shipment was defective? Round to the nearest hundredth of a percent.

40. Jackie earned $2122.40 in interest after investing $24,000 for one year. She placed part of the money in an account earning 7% simple interest. She took a risk on some stock options with the rest of the savings, which yielded her 10.5%. How much was invested in the stock options?

7.5 Graphing Linear Inequalities

In Section 2.7 we discussed inequalities in one variable. Look at the inequality $x < -2$ (x is less than -2). Some of the solutions to the inequality are $-3, -5,$ and $-5\frac{1}{2}$. In fact all numbers to the left of -2 on the number line are solutions. The graph of the inequality is given in the following figure. Notice that the open circle at -2 indicates that -2 is *not* a solution.

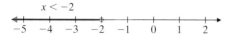

$x < -2$

Now we will extend our discussion to consider linear inequalities in two variables.

Student Learning Objectives

After studying this section, you will be able to:

① Graph linear inequalities in two variables.

SSM   PH TUTOR CENTER   CD & VIDEO   MATH PRO   WEB

① Graphing Linear Inequalities in Two Variables

Consider the inequality $y \geq x$. The solution of the inequality is the set of all possible ordered pairs that when substituted into the inequality will yield a true statement. Which ordered pairs will make the statement $y \geq x$ true? Let's try some.

$(0, 6)$	$(-2, 1)$	$(1, -2)$	$(3, 5)$	$(4, 4)$
$6 \geq 0$, true	$1 \geq -2$, true	$-2 \geq 1$, false	$5 \geq 3$, true	$4 \geq 4$, true

$(0, 6), (-2, 1), (3, 5),$ and $(4, 4)$ are solutions to the inequality $y \geq x$. In fact, every point at which the y-coordinate is greater than or equal to the x-coordinate is a solution to the inequality. This is shown by the solid line and the shaded region in the graph at the right.

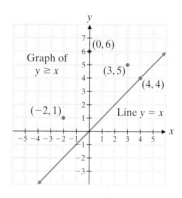

Is there an easier way to graph a linear inequality in two variables? It turns out that we can graph such an inequality by first graphing the associated linear equation and then testing one point that is not on that line. That is, we can change the inequality symbol to an equal sign and graph the equation. If the inequality symbol is $\geq$ or $\leq$, we use a solid line to indicate that the points on the line are included in the solution of the inequality. If the inequality symbol is $>$ or $<$, we use a dashed line to indicate that the points on the line are not included in the solution of the inequality. Then we test one point that is not on the line. If the point is a solution to the inequality, we shade the region on the side of the line that includes the point. If the point is not a solution, we shade the region on the other side of the line.

EXAMPLE 1 Graph $5x + 3y > 15$. Use the given coordinate system.

We begin by graphing the line $5x + 3y = 15$. You may use any method discussed previously to graph the line. Since there is no equal sign in the inequality, we will draw a dashed line to indicate that the line is *not* part of the solution set.

Look for a test point. The easiest point to test is $(0, 0)$. Substitute $(0, 0)$ for (x, y) in the inequality.

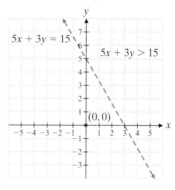

$$5x + 3y > 15$$
$$5(0) + 3(0) > 15$$
$$0 > 15 \quad \text{false}$$

$(0, 0)$ is not a solution. Shade the side of the line that does *not* include $(0, 0)$.

Practice Problem 1 Graph $x - y \geq -10$. Use the coordinate system in the margin.

PRACTICE PROBLEM 1

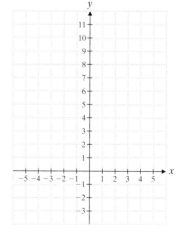

Graphing Linear Inequalities

1. Replace the inequality symbol by an equality symbol. Graph the line.
   (a) The line will be solid if the inequality is $\geq$ or $\leq$.
   (b) The line will be dashed if the inequality is $>$ or $<$.
2. Test the point $(0, 0)$ in the inequality if $(0, 0)$ does not lie on the graphed line in step 1.
   (a) If the inequality is true, shade the side of the line that includes $(0, 0)$.
   (b) If the inequality is false, shade the side of the line that does not include $(0, 0)$.
3. If the point $(0, 0)$ is a point on the line, choose another test point and proceed accordingly.

PRACTICE PROBLEM 2

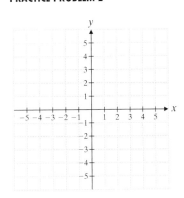

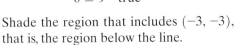 **EXAMPLE 2** Graph $2y \leq -3x$.

Step 1 Graph $2y = -3x$. Since $\leq$ is used, the line should be a solid line.

Step 2 We see that the line passes through $(0, 0)$.

Step 3 Choose another test point. We will choose $(-3, -3)$.

$$2y \leq -3x$$
$$2(-3) \leq -3(-3)$$
$$-6 \leq 9 \quad \text{true}$$

Shade the region that includes $(-3, -3)$, that is, the region below the line.

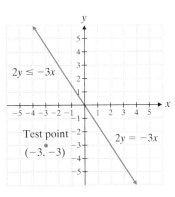

Practice Problem 2 Graph $y > \frac{1}{2}x$ on the coordinate system in the margin.

If we are graphing the inequality $x < -2$ on the coordinate plane, the solution will be a region. Notice that this is very different from the solution $x < -2$ on the number line discussed earlier.

PRACTICE PROBLEM 3

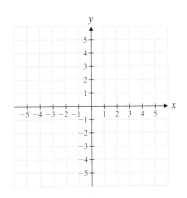

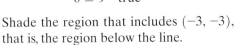 **EXAMPLE 3** Graph $x < -2$.

Step 1 Graph $x = -2$. Since $<$ is used, the line should be dashed.

Step 2 Test $(0, 0)$ in the inequality.

$$x < -2$$
$$0 < -2 \quad \text{false}$$

Shade the region that does not include $(0, 0)$, that is, the region to the left of the line $x = -2$. Observe that every point in the shaded region has an x-value that is less than -2.

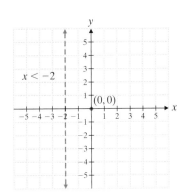

Practice Problem 3 Graph $y \geq -3$ on the coordinate system in the margin.

Verbal and Writing Skills

1. Does it matter what point you use as your test point? Justify your response.

2. Explain when to use a solid line or a dashed line when graphing linear inequalities in two variables.

Graph the region described by the inequality.

3. $y > 2 - 3x$

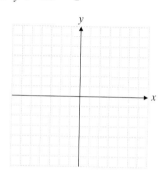

4. $y > 3x - 1$

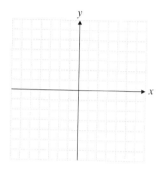

5. $2x - 3y < 6$

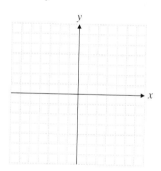

6. $3x + 2y < -6$

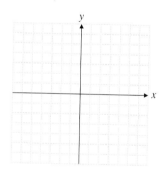

7. $2x - y \geq 3$

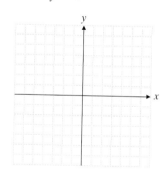

8. $3x - y \geq 4$

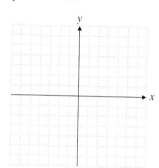

9. $y \geq 3x$

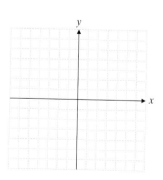

10. $y \geq -4x$

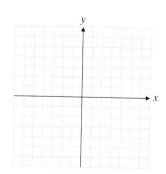

11. $y < -\dfrac{1}{2}x$

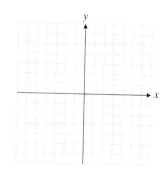

12. $y > \dfrac{1}{5}x$

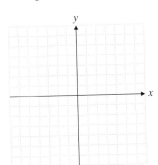

13. $x \geq 2$

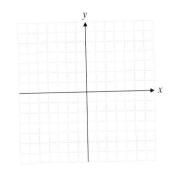

14. $y \leq -2$

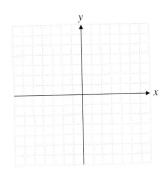

15. $3x - y + 1 \geq 0$

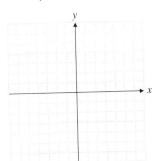

16. $2x + y - 5 \leq 0$

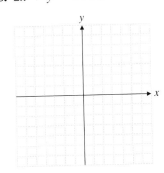

17. $2x > -3y$

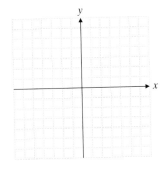

18. $3x \leq -2y$

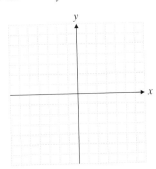

19. $2x > 3 - y$

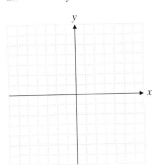

20. $3x > 1 + y$

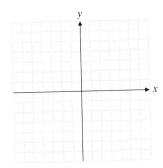

21. $x > -2y$

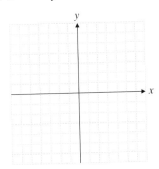

22. $x < -3y$

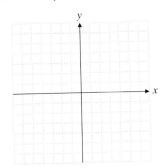

To Think About

23. Graph the inequality $3x + 3 > 5x - 3$ by graphing the lines $y = 3x + 3$ and $y = 5x - 3$ on the same coordinate plane.

(a) For which values of x does the graph of $y = 3x + 3$ lie above the graph of $y = 5x - 3$?

(b) For which values of x is $3x + 3 > 5x - 3$ true?

(c) Explain how the answer to (a) can help you to find the answer to (b).

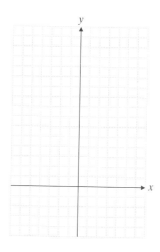

Cumulative Review Problems

24. Solve for x. $2x^2 + 7x + 6 = 0$

25. What is the slope–intercept form of the equation for the line through $(3, 0)$ and $(6, -7)$?

26. Angela has a new sales job that allows her to lease a car. Her lease budget is $8400 for a 3-year period. She wants to lease a new Chrysler Cirrus at $195 per month for the next 36 months. However, all mileage in excess of 36,000 miles driven over this 3-year period will be charged at $0.15 per mile. How many miles above the 36,000-mile limit can Angela drive and still not exceed the leasing budget?

27. Samantha is a great telemarketer. She has great success in selling subscriptions to a travel club. She has an average of 0.492 after having made 1000 telephone calls. (She sells a subscription 49.2% of the time.) Assuming that she makes a sale with each of her next telephone calls, what is the minimum number of calls she needs to make in order to raise her average to 0.500?

Student Learning Objectives

After studying this section, you will be able to:

1. Understand and use the definitions of a relation and a function.

2. Graph simple nonlinear equations.

3. Determine whether a graph represents a function.

4. Use function notation.

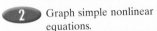

SSM
PH TUTOR   CD & VIDEO   MATH PRO   WEB
CENTER

1 Understanding and Using the Definitions of a Relation and a Function

Thus far you have studied linear equations in two variables. You have seen that such an equation can be represented by a table of values, by the algebraic equation itself, and by a graph.

x	0	1	3
y	4	1	−5

$$y = -3x + 4$$

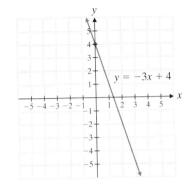

$y = -3x + 4$

The solutions to the linear equation are all the ordered pairs that satisfy the equation (make the equation true). They are all the points that lie on the graph of the line. These ordered pairs can be represented in a table of values. Notice the relationship between the ordered pairs. We can choose any value for x. But once we have chosen a value for x, the value of y is determined. For the preceding equation, if x is 0, then y must be 4. We say that x is the **independent variable** and that y is the **dependent variable**.

Mathematicians call such a pairing of two values a *relation*.

Definition of a Relation

A **relation** is any set of ordered pairs.

All the first coordinates in all of the ordered pairs of the relation make up the **domain** of the relation. All the second coordinates in all of the ordered pairs make up the **range** of the relation. Notice that the definition of a relation is very broad. Some relations cannot be described by an equation. These relations may simply be a set of discrete ordered pairs.

EXAMPLE 1 State the domain and range of the relation.

$$\{(5, 7), (9, 11), (10, 7), (12, 14)\}$$

The domain consists of all the first coordinates in the ordered pairs.

Domain

$$\{(5, 7), (9, 11), (10, 7), (12, 14)\}$$

Range

The range consists of all of the second coordinates in the ordered pairs.

The domain is $\{5, 9, 10, 12\}$.

The range is $\{7, 11, 14\}$.   We list 7 only once.

Practice Problem 1   State the domain and range of the relation.

$$\{(-3, -5), (3, 5), (0, -5), (20, 5)\}$$

Some relations have the special property that no two different ordered pairs have the same first coordinate. Such relations are called **functions**. The relation $y = -3x + 4$ is a function. If we substitute a value for x, we get just one value for y. Thus no two ordered pairs will have the same x-coordinate and different y-coordinates.

Definition of a Function

A **function** is a relation in which no two different ordered pairs have the same first coordinate.

EXAMPLE 2   Determine whether the relation is a function.

(a) $\{(3, 9), (4, 16), (5, 9), (6, 36)\}$ **(b)** $\{(7, 8), (9, 10), (12, 13), (7, 14)\}$

(a) Look at the ordered pairs. No two ordered pairs have the same first coordinate. Thus this set of ordered pairs defines a function. Note that the ordered pairs $(3, 9)$ and $(5, 9)$ have the same second coordinate, but the relation is still a function. It is the first coordinates that cannot be the same.

(b) Look at the ordered pairs. Two different ordered pairs, $(7, 8)$ and $(7, 14)$, have the same first coordinate. Thus this relation is *not* a function.

Practice Problem 2   Determine whether the relation is a function.

(a) $\{(-5, -6), (9, 30), (-3, -3), (8, 30)\}$
(b) $\{(60, 30), (40, 20), (20, 10), (60, 120)\}$

A functional relationship is often what we find when we analyze two sets of data. Look at the following table of values, which compares Celsius temperature with Fahrenheit temperature. Is there a relationship between degrees Fahrenheit and degrees Celsius? Is the relation a function?

Temperature

°F	23	32	41	50
°C	-5	0	5	10

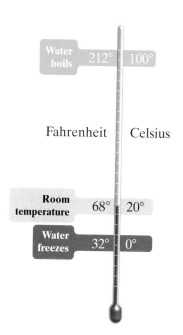

Since every Fahrenheit temperature produces a unique Celsius temperature, we would expect this to be a function. We can verify our assumption by looking at the formula $C = \frac{5}{9}(F - 32)$ and its graph. The formula is a linear equation, and its graph is a line with slope $\frac{5}{9}$ and y-intercept at about -17.8. The relation is a function. In the equation given here, notice that the *dependent variable* is C, since the value of C depends on the value of F. We say that F is the *independent variable*. The *domain* can be described as the set of possible values of the independent variable. The *range* is the set of corresponding values of the dependent variable. Scientists believe that the coldest temperature possible is approximately $-273°C$. They call this temperature **absolute zero**. Thus,

Domain = {all possible Fahrenheit temperatures from absolute zero to infinity}
   Range = {all corresponding Celsius temperatures from $-273°$ C to infinity}

● **EXAMPLE 3** Each of the following tables contains some data pertaining to a relation. Determine whether the relation suggested by the table is a function. If it is a function, identify the domain and range.

(a)
Circle

Radius	1	2	3	4	5
Area	3.14	12.56	28.26	50.24	78.5

(b)
$4000 Loan at 8% for a Minimum of One Year

Time (yr)	1	2	3	4	5
Interest	$320	$665.60	$1038.85	$1441.96	$1877.31

(a) Looking at the table, we see that no two different ordered pairs have the same first coordinate. The area of a circle is a function of the length of the radius.

Next we need to identify the independent variable to determine the domain. Sometimes it is easier to identify the dependent variable. Here we notice that the area of the circle depends on the length of the radius. Thus radius is the independent variable. Since a negative length does not make sense, the radius cannot be a negative number. Although only integer radius values are listed in the table, the radius of a circle can be any nonnegative real number.

Domain = {all nonnegative real numbers}
Range = {all nonnegative real numbers}

(b) No two different ordered pairs have the same first coordinate. Interest is a function of time.

Since the amount of interest paid on a loan depends on the number of years (term of the loan), interest is the dependent variable and time is the independent variable. Negative numbers do not apply in this situation. Although the table includes only integer values for the time, the length of a loan in years can be any real number that is greater than or equal to 1.

Domain = {all real numbers greater than or equal to 1}
Range = {all positive real numbers greater than or equal to $320}

Practice Problem 3 Determine whether the relation suggested by the table is a function. If it is a function, identify the domain and the range.

(a)
28 Mpg at $1.20 per Gallon

Distance	0	28	42	56	70
Cost	0	1.20	1.80	2.40	3.00

(b)
Store's Inventory of Shirts

Number of Shirts	5	10	5	2	8
Price of Shirt	$20	$25	$30	$45	$50

To Think About Look at the following bus schedule. Determine whether the relation is a function. Which is the independent variable? Explain your choice.

Bus Schedule

Bus Stop	Main St.	8th Ave.	42nd St.	Sunset Blvd.	Cedar Lane
Time	7:00	7:10	7:15	7:30	7:39

2 *Graphing Simple Nonlinear Equations*

Thus far in this chapter we have graphed linear equations in two variables. We now turn to graphing a few nonlinear equations. We will need to plot more than three points to get a good idea of what the graph of a nonlinear equation will look like.

EXAMPLE 4 Graph $y = x^2$.

Begin by constructing a table of values. We select values for x and then determine by the equation the corresponding values of y. We will include negative values for x as well as positive values. We then plot the ordered pairs and connect the points with a smooth curve.

x	$y = x^2$	y
-2	$y = (-2)^2 = 4$	4
-1	$y = (-1)^2 = 1$	1
0	$y = (0)^2 = 0$	0
1	$y = (1)^2 = 1$	1
2	$y = (2)^2 = 4$	4

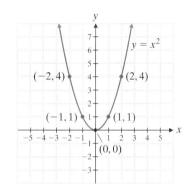

This type of curve is called a *parabola*. We will study this type of curve more extensively in Chapter 10.

Practice Problem 4 Graph $y = x^2 - 2$ on the coordinate system in the margin.

Some equations are solved for x. Usually, in those cases we pick values of y and then obtain the corresponding values of x from the equation.

EXAMPLE 5 Graph $x = y^2 + 2$.

We will select a value of y and then substitute it into the equation to obtain x. For convenience in graphing, we will repeat the y column at the end so that it is easy to write the ordered pairs (x, y).

y	$x = y^2 + 2$	x	y
-2	$x = (-2)^2 + 2 = 4 + 2 = 6$	6	-2
-1	$x = (-1)^2 + 2 = 1 + 2 = 3$	3	-1
0	$x = (0)^2 + 2 = 0 + 2 = 2$	2	0
1	$x = (1)^2 + 2 = 1 + 2 = 3$	3	1
2	$x = (2)^2 + 2 = 4 + 2 = 6$	6	2

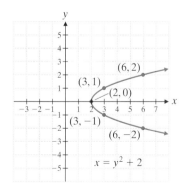

Practice Problem 5 Graph $x = y^2 - 1$ on the coordinate system in the margin.

If the equation involves fractions with variables in the denominator, we must use extra caution. Remember that you may never divide by zero.

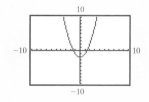
PRACTICE PROBLEM 4

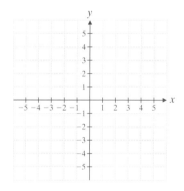

PRACTICE PROBLEM 5

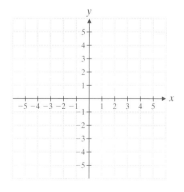

EXAMPLE 6 Graph $y = \dfrac{4}{x}$.

It is important to note that x cannot be zero because division by zero is not defined. $y = \frac{4}{0}$ is not allowed! Observe that when we draw the graph we get two separate branches that do not touch.

x	$y = \dfrac{4}{x}$	y
-4	$y = \dfrac{4}{-4} = -1$	-1
-2	$y = \dfrac{4}{-2} = -2$	-2
-1	$y = \dfrac{4}{-1} = -4$	-4
0	We cannot divide by zero.	There is no value.
1	$y = \dfrac{4}{1} = 4$	4
2	$y = \dfrac{4}{2} = 2$	2
4	$y = \dfrac{4}{4} = 1$	1

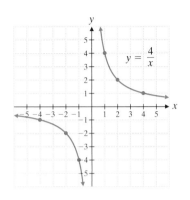

PRACTICE PROBLEM 6

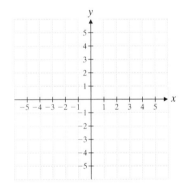

Practice Problem 6 Graph $y = \dfrac{6}{x}$ on the coordinate system in the margin.

③ Determining Whether a Graph Represents a Function

Can we tell whether a graph represents a function? Recall that a function cannot have two different ordered pairs with the same first coordinate. That is, each value of x must have a separate unique value of y. Look at the graph of the function $y = x^2$ in Example 4. Each x-value has a unique y-value. Look at the graph of $x = y^2 + 2$ in Example 5. At $x = 3$ there are two y-values, 1 and -1. In fact, for every x-value greater than 2 there are two y-values. $x = y^2 + 2$ is not a function.

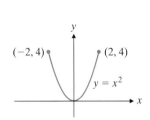

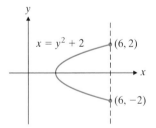

Observe that we can draw a vertical line through $(6, 2)$ and $(6, -2)$. Any graph that is not a function will have at least one region in which a vertical line will cross the graph more than once.

Vertical Line Test

If a vertical line can intersect the graph of a relation more than once, the relation is not a function. If no such line can be drawn, then the relation is a function.

EXAMPLE 7   Determine whether each of the following is the graph of a function.

(a)
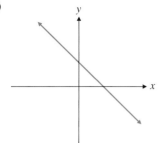

(b)
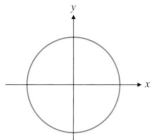

(c)
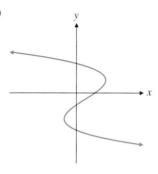

(a) The graph of the straight line is a function. Any vertical line will cross this straight line in only one location.

(b) and **(c)** Each of these graphs is not the graph of a function. In each case there exists a vertical line that will cross the curve in more than one place.

(a)
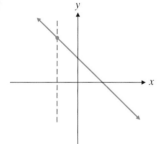

(b)
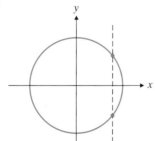

(c)
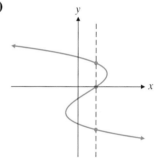

Practice Problem 7   Determine whether each of the following is the graph of a function.

(a)
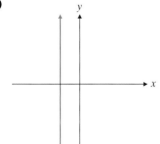

(b)
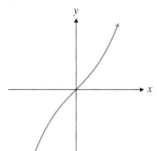

(c)
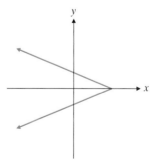

4 Using Function Notation

We have seen that an equation like $y = 2x + 7$ is a function. For each value of x, the equation assigns a unique value to y. We could say "y is a function of x." If we name the function f, this statement can be symbolized by using the **function notation** $y = f(x)$. Many times we avoid using the y-variable completely and write the function as $f(x) = 2x + 7$.

🚫 **WARNING** Be careful. The notation $f(x)$ does not mean f multiplied by x.

EXAMPLE 8 If $f(x) = 3x^2 - 4x + 5$, find each of the following.

(a) $f(-2)$ **(b)** $f(4)$ **(c)** $f(0)$

(a) $f(-2) = 3(-2)^2 - 4(-2) + 5 = 3(4) - 4(-2) + 5 = 12 + 8 + 5 = 25$

(b) $f(4) = 3(4)^2 - 4(4) + 5 = 3(16) - 4(4) + 5 = 48 - 16 + 5 = 37$

(c) $f(0) = 3(0)^2 - 4(0) + 5 = 3(0) - 4(0) + 5 = 0 - 0 + 5 = 5$

Practice Problem 8 If $f(x) = -2x^2 + 3x - 8$, find each of the following.

(a) $f(2)$ **(b)** $f(-3)$ **(c)** $f(0)$

When evaluating a function, it is helpful to place parentheses around the value that is being substituted for x. Taking the time to do this will minimize sign errors in your work.

Some functions are useful in medicine. For example, the approximate width of a man's elbow is given by the function $e(x) = 0.03x + 0.6$, where x is the height of the man in inches. If a man is 68 inches tall, $e(68) = 0.03(68) + 0.6 = 2.64$. A man 68 inches tall would have an elbow width of approximately 2.64 inches.

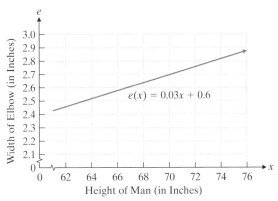

Source: U.S. National Center for Health Statistics

Verbal and Writing Skills

1. What are the three ways you can describe a function?

2. What is the difference between a function and a relation?

3. The domain of a function is the set of _____ _____ of the _____ variable.

4. The range of a function is the set of _____ _____ of the _____ variable.

5. How can we tell whether a graph is the graph of a function?

Find the domain and range of the relation. Determine whether the relation is a function.

6. $\left\{\left(\frac{1}{2}, \frac{1}{2}\right), \left(-10, \frac{1}{2}\right), \left(7, \frac{1}{4}\right), \left(\frac{1}{2}, \frac{1}{4}\right)\right\}$

7. $\left\{\left(\frac{1}{2}, 5\right), \left(\frac{1}{4}, 10\right), \left(\frac{3}{4}, 6\right), \left(\frac{1}{2}, 6\right)\right\}$

8. $\left\{(7, 3.1), (5, 0), (7, 2.3)\right\}$

9. $\left\{(7.3, 1), (0, 8), (2, 1)\right\}$

10. $\left\{(12, 1), (14, 3), (1, 12), (9, 12)\right\}$

11. $\left\{(5.6, 8), (5.8, 6), (6, 5.8), (5, 6)\right\}$

12. $\left\{(3, 75), (5, 95), (3, 85), (7, 100)\right\}$

13. $\left\{(85, 3), (95, 11), (110, 15), (110, 20)\right\}$

Graph the equation.

14. $y = x^2 + 3$

15. $y = x^2 - 1$

16. $y = 2x^2$

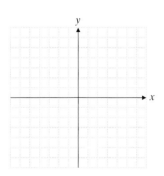
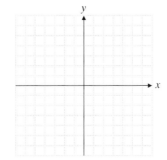
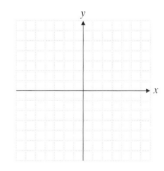

17. $y = \dfrac{1}{2}x^2$

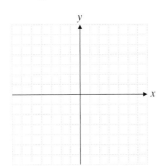

18. $x = -2y^2$

19. $x = \dfrac{1}{2}y^2$

20. $x = y^2 + 1$

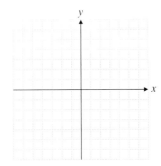

21. $x = 2y^2$

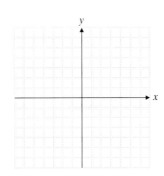

22. $y = \dfrac{2}{x}$

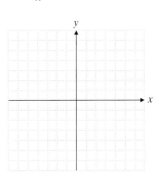

23. $y = -\dfrac{2}{x}$

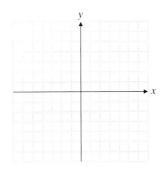

24. $y = \dfrac{4}{x^2}$

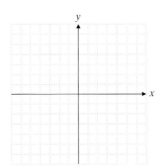

25. $y = -\dfrac{6}{x^2}$

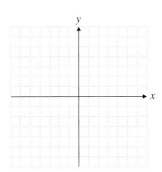

26. $y = (x + 1)^2$

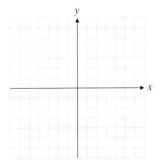

27. $x = (y - 2)^2$

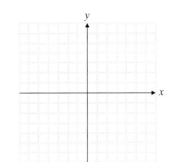

28. $y = \dfrac{4}{x - 2}$

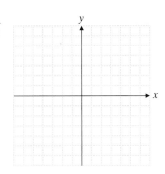

29. $x = \dfrac{2}{y + 1}$

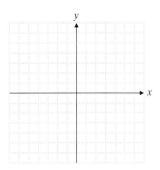

Determine whether each relation is a function.

30.

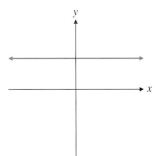

31.

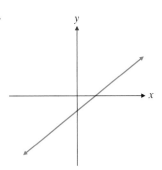

32.

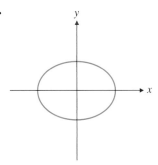

33.

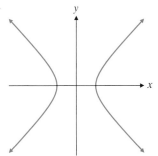

34.

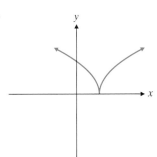

35.

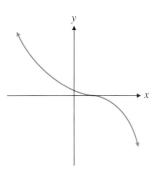

36.

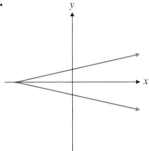

37.

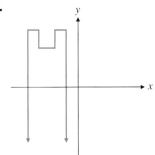

Given the following functions, find the indicated values.

38. $f(x) = 2 - 3x$
   (a) $f(-8)$ **(b)** $f(0)$ **(c)** $f(2)$

39. $f(x) = 3 - 5x$
   (a) $f(-1)$ **(b)** $f(1)$ **(c)** $f(2)$

40. $f(x) = x^2 + 3x + 4$
   (a) $f(2)$ **(b)** $f(0)$ **(c)** $f(-3)$

41. $f(x) = x^2 + 2x - 3$
   (a) $f(-1)$ **(b)** $f(0)$ **(c)** $f(3)$

42. $f(x) = 2x + 3 - \dfrac{12}{x}$

   (a) $f(-2)$ **(b)** $f(1)$ **(c)** $f(4)$

43. $f(x) = x - 2 + \dfrac{8}{x}$

   (a) $f(-2)$ **(b)** $f(1)$ **(c)** $f(4)$

Applications

The accompanying graph represents the approximate percent of people in the United States who smoked during selected years from 1980 to 2000. Source: National Center for Health Statistics.

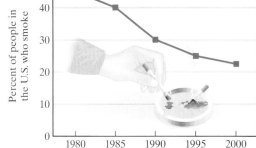

Percent of People Smoking

44. Does the graph represent a function?

45. Find $f(1990)$.

46. If $f(x) = 45\%$, what year is represented by x?

47. Between what two years is the difference in function values equal to 10%?

48. During a recent population growth period in Nevada from 1980 to 1990, the approximate population of the state measured in thousands could be predicted by the function $f(x) = 2x^2 + 20x + 800$, where x is the number of years since 1980. (*Source:* U.S. Census Bureau.) Find $f(0), f(3), f(6)$, and $f(10)$. Graph the function. What pattern do you observe?

49. During a recent population growth period in Alaska from 1980 to 2000, the approximate population of the state measured in thousands could be predicted by the function $f(x) = -0.6x^2 + 22x + 411$, where x is the number of years since 1980. (*Source:* U.S. Census Bureau.) Find $f(0), f(4), f(10)$, and $f(20)$. Graph the function. What pattern do you observe?

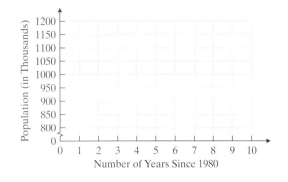

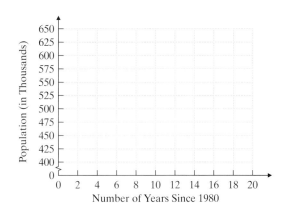

Cumulative Review Problems

Simplify.

50. $(5x^2 - 6x + 2) + (-7x^2 - 8x + 4)$

51. $(-2x^2 - 3x + 8) - (x^2 + 4x - 3)$

52. $(4x^2 - 2x + 3)(x + 2)$

53. $(6x^2 + 10x - 5) \div (3x - 1)$

Putting Your Skills to Work

Graphing the Size of Rainforests

Most of us are aware that the world's rainforests are disappearing, but do you realize how quickly? Rainforests cover only 2% of the Earth's surface, but are home to almost half of all life forms (plants and animals) on the planet. As deforestation continues, many species are becoming extinct and the world's climate is being negatively affected (a phenomenon known as the greenhouse effect).

In 1990, there were 1756 million hectares of rainforest on the planet. Since that time, rainforest destruction has occurred at 2.47 acres per second. (This area is about as large as two football fields.)

Problems for Individual Investigation and Study

1. How many acres of rainforest are destroyed each day? Each year? (Round to the nearest million.)

2. Approximately how many acres of rainforest were there in 1990? (1 hectare = 2.47 acres)

3. Use your answers from questions 1 and 2 to determine how many acres of rainforest remained in 1992. How many acres of rainforest were there in 1995?

Problems for Cooperative Group Activity

Suppose that the current rate of rainforest destruction continues without intervention. How many acres of rainforest will there be 20 years from now? 35 years from now? In what year will there be no rainforests left in the world? Answer the following questions to find out.

4. Write a function that gives the number of acres of rainforest r (in millions of acres) that remain t years after 1990.

5. Complete the following table.

Number of Years after 1990	Amount of Rainforest Remaining on Earth (in Millions of Acres)
0	
5	
10	
15	
20	

6. Graph the information from question 5. What is the dependent variable? What is the independent variable? Be sure to use appropriate scales on the axes.

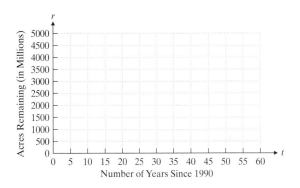

7. Using the graph, determine how many acres of rainforest will remain in the year 2035.

8. Using the graph, determine when there will be no rainforests left on Earth.

Internet Connections

 Netsite: http://www.prenhall.com/tobey_beginning

Site: Food and Agricultural Organization of the United Nations

9. Comparing Asia, Latin America, and Africa, which of these is projected to have the greatest absolute growth in population in the period 2000–2050? Can you tell just by looking at the slopes of the three plots which of these areas will experience the highest *rate* of growth in that period?

10. What is the relative growth of urban population as a percent of total population between 1950 and 2030 in Asia, Latin America/Caribbean, and Africa, respectively?

461

Math in the Media

Automobiles — Costs Beyond the Purchase Price

By Jim Mateja

An article in the *Chicago Tribune-Northern Light*, August 25, 2000, reported, "The Saturn sedan and Chevy Metro 2000 models cost less than half as much to own and operate as the Cadillac DeVille or Lincoln Town Car. They were less than one-third as much as the Mercedes-Benz 500S, according to Runzheiner International, the Wisconsin-based management consulting firm."

The analysis considered the Total Annual Costs of each model.

The Total Annual Costs = annual operating costs + fixed costs.

It is important to consider all the costs associated with a car when making a decision to purchase. The questions that follow give you the chance to compare costs in a similar way using four different vehicle models.

EXERCISES

The chart below shows the Total Annual Costs of operating four different motor vehicles, including the costs associated with gasoline and oil, maintenance, tires, insurance, license, registration, taxes, depreciation, and finance charges.

Total Annual Costs in Dollars

Miles Driven Per Year	Vehicle #1	Vehicle #2	Vehicle #3	Vehicle #4
10,000 miles per year	3,800	5,500	5,800	6,700
15,000 miles per year	5,100	7,500	6,800	7,600
20,000 miles per year	7,300	9,200	8,700	9,900

1. Draw a graph with *Miles per Year* on the *x*-axis and *Total Annual Cost* on the *y*-axis. Use a scale of 0 to 20,000 on the *x*-axis and a scale of 0 to 10,000 on the *y*-axis. Plot three points for each of the vehicle types and, for each vehicle, connect each pair of ordered pairs with straight lines.
   a. Based on your graph, which vehicle type had the highest dollar increase in cost per year comparing 10,000 miles/year usage with 15,000 miles/year usage? Comparing 15,000 miles/year usage with 20,000 miles/year usage?
   b. Based on your graph, which vehicle type had the lowest percent increase in cost per mile comparing 10,000 miles/year usage with 15,000 miles/year usage? Comparing 15,000 miles/year usage with 20,000 miles/year usage?

2. What is being measured by the slope between the points plotted for any given vehicle on the graph you constructed?

Chapter 7 Organizer

Topic	Procedure	Examples
Plotting points, p. 407.	To plot (x, y): **1.** Begin at the origin. **2.** If x is positive, move to the right along the x-axis. If x is negative, move to the left along the x-axis. **3.** If y is positive, move up. If y is negative, move down.	To plot $(-2, 3)$:
Graphing straight lines, p. 416.	An equation of the form $$Ax + By = C$$ has a graph that is a straight line. To graph such an equation, plot any three points; two give the line and the third checks it. (Where possible, use the x- and y-intercepts.)	Graph $3x + 2y = 6$. | x	y
---	---
0	3
4	-3
2	0	
Finding the slope given two points, p. 427.	Nonvertical lines passing through distinct points (x_1, y_1) and (x_2, y_2) have slope $$m = \frac{y_2 - y_1}{x_2 - x_1}.$$ The slope of a horizontal line is 0. The slope of a vertical line is undefined.	What is the slope of the line through $(2, 8)$ and $(5, 1)$? $$m = \frac{1 - 8}{5 - 2} = -\frac{7}{3}$$
Finding the slope and y-intercept of a line given the equation, p. 430.	**1.** Rewrite the equation in the form $y = mx + b$. **2.** The slope is m. **3.** The y-intercept is $(0, b)$.	Find the slope and y-intercept. $$3x - 4y = 8$$ $$-4y = -3x + 8$$ $$y = \tfrac{3}{4}x - 2$$ The slope is $\frac{3}{4}$. The y-intercept is $(0, -2)$.
Finding the equation of a line given the slope and y-intercept, p. 430.	The slope–intercept form of the equation of a line is $$y = mx + b$$ The slope is m and the y-intercept is $(0, b)$.	Find the equation of the line with y-intercept $(0, 7)$ and with slope $m = 3$. $$y = 3x + 7$$
Graphing a line using slope and y-intercept, p. 431.	**1.** Plot the y-intercept. **2.** Starting from $(0, b)$, plot a second point using the slope. $$\text{slope} = \frac{\text{rise}}{\text{run}}$$ **3.** Draw a line that connects the two points.	Graph $y = -4x + 1$. First plot the y-intercept at $(0, 1)$. Slope $= -4$ or $\frac{-4}{1}$
Finding the slope of parallel and perpendicular lines, p. 432.	Parallel lines have the same slope. Perpendicular lines have slopes whose product is -1.	Line q has a slope of 2. The slope of a line parallel to q is 2. The slope of a line perpendicular to q is $-\frac{1}{2}$.

Topic	Procedure	Examples
Finding the equation of a line through a point with a given slope, p. 440.	1. Substitute the known values in the equation $y = mx + b$. 2. Solve for b. 3. Use the values of m and b to write the general equation.	Find the equation of the line through $(3, 2)$ with slope $m = \frac{4}{5}$. $$y = mx + b$$ $$2 = \frac{4}{5}(3) + b$$ $$2 = \frac{12}{5} + b$$ $$-\frac{2}{5} = b$$ The equation is $y = \frac{4}{5}x - \frac{2}{5}$.
Finding the equation of a line through two points, p. 440.	1. Find the slope. 2. Use the procedure when given a point and the slope.	Find the equation of the line through $(3, 2)$ and $(13, 10)$. $$m = \frac{y_2 - y_1}{x_2 - x_1} = \frac{10 - 2}{13 - 3} = \frac{8}{10} = \frac{4}{5}$$ We choose the point $(3, 2)$. $$y = mx + b$$ $$2 = \frac{4}{5}(3) + b$$ $$2 = \frac{12}{5} + b$$ $$-\frac{2}{5} = b$$ The equation is $y = \frac{4}{5}x - \frac{2}{5}$.
Writing the equation of a line given the graph, p. 441.	1. Identify the y-intercept. 2. Find the slope. $$\text{slope} = \frac{\text{change in } y}{\text{change in } x}$$	 The equation is $y = -\frac{5}{4}x - 2$.
Graphing linear inequalities, p. 445.	1. Graph as if it were an equation. If the inequality symbol is $>$ or $<$, use a dashed line. If the inequality symbol is $\geq$ or $\leq$, use a solid line. 2. Look for a test point. The easiest test point is $(0, 0)$, unless the line passes through $(0, 0)$. In that case, choose another test point. 3. Substitute the coordinates of the test point into the inequality. 4. If it is a true statement, shade the side of the line containing the test point. If it is a false statement, shade the side of the line that does *not* contain the test point.	Graph $y \geq 3x + 2$. Graph the line $y = 3x + 2$. Use a solid line. $$\text{Test } (0, 0). \quad 0 \geq 3(0) + 2$$ $$0 \geq 2 \quad \text{false}$$ Shade the side of the line that does not contain $(0, 0)$.

Topic	Procedure	Examples
Determining whether a relation is a function, p. 451.	A function is a relation in which no two different ordered pairs have the same first coordinate.	Is this relation a function? $\{(5,7), (3,8), (5,10)\}$ It is *not* a function since $(5, 7)$ and $(5, 10)$ are two different ordered pairs with the same x-coordinate, 5.
Determining whether a graph represents a function, p. 454.	If a vertical line can intersect the graph of a relation more than once, the relation is not a function. If no such line exists, the relation is a function.	Is this graph a function? Yes. Any vertical line will intersect it at most once.

Chapter 7 Review Problems

7.1

1. Plot and label the following points.

A: $(2, -3)$ B: $(-1, 0)$ C: $(3, 2)$ D: $(-2, -3)$

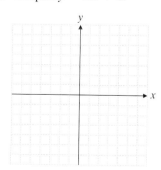

2. Give the coordinates of each point.

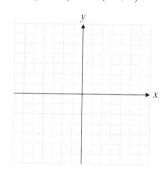

Complete the ordered pairs so that each is a solution to the given equation.

3. $y = 3x - 5$

   (a) $(0, \quad)$ **(b)** $(3, \quad)$

4. $2x + 5y = 12$

   (a) $(1, \quad)$ **(b)** $(\quad, 4)$

5. $x = 6$

   (a) $(\quad, -1)$ **(b)** $(\quad, 3)$

7.2

6. Graph $3y = 2x + 6$.

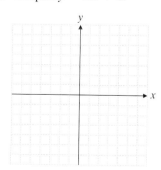

7. Graph $5y + x = -15$.

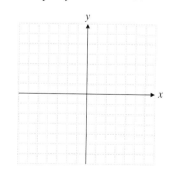

8. Graph $3x + 5y = 15 + 3x$.

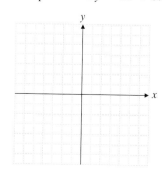

7.3

9. Find the slope of the line passing through $(5, -3)$ and $(2, -\frac{1}{2})$.

10. Find the slope and y-intercept of the line $9x - 11y + 15 = 0$.

11. Write an equation of the line with slope $-\frac{1}{2}$ and y-intercept $(0, 3)$.

12. A line has a slope of $-\frac{2}{3}$. What is the slope of a line perpendicular to that line?

13. Graph $y = -\dfrac{1}{2}x + 3$.

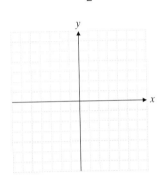

14. Graph $2x - 3y = -12$.

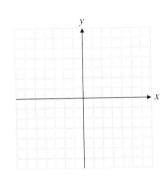

15. Graph $5x + 2y = 20 + 2y$.

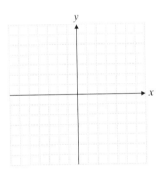

7.4

16. Write an equation of the line passing through $(5, 6)$ having a slope of 2.

17. Write an equation of the line passing through $(3, -4)$ having a slope of -6.

18. Write an equation of the line passing through $(6, -3)$ having a slope of $\frac{1}{3}$.

19. Write an equation of the line passing through $(3, 7)$ and $(-6, 7)$.

Write an equation of the graph.

20.

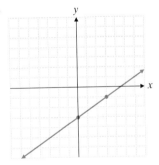

21.

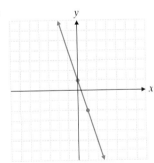

22.

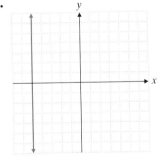

7.5

23. Graph $y < \frac{1}{3}x + 2$.

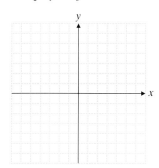

24. Graph $3y + 2x \geq 12$.

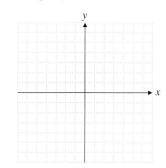

25. Graph $y \geq -2$.

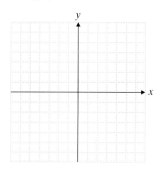

7.6

Determine the domain and range of the relation. Determine whether the relation is a function.

26. $\{(5, -6), (-6, 5), (-5, 5), (-6, -6)\}$

27. $\{(3, -7), (-7, 3), (-3, 7), (7, -3)\}$

In exercises 28–30, determine whether the graphs represent a function.

28.

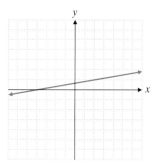

29.

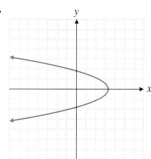

30.

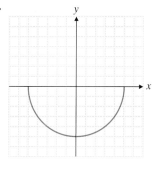

31. Graph $y = x^2 - 5$.

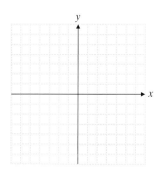

32. Graph $x = y^2 + 3$.

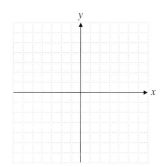

33. Graph $y = (x - 3)^2$.

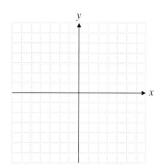

Given the following functions, find the indicated values.

34. $f(x) = 7 - 6x$ **(a)** $f(0)$ **(b)** $f(-4)$

35. $g(x) = -2x^2 + 3x + 4$ **(a)** $g(-1)$ **(b)** $g(3)$

36. $p(x) = \dfrac{-3}{x}$ **(a)** $p(-3)$ **(b)** $p(0.5)$

37. $f(x) = \dfrac{2}{x + 4}$ **(a)** $f(-2)$ **(b)** $f(6)$

38. $f(x) = \dfrac{x + 6}{x - 5}$ **(a)** $f(0)$ **(b)** $f(-2)$

Applications

Bob and Evelyn Hanson have found that when they travel across the country, they can estimate the cost of their trip with the equation $y = 150 + 0.15x$. In this equation y represents the cost in dollars and x represents the number of miles traveled.

39. What is the approximate cost to travel 1000 miles?

40. What is the approximate cost to travel 3000 miles?

41. Write the equation in the form $y = mx + b$, and determine the numerical value of the slope.

42. What is the significance of the slope? What does it tell us about the cross-country costs?

43. Bob Hanson estimated that the total budget for a trip was $540. With that limit, how many miles can they travel?

44. Evelyn Hanson estimated that the total budget for a trip was $750. With that limit, how many miles can they travel?

Russ and Norma Camp found that their monthly electric bill could be calculated by the equation $y = 30 + 0.09x$. In this equation y represents the amount of the monthly bill in dollars and x represents the number of kilowatt-hours used during the month.

45. What would be their monthly bill if they used 2000 kilowatt-hours of electricity?

46. What would be their monthly bill if they used 1600 kilowatt-hours of electricity?

47. Write the equation in the form $y = mx + b$, and determine the numerical value of the y-intercept. What is the significance of this y-intercept? What does it tell us?

48. If the equation is placed in the form $y = mx + b$, what is the numerical value of the slope? What is the significance of this slope? What does it tell us?

49. If Russ and Norma have a monthly bill of $147, how many kilowatt-hours of electricity did they use?

50. If Russ and Norma have a monthly bill of $246, how many kilowatt-hours of electricity did they use?

Chapter 7 Test

1. Plot and label the following points.
$B: (6, 1)$ $C: (-4, -3)$
$D: (-3, 0)$ $E: (5, -2)$

2. Graph the line $6x - 3 = 5x - 2y$.

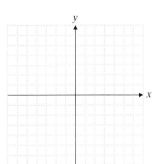

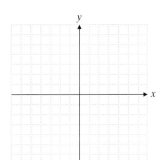

3. Graph the line $12x - 3y = 6$.

4. Graph $y = \frac{2}{3}x - 4$.

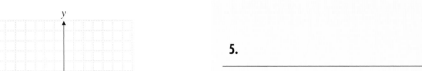

5. What is the slope and the y-intercept of the line $3x + 2y - 5 = 0$?

6. Find the slope of the line that passes through $(8, 6)$ and $(-3, -5)$.

7. Write an equation for the line that passes through $(4, -2)$ and has a slope of $\frac{1}{2}$.

8. Find the slope of the line through $(-3, 11)$ and $(6, 11)$.

9. Find an equation for the line passing through $(2, 5)$ and $(8, 3)$.

10. Write an equation for the line through $(2, 7)$ and $(2, -2)$. What is the slope of this line?

1. _____

2. _____

3. _____

4. _____

5. _____

6. _____

7. _____

8. _____

9. _____

10. _____

469

11. _____

12. _____

13. _____

14. _____

15. _____

16. (a) _____

(b) _____

17. (a) _____

(b) _____

11. Graph the region described by $4y \leq 3x$.

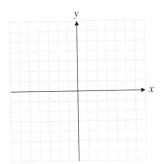

12. Graph the region described by $-3x - 2y > 10$.

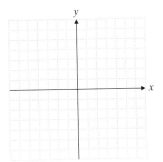

13. Is this relation a function? $\{(-8, 2), (3, 2), (5, -2)\}$ Why?

14. Look at the relation graphed below. Is this relation a function? Why?

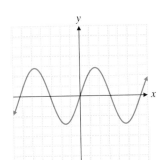

15. Graph $y = 2x^2 - 3$.

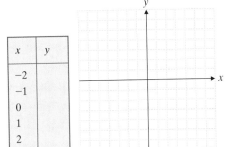

x	y
-2	
-1	
0	
1	
2	

16. For $f(x) = 2x^2 + 3x - 4$:

   (a) Find $f(0)$.

   (b) Find $f(-2)$.

17. For $g(x) = \dfrac{3}{x - 4}$:

   (a) Find $g(3)$.

   (b) Find $g(-11)$.

Approximately one-half of this test covers the content of Chapters 0–6. The remainder covers the content of Chapter 7.

1. Simplify. $\left(-3x^3y^4z\right)^3$

2. Simplify. $-\dfrac{18a^3b^4}{24ab^6}$

Solve.

3. $2 - 3(4 - x) = x - (3 - x)$

4. $\dfrac{600}{R} = \dfrac{600}{R + 50} + \dfrac{500(R + 20)}{R^2 + 50R}$

5. $x = 2^3 - 6 \div 3 - 1$

6. Solve $A = lw + lh$ for w.

7. Factor. $50a^2 - 98b^2$

8. Factor. $3x^2 - 2x - 21$

9. Simplify. $\dfrac{x - \dfrac{1}{x}}{\dfrac{1}{2} + \dfrac{1}{2x}}$

10. Find an equation of the line through $(6, 8)$ and $(7, 11)$.

In questions 11 and 12, give the equation of the line that fits each description.

11. Passes through $(7, -4)$ and is vertical.

12. Passes through $(-2, 3)$ with slope $\frac{1}{3}$.

13. Find the slope of the line through $(-8, -3)$ and $(11, -3)$.

14. What is the slope of the line $3x - 7y = -2$?

1. _____

2. _____

3. _____

4. _____

5. _____

6. _____

7. _____

8. _____

9. _____

10. _____

11. _____

12. _____

13. _____

14. _____

15. _____

16. _____

17. _____

18. _____

19. _____

20. (a) _____

(b) _____

15. Graph $y = \frac{2}{3}x - 4$.

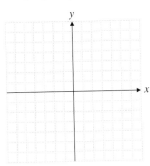

16. Graph $3x + 8 = 5x$.

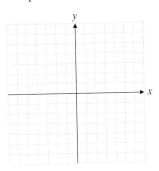

17. Graph the region $2x + 5y \leq -10$.

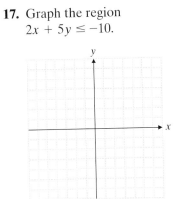

18. Graph $x = y^2 + 3$. Is this a function?

x	y
-2	
-1	
0	
1	
2	

19. Is this relation a function?

$\{(3, 10), (10, -3), (0, -3), (5, 10)\}$

20. For $f(x) = -2x^2 - 3x + 5$:

(a) Find $f(4)$.   **(b)** Find $f(-2)$.

Systems of Equations

When we think of automobile production, we envision an assembly line of thousands of cars. However, some auto companies still produce cars by hand. One such company is the Morgan Motor Company of England. Their autos, of course, are quite expensive and available only to wealthy individuals. They have been producing essentially the same hand-built racing sports car since 1936. Do you think you can use your knowledge of systems of equations to solve some problems relating to the sales of these cars? Turn to the Putting Your Skills to Work problems on page 496 to find out.

Pretest Chapter 8

1. _____

2. _____

3. _____

4. _____

5. _____

6. _____

7. _____

8. _____

9. _____

10. _____

11. _____

12. _____

13. _____

14. _____

If you are familiar with the topics in this chapter, take this test now. Check your answers with those in the back of the book. If an answer is wrong or you can't answer a question, study the appropriate section of the chapter.

If you are not familiar with the topics in this chapter, don't take this test now. Instead, study the examples, work the practice problems, and then take the test.

This test will help you identify which concepts you have mastered and which you need to study further.

Section 8.1

Solve by the graphing method. If there is no solution, state why.

1. $4x + 2y = -8$
$-2x + 3y = 12$

2. $x - 2y = -2$
$3x - 6y = 12$

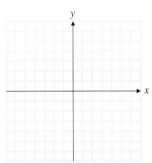
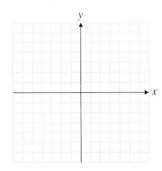

Section 8.2

Solve by the substitution method.

3. $3x + y = -11$
$4x - 3y = -6$

4. $\frac{2}{3}x + y = 1$
$3x + 2y = 3$

Section 8.3

In questions 5–7, solve by the addition method.

5. $5x - 3y = 10$
$4x + 7y = 8$

6. $\frac{1}{2}x - \frac{3}{4}y = \frac{3}{2}$
$x + y = 8$

7. $0.8x - 0.3y = 0.7$
$1.2x + 0.6y = 4.2$

8. If a system of linear equations has one solution, the graphs of the equations will be lines that _____.

9. If a system of linear equations has no solutions, the graphs of the equations will be lines that _____.

10. If a system of linear equations has an infinite number of solutions, the graphs of the equations will be lines that _____.

Section 8.4

Solve by any method.

11. $x + 8y = 7$
$5x - 2y = 14$

12. $3x - y = 5$
$-5x + 2y = -10$

13. $x + y = 10$
$0.05x + 0.25y = 2$

Section 8.5

14. James and Michael work sorting letters for the post office. Yesterday James sorted letters for 3 hours and Michael sorted for 2 hours. Together they sorted 2250 letters. Today James sorted letters for 2 hours and Michael for 3 hours. Together they sorted 2650 letters. How many letters per hour can be sorted by James? By Michael?

In Chapter 7 we examined linear equations and inequalities in two variables. In this chapter we will examine systems of linear equations and inequalities. Two or more equations or inequalities in several variables that are considered simultaneously are called a **system of equations** or a **system of inequalities**. Recall that the graph of a linear equation is a straight line. If we have the graphs of two linear equations on one coordinate plane, how might they relate to one another?

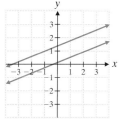

The lines may be parallel.

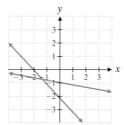

The lines may intersect.

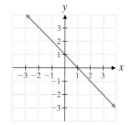

The lines may coincide.

The solutions to a system of linear equations are those points that both lines have in common. Note that parallel lines have no points in common. Thus the first system has no solution. Intersecting lines intersect at one point. Thus the second system has one solution. Lines that coincide have an infinite number of points in common. Thus the third system has an infinite number of solutions. You can determine the solution to a system of equations by graphing.

1 **Solving a System of Linear Equations That Has a Unique Solution by Graphing**

To solve a system of equations by graphing, graph both equations on the same coordinate system. If the lines intersect, the system has a unique solution. The ordered pair of coordinates of the point of intersection is the solution to the system of equations.

EXAMPLE 1 Solve by graphing.

$$-2x + 3y = 6$$

$$2x + 3y = 18$$

Graph both equations on the same coordinate plane and determine the point of intersection. The graph of the system is shown in the figure at the right. The lines intersect at the point for which $x = 3$ and $y = 4$. Thus the unique solution to the system of equations is $(3, 4)$. Always check your answer.

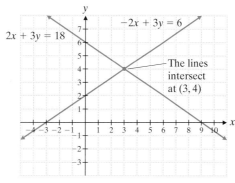

$$-2(3) + 3(4) \overset{?}{=} 6 \qquad 2(3) + 3(4) \overset{?}{=} 18$$

$$6 = 6 \qquad\qquad 18 = 18 \quad \checkmark$$

A linear system of equations that has one solution is said to be **consistent**.

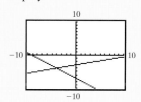

PRACTICE PROBLEM 1

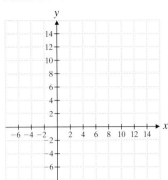

Practice Problem 1   Solve by graphing. $x + y = 12$
$$-x + y = 4$$

2 Graphing a System of Linear Equations to Determine the Type of Solution

Two lines in a plane may intersect, may never intersect (be parallel lines), or may be the same line. The corresponding system of equations will have one solution, no solution, or an infinite number of solutions, respectively. Thus far we have focused on systems of linear equations that have a unique solution. We now draw your attention to the other two cases.

EXAMPLE 2   Solve by graphing.
$$3x - y = 1$$
$$3x - y = -7$$

We graph both equations on the same coordinate plane. The graph is at the right. Notice that the lines are parallel. They do not intersect. Hence there is no solution to this system of equations.
A system of linear equations that has no solution is called **inconsistent**.

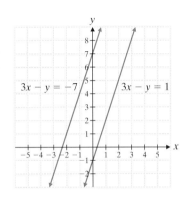

PRACTICE PROBLEM 2

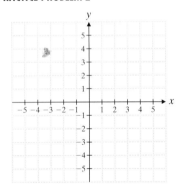

Practice Problem 2   Solve by graphing.
$$4x + 2y = 8$$
$$-6x - 3y = 6$$

To Think About   Without graphing, how could you tell that the system in Example 2 represented parallel lines? Without graphing, determine whether each of the following systems is inconsistent (has no solution) or consistent (has a solution).

(a) $2x - 3y = 6$ **(b)** $3x + y = 2$ **(c)** $x - 2y = 1$
       $6y = 4x + 1$ $2x + y = 3$ $3x - 6y = 3$

EXAMPLE 3   Solve by graphing.
$$x + y = 4$$
$$3x + 3y = 12$$

We graph both equations on the same coordinate plane at the right. Notice that the equations represent the same line (coincide). The coordinates of every point on the line will satisfy both equations. That is, every point on the line represents a solution to $x + y = 4$ and to $3x + 3y = 12$. Thus there is *an infinite number of solutions* to this system.
Such equations are said to be **dependent**.

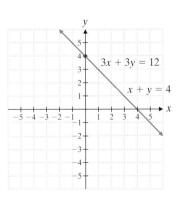

PRACTICE PROBLEM 3

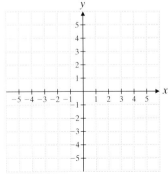

Practice Problem 3   Solve by graphing.
$$3x - 9y = 18$$
$$-4x + 12y = -24$$

❸ Solving Applied Problems by Graphing

⬭ EXAMPLE 4   Walter and Barbara need some plumbing repairs done at their house. They called two plumbing companies for estimates of the work that needed to be done. Roberts Plumbing and Heating charges $40 for a house call and then $35 per hour for labor. Instant Plumbing Repairs charges $70 for a house call and then $25 per hour for labor.

(a) Create a cost equation for each company where y is the total cost of plumbing repairs and x is the number of hours of labor. Write a system of equations.

(b) Graph the two equations using the values $x = 0, 3$, and 6.

(c) Determine from your graph how many hours of plumbing repairs would be required for the two companies to charge the same.

(d) Determine from your graph which company charges less if the estimated amount of time to complete the plumbing repairs is 4 hours.

(a)   For each company we will obtain an equation of the form

$$y = (\text{cost of house call}) + (\text{cost per hour})x.$$

Roberts Plumbing and Heating charges $40 for a house call and $35 per hour.
    Thus we obtain the first equation, $y = 40 + 35x$.
Instant Plumbing Repairs charges $70 for a house call and $25 per hour.
    Thus we obtain the second equation, $y = 70 + 25x$.
This yields the following system of equations:

$$y = 40 + 35x$$
$$y = 70 + 25x.$$

(b)   We will graph the system by first obtaining a table of values for each equation.

Roberts Plumbing and Heating $y = 40 + 35x$

Let $x = 0$ $y = 40 + 35(0) = 40$
Let $x = 3$ $y = 40 + 35(3) = 145$
Let $x = 6$ $y = 40 + 35(6) = 250$

x	y
0	40
3	145
6	250

Instant Plumbing Repairs $y = 70 + 25x$

Let $x = 0$ $y = 70 + 25(0) = 70$
Let $x = 3$ $y = 70 + 25(3) = 145$
Let $x = 6$ $y = 70 + 25(6) = 220$

The graph of the system follows.

x	y
0	70
3	145
6	220

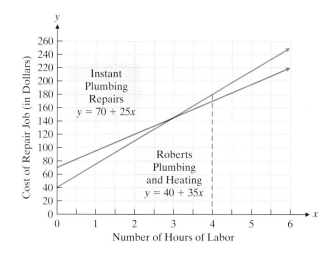

(c) We see that the graphs of the two lines intersect at $(3, 145)$. Thus the two companies will charge the same if 3 hours of plumbing repairs are required.

(d) We draw a dashed line at $x = 4$. We see that the blue line representing Roberts Plumbing and Heating is higher than the red line representing Instant Plumbing Repairs after 3 hours. Thus the cost would be less if we use Instant Plumbing Repairs for 4 hours of work.

Practice Problem 4   Jonathan and Joanne Wells need some electrical work done on their new house. They obtained estimates from two companies. Bill Tupper's Electrical Service charges $100 for a house call and $30 per hour for labor. Wire for Hire charges $50 for a house call and $40 per hour for labor.

(a) Create a cost equation for each company where y is the total cost of the electrical work and x is the number of hours of labor. Write a system of equations.

(b) Graph the two equations using the values $x = 0, 4,$ and 8.

(c) Determine from your graph how many hours of electrical repairs would be required for the two companies to charge the same.

(d) Determine from your graph which company charges less if the estimated amount of time to complete the electrical repairs to the house is 6 hours.

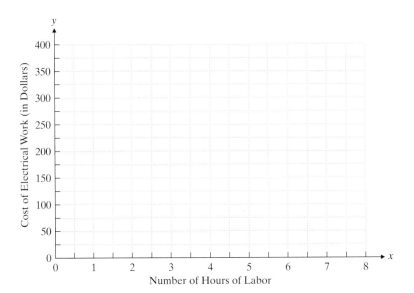

Solve by graphing. If there isn't a unique solution to the system, state the reason.

1. $x - y = 3$
$x + y = 5$

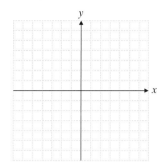

2. $x - y = 4$
$-x - y = 6$

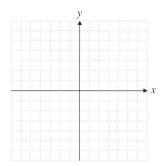

3. $2x + y = 0$
$-3x + y = 5$

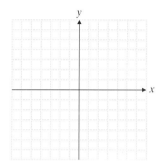

4. $x + 2y = 5$
$-3x + 2y = 17$

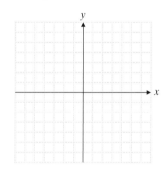

5. $2x + y = 6$
$-2x + y = 2$

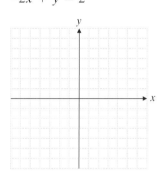

6. $x - 3y = 6$
$4x + 3y = 9$

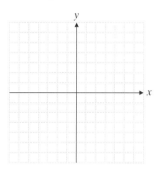

7. $-2x + y - 3 = 0$
$4x + y + 3 = 0$

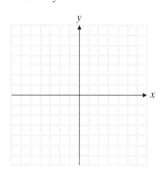

8. $x - 2y - 10 = 0$
$2x + 3y - 6 = 0$

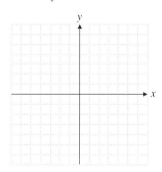

9. $3x - 2y = -18$
$2x + 3y = 14$

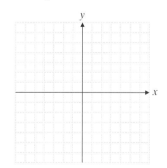

10. $3x + 2y = -10$
$-2x + 3y = 24$

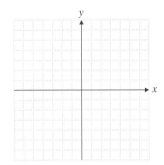

11. $y = \dfrac{3}{4}x + 7$

$y = -\dfrac{1}{2}x + 2$

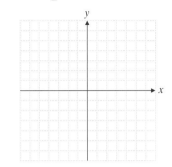

12. $y = \dfrac{5}{7}x - 2$

$y = \dfrac{1}{3}x + \dfrac{2}{3}$

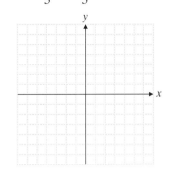

13. $3x - 2y = -4$
$-9x + 6y = -9$

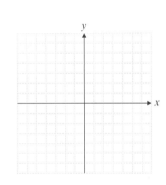

14. $4x - 6y = 8$
$-2x + 3y = -4$

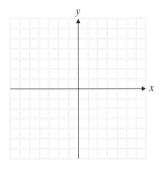

15. $y - 2x - 6 = 0$

$\dfrac{1}{2}y - 3 = x$

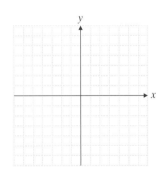

16. $2y + x - 6 = 0$

$y + \dfrac{1}{2}x = 4$

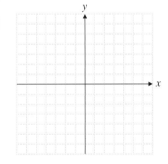

Verbal and Writing Skills

17. In the system $y = 2x - 5$ and $y = 2x + 6$, the lines have the same slope but different y-intercepts. What does this tell you about the graph of this system?

18. In the system $y = -3x + 4$ and $y = -3x + 4$, the lines have the same slope and the same y-intercepts. What does this tell you about the graph of the system?

19. Before you graph a system of equations, if you notice that the lines have different slopes, you can conclude that they _____ and that the system has _____ solution.

20. If lines have the same slope, you cannot conclude that the lines are parallel. Why? How else might the lines relate to each other?

21. Two lines have different slopes but the same y-intercept. What does this tell you about the graphs of the lines and about the solution of the system?

22. Fred Lohnes needs to have some transmission repair work done on his Chrysler Concorde. Cityside Chrysler charges $100 for an initial examination and $40 an hour for labor costs for the mechanic. Transmission World charges $30 for the initial examination and $50 an hour for labor costs for the mechanic.
   (a) Create a cost equation for each company where y is the total cost of the transmission work and x is the number of hours of labor. Write a system of equations.
   (b) Graph the two equations using the values $x = 0, 7,$ and 10.
   (c) Determine from your graph how many hours of transmission repairs would be required for the two companies to charge the same.
   (d) Determine from your graph which company charges less to perform the transmission repairs if the number of hours of labor is 5 hours.

23. Fred and Amy want to have shrubs planted in front of their new house. They have obtained estimates from two landscaping companies. Camp Property Care charges $100 for an initial consultation and $60 an hour for labor. Manchester Landscape Designs charges $200 for an initial consultation and $40 an hour for labor.
   (a) Create a cost equation for each company where y is the total cost of the landscaping work and x is the number of hours of labor. Write a system of equations.
   (b) Graph the two equations using the values $x = 0, 5,$ and 10.
   (c) Determine from your graph how many hours of labor would be required for the cost of landscaping from each of the two companies to be the same.
   (d) Determine from your graph which company would charge less if 7 hours of landscaping are required.

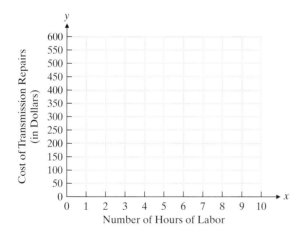

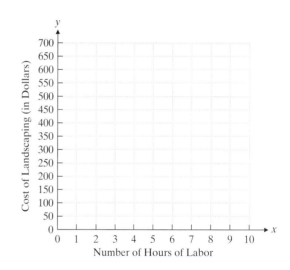

Graphing Calculator Exercises (Optional)

If you have a graphing calculator, solve each system by graphing the equations and estimating the point of intersection to the nearest hundredth. You will need to adjust the window for the appropriate x- and y-values in order to see the region where the straight lines intersect.

24. $y = 56x + 1808$
$y = -62x - 2086$

25. $y = 47x - 960$
$y = -36x + 783$

26. $88x + 57y = 683.10$
$95x - 48y = 7460.64$

27. $64x + 99y = -8975.52$
$58x - 73y = 3503.58$

Cumulative Review Problems

28. Solve. $7 - 3(4 - x) \le 11x - 6$

29. Solve for d. $3dx - 5y = 2(dx + 8)$

30. Multiply. $(3x - 7)^2$

31. Factor. $25x^2 - 60x + 36$

In 1999, Machu Picchu, the 15th-century Incan citadel perched between the Andes and the Amazon jungle, drew a record 360,000 visitors. This was 20% more than the year before and four times higher than in 1991. According to the government, the ideal number of tourists would be just under two million a year. Critics state that the ruins can sustain only 1000 to 1500 visitors a day without suffering further deterioration. The general manager of a company that plans to install and operate a tram there believes that 400,000 to 500,000 visitors annually would be a better target than two million. (Source: Peru Department of Tourism)

32. How many people visited Machu Picchu in 1991?

33. How many people visited Machu Picchu in 1998?

34. In the view of the critics, what is the maximum number of people per year that should be allowed to visit Machu Picchu?

35. If the target of 500,000 visitors offered by the general manager of the tram company is achieved, what will be the percent of increase in the number of visitors since 1999?

8.2 Solving a System of Equations by the Substitution Method

1 *Solving a System of Two Linear Equations with Integer Coefficients by the Substitution Method*

Since the solution to a system of two linear equations may be a point where the two lines intersect, we would expect the solution to be an ordered pair. Since the point must lie on both lines, the ordered pair (x, y) must satisfy both equations. For example, the solution to the system $3x + 2y = 6$ and $x + y = 3$ is $(0, 3)$. Let's check.

$$3x + 2y = 6 \qquad\qquad x + y = 3$$
$$3(0) + 2(3) \overset{?}{=} 6 \qquad\qquad 0 + 3 \overset{?}{=} 3$$
$$6 = 6 \;\checkmark \qquad\qquad\qquad 3 = 3 \;\checkmark$$

Given a system of linear equations, we can develop algebraic methods for finding a solution. These methods reduce the system to one equation in one variable, which we already know how to solve. Once we have solved for this variable, we can substitute this known value into any one of the original equations and solve for the second variable.

The first method we will discuss is the **substitution method**.

Student Learning Objectives

After studying this section, you will be able to:

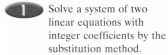

1 Solve a system of two linear equations with integer coefficients by the substitution method.

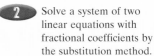

2 Solve a system of two linear equations with fractional coefficients by the substitution method.

SSM   PH TUTOR CENTER   CD & VIDEO   MATH PRO   WEB

Procedure for Solving a System of Equations by the Substitution Method

1. Solve one of the two equations for one variable.
2. Substitute the expression you obtain in step 1 for this variable in the *other* equation.
3. You now have one equation with one variable. Solve this equation to find the value for that one variable.
4. Substitute this value for the variable into one of the original equations to obtain a value for the other variable.
5. Check the solution in each equation to verify your results.

EXAMPLE 1 Find the solution.

$$x - 2y = 7$$
$$-5x + 4y = -5$$

Step 1 Solve one equation for one variable.

$$x - 2y = 7 \qquad \text{The first equation is the easiest one in which to isolate a variable.}$$

$$x = 7 + 2y \qquad \text{Add } 2y \text{ to both sides to solve for } x.$$

Step 2 Substitute the expression into the other equation.

$$-5x + 4y = -5 \qquad \text{This is the original second equation.}$$
$$-5(7 + 2y) + 4y = -5 \qquad \text{Substitute the value } 7 + 2y \text{ for } x \text{ in this equation.}$$

Step 3 Solve this equation.

$$-35 - 10y + 4y = -5 \qquad \text{Remove the parentheses.}$$
$$-35 - 6y = -5 \qquad \text{Collect like terms.}$$
$$-6y = 30 \qquad \text{Add 35 to both sides.}$$
$$y = -5 \qquad \text{Divide both sides by } -6.$$

483

Step 4 Obtain the value of the other variable. We will now use the value $y = -5$ in one of the equations that contains both variables to find the value for x.

$$x = 7 + 2y \qquad \text{The easiest equation to use is the one that we solved for } x.$$
$$= 7 + 2(-5) \quad \text{Replace } y \text{ by } -5.$$
$$= 7 - 10$$
$$= -3$$

The solution is $(-3, -5)$.

Step 5 Check. To be sure that we have the correct solution, we will need to check that the values obtained for x and y can be substituted into *both original* equations to obtain true mathematical statements.

$$\begin{array}{ll} x - 2y = 7 & -5x + 4y = -5 \\ -3 - 2(-5) \stackrel{?}{=} 7 & -5(-3) + 4(-5) \stackrel{?}{=} -5 \\ -3 + 10 \stackrel{?}{=} 7 & 15 - 20 \stackrel{?}{=} -5 \\ 7 = 7 \quad \checkmark & -5 = -5 \quad \checkmark \end{array}$$

Practice Problem 1 Find the solution.

$$5x + 3y = 19$$
$$2x - y = 12$$

2 Solving a System of Two Linear Equations with Fractional Coefficients by the Substitution Method

If a system of equations contains fractions, clear the equations of fractions *before* performing any other steps.

EXAMPLE 2 Find the solution.

$$\frac{3}{2}x + y = \frac{5}{2}$$
$$-y + 2x = -1$$

We want to clear the first equation of fractions. We observe that the LCD of the fractions is 2.

$$2\left(\frac{3}{2}x\right) + 2(y) = 2\left(\frac{5}{2}\right) \qquad \text{Multiply each term of the equation by the LCD.}$$

$$3x + 2y = 5 \qquad \text{This is equivalent to } \frac{3}{2}x + y = \frac{5}{2}.$$

Now follow the five-step procedure.

Step 1 Solve for one variable. We will use the second equation.

$$-y + 2x = -1$$
$$-y = -1 - 2x \qquad \text{Add } -2x \text{ to both sides.}$$
$$y = 1 + 2x \qquad \text{Multiply each term by } -1.$$

Step 2 Substitute the resulting expression into the other equation.

$$3x + 2(1 + 2x) = 5$$

Step 3 Solve this equation for the variable.

$$3x + 2 + 4x = 5 \qquad \text{Remove parentheses.}$$
$$7x + 2 = 5$$
$$7x = 3$$
$$x = \frac{3}{7}$$

Step 4 Find the value of the second variable. We will use the second equation.

$$-y + 2x = -1$$
$$-y + 2\left(\frac{3}{7}\right) = -1$$
$$-y + \frac{6}{7} = -1$$
$$y - \frac{6}{7} = 1$$
$$y = 1 + \frac{6}{7} = \frac{13}{7}$$

The solution to the system is $\left(\dfrac{3}{7}, \dfrac{13}{7}\right)$.

Step 5 Check.

$$\frac{3}{2}x + y = \frac{5}{2} \qquad\qquad\qquad -y + 2x = -1$$

$$\frac{3}{2}\left(\frac{3}{7}\right) + \left(\frac{13}{7}\right) \overset{?}{=} \frac{5}{2} \qquad -\frac{13}{7} + 2\left(\frac{3}{7}\right) \overset{?}{=} -1$$

$$\frac{9}{14} + \frac{13}{7} \overset{?}{=} \frac{5}{2} \qquad\qquad -\frac{13}{7} + \frac{6}{7} \overset{?}{=} -\frac{7}{7}$$

$$\frac{9}{14} + \frac{26}{14} \overset{?}{=} \frac{35}{14} \qquad\qquad -\frac{7}{7} = -\frac{7}{7} \checkmark$$

$$\frac{35}{14} = \frac{35}{14} \quad \checkmark$$

Practice Problem 2 Find the solution. Be sure to check your answer.

$$\frac{x}{3} - \frac{y}{2} = 1$$
$$x + 4y = -8$$

Graphing Calculator

Finding an Approximate Solution to a System of Equations

We can solve systems of equations graphically by using a graphing calculator. For example, to solve the system of equations in Example 2, first rewrite each equation in slope–intercept form.

$$y = -\tfrac{3}{2}x + \tfrac{5}{2}$$
$$y = 2x + 1$$

Then graph $y_1 = -\tfrac{3}{2}x + \tfrac{5}{2}$ and $y_2 = 2x + 1$ on the same screen. Display:

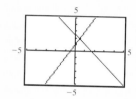

Next you can use the Trace and Zoom features to find the intersection of the two lines. Some graphing calculators have a command to find and calculate the intersection. Display:

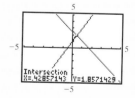

Rounded to four decimal places, the solution is (0.4286, 1.8571). Observe that $\dfrac{3}{7} = 0.4285714\ldots$ and $\dfrac{13}{7} = 1.8571428\ldots$, so the answer agrees with the answer found in Example 2.

Find the solution to each system of equations by the substitution method. Check your answers. Place your solution in the form $(x, y), (a, b), (p, q),$ *or* (s, t).

1. $4x + 3y = 9$
$x - 3y = 6$

2. $x + 4y = 4$
$-x + 2y = 2$

3. $2x + y = 4$
$2x - y = 0$

4. $7x - 3y = -10$
$x + 3y = 2$

5. $5x + 2y = 5$
$3x + y = 4$

6. $4x + 2y = 4$
$3x + y = 4$

7. $3x - 2y = -8$
$x + y = 4$

8. $-x + 3y = -6$
$2x - y = 2$

9. $3a - 5b = 2$
$3a + b = 32$

10. $a - 5b = 21$
$3a + 4b = -13$

11. $3x - 4y = 6$
$-2x - y = -4$

12. $-2x - 3y = -1$
$3x - y = 7$

13. $p + 2q - 4 = 0$
$7p - q - 3 = 0$

14. $7s + 2t - 10 = 0$
$5s - t + 5 = 0$

15. $3x - y - 9 = 0$
$8x + 5y - 1 = 0$

16. $8x + 2y - 7 = 0$
$-2x - y + 2 = 0$

17. $\dfrac{5}{3}x + \dfrac{1}{3}y = -3$
$-2x + 3y = 24$

18. $-x + y = -4$
$\dfrac{3x}{7} + \dfrac{2y}{3} = 5$

19. $4x + 5y = 2$
$\dfrac{1}{5}x + y = -\dfrac{7}{5}$

20. $x + 2y = -6$
$\dfrac{2}{3}x + \dfrac{1}{3}y = -3$

21. $\dfrac{3}{5}x - \dfrac{2}{5}y = 5$
$x + 4y = -1$

22. $-\dfrac{1}{3}x + \dfrac{1}{2}y = 1$
$2x - y = 2$

23. $2x = 4(2y + 2)$
$3(x - 3y) + 2 = 17$

24. $2(2x - y - 11) = 1 - y$
$3(3x + y) - 17 = 2(4x + y)$

Verbal and Writing Skills

25. How many equations do you think you would need to solve a system with three unknowns? With seven unknowns? Explain your reasoning.

26. The point where the graphs of two linear equations intersect is the _____ to the system of linear equations.

27. The solution to a system of two equations must satisfy _____ equations.

28. How many solutions will a system of equations have if the graphs of the lines intersect? Justify your answer.

29. How many solution(s) will a system of equations have if the graphs of the equations are parallel lines? Why?

To Think About

30. Tim Martinez is involved in a large construction project in the city. He needs to rent a heavy construction crane for several weeks. He is considering renting from one of two companies. Boston Construction will rent a crane for an initial delivery charge of $1500 and a rental fee of $900 per week. North End Contractors will rent a crane for an initial delivery charge of $500 and a rental fee of $1000 per week.

(a) Create a cost equation for each company where y is the total cost of renting a crane and x is the number of weeks the crane is rented. Write a system of equations.

(b) Solve the system of equations by the substitution method to find out how many weeks would be required for the two companies to charge the same. What would the cost be for each company?

(c) Tim remembers that last year he considered renting from the same two companies. For the number of weeks he needed the crane, the cost of renting from one company was $4,000 more than the other. How many weeks was the rental last year? Which company was less expensive for that period last year?

31. Walter's mother is planning to move to a retirement community. She is trying to decide between Sunset Acres and Maple Tree Center. A senior citizen moving to Sunset Acres has to pay an initial fee of $5000 and then a monthly rental fee of $2000 per month. At Maple Tree Center the initial fee is $3000 and the monthly fee is $2200.

(a) Create a cost equation for each retirement community where y is the total cost of living there and x is the number of months. Write a system of equations.

(b) Solve the system of equations by the substitution method to find out how many months would be required for the cost of each retirement community to be the same. How many months would this be? What would be the cost for each retirement community?

(c) Walter's mother commented that if she picked her favorite of the two retirement communities, she would save $6000 if she stayed for "several months." How many months would she have to stay for the difference in cost for the two retirement communities to equal $6000? Which retirement community would be less expensive for that period of time?

Cumulative Review Problems

32. Find the slope and y-intercept of $7x + 11y = 19$.

33. Graph $3x - y < 2$.

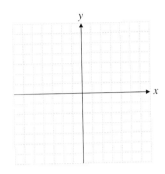

34. Find the running speed r of a 85-kilogram runner who is using up energy at the rate of 35 kilocalories per minute. Use the equation $85(dr - g) = 35$, where d is 0.0011, g is 0.026, and r is measured in meters per minute. Round to the nearest whole number.

35. West Coast Scholastic Construction has been given a contract to build the new wing of the Benson Memorial Middle School for $3.8 million. The construction company has been offered a 3% bonus if the work is completed 3 months ahead of schedule. What will be the total cost of the construction if the work is completed 3 months ahead of schedule?

36. The average American leisure traveler usually vacations with three other people for three days and three nights. If each week the average travel agency books lodgings for 14 leisure travelers (who are making arrangements for their traveling companions) and all of the travelers request single rooms, how many nights in rooms will the travel agency have to book in one year?

37. The average American family that owns a car will take five major family trips by car per year. The average car trip amounts to 360 miles. The average American car owned by a family gets about 20 miles per gallon. If gas costs $1.80 per gallon, what is the total cost of gasoline for family trips for the average American family in one year?

8.3 Solving a System of Equations by the Addition Method

Using the Addition Method to Solve a System of Two Linear Equations with Integer Coefficients

The substitution method is useful for solving a system of equations when the coefficient of one variable is 1. In the following system, the variable x in the first equation has a coefficient of 1, and we can easily solve for x.

$$x - 2y = 7 \Rightarrow x = 7 + 2y$$
$$-5x + 4y = -5$$

This makes the substitution method a natural choice for solving this system. But we may not always have 1 as a coefficient of a variable. We need a method of solving systems of equations that will work for integer coefficients, fractional coefficients, and decimal coefficients. One such method is the **addition method**.

EXAMPLE 1   Solve by addition.

$$5x + 2y = 7 \quad \textbf{(1)}$$
$$3x - y = 13 \quad \textbf{(2)}$$

We would like the coefficients of the y-terms to be opposites. One should be $+2y$ and the other $-2y$. Therefore, we multiply each term of equation **(2)** by 2.

$2(3x) - 2(y) = 2(13)$	Multiply each term of equation **(2)** by 2.
$5x + 2y = 7$	We now have an equivalent system of equations.
$\underline{6x - 2y = 26}$	Add the two equations. This will eliminate the y-variable.
$11x \qquad = 33$	
$x = 3$	Divide both sides by 11.
$5(3) + 2y = 7$	Substitute $x = 3$ into one of the original equations. Arbitrarily, we pick equation **(1)**.
$15 + 2y = 7$	Remove parentheses.
$2y = -8$	Subtract 15 from both sides.
$y = -4$	Divide both sides by 2.

The solution is $(3, -4)$.

   Check: Replace x by 3 and y by -4 in *both* original equations.

$5x + 2y = 7$	$3x - y = 13$
$5(3) + 2(-4) \stackrel{?}{=} 7$	$3(3) - (-4) \stackrel{?}{=} 13$
$15 - 8 \stackrel{?}{=} 7$	$9 + 4 \stackrel{?}{=} 13$
$7 = 7 \quad ✓$	$13 = 13 \quad ✓$

Practice Problem 1   Solve by addition.

$$3x + y = 7$$
$$5x - 2y = 8$$

   Notice that when we added the two equations together, one variable was eliminated. The addition method is therefore often called the **elimination method**.

   For convenience, we will make a list of the steps we use to solve a system of equations by the addition method.

Student Learning Objectives

After studying this section, you will be able to:

 Use the addition method to solve a system of two linear equations with integer coefficients.

 Use the addition method to solve a system of two linear equations with fractional coefficients.

 Use the addition method to solve a system of two linear equations with decimal coefficients.

SSM   PH TUTOR   CD & VIDEO   MATH PRO   WEB
       CENTER

Procedure for Solving a System of Equations by the Addition Method

1. Multiply each term of one or both equations by some nonzero integer so that the coefficients of one of the variables are opposites.
2. Add the equations of this new system so that one variable is eliminated.
3. Solve the resulting equation for the remaining variable.
4. Substitute the value found in step 3 into either of the original equations to find the value of the other variable.
5. Check your solution in both of the original equations.

Care should be used when the solution of a system contains fractions.

EXAMPLE 2   Solve by addition.

$$3x + 4y = 17 \qquad \textbf{(1)}$$
$$2x + 7y = 19 \qquad \textbf{(2)}$$

Suppose we want the x-terms to have opposite coefficients. One equation could have $+6x$ and the other $-6x$. This would happen if we multiply equation **(1)** by $+2$ and equation **(2)** by -3.

$$(2)(3x) + (2)(4y) = (2)(17) \qquad \text{Multiply each term of equation \textbf{(1)} by 2.}$$

$$(-3)(2x) + (-3)(7y) = (-3)(19) \qquad \text{Multiply each term of equation \textbf{(2)} by } -3.$$

(Often this multiplication step can be done mentally.)

$$
\begin{array}{rl}
6x + 8y = 34 & \text{We now have an equivalent system of equations.} \\
\underline{-6x - 21y = -57} & \text{The coefficients of the } x\text{-terms are opposites.} \\
-13y = -23 & \text{Add the two equations to eliminate the variable } x. \\
y = \dfrac{23}{13} & \text{Divide both sides by } -13.
\end{array}
$$

Substitute $y = \dfrac{23}{13}$ into one of the original equations.

$$3x + 4\left(\frac{23}{13}\right) = 17$$

$$3x + \frac{92}{13} = 17 \qquad \text{Solve for } x.$$

$$13(3x) + 13\left(\frac{92}{13}\right) = 13(17) \qquad \text{Multiply both sides by 13 to clear the fraction.}$$

$$39x + 92 = 221 \qquad \text{Remove the parentheses.}$$

$$39x = 129 \qquad \text{Subtract 92 from both sides.}$$

$$x = \frac{43}{13} \qquad \text{Divide both sides by 39. The solution is } \left(\frac{43}{13}, \frac{23}{13}\right).$$

Alternative solution: You could also have eliminated the y-variable. For example, if you multiply equation **(1)** by 7 and equation **(2)** by -4, you obtain the equivalent system shown here.

$$21x + 28y = 119$$
$$-8x - 28y = -76$$

You can add these equations to eliminate the y-variable. Since the numbers involved in this approach are somewhat larger, it is probably wiser to eliminate the x-variable in this example.

Note: A common error occurs when students forget that we want to obtain two terms with *opposite signs* that when added will equal zero. If all the coefficients in the equations are positive, such as in Example 2, it will be necessary to multiply *one* of the equations by a negative number.

Practice Problem 2   Solve by addition.

$$4x + 5y = 17$$
$$3x + 7y = 12$$

2 *Using the Addition Method to Solve a System of Two Linear Equations with Fractional Coefficients*

If the system of equations has fractional coefficients, you should first clear each equation of the fractions. To do so, you will need to multiply each term in the equation by the LCD of the fractions.

EXAMPLE 3   Solve.

$$x - \frac{5}{2}y = \frac{5}{2} \quad \textbf{(1)}$$

$$\frac{4}{3}x + y = \frac{23}{3} \quad \textbf{(2)}$$

$$2(x) - 2\left(\frac{5}{2}y\right) = 2\left(\frac{5}{2}\right) \qquad \text{Multiply each term of equation \textbf{(1)} by 2.}$$

$$2x - 5y = 5$$

$$3\left(\frac{4}{3}x\right) + 3(y) = 3\left(\frac{23}{3}\right) \qquad \text{Multiply each term of equation \textbf{(2)} by 3.}$$

$$4x + 3y = 23$$

We now have an equivalent system of equations that does not contain fractions.

$$2x - 5y = 5 \quad \textbf{(3)}$$
$$4x + 3y = 23 \quad \textbf{(4)}$$

Let us eliminate the *x*-variable. We want the coefficients of *x* to be opposites.

$$(-2)(2x) - (-2)5y = (-2)(5) \qquad \text{Multiply each term of equation \textbf{(3)} by } -2.$$

$$\begin{array}{rl} -4x + 10y = -10 & \text{We now have an equivalent system of equations.} \\ 4x + 3y = 23 & \text{The coefficients of the } x\text{-terms are opposites.} \\ \hline 13y = 13 & \text{Add the two equations to eliminate the variable } x. \\ y = 1 & \text{Divide both sides by 13.} \\ 4x + 3(1) = 23 & \text{Substitute } y = 1 \text{ into equation \textbf{(4)}.} \\ 4x + 3 = 23 \\ 4x = 20 \\ x = 5 \end{array}$$

The solution is $(5, 1)$.

 Check: Check the solution in both of the *original* equations.

Practice Problem 3   Solve.

$$\frac{2}{3}x - \frac{3}{4}y = 3$$

$$-2x + y = 6$$

3 Using the Addition Method to Solve a System of Two Linear Equations with Decimal Coefficients

Some linear equations will have decimal coefficients. It will be easier to work with the equations if we change the decimal coefficients to integer coefficients. To do so, we will multiply each term of the equation by a power of 10.

EXAMPLE 4   Solve.

$$0.12x + 0.05y = -0.02 \quad \textbf{(1)}$$
$$0.08x - 0.03y = -0.14 \quad \textbf{(2)}$$

Since the decimals are hundredths, we will multiply each term of both equations by 100.

$$100(0.12x) + 100(0.05y) = 100(-0.02)$$
$$100(0.08x) + 100(-0.03y) = 100(-0.14)$$

$$12x + 5y = -2 \quad \textbf{(3)} \qquad \text{We now have an equivalent system of equations that has integer coefficients.}$$

$$8x - 3y = -14 \quad \textbf{(4)}$$

We will eliminate the variable y. We want the coefficients of the y-terms to be opposites.

$$
\begin{array}{ll}
36x + 15y = -6 & \text{Multiply equation \textbf{(3)} by 3 and equation \textbf{(4)} by 5.} \\
\underline{40x - 15y = -70} & \\
76x = -76 & \text{Add the equations.} \\
x = -1 & \text{Solve for } x. \\
12(-1) + 5y = -2 & \text{Replace } x \text{ by } -1 \text{ in equation \textbf{(3)}, and solve for } y. \\
-12 + 5y = -2 & \\
5y = -2 + 12 & \\
5y = 10 & \\
y = 2 &
\end{array}
$$

The solution is $(-1, 2)$.

Check: We substitute $x = -1$ and $y = 2$ into the original equations. Most students probably would rather not use equation **(1)** and equation **(2)**! However, *we must check the solutions in the original equations* if we want to be sure of our answers. It is possible to make an error going from equations **(1)** and **(2)** to equations **(3)** and **(4)**. The solutions could satisfy equations **(3)** and **(4)**, but might not satisfy equations **(1)** and **(2)**.

$$
\begin{array}{ll}
0.12x + 0.05y = -0.02 & 0.08x - 0.03y = -0.14 \\
0.12(-1) + 0.05(2) \overset{?}{=} -0.02 & 0.08(-1) - 0.03(2) \overset{?}{=} -0.14 \\
-0.12 + 0.10 \overset{?}{=} -0.02 & -0.08 - 0.06 \overset{?}{=} -0.14 \\
-0.02 = -0.02 \ \checkmark & -0.14 = -0.14 \ \checkmark
\end{array}
$$

Practice Problem 4   Solve. $0.2x + 0.3y = -0.1$
$$0.5x - 0.1y = -1.1$$

Warning

A common error when solving a system is to find the value for x and then stop. You have not solved a system in x and y until you find both the x- and y-values that make both equations true.

8.3 Exercises

Verbal and Writing Skills

1. Look at the following systems of equations. Decide which variable to eliminate in each system, and explain how you would eliminate that variable.

   (a) $3x + 2y = 5$
   $5x - y = 3$

   (b) $2x - 9y = 1$
   $2x + 3y = 2$

   (c) $4x - 3y = 10$
   $5x + 4y = 0$

Find the solution by the addition method. Check your answers.

2. $-x + y = -3$
   $-2x - y = 6$

3. $-x + y = 2$
   $x + y = 4$

4. $2x + 3y = 1$
   $x - 2y = 4$

5. $2x + y = 4$
   $3x - 2y = -1$

6. $x + 5y = 2$
   $2x + 3y = -3$

7. $x + y = 5$
   $2x - y = -5$

8. $5x - 15y = 9$
   $-x + 10y = 1$

9. $-4x + 5y = -16$
   $8x + y = -1$

10. $2x + 5y = 2$
    $3x + y = 3$

11. $5x - 3y = 14$
    $2x - y = 6$

12. $8x + 6y = -2$
    $10x - 9y = -8$

13. $4x + 9y = 0$
    $8x - 5y = -23$

14. $2x + 3y = -8$
    $5x + 4y = -34$

15. $9x - 2y = 14$
    $4x + 3y = 14$

16. $5x - 2y = 6$
    $6x + 7y = 26$

17. $-5x + 4y = -13$
    $11x + 6y = -1$

18. $12x - 6y = -2$
    $-9x - 7y = -10$

19. $2x - 4y = -22$
    $-6x + 3y = 3$

Verbal and Writing Skills

20. Before using the addition method to solve each system, you will need to simplify the equation(s). Explain how you would change the fractional coefficients to integer coefficients in each system.

    (a) $\dfrac{1}{4}x - \dfrac{3}{4}y = 2$
    $2x + 3y = 1$

    (b) $5x + 2y = 9$
    $\dfrac{4}{9}x - \dfrac{2}{3}y = 4$

    (c) $\dfrac{1}{2}x + \dfrac{2}{3}y = \dfrac{1}{3}$
    $\dfrac{3}{4}x - \dfrac{4}{5}y = 2$

Find the solution. Check your answers.

21. $x + \dfrac{5}{4}y = \dfrac{9}{4}$
    $\dfrac{2}{5}x - y = \dfrac{3}{5}$

22. $\dfrac{2}{3}x + y = 2$
    $x + \dfrac{1}{2}y = 7$

23. $\dfrac{x}{6} + \dfrac{y}{2} = -\dfrac{1}{2}$
    $x - 9y = 21$

24. $\dfrac{3}{2}x - \dfrac{y}{8} = -1$
    $16x + 3y = -28$

25. $\dfrac{5}{6}x + y = -\dfrac{1}{3}$

$-8x + 9y = 28$

26. $\dfrac{2}{9}x - \dfrac{1}{3}y = 1$

$4x + 9y = -2$

27. $\dfrac{1}{2}x - \dfrac{3}{4}y = -\dfrac{1}{2}$

$\dfrac{1}{6}x + \dfrac{1}{2}y = -\dfrac{5}{3}$

28. $\dfrac{x}{5} - \dfrac{y}{3} = \dfrac{4}{5}$

$-\dfrac{x}{7} + \dfrac{y}{3} = -\dfrac{8}{7}$

Verbal and Writing Skills

29. Explain how you would change each system to an equivalent system with integer coefficients. Write the equivalent system.

(a) $0.5x - 0.3y = 0.1$

$5x + 3y = 6$

(b) $0.08x + y = 0.05$

$2x - 0.1y = 3$

(c) $4x + 0.5y = 9$

$0.2x - 0.05y = 1$

Solve for x and y using the addition method.

30. $0.2x - 0.3y = 0.4$

$0.3x - 0.4y = 0.9$

31. $0.5x - 0.2y = 0.5$

$0.4x + 0.7y = 0.4$

32. $0.02x - 0.04y = 0.26$

$0.07x - 0.09y = 0.66$

33. $0.04x - 0.03y = 0.05$

$0.05x + 0.08y = -0.76$

34. $0.4x - 5y = -1.2$

$-0.03x + 0.5y = 0.14$

35. $-0.6x - 0.08y = -4$

$3x + 2y = 4$

To Think About

Find the solution.

36. $5(3x - y) = 2 - (y - x)$

$9(2x - 2) = y - 3x$

37. $4(x - y) = 3(1 - 2x) - y$

$1 = -3(5x + 4y)$

38. $5(x + 5) = 2(y + 13)$

$3x + 11 = 2(17 - y)$

39. $3(x - 6) = -2(1 + y)$

$7(x + 3) = 2y + 25$

Cumulative Review Problems

40. At any given moment of the day, between 5000 and 7000 commercial aircraft are flying in U.S. airspace. If 89% of the air traffic is flying over the contiguous states, how many commercial airplanes are flying over Alaska and Hawaii?

41. An international study of laughter found that Germans laugh an average of 6 minutes a day; the British, 15; the French, 18; and Italians, 19. If Americans laugh $\frac{1}{3}$ times longer than the Germans, what is the average length of time per day that Americans laugh?

42. Carlos and Bobby are professional race car drivers. They are in a 100-lap race on an oval track that is 4.75 miles long. Carlos makes 4 complete laps during the time that Bobby is out of the race for repairs to his car. If Carlos maintains an average speed of 95 miles per hour for the entire race, what average speed will Bobby have to achieve in order to beat Carlos? (Round to the nearest whole number.)

43. At one point during a recent season, the Boston Red Sox had won 45 baseball games and lost 55 games. Tricia and Mindy were discussing the potential for the Red Sox to win 50 percent of all their games for the season. If the team had 62 games remaining at that point, how many of the future games would they need to win in order to achieve that percent?

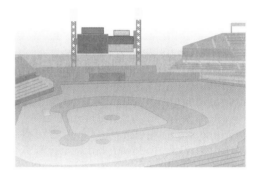

44. Factor $75x^2 - 3$ completely.

45. Factor. $6x^2 - 25x + 25$

Putting Your Skills to Work

The Mathematics of Sports Cars

The Morgan Motor Company in England specializes in handmade sports cars. They emphasize craftsmanship and refined and improved their cars slowly and carefully. After 1936, they did not offer a new model car until they introduced the "Aero 8" and "Plus 8" in 2000. Most of the car dealerships for Morgan Cars are in England. See if you can solve the following problems.

Problems for Individual Investigation and Analysis

1. Determine the prices for the Aero 8 and the Plus 8 for 2000 from the following information. A dealership in Bristol sold three Aero 8s and two Plus 8s in January, for total sales of $349,700. A dealership in Lancaster sold four Aero 8s and three Plus 8s in January, for total sales of $483,800. What was the cost of an Aero 8 and the cost of a Plus 8 in January 2000?

2. The prices were raised slightly for 2001. A dealership in Chichester sold four Aero 8s and five Plus 8s in February 2001, for total sales of $602,000. A dealership in Truro sold seven Aero 8s and six Plus 8s in February 2001, for total sales of $905,000. What was the cost of an Aero 8 and the cost of a Plus 8 in February 2001?

Problems for Group Investigation and Cooperative Learning

3. Suppose a car dealership in Malvern makes a profit of $8500 on each car sold and has annual fixed costs of $256,000. Suppose a car dealership in Southport makes a profit of $9200 on each car sold and has annual fixed costs of $326,000. How many cars does each dealership need to sell in a year in order to make a profit?

4. Suppose the dealership in Malvern and the dealership in Southport sold the same number of cars last year and made exactly the same profit. Using the information in question 3, find out: **(a)** how many cars were sold by each dealer, and **(b)** what each dealer's profit for the year was.

Internet Connections

  Netsite: http://www.prenhall.com/tobey_beginning

Site: Morgan Motor Company

The hand-built cars of the Morgan Motor Company can be ordered in any color you desire. You may custom select a color on your computer screen by combining various percentages (between 0% and 100%) of each of the following four colors: red, green, blue, and black. Go to the color selection page. Use the sliding scales to answer the following questions.

5. What color do you obtain if you combine 25% red, 50% green, 75% blue, and 0% black?

6. What color do you obtain if you combine 25% red, 25% green, 0% blue, and 50% black?

8.4 Review of Methods for Solving Systems of Equations

1 Choosing an Appropriate Method to Solve a System of Equations Algebraically

Student Learning Objectives

After studying this section, you will be able to:

1 Choose an appropriate method to solve a system of equations algebraically.

2 Identify inconsistent and dependent systems by algebraic methods.

At this point we will review the algebraic methods for solving systems of linear equations and discuss the advantages and disadvantages of each method.

Method	Advantage	Disadvantage
Substitution	Works well if one or more variable has a coefficient of 1 or −1.	Often becomes difficult to use if no variable has a coefficient of 1 or −1.
Addition	Works well if equations have fractional or decimal coefficients. Works well if no variable has a coefficient of 1 or −1.	None

SSM · PH TUTOR CENTER · CD & VIDEO · MATH PRO · WEB

EXAMPLE 1 Select a method and solve each system of equations.

(a) $x + y = 3080$
$2x + 3y = 8740$

(b) $5x - 2y = 19$
$-3x + 7y = 35$

(a) Since there are x- and y-values that have coefficients of 1, we will select the substitution method.

$$y = 3080 - x \qquad \text{Solve the first equation for } y.$$
$$2x + 3(3080 - x) = 8740 \qquad \text{Substitute the expression into the second equation.}$$
$$2x + 9240 - 3x = 8740 \qquad \text{Remove parentheses.}$$
$$-1x = -500 \qquad \text{Simplify.}$$
$$x = 500 \qquad \text{Divide each side by } -1.$$
$$y = 3080 - x$$
$$= 3080 - 500 \qquad \text{Substitute 500 for } x.$$
$$= 2580 \qquad \text{Simplify.}$$

The solution is $(500, 2580)$.

(b) Because none of the x- and y-variables have a coefficient of 1 or −1, we select the addition method. We choose to eliminate the y-variable. Thus, we would like the coefficients of y to be −14 and 14.

$$7(5x) - 7(2y) = 7(19) \qquad \text{Multiply each term of the first equation by 7.}$$
$$2(-3x) + 2(7y) = 2(35) \qquad \text{Multiply each term of the second equation by 2.}$$
$$35x - 14y = 133 \qquad \text{We now have an equivalent system of equations.}$$
$$\underline{-6x + 14y = 70}$$
$$29x = 203 \qquad \text{Add the two equations.}$$
$$x = 7 \qquad \text{Divide each side by 29.}$$

Substitute $x = 7$ into one of the original equations.

$$5(7) - 2y = 19$$
$$35 - 2y = 19 \qquad \text{Solve for } y.$$
$$-2y = -16$$
$$y = 8$$

The solution is $(7, 8)$.

Practice Problem 1   Select a method and solve each system of equations.

(a) $3x + 5y = 1485$
          $x + 2y = \ \ 564$

(b) $7x + 6y = 45$
          $6x - 5y = -2$

2 Identifying Inconsistent and Dependent Systems by Algebraic Methods

Recall that an inconsistent system of linear equations is a system of parallel lines. Since parallel lines never intersect, the system has no solution. Can we determine this algebraically?

EXAMPLE 2   Solve algebraically.

$$3x - y = -1$$
$$3x - y = -7$$

Clearly, the addition method would be very convenient in this case.

$3x - y = -1$ Keep the first equation unchanged.
$\underline{-3x + y = +7}$ Multiply each term in the second equation by -1.
$\ \ \ \ \ \ 0 = 6$ Add the two equations.

Notice we have $0 = 6$, which we know is not true. The statement $0 = 6$ is inconsistent with known mathematical facts. No possible x- and y-values can make this equation true. Thus there is no solution to this system of equations.

   This example shows us that, if we obtain a mathematical statement that is not true (inconsistent with known facts), we can identify the entire system as inconsistent. If graphed, these lines would be parallel.

Practice Problem 2   Solve algebraically.

$$4x + 2y = 2$$
$$-6x - 3y = 6$$

   What happens if we try to solve a dependent system of equations algebraically? Recall that a dependent system consists of two lines that coincide (are the same line).

EXAMPLE 3   Solve algebraically.

$$x + \ \ y = 4$$
$$3x + 3y = 12$$

Let us use the substitution method.

$y = 4 - x$ Solve the first equation for y.
$3x + 3(4 - x) = 12$ Substitute $4 - x$ for y in the second equation.
$3x + 12 - 3x = 12$ Remove parentheses.
$12 = 12$

Notice we have $12 = 12$, which is always true. Thus we obtain an equation that is true for any value of x. An equation that is always true is called an **identity**. All the solutions of one equation of the system are also solutions of the other equation. Thus the lines coincide and there are an infinite number of solutions.

   This example shows us that, if we obtain a mathematical statement that is always true (an identity), we can identify the equations as dependent. There are an unlimited number of solutions to a system that has dependent equations.

Practice Problem 3 Solve algebraically.

$$3x - 9y = 18$$
$$-4x + 12y = -24$$

Two Linear Equations with Two Variables

Now is a good time to look back over what we have learned. When you graph a system of two linear equations, what possible kinds of graphs will you obtain?

What will happen when you try to solve a system of two linear equations using algebraic methods? How many solutions are possible in each case? The following chart may help you to organize your answers to these questions.

Graph	Number of Solutions	Algebraic Interpretation
Two lines intersect at one point $(6, -3)$	One unique solution	You obtain one value for x and one value for y. For example, $x = 6, y = -3$.
Parallel lines	No solution	You obtain an equation that is inconsistent with known facts. For example, $0 = 6$.
Lines coincide	Infinite number of solutions	You obtain an equation that is always true. For example, $8 = 8$.

8.4 Exercises

Verbal and Writing Skills

1. If there is no solution to a system of linear equations, the graphs of the equations are _____. Solving the system algebraically, you will obtain an equation that is _____ with known facts.

2. If an algebraic attempt at solving a system of linear equations results in an identity, the system is said to be _____. There are an _____ number of solutions, and the graphs of the lines _____.

3. If there is exactly one solution, the graphs of the equations _____. This system is said to be _____ and _____.

4. A student solves a system of equations and obtains the equation $-36x = 0$. He is unsure of what to do next. What should he do next? What should he conclude about the system of equations?

If possible, solve by an algebraic method (substitution or addition method, without the use of graphing). Otherwise, state that the problem has no solution or an infinite number of solutions.

5. $-2x - 3y = 15$
   $5x + 2y = 1$

6. $2x - 4y = 5$
   $-4x + 8y = 9$

7. $3x + 6y = 12$
   $x + 2y = 7$

8. $-5x + 2y = 2$
   $15x - 6y = -6$

9. $5x - 3y = 12$
   $7x + 2y = -5$

10. $-4x + 5y = 10$
    $6x - 7y = 8$

11. $3x - 2y = 70$
    $0.6x + 0.5y = 50$

12. $5x - 3y = 10$
    $0.9x + 0.4y = 30$

13. $0.3x - 0.5y = 0.4$
    $0.6x + 1.0y = 2.8$

14. $-0.05x + 0.02y = -0.01$
    $0.10x + 0.04y = -0.22$

15. $\frac{4}{3}x + \frac{1}{2}y = 1$
    $\frac{1}{3}x - y = -\frac{1}{2}$

16. $\frac{2}{3}x + \frac{1}{6}y = 2$
    $\frac{1}{4}x - \frac{1}{2}y = -\frac{3}{4}$

To Think About

Remove parentheses and solve the system for (a, b).

17. $2(a + 3) = b + 1$
$3(a - b) = a + 1$

18. $3(a + b) - 2 = b + 2$
$a - 2(b - 1) = 3a - 4$

Cumulative Review Problems

Solve for x.

19. $20x + 12\left(\dfrac{9}{2} - x\right) = 70$

20. $0.3(150) + x = 0.4(x + 150)$

In 1999, a federal income tax rate of 15% applied to single people who had a taxable income of $23,350 or less and to married people filing jointly who had a combined taxable income of $39,000 or less.

21. What was Robert's taxable income if he paid an income tax of $3225 for 1999 and he filed as a single person?

22. What was the combined taxable income of Alberto and Maria Lazano if they paid an income tax of $5775 for 1999 and they filed as a married couple?

23. On a given day, approximately 92,500 auto drivers in the United States are stopped by police officers and given tickets. (*Source:* National Highway Traffic Safety Administration.) If New York, New Jersey, Maryland, California, Texas, and Illinois as a group account for 40% of the tickets, what is the average number of tickets written per day in each of the remaining states? (Round to the nearest whole number.)

24. The cost of owning and operating a midsize car in 1960 was about 10 cents per mile. Today, it is 45.9 cents per mile. If fuel prices increase by 47% next year and if 30% of the cost of operating a midsize car is based on fuel, how much will it cost to run a car next year if other car maintenance costs remain the same?

8.5 Solving Word Problems Using a System of Equations

1 Using a System of Equations to Solve Word Problems

Word problems can be solved in a variety of ways. In Chapter 3 and throughout the text, you have used one equation with one unknown to describe situations found in real-life applications. Some of these same problems can be solved using two equations in two unknowns. The method you use will depend on what you find works best for you. Become familiar with both methods before you decide. Generally, you will want to use an equation to represent each fact provided in the problem situation.

EXAMPLE 1 A worker in a large post office is trying to verify the rate at which two electronic card-sorting machines operate. Yesterday the first machine sorted for 3 minutes and the second machine sorted for 4 minutes. The total workload both machines processed during that time period was 10,300 cards. Two days ago the first machine sorted for 2 minutes and the second machine for 3 minutes. The total workload both machines processed during that time period was 7400 cards. Can you determine the number of cards per minute sorted by each machine?

1. **Understand the problem.**

   The number of cards processed by two different machines on two different days provides the basis for two linear equations with two unknowns.

   The two equations will represent what occurred on two different days:

   (1) yesterday and

   (2) two days ago.

   The number of cards sorted by each machine is what we need to find.
   Let x = the number of cards per minute sorted by the first machine and
   y = the number of cards per minute sorted by the second machine.

2. **Write the equation(s).**

   Yesterday the first machine sorted for 3 minutes and the second machine sorted for 4 minutes, processing a total of 10,300 cards.

first machine		second machine		total no. of cards
3 (no. of cards per minute)		4 (no. of cards per minute)		
$3x$	$+$	$4y$	$=$	$10{,}300$ **(1)**

   Two days ago the first machine sorted for 2 minutes and the second machine sorted for 3 minutes, processing a total of 7400 cards.

first machine		second machine		total no. of cards
2 (no. of cards per minute)		3 (no. of cards per minute)		
$2x$	$+$	$3y$	$=$	7400 **(2)**

3. **Solve and state the answer.**

   We will use the addition method to solve this system.

   $$3x + 4y = 10{,}300 \quad \textbf{(1)}$$
   $$2x + 3y = 7400 \quad \textbf{(2)}$$

   $6x + 8y = 20{,}600$ Multiply equation **(1)** by 2.
   $\underline{-6x - 9y = -22{,}200}$ Multiply equation **(2)** by -3.
   $-1y = -1600$ Add the two equations to eliminate x and solve for y.
   $y = 1600$

   $2x + 3(1600) = 7400$ Substitute the value of y into equation **(2)**.
   $2x + 4800 = 7400$ Simplify and solve for x.
   $2x = 2600$
   $x = 1300$

Thus, the first machine sorts 1300 cards per minute and the second machine sorts 1600 cards per minute.

4. **Check.**

We can verify each statement in the problem.

Yesterday: Were 10,300 cards sorted?

$$3(1300) + 4(1600) \overset{?}{=} 10{,}300$$
$$3900 + 6400 \overset{?}{=} 10{,}300$$
$$10{,}300 = 10{,}300 \quad \checkmark$$

Two days ago: Were 7400 cards sorted?

$$2(1300) + 3(1600) \overset{?}{=} 7400$$
$$2600 + 4800 \overset{?}{=} 7400$$
$$7400 = 7400 \quad \checkmark$$

Practice Problem 1   After a recent disaster, the Red Cross brought in a pump to evacuate water from a basement. After 3 hours a larger pump was brought in and the two ran for an additional 5 hours, removing 49,000 gallons of water. The next day, after running both pumps for 3 hours, the larger one was taken to another site. The smaller pump finished the job after another 2 hours. If on the second day 30,000 gallons were pumped, how much water does each pump remove per hour?

EXAMPLE 2   Fred is considering working for a company that sells encyclopedias. Fred found out that all starting representatives receive the same annual base salary and a standard commission of a certain percentage of the sales they make during the first year. He has been told that one representative sold $50,000 worth of encyclopedias her first year and that she earned $14,000. He was able to find out that another representative sold $80,000 worth of encyclopedias and that he earned $17,600. Determine the base salary and the commission rate of a beginning sales representative.

The earnings of the two different sales representatives will represent the two equations, each of the form

base salary + commission = total earnings.

What is unknown? The base salary and the commission rate are what we need to find.

Let b = the amount of base salary and

c = the commission rate.

$b + 50{,}000c = 14{,}000$ **(1)**   Earnings for the first sales representative.
$b + 80{,}000c = 17{,}600$ **(2)**   Earnings for the second sales representative.

We will use the addition method to solve these equations.

$$
\begin{aligned}
-b - 50{,}000c &= -14{,}000 \qquad \text{Multiply equation (1) by } -1. \\
\underline{b + 80{,}000c} &= \underline{\;\;17{,}600} \qquad \text{Leave equation (2) unchanged.} \\
30{,}000c &= \;\;\;3600 \qquad \text{Add the two equations and solve for } c.
\end{aligned}
$$

$$c = \frac{3600}{30{,}000}$$

$$c = 0.12 \qquad \text{The commission rate is 12\%.}$$

$$b + (50{,}000)(0.12) = 14{,}000 \qquad \text{Substitute this value into equation (1).}$$

$$b + 6000 = 14{,}000 \qquad \text{Simplify and solve for } b.$$

$$b = 8000 \qquad \text{The base salary is \$8000 per year.}$$

Thus the base salary is $8000 per year, and the commission rate is 12%.

Practice Problem 2 Last week Rick bought 7 gallons of unleaded premium and 8 gallons of unleaded regular gasoline and paid $27.22. This week he purchased 8 gallons of unleaded premium and 4 gallons of unleaded regular and paid $22.16. He forgot to record how much each type of fuel cost per gallon. Can you determine these values?

EXAMPLE 3 A lab technician is required to prepare 200 liters of a solution. The prepared solution must contain 42% fungicide. The technician wishes to combine a solution that contains 30% fungicide with a solution that contains 50% fungicide. How much of each solution should he use?

Understand the problem. The technician needs to make one solution from two different solutions. We need to find the amount (number of liters) of each solution that will give us 200 liters of a 42% solution. The two unknowns are easy to identify.

Let x = the number of liters of the 30% solution needed and

y = the number of liters of the 50% solution needed.

Now, what are the two equations? We know that the 200 liters of the final solution will be made by combining x and y. $x + y = 200$ **(1)**

The second piece of information concerns the percent of fungicide in each solution. See the picture at left. $0.3x + 0.5y = 0.42\,(200)$ **(2)**.

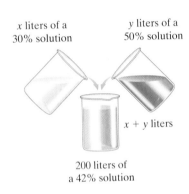

x liters of a
30% solution

y liters of a
50% solution

$x + y$ liters

200 liters of
a 42% solution

Thus our system of equations is

$$x + y = 200 \quad \textbf{(1)}$$
$$0.3x + 0.5y = 84 \quad \textbf{(2)}$$

We will solve this system by the addition method.

$$\begin{aligned} 1.0x + 1.0y &= 200 \\ -0.6x - 1.0y &= -168 \\ \hline 0.4x \phantom{{}- 1.0y} &= 32 \end{aligned}$$

Rewrite equation **(1)** in an equivalent form.

Multiply equation **(2)** by -2.

Solve for x.

$$\frac{0.4x}{0.4} = \frac{32}{0.4}$$

$$x = 80$$

$$80 + y = 200$$
$$y = 120$$

Substitute the value $x = 80$ into one of the original equations and solve for y.

The technician should use 80 liters of the 30% fungicide and 120 liters of the 50% fungicide to obtain the required solution.

Check. Do the amounts total 200 liters?

$$80 + 120 = 200 \ \checkmark$$

Do the amounts yield a 42% strength mixture?

$$0.30x + 0.50y = 0.42(200)$$
$$0.30(80) + 0.50(120) \overset{?}{=} 84$$
$$24 + 60 = 84 \ \checkmark$$

Practice Problem 3 Dwayne Laboratories needs to ship 4000 liters of H_2SO_4 (sulfuric acid). Dwayne keeps in stock solutions that are 20% H_2SO_4 and 80% H_2SO_4. If the strength of the solution shipped is to be 65%, how should Dwayne's stock be mixed?

EXAMPLE 4   Mike recently rode his boat on Lazy River. He took a 48-mile trip up the river traveling against the current in exactly 3 hours. He refueled and made the return trip in exactly 2 hours. What was the speed of his boat in still water and the speed of the current on the river?

$$\text{Let } x = \text{ the speed of the boat in still water in miles/hour and}$$
$$y = \text{ the speed of the river current in miles/hour.}$$

You may want to draw a picture to show the boat traveling in each direction.

Against the current boat takes 3 hours

With the current boat takes 2 hours

Traveling *against* the current, the boat took 3 hours to travel 48 miles. We can calculate the rate (speed) the boat traveled upstream. Use distance = rate · time or, $d = r \cdot t$.

$$48 = 3 \cdot r \qquad r = 16 \text{ miles/hour}$$

The boat was traveling *against* the current. Thus the speed of the boat in still water minus the speed of the current is 16.

$$x - y = 16 \quad \textbf{(1)}$$

Traveling with the current, the boat took 2 hours to travel 48 miles. We can calculate the rate (speed) the boat traveled downstream. Use $d = r \cdot t$.

$$48 = 2 \cdot r \qquad r = 24 \text{ miles/hour}$$

The boat was traveling *with* the current. Thus the speed of the boat in still water plus the speed of the current is 24.

$$x + y = 24 \quad \textbf{(2)}$$

This system is solved most easily by the addition method.

$$
\begin{array}{ll}
x - y = 16 & \textbf{(1)} \\
\underline{x + y = 24} & \textbf{(2)} \\
2x = 40 & \text{Add the two equations.} \\
x = 20 & \text{Solve for } x \text{ by dividing both sides by 2.} \\
x + y = 24 \quad \textbf{(2)} & \text{Substitute the value of } x = 20 \text{ into one of the original} \\
& \text{equations and solve for } y. \text{ We'll use equation } \textbf{(2)}. \\
20 + y = 24 & \\
y = 4 &
\end{array}
$$

Thus the speed of Mike's boat in still water was 20 miles/hour and the speed of the current in Lazy River was 4 miles/hour.

   Check.   Can you verify these answers?

Practice Problem 4   Mr. Caminetti traveled downstream in his boat a distance of 72 miles in a time of 3 hours. His return trip up the river against the current took 4 hours. Find the speed of the boat in still water. Find the speed of the current in the river.

Applications

Solve using two equations with two variables.

1. A restaurant chef is shopping around for the best price for sourdough bread. The Meilleur Pain company has bread priced $0.12 per loaf more than the Corner Loaf company. If you average the two prices for the sourdough bread, you obtain $0.36 per loaf. Find the price of the bread at each company.

▲ 2. In any isosceles triangle, the base angles have the same measure. In a certain isosceles triangle, the measure of the third angle is 60° more than the measure of a base angle. Find the measure of each angle.

3. An art dealer bought 12 etchings for $375. He sold some of them at $35 each and the rest at $40 each, making a profit of $80 on the transaction. How many etchings did he sell at each price?

4. El Segundo High School put on their annual musical. The students sold 650 tickets for a value of $4375. If orchestra seats cost $7.50 and balcony seats cost $3.50, how many of each kind of seat were sold?

▲ 5. KidTime Daycare has plans to expand its playground. Presently it takes 1050 feet of fencing to enclose its rectangular playground. The expansion plans call for the daycare to triple the width of the playground and to double the length. Then the center will require 2550 feet of fencing. What is the length and width of the current playground?

6. Bill is the busiest psychotherapist at a local clinic. Last week he treated 50 patients. The head of the clinic stated that the rest of the staff (eight other psychotherapists) saw four times as many men and three times as many women as Bill saw during the week. This amounted to 179 patients. How many men and how many women did Bill treat?

7. During her beginning mechanics course, Rose learned that her leaking radiator holds 16 quarts of water. She temporarily patched the leak when the radiator was partially full. At that time it was 50% antifreeze. After she filled the remaining space in the radiator with an 80% antifreeze solution and waited a little while, she found that it was 65% antifreeze. How many quarts of solution were in the radiator just before Rose added antifreeze? How many quarts of 80% antifreeze solution did she add?

8. A candy company wants to produce 20 kilograms of a special 45% fat content chocolate to conform with customer dietary demands. To obtain this, a 50% fat content Hawaiian chocolate is combined with a 30% fat content domestic chocolate in a special melting process. How many kilograms of the 50% fat content chocolate and how many kilograms of the 30% fat content chocolate are used to make the required 20 kilograms?

9. The hiking club wants to sell 50 pounds of an energy snack food that will cost them only $1.80 per pound. The mixture will be made up of nuts that cost $2.00 per pound and raisins that cost $1.50 per pound. How much of each should they buy to obtain 50 pounds of the desired mixture?

10. How many bunches of local daisies that cost $2.50 per bunch should be mixed with imported daisies that cost $3.50 per bunch to obtain 100 bunches of daisies that will cost $2.90 per bunch?

11. An airplane traveled between two cities that are 3000 kilometers apart. The trip against the wind took 6 hours. The return trip with the benefit of the wind was 5 hours. What is the wind speed in kilometers per hour? What is the speed of the plane in still air?

12. An airplane flies between Boston and Cleveland. During a round-trip flight, the plane flew a distance of 630 miles each way. The trip with a tailwind (the wind traveling the same direction as the plane) took 3 hours. The return trip traveling against the wind took 3.5 hours. Find the speed of the wind. Find the speed of the airplane in still air.

13. In 1997 the number of full-time state employees in Oklahoma was 72,000. Over a seven-year period, it was observed that the number of state employees was growing. In contrast, the number of full-time state employees in 1997 in Illinois was 141,000. Over a seven-year period, it was observed that this number was decreasing. A relationship between the number of state employees measured in thousands (y) and the number of years since 1990 (x) can be used to predict the number of employees for each state in the future. For Oklahoma, the equation is $x - y = -65$. For Illinois, the equation is $3x + 5y = 725$. (*Source:* U.S. Census Bureau.) One politician joked that if we wait enough years, Illinois and Oklahoma will have the same number of full-time state employees. Solve the system of equations and determine what year that will be. What will be the number of state employees in each state at that time?

14. In 1997 the number of full-time state employees in Indiana was 87,000. Over a seven-year period, it was observed that the number of state employees was decreasing. In contrast, the number of full-time state employees in Nevada was 23,000. Over a seven-year period, it was observed that this number was increasing. A relationship between the number of state employees measured in thousands (y) and the number of years since 1990 (x) can be used to predict the number of employees for each state in the future. For Indiana, the equation is $y = -0.3x + 90$. For Nevada, the equation is $6x - 10y = -180$. (*Source:* U.S. Census Bureau.) One politician joked that if we wait enough years, Indiana and Nevada will have the same number of full-time state employees. Solve the system of equations and determine what year that will be. What will be the number of state employees in each state at that time?

15. In 2000 the population of Maryland was 5,275,000 and is projected to grow moderately. In the same year, the population of Missouri was 5,540,000 and is projected to grow at a slower rate. The relationship between the population of Maryland in thousands (y) and the number of years (x) since 2000 can be approximated by $40x - y = -5275$. A similar relationship for Missouri can be approximated by $28x - 2y = -11,078$. (*Source:* U.S. Census Bureau.) Solve the system of equations to determine the approximate year that the two states will have the same population. What will be the population of each state during that year?

16. In 2000 the population of Iowa was 2,900,000 and was projected to grow at a slow rate. In the same year, the population of Kansas was 2,668,000 and was projected to grow more rapidly. The relationship between the population of Iowa in thousands (y) and the number of years (x) since 2000 can be approximated by $56x - 10y = -29,500$. A similar relationship for Kansas can by approximated by $176x - 10y = -27,100$. (*Source:* U.S. Census Bureau.) Solve the system of equations to determine the approximate year that the two states will have the same population. What will be the population of each state during that year?

17. Sam's commission is a given percent of his total sales of surgical transplant equipment. He receives an 8% commission for selling heart transplant kits and a 6% commission for selling kidney transplant kits. Last month he earned $4480 in commissions. This month he sold the same quantity of heart and kidney kits; however, his company gave him a raise, and he now earns a 10% commission for selling heart kits and an 8% commission for selling kidney kits. This month he earned $5780. What is the total value of the heart kits that he sold last month? What is the total value of the kidney kits?

18. The present population of the town of Springfield is 52,500. The town is growing at the rate of 500 people per year. The population of the city of Essex is decreasing at the rate of 600 people per year. There are presently 80,000 people in Essex. How many years will it be until the populations of Springfield and Essex are the same? What will each population be?

Cumulative Review Problems

19. Add. $\dfrac{1}{x-3} - \dfrac{1}{x+3} + 2$

20. Evaluate when $x = -2$ and $y = 3$.
$2x^2 - 3y + 4y^2$

21. Simplify. $\left(-4x^3y^2\right)^3$

22. Jillian inspected a 32-ounce bottle of cranberry-raspberry drink. She discovered that it contained only 11.2 ounces of pure fruit juice. What percent of the contents of the bottle is pure fruit juice?

Putting Your Skills to Work

Using Mathematics to Predict the Number of Pets

More U.S. households have dogs than have cats. However, many experts in the field say that with increased urbanization, the number of households with cats is increasing at a greater rate than the number of households with dogs.

**Percentage of U.S. Households in 2000
That Have Certain Pets**

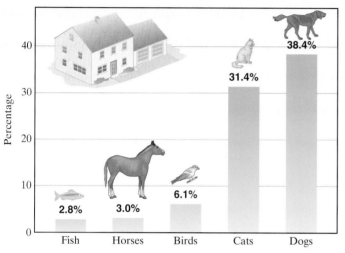

Source: American Veterinary Medical Association

Problems for Individual Investigation and Analysis

Let P = the percentage of households that have one or more pets and x = the number of years since 2000.

1. As the graph shows, 38.4% of households had one or more pet dogs in 2000. Suppose that this percentage has been increasing at a rate of 0.2% per year since then. Write an equation that approximates the percentage x years after 2000.

2. As the graph shows, 31.4% of households had one or more pet cats in 2000. Suppose that this percentage has been increasing at a rate of 0.3% per year since then. Write an equation that approximates the percentage x years after 2000.

Problems for Cooperative Investigation and Group Activities

Together with some other members of your class, see if you can answer the following questions.

3. Solve the equations from questions 1 and 2 simultaneously to determine how many years after 2000 the percentages of households with cats and dogs will be the same.

4. What will be the percentage of households with each pet when this equality is achieved? (Round to the nearest tenth of a percent.)

Internet Connections

  Netsite: http://www.prenhall.com/tobey_beginning

Site: Companion Animal Overpopulation: The Facts & Figures (Doris Day Animal League) or a related site

This site gives facts and figures related to pet overpopulation. Use the information to answer the following questions.

5. Suppose that a neighborhood has 12 female cats. How many kittens can these cats produce in 7 years?

6. Let x be the number of dogs that enter animal shelters in a year. Write an equation of the form $K = ax$ that estimates the number of these dogs that are purebreds. What is the value of a?

7. Let x be the number of cats that enter animal shelters in a year. Write an equation of the form $K = bx$ that estimates the number of these cats that are killed. What is the value of b?

Math in the Media

Using Home Computers to Search for Evidence of Extraterrestrial Life

Sources: Guinness World Records.com; United Devices and SETI@home webpages

On May 17, 1999, the University of California at Berkeley launched SETI@home. The project is the largest Internet-distributed computing project to date. It employs the power of hundreds of thousands of Internet-connected computers to perform the search for radio signals from extraterrestrial civilizations.

At launch, the project received more data than assigned computers could process. So, the call went out to volunteers to donate idle processing time on their home computers. Volunteers download the program that runs as a background task on the computer. The program downloads and analyzes the radio telescope data collected by the Arecibo Radio Telescope in Puerto Rico. When analysis is complete, the computer transmits results to SETI@home which sends users more data for analysis.

According to Guinness World Records.com, SETI@home had over 2, 190,000 users as of July 2000. The hope is that one of them may be the lucky individual whose computer detects evidence of extraterrestrial life.

SETI@home earned its place in the Guinness World Records for the largest computation. This remarkable network of computers reached 3×10^{20} floating point operations as of July 2000. It performs an average of 14 trillion floating point operations per second. Over 500,000 years of processing time were gained over the course of 18 months.

EXERCISES

The exercises that follow challenge you to use your calculating and reasoning skills. A calculator is needed for exercises 1 and 3.

1. How long in seconds does it take to perform one floating point operation with Seti@home? Express your answer in scientific notation. (Carry out to 6 decimal places.)

2. What benefits can you see to performing the research as a distributed computing project vs. the use of 1 or 2 supercomputers?

3. If the number of users of SETI@home increases at the rate of 12% each year, approximately how many users will there be in July 2003?

Chapter 8 Organizer

Topic	Procedure	Examples
Solving a system of equations by graphing, p. 475.	Graph both equations on the same coordinate system. One of the following will be true. **1.** The lines intersect. The point of intersection is the solution to the system. Such a system is consistent. **2.** The lines are parallel. There is no point of intersection. The system has no solution. Such a system is inconsistent. **3.** The lines coincide. Every point on the line represents a solution. There are an infinite number of solutions. Such a system has dependent equations.	**1.** **2.** **3.**
Solving a system of equations: substitution method, p. 483.	**1.** Solve for one variable in terms of the other variable in one of the two equations. **2.** Substitute the expression you obtain for this variable into the other equation. **3.** Solve this equation to find the value of the variable. **4.** Substitute this value for the variable into one of the equations to obtain a value for the other variable. **5.** Check the solution in both equations.	Solve: $$2x + 3y = -6$$ $$3x - y = 13$$ Solve the second equation for y. $-y = -3x + 13$ $$y = 3x - 13$$ Substitute this expression into the first equation. $$2x + 3(3x - 13) = -6$$ $$2x + 9x - 39 = -6$$ $$11x - 39 = -6$$ $$x = 3$$ Substitute $x = 3$ into $y = 3x - 13$. $$y = 3(3) - 13 = 9 - 13 = -4$$ *Check:* $2(3) + 3(-4) \overset{?}{=} -6 \qquad 3(3) - (-4) \overset{?}{=} 13$ $6 - 12 \overset{?}{=} -6 \qquad\qquad 9 - (-4) \overset{?}{=} 13$ $-6 = -6 \ \checkmark \qquad\qquad 9 + 4 = 13 \ \checkmark$ The solution is $(3, -4)$.

Topic	Procedure	Examples
Solving a system of equations: addition method, p. 489.	1. Multiply each term of one or both equations by some nonzero integer so that the coefficients of one of the variables are opposites. 2. Add the equations of this new system so that one variable is eliminated. 3. Solve the resulting equation for the remaining variable. 4. Substitute the value found into either of the original equations to find the value of the other variable. 5. Check your solution in both of the original equations.	Solve: $\begin{aligned} 3x - 2y &= 0 \quad \textbf{(1)} \\ -4x + 3y &= 1 \quad \textbf{(2)} \end{aligned}$ Eliminate the x's: a common multiple of 3 and 4 is 12. Multiply equation **(1)** by 4 and equation **(2)** by 3; then add. $$\begin{aligned} 12x - 8y &= 0 \\ \underline{-12x + 9y} &= \underline{3} \\ y &= 3 \end{aligned}$$ Substitute $y = 3$ into either equation and solve for x. Let us pick $3x - 2y = 0$. $3x - 2(3) = 0 \qquad 3x - 6 = 0, \qquad x = 2$ *Check:* $3(2) - 2(3) \overset{?}{=} 0 \qquad -4(2) + 3(3) \overset{?}{=} 1$ $ 6 - 6 \overset{?}{=} 0 \qquad -8 + 9 \overset{?}{=} 1$ $ 0 = 0 \ \checkmark \qquad 1 = 1 \ \checkmark$ The solution is $(2, 3)$.
Choosing a method to solve a system of equations, p. 497.	1. Substitution works well if at least one variable has a coefficient of 1 or -1. 2. Addition works well for integer, fractional, or decimal coefficients. 3. Graphing gives you a picture of the system. It works well if the system has integer solutions or if the system is inconsistent or dependent.	Use substitution. $$\begin{aligned} x + y &= 8 \\ 2x - 3y &= 9 \end{aligned}$$ Use addition. $$\frac{3}{5}x - \frac{1}{4}y = 4$$ $$\frac{1}{5}x + \frac{3}{4}y = 8$$ Graph to get a picture of the system. If the system is consistent and if it is difficult to determine the solutions because they are fractions, use an algebraic method.
To identify inconsistent and dependent systems algebraically, p. 498.	1. If you obtain an equation that is inconsistent with known facts, such as $0 = 2$, the system is inconsistent. There is no solution. 2. If you obtain an equation that is always true, such as $0 = 0$, the system is dependent. We often say that in this case the equations are dependent. There are an infinite number of solutions.	Solve: $\begin{aligned} x + 2y &= 1 \\ 3x + 6y &= 12 \end{aligned}$ Multiply the first equation by -3. $$\begin{aligned} -3x - 6y &= -3 \\ \underline{3x + 6y} &= \underline{12} \\ 0 &= 9 \end{aligned}$$ 0 is not equal to 9, so there is *no solution*. The system is inconsistent. Solve: $\begin{aligned} 2x + y &= 1 \\ 6x + 3y &= 3 \end{aligned}$ Multiply the first equation by -3. $$\begin{aligned} -6x - 3y &= -3 \\ \underline{6x + 3y} &= \underline{3} \\ 0 &= 0 \end{aligned}$$ Since 0 is always equal to 0, the system is dependent. There are an infinite number of solutions.

Topic	Procedure	Examples
Solving word problems with a system of equations, p. 502.	1. Understand the problem. Choose a variable to represent each unknown quantity. 2. Write a system of equations in two variables. 3. Solve the system of equations and state the answer. 4. Check the answer.	Apples sell for \$0.35 per pound. Oranges sell for \$0.40 per pound. Nancy bought 12 pounds of apples and oranges for \$4.45. How many pounds of each fruit did she buy? Let x = number of pounds of apples and y = number of pounds of oranges. 12 pounds in all were purchased. $x + y = 12$ The purchase cost \$4.45. $0.35x + 0.40y = 4.45$ Multiply the second equation by 100 to obtain the following system.$$\begin{aligned} x + y &= 12 \\ 35x + 40y &= 445 \end{aligned}$$Multiply the top equation by -35 and add the equations.$$\begin{aligned} -35x - 35y &= -420 \\ \underline{35x + 40y} &= \underline{445} \\ 5y &= 25 \\ y &= 5 \end{aligned}$$Substitute $y = 5$ into $x + y = 12$.$$\begin{aligned} x + 5 &= 12 \\ x &= 7 \end{aligned}$$Nancy purchased 7 pounds of apples and 5 pounds of oranges. *Check:* Are there 12 pounds of apples and oranges?$$7 + 5 = 12 \;\checkmark$$Would this purchase cost \$4.45?$$\begin{aligned} 0.35(7) + 0.40(5) &\overset{?}{=} 4.45 \\ 2.45 + 2.00 &\overset{?}{=} 4.45 \\ 4.45 &= 4.45 \;\checkmark \end{aligned}$$

Chapter 8 Review Problems

8.1 *Solve by graphing the lines and finding the point of intersection.*

1. $3x + y = 4$
$3x - 2y = 10$

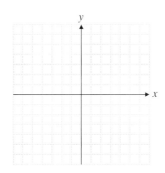

2. $-3x + y = -2$
$-2x - y = -8$

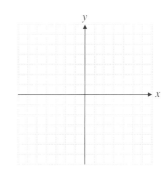

3. $-4x + 3y = 15$
$2x + y = -5$

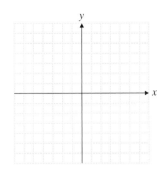

4. $2x - y = 1$
$3x + y = -6$

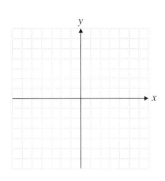

8.2 *Solve by the substitution method. Check your solution.*

5. $x + y = 6$
$-2x + y = -3$

6. $x + 3y = 18$
$2x + y = 11$

7. $5x - y = 4$
$2x - \dfrac{1}{3}y = 6$

8. $3x + 2y = 0$
$2x - y = 7$

8.3 *Solve by the addition method. Check your solution.*

9. $6x - 2y = 10$
$2x + 3y = 7$

10. $4x - 5y = -4$
$-3x + 2y = 3$

11. $2x + 4y = 10$
$4x - 3y = 9$

12. $-5x + 8y = 4$
$2x - 7y = 6$

8.4 *Solve by any appropriate method. If it is not possible to obtain one solution, state the reason. Place the solution in the form (x, y), (a, b), (s, t) or (m, n).*

13. $5x - 3y = 32$
$5x + 7y = -8$

14. $4x + 3y = 46$
$-2x + 3y = -14$

15. $7x + 3y = 2$
$-8x - 7y = 2$

16. $7x + 5y = 7$
$-3x - 4y = 1$

17. $5x - 11y = -4$
$6x - 8y = -10$

18. $3x + 5y = -9$
$4x - 3y = 17$

19. $2x - 4y = 6$
$-3x + 6y = 7$

20. $4x - 7y = 8$
$5x + 9y = 81$

21. $2x - 9y = 0$
$3x + 5 = 6y$

22. $2x + 10y = 1$
$-4x - 20y = -2$

23. $5x - 8y = 3x + 12$
$7x + y = 6y - 4$

24. $1 + x - y = y + 4$
$4(x - y) = 3 - x$

25. $3x + y = 9$
$x - 2y = 10$

26. $x + 12y = 0$
$3x - 5y = -2$

27. $2(x + 3) = y + 4$
$4x - 2y = -4$

28. $x + y = 3000$
$x - 2y = -120$

29. $5x + 4y + 3 = 23$
$8x - 3y - 4 = 75$

30. $4x - 3y + 1 = 6$
$5x + 8y + 2 = -74$

31. $\dfrac{2x}{3} - \dfrac{3y}{4} = \dfrac{7}{12}$
$8x + 5y = 9$

32. $\dfrac{x}{2} + \dfrac{y}{5} = 4$
$\dfrac{x}{3} + \dfrac{y}{5} = \dfrac{10}{3}$

33. $\dfrac{1}{5}a + \dfrac{1}{2}b = 6$
$\dfrac{3}{5}a - \dfrac{1}{2}b = 2$

34. $\dfrac{2}{3}a + \dfrac{3}{5}b = -17$
$\dfrac{1}{2}a - \dfrac{1}{3}b = -1$

35. $0.2s - 0.3t = 0.3$
$0.4s + 0.6t = -0.2$

36. $0.3s + 0.2t = 0.9$
$0.2s - 0.3t = -0.6$

37. $3m + 2n = 5$
$4n = 10 - 6m$

38. $12n - 8m = 6$
$4m + 3 = 6n$

39. $3(x + 2) = -2 - (x + 3y)$
$3(x + y) = 3 - 2(y - 1)$

40. $13 - x = 3(x + y) + 1$
$14 + 2x = 5(x + y) + 3x$

41. $0.2b = 1.4 - 0.3a$
$0.1b + 0.6 = 0.5a$

42. $0.3a = 1.1 - 0.2b$
$0.3b = 0.4a - 0.9$

43. $\dfrac{b}{5} = \dfrac{2}{5} - \dfrac{a - 3}{2}$
$4(a - b) = 3b - 2(a - 2)$

44. $9(b + 4) = 2(2a + 5b)$
$\dfrac{b}{5} + \dfrac{a}{2} = \dfrac{18}{5}$

8.5 *Solve.*

45. During a concert, $3280 worth of tickets were sold. A total of 760 students attended the concert. Reserved seats were $6 and general admission seats were $4. How many of each kind of ticket were sold?

46. In a local fruit market, apples are on sale for $0.25 per pound. Oranges are on sale for $0.39 per pound. Coach Smith purchased 12 pounds of a mixture of apples and oranges before the basketball team left on a bus trip to play at Center City. If the coach paid $3.70, how many pounds of each fruit did he get?

47. A plane travels 1500 miles in 5 hours with the benefit of a tailwind. On the return trip it requires 6 hours to fly against the wind. Can you find the speed of the plane in still air and the wind speed?

48. A chemist needs to combine two solutions of a certain acid. How many liters of a 30% acid solution should be added to 40 liters of a 12% acid solution to obtain a final mixture that is 20% acid?

49. The girls and boys sponsoring the Scouts Car Wash had 86 customers on Saturday. They charged $4 to wash regular-sized cars and $3.50 to wash compact cars. The gross receipts for the day were $323. How many cars of each type were washed?

50. Carl manages a highway department garage in Maine. A mixture of salt and sand is stored in the garage for use during the winter. Carl needs 24 tons of a salt/sand mixture that is 25% salt. He will combine shipments of 15% salt and shipments of 30% salt to achieve the desired 24 tons. How much of each type should he use?

51. A water ski boat traveled 23 kilometers/hour going with the current. It went back in the opposite direction against the current and traveled only 15 kilometers/hour. What was the speed of the boat? What was the speed of the current?

52. Fred is analyzing the cost of producing two different items at an electronics company. An electrical sensing device uses 5 grams of copper and requires 3 hours to assemble. A smaller sensing device made by the same company uses 4 grams of copper but requires 5 hours to assemble. The first device has a production cost of $27. The second device has a production cost of $32. How much does it cost the company for a gram of this type of copper? What is the hourly labor cost at this company? (Assume that production cost is obtained by adding the copper cost and the labor cost.)

Use the following information for exercises 53 and 54.

Fenway Park Tickets for Boston Red Sox 2000 Season			
Field Boxes	$45	Outfield Grandstand Seats	$20
Loge Boxes	$40	Upper Bleacher Seats	$14
Right Field Roof Seats	$27	Lower Bleacher Seats	$16

53. Connie and Dan Lacorazza took the youth group from their church to a Red Sox game. They purchased 23 tickets. Some of the tickets were outfield grandstand seats, while the rest were right field roof seats. The cost of all the tickets was $523. How many of each type of ticket did they buy?

54. Russ and Norma Camp took a group of senior citizens from Hamilton to a Red Sox game. They purchased 27 tickets. Some of the tickets were upper bleacher seats, while the rest were lower bleacher seats. The cost of all the tickets was $406. How many of each type of ticket did they buy?

Putting Your Skills to Work

The Mathematics of the Break-Even Point

Systems of equations are used in the business world to determine the **break-even point**. The break-even point occurs when the revenue received by a company equals the costs incurred by the company. Below the break-even point, a business is operating at a loss. Above the break-even point, a business is operating at a profit. Dave's Candy Company has fixed costs from salaries, rent, light, heat, and telephone of $300 per day. Each pound of candy produced costs $2 per pound for ingredients. Each pound of candy sold yields $3 in revenue.

Let x = the number of pounds of candy,

C = cost per day, and

R = revenue per day.

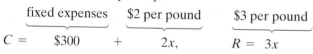

$$C = \$300 + 2x, \qquad R = 3x$$

The break-even point is where cost equals revenue, the intersection of the two lines. This occurs when $x = 300$ pounds of candy.

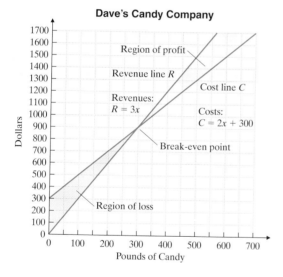

Dave's Candy Company

Problems for Individual Investigation

1. 200 pounds of candy are made and sold in one day. Is this a loss or a profit? Why?

2. If you manufacture and sell 625 pounds of candy daily, how much is the daily profit or loss?

Problems for Cooperative Group Activity and Joint Investigation

Together with other members of your class answer these questions. Campus T-Shirts, Inc., is a local student-run enterprise. It has daily fixed costs of $400. Every T-shirt costs the company $5 to make. Every T-shirt sold yields $7 in revenue.

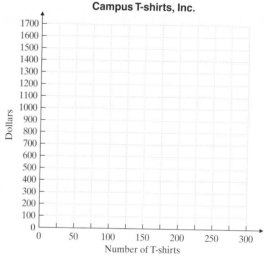

Campus T-shirts, Inc.

3. Write a system of equations to describe the cost and revenue for Campus T-Shirts, Inc., and draw a graph to illustrate the profit and loss for the company.

4. Determine the break-even point by solving the system of equations.

5. What is the daily profit or loss if 320 shirts are made and sold each day?

Internet Connections

 Netsite: http://www.prenhall.com/tobey_beginning

Site: A break-even calculator.

The break-even calculator allows you to vary cost and revenue levels and immediately see the effect on the break-even point.

6. Enter the cost and revenue figures from the example at the beginning of this section. Confirm the break-even point. Now increase the variable cost by 10%. What is the new break-even point? What would it have been if you had decreased the variable cost by 10%? Graph both situations.

7. Figure out three different ways of lowering the break-even point by 20% by either (a) changing the unit cost, (b) changing the fixed cost, or (c) changing the income per unit. What is the new break-even point in each case? What might be the business consequence of each approach in the real world?

517

Chapter 8 Test

1. _____

2. _____

3. _____

4. _____

5. _____

6. _____

7. _____

8. _____

9. _____

10. _____

11. _____

12. _____

13. _____

14. _____

15. _____

16. _____

17. _____

Solve by the method specified.

1. Substitution method
$$3x - y = -5$$
$$-2x + 5y = -14$$

2. Addition method
$$3x + 4y = 7$$
$$2x + 3y = 6$$

3. Graphing method
$$2x - y = 4$$
$$4x + y = 2$$

4. Any method
$$x + 3y = 12$$
$$2x - 4y = 4$$

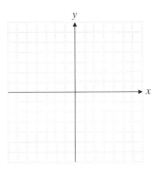

Solve by any method. If there is not one solution to a system, state why.

5. $2x - y = 5$
$-x + 3y = 5$

6. $2x + 3y = 13$
$3x - 5y = 10$

7. $\dfrac{2}{3}x - \dfrac{1}{5}y = 2$
$\dfrac{4}{3}x + 4y = 4$

8. $3x - 6y = 5$
$-\dfrac{1}{2}x + y = \dfrac{7}{2}$

9. $5x - 2 = y$
$10x = 4 + 2y$

10. $0.3x + 0.2y = 0$
$1.0x + 0.5y = -0.5$

11. $2(x + y) = 2(1 - y)$
$5(-x + y) = 2(23 - y)$

12. $3(x - y) = 12 + 2y$
$8(y + 1) = 6x - 7$

Solve.

13. Twice one number plus three times a second is one. Twice the second plus three times the first is nine. Find the numbers.

14. Both $8 and $12 tickets were sold for a basketball game. In all, 30,500 people paid for admission, resulting in a "gate" of $308,000. How many of each type of ticket were sold?

15. Five shirts and three pairs of slacks cost $172. Three of the same shirts and four pairs of the same slacks cost $156. How much does each shirt cost? How much is a pair of slacks?

16. Two machines place address labels on business envelopes. The first machine operated for 8 hours and the second machine for 3 hours. That day 5200 envelopes were affixed with address labels. The next day 4400 envelopes were addressed when the first machine was operated for 4 hours and the second machine for 6 hours. How many envelopes are addressed per hour by each machine?

17. A jet plane flew 2000 kilometers against the wind in a time of 5 hours. It refueled and flew back the same distance with the wind in 4 hours. Find the wind speed in kilometers per hour. Find the speed of the jet in still air in kilometers per hour.

Approximately one-half of this test covers the content of Chapters 0–7. The remainder covers the content of Chapter 8.

Simplify.

1. 11% of $37.20

(Round to the nearest cent.)

2. $\dfrac{-2^4 \cdot \left(2^2\right)}{2^3}$

3. $\left(3x^2 - 2x + 1\right)(5x - 3)$

4. $\dfrac{x^2 - 7x + 12}{x^2 - 9} \div \left(x^2 - 8x + 16\right)$

In questions 5 and 6, solve.

5. $5(3 - x) \geq 6$

6. $6x^2 + x - 2 = 0$

7. 12% of what number is 360?

8. Solve for x. $\dfrac{1}{10}x + \dfrac{1}{3} = \dfrac{2}{5}x - \dfrac{1}{6}$

9. Factor. $5a^3 - 5a^2 - 60a$

10. Solve for x. $4x = 2(12 - 2x)$

Solve by any method. If there is not one solution to a system, state why.

11. $2x + y = 8$
$3x + 4y = -8$

12. $\dfrac{1}{3}x - \dfrac{1}{2}y = 4$
$\dfrac{3}{8}x - \dfrac{1}{4}y = 7$

13. $1.3x - 0.7y = 0.4$
$-3.9x + 2.1y = -1.2$

14. $10x - 5y = 45$
$3x - 8y = 7$

Solve.

15. The equations in a system are graphed and parallel lines are obtained. What is this system called? What does that mean?

16. A mixture of 8% and 20% solutions yields 12,000 liters of a 16% solution. What quantities of the original solutions were used?

17. A boat trip 6 miles upstream takes 3 hours. The return takes 90 minutes. How fast is the stream?

18. Two printers were used for a 5-hour job of printing 15,000 labels. The next day, a second run of the same labels took 7 hours because one printer broke down after 2 hours. How many labels can each printer process per hour?

1. _____

2. _____

3. _____

4. _____

5. _____

6. _____

7. _____

8. _____

9. _____

10. _____

11. _____

12. _____

13. _____

14. _____

15. _____

16. _____

17. _____

18. _____

Radicals

Did you know that if you jumped out of a plane and did not deploy your parachute for twelve seconds, you would fall 2304 feet during that time? How many seconds do you think it would take to fall a distance of one mile? If you think you can solve problems like this, please turn to the Putting Your Skills to Work problems on page 571.

1. _____

2. _____

3. _____

4. _____

5. _____

6. _____

7. _____

8. _____

9. _____

10. _____

11. _____

12. _____

13. _____

14. _____

15. _____

16. _____

17. _____

If you are familiar with the topics in this chapter, take this test now. Check your answers with those in the back of the book. If an answer is wrong or you can't answer a question, study the appropriate section of the chapter.

If you are not familiar with the topics in this chapter, don't take this test now. Instead, study the examples, work the practice problems, and then take the test. This test will help you identify which concepts you have mastered and which you need to study further.

Assume all variables in the radicand represent positive real numbers.

Section 9.1

Find the square roots, if possible. (If it is impossible to obtain a real number for an answer, say so.) Approximate to the nearest thousandth.

1. $\sqrt{81}$ **2.** $\sqrt{\dfrac{25}{64}}$ **3.** $\sqrt{144}$

4. $-\sqrt{100}$ **5.** $\sqrt{5}$ **6.** $-\sqrt{-1}$

Section 9.2

Simplify.

7. $\sqrt{x^4}$ **8.** $\sqrt{25x^4y^6}$ **9.** $\sqrt{40}$

10. $\sqrt{x^5}$ **11.** $\sqrt{36x^3}$ **12.** $\sqrt{8a^4b^3}$

Section 9.3

Combine, if possible.

13. $12\sqrt{7} - 3\sqrt{7}$ **14.** $\sqrt{98} + \sqrt{128}$

15. $3\sqrt{2} - \sqrt{8} + \sqrt{18}$ **16.** $2\sqrt{28} + 3\sqrt{63} - \sqrt{49}$

17. $3\sqrt{8} + \sqrt{12} + \sqrt{50} - 4\sqrt{75}$

Section 9.4

Multiply. Simplify your answer.

18. $(3\sqrt{5})(2\sqrt{10})$

19. $\sqrt{2}(\sqrt{6} + 3\sqrt{2})$

20. $(\sqrt{7} + 4)^2$

21. $(\sqrt{11} - \sqrt{10})(\sqrt{11} + \sqrt{10})$

22. $(\sqrt{m} + \sqrt{m+1})(\sqrt{m} - \sqrt{m+1})$

23. $(2\sqrt{3} - \sqrt{5})^2$

Section 9.5

Simplify. Rationalize all denominators.

24. $\sqrt{\dfrac{36}{x^2}}$

25. $\dfrac{5}{\sqrt{3}}$

26. $\dfrac{3}{\sqrt{5} + 2}$

27. $\dfrac{\sqrt{7} - \sqrt{6}}{\sqrt{7} + \sqrt{6}}$

Section 9.6

Use the Pythagorean theorem to find the length x of the third side of the right triangle.

28.

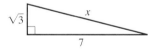

29.

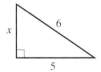

Solve and verify your answer.

30. $2 + \sqrt{3x + 4} = x$

31. $\sqrt{3x + 1} = 6$

32. $\sqrt{11x - 5} = \sqrt{4 - 7x}$

Section 9.7

33. If y varies inversely as x, and $y = \frac{2}{3}$ when $x = 21$, find the value of y if $x = 2$.

18. _____

19. _____

20. _____

21. _____

22. _____

23. _____

24. _____

25. _____

26. _____

27. _____

28. _____

29. _____

30. _____

31. _____

32. _____

33. _____

1 *Evaluating the Square Root of a Perfect Square*

How long is the side of a square whose area is 4? Recall the formula for the area of a square.

$$\text{area of a square} = s^2 \quad \text{Our question then becomes, what number times itself is 4?}$$

$$s^2 = 4$$

$$s = 2 \quad \text{because } (2)(2) = 4$$

$$s = -2 \quad \text{because } (-2)(-2) = 4$$

We say that 2 is a **square root** of 4 because $(2)(2) = 4$. We can also say that -2 is a square root of 4 because $(-2)(-2) = 4$. Note that 4 is a **perfect square**. A square root of a perfect square is an integer.

The symbol $\sqrt{\ }$ is used in mathematics for the **principal square root** of a number, which is the nonnegative square root. The symbol itself $\sqrt{\ }$ is called the **radical sign**. If we want to find the negative square root of a number, we use the symbol $-\sqrt{\ }$.

Definition of Principal Square Root

For all nonnegative numbers N, the principal square root of N (written $\sqrt{N}$) is defined to be the nonnegative number a if and only if $a^2 = N$.

Notice in the definition that we did not use the words "for all *positive* numbers N" because we also want to include the square root of 0. $\sqrt{0} = 0$ since $0^2 = 0$.

Let us now examine the use of this definition with a few examples.

EXAMPLE 1 Find. **(a)** $\sqrt{144}$ **(b)** $-\sqrt{9}$

(a) $\sqrt{144} = 12$ since $(12)^2 = (12)(12) = 144$ and $12 \geq 0$.

(b) The symbol $-\sqrt{9}$ is read "the opposite of the square root of 9."

Because $\sqrt{9} = 3$, we have $-\sqrt{9} = -3$.

Practice Problem 1 Find. **(a)** $\sqrt{64}$ **(b)** $-\sqrt{121}$

The number beneath the radical sign is called the **radicand**. The radicand will not always be an integer.

EXAMPLE 2 Find. **(a)** $\sqrt{\dfrac{1}{4}}$ **(b)** $-\sqrt{\dfrac{4}{9}}$

(a) $\sqrt{\dfrac{1}{4}} = \dfrac{1}{2}$ since $\left(\dfrac{1}{2}\right)^2 = \left(\dfrac{1}{2}\right)\left(\dfrac{1}{2}\right) = \dfrac{1}{4}$.

(b) $-\sqrt{\dfrac{4}{9}} = -\dfrac{2}{3}$ since $\sqrt{\dfrac{4}{9}} = \dfrac{2}{3}$.

Practice Problem 2 Find. **(a)** $\sqrt{\dfrac{9}{25}}$ **(b)** $-\sqrt{\dfrac{121}{169}}$

EXAMPLE 3 Find. **(a)** $\sqrt{0.09}$ **(b)** $\sqrt{1600}$ **(c)** $\sqrt{225}$

(a) $\sqrt{0.09} = 0.3$ since $(0.3)(0.3) = 0.09$.

Notice that $\sqrt{0.09}$ is *not* 0.03 because $(0.03)(0.03) = 0.0009$. Remember to count the decimal places when you multiply decimals. When finding the square root of a decimal, you should multiply to check your answer.

(b) $\sqrt{1600} = 40$ since $(40)(40) = 1600$.

(c) $\sqrt{225} = 15$

Practice Problem 3 Find. **(a)** $-\sqrt{0.0036}$ **(b)** $\sqrt{2500}$

2 *Approximating a Square Root*

Not all numbers are perfect squares. How can we find the square root of 2? That is, what number times itself equals 2? $\sqrt{2}$ is an irrational number. It cannot be written as $\dfrac{p}{q}$, where p and q are integers and $q \neq 0$. $\sqrt{2}$ is a nonterminating, nonrepeating decimal, and the best we can do is approximate its value. To do so, we can use the square root key on a calculator.

To use a calculator, enter 2 and push the $\boxed{\sqrt{x}}$ or $\boxed{\sqrt{}}$ key.

$$\boxed{2}\ \boxed{\sqrt{x}}\ \boxed{1.4142136}$$

Remember, this is only an approximation. Most calculators are limited to an eight-digit display.

To use the table of square roots, look under the x column until you find the number 2. Then look across the row to the $\sqrt{x}$ column. The value there is rounded to the nearest thousandth.

$$\sqrt{2} \approx 1.414$$

x	$\sqrt{x}$
1	1.000
2	1.414
3	1.732
4	2.000
5	2.236
6	2.449
7	2.646

EXAMPLE 4 Use a calculator or a table to approximate. Round to the nearest thousandth.

(a) $\sqrt{28}$ **(b)** $\sqrt{191}$

(a) Using a calculator, we have 28 $\boxed{\sqrt{x}}$ 5.291502622. Thus $\sqrt{28} \approx 5.292$.

(b) Using a calculator, we have 191 $\boxed{\sqrt{x}}$ 13.82027496. Thus $\sqrt{191} \approx 13.820$.

Practice Problem 4 Approximate.

(a) $\sqrt{3}$ **(b)** $\sqrt{35}$ **(c)** $\sqrt{127}$

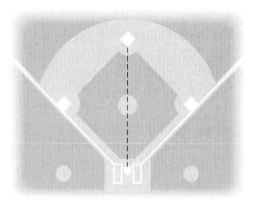

On a professional softball diamond, the distance from home plate to second base is exactly $\sqrt{7200}$ feet. Approximate this value using a calculator.

A Topic for Further Thought

Let's look at the radicand carefully. Does the radicand need to be a nonnegative number? What if we write $\sqrt{-4}$? Is there a number that you can square to get -4?

Obviously, there is *no real number* that you can square to get -4. We know that $(2)^2 = 4$ and $(-2)^2 = 4$. Any real number that is squared will be nonnegative. We therefore conclude that $\sqrt{-4}$ does not represent a real number. Because we want to work with real numbers, our definition of $\sqrt{n}$ requires that n be nonnegative. Thus $\sqrt{-4}$ is *not a real number*.

You may encounter a more sophisticated number system called **complex numbers** in a higher-level mathematics course. In the complex number system, negative numbers have square roots.

9.1 Exercises

Verbal and Writing Skills

1. Define a principal square root in your own words.

2. Is $\sqrt{576} = 24$? Why or why not?

3. Is $\sqrt{0.9} = 0.3$? Why or why not?

4. How would you find $\sqrt{90}$?

Find the two square roots of each number.

5. 9

6. 4

7. 49

8. 36

Find the square root. Do not use a calculator or a table of square roots.

9. $\sqrt{36}$

10. $\sqrt{25}$

11. $\sqrt{81}$

12. $\sqrt{64}$

13. $-\sqrt{49}$

14. $-\sqrt{16}$

15. $\sqrt{0.81}$

16. $\sqrt{0.64}$

17. $\sqrt{\dfrac{36}{121}}$

18. $\sqrt{\dfrac{25}{81}}$

19. $\sqrt{\dfrac{49}{64}}$

20. $\sqrt{\dfrac{49}{100}}$

21. $\sqrt{900}$

22. $\sqrt{2500}$

23. $-\sqrt{10,000}$

24. $\sqrt{40,000}$

25. $\sqrt{144}$

26. $\sqrt{256}$

27. $-\sqrt{\dfrac{1}{64}}$

28. $\sqrt{\dfrac{81}{100}}$

29. $\sqrt{\dfrac{9}{16}}$

30. $-\sqrt{\dfrac{16}{49}}$

31. $\sqrt{0.0049}$

32. $\sqrt{0.0025}$

33. $\sqrt{16,900}$

34. $\sqrt{22,500}$

▲ **35.** A tree is 12 feet tall. A wire that stretches from a point that is 5 feet from the base of the tree to the top of the tree is $\sqrt{169}$ feet long. Exactly how many feet long is the wire?

▲ **36.** A tree is 24 feet tall. A wire that stretches from a point that is 10 feet from the base of the tree to the top of the tree is $\sqrt{676}$ feet long. Exactly how many feet long is the wire?

Use a calculator or the square root table to approximate to the nearest thousandth.

37. $\sqrt{42}$

38. $\sqrt{46}$

39. $\sqrt{74}$

40. $\sqrt{79}$

41. $-\sqrt{133}$

42. $-\sqrt{146}$

43. $-\sqrt{195}$

44. $-\sqrt{182}$

Applications

If we ignore wind resistance, the time in seconds it takes for an object to fall s feet is given by $t = \frac{1}{4}\sqrt{s}$. Thus an object will fall 100 feet in $t = \frac{1}{4}\sqrt{100} = \frac{1}{4}(10) = 2.5$ seconds.

45. How long would it take an object dropped off a building to fall 900 feet?

46. How long would it take an object dropped out of a tower to fall 1600 feet?

47. The Sears Tower in Chicago is one of the world's tallest buildings. It is 1454 feet tall. If a metal frame fell off the building from a height of 1444 feet, how long would it take the metal frame to hit the ground?

48. The CN Tower in Toronto, Ontario, Canada, is the world's tallest self-supporting structure. It is 1821 feet tall. If a worker installing an antenna for cellular phone reception dropped a bolt from a height of 1764 feet, how long would it take the bolt to hit the ground?

Finding Higher Roots

In addition to finding the square root of a number, we can also find the cube root of a number. To find the cube root of N (which is written as $\sqrt[3]{N}$), we find a number that, when cubed, will equal N.

For example, $\sqrt[3]{8} = 2$ since $2^3 = 8$. We can also find the cube root of a negative number—for example, $\sqrt[3]{-8} = -2$ since $(-2)^3 = -8$. The cube root of a positive number is positive. The cube root of a negative number is negative. Therefore each real number has only one real cube root.

The concept of roots can be extended to fourth and fifth roots. For example, $\sqrt[4]{16} = 2$ since $2^4 = 16$. Also, $\sqrt[5]{32} = 2$ since $2^5 = 32$.

Find each higher root.

49. $\sqrt[3]{27}$

50. $\sqrt[3]{64}$

51. $\sqrt[3]{-64}$

52. $\sqrt[3]{-27}$

53. $\sqrt[4]{81}$

54. $\sqrt[4]{625}$

55. $\sqrt[5]{243}$

56. $\sqrt[5]{1024}$

57. Is there a real number that equals $\sqrt[4]{-16}$? Why?

58. Is there a real number that equals $\sqrt[4]{-81}$? Why?

Cumulative Review Problems

59. Solve. $3x + 2y = 8$
$ 7x - 3y = 11$

60. Solve if possible. $2x - 3y = 1$
$ -8x + 12y = 4$

61. A snowboard racer bought three new snowboards and two pairs of racing goggles last month for $850. This month, she bought four snowboards and three pairs of goggles for $1150. How much did each snowboard cost? How much did each pair of goggles cost?

62. With a tailwind, the Qantas flight from Los Angeles to Sydney, Australia, covered 7280 miles in 14 hours, nonstop. The return flight of 7140 miles took 17 hours because of a headwind. If the airspeed of the plane remained constant, and if the return flight encountered a constant headwind equal to the earlier tailwind, what was the plane's airspeed in still air?

NASA obtained $14 billion in federal funds for 2001, a 3.2% increase over the amount obtained in 2000. NASA is seeking $15.6 billion for 2005. Of these amounts $3.2 billion will go toward operations and research for the space shuttle each year. Source: Office of Management and Budget.

63. If NASA obtains the funds it is requesting for 2005, what will the percent of increase in its budget be from 2001 to 2005?

64. What percent of the proposed 2005 NASA budget will go toward the space shuttle?

9.2 Simplifying Radicals

 Simplifying a Radical Expression with a Radicand That Is a Perfect Square

We know that $\sqrt{25} = \sqrt{5^2} = 5$. $\sqrt{9} = \sqrt{3^2} = 3$ and $\sqrt{36} = \sqrt{6^2} = 6$. If we write the radicand as a square of a nonnegative number, the square root is this nonnegative base. That is, $\sqrt{x^2} = x$ when x is nonnegative. We can use this idea to simplify radicals.

EXAMPLE 1 Find. **(a)** $\sqrt{9^4}$ **(b)** $\sqrt{126^2}$ **(c)** $\sqrt{17^6}$

(a) Using the law of exponents, we rewrite 9^4 as a square.

$$\sqrt{9^4} = \sqrt{(9^2)^2} \qquad \text{Raise a power to a power: } (9^2)^2 = 9^4.$$
$$= 9^2 \qquad \text{To check, multiply: } (9^2)(9^2) = 9^4.$$

(b) $\sqrt{126^2} = 126$

(c) $\sqrt{17^6} = \sqrt{(17^3)^2} \qquad \text{Raise a power to a power: } (17^3)^2 = 17^6.$
$$= 17^3 \qquad \text{To check, multiply: } (17^3)(17^3) = 17^6.$$

Practice Problem 1 Find. **(a)** $\sqrt{6^2}$ **(b)** $\sqrt{13^4}$ **(c)** $\sqrt{18^{12}}$

This same concept can be used with variable expressions. We will restrict the values of each variable to be nonnegative if the variable is under the radical sign. Thus *all variable radicands in this chapter* are assumed to *represent positive numbers or zero*.

EXAMPLE 2 Find. **(a)** $\sqrt{x^6}$ **(b)** $\sqrt{y^{24}}$

(a) $\sqrt{x^6} = \sqrt{(x^3)^2} \qquad \text{By the law of exponents, } (x^3)^2 = x^6.$
$$= x^3$$

(b) $\sqrt{y^{24}} = y^{12}$

Practice Problem 2 Find. **(a)** $\sqrt{y^{18}}$ **(b)** $\sqrt{x^{30}}$

In each of these examples we found the square root of a perfect square, and thus the radical sign disappeared. When you find the square root of a perfect square, do *not* leave a radical sign in your answer.

Often the coefficient of a variable radical will not be written in exponent form. You will want to be able to recognize the squares of numbers from 1 to 15, 20, and 25. If you do, you will be able to find many square roots faster mentally than by using a calculator or a table of square roots.

Before you begin, you need to know the multiplication rule for square roots.

Multiplication Rule for Square Roots

For all nonnegative numbers a and b,

$$\sqrt{a} \cdot \sqrt{b} = \sqrt{ab} \quad \text{and} \quad \sqrt{ab} = \sqrt{a} \cdot \sqrt{b}.$$

Student Learning Objectives

After studying this section, you will be able to:

1 Simplify a radical expression with a radicand that is a perfect square.

2 Simplify a radical expression with a radicand that is not a perfect square.

SSM PH TUTOR CD & VIDEO MATH PRO WEB
CENTER

⬭ **EXAMPLE 3** Find. **(a)** $\sqrt{225x^2}$ **(b)** $\sqrt{x^6y^{14}}$ **(c)** $\sqrt{169x^8y^{10}}$

(a) $\sqrt{225x^2} = 15x$ Using the multiplication rule for square roots,
$$\sqrt{225x^2} = \sqrt{225}\,\sqrt{x^2} = 15x.$$

(b) $\sqrt{x^6y^{14}} = x^3y^7$

(c) $\sqrt{169x^8y^{10}} = 13x^4y^5$

To check each answer, square it. The result should be the expression under the radical sign.

Practice Problem 3 Find.

(a) $\sqrt{625y^4}$ **(b)** $\sqrt{x^{16}y^{22}}$ **(c)** $\sqrt{121x^{12}y^6}$

② Simplifying a Radical Expression with a Radicand That Is Not a Perfect Square

Most of the time when we encounter a square root, the radicand is not a perfect square. Thus, when we simplify the square root algebraically, we must retain the radical sign. Still, being able to simplify such radical expressions is a useful skill because it allows us to combine them, as we shall see in Section 9.3.

⬭ **EXAMPLE 4** Simplify. **(a)** $\sqrt{20}$ **(b)** $\sqrt{50}$ **(c)** $\sqrt{48}$

To begin, when you look at a radicand, look for perfect squares.

(a) $\sqrt{20} = \sqrt{4\cdot 5}$ Note that 4 is a perfect square, and we can write 20 as $4\cdot 5$.
$= \sqrt{4}\,\sqrt{5}$ Recall that $\sqrt{4}=2$.
$= 2\sqrt{5}$

(b) $\sqrt{50} = \sqrt{25\cdot 2}$ **(c)** $\sqrt{48} = \sqrt{16\cdot 3}$
$= \sqrt{25}\,\sqrt{2}$ $= \sqrt{16}\,\sqrt{3}$
$= 5\sqrt{2}$ $= 4\sqrt{3}$

Practice Problem 4 Simplify. **(a)** $\sqrt{98}$ **(b)** $\sqrt{12}$ **(c)** $\sqrt{75}$

The same procedure can be used with square root radical expressions containing variables. The key is to think of squares. That is, think, "How can I rewrite the expression so that it contains a perfect square?"

EXAMPLE 5 Simplify. **(a)** $\sqrt{x^3}$ **(b)** $\sqrt{x^7y^9}$

(a) Recall that the law of exponents tells us to add the exponents when multiplying two expressions. Because we want one of these expressions to be a perfect square, we think of the exponent 3 as a sum of an even number and 1: $3 = 2 + 1$.

$\sqrt{x^3} = \sqrt{x^2}\sqrt{x}$ Because $(x^2)(x) = x^3$ by the law of exponents.
$= x\sqrt{x}$

In general, to simplify a square root that has a variable with an exponent in the radicand, first write the square root as a product of the form $\sqrt{x^n}\sqrt{x}$, where n is the largest possible even exponent.

(b) $\sqrt{x^7y^9} = \sqrt{x^6}\sqrt{x}\sqrt{y^8}\sqrt{y}$
$= x^3\sqrt{x}y^4\sqrt{y}$
$= x^3y^4\sqrt{xy}$

In the final simplified form, we place the variable factors with exponents first and the radical factor second.

Practice Problem 5 Simplify. **(a)** $\sqrt{x^{11}}$ **(b)** $\sqrt{x^5y^3}$

In summary, to simplify a square root radical, you factor each part of the expression under the radical sign and simplify the perfect squares.

EXAMPLE 6 Simplify. **(a)** $\sqrt{12y^5}$ **(b)** $\sqrt{18x^3y^7w^{10}}$

(a) $\sqrt{12y^5} = \sqrt{4 \cdot 3 \cdot y^4 \cdot y}$
$= 2y^2\sqrt{3y}$ Remember, $y^4 = (y^2)^2$ and $\sqrt{(y^2)^2} = y^2$.

(b) $\sqrt{18x^3y^7w^{10}} = \sqrt{9 \cdot 2 \cdot x^2 \cdot x \cdot y^6 \cdot y \cdot w^{10}}$
$= 3xy^3w^5\sqrt{2xy}$

Practice Problem 6 Simplify. **(a)** $\sqrt{48x^{11}}$ **(b)** $\sqrt{121x^6y^7z^8}$

9.2 Exercises

Simplify. Leave the answer in exponent form.

1. $\sqrt{8^2}$
2. $\sqrt{2^2}$
3. $\sqrt{10^4}$
4. $\sqrt{8^4}$
5. $\sqrt{9^6}$

6. $\sqrt{7^{10}}$
7. $\sqrt{33^8}$
8. $\sqrt{39^8}$
9. $\sqrt{5^{140}}$
10. $\sqrt{7^{160}}$

Simplify. Assume that all variables represent positive numbers.

11. $\sqrt{x^{14}}$
12. $\sqrt{x^{20}}$
13. $\sqrt{t^{18}}$
14. $\sqrt{w^{22}}$
15. $\sqrt{y^{26}}$

16. $\sqrt{w^{14}}$
17. $\sqrt{36x^8}$
18. $\sqrt{64x^6}$
19. $\sqrt{144x^2}$
20. $\sqrt{196y^{10}}$

21. $\sqrt{x^6y^4}$
22. $\sqrt{x^2y^{18}}$
23. $\sqrt{16x^2y^{20}}$
24. $\sqrt{64x^2y^4}$
25. $\sqrt{100x^{12}y^8}$

26. $\sqrt{20}$
27. $\sqrt{12}$
28. $\sqrt{45}$
29. $\sqrt{75}$
30. $\sqrt{18}$

31. $\sqrt{32}$
32. $\sqrt{72}$
33. $\sqrt{60}$
34. $\sqrt{90}$
35. $\sqrt{125}$

36. $\sqrt{98}$
37. $\sqrt{54}$
38. $\sqrt{8x^3}$
39. $\sqrt{12y^5}$
40. $\sqrt{27w^5}$

41. $\sqrt{25x^5}$
42. $\sqrt{32z^9}$
43. $\sqrt{49x^5}$
44. $\sqrt{75x^3y^5}$
45. $\sqrt{50x^7y}$

46. $\sqrt{48y^3w}$
47. $\sqrt{12x^2y^3}$
48. $\sqrt{75x^2y^3}$
49. $\sqrt{27a^8y^7}$
50. $\sqrt{135x^5y^7}$

51. $\sqrt{180a^5bc^2}$
52. $\sqrt{80a^3b^3c^6}$
53. $\sqrt{63a^4b^6c^3}$
54. $\sqrt{28a^2b^5c^8}$
55. $\sqrt{81x^{12}y^{11}w^5}$

To Think About

Simplify. Assume that all variables represent positive numbers.

56. $\sqrt{(x+5)^2}$
57. $\sqrt{x^2 + 4x + 4}$
58. $\sqrt{16x^2 + 8x + 1}$

59. $\sqrt{4x^2 + 20x + 25}$
60. $\sqrt{x^2y^2 + 14xy^2 + 49y^2}$
61. $\sqrt{16x^2 + 16x + 4}$

Applications

62. In a research experiment at Lucent Technologies, Maria found that the rectangular viewing area of her electron microscope measured $\sqrt{75}$ square millimeters. Write this answer in simplified radical form.

63. In a research project at North Shore Biolabs, Charlene found that the rectangular area of a test slide measured $\sqrt{147}$ square millimeters. Write this answer in simplified radical form.

64. When Maria adjusted the microscope discussed in exercise 62 to a lower magnification, she increased the rectangular viewing area to $\sqrt{1200}$ square millimeters.

(a) Write this answer in simplified radical form.

(b) How many times larger is this area than the original area found in exercise 62?

65. When Charlene conducted the next phase of the experiment discussed in exercise 63, her test slide measured $\sqrt{3675}$ square millimeters.

(a) Write this answer in simplified radical form.

(b) How many times larger is this area than the original area found in exercise 63?

Cumulative Review Problems

66. Solve. $\begin{aligned} 3x - 2y &= 14 \\ y &= x - 5 \end{aligned}$

67. Solve. $\frac{1}{3}(2x - 4) = \frac{1}{4}x - 3$

68. Surprisingly, Asia and Australia lead the world in nuclear energy consumption. In 1984, they used 310 million tons of equivalent energy. In 1993, they used 560 million tons. Assuming the same annual rate of increase in energy consumption, estimate how much energy will be used in 2002.

69. In 1993, the United States consumed 84 quadrillion Btu of energy. This was 265% of the energy consumed by China in that same year. How many quadrillion Btu of energy were consumed by China?

Asia and Australia Nuclear Energy Consumption

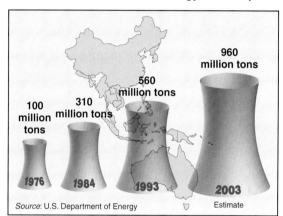

70. What is the expected percent of increase of consumption of nuclear energy for Asia and Australia from 1993 to 2003? Use the data in the Energy Consumption chart above to determine your answer.

1 Adding and Subtracting Radical Expressions with Like Radicals

Recall that when you simplify an algebraic expression you combine like terms. $8b + 5b = 13b$ because both terms contain the same variable b, and you can add the coefficients. This same idea is used when combining radicals. To add or subtract square root radicals, the numbers or the algebraic expressions under the radical signs must be the same. That is, the radicals must be **like radicals**.

EXAMPLE 1   Combine.   **(a)** $5\sqrt{2} - 8\sqrt{2}$ **(b)** $7\sqrt{a} + 3\sqrt{a} - 5\sqrt{a}$

First check to see if the radicands are the same. If they are, you can combine the radicals.

(a) $5\sqrt{2} - 8\sqrt{2} = (5 - 8)\sqrt{2} = -3\sqrt{2}$

(b) $7\sqrt{a} + 3\sqrt{a} - 5\sqrt{a} = (7 + 3 - 5)\sqrt{a} = 5\sqrt{a}$

Practice Problem 1   Combine.   **(a)** $7\sqrt{11} + 4\sqrt{11}$
(b) $4\sqrt{t} - 7\sqrt{t} + 6\sqrt{t} - 2\sqrt{t}$

Be careful. Not all the terms in an expression will have like radicals.

EXAMPLE 2   Combine. $5\sqrt{2a} + 3\sqrt{a} - 7\sqrt{2} + 3\sqrt{2a}$

The only terms that have the same radicand are $5\sqrt{2a}$ and $3\sqrt{2a}$. These terms may be combined. All other terms stay the same.

$$5\sqrt{2a} + 3\sqrt{a} - 7\sqrt{2} + 3\sqrt{2a} = 8\sqrt{2a} + 3\sqrt{a} - 7\sqrt{2}$$

Practice Problem 2   Combine. $3\sqrt{x} - 2\sqrt{xy} - 5\sqrt{y} + 7\sqrt{xy}$

2 Adding and Subtracting Radical Expressions That Must First Be Simplified

Sometimes it is necessary to simplify one or more radicals before their terms can be combined. Be sure to combine only like radicals.

EXAMPLE 3   Combine.   **(a)** $2\sqrt{3} + \sqrt{12}$ **(b)** $\sqrt{12} - \sqrt{27} + \sqrt{50}$

(a) $2\sqrt{3} + \sqrt{12} = 2\sqrt{3} + \sqrt{4 \cdot 3}$ Look for perfect square factors.

$\qquad\qquad\quad = 2\sqrt{3} + 2\sqrt{3}$ Simplify and combine like radicals.

$\qquad\qquad\quad = 4\sqrt{3}$

(b) $\sqrt{12} - \sqrt{27} + \sqrt{50} = \sqrt{4 \cdot 3} - \sqrt{9 \cdot 3} + \sqrt{25 \cdot 2}$

$\qquad\qquad\qquad\qquad = 2\sqrt{3} - 3\sqrt{3} + 5\sqrt{2}$ Only $2\sqrt{3}$ and $-3\sqrt{3}$ can be combined.

$\qquad\qquad\qquad\qquad = -\sqrt{3} + 5\sqrt{2}$ or $5\sqrt{2} - \sqrt{3}$

Practice Problem 3   Combine.   **(a)** $\sqrt{50} - \sqrt{18} + \sqrt{98}$
(b) $\sqrt{12} + \sqrt{18} - \sqrt{50} + \sqrt{27}$

EXAMPLE 4   Combine. $\sqrt{2a} + \sqrt{8a} + \sqrt{27a}$

$$\sqrt{2a} + \sqrt{8a} + \sqrt{27a} = \sqrt{2a} + \sqrt{4 \cdot 2a} + \sqrt{9 \cdot 3a}$$

Look for perfect square factors.

$$= \sqrt{2a} + 2\sqrt{2a} + 3\sqrt{3a}$$

Be careful. $\sqrt{2a}$ and $\sqrt{3a}$ are **not** like radicals. The expressions under the radical signs must be the same.

$$= 3\sqrt{2a} + 3\sqrt{3a}$$

Practice Problem 4   Combine. $\sqrt{9x} + \sqrt{8x} - \sqrt{4x} + \sqrt{50x}$

Special care should be taken if the original radical has a numerical coefficient. If the radical can be simplified, the two resulting numerical coefficients should be multiplied.

EXAMPLE 5   Combine. $2\sqrt{20} + 3\sqrt{45} - 4\sqrt{80}$

$$2\sqrt{20} + 3\sqrt{45} - 4\sqrt{80} = 2 \cdot \sqrt{4} \cdot \sqrt{5} + 3 \cdot \sqrt{9} \cdot \sqrt{5} - 4 \cdot \sqrt{16} \cdot \sqrt{5}$$

$$= 2 \cdot 2 \cdot \sqrt{5} + 3 \cdot 3 \cdot \sqrt{5} - 4 \cdot 4 \cdot \sqrt{5}$$

$$= 4\sqrt{5} + 9\sqrt{5} - 16\sqrt{5}$$

$$= -3\sqrt{5}$$

Practice Problem 5   Combine. $\sqrt{27} - 4\sqrt{3} + 2\sqrt{75}$

EXAMPLE 6   Combine. $3a\sqrt{8a} + 2\sqrt{50a^3}$

$$3a\sqrt{8a} + 2\sqrt{50a^3} = 3a\sqrt{4}\sqrt{2a} + 2\sqrt{25a^2}\sqrt{2a}$$

$$= 3a \cdot 2 \cdot \sqrt{2a} + 2 \cdot 5a \cdot \sqrt{2a}$$

$$= 6a\sqrt{2a} + 10a\sqrt{2a}$$

$$= 16a\sqrt{2a}$$

(*Note:* If you are unsure of the last step, show the use of the distributive property in performing the addition.)

$$6a\sqrt{2a} + 10a\sqrt{2a} = (6 + 10)a\sqrt{2a}$$

$$= 16a\sqrt{2a}$$

Practice Problem 6   Combine. $2\sqrt{12x} - 3\sqrt{45x} - 3\sqrt{27x} + \sqrt{20}$

Verbal and Writing Skills

1. List the steps involved in simplifying an expression such as $\sqrt{75x} + \sqrt{48x} + \sqrt{16x^2}$.

2. Julio said that, since $\sqrt{8}$ and $\sqrt{2}$ are not like radicals, $\sqrt{8} + \sqrt{2}$ cannot be simplified. What do you think?

Combine, if possible. Do not use a calculator or a table of square roots.

3. $3\sqrt{5} - \sqrt{5} + 4\sqrt{5}$

4. $\sqrt{7} + 2\sqrt{7} - 6\sqrt{7}$

5. $\sqrt{2} + 8\sqrt{3} - 5\sqrt{3} + 4\sqrt{2}$

6. $\sqrt{5} - \sqrt{6} + 3\sqrt{5} - 2\sqrt{6}$

7. $\sqrt{5} + \sqrt{45}$

8. $\sqrt{7} - \sqrt{63}$

9. $\sqrt{50} + 3\sqrt{32}$

10. $2\sqrt{12} + \sqrt{48}$

11. $2\sqrt{8} - 3\sqrt{2}$

12. $2\sqrt{27} - 4\sqrt{3}$

13. $\sqrt{18} - \sqrt{2} + \sqrt{98}$

14. $\sqrt{24} + 5\sqrt{6} - \sqrt{54}$

15. $2\sqrt{12} + \sqrt{20} + \sqrt{36}$

16. $\sqrt{24} - 2\sqrt{54} + 2\sqrt{18}$

17. $2\sqrt{48} - 3\sqrt{8} + \sqrt{50}$

18. $\sqrt{50} + \sqrt{28} - \sqrt{9}$

19. $8\sqrt{2x} + \sqrt{50x}$

20. $6\sqrt{5y} - \sqrt{20y}$

21. $1.2\sqrt{3x} - 0.5\sqrt{12x}$

22. $-1.5\sqrt{2a} + 0.2\sqrt{8a}$

23. $\sqrt{20y} + 2\sqrt{45y} - \sqrt{5y}$

24. $\sqrt{72w} - 3\sqrt{2w} - \sqrt{50w}$

25. $3\sqrt{y^3} + 2y\sqrt{y}$

26. $4x\sqrt{x} - \sqrt{16x^3}$

27. $3\sqrt{28x} - 5x\sqrt{63x}$

28. $-2b\sqrt{45b} + 5\sqrt{80b}$

29. $5\sqrt{8x^3} - 3x\sqrt{50x}$

30. $-2\sqrt{27y^3} + y\sqrt{12y}$

31. $3\sqrt{27x^2} - 2\sqrt{48x^2}$

32. $-5\sqrt{8y^2} + 2\sqrt{32y^2}$

33. $2\sqrt{6y^3} - 2y\sqrt{54}$

34. $3\sqrt{20x^3} - 4x\sqrt{5x^2}$

35. $5x\sqrt{8x} - 24\sqrt{50x^3}$

Applications

▲ **36.** A planned unmanned mission to Mars may possibly require a small vehicle to make scientific measurements as it travels along the perimeter of a region like the one shown below. If the region is rectangular and the dimensions are as labeled, find the exact distance to be traveled by the vehicle in one trip around the perimeter.

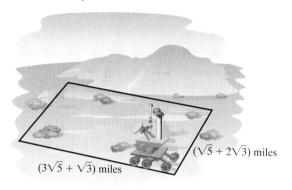

$(\sqrt{5} + 2\sqrt{3})$ miles

$(3\sqrt{5} + \sqrt{3})$ miles

▲ **37.** Suppose that on the second day of the mission to Mars, the vehicle must travel along the perimeter of a right triangle with dimensions as labeled below. Find the exact distance to be traveled by the vehicle in one trip around the perimeter.

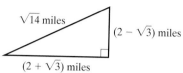

$\sqrt{14}$ miles

$(2 - \sqrt{3})$ miles

$(2 + \sqrt{3})$ miles

Cumulative Review Problems

38. Solve and graph on a number line. $7 - 3x \le 11$

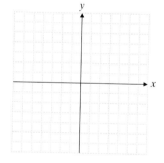

39. Graph on a coordinate system. $3y - 2x \le 6$

40. Enrico went to the local music store and bought 10 CDs for $107.90. Some of the CDs were $7.99 and some were $11.99. How many of each did Enrico buy?

41. Shelly invested $8000 in two places. She bought a certificate of deposit with a return of 7% simple interest, and she invested the rest of the money at 5% simple interest in a money market fund. At the end of the year she earned a total of $432 from these two places. How much did she invest in each place?

42. A real estate developer is planning to build a residential high-rise tower next to the United Nations. This tower is $\frac{1}{6}$ mile tall. If the rooms have 11-foot ceilings, the apartments are separated by 6-foot-thick floors, the lobby has a 31-foot ceiling, and there is a 45-foot space between the highest apartment and the roof (for central air, wiring, cable, and crawl space), how many levels can this $\frac{1}{6}$-mile tower have? Round to the nearest whole number.

1 **Multiplying Monomial Radical Expressions**

Recall the basic rule for multiplication of square root radicals,

$$\sqrt{a}\,\sqrt{b} = \sqrt{ab}.$$

We will use this concept to multiply square root radical expressions. Note that the direction "multiply" means to express the product in simplest form.

EXAMPLE 1 Multiply. $\sqrt{7}\,\sqrt{14x}$

$$\sqrt{7}\,\sqrt{14x} = \sqrt{98x}$$

We do *not* stop here, because the radical $\sqrt{98x}$ can be simplified.

$$\sqrt{98x} = \sqrt{49 \cdot 2x} = \sqrt{49}\,\sqrt{2x} = 7\sqrt{2x}$$

Practice Problem 1 Multiply. $\sqrt{3a}\,\sqrt{6a}$

EXAMPLE 2 Multiply.

(a) $(2\sqrt{3})(5\sqrt{7})$ **(b)** $(2a\sqrt{3a})(3\sqrt{6a})$

(a) $(2\sqrt{3})(5\sqrt{7}) = 10\sqrt{21}$ Multiply the coefficients: $(2)(5) = 10$.
Multiply the radicals: $\sqrt{3}\,\sqrt{7} = \sqrt{21}$.

(b) $(2a\sqrt{3a})(3\sqrt{6a}) = 6a\sqrt{18a^2}$

$$= 6a\sqrt{9}\,\sqrt{2}\,\sqrt{a^2}\quad \text{Simplify } \sqrt{18a^2}.$$

$$= 18a^2\sqrt{2}\qquad\qquad \text{Multiply } (6a)(3)(a).$$

Practice Problem 2 Multiply.

(a) $(2\sqrt{3})(5\sqrt{5})$ **(b)** $(4\sqrt{3x})(2x\sqrt{6x})$

EXAMPLE 3 Find the area of a rectangular computer screen that measures $\sqrt{200}$ inches tall and $\sqrt{120}$ inches wide. Express your answer as a simplified radical.

We multiply.

$$(\sqrt{200})(\sqrt{120}) = (10\sqrt{2})(2\sqrt{30}) = 20\sqrt{60}$$

$$= 20(2)\sqrt{15} = 40\sqrt{15} \text{ square inches.}$$

Practice Problem 3 Find the area of a computer chip that measures $\sqrt{180}$ millimeters long and $\sqrt{150}$ millimeters wide.

2 Multiplying a Monomial Radical Expression by a Polynomial

Recall that a binomial consists of two terms. $\sqrt{2} + \sqrt{5}$ is a binomial. We use the distributive property when multiplying a binomial by another factor.

EXAMPLE 4   Multiply. $\sqrt{5}(\sqrt{2} + 3\sqrt{7})$

$$= \sqrt{5}\,\sqrt{2} + 3\sqrt{5}\,\sqrt{7}$$
$$= \sqrt{10} + 3\sqrt{35}$$

In similar fashion, we can multiply a trinomial by another factor.

Practice Problem 4   Simplify. $2\sqrt{3}(\sqrt{3} + \sqrt{5} - \sqrt{12})$

Note that $\sqrt{3} \cdot \sqrt{3} = 3$. This is a specific case of the multiplication rule for square root radicals.

> For any nonnegative real number a,
> $$\sqrt{a} \cdot \sqrt{a} = a.$$

EXAMPLE 5   Multiply. $\sqrt{a}(3\sqrt{a} - 2\sqrt{5})$

$$= 3\sqrt{a}\,\sqrt{a} - 2\sqrt{5}\,\sqrt{a} \quad \text{Use the distributive property.}$$
$$= 3a - 2\sqrt{5a}$$

Practice Problem 5   Multiply. $2\sqrt{x}(4\sqrt{x} - x\sqrt{2})$

Be sure to simplify all radicals after multiplying.

3 Multiplying Two Polynomial Radical Expressions

Recall the FOIL method used to multiply two binomials. The same method can be used for radical expressions.

Algebraic Expressions

$$(2x + y)(x - 2y) = 2x^2 - 4xy + xy - 2y^2$$
$$= 2x^2 - 3xy - 2y^2$$

Radical Expressions

$$(2\sqrt{3} + \sqrt{5})(\sqrt{3} - 2\sqrt{5}) = 2\sqrt{9} - 4\sqrt{15} + \sqrt{15} - 2\sqrt{25}$$
$$= 2(3) - 3\sqrt{15} - 2(5)$$
$$= 6 - 3\sqrt{15} - 10$$
$$= -4 - 3\sqrt{15}$$

Let's look at this procedure more closely.

EXAMPLE 6 Multiply. $\left(\sqrt{2} + 5\right)\left(\sqrt{2} - 3\right)$

$$\left(\sqrt{2} + 5\right)\left(\sqrt{2} - 3\right)$$

$$= \sqrt{4} - 3\sqrt{2} + 5\sqrt{2} - 15 \qquad \text{Multiply to obtain the four products.}$$
$$= 2 + 2\sqrt{2} - 15 \qquad\qquad\quad \text{Combine the middle terms.}$$
$$= -13 + 2\sqrt{2}$$

Practice Problem 6 Multiply. $\left(\sqrt{2} + \sqrt{6}\right)\left(2\sqrt{2} - \sqrt{6}\right)$

EXAMPLE 7 Multiply. $\left(\sqrt{2} - 3\sqrt{6}\right)\left(\sqrt{2} + \sqrt{6}\right)$

$$\left(\sqrt{2} - 3\sqrt{6}\right)\left(\sqrt{2} + \sqrt{6}\right)$$

$$= \sqrt{4} + \sqrt{12} - 3\sqrt{12} - 3\sqrt{36} \qquad \text{Multiply to obtain the four products.}$$
$$= 2 - 2\sqrt{12} - 18 \qquad\qquad\qquad\quad \text{Combine the middle terms.}$$
$$= -16 - 4\sqrt{3} \qquad\qquad\qquad\qquad\; \text{Simplify } -2\sqrt{12}.$$

Practice Problem 7 Multiply. $\left(\sqrt{6} + \sqrt{5}\right)\left(\sqrt{2} + 2\sqrt{5}\right)$

If an expression with radicals is squared, write the expression as the product of two binomials and multiply.

EXAMPLE 8 Multiply. $\left(2\sqrt{3} - \sqrt{6}\right)^2$

$$= \left(2\sqrt{3} - \sqrt{6}\right)\left(2\sqrt{3} - \sqrt{6}\right)$$
$$= 4\sqrt{9} - 2\sqrt{18} - 2\sqrt{18} + \sqrt{36}$$
$$= 12 - 4\sqrt{18} + 6$$
$$= 18 - 12\sqrt{2}$$

Practice Problem 8 Multiply. $\left(3\sqrt{5} - \sqrt{10}\right)^2$

Multiply. Be sure to simplify any radicals in your answer. Do not use a calculator or a table of square roots.

1. $\sqrt{3}\,\sqrt{10}$

2. $\sqrt{2}\,\sqrt{22}$

3. $\sqrt{2a}\,\sqrt{6a}$

4. $\sqrt{8}\,\sqrt{6b}$

5. $(3\sqrt{5})(2\sqrt{6})$

6. $(2\sqrt{14x})(3x\sqrt{7x})$

7. $(-4a\sqrt{8a})(-a\sqrt{3a})$

8. $(4\sqrt{12})(2\sqrt{6})$

9. $(-3\sqrt{ab})(2\sqrt{b})$

10. $(2\sqrt{x})(5\sqrt{xy})$

11. $\sqrt{3}(\sqrt{7}+5\sqrt{3})$

12. $\sqrt{7}(2\sqrt{3}-3\sqrt{2})$

13. $\sqrt{3}(2\sqrt{6}+5\sqrt{15})$

14. $\sqrt{6}(5\sqrt{12}-4\sqrt{3})$

15. $2\sqrt{x}(\sqrt{x}-8\sqrt{5})$

16. $-3\sqrt{b}(2\sqrt{a}+3\sqrt{b})$

17. $\sqrt{6}(\sqrt{2}-3\sqrt{6}+2\sqrt{10})$

18. $\sqrt{10}(\sqrt{5}-3\sqrt{10}+5\sqrt{2})$

19. $2\sqrt{a}(3\sqrt{b}+\sqrt{ab}-2\sqrt{a})$

20. $3\sqrt{x}(\sqrt{y}+2\sqrt{xy}-4\sqrt{x})$

21. $(2\sqrt{3}+\sqrt{6})(\sqrt{3}-2\sqrt{6})$

22. $(3\sqrt{5}+\sqrt{3})(\sqrt{5}+\sqrt{3})$

23. $(5+3\sqrt{2})(3+\sqrt{2})$

24. $(7-2\sqrt{3})(4+\sqrt{3})$

25. $(2\sqrt{7}-3\sqrt{3})(\sqrt{7}+\sqrt{3})$

26. $(5\sqrt{6}-\sqrt{2})(\sqrt{6}+3\sqrt{2})$

27. $(\sqrt{3}+2\sqrt{6})(2\sqrt{3}-\sqrt{6})$

28. $(\sqrt{2}+3\sqrt{10})(2\sqrt{2}-\sqrt{10})$

29. $(3\sqrt{7}-\sqrt{8})(\sqrt{8}+2\sqrt{7})$

30. $(\sqrt{12}-\sqrt{5})(\sqrt{5}+2\sqrt{12})$

31. $(2\sqrt{5}-3)^2$

32. $(3\sqrt{2}+5)^2$

33. $(\sqrt{3}+5\sqrt{2})^2$

34. $(\sqrt{3}-2\sqrt{7})^2$

35. $(3\sqrt{5}-\sqrt{3})^2$

36. $(2\sqrt{3}+5\sqrt{5})^2$

37. $(3x\sqrt{y}+\sqrt{5})(3x\sqrt{y}-\sqrt{5})$

38. $(5a\sqrt{3}-3\sqrt{2})(5a\sqrt{3}+3\sqrt{2})$

To Think About

39. **(a)** If a can be any number, can you think of an example of when it is not true that $\sqrt{a} \cdot \sqrt{a} = a$?

   (b) Why is it necessary to state the restriction "for any nonnegative real number a"?

40. Hank was trying to evaluate the expression $\left(\sqrt{a} + \sqrt{b}\right)^2$. He wrote $\left(\sqrt{a} + \sqrt{b}\right)^2 = \left(\sqrt{a}\right)^2 + \left(\sqrt{b}\right)^2 = a + b$. This was a serious error! What is wrong with his work? How should he have done the problem?

A wheelchair access ramp is constructed with the dimensions as listed on the diagram. The gray shaded triangle is a right triangle.

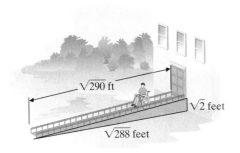

$\sqrt{290}$ ft

$\sqrt{2}$ feet

$\sqrt{288}$ feet

▲ **41.** Find the area of the shaded portion of the wheelchair ramp.

▲ **42.** Find the perimeter of the shaded portion of the wheelchair ramp. Leave your answer as a radical in simplified form.

Cumulative Review Problems

43. Factor. $36x^2 - 49y^2$

44. Factor. $6x^2 + 11x - 10$

45. Ships often measure their speed in knots (or nautical miles per hour). The regular mile (statute mile) that we are more familiar with is 5280 feet. A nautical mile is about 6076 feet. Write a formula of the form $m = ak$ where m is the number of statute miles, k is the number of nautical miles, and a is a constant number (in this case a fraction). Use the formula to find out how fast a Coast Guard Cutter is going in miles per hour if it is traveling at 35 knots.

46. Phil and Melissa LaBelle found that the collision and theft portion of their car insurance could be reduced by 15% if they purchased an auto security device. They purchased and had such a device installed for $350. The yearly collision and theft portion of their car insurance is $280. How many years will it take for the security device to pay for itself?

9.5 Division of Radicals

① Simplifying a Fraction Involving Radicals Using the Quotient Rule for Square Roots

Just as there is a multiplication rule for square roots, there is a quotient rule for square roots. The rules are similar.

Quotient Rule for Square Roots

For all positive numbers a and b,

$$\frac{\sqrt{a}}{\sqrt{b}} = \sqrt{\frac{a}{b}} \quad \text{and} \quad \sqrt{\frac{a}{b}} = \frac{\sqrt{a}}{\sqrt{b}}.$$

This can be used to divide square root radicals or to simplify square root radical expressions involving division. We will be using both parts of the quotient rule.

EXAMPLE 1 Simplify. **(a)** $\frac{\sqrt{75}}{\sqrt{3}}$ **(b)** $\sqrt{\frac{25}{36}}$

(a) We notice that 3 is a factor of 75. We use the quotient rule to rewrite the expression.

$$\frac{\sqrt{75}}{\sqrt{3}} = \sqrt{\frac{75}{3}}$$
$$= \sqrt{25} \quad \text{Divide.}$$
$$= 5 \quad \text{Simplify.}$$

(b) Since both 25 and 36 are perfect squares, we will rewrite this as the quotient of square roots.

$$\sqrt{\frac{25}{36}} = \frac{\sqrt{25}}{\sqrt{36}}$$
$$= \frac{5}{6}$$

Practice Problem 1 Simplify. **(a)** $\sqrt{\frac{x^3}{x}}$ **(b)** $\sqrt{\frac{49}{x^2}}$

EXAMPLE 2 Simplify. $\sqrt{\frac{20}{x^6}}$

$$\sqrt{\frac{20}{x^6}} = \frac{\sqrt{20}}{\sqrt{x^6}} = \frac{\sqrt{4}\sqrt{5}}{x^3} = \frac{2\sqrt{5}}{x^3} \quad \text{Don't forget to simplify } \sqrt{20} \text{ as } 2\sqrt{5}.$$

Practice Problem 2 Simplify. $\sqrt{\frac{50}{a^4}}$

② Rationalizing the Denominator of a Fraction with a Square Root in the Denominator

Sometimes, when calculating with fractions that contain radicals, it is advantageous to have an integer in the denominator. If a fraction has a radical in the denominator, we **rationalize the denominator**. That is, we multiply to change the fraction to an equivalent one that has an integer in the denominator. Remember, we do not want to change the value of the fraction. Thus we will multiply the numerator and the denominator by the same number.

After studying this section, you will be able to:

① Simplify a fraction involving radicals using the quotient rule for square roots.

② Rationalize the denominator of a fraction with a square root in the denominator.

③ Rationalize the denominator of a fraction with a binomial in the denominator containing at least one square root.

SSM | PH TUTOR CENTER | CD & VIDEO | MATH PRO | WEB

543

EXAMPLE 3 Simplify. $\dfrac{3}{\sqrt{2}}$

Think, "What times 2 will make a perfect square?"

$$\frac{3}{\sqrt{2}} = \frac{3}{\sqrt{2}} \times 1 = \frac{3}{\sqrt{2}} \times \frac{\sqrt{2}}{\sqrt{2}} = \frac{3\sqrt{2}}{\sqrt{4}} = \frac{3\sqrt{2}}{2}$$

Practice Problem 3 Simplify. $\dfrac{9}{\sqrt{7}}$

(Note that a fraction is not simplified unless the denominator is rationalized.)

When rationalizing a denominator containing a square root radical, we will want to use the smallest possible radical that will yield the square root of a perfect square. Very often we will not use the radical that is in the denominator.

EXAMPLE 4 Simplify. **(a)** $\dfrac{\sqrt{7}}{\sqrt{8}}$ **(b)** $\dfrac{3}{\sqrt{x^3}}$

(a) Think, "What times 8 will make a perfect square?"

$$\frac{\sqrt{7}}{\sqrt{8}} = \frac{\sqrt{7}}{\sqrt{8}} \times \frac{\sqrt{2}}{\sqrt{2}} = \frac{\sqrt{14}}{\sqrt{16}} = \frac{\sqrt{14}}{4}$$

(b) Think, "What times x^3 will give an even exponent?"

$$\frac{3}{\sqrt{x^3}} \times \frac{\sqrt{x}}{\sqrt{x}} = \frac{3\sqrt{x}}{\sqrt{x^4}} = \frac{3\sqrt{x}}{x^2}$$

We could have simplified $\sqrt{x^3}$ before we rationalized the denominator.

$$\frac{3}{\sqrt{x^3}} = \frac{3}{x\sqrt{x}} \times \frac{\sqrt{x}}{\sqrt{x}} = \frac{3\sqrt{x}}{x\sqrt{x^2}} = \frac{3\sqrt{x}}{(x)(x)} = \frac{3\sqrt{x}}{x^2}$$

Practice Problem 4 Simplify. **(a)** $\dfrac{\sqrt{2}}{\sqrt{12}}$ **(b)** $\dfrac{6a}{\sqrt{a^7}}$

EXAMPLE 5 Simplify. $\dfrac{\sqrt{2}}{\sqrt{27x}}$

Since it is not apparent what we should multiply 27 by to obtain a perfect square, we will begin by simplifying the denominator.

$$\frac{\sqrt{2}}{\sqrt{27x}} = \frac{\sqrt{2}}{3\sqrt{3x}}$$

Now it is easy to see that we multiply by $\dfrac{\sqrt{3x}}{\sqrt{3x}}$ to rationalize the denominator.

$$= \frac{\sqrt{2}}{3\sqrt{3x}} \times \frac{\sqrt{3x}}{\sqrt{3x}}$$

$$= \frac{\sqrt{6x}}{3\sqrt{9x^2}}$$

$$= \frac{\sqrt{6x}}{9x}$$

Practice Problem 5 Simplify. $\dfrac{\sqrt{2x}}{\sqrt{8x}}$

3 *Rationalizing the Denominator of a Fraction with a Binomial Denominator Containing at Least One Square Root*

Sometimes the denominator of a fraction is a binomial with a square root radical term. $\dfrac{1}{5 - 3\sqrt{2}}$ is such a fraction. How can we eliminate the radical term in the denominator? Recall that when we multiply $(a + b)(a - b)$, we obtain the square of a minus the square of b: $a^2 - b^2$. For example, $(x + 3)(x - 3) = x^2 - 9$. The resulting terms are perfect squares. We can use this idea to eliminate the radical in $5 - 3\sqrt{2}$.

$$(5 - 3\sqrt{2})(5 + 3\sqrt{2}) = 5^2 + 15\sqrt{2} - 15\sqrt{2} - (3\sqrt{2})^2 = 5^2 - (3\sqrt{2})^2 = 25 - 18 = 7$$

Expressions like $(5 - 3\sqrt{2})$ and $(5 + 3\sqrt{2})$ are called **conjugates**.

EXAMPLE 6 Simplify. **(a)** $\dfrac{2}{\sqrt{3} - 4}$ **(b)** $\dfrac{\sqrt{x}}{\sqrt{5} + \sqrt{3}}$

(a) The conjugate of $\sqrt{3} - 4$ is $\sqrt{3} + 4$.

$$\frac{2}{(\sqrt{3} - 4)} \cdot \frac{(\sqrt{3} + 4)}{(\sqrt{3} + 4)} = \frac{2\sqrt{3} + 8}{(\sqrt{3})^2 + 4\sqrt{3} - 4\sqrt{3} - 4^2}$$

$$= \frac{2\sqrt{3} + 8}{3 - 16}$$

$$= \frac{2\sqrt{3} + 8}{-13}$$

$$= -\frac{2\sqrt{3} + 8}{13}$$

(b) The conjugate of $\sqrt{5} + \sqrt{3}$ is $\sqrt{5} - \sqrt{3}$.

$$\frac{\sqrt{x}}{(\sqrt{5} + \sqrt{3})} \cdot \frac{(\sqrt{5} - \sqrt{3})}{(\sqrt{5} - \sqrt{3})} = \frac{\sqrt{5x} - \sqrt{3x}}{(\sqrt{5})^2 - \sqrt{15} + \sqrt{15} - (\sqrt{3})^2}$$

$$= \frac{\sqrt{5x} - \sqrt{3x}}{5 - 3}$$

$$= \frac{\sqrt{5x} - \sqrt{3x}}{2}$$

Be careful not to combine $\sqrt{5x}$ and $\sqrt{3x}$ in the numerator. They are not like radicals.

Practice Problem 6 Simplify. **(a)** $\dfrac{4}{\sqrt{3} + \sqrt{5}}$ **(b)** $\dfrac{\sqrt{a}}{\sqrt{10} - 3}$

EXAMPLE 7 Rationalize the denominator. $\dfrac{\sqrt{3} + \sqrt{2}}{\sqrt{3} - \sqrt{2}}$

The conjugate of $\sqrt{3} - \sqrt{2}$ is $\sqrt{3} + \sqrt{2}$.

$$\frac{\left(\sqrt{3} + \sqrt{2}\right)}{\left(\sqrt{3} - \sqrt{2}\right)} \cdot \frac{\left(\sqrt{3} + \sqrt{2}\right)}{\left(\sqrt{3} + \sqrt{2}\right)}$$

$$= \frac{\sqrt{9} + \sqrt{6} + \sqrt{6} + \sqrt{4}}{\left(\sqrt{3}\right)^2 - \left(\sqrt{2}\right)^2} \qquad \text{Multiply.}$$

$$= \frac{3 + 2\sqrt{6} + 2}{3 - 2} \qquad \text{Simplify and combine like terms.}$$

$$= \frac{5 + 2\sqrt{6}}{1}$$

$$= 5 + 2\sqrt{6}$$

Practice Problem 7 Rationalize the denominator. $\dfrac{\sqrt{7} - \sqrt{x}}{\sqrt{7} + \sqrt{x}}$

Simplify. Be sure to rationalize all denominators. Do not use a calculator or a table of square roots.

1. $\dfrac{\sqrt{12}}{\sqrt{3}}$ **2.** $\dfrac{\sqrt{3}}{\sqrt{27}}$ **3.** $\dfrac{\sqrt{5}}{\sqrt{20}}$ **4.** $\dfrac{\sqrt{24}}{\sqrt{6}}$ **5.** $\dfrac{\sqrt{6}}{\sqrt{x^4}}$ **6.** $\dfrac{\sqrt{12}}{\sqrt{x^2}}$

7. $\dfrac{\sqrt{75}}{\sqrt{3}}$ **8.** $\dfrac{\sqrt{3}}{\sqrt{48}}$ **9.** $\dfrac{\sqrt{18}}{\sqrt{a^4}}$ **10.** $\dfrac{\sqrt{24}}{\sqrt{b^6}}$ **11.** $\dfrac{3}{\sqrt{7}}$ **12.** $\dfrac{2}{\sqrt{10}}$

13. $\dfrac{x\sqrt{x}}{\sqrt{2}}$ **14.** $\dfrac{\sqrt{3y}}{\sqrt{6}}$ **15.** $\dfrac{\sqrt{8}}{\sqrt{x}}$ **16.** $\dfrac{\sqrt{18}}{\sqrt{y}}$ **17.** $\dfrac{3}{\sqrt{12}}$ **18.** $\dfrac{4}{\sqrt{20}}$

19. $\dfrac{7}{\sqrt{a^3}}$ **20.** $\dfrac{5}{\sqrt{b^5}}$ **21.** $\dfrac{x}{\sqrt{2x^5}}$ **22.** $\dfrac{y}{\sqrt{5x^3}}$ **23.** $\dfrac{\sqrt{18}}{\sqrt{2x^3}}$ **24.** $\dfrac{\sqrt{24}}{\sqrt{6x^7}}$

25. $\sqrt{\dfrac{3}{5}}$ **26.** $\sqrt{\dfrac{7}{11}}$ **27.** $\dfrac{9}{\sqrt{32x}}$ **28.** $\dfrac{3}{\sqrt{50x}}$ **29.** $\dfrac{4}{\sqrt{3}-1}$

30. $\dfrac{5}{\sqrt{5}+1}$ **31.** $\dfrac{3}{\sqrt{2}+\sqrt{5}}$ **32.** $\dfrac{4}{\sqrt{7}-\sqrt{2}}$ **33.** $\dfrac{\sqrt{6}}{\sqrt{6}-\sqrt{3}}$ **34.** $\dfrac{\sqrt{3}}{\sqrt{5}+\sqrt{3}}$

35. $\dfrac{3x}{2\sqrt{2}-\sqrt{5}}$ **36.** $\dfrac{4x}{2\sqrt{7}+2\sqrt{6}}$ **37.** $\dfrac{\sqrt{7}}{\sqrt{8}+\sqrt{7}}$ **38.** $\dfrac{\sqrt{5}}{\sqrt{5}+\sqrt{6}}$ **39.** $\dfrac{\sqrt{5}-\sqrt{2}}{\sqrt{5}+\sqrt{2}}$

40. $\dfrac{\sqrt{7}+\sqrt{3}}{\sqrt{7}-\sqrt{3}}$ **41.** $\dfrac{2\sqrt{3}+\sqrt{6}}{\sqrt{6}-\sqrt{3}}$ **42.** $\dfrac{2\sqrt{5}+3}{\sqrt{5}+\sqrt{3}}$ **43.** $\dfrac{4\sqrt{3}+2}{\sqrt{8}-\sqrt{6}}$ **44.** $\dfrac{3\sqrt{5}+4}{\sqrt{15}-\sqrt{3}}$

45. $\dfrac{x-25}{\sqrt{x}+5}$ **46.** $\dfrac{x-36}{\sqrt{x}-6}$

To Think About

47. (a) Multiply $\left(x + \sqrt{2}\right)$ by its conjugate.
(b) Use the result of part (a) to show that while $x^2 - 2$ cannot be factored by earlier methods, it can be factored using radicals.

(c) Factor using radicals. $x^2 - 12$

 48. When rationalized, $\dfrac{3}{\sqrt{11}}$ becomes $\dfrac{3\sqrt{11}}{11}$. Evaluate each expression with a calculator. What do you notice? Why?

 49. When rationalized, $\dfrac{7}{\sqrt{35}}$ becomes $\dfrac{\sqrt{35}}{5}$. Evaluate each expression with a calculator. What do you notice? Why?

Applications

 50. A cylinder has a volume V and a height h. The radius of the cylinder is $r = \sqrt{\dfrac{V}{\pi h}}$.

(a) Simplify this expression by rationalizing the denominator.

(b) Find the radius of a cylinder of volume 81 cubic feet and height π feet.

 51. A reflector sheet for a microwave relay tower is shaped like a rectangle. The area of the rectangle is 3 square meters. Ideally the length should be exactly $\left(\sqrt{5} + 2\right)$ meters long. What should be the width of the rectangle? Express the answer exactly using radical expressions.

 52. A computer chip for a new computer is shaped like a rectangle. The area of the rectangle is 4 square millimeters. Ideally the width should be exactly $\left(\sqrt{5} - \sqrt{3}\right)$ millimeters. What should be the length of the rectangle? Express the answer exactly using radical expressions.

 53. An engineer needs the approximate dimensions of the rectangle in exercise 51. Approximate the length and the width of that rectangle to the nearest thousandth.

 54. An engineer needs the approximate dimensions of the computer chip in exercise 52. Approximate the length and the width of that computer chip to the nearest thousandth.

Cumulative Review Problems

55. Solve for x (round to two decimal places).
$3(2x - 6) = 5x - 7(x - 2) + 27$

56. Evaluate $3x^2 - x \div y + y^2 - y$ for $x = 12$ and $y = -3$.

57. Dr. Frank Day, a chemist, has determined that the types of percolated coffee he buys have 70% more caffeine per cup than the types of instant coffee he usually buys. He determined that the percolated coffee has 120 mg of caffeine per cup. Approximately how much caffeine is there per cup of instant coffee?

58. Wally and Mary Coyle are considering two family medical insurance plans that are quite similar. The Gold Plan costs $5000 per year and covers all doctor visits and all prescription drugs after a $50 deductible per doctor visit and a $30 deductible per prescription. The Master Plan costs $6000 per year and covers all doctor visits and all prescription drugs. Wally and Mary have discovered that half of all their doctor visits result in a written prescription. How many doctor visits per year would they need to average for the Master Plan to be more economical?

9.6 **The Pythagorean Theorem and Radical Equations**

① *Using the Pythagorean Theorem*

In ancient Greece, mathematicians studied a number of properties of right triangles. They proved that the square of the longest side of a right triangle is equal to the sum of the squares of the other two sides. This property is called the Pythagorean theorem, in honor of the Greek mathematician Pythagoras (ca. 590 B.C.). The shorter sides, a and b, are referred to as the **legs** of the right triangle. The longest side, c, is called the **hypotenuse** of the right triangle. We state the theorem as follows.

Student Learning Objectives

After studying this section, you will be able to:

① Use the Pythagorean theorem.

② Solve radical equations.

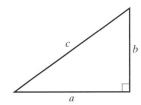

SSM
PH TUTOR CENTER CD & VIDEO MATH PRO WEB

Pythagorean Theorem

In any right triangle, if c is the length of the hypotenuse and a and b are the lengths of the two legs, then

$$c^2 = a^2 + b^2.$$

If we know any two sides of a right triangle, we can find the third side using this theorem.

Recall that in Section 9.1 we showed that if $s^2 = 4$ then $s = 2$ or -2. In general, if $x^2 = a$, then $x = \pm\sqrt{a}$. This is sometimes called "taking the square root of each side of an equation." The abbreviation $\pm\sqrt{a}$ means $+\sqrt{a}$ or $-\sqrt{a}$. It is read "plus or minus $\sqrt{a}$."

▲ ⬤ **EXAMPLE 1** The ramp to Tony Pithin's barn rises 5 feet over a horizontal distance of 12 feet. How long is the ramp? That is, find the length of the hypotenuse of a right triangle whose legs are 5 feet and 12 feet.

1. ***Understand the problem.***
   Draw and label a diagram.

   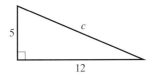

2. ***Write an equation.***
   Write the Pythagorean Theorem. Substitute the known values into the equation.
   $$c^2 = a^2 + b^2$$
   $$c^2 = 5^2 + 12^2$$

3. ***Solve and state the answer.***
   $$c^2 = 25 + 144$$
   $$c^2 = 169 \qquad \text{Take the square root of each side of the equation.}$$
   $$c = \pm\sqrt{169}$$
   $$= \pm 13$$

   The hypotenuse is 13 feet. We do not use -13 because length is not negative. Thus the length of the ramp is 13 feet.

4. ***Check.***
   Substitute the values for a, b, and c in the Pythagorean Theorem and evaluate.
   Does $13^2 = 5^2 + 12^2$? Yes. The solution checks. ✓

▲ **Practice Problem 1** Find the length of the hypotenuse of a right triangle with legs of 9 centimeters and 12 centimeters. ⬤

Sometimes you will need to find one of the legs of a right triangle given the hypotenuse and the other leg.

549

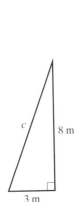

▲ ● **EXAMPLE 2** Find the unknown leg of a right triangle that has a hypotenuse of 10 yards and one leg of $\sqrt{19}$ yards.

Draw a diagram. The diagram is in the margin on the left.

$c^2 = a^2 + b^2$	Write the Pythagorean Theorem.
$10^2 = \left(\sqrt{19}\right)^2 + b^2$	Substitute the known values into the equation.
$100 = 19 + b^2$	
$81 = b^2$	
$\pm 9 = b$	Take the square root of each side of the equation.

The leg is 9 yards.

▲ **Practice Problem 2** The hypotenuse of a right triangle is $\sqrt{17}$ meters. One leg is 1 meter long. Find the length of the other leg.

Sometimes the answer to a problem will be an irrational number such as $\sqrt{33}$. It is best to leave the answer in radical form unless you are asked for an approximate answer. To find the approximate value, use a calculator or the square root table. $\sqrt{33}$ is an exact answer.

▲ ● **EXAMPLE 3** A 25-foot ladder is placed against a building. The foot of the ladder is 8 feet from the wall. At approximately what height does the top of the ladder touch the building? Round to the nearest tenth.

$$c^2 = a^2 + b^2$$
$$25^2 = 8^2 + b^2$$
$$625 = 64 + b^2$$
$$561 = b^2$$
$$\pm\sqrt{561} = b$$

We want only the positive value for the distance, so $b = \sqrt{561}$. Using a calculator, we have 561 $\boxed{\sqrt{}}$ 23.68543856. Rounding, we obtain $b \approx 23.7$.

If you do not have a calculator, you will need to use the square root table in Appendix A. However, this square root table goes only as far as $\sqrt{200}$! To approximate larger radicals with the square root table, you will have to write the radical in a different fashion.

$b = \sqrt{3}\sqrt{187}$	Since $\sqrt{ab} = \sqrt{a}\sqrt{b}$ and $3 \cdot 187 = 561$.
$b \approx (1.732)(13.675)$	Replace each radical by a decimal approximation from the table.
$b \approx 23.7$	Multiply and round to the nearest tenth.

The ladder touches the building at a height of approximately 23.7 feet.

▲ **Practice Problem 3** A support line is placed 3 meters away from the base of an 8-meter pole. If the support line is attached to the top of the pole and pulled tight (assume that it is a straight line), how long is the support line from the ground to the pole? Round to the nearest tenth.

2 Solving Radical Equations

A **radical equation** is an equation with a variable in one or more of the radicands. A square root radical equation can be simplified by squaring each side of the equation. The apparent solution to a radical equation *must* be checked by substitution into the *original* equation.

For some square root radical equations, you will need to *isolate the radical before* you square each side of the equation.

EXAMPLE 4 Solve and check. $1 + \sqrt{5x - 4} = 5$

$$1 + \sqrt{5x - 4} = 5 \qquad \text{We want to isolate the radical first.}$$
$$\sqrt{5x - 4} = 4 \qquad \text{Subtract 1 from each side to isolate the radical.}$$
$$(\sqrt{5x - 4})^2 = (4)^2 \qquad \text{Square each side.}$$
$$5x - 4 = 16 \qquad \text{Simplify and solve for } x.$$
$$5x = 20$$
$$x = 4$$

Check.
$$1 + \sqrt{5(4) - 4} \stackrel{?}{=} 5$$
$$1 + \sqrt{20 - 4} \stackrel{?}{=} 5$$
$$1 + \sqrt{16} \stackrel{?}{=} 5$$
$$1 + 4 = 5 \checkmark$$

Thus 4 is the solution.

Practice Problem 4 Solve and check. $\sqrt{3x - 2} - 7 = 0$

Some problems will have an apparent solution, but that solution does not always check. An apparent solution that does not satisfy the original equation is called an **extraneous root**.

EXAMPLE 5 Solve and check. $\sqrt{x + 3} = -7$

$$\sqrt{x + 3} = -7$$
$$(\sqrt{x + 3})^2 = (-7)^2 \qquad \text{Square each side.}$$
$$x + 3 = 49 \qquad \text{Simplify and solve for } x.$$
$$x = 46$$

Check.
$$\sqrt{46 + 3} \stackrel{?}{=} -7$$
$$\sqrt{49} \stackrel{?}{=} -7$$
$$7 \neq -7! \qquad \text{Does not check! The apparent solution is an extraneous root.}$$

There is no solution to Example 5.

Practice Problem 5 Solve and check. $\sqrt{5x + 4} + 2 = 0$

Wait a minute! Is the original equation in Example 5 possible? No. $\sqrt{x + 3} = -7$ is an impossible statement. We defined the $\sqrt{\ }$ symbol to mean the positive square root of the radicand. Thus, it cannot be equal to a negative number. If we had noticed this in the first step, we could have written down "no solution" immediately.

When you square both sides of an equation you may obtain a quadratic equation. In such cases, you may obtain two apparent solutions, and you must check both of them.

EXAMPLE 6 Solve and check. $\sqrt{3x+1} = x+1$.

$$\sqrt{3x+1} = x+1$$

$$\left(\sqrt{3x+1}\right)^2 = (x+1)^2 \qquad \text{Square each side.}$$

$$3x+1 = x^2 + 2x + 1 \qquad \text{Simplify and set the equation equal to } 0.$$

$$0 = x^2 - x$$

$$0 = x(x-1) \qquad \text{Factor the quadratic equation.}$$

$$x = 0 \qquad x - 1 = 0 \qquad \text{Set each factor equal to } 0 \text{ and solve.}$$

$$x = 1$$

Check. If $x = 0$: $\sqrt{3(0)+1} \overset{?}{=} 0 + 1$ If $x = 1$: $\sqrt{3(1)+1} \overset{?}{=} 1 + 1$

$$\sqrt{0+1} \overset{?}{=} 0 + 1 \qquad\qquad\qquad \sqrt{3+1} \overset{?}{=} 2$$

$$\sqrt{1} \overset{?}{=} 1 \qquad\qquad\qquad\qquad \sqrt{4} \overset{?}{=} 2$$

$$1 = 1 \;\checkmark \qquad\qquad\qquad\qquad\quad 2 = 2 \;\checkmark$$

Thus 0 and 1 are both solutions.

It is possible in these cases for both apparent solutions to check, for both to be extraneous, or for one to check and the other to be extraneous.

Practice Problem 6 Solve and check. $-2 + \sqrt{6x-1} = 3x - 2$

EXAMPLE 7 Solve and check. $\sqrt{2x-1} = x - 2$

$$\sqrt{2x-1} = x - 2$$

$$\left(\sqrt{2x-1}\right)^2 = (x-2)^2 \qquad \text{Square each side.}$$

$$2x - 1 = x^2 - 4x + 4$$

$$0 = x^2 - 6x + 5 \qquad \text{Simplify and set the equation to } 0.$$

$$0 = (x-5)(x-1) \qquad \text{Solve for } x.$$

$$x - 5 = 0 \qquad x - 1 = 0$$

$$x = 5 \qquad\quad x = 1$$

Check. If $x = 5$: $\sqrt{2(5)-1} \overset{?}{=} 5 - 2$ If $x = 1$: $\sqrt{2(1)-1} \overset{?}{=} 1 - 2$

$$\sqrt{10-1} \overset{?}{=} 3 \qquad\qquad\qquad \sqrt{2-1} \overset{?}{=} -1$$

$$\sqrt{9} \overset{?}{=} 3 \qquad\qquad\qquad\qquad \sqrt{1} \overset{?}{=} -1$$

$$3 = 3 \;\checkmark \qquad\qquad\qquad\qquad 1 \neq -1$$

It does not check. In this case 1 is an extraneous root.

Thus only 5 is a solution to this equation.

Practice Problem 7 Solve and check. $2 - x + \sqrt{x+4} = 0$

EXAMPLE 8 Solve and check. $\sqrt{3x+3} = \sqrt{5x-1}$

Here there are two radicals. Each radical is already isolated.

$$\sqrt{3x+3} = \sqrt{5x-1}$$

$$\left(\sqrt{3x+3}\right)^2 = \left(\sqrt{5x-1}\right)^2 \qquad \text{Square each side.}$$

$$3x + 3 = 5x - 1 \qquad\qquad \text{Simplify and solve for } x.$$

$$3 = 2x - 1$$

$$4 = 2x$$

$$2 = x$$

Check. $\sqrt{3(2)+3} \overset{?}{=} \sqrt{5(2)-1}$

$$\sqrt{6+3} \overset{?}{=} \sqrt{10-1}$$

$$\sqrt{9} = \sqrt{9} \;\checkmark$$

Thus 2 is a solution.

Practice Problem 8 Solve and check. $\sqrt{2x+1} = \sqrt{x-10}$

9.6 Exercises

Use the Pythagorean Theorem to find the length of the third side of each right triangle. Leave any irrational answers in radical form.

▲ **1.**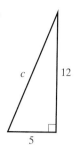

▲ **2.**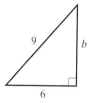

▲ **3.**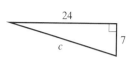

▲ **4.**

Exercises 5–14 refer to a right triangle with legs a and b and hypotenuse c. Find the exact length of the missing side.

▲ **5.** $a = 5, b = 7$
Find c.

▲ **6.** $a = 6, b = 9$
Find c.

▲ **7.** $a = \sqrt{14}, b = 7$
Find c.

▲ **8.** $a = \sqrt{21}, b = 5$
Find c.

▲ **9.** $c = 20, b = 18$
Find a.

▲ **10.** $c = 17, b = 16$
Find a.

▲ **11.** $c = \sqrt{82}, a = 5$
Find b.

▲ **12.** $a = \sqrt{5}, c = 7$
Find b.

▲ **13.** $c = 12.96, b = 8.35$ Find a to the nearest hundredth.

▲ **14.** $a = 7.61, b = 5.38$ Find c to the nearest hundredth.

Applications

Draw a diagram and use the Pythagorean Theorem to solve. Round to the nearest tenth.

▲ **15.** A ladder is 18 feet long. The top of it touches a wall at a height of 15 feet. How far is the base of the ladder from the wall?

▲ **16.** A baseball diamond is a square. Each side of the square is 90 feet long. How far is it from home plate to second base? *Hint:* Draw the diagonal.

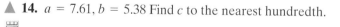

553

▲ **17.** A kite is flying on a string that is 100 feet long and is fastened to the ground at the other end. Assume that the string is a straight line. How high is the kite if it is flying above a point that is 20 feet away from where it is fastened on the ground?

▲ **18.** A boat must be moored 20 feet away from a dock. The boat will be 10 feet below the level of the dock at low tide. What is the minimum length of rope needed to go from the boat to the dock at low tide?

To estimate the cost of erecting a radio tower, we need to know how many feet of support cable are needed.

▲ **19.** Ground anchor #2 is exactly 54 feet from the base of the tower (dashed line in diagram). The bottom ground support cable from level 6 to ground anchor #2 is attached to the tower at a height of 130 feet. The top ground support cable from level 6 to ground anchor #2 is attached to the tower at a height of 135 feet. How long is each support cable to level 6? (Round to the nearest hundredth of a foot.)

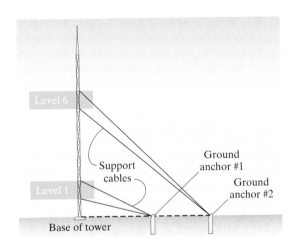

▲ **20.** Ground anchor #1 is exactly 28 feet from the base of the tower (dashed line in diagram). The bottom ground support cable from level 1 to ground anchor #1 is attached to the tower at a height 42 feet. The top ground support cable from level 1 to ground anchor #1 is attached to the tower at a height 47 feet. How long is each support cable to level 1? (Round to the nearest hundredth of a foot.

Solve for the variable. Check your solutions.

21. $\sqrt{x + 2} = 3$

22. $\sqrt{x - 2} = 6$

23. $\sqrt{3x + 6} = 2$

24. $\sqrt{3x - 8} = 4$

25. $\sqrt{2x + 2} = \sqrt{3x - 5}$

26. $\sqrt{5x - 5} = \sqrt{4x + 1}$

27. $\sqrt{2x} - 5 = 4$

28. $\sqrt{13x} - 2 = 5$

29. $\sqrt{3x + 10} = x$

30. $\sqrt{5x - 6} = x$

31. $\sqrt{5y + 1} = y + 1$

32. $\sqrt{2y + 9} = y + 3$

33. $\sqrt{x + 3} = 3x - 1$

34. $\sqrt{2x + 3} = 2x - 9$

35. $\sqrt{3y + 1} - y = 1$

36. $\sqrt{3y - 8} + 2 = y$

37. $\sqrt{6y + 1} - 3y = y$

38. $\sqrt{2x + 3} - 2 = x$

To Think About

Solve for x.

39. $\sqrt{2x + 5} = 2\sqrt{2x} + 1$

40. $\sqrt{x - 2} + 3 = \sqrt{4x + 1}$

Cumulative Review Problems

41. Solve. $\dfrac{5x}{x - 4} = 5 + \dfrac{4x}{x - 4}$

42. Multiply mentally. $(5x - 7)^2$

43. $f(x) = 2x^2 - 3x + 6$
Find $f(-2)$.

44. Is $x^2 + y^2 = 9$ a relation or a function?

Peter J. Dubacher and his wife, Betty Ann, have run a bird sanctuary for 25 years. This sanctuary costs $40,000 per year to run. Peter and Betty Ann care for 1700 birds, including 15 eagles, 2 peregrine falcons, and 19 emus.

45. If the number of birds has grown by 50 every year for the last 25 years, how many birds did they start with in year 1?

46. If you average the cost to care for the birds over the 25 years, how much was spent per bird per year?

1 Solving Problems Involving Direct Variation

Often in daily life there is a relationship between two measurable quantities. For example, we turn up the thermostat and the heating bill increases. We say that the heating bill varies directly as the temperature on the thermostat. If y **varies directly as** x, then $y = kx$, where k is a constant. The constant is often called the **constant of variation**. Consider the following example.

EXAMPLE 1 Cliff works part-time in a local supermarket while going to college. His salary varies directly as the number of hours worked. Last week he earned $33.60 for working 7 hours. This week he earned $52.80. How many hours did he work?

Let S = his salary,
      h = the number of hours he worked, and
      k = the constant of variation.

Since his salary varies directly as the number of hours he worked, we write

$$S = k \cdot h.$$

We can find the constant k by substituting the known values of $S = 33.60$ and $h = 7$.

$$33.60 = k \cdot 7 = 7k$$

$$\frac{33.60}{7} = \frac{7k}{7} \qquad \text{Solve for } k.$$

$$4.80 = k \qquad \text{The constant of variation is 4.80.}$$

$$S = 4.80h \qquad \text{Replace } k \text{ in the variation equation by 4.80.}$$

How many hours did he work to earn $52.80?

$$S = 4.80h \qquad \text{The direct variation equation with } k = 4.80.$$

$$52.80 = 4.80h \qquad \text{Substitute 52.80 for } S \text{ and solve for } h.$$

$$\frac{52.80}{4.80} = \frac{4.80h}{4.80}$$

$$11 = h$$

Cliff worked 11 hours this week.

Practice Problem 1 The *change* in temperature measured on the Celsius scale varies directly as the change measured on the Fahrenheit scale. The change in temperature from freezing water to boiling water is 100° on the Celsius scale and 180° on the Fahrenheit scale. If the Fahrenheit temperature drops 20°, what will be the change in temperature on the Celsius scale?

Solving a Direct Variation Problem

1. Write the direct variation equation.
2. Solve for the constant k by substituting in known values.
3. Replace k in the direct variation equation by the value obtained in step 2.
4. Solve for the desired value.

EXAMPLE 2 If y varies directly as x, and $y = 20$ when $x = 3$, find the value of y when $x = 21$.

Step 1 $y = kx$

Step 2 To find k, we substitute $y = 20$ and $x = 3$.

$$20 = k(3)$$
$$20 = 3k$$
$$\frac{20}{3} = k$$

Step 3 We now write the variation equation with k replaced by $\frac{20}{3}$.

$$y = \frac{20}{3}x$$

Step 4 We replace x by 21 and find y.

$$y = \left(\frac{20}{\cancel{3}}\right)(\cancel{21}^{7})$$
$$= (20)(7) = 140$$

Thus $y = 140$ when $x = 21$.

Practice Problem 2 If y varies directly as x, and $y = 18$ when $x = 5$, find the value of y when $x = \frac{20}{23}$.

A number of real-life situations can be described by direct variation equations. The following table shows some of the more common forms of the direct variation equation. In each case, $k =$ the constant of variation.

Sample Direct Variation Situations

Verbal Description	*Variation Equation*
y varies directly as x	$y = kx$
b varies directly as the square of c	$b = kc^2$
l varies directly as the cube of m	$l = km^3$
V varies directly as the square root of h	$V = k\sqrt{h}$

EXAMPLE 3 In a certain class of racing cars, the maximum speed varies directly as the square root of the horsepower of the engine. If a car with 225 horsepower can achieve a maximum speed of 120 mph, what speed could it achieve with 256 horsepower?

Let $V =$ the maximum speed,
$h =$ the horsepower of the engine, and
$k =$ the constant of variation.

Step 1 Since the maximum speed (V) varies directly as the square root of the horsepower of the engine,

$$V = k\sqrt{h}.$$

Step 2 $120 = k\sqrt{225}$ Substitute known values of V and h.
$120 = k \cdot 15$ Solve for k.
$8 = k$

Step 3 Now we can write the direct variation equation with the known value for k.

$$V = 8\sqrt{h}$$

Step 4 $V = 8\sqrt{256}$ Substitute the value of $h = 256$.

$= (8)(16)$ Solve for V.

$= 128$

Thus a car with 256 horsepower could achieve a maximum speed of 128 mph.

Practice Problem 3   A certain type of car has a stopping distance that varies directly as the square of its speed. If this car is traveling at a speed of 20 mph on an ice-covered road, it can stop in 60 feet. If the car is traveling at 40 mph on an icy road, what will its stopping distance be?

2 Solving Problems Involving Inverse Variation

If one variable is a constant multiple of the reciprocal of another, the two variables are said to **vary inversely**. If y varies inversely as x, we express this by the equation $y = \dfrac{k}{x}$, where k is the constant of variation. Inverse variation problems can be solved by a four-step procedure similar to that used for direct variation problems.

Solving an Inverse Variation Problem

1. Write the inverse variation equation.
2. Solve for the constant k by substituting in known values.
3. Replace k in the inverse variation equation by the value obtained in step 2.
4. Solve for the desired value.

EXAMPLE 4   If y varies inversely as x, and $y = 12$ when $x = 7$, find the value of y when $x = \dfrac{2}{3}$.

Step 1 $y = \dfrac{k}{x}$

Step 2 $12 = \dfrac{k}{7}$ Substitute known values of x and y to find k.

$84 = k$

Step 3 $y = \dfrac{84}{x}$

Step 4   To find y when $x = \dfrac{2}{3}$, we substitute.

$$y = \dfrac{84}{\dfrac{2}{3}}$$

$$= \dfrac{84}{1} \div \dfrac{2}{3}$$

$$= \dfrac{\overset{42}{\cancel{84}}}{1} \cdot \dfrac{3}{\cancel{2}}$$

$$= 42 \cdot 3 = 126$$

Thus $y = 126$ when $x = \dfrac{2}{3}$.

Practice Problem 4   If y varies inversely as x, and $y = 8$ when $x = 15$, find the value of y when $x = \dfrac{3}{5}$.

EXAMPLE 5   A car manufacturer is thinking of reducing the size of the wheel used in a subcompact car. The number of times a car wheel must turn to cover a given distance varies inversely as the radius of the wheel. (Notice that this says that the smaller the wheel, the more times it must turn to cover a given distance.) A wheel with a radius of 0.35 meter must turn 400 times to cover a specified distance on a test track. How many times would it have to turn if the radius were reduced to 0.30 meter (see the sketch)?

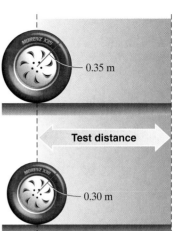

Let n = the number of times the car wheel turns,

r = the radius of the wheel, and

k = the constant of variation.

Step 1   Since the number of turns varies inversely as the radius, we can write the following.

$$n = \frac{k}{r} \quad \text{Write the variation equation.}$$

Step 2 $400 = \dfrac{k}{0.35}$ Substitute known values of n and r to find k.

$140 = k$

Step 3 $n = \dfrac{140}{r}$ Use the variation equation where k is known.

How many times must the wheel turn if the radius is 0.30 meter?

Step 4 $n = \dfrac{140}{0.30}$ Substitute 0.30 for r.

$n = 466\dfrac{2}{3}$

The wheel would have to turn $466\dfrac{2}{3}$ times to cover the same distance if the radius were only 0.30 meter.

Practice Problem 5   Over the last three years, the market research division of a calculator company found that the volume of sales of scientific calculators varies inversely as the price of the calculator. One year 120,000 calculators were sold at $30 each. How many calculators were sold the next year when the price was $24 for each calculator?

The following table contains various forms of the inverse variation equation. In each case, k = the constant of variation.

Sample Inverse Variation Situations

Verbal Description	Variation Equation
y varies inversely as x	$y = \dfrac{k}{x}$
b varies inversely as the square of c	$b = \dfrac{k}{c^2}$
l varies inversely as the cube of m	$l = \dfrac{k}{m^3}$
d varies inversely as the square root of t	$d = \dfrac{k}{\sqrt{t}}$

⬤ **EXAMPLE 6**   The illumination of a light source varies inversely as the square of the distance from the source. The illumination measures 25 candlepower when a certain light is 4 meters away. Find the illumination when the light is 8 meters away (see figure).

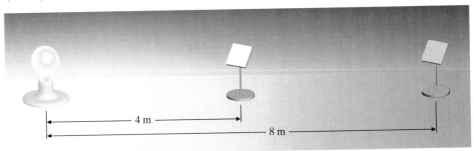

Let I = the measurement of illumination,
   d = the distance from the light source, and
   k = the constant of variation.

Step 1   Since the illumination varies inversely as the square of the distance,

$$I = \frac{k}{d^2}.$$

Step 2   We evaluate the constant by substituting the given values.

$$25 = \frac{k}{4^2} \quad \text{Substitute } I = 25 \text{ and } d = 4.$$

$$25 = \frac{k}{16} \quad \text{Simplify and solve for } k.$$

$$400 = k$$

Step 3   We may now write the variation equation with the constant evaluated.

$$I = \frac{400}{d^2}$$

Step 4 $I = \dfrac{400}{8^2}$ \qquad Substitute a distance of 8 meters.

$$= \frac{400}{64} \qquad \text{Square 8.}$$

$$= \frac{25}{4} = 6.25$$

The illumination is 6.25 candlepower when the light source is 8 meters away.

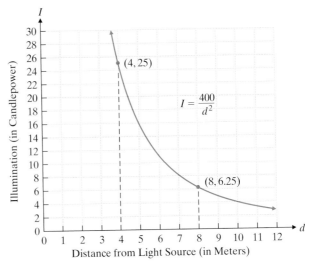

Practice Problem 6   If the amount of power in an electrical circuit is held constant, the resistance in the circuit varies inversely as the square of the amount of current. If the amount of current is 0.01 ampere, the resistance is 800 ohms. What is the resistance if the amount of current is 0.02 ampere?

Graphs of Variation Equations

The graphs of direct variation equations are shown in the following chart. Notice that as x increases, y also increases.

Variation Statement	Equation	Graph
1. y varies directly as x	$y = kx$	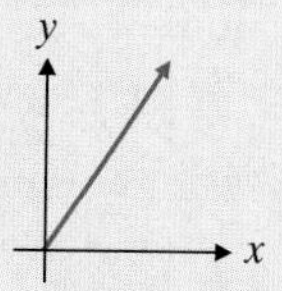
2. y varies directly as x^2	$y = kx^2$	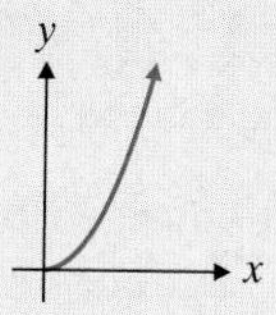
3. y varies directly as x^3	$y = kx^3$	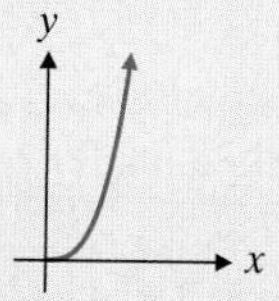
4. y varies directly as the square root of x	$y = k\sqrt{x}$	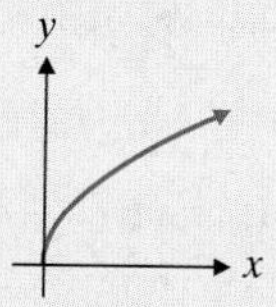

The graphs of inverse variation equations are shown in the following chart. Notice that as x increases, y decreases.

Variation Statement	Equation	Graph
5. y varies inversely as x	$y = \dfrac{k}{x}$	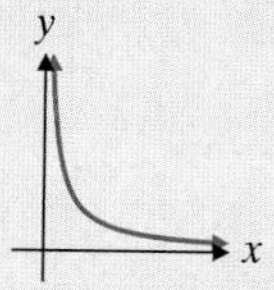
6. y varies inversely as x^2	$y = \dfrac{k}{x^2}$	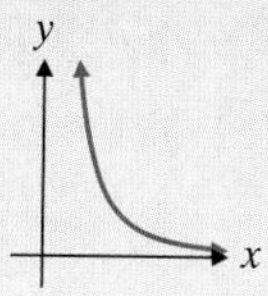

1. If y varies directly as x, and $y = 9$ when $x = 2$, find y when $x = 16$.

2. If y varies directly as x, and $y = 5$ when $x = 3$, find y when $x = 18$.

3. If y varies directly as the cube of x, and $y = 12$ when $x = 2$, find y when $x = 7$.

4. If y varies directly as the square root of x, and $y = 7$ when $x = 4$, find y when $x = 25$.

5. If y varies directly as the square of x, and $y = 900$ when $x = 25$, find y when $x = 30$.

6. If y varies directly as the square of x, and $y = 8000$ when $x = 16$, find y when $x = 5$.

Applications

7. The pressure of water on an object submerged beneath the surface varies directly as the distance beneath the surface. A submarine experiences a pressure of 26 pounds per square inch at 60 feet below the surface. How much pressure will the submarine experience at 390 feet below the surface?

8. The revenue from sales of pizza at College Pizza is directly proportional to the advertising budget. When the owners spent $3000 per month to advertise on local radio and television stations, the monthly gross revenue at College Pizza was $120,000. If the owners increase their advertising budget to $5000 per month, what can they expect for a monthly gross revenue?

▲ 9. The time it takes to fill a storage cube with sand varies directly as the cube of the side of the box. A storage cube that is 2.0 meters on each side (inside dimensions) can be filled in 7 minutes by a sand loader. How long will it take to fill a storage cube that is 4.0 meters on each side (inside dimensions)?

10. The weight of an object on the surface of the moon varies directly as the weight of the object on the surface of Earth. An astronaut with his protective suit weighs 80 kilograms on Earth's surface; on the moon and wearing the same suit, he will weigh 12.8 kilograms. If his moon rover vehicle weighs 900 kilograms on Earth, how much will it weigh on the moon?

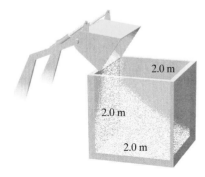

2.0 m

2.0 m

2.0 m

11. If y varies inversely as x, and $y = 12$ when $x = 4$, find y when $x = 7$.

12. If y varies inversely as x, and $y = 39$ when $x = 3$, find y when $x = 11$.

13. If y varies inversely as x, and $y = \dfrac{1}{4}$ when $x = 8$, find y when $x = 1$.

14. If y varies inversely as x, and $y = \dfrac{1}{5}$ when $x = 20$, find y when $x = 2$.

15. If y varies inversely as the square of x, and $y = 30$ when $x = 2$, find y when $x = 9$.

16. If y varies inversely as the cube of x, and $y = \dfrac{1}{9}$ when $x = 3$, find y when $x = \dfrac{1}{2}$.

17. The amount of time in minutes that it takes for an ice cube to melt varies inversely as the temperature of the water that the ice cube is placed in. When an ice cube is placed in 60°F water, it takes 2.3 minutes to melt. How long would it take for an ice cube of this size to melt if it were placed in 40°F water?

18. The textile mills in Lowell, Massachusetts, were operated with water power. The Merrimack River and its canals powered huge turbines that generated the electricity for the mills. In each mill room, a single driveshaft supplied power to all machines by means of belts that ran to pulleys on the machines. The driveshaft turned at a constant speed, but the pulleys were different diameters, which provided for the different speeds of each machine. The speed of each machine varied inversely as the diameter of its pulley. If a machine with a 60-centimeter pulley turned at 1000 rpm, what was the speed of a machine with an 80-centimeter pulley?

19. The weight of an object near Earth's surface varies inversely as the square of its distance from the center of Earth. An object weighs 1000 pounds on Earth's surface. This is approximately 4000 miles from the center of Earth. How much will the object weigh 6000 miles from the center of Earth?

20. The illumination provided from a table lamp varies inversely with the square of the distance from the lamp. A manufacturer rates his lamps so that the standard illumination is measured for someone sitting 3 feet from the lamp. How close to this lamp must one be to get twice as much light?

21. A person on the mast of a sailing ship 81 meters above the water can see approximately 12.3 kilometers out to sea. In general, the distance, d, you can see varies directly as the square root of your height, h, above the water. In this case, the equation is $d = \left(\frac{15}{11}\right)\sqrt{h}$, where d is measured in kilometers and h is measured in meters.

   (a) Find the distances you can see when $h = 121$, 36, and 4. Round to the nearest tenth.

   (b) Graph the equation using values of h of 121, 81, 36, and 4.

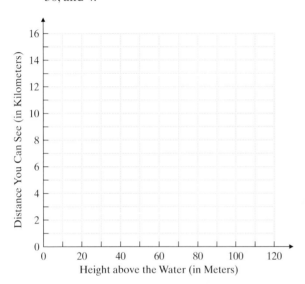

22. Suppose a computer has a hard drive that can hold 2000 files if each file is 100,000 bytes long. The number of files, f, the hard drive can hold varies inversely as the size, s, of a file. In this case, the equation is $f = \dfrac{200,000,000}{s}$, where f is the number of files and s is the size of a file in bytes.

   (a) Find the number of files that can be held if $s = 400,000$, $s = 200,000$, and $s = 50,000$.

   (b) Graph the equation using the values $s = 400,000$, $s = 200,000$, $s = 100,000$, and $s = 50,000$.

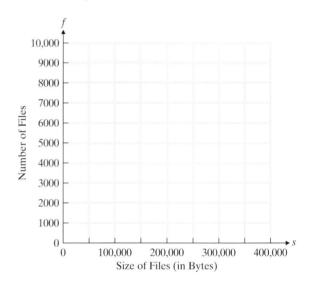

To Think About

Write an equation to describe the variation and solve.

23. The amount of heat that passes through a wall depends on various factors: the temperature difference between the two sides of the wall, the area of the wall, and the thickness of the wall. The heat transfer varies directly as the product of the area of the wall and the temperature difference between the two sides of the wall and inversely as the thickness of the wall. In a certain house, 6000 Btu/hour was transferred by a wall with an area of 80 square feet and a thickness of 0.5 foot when the outside temperature was 40°F and the inside temperature was 70°F. Suppose that the wall had been exactly 1.0 foot thick and the outside temperature had been 20°F. Assuming the same wall area and inside temperature, how much heat would have been transferred?

24. An important element of sailboat design is the calculation of the force of wind on the sail. The force varies directly with the square of the wind speed and directly with the area of the sail. For a certain class of racing boat, the force on 100 square feet of sail is 280 pounds when the wind speed is 6 mph. What would be the force on the sail if it were rigged for a storm (the amount of sail exposed to the wind is reduced by lashing part of the sail to the boom) so that only 60 square feet of sail were exposed and the wind speed were 40 mph?

Cumulative Review Problems

25. Solve for a. $\dfrac{80,000}{320} = \dfrac{120,000}{320 + a}$

26. Graph the equation $S = kh$, where $k = 4.80$. Use h for the horizontal axis and S for the vertical. Assume that h and S are nonnegative.

27. Factor. $12x^2 - 20x - 48$

28. Solve. $2x - (5 - 3x) = 6(x - 2)$

Math in the Media

The Pythagorean Winning Percentage Formula

The Pythagorean winning percentage formula, developed by Bill James, is used to predict the record of a team based on its runs scored and runs allowed. The formula is:

$$\frac{(\text{Runs Scored})^2}{(\text{Runs Scored})^2 + (\text{Runs Allowed})^2}$$

In each part of the formula, an exponent of 1.83 is sometimes used instead of 2 because it provides a slightly more precise calculation.

The Pythagorean winning percentage formula was put to just such a use in a recent article, *So Who Would Win?* by Devin Clancy in USATODAY.com.

To get some hands on experience with this formula try the following exercises.

EXERCISES

1. The Cincinnati Reds scored 865 runs and allowed 711 runs in the 1999 season. Using the Pythagorean winning percentage formula (with the exponent of 2), calculate the team's predicted winning percent. Round to 3 decimal places.

   exponent of 2 in the winning percentage formula. What is the predicted winning percent? Did the accuracy of the prediction improve?

   2000 season. Using the Pythagorean winning percentage formula, with an exponent of 2, calculate the team's predicted winning percentage. Repeat the formula with an exponent of 1.83. Which calculation is closer to the exact value?

2. The Cincinnati Reds ended the 1999 season with a 0.589 win percentage. Calculate exercise 1 again and use 1.83 rather than an

3. The New York Yankees ended the 2000 season with a 0.540 win percentage. The team scored 871 runs and allowed 814 runs in the

Chapter 9 Organizer

Topic	Procedure	Examples
Square roots, p. 524.	**1.** The square roots of perfect squares up to $\sqrt{225} = 15$ (or so) should be memorized. Approximate values of others can be found in tables or by calculator. **2.** Negative numbers do not have real number square roots.	$\sqrt{196} = 14 \qquad \sqrt{169} = 13$ $\sqrt{200} \approx 14.142 \qquad \sqrt{13} \approx 3.606$ $\sqrt{-5}$ This is not a real number.
Simplifying radicals, p. 529.	$\sqrt{ab} = \sqrt{a}\,\sqrt{b}$ Square root radicals are simplified by taking the square roots of all factors that are perfect squares and leaving all other factors under the radical sign.	$\sqrt{48} = \sqrt{16}\,\sqrt{3} = 4\sqrt{3}$ $\sqrt{12x^2y^3z} = \sqrt{2^2x^2y^2}\,\sqrt{3yz}$ $\qquad = 2xy\sqrt{3yz}$ $\sqrt{a^{11}} = \sqrt{a^{10} \cdot a} = \sqrt{a^{10}}\,\sqrt{a} = a^5\sqrt{a}$
Adding and subtracting radicals, p. 534.	Simplify all radicals and then add or subtract like radicals.	$\sqrt{12} + \sqrt{18} + \sqrt{27} + \sqrt{50}$ $= \sqrt{4}\,\sqrt{3} + \sqrt{9}\,\sqrt{2} + \sqrt{9}\,\sqrt{3} + \sqrt{25}\,\sqrt{2}$ $= 2\sqrt{3} + 3\sqrt{2} + 3\sqrt{3} + 5\sqrt{2}$ $= 5\sqrt{3} + 8\sqrt{2}$
Multiplying radicals, p. 538.	Square root radicals are multiplied like polynomials. Apply the rule $\sqrt{a}\,\sqrt{b} = \sqrt{ab}$ and simplify all results.	$(2\sqrt{6} + \sqrt{3})(\sqrt{6} - 3\sqrt{3})$ $= 2\sqrt{36} - 6\sqrt{18} + \sqrt{18} - 3\sqrt{9}$ $= 2(6) - 5\sqrt{18} - 3(3)$ $= 12 - 5\sqrt{9}\,\sqrt{2} - 9$ $= 12 - 5(3)\sqrt{2} - 9$ $= 3 - 15\sqrt{2}$
Dividing radicals, p. 543.	Division of radical expressions is accomplished by writing the division problem as a fraction and rationalizing the denominator.	Divide. $\sqrt{x} \div \sqrt{3}$ $\dfrac{\sqrt{x}}{\sqrt{3}} \cdot \dfrac{\sqrt{3}}{\sqrt{3}} = \dfrac{\sqrt{3x}}{\sqrt{9}} = \dfrac{\sqrt{3x}}{3}$
Conjugates, p. 545.	Note that the product of the sum and difference of two square root radicals is equal to the difference of their squares and thus contains no radicals. This sum and difference are called conjugates.	$(3\sqrt{6} + \sqrt{10})$ and $(3\sqrt{6} - \sqrt{10})$ are conjugates. $(3\sqrt{6} + \sqrt{10})(3\sqrt{6} - \sqrt{10})$ $= 9\sqrt{6}^2 - \sqrt{10}^2$ $= 9 \cdot 6 - 10 = 54 - 10 = 44$
Rationalizing denominators, p. 545.	To simplify a fraction with a square root radical in the denominator, the radical (in the denominator) must be removed. **1.** Multiply the denominator by itself if it is a monomial. Multiply the numerator by the same quantity **2.** Multiply the denominator by its conjugate if it is a binomial. Multiply the numerator by the same quantity.	$\dfrac{\sqrt{10}}{\sqrt{6}} = \dfrac{\sqrt{10} \cdot \sqrt{6}}{\sqrt{6} \cdot \sqrt{6}} = \dfrac{2\sqrt{15}}{6} = \dfrac{\sqrt{15}}{3}$ $\dfrac{\sqrt{3}}{\sqrt{3} - \sqrt{2}} = \dfrac{\sqrt{3}(\sqrt{3} + \sqrt{2})}{(\sqrt{3} - \sqrt{2})(\sqrt{3} + \sqrt{2})}$ $= \dfrac{\sqrt{9} + \sqrt{6}}{3 - 2} = 3 + \sqrt{6}$
Radicals with fractions, p. 543.	Note that $\sqrt{\dfrac{5}{3}}$ is the same as $\dfrac{\sqrt{5}}{\sqrt{3}}$ and so must be rationalized.	$\sqrt{\dfrac{5}{3}} = \dfrac{\sqrt{5}}{\sqrt{3}} \cdot \dfrac{\sqrt{3}}{\sqrt{3}} = \dfrac{\sqrt{15}}{3}$

Topic	Procedure	Examples
Pythagorean Theorem, p. 549.	In any right triangle with hypotenuse of length c and legs of lengths a and b, $$c^2 = a^2 + b^2.$$	Find c to the nearest tenth if $a = 7$ and $b = 9$. $$c^2 = 7^2 + 9^2 = 49 + 81 = 130$$ $$c = \sqrt{130} \approx 11.4$$
Radical equations, p. 551.	To solve an equation containing a square root radical: 1. Isolate the radical by itself on one side of the equation. 2. Square both sides. 3. Solve the resulting equation. 4. Check all apparent solutions. Extraneous roots may have been introduced in step 2.	Solve. $\sqrt{3x + 2} - 4 = 8$ $$\sqrt{3x + 2} = 12$$ $$\left(\sqrt{3x + 2}\right)^2 = 12^2$$ $$3x + 2 = 144$$ $$3x = 142$$ $$x = \frac{142}{3}$$ $$\sqrt{3 \cdot \frac{142}{3} + 2} \overset{?}{=} 12$$ $$\sqrt{144} = 12 \quad \checkmark$$ So $\quad x = \frac{142}{3}$.
Variation: direct, p. 556.	If y varies directly with x, there is a constant of variation k such that $y = kx$. Once k is determined, the value of y or x can easily be computed.	y varies directly with x. When $x = 2$, $y = 7$. $$y = kx$$ $$7 = k(2) \quad \text{Substituting.}$$ $$k = \frac{7}{2} \quad \text{Solving.}$$ $$y = \frac{7}{2}x$$ What is y when $x = 8$? $$y = \frac{7}{2}x = \frac{7}{2} \cdot 8 = 28$$
Variation: inverse, p. 558.	If y varies inversely with x, there is a constant of variation k such that $$y = \frac{k}{x}.$$	y varies inversely with x. When x is 5, y is 12. What is y when x is 15? $$y = \frac{k}{x}$$ $$12 = \frac{k}{5} \quad \text{Substituting.}$$ $$k = 60 \quad \text{Solving.}$$ $$y = \frac{60}{x} \quad \text{Substituting.}$$ $$x = 15 \quad y = \frac{60}{15} = 4$$

Chapter 9 Review Problems

9.1 *Simplify or evaluate, if possible.*

1. $\sqrt{121}$ **2.** $\sqrt{144}$ **3.** $\sqrt{169}$ **4.** $\sqrt{196}$ **5.** $\sqrt{-81}$

6. $\sqrt{-64}$ **7.** $\sqrt{49}$ **8.** $\sqrt{256}$ **9.** $\sqrt{225}$ **10.** $\sqrt{289}$

11. $\sqrt{0.81}$ **12.** $\sqrt{0.36}$ **13.** $\sqrt{\dfrac{1}{25}}$ **14.** $\sqrt{\dfrac{36}{49}}$

Approximate, using the square root table in Appendix A or a calculator. Round to the nearest thousandth, if necessary.

15. $\sqrt{105}$ **16.** $\sqrt{198}$ **17.** $\sqrt{77}$ **18.** $\sqrt{88}$

9.2 *Simplify.*

19. $\sqrt{50}$ **20.** $\sqrt{48}$ **21.** $\sqrt{98}$ **22.** $\sqrt{72}$

23. $\sqrt{40}$ **24.** $\sqrt{80}$ **25.** $\sqrt{x^8}$ **26.** $\sqrt{y^{10}}$

27. $\sqrt{x^5 y^6}$ **28.** $\sqrt{a^3 b^4}$ **29.** $\sqrt{16x^3 y^5}$ **30.** $\sqrt{98x^4 y^6}$

31. $\sqrt{12x^5}$ **32.** $\sqrt{27x^7}$ **33.** $\sqrt{75x^{10}}$ **34.** $\sqrt{125x^{12}}$

35. $\sqrt{120a^3 b^4 c^5}$ **36.** $\sqrt{121a^6 b^4 c}$ **37.** $\sqrt{56x^7 y^9}$ **38.** $\sqrt{99x^{13} y^7}$

9.3 *Simplify.*

39. $\sqrt{50} + \sqrt{2} - \sqrt{8}$ **40.** $2\sqrt{5} + 3\sqrt{20} + \sqrt{45}$

41. $x\sqrt{3} + 3x\sqrt{3} + \sqrt{27x^2}$ **42.** $a\sqrt{2} + \sqrt{12a^2} + a\sqrt{98}$

43. $5\sqrt{5} - 6\sqrt{20} + 2\sqrt{10}$ **44.** $3\sqrt{6} - 5\sqrt{18} + 3\sqrt{24}$

9.4 *Simplify.*

45. $(2\sqrt{x})(3\sqrt{x^3})$ **46.** $(-5\sqrt{a})(2\sqrt{ab})$

47. $(\sqrt{2a^3})(\sqrt{8b^2})$ **48.** $(5x\sqrt{x})(-3x^2\sqrt{x})$

49. $\sqrt{5}(3\sqrt{5} - \sqrt{20})$ **50.** $2\sqrt{3}(\sqrt{27} - 6\sqrt{3})$

51. $\sqrt{2}(\sqrt{5} - \sqrt{3} - 2\sqrt{2})$ **52.** $\sqrt{5}(\sqrt{6} - 2\sqrt{5} + \sqrt{10})$

53. $(\sqrt{11} + 2)(2\sqrt{11} - 1)$ **54.** $(\sqrt{10} + 3)(3\sqrt{10} - 1)$

55. $(2 + 3\sqrt{6})(4 - 2\sqrt{3})$ **56.** $(5 - \sqrt{2})(3 - \sqrt{12})$

57. $(2\sqrt{3} + 3\sqrt{6})^2$ **58.** $(5\sqrt{2} - 2\sqrt{6})^2$

9.5 *Rationalize the denominator.*

59. $\dfrac{1}{\sqrt{3x}}$

60. $\dfrac{2y}{\sqrt{5}}$

61. $\dfrac{x^2 y}{\sqrt{8}}$

62. $\dfrac{3ab}{\sqrt{2b}}$

63. $\sqrt{\dfrac{3}{7}}$

64. $\sqrt{\dfrac{2}{9}}$

65. $\dfrac{\sqrt{a^5}}{\sqrt{2a}}$

66. $\dfrac{\sqrt{x^3}}{\sqrt{3x}}$

67. $\dfrac{3}{\sqrt{5} + \sqrt{2}}$

68. $\dfrac{2}{\sqrt{6} - \sqrt{3}}$

69. $\dfrac{1 - \sqrt{5}}{2 + \sqrt{5}}$

70. $\dfrac{1 - \sqrt{3}}{3 + \sqrt{3}}$

9.6 *In exercises 71–77, use the Pythagorean Theorem to find the length of the missing side of a right triangle with legs a and b and hypotenuse c.*

△ 71. $a = 5, b = 8$

△ 72. $c = \sqrt{11}, b = 3$

△ 73. $c = 5, a = 3.5$

△ 74. If $a = 2.400$ and $b = 2.000$, find c and round to the nearest thousandth.

△ 75. A flagpole is 24 meters tall. A man stands 18 meters from the base of the pole. How far is it from the feet of the man to the top of the pole?

△ 76. A city is 20.0 miles east of a major airport. A small town is directly south of the city. The town is 50.0 miles from the airport. How far is it from the town to the city? (Round to the nearest tenth of a mile.)

△ 77. The diagonal measurement of a small color television screen is 10 inches. The screen is square. How long is each side of the screen? (Round to the nearest tenth of an inch.)

Solve. Be sure to verify your answers.

78. $\sqrt{x + 3} = 4$

79. $\sqrt{2x - 3} = 9$

80. $\sqrt{1 - 3x} = \sqrt{5 + x}$

81. $\sqrt{-5 + 2x} = \sqrt{1 + x}$

82. $\sqrt{10x + 9} = -1 + 2x$

83. $\sqrt{2x - 5} = 10 - x$

84. $6 - \sqrt{5x - 1} = x + 1$

85. $4x + \sqrt{x + 2} = 5x - 4$

9.7

86. If y varies directly with the square root of x, and $y = 35$ when $x = 25$, find the value of y when $x = 121$.

87. If y varies inversely with the square of x, and $y = \dfrac{6}{5}$ when $x = 5$, find the value of y with $x = 15$.

88. If y varies inversely with the cube of x, and $y = 4$ when $x = 2$, find the value of y when $x = 4$.

89. The insect population in a potato field varies inversely with the amount of pesticide used. When 40 pounds of pesticide was used, the insect population of the field was estimated to be 1000 bugs. How many pounds of pesticide are required to reduce the number of bugs in the field to 100?

90. When an automobile driver slams on the brakes, the length of skid marks on the road varies directly with the square of the speed of the car. At 30 mph a certain car had skid marks 40 feet long. How long will the skid marks be if the car travels at 55 mph?

91. The horsepower that is needed to drive a racing boat through water varies directly with the cube of the speed of the boat. What will happen to the horsepower requirement if someone wants to double the maximum speed of a given boat?

Putting Your Skills to Work

Determining Free-Fall Times

When people jump out of a plane to parachute to Earth, they are in free-fall until they deploy their parachutes. If we ignore wind resistance, we can approximate the amount of time in seconds it takes to fall x feet by the equation $t = 0.25\sqrt{x}$.

Problems for Individual Investigation and Analysis

1. Using the preceding equation, complete the following table. Round all values to the nearest tenth. Use a calculator or square root table when necessary.

x	0	64	100	144	256	400	784
t							

2. Using the preceding table of values, graph the equation $t = 0.25\sqrt{x}$. From your graph, estimate the number of seconds it would take to drop 900 feet.

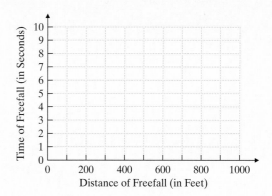

Problems for Group Investigation and Cooperative Learning

The speed v in feet per second at which a person in free-fall is traveling after x seconds is given by the equation $v = 32x$. For example, after 5 seconds a person is falling at a rate of $32(5) = 160$ feet per second.

3. How fast is a person in free-fall traveling when he has fallen a distance of 576 feet?

4. How fast is a person in free-fall traveling when she has fallen a distance of 1024 feet?

Internet Connections

  Netsite: http://www.prenhall.com/tobey_beginning

Site: Records of Parachute Events

5. What is the world's record for the largest number of people who jumped out of an airplane, made contact with each other, and then successfully parachuted to Earth?

6. Disregarding air resistance, if it had taken 27 seconds to achieve the link, approximately how many feet would they have been from the ground when the link was broken?

Chapter 9 Test

1. _____

2. _____

3. _____

4. _____

5. _____

6. _____

7. _____

8. _____

9. _____

10. _____

11. _____

12. _____

13. _____

14. _____

15. _____

16. _____

17. _____

18. _____

19. _____

20. _____

21. _____

22. _____

Evaluate.

1. $\sqrt{121}$

2. $\sqrt{\dfrac{9}{100}}$

Simplify.

3. $\sqrt{48x^2y^7}$

4. $\sqrt{100x^3yz^4}$

Combine and simplify.

5. $\sqrt{5} + 3\sqrt{20} - 2\sqrt{45}$

6. $\sqrt{4a} + \sqrt{8a} + \sqrt{36a} + \sqrt{18a}$

Multiply.

7. $(2\sqrt{a})(3\sqrt{b})(2\sqrt{ab})$

8. $\sqrt{3}(\sqrt{6} - \sqrt{2} + 5\sqrt{27})$

9. $(2 - \sqrt{6})^2$

10. $(4\sqrt{2} - \sqrt{5})(3\sqrt{2} + \sqrt{5})$

In questions 11–14, simplify.

11. $\sqrt{\dfrac{x}{5}}$

12. $\dfrac{3}{\sqrt{12}}$

13. $\dfrac{3 - \sqrt{2}}{\sqrt{2} - 6}$

14. $\dfrac{3a}{\sqrt{5} + \sqrt{2}}$

15. Use the table of square roots or a calculator to approximate $\sqrt{156}$. Round your answer to the nearest hundredth.

Find the missing side by using the Pythagorean Theorem.

▲ **16.**

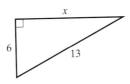

▲ **17.**

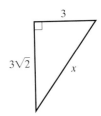

Solve. Verify your solutions.

18. $6 - \sqrt{2x + 1} = 0$

19. $x = 5 + \sqrt{x + 7}$

Solve.

20. The intensity of illumination from a light source varies inversely with the square of the distance from the light source. A photoelectric cell is placed 8 inches from a light source. Then it is moved so that it receives only one-fourth as much illumination. How far is it from the light source?

21. Sales tax varies directly with the cost of the sale. If $26.10 is the tax on a purchase of $360, what will the tax be on a purchase of $12,580?

22. The area of an equilateral triangle varies directly as the square of its perimeter. If the perimeter is 12 centimeters, the area is 6.93 square centimeters. What is the area of an equilateral triangle whose perimeter is 21 centimeters? Round to the nearest tenth.

Approximately one-half of this test covers the content of Chapters 0–8. The remainder covers the content of Chapter 9.

In questions 1–3, simplify.

1. $\dfrac{12}{15}$

2. $3\dfrac{1}{4} + 5\dfrac{2}{3}$

3. $6\dfrac{1}{4} \div 6\dfrac{2}{3}$

4. Write 0.07% as a decimal.

Simplify.

5. $-11 - 16 + 8 + 4 - 13 + 31$

6. $(-3)^2 \cdot 4 - 8 \div 2 - (3 - 2)^3$

7. $3x^3yz^2 + 4xyz - 5x^3yz^2$

8. $(3x - 2)^3$

9. $(4x - 5)^2$

10. $(3x - 11)(3x + 11)$

11. $\dfrac{4x}{x + 2} + \dfrac{4}{x + 2}$

12. $\dfrac{x^2 - 5x + 6}{x^2 - 5x - 6} \div \dfrac{x^2 - 4}{x^2 + 2x + 1}$

Simplify, if possible.

13. $\sqrt{98x^5y^6}$

14. $\sqrt{12} - \sqrt{27} + 3\sqrt{75}$

15. $\left(\sqrt{6} - \sqrt{3}\right)^2$

16. $\dfrac{\sqrt{3} + \sqrt{2}}{\sqrt{3} - \sqrt{2}}$

17. $\left(3\sqrt{6} - \sqrt{2}\right)\left(\sqrt{6} + 4\sqrt{2}\right)$

18. $\dfrac{1}{3 - \sqrt{2}}$

19. $-\sqrt{16}$

20. $\sqrt{-16}$

Use the Pythagorean Theorem to find the missing side of each triangle. Be sure to simplify your answer.

▲ **21.** $a = 4, b = \sqrt{7}$

▲ **22.** $a = 19, c = 21$

In questions 23 and 24, solve for the variable. Be sure to verify your answer(s).

23. $\sqrt{4x + 5} = x$

24. $\sqrt{3y - 2} + 2 = y$

25. If y varies inversely as x, and $y = 4$ when $x = 5$, find the value of y when $x = 2$.

26. The surface area of a balloon varies directly with the square of the cross-sectional radius. When the radius is 2 meters, the surface area is 1256 square meters. Find the surface area if the radius is 0.5 meter.

1. _____

2. _____

3. _____

4. _____

5. _____

6. _____

7. _____

8. _____

9. _____

10. _____

11. _____

12. _____

13. _____

14. _____

15. _____

16. _____

17. _____

18. _____

19. _____

20. _____

21. _____

22. _____

23. _____

24. _____

25. _____

26. _____

Quadratic Equations

T he beauty of the Taj Mahal is noticed by every observer. But how much of this beauty is due to the cleverly designed reflecting pool in front of the Taj Mahal? Do you know the reasons for the choice of the dimensions of this pool? Please turn to the Putting Your Skills to Work problems on page 619 to find out.

Pretest Chapter 10

1. _____

2. _____

3. _____

4. _____

5. _____

6. _____

7. _____

8. _____

9. _____

10. _____

11. _____

12. _____

13. _____

14. _____

15. _____

If you are familiar with the topics in this chapter, take this test now. Check your answers with those in the back of the book. If an answer is wrong or you can't answer a question, study the appropriate section of the chapter.

If you are not familiar with the topics in this chapter, don't take this test now. Instead, study the examples, work the practice problems, and then take the test.

This test will help you to identify those concepts that you have mastered and those that need more study.

Section 10.1

Solve by factoring.

1. $x^2 - 13x - 48 = 0$

2. $5x^2 + 7x = 14x$

3. $5x^2 = 22x - 8$

4. $-2x + 1 = 8x^2$

Section 10.2

Solve by taking the square root of each side of the equation.

5. $x^2 - 18 = 0$

6. $3x^2 + 1 = 76$

Solve by completing the square.

7. $x^2 + 8x + 5 = 0$

8. $2x^2 + 3x - 7 = 0$

Section 10.3

Solve if possible using the quadratic formula.

9. $2x^2 + 4x - 5 = 0$

10. $2x^2 = 7x - 4$

11. $3x^2 + 8x + 1 = 0$

12. $5x^2 + 3 = 4x$

Section 10.4

Graph. Locate the vertex.

13. $y = 5 - 3x^2$

14. $y = 2x^2 - 4x - 1$

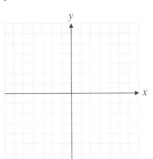

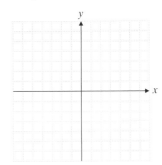

Section 10.5

▲ **15.** A triangle has an area of 39 square centimeters. The altitude of the triangle is 1 centimeter longer than double the length of the base. Find the dimensions of the triangle.

10.1 Introduction to Quadratic Equations

① Placing a Quadratic Equation in Standard Form

In Section 5.7 we introduced quadratic equations. A **quadratic equation** is a polynomial equation of degree two. For example,

$$3x^2 + 5x - 7 = 0, \quad \frac{1}{2}x^2 - 5x = 0, \quad 2x^2 = 5x + 9, \quad \text{and} \quad 3x^2 = 8$$

are all quadratic equations. It is particularly useful to place quadratic equations in standard form.

> The **standard form of a quadratic equation** is $ax^2 + bx + c = 0$, where a, b, and c are real numbers, and a is a positive real number.

You will need to be able to recognize the real numbers that represent a, b, and c in specific equations. It will be easier to do so if these equations are placed in standard form.

EXAMPLE 1 Place each quadratic equation in standard form and identify the real numbers a, b, and c.

(a) $5x^2 - 6x + 3 = 0$ This equation is in standard form.

$\quad ax^2 + bx + c = 0$ Match each term to the standard form.

$\quad a = 5, \quad b = -6, \quad c = 3$

(b) $2x^2 + 5x = 4$ The right-hand side is not zero. It is not in standard form.

$\quad 2x^2 + 5x - 4 = 0$ Add -4 to each side of the equation.

$\quad ax^2 + bx + c = 0$ Match each term to the standard form.

$\quad a = 2, \quad b = 5, \quad c = -4$

(c) $-2x^2 + 15x + 4 = 0$ This is not in standard form since the coefficient of x^2 is negative.

$\quad 2x^2 - 15x - 4 = 0$ Multiply each term on both sides of the equation by -1.

$\quad ax^2 + bx + c = 0$

$\quad a = 2, \quad b = -15, \quad c = -4$

(d) $7x^2 - 9x = 0$ This equation is in standard form.

$\quad ax^2 + bx + c = 0$ Note that the constant term is missing, so we know $c = 0$.

$\quad a = 7, \quad b = -9, \quad c = 0$

Practice Problem 1 Place each quadratic equation in standard form and identify the real numbers a, b, and c. Standard form requires that a be a positive number. If $a < 0$, you can multiply each term of the equation by -1 to obtain an equivalent equation.

(a) $2x^2 + 12x - 9 = 0$ **(b)** $7x^2 = 6x - 8$

(c) $-x^2 - 6x + 3 = 0$ **(d)** $10x^2 - 12x = 0$

② Solving Quadratic Equations of the Form $ax^2 + bx = 0$ by Factoring

Notice that the terms in a quadratic equation of the form $ax^2 + bx = 0$ both have x as a factor. To solve such equations, begin by factoring out the x and any common numerical factor. Thus you may be able to factor out an x or an expression such as $5x$.

Student Learning Objectives

After studying this section, you will be able to:

① Place a quadratic equation in standard form.

② Solve quadratic equations of the form $ax^2 + bx = 0$ by factoring.

③ Solve quadratic equations of the form $ax^2 + bx + c = 0$ by factoring.

④ Solve applied problems that require the use of quadratic equations.

SSM   PH TUTOR   CD & VIDEO   MATH PRO   WEB
CENTER

Remember that you want to remove the *greatest common factor*. So if $5x$ is a common factor of each term, be sure to factor it out. Then use the zero factor property discussed in Section 5.7. This property states that if $a \cdot b = 0$, then either $a = 0$ or $b = 0$. Once each factor is set equal to 0, solve for x.

EXAMPLE 2 Solve. $7x^2 + 9x - 2 = -8x - 2$

$$7x^2 + 9x - 2 = -8x - 2 \qquad \text{The equation is not in standard form.}$$
$$7x^2 + 9x - 2 + 8x + 2 = 0 \qquad \text{Add } 8x + 2 \text{ to each side.}$$
$$7x^2 + 17x = 0 \qquad \text{Collect like terms.}$$
$$x(7x + 17) = 0 \qquad \text{Factor.}$$
$$x = 0 \qquad 7x + 17 = 0 \qquad \text{Set each factor equal to zero.}$$
$$7x = -17$$
$$x = -\frac{17}{7}$$

One of the solutions to the preceding equation is 0. This will always be true of equations of this form. One root will be zero. The other root will be a nonzero real number.

Practice Problem 2 Solve. $2x^2 - 7x - 6 = 4x - 6$

3 Solving Quadratic Equations of the Form $ax^2 + bx + c = 0$ by Factoring

If an equation you are trying to solve contains fractions, clear the fractions by multiplying each term by the least common denominator of all the fractions in the equation. Sometimes an equation does not look like a quadratic equation. However, it takes the quadratic form once the fractions have been cleared. A note of caution: Always check a possible solution in the original equation. Any apparent solution that would make the denominator of any fraction in the original equation 0 is not a valid solution.

EXAMPLE 3 Solve and check. $8x - 6 + \dfrac{1}{x} = 0$

$$8x - 6 + \frac{1}{x} = 0 \qquad \text{The equation has a fractional term.}$$

$$x(8x) - (x)(6) + x\left(\frac{1}{x}\right) = x(0) \qquad \text{Multiply each term by the LCD which is } x.$$

$$8x^2 - 6x + 1 = 0 \qquad \text{Simplify. The equation is now in standard form.}$$

$$(4x - 1)(2x - 1) = 0 \qquad \text{Factor.}$$
$$4x - 1 = 0 \qquad 2x - 1 = 0 \qquad \text{Set each factor equal to 0.}$$
$$4x = 1 \qquad 2x = 1 \qquad \text{Solve each equation for } x.$$
$$x = \frac{1}{4} \qquad x = \frac{1}{2}$$

Check. Checking fractional roots is more difficult, but you should be able to do it if you work carefully.

$$\text{If } x = \frac{1}{4}: \quad 8\left(\frac{1}{4}\right) - 6 + \frac{1}{\frac{1}{4}} = 2 - 6 + 4 = 0 \quad \left(\text{since } 1 \div \frac{1}{4} = 1 \cdot \frac{4}{1} = 4\right)$$

$$0 = 0 \checkmark$$

If $x = \dfrac{1}{2}$: $\quad 8\left(\dfrac{1}{2}\right) - 6 + \dfrac{1}{\frac{1}{2}} = 4 - 6 + 2 = 0 \quad \left(\text{since } 1 \div \dfrac{1}{2} = 1 \cdot \dfrac{2}{1} = 2\right)$

$$0 = 0 \quad \checkmark$$

Both roots check, so $\dfrac{1}{2}$ and $\dfrac{1}{4}$ are the two roots that satisfy the original equation

$$8x - 6 + \dfrac{1}{x} = 0.$$

It is also correct to write the answers in decimal form as 0.5 and 0.25. This will speed up the checking process, especially if you use a scientific calculator when performing the check.

Practice Problem 3 Solve and check. $\quad 3x + 10 - \dfrac{8}{x} = 0$

If a quadratic equation is given to us in standard form, we examine it to see if it can be factored. If it is possible to factor the quadratic equation, then we use the zero factor property to find the solutions.

EXAMPLE 4 Solve and check. $\quad 6x^2 + 7x - 10 = 0$

$\qquad (6x - 5)(x + 2) = 0 \qquad$ Factor the quadratic equation.

$6x - 5 = 0 \qquad x + 2 = 0 \qquad$ Set each factor equal to zero and solve for x.

$\qquad 6x = 5 \qquad\qquad x = -2$

$\qquad x = \dfrac{5}{6}$

Thus the solutions to the equation are $\dfrac{5}{6}$ and -2.

Check each solution in the original equation.

$$6\left(\dfrac{5}{6}\right)^2 + 7\left(\dfrac{5}{6}\right) - 10 \overset{?}{=} 0 \qquad\qquad 6(-2)^2 + 7(-2) - 10 \overset{?}{=} 0$$

$$6\left(\dfrac{25}{36}\right) + 7\left(\dfrac{5}{6}\right) - 10 \overset{?}{=} 0 \qquad\qquad 6(4) + 7(-2) - 10 \overset{?}{=} 0$$

$$\dfrac{25}{6} + \dfrac{35}{6} - 10 \overset{?}{=} 0 \qquad\qquad 24 + (-14) - 10 \overset{?}{=} 0$$

$$\dfrac{60}{6} - 10 \overset{?}{=} 0 \qquad\qquad 10 - 10 = 0 \quad \checkmark$$

$$10 - 10 = 0 \quad \checkmark$$

Practice Problem 4 Solve and check. $2x^2 + 9x - 18 = 0$

It is important to place a quadratic equation in standard form before factoring. This will sometimes involve removing parentheses and collecting like terms.

EXAMPLE 5 Solve. $\quad x + (x - 6)(x - 2) = 2(x - 1)$

$\qquad x + x^2 - 6x - 2x + 12 = 2x - 2 \qquad$ Remove parentheses.

$\qquad\qquad x^2 - 7x + 12 = 2x - 2 \qquad$ Collect like terms.

$\qquad\qquad x^2 - 9x + 14 = 0 \qquad$ Add $-2x + 2$ to each side.

$\qquad\qquad (x - 7)(x - 2) = 0 \qquad$ Factor.

$\qquad\qquad x = 7, \quad x = 2 \qquad$ Use the zero property.

The solutions to the equation are 7 and 2. The check is left to the student.

Practice Problem 5 Solve. $-4x + (x - 5)(x + 1) = 5(1 - x)$

EXAMPLE 6 Solve. $\dfrac{8}{x+1} - \dfrac{x}{x-1} = \dfrac{2}{x^2-1}$

The LCD $= (x+1)(x-1)$. We multiply each term by the LCD.

$$(x+1)(x-1)\left(\frac{8}{x+1}\right) - (x+1)(x-1)\left(\frac{x}{x-1}\right) = (x+1)(x-1)\left[\frac{2}{(x+1)(x-1)}\right]$$

$8(x-1) - x(x+1) = 2$	Simplify.
$8x - 8 - x^2 - x = 2$	Remove parentheses.
$-x^2 + 7x - 8 = 2$	Collect like terms.
$-x^2 + 7x - 8 - 2 = 0$	Add -2 to each side.
$-x^2 + 7x - 10 = 0$	Simplify.
$x^2 - 7x + 10 = 0$	Multiply each term by -1.
$(x-5)(x-2) = 0$	Factor and solve for x.
$x = 5, \quad x = 2$	Use the zero factor property.

Check. The check is left to the student.

Practice Problem 6 Solve. $\dfrac{10x+18}{x^2+x-2} = \dfrac{3x}{x+2} + \dfrac{2}{x-1}$

4 **Solving Applied Problems That Require the Use of Quadratic Equations**

A manager of a truck delivery service is determining the possible routes he may have to send trucks on during a given day. His delivery service must have the capability to send trucks between any two of the following Texas cities: Austin, Dallas, Tyler, Lufkin, and Houston. He made a rough map of the cities.

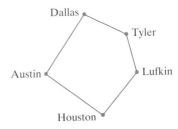

What is the maximum number of routes he may have to send trucks on in a given day, if there is a route between every two cities? One possible way to figure this out is to draw a straight line between every two cities. Let us assume that no three cities lie on a straight line. Therefore, the route between any two cities is distinct from all other routes. The following figure shows that we can draw exactly 10 lines, so the answer is that there are 10 truck routes. What if the manager has to service 18 cities? How many truck routes will there be then?

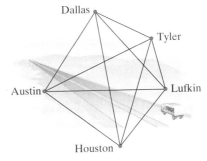

As the number of cities increases, it becomes quite difficult (and time consuming) to use a drawing to determine the number of routes. In a more advanced math course, it can be proved that the maximum number of truck routes (t) can be found by using the number of cities (n) in the following equation.

$$t = \frac{n^2 - n}{2}$$

Thus, to find the number of truck routes for 18 cities, we merely substitute 18 for n.

$$t = \frac{(18)^2 - (18)}{2}$$

$$= \frac{324 - 18}{2}$$

$$= \frac{306}{2}$$

$$= 153$$

Thus 153 truck routes are needed to service 18 cities if no three of those cities lie on a straight line.

EXAMPLE 7 A truck delivery company can handle a maximum of 36 truck routes in one day, and every two cities have a distinct truck route between them. How many separate cities can the truck company service?

$$t = \frac{n^2 - n}{2}$$

$$36 = \frac{n^2 - n}{2} \qquad \text{Substitute 36 for } t, \text{ the number of truck routes.}$$

$$72 = n^2 - n \qquad \text{Multiply both sides of the equation by 2.}$$

$$0 = n^2 - n - 72 \qquad \text{Add } -72 \text{ to each side to obtain the standard form of a quadratic equation.}$$

$$0 = (n - 9)(n + 8) \qquad \text{Factor.}$$

$$n - 9 = 0 \quad n + 8 = 0 \qquad \text{Set each factor equal to zero.}$$

$$n = 9 \qquad n = -8 \qquad \text{Solve for } n.$$

We reject -8 as a meaningless solution for this problem. We cannot have a negative number of cities. Thus our answer is that the company can service 9 cities.

Check. If we service $n = 9$ cities, does that really result in a maximum number of $t = 36$ truck routes?

$$t = \frac{n^2 - n}{2} \qquad \text{The equation in which } t = \text{ the number of truck routes and } n = \text{ the number of cities to be connected.}$$

$$36 \overset{?}{=} \frac{(9)^2 - (9)}{2} \qquad \text{Substitute } t = 36 \text{ and } n = 9.$$

$$36 \overset{?}{=} \frac{81 - 9}{2}$$

$$36 \overset{?}{=} \frac{72}{2}$$

$$36 = 36 \quad \checkmark$$

Practice Problem 7 How many cities can the company service with 28 truck routes?

Write in standard form. Determine the values of a, b, and c.

1. $x^2 + 8x + 7 = 0$

2. $x^2 + 6x - 5 = 0$

3. $8x^2 - 11x = 0$

4. $4x^2 + 20x = 0$

5. $x^2 + 15x - 7 = 12x + 8$

6. $3x^2 + 10x + 8 = -17x - 2$

Using the factoring method, solve for the roots of each quadratic equation. Be sure to place your equation in standard form before factoring.

7. $27x^2 - 9x = 0$

8. $6x + 18x^2 = 0$

9. $7x^2 - 3x = 5x$

10. $9x^2 + 2x = -3x$

11. $11x^2 - 13x = 8x - 3x^2$

12. $60x^2 + 14x = 9x^2 - 20x$

13. $x^2 - 3x - 28 = 0$

14. $x^2 + 3x - 40 = 0$

15. $2x^2 - 15x - 8 = 0$

16. $2x^2 + 13x - 7 = 0$

17. $3x^2 - 16x + 5 = 0$

18. $5x^2 - 21x + 4 = 0$

19. $x^2 = 3x + 18$

20. $x^2 = 10x - 24$

21. $15x^2 + 31x + 10 = 0$

22. $10x^2 + 31x + 15 = 0$

23. $m^2 + 13m + 24 = 2m - 6$

24. $p^2 + 15p + 25 = 6p + 5$

25. $8y^2 = 14y - 3$

26. $6y^2 = 11y - 3$

27. $25x^2 - 60x + 36 = 0$

28. $9x^2 - 24x + 16 = 0$ **29.** $(x - 5)(x + 2) = -10$ **30.** $(x - 6)(x + 1) = 8$

31. $3x^2 + 8x - 10 = -6x - 10$ **32.** $5x^2 - 7x + 12 = -8x + 12$ **33.** $n^2 + 2n - 11 = 5n + 7$

34. $x(x - 4) = 3(x - 2)$ **35.** $(x - 6)(x - 4) = 3$ **36.** $x(x + 9) = 4(x + 6)$

37. $2x(x + 3) = (3x + 1)(x + 1)$ **38.** $2(2x + 1)(x + 1) = x(7x + 1)$

Solve. Check your solutions.

39. $\dfrac{4}{x} + \dfrac{3}{x + 5} = 2$ **40.** $\dfrac{5}{x + 2} = \dfrac{2x - 1}{5}$

41. $\dfrac{12}{x - 2} = \dfrac{2x - 3}{3}$ **42.** $\dfrac{7}{x} + \dfrac{6}{x - 1} = 2$

43. $\dfrac{x}{4} - \dfrac{7}{x} = -\dfrac{3}{4}$ **44.** $\dfrac{x}{3} + \dfrac{2}{3} = \dfrac{5}{x}$

45. $x + \dfrac{8}{x} = 6$ **46.** $x - 1 = \dfrac{20}{x}$

47. $\dfrac{5x + 7}{x^2 - 9} = 1 + \dfrac{2x - 8}{x - 3}$

48. $\dfrac{24}{x^2 - 4} = 1 + \dfrac{2x - 6}{x - 2}$

To Think About

49. Why can a quadratic equation in standard form with $c = 0$ always be solved?

50. Martha solved $(x + 3)(x - 2) = 14$ as follows.

$x + 3 = 14 \qquad x - 2 = 14$

$\qquad x = 11 \qquad\qquad x = 16$

Josette said this had to be wrong because these values do not check. Explain what is wrong with Martha's method.

51. The equation $ax^2 - 7x + c = 0$ has the roots $-\dfrac{3}{2}$ and 6. Find the values of a and c.

Applications

The maximum number of truck routes, t, that are needed to provide direct service to n cities, where no three cities lie in a straight line, is given by the equation

$$t = \frac{n^2 - n}{2}.$$

Use this equation to find the solutions to exercises 52–57.

52. How many truck routes will be needed to run service between 22 cities?

53. How many truck routes will be needed to run service between 20 cities?

54. A company can handle a maximum of 15 truck routes in one day. How many separate cities can be serviced?

55. A company can handle a maximum of 21 truck routes in one day. How many separate cities can be serviced?

56. A company can handle a maximum of 45 truck routes in one day. How many separate cities can be serviced?

57. A company can handle a maximum of 55 truck routes in one day. How many separate cities can be serviced?

The cost c in dollars to produce x digital telephone answering machines is given by the equation
$c = 0.2(x^2 - 300x + 45,000)$.

Use this equation to find the solutions to exercises 58–63.

58. If the cost to produce the machines is $4680, what are the two values of x (the two different numbers of machines) that may be produced?

59. If the cost to produce the machines is $4520, what are the two values of x (the two different numbers of machines) that may be produced?

60. What is the average of the two answers you obtained in exercise 58? Use the cost equation to find the cost of producing that many machines. What is significant about your answer?

61. What is the average of the two answers you obtained in exercise 59? Use the cost equation to find the cost of producing that many machines. What is significant about your answer?

62. Find the cost to produce 149 machines and the cost to produce 151 machines. Compare these two costs with your results for exercises 58 and 60. What can you conclude?

63. Find the cost to produce 148 machines and the cost to produce 152 machines. Compare these two costs with your results for exercises 59 and 61. What can you conclude?

Cumulative Review Problems

64. *Simplify.* $\dfrac{\dfrac{1}{x} + \dfrac{3}{x-2}}{\dfrac{4}{x^2-4}}$

65. *Add the fractions.* $\dfrac{2x}{3x-5} + \dfrac{2}{3x^2-11x+10}$

Cars kick up a lot of dust. In Los Angeles, the dust consists of 20 known allergens, including molds, pollen, and animal dander. In two residential areas of Los Angeles, road dust contributed 12% of the allergens in the air. In contrast, in downtown Los Angeles, road dust is responsible for about 2% of the airborne allergens.

66. If a 36-mg sample of dust were taken from the two residential areas of Los Angeles mentioned, how many milligrams of it would be road dust?

67. A sample of dust containing 2.6 mg of road dust was collected from downtown Los Angeles. What was the size of the sample of dust?

10.2 Finding Solutions Using the Square Root Property and by Completing the Square

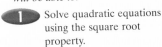

1 Solving Quadratic Equations Using the Square Root Property

Recall that the quadratic equation $x^2 = a$ has two possible solutions, $x = \sqrt{a}$ or $x = -\sqrt{a}$, where a is a nonnegative real number. That is $x = \pm\sqrt{a}$. This basic idea is called the **square root property**.

Square Root Property

If $x^2 = a$, then $x = \sqrt{a}$ or $x = -\sqrt{a}$, for all nonnegative real numbers a.

EXAMPLE 1 Solve. **(a)** $x^2 = 49$ **(b)** $x^2 = 20$ **(c)** $5x^2 = 125$

(a) $x^2 = 49$

$\quad x = \pm\sqrt{49}$

$\quad x = \pm 7$

(b) $x^2 = 20$

$\quad x = \pm\sqrt{20}$

$\quad x = \pm 2\sqrt{5}$

(c) $5x^2 = 125$ Divide both sides

$\quad x^2 = 25$ by 5 before taking

$\quad x = \pm\sqrt{25}$ the square root.

$\quad x = \pm 5$

Practice Problem 1 Solve. **(a)** $x^2 = 1$ **(b)** $x^2 - 50 = 0$ **(c)** $3x^2 = 81$

EXAMPLE 2 Solve. $3x^2 + 5x = 18 + 5x + x^2$

Simplify the equation by placing all the variable terms on the left and the constants on the right.

$$3x^2 + 5x = 18 + 5x + x^2$$
$$2x^2 = 18$$
$$x^2 = 9$$
$$x = \pm 3$$

Practice Problem 2 Solve. $4x^2 - 5 = 319$

The square root property can also be used if the term that is squared is a binomial.

Sometimes when we use the square root property we obtain an irrational number. This will occur if the number on the right side of the equation is not a perfect square.

EXAMPLE 3 Solve. $(3x + 1)^2 = 8$

$$(3x + 1)^2 = 8$$
$$3x + 1 = \pm\sqrt{8} \qquad \text{Take the square root of both sides.}$$
$$3x + 1 = \pm 2\sqrt{2} \qquad \text{Simplify as much as possible.}$$

Now we must solve the two equations expressed by the plus or minus statement.

$$3x + 1 = +2\sqrt{2} \qquad\qquad 3x + 1 = -2\sqrt{2}$$
$$3x = -1 + 2\sqrt{2} \qquad\qquad 3x = -1 - 2\sqrt{2}$$
$$x = \frac{-1 + 2\sqrt{2}}{3} \qquad\qquad x = \frac{-1 - 2\sqrt{2}}{3}$$

The roots of this quadratic equation are irrational numbers. They are

$$\frac{-1 + 2\sqrt{2}}{3} \quad \text{and} \quad \frac{-1 - 2\sqrt{2}}{3}.$$

We cannot simplify these roots further, so we leave them in this form.

Practice Problem 3 Solve. $(2x - 3)^2 = 12$

2 *Solving Quadratic Equations by Completing the Square*

If a quadratic equation is not in a form where we can use the square root property, we can rewrite the equation by a method called **completing the square** to make it so. That is, we rewrite the equation with a perfect square on the left side so that we have an equation of the form $(ax + b)^2 = c$.

EXAMPLE 4 Solve. $x^2 + 12x = 4$

Draw a picture of $x^2 + 12x = 4$. Fill in the pieces you know.

$$x^2 + 12x \qquad = 4$$

$$(x + 6)(x + 6) = 4 + 36 \qquad \text{We need a +36 to complete the square.}$$

$$x^2 + 12x + 36 = 4 + 36 \qquad \text{Add +36 to both sides of the equation.}$$
$$(x + 6)^2 = 40 \qquad \text{Write the left side as a perfect square.}$$
$$x + 6 = \pm\sqrt{40} \qquad \text{Solve for } x.$$

$$x + 6 = +2\sqrt{10} \qquad x + 6 = -2\sqrt{10}$$
$$x = -6 + 2\sqrt{10} \qquad x = -6 - 2\sqrt{10}$$

The two roots are $-6 + 2\sqrt{10}$ and $-6 - 2\sqrt{10}$.

Practice Problem 4 Solve. $x^2 + 10x = 3$

Completing the square is a little more difficult when the coefficient of x is an odd number. Let's see what happens.

EXAMPLE 5 Solve. $x^2 - 3x - 1 = 0$

Write the equation so that all the x-terms are on the left and the constants are on the right.

$$x^2 - 3x \qquad = 1$$

$$(x \qquad)(x \qquad) \qquad \text{The missing number when added to itself must be } -3.$$
$$\text{Since } -\frac{3}{2} + \left(-\frac{3}{2}\right) = -\frac{6}{2} = -3, \text{ the missing number is } -\frac{3}{2}.$$

$$\left(x - \frac{3}{2}\right)\left(x - \frac{3}{2}\right) \qquad \text{Complete the picture.}$$

$$x^2 - 3x + \frac{9}{4} = 1 + \frac{9}{4} \qquad \text{Complete the square by adding } \left(-\frac{3}{2}\right)\left(-\frac{3}{2}\right) = \frac{9}{4}$$
$$\text{to each side.}$$

$$\left(x - \frac{3}{2}\right)^2 = \frac{13}{4} \qquad \text{Solve for } x.$$

$$x - \frac{3}{2} = \pm\sqrt{\frac{13}{4}}$$

$$x - \frac{3}{2} = \pm\frac{\sqrt{13}}{2}$$

$$x - \frac{3}{2} = +\frac{\sqrt{13}}{2} \qquad x - \frac{3}{2} = -\frac{\sqrt{13}}{2}$$

$$x = \frac{3}{2} + \frac{\sqrt{13}}{2} \qquad x = \frac{3}{2} - \frac{\sqrt{13}}{2}$$

$$x = \frac{3 + \sqrt{13}}{2} \qquad x = \frac{3 - \sqrt{13}}{2}$$

The two roots are $\dfrac{3 \pm \sqrt{13}}{2}$.

Practice Problem 5 Solve. $x^2 - 5x = 7$

If a, the coefficient of the squared variable, is not 1, we divide all terms of the equation by a so that the coefficient of the squared variable will be 1.

Let us summarize for future reference the steps we have performed in order to solve a quadratic equation by completing the square.

Completing the Square

1. Put the equation in the form $ax^2 + bx = c$. If $a \neq 1$, divide each term by a.

2. Square $\dfrac{b}{2}$ and add the result to both sides of the equation.

3. Factor the left side (a perfect square trinomial).

4. Use the square root property.

5. Solve the equations.

6. Check the solutions in the original equation.

EXAMPLE 6 Solve. $4x^2 + 4x - 3 = 0$

Step 1 $\quad 4x^2 + 4x = 3 \qquad$ Place the constant on the right. This puts the equation in the form $ax^2 + bx = c$.

$$x^2 + x = \frac{3}{4} \qquad \text{Divide all terms by 4.}$$

Step 2 $\quad x^2 + 1x = \dfrac{3}{4} \qquad$ Take one-half the coefficient of x and square it. $\left(\dfrac{1}{2}\right)^2 = \dfrac{1}{4}$.

$$x^2 + x + \frac{1}{4} = \frac{3}{4} + \frac{1}{4} \quad \text{Add } \frac{1}{4} \text{ to each side.}$$

Step 3 $\quad \left(x + \dfrac{1}{2}\right)^2 = 1 \qquad$ Factor the left side.

Step 4 $\quad x + \dfrac{1}{2} = \pm\sqrt{1} \quad$ Use the square root property.

Step 5 $\quad x + \dfrac{1}{2} = \pm 1$

$$x + \frac{1}{2} = 1 \qquad\qquad x + \frac{1}{2} = -1$$

$$x = \frac{2}{2} - \frac{1}{2} \qquad\qquad x = -\frac{2}{2} - \frac{1}{2}$$

$$x = \frac{1}{2} \qquad\qquad x = -\frac{3}{2}$$

Thus the two roots are $\dfrac{1}{2}$ and $-\dfrac{3}{2}$.

Step 6 $\qquad 4\left(\dfrac{1}{2}\right)^2 + 4\left(\dfrac{1}{2}\right) - 3 \overset{?}{=} 0 \qquad$ Check.

$$4\left(\dfrac{1}{4}\right) + 4\left(\dfrac{1}{2}\right) - 3 \overset{?}{=} 0$$

$$1 + 2 - 3 = 0 \quad \checkmark$$

$$4\left(-\dfrac{3}{2}\right)^2 + 4\left(-\dfrac{3}{2}\right) - 3 \overset{?}{=} 0$$

$$4\left(\dfrac{9}{4}\right) + 4\left(-\dfrac{3}{2}\right) - 3 \overset{?}{=} 0$$

$$9 - 6 - 3 = 0 \quad \checkmark$$

Both values check.

This method of completing the square will enable us to solve any quadratic equation that has real number roots. It is usually faster, however, to factor the quadratic equation if the polynomial is factorable.

Practice Problem 6   Solve by factoring. $4y^2 + 4y - 3 = 0$

Solve using the square root property.

1. $x^2 = 64$

2. $x^2 = 36$

3. $x^2 = 98$

4. $x^2 = 72$

5. $x^2 - 40 = 0$

6. $x^2 - 75 = 0$

7. $5x^2 = 45$

8. $3x^2 = 48$

9. $6x^2 = 120$

10. $3x^2 = 150$

11. $5x^2 - 140 = 0$

12. $2x^2 - 196 = 0$

13. $5x^2 + 13 = 73$

14. $6x^2 + 5 = 53$

15. $13x^2 + 17 = 82$

16. $11x^2 + 14 = 47$

17. $(x - 3)^2 = 5$

18. $(x + 7)^2 = 5$

19. $(x + 4)^2 = 6$

20. $(x - 5)^2 = 36$

21. $(2x + 5)^2 = 2$

22. $(3x - 4)^2 = 6$

23. $(3x - 1)^2 = 7$

24. $(2x + 3)^2 = 10$

25. $(7x + 2)^2 = 12$

26. $(8x + 5)^2 = 20$

27. $(5x - 3)^2 = 75$

28. $(7x - 2)^2 = 98$

Solve by completing the square.

29. $x^2 - 6x = 11$

30. $x^2 - 2x = 4$

31. $x^2 + 6x + 7 = 0$

32. $x^2 - 12x - 4 = 0$

33. $x^2 - 12x - 5 = 0$

34. $x^2 + 16x + 30 = 0$

35. $x^2 + 3x = 0$

36. $x^2 + 5x = 3$

37. $4x^2 - 8x + 3 = 0$

38. $2x^2 + 5x = 3$

39. $2x^2 - 7x = 9$

To Think About

40. Let b represent a constant value. Solve $x^2 + bx - 7 = 0$ for x by completing the square.

41. Solve $x^2 + bx + c = 0$ for x in terms of b and c.

Cumulative Review Problems

42. Solve. $3a - 5b = 8$
$\qquad\quad 5a - 7b = 8$

43. Solve. $\dfrac{x-4}{x-2} - \dfrac{1}{x} = \dfrac{x-3}{x^2 - 2x}$

44. While hiking in the Grand Canyon, Will had to use his hunting knife to clear a fallen branch from the path. Find the pressure on the blade of his knife if the knife is 8 in. long and sharpened to an edge 0.01 in. wide and he applies a force of 16 pounds. Use the formula $P = \frac{F}{A}$, where P is the pressure on the blade of the knife, F is the force in pounds placed on the knife, and A is the area exposed to an object by the sharpened edge of the knife.

45. Darlene inspected a sample of 100 new steel-belted radial tires. 13 of the tires had defects in workmanship, while 9 of the tires had defects in materials. Of those defective tires, 6 had defects in both workmanship and materials. How many of the 100 tires had any kind of defect? (Assume that the only defects found were those in materials or workmanship.)

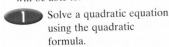

Student Learning Objectives

After studying this section, you will be able to:

1 Solve a quadratic equation using the quadratic formula.

2 Find a decimal approximation to the real roots of a quadratic equation.

3 Determine whether a quadratic equation has no real solutions.

SSM PH TUTOR CD & VIDEO MATH PRO WEB
CENTER

1 *Solving a Quadratic Equation Using the Quadratic Formula*

To find solutions of a quadratic equation, you can factor or you can complete the square when the trinomial is not factorable. If we use the method of completing the square on the general equation $ax^2 + bx + c = 0$, we can derive a general formula for the solution of any quadratic equation. This is called the **quadratic formula**.

Quadratic Formula

The roots of any quadratic equation of the form $ax^2 + bx + c = 0$, where a, b, and c are real numbers and $a \neq 0$, are

$$x = \frac{-b \pm \sqrt{b^2 - 4ac}}{2a}.$$

All you will need to know are the values of a, b, and c to solve a quadratic equation using the quadratic formula.

EXAMPLE 1 Solve using the quadratic formula. $3x^2 + 10x + 7 = 0$

In our given equation, $3x^2 + 10x + 7 = 0$, we have

$a = 3$ (the coefficient of x^2), $b = 10$ (the coefficient of x), and
$c = 7$ (the constant term).

$$x = \frac{-b \pm \sqrt{b^2 - 4ac}}{2a} = \frac{-10 \pm \sqrt{(10)^2 - 4(3)(7)}}{2(3)}$$ Write the quadratic formula and substitute values for a, b, and c.

$$= \frac{-10 \pm \sqrt{100 - 84}}{6}$$ Simplify.

$$= \frac{-10 \pm \sqrt{16}}{6} = \frac{-10 \pm 4}{6}$$

$$x = \frac{-10 + 4}{6} = \frac{-6}{6} = -1$$ Using the positive sign.

$$x = \frac{-10 - 4}{6} = \frac{-14}{6} = -\frac{7}{3}$$ Using the negative sign.

Thus the two solutions are -1 and $-\dfrac{7}{3}$.

[*Note*: Here we obtain rational roots. We would obtain the same answer by factoring $3x^2 + 10x + 7 = 0$ into $(3x + 7)(x + 1) = 0$ and setting each factor $= 0$.]

Practice Problem 1 Solve using the quadratic formula. $x^2 - 7x + 6 = 0$

Often the roots will be irrational numbers. Sometimes these roots can be simplified. You should always leave your answer in simplest form.

EXAMPLE 2 Solve. $2x^2 - 4x - 9 = 0$

$$a = 2, \quad b = -4, \quad c = -9$$

$$x = \frac{-b \pm \sqrt{b^2 - 4ac}}{2a} = \frac{-(-4) \pm \sqrt{(-4)^2 - 4(2)(-9)}}{2(2)}$$

$$= \frac{4 \pm \sqrt{16 + 72}}{4} = \frac{4 \pm \sqrt{88}}{4}$$ Do not stop here!

Notice that we can simplify $\sqrt{88}$.

$$x = \frac{4 \pm 2\sqrt{22}}{4} = \frac{2(2 \pm \sqrt{22})}{4} = \frac{\overset{1}{2}(2 \pm \sqrt{22})}{\underset{2}{\cancel{4}}} = \frac{2 \pm \sqrt{22}}{2}$$

Be careful. Here we were able to divide the numerator and denominator by 2 because 2 was a factor of every term.

Practice Problem 2   Solve. $3x^2 - 8x + 3 = 0$

A quadratic equation *must* be written in the standard form $ax^2 + bx + c = 0$ *before* the quadratic formula can be used. Several algebraic steps may be needed to accomplish this objective. Also, since it is much easier to use the formula if a, b, and c are integers, we will avoid using fractional values.

EXAMPLE 3   Solve. $x^2 = 5 - \dfrac{3}{4}x$

First we obtain an equivalent equation that does not have fractions.

$$x^2 = 5 - \frac{3}{4}x$$

$$4(x^2) = 4(5) - 4\left(\frac{3}{4}x\right) \qquad \text{Multiply each term by the LCD of 4.}$$

$$4x^2 = 20 - 3x \qquad \text{Simplify.}$$

$$4x^2 + 3x - 20 = 0 \qquad \text{Add } 3x - 20 \text{ to each side.}$$

$$a = 4, \quad b = 3, \quad c = -20 \qquad \text{Substitute } a = 4, b = 3, \text{ and } c = -20 \text{ into the quadratic formula.}$$

$$x = \frac{-3 \pm \sqrt{(3)^2 - 4(4)(-20)}}{2(4)}$$

$$x = \frac{-3 \pm \sqrt{9 + 320}}{8}$$

$$x = \frac{-3 \pm \sqrt{329}}{8}$$

Practice Problem 3   Solve. $x^2 = 7 - \dfrac{3}{5}x$

2   *Finding a Decimal Approximation to the Real Roots of a Quadratic Equation*

EXAMPLE 4   Find the roots of $3x^2 - 5x = 7$. Approximate to the nearest thousandth.

$$3x^2 - 5x = 7$$

$$3x^2 - 5x - 7 = 0 \qquad \text{Place the equation in standard form.}$$

$$a = 3, \quad b = -5, \quad c = -7$$

$$x = \frac{-(-5) \pm \sqrt{(-5)^2 - 4(3)(-7)}}{2(3)} = \frac{5 \pm \sqrt{25 + 84}}{6}$$

$$= \frac{5 \pm \sqrt{109}}{6}$$

Using most scientific calculators, we can find $\dfrac{5 + \sqrt{109}}{6}$ with these keystrokes.

$$5 \boxed{+} \; 109 \boxed{\sqrt{}} \boxed{=} \boxed{\div} \; 6 \boxed{=} \; 2.5733844$$

Rounding to the nearest thousandth, we have $x \approx 2.573$.

To find $\dfrac{5 - \sqrt{109}}{6}$, we use the following keystrokes.

$5 \boxed{-} \ 109 \ \boxed{\sqrt{\ }} \ \boxed{=} \ \boxed{\div} \ 6 \ \boxed{=} \ -0.9067178$

Rounding to the nearest thousandth, we have $x \approx -0.907$.
If you do not have a calculator with a square root key, look up $\sqrt{109}$ in the square root table. $\sqrt{109} \approx 10.440$. Using this result, we have

$$x \approx \frac{5 + 10.440}{6} = \frac{15.440}{6} \approx 2.573 \quad (\text{rounded to the nearest thousandth})$$

and

$$x \approx \frac{5 - 10.440}{6} = \frac{-5.440}{6} \approx -0.907 \quad (\text{rounded to the nearest thousandth}).$$

If you have a graphing calculator, there is an alternate method for finding the decimal approximation of the roots of a quadratic equation. In Section 10.4, you will cover graphing quadratic equations. You will discover that finding the roots of a quadratic equation of the form $ax^2 + bx + c = 0$ is equivalent to finding the values of x on the graph of $y = ax^2 + bx + c$ where $y = 0$. These may be approximated quite quickly with a graphing calculator.

Practice Problem 4   Find the roots of $2x^2 = 13x + 5$. Approximate to the nearest thousandth.

3 ▸ Determining Whether a Quadratic Equation Has No Real Solutions

EXAMPLE 5   Solve. $2x^2 + 5 = -3x$

The equation is not in standard form.

$\qquad 2x^2 + 5 = -3x \qquad$ Add $3x$ to both sides.

$2x^2 + 3x + 5 = 0 \qquad$ We can now find a, b, and c.

$a = 2, \quad b = 3, \quad c = 5 \qquad$ Substitute these values into the quadratic formula.

$$x = \frac{-3 \pm \sqrt{(3)^2 - 4(2)(5)}}{2(2)} = \frac{-3 \pm \sqrt{9 - 40}}{4} = \frac{-3 \pm \sqrt{-31}}{4}$$

There is no real number that is $\sqrt{-31}$. Since we are using only real numbers in this book, we say there is no solution to the problem. (*Note*: In more advanced math courses, these types of numbers, called complex numbers, will be studied.)

Practice Problem 5   Solve. $5x^2 + 2x = -3$

We can tell whether the roots of any given quadratic equation are real numbers. Look at the quadratic formula.

$$\frac{-b \pm \sqrt{b^2 - 4ac}}{2a}$$

The expression under the radical sign is called the **discriminant**.

$$b^2 - 4ac \text{ is the discriminant.}$$

If the discriminant is a negative number, the roots are not real numbers, and there is no real number solution to the equation.

EXAMPLE 6 Determine whether $3x^2 = 5x - 4$ has real number solution(s).

First, we place the equation in standard form. Then we need only check the discriminant.

$$3x^2 = 5x - 4$$

$$3x^2 - 5x + 4 = 0$$

$a = 3, \quad b = -5, \quad c = 4$ Substitute the values for a, b, and c into the discriminant and evaluate.

$$b^2 - 4ac = (-5)^2 - 4(3)(4) = 25 - 48 = -23$$

The discriminant is negative. Thus $3x^2 = 5x - 4$ has no real number solution(s).

Practice Problem 6 Determine whether each equation has real number solution(s).

(a) $5x^2 = 3x + 2$ **(b)** $2x^2 - 5 = 4x$

Solve using the quadratic formula. If there are no real roots, say so.

1. $x^2 + 3x - 5 = 0$

2. $x^2 - 5x + 3 = 0$

3. $x^2 - 3x - 8 = 0$

4. $x^2 + 9x + 7 = 0$

5. $4x^2 - 5x - 6 = 0$

6. $3x^2 + 7x - 6 = 0$

7. $5x - 1 = 2x^2$

8. $5x^2 = -3x + 2$

9. $6x^2 - 3x = 1$

10. $3 = 7x - 4x^2$

11. $x + \dfrac{3}{2} = 3x^2$

12. $\dfrac{2}{3}x^2 - x = \dfrac{5}{2}$

13. $5x^2 + 6x + 2 = 0$

14. $\dfrac{x}{2} + \dfrac{2}{x} = \dfrac{5}{2}$

15. $\dfrac{x}{3} + \dfrac{2}{x} = \dfrac{7}{3}$

16. $3x^2 - 4x + 2 = 0$

17. $5y^2 = 3 - 7y$

18. $5y - 6 = -2y^2$

19. $\dfrac{d^2}{2} + \dfrac{5d}{6} - 2 = 0$

20. $\dfrac{k^2}{4} - \dfrac{5k}{6} = \dfrac{2}{3}$

Use the quadratic formula to find the roots. Find a decimal approximation to the nearest thousandth.

21. $x^2 + 5x - 2 = 0$

22. $x^2 + 3x - 7 = 0$

23. $2x^2 + 3x - 3 = 0$

24. $3x^2 - 2x - 3 = 0$

25. $5x^2 + 10x + 1 = 0$

26. $2x^2 + 9x + 2 = 0$

Mixed Practice

Solve for the variable. Choose the method you feel will work best.

27. $t^2 + 1 = \dfrac{13}{6}t$

28. $2t^2 - 1 = \dfrac{7}{3}t$

29. $3(s^2 + 1) = 10s$

30. $3 + p(p + 2) = 18$

31. $(t + 5)(t - 3) = 7$

32. $(b + 4)(b - 7) = 4$

33. $y^2 - \dfrac{2}{5}y = 2$

34. $y^2 - 3 = \dfrac{8}{3}y$

35. $5x^2 - 11 = 0$

36. $4x^2 + 15x = 0$

37. $x(x - 2) = 7$

38. $(3x + 1)^2 = 12$

To Think About

▲ **39.** John and Chris Maney have designed a new in-ground swimming pool for their backyard. The pool is rectangular and measures 30 feet by 20 feet. They have also designed a tile border to go around the pool. In the diagram at right, this border is x feet wide. The tile border covers 216 square feet in area. What is the width x of this tile border? (*Hint*: Find an expression for the area of the large rectangle and subtract an expression for the area of the small rectangle.)

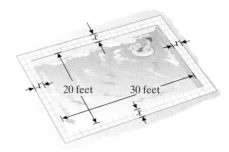

▲ **40.** The Alumni Association of North Shore Community College has designed a new garden for the front entrance of the Lynn campus. It is rectangular and measures 40 feet by 30 feet. The association has designed a cement walkway to go around this rectangular garden. In the diagram at right, this cement walkway is x feet wide. The entire cement walkway covers 456 square feet in area. What is the width x of this cement walkway? (*Hint*: Find an expression for the area of the large rectangle and subtract an expression for the area of the small rectangle.)

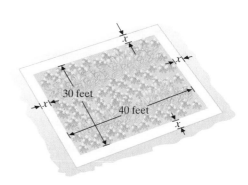

Cumulative Review Problems

41. Simplify. $\dfrac{3x}{x-2} + \dfrac{4}{x+2} - \dfrac{x+22}{x^2-4}$

42. Give the equation of the line passing through $(-8, 2)$ and $(5, -3)$ in **(a)** slope–intercept form and **(b)** standard form with integer coefficients.

43. Divide. $(x^3 + 8x^2 + 17x + 6) \div (x + 3)$

44. Solve for g. $2p = 3gf - 7p + 20$

▲ **45.** A very large kaleidoscope is produced for a children's science museum. It is so big that it must be supported by a specially made stand. The kaleidoscope is in the shape of a cylinder, measures 5.35 feet in length, and has a radius of 2.40 feet. To prevent contamination by dust, the museum keeps the interior of the cylinder in a vacuum. Find the volume of air in cubic feet that must be removed from the cylinder to establish a vacuum. Use $\pi \approx 3.14$. Round to the nearest hundredth.

▲ **46.** An archery bull's-eye is set up for the purpose of training Olympic athletes. There are 5 concentric circles within the largest circle. Each circle is 2 inches from the adjacent circle. The smallest circle has an area of 12.56 square inches. What is the area between the largest circle and the circle that is closest to the largest circle? Use $\pi \approx 3.14$. Round to the nearest hundredth.

10.4 Graphing Quadratic Equations

1 Graphing Equations of the Form $y = ax^2 + bx + c$ Using Ordered Pairs

Recall that the graph of a linear equation is always a straight line. What is the graph of a quadratic equation? What does the graph look like if one of the variable terms is squared? We will use point plotting to look for a pattern. We begin with a table of values.

EXAMPLE 1 Graph. **(a)** $y = x^2$ **(b)** $y = -x^2$

Student Learning Objectives

After studying this section, you will be able to:

1 Graph equations of the form $y = ax^2 + bx + c$ using ordered pairs.

2 Graph equations of the form $y = ax^2 + bx + c$ using the vertex.

SSM
PH TUTOR   CD & VIDEO   MATH PRO   WEB
CENTER

(a) $y = x^2$

x	y
-3	9
-2	4
-1	1
0	0
1	1
2	4
3	9

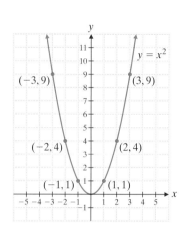

(b) $y = -x^2$

x	y
-3	-9
-2	-4
-1	-1
0	0
1	-1
2	-4
3	-9

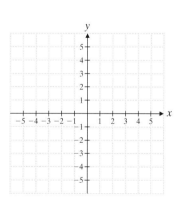

Practice Problem 1 Graph. **(a)** $y = 2x^2$ **(b)** $y = -2x^2$

$y = 2x^2$

x	y
-2	
-1	
0	
1	
2	

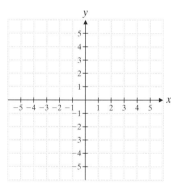

$y = -2x^2$

x	y
-2	
-1	
0	
1	
2	

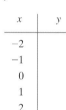

To Think About What happens to the graph when the coefficient of x^2 is negative? What happens when the coefficient of x^2 is an integer greater than 1 or less than -1? Graph $y = 3x^2$ and $y = -3x^2$. What happens when $-1 < a < 1$? Graph $y = \frac{1}{3}x^2$ and $y = -\frac{1}{3}x^2$.

The curves that you have been graphing are called **parabolas**. What happens to the graph of a parabola as the quadratic equation that describes it changes?

In the preceding examples, the equations we worked with were relatively simple. They provided us with an understanding of the general shape of the graph of a quadratic equation. We have seen how changing the equation affects the graph of that equation. Let's look at an equation that is slightly more difficult to graph.

The **vertex** is the lowest point on a parabola that opens upward, or the highest point on a parabola that opens downward.

EXAMPLE 2   Graph $y = x^2 - 2x$. Identify the coordinates of the vertex.

x	y
-2	8
-1	3
0	0
1	-1 vertex
2	0
3	3
4	8

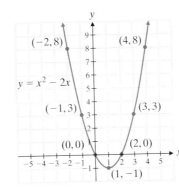

The vertex is at $(1, -1)$.

Notice that the x-intercepts are at $(0,0)$ and $(2,0)$. The x-coordinates are the solutions to the equation $x^2 - 2x = 0$.

$$x^2 - 2x = 0$$
$$x(x - 2) = 0$$
$$x = 0, \qquad x = 2$$

Practice Problem 2   Graph.   For each table of values, use x-values from -4 to 2. Identify the coordinates of the vertex.   **(a)** $y = x^2 + 2x$ **(b)** $y = -x^2 - 2x$

$y = x^2 + 2x$

x	y
-4	
-3	
-2	
-1	
0	
1	
2	

$y = -x^2 - 2x$

x	y
-4	
-3	
-2	
-1	
0	
1	
2	

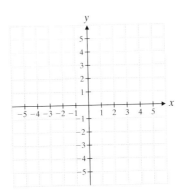

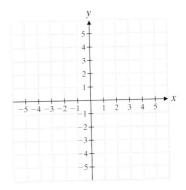

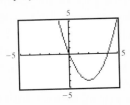

Graphing Calculator

Graphing Quadratic Equations

Graph $y = x^2 - 4x$ on a graphing calculator using an appropriate window. Display:

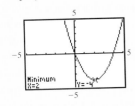

Next you can use the Trace and Zoom features of your calculator to determine the vertex. Some calculators have a feature that will calculate the maximum or minimum point on the graph. Use the feature that calculates the minimum point to find the vertex. Display:

Thus the vertex is at $(2, -4)$.

2 Graphing Equations of the Form $y = ax^2 + bx + c$ Using the Vertex

Finding the vertex is the key to graphing a parabola. Let's take a look at some of the previous graphs and identify the vertex of each parabola.

Equation: $y = x^2$ $y = x^2 - 3$ $y = (x - 1)^2$ $y = (x + 2)^2$ $y = x^2 - 2x$
                                                        $y = x^2 - 2x + 1$ $y = x^2 + 4x + 4$

Vertex: $(0, 0)$ $(0, -3)$ $(1, 0)$ $(-2, 0)$ $(1, -1)$

Notice that the x-coordinate of the vertex of the parabola for each of the first two equations is 0. You may also notice that these two equations do not have middle terms. That is, for $y = ax^2 + bx + c, b = 0$. This is not true of the last three equations. As you may have already guessed, there is a relationship between the quadratic equation and the x-coordinate of the vertex of the parabola it describes.

Vertex of a Parabola

The x-coordinate of the vertex of the parabola described by $y = ax^2 + bx + c$ is

$$x = \frac{-b}{2a}.$$

Once we have the x-coordinate, we can find the y-coordinate using the equation. We can plot the point and draw a general sketch of the curve. The sign of the squared term will tell us whether the graph opens upward or downward.

◖ EXAMPLE 3 $y = -x^2 - 2x + 3$. Determine the vertex and the x-intercepts. Then sketch the graph.

We will first determine the coordinates of the vertex. We begin by finding the x-coordinate.

$$x = \frac{-b}{2a} = \frac{-(-2)}{2(-1)} = \frac{2}{-2} = -1$$

$$y = -(-1)^2 - 2(-1) + 3 \qquad \text{Determine } y \text{ when } x \text{ is } -1.$$
$$= 4 \qquad \qquad \text{The point } (-1, 4) \text{ is the vertex of the parabola.}$$

Now we will determine the x-intercepts, those points where the graph crosses the x-axis. This occurs when $y = 0$. These are the solutions to the equation $-x^2 - 2x + 3 = 0$. Note that this equation is factorable.

$$-x^2 - 2x + 3 = 0 \qquad \text{Find the solutions to the equation.}$$
$$(-x + 1)(x + 3) = 0 \qquad \text{Factor.}$$
$$-x + 1 = 0 \qquad x + 3 = 0 \qquad \text{Solve for } x.$$
$$x = 1 \qquad \qquad x = -3 \qquad \text{The } x\text{-intercepts are at } (1, 0) \text{ and at } (-3, 0).$$

We now have enough information to sketch the graph. Since $a = -1$, the parabola opens downward and the vertex $(-1, 4)$ is the highest point. The graph crosses the x-axis at $(-3, 0)$ and at $(1, 0)$.

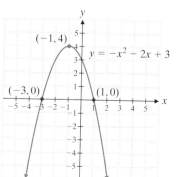

Practice Problem 3 $y = x^2 - 4x - 5$. Determine the vertex and the x-intercepts. Then sketch the graph.

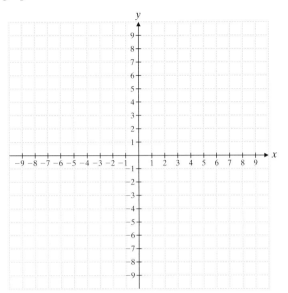

 EXAMPLE 4   A person throws a ball into the air and its distance from the ground in feet is given by the equation $h = -16t^2 + 64t + 5$, where t is measured in seconds.

(a) Graph the equation $h = -16t^2 + 64t + 5$.

(b) What physical significance does the vertex of the graph have?

(c) After how many seconds will the ball hit the ground?

(a) Since the equation is a quadratic equation, the graph is a parabola. a is negative. Thus the parabola opens downward and the vertex is the highest point. We begin by finding the coordinates of the vertex. Our equation has ordered pairs (t, h) instead of (x, y). The t-coordinate of the vertex is

$$t = \frac{-b}{2a} = \frac{-64}{2(-16)} = \frac{-64}{-32} = 2.$$

Substitute $t = 2$ into the equation to find the h-coordinate of the vertex.

$$h = -16(2)^2 + 64(2) + 5 = -64 + 128 + 5 = 69$$

The vertex (t, h) is $(2, 69)$.

To sketch this graph, it would be helpful to know the h-intercept, that is, the point where $t = 0$.

$$h = -16(0)^2 + 64(0) + 5$$
$$= 5$$

The h-intercept is at $(0, 5)$.

The t-intercept is where $h = 0$. We will use the quadratic formula to find t when h is 0, that is, to find the roots of the quadratic equation $-16t^2 + 64t + 5 = 0$.

$$-16t^2 + 64t + 5 = 0$$

$$t = \frac{-64 \pm \sqrt{(64)^2 - 4(-16)(5)}}{2(-16)}$$

$$t = -\frac{64 \pm \sqrt{4096 + 320}}{-32} = -\frac{64 \pm \sqrt{4416}}{-32} = \frac{8 \pm \sqrt{69}}{4}$$

Using a calculator or a square root table, we have the following.

$$t = \frac{8 + 8.307}{4} \approx 4.077, \qquad t \approx 4.1 \quad \text{(to the nearest tenth)}$$

$$t = \frac{8 - 8.307}{4} \approx -0.077$$

We will not use negative t-values. We are studying the height of the ball from the time it is thrown ($t = 0$ seconds) to the time that it hits the ground (approximately $t = 4.1$ seconds). It is not useful to find points with negative values of t because they have no meaning in this situation.

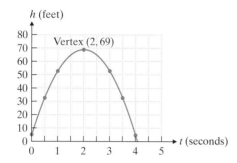

(b) The vertex represents the greatest height of the ball. The ball is 69 feet above the ground 2 seconds after it is thrown.

(c) The ball will hit the ground approximately 4.1 seconds after it is thrown. This is the point at which the graph intersects the t-axis, the positive root of the equation.

Practice Problem 4   A ball is thrown upward with a speed of 32 feet per second from a distance 10 feet above the ground. The height of the ball in feet is given by the equation $h = -16t^2 + 32t + 10$, where t is measured in seconds.

(a) Graph the equation $h = -16t^2 + 32t + 10$.

(b) What is the greatest height of the ball?

(c) After how many seconds will the ball hit the ground?

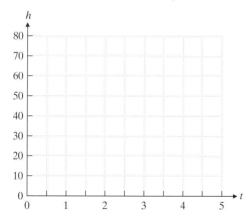

Graph each equation.

1. $y = x^2 + 2$

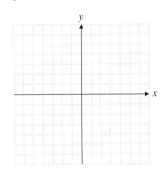

2. $y = 2x^2 - 1$

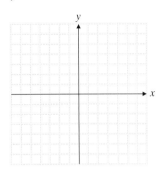

3. $y = -\frac{1}{3}x^2$

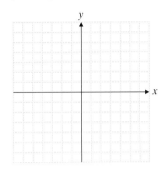

4. $y = -\frac{1}{2}x^2$

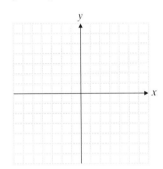

5. $y = x^2 - 5$

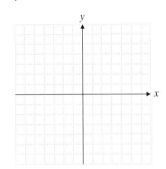

6. $y = 3x^2 + 2$

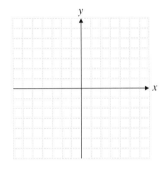

7. $y = (x - 2)^2$

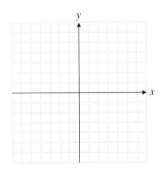

8. $y = (x + 1)^2$

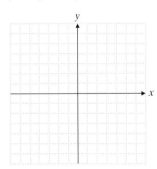

9. $y = -\frac{1}{2}x^2 + 4$

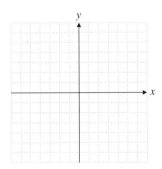

10. $y = -\frac{1}{3}x^2 + 6$

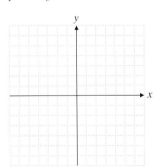

11. $y = \frac{1}{2}(x - 3)^2$

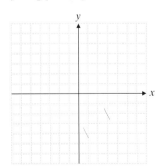

12. $y = \frac{1}{2}(x + 2)^2$

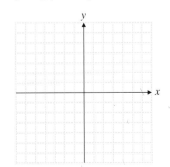

Graph each equation. Identify the coordinates of the vertex.

13. $y = x^2 + 4x$

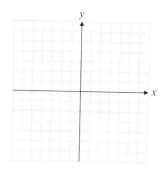

14. $y = -x^2 + 4x$

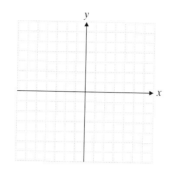

15. $y = -x^2 - 4x$

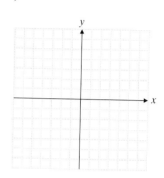

16. $y = x^2 - 4x$

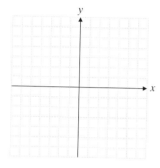

17. $y = -2x^2 + 8x$

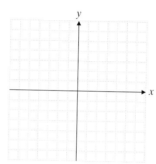

18. $y = x^2 + 6x$

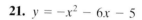

Determine the vertex and the x-intercepts. Then sketch the graph.

19. $y = x^2 + 2x - 3$

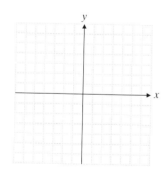

20. $y = x^2 - 4x + 3$

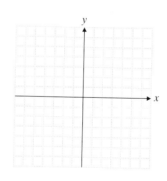

21. $y = -x^2 - 6x - 5$

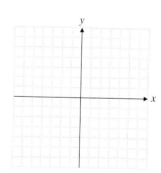

22. $y = -x^2 + 2x + 3$

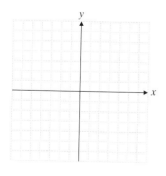

23. $y = x^2 + 6x + 9$

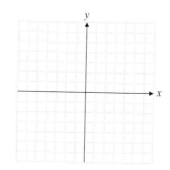

24. $y = -x^2 + 2x - 1$

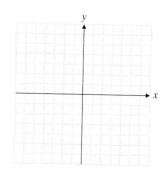

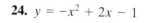

Applications

25. A miniature rocket is launched so that its height h in meters after t seconds is given by the equation
$h = -4.9t^2 + 19.6t + 10$.

(a) Draw a graph of the equation.

(b) How high is the rocket after 3 seconds?

(c) What is the maximum height the rocket will attain?

(d) How long after the launch will the rocket strike the Earth?

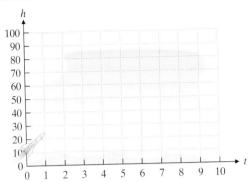

26. A miniature rocket is launched so that its height h in meters after t seconds is given by the equation
$h = -4.9t^2 + 39.2t + 4$.

(a) Draw a graph of the equation.

(b) How high is the rocket after 2 seconds?

(c) What is the maximum height the rocket will attain?

(d) How long after the launch will the rocket strike the Earth?

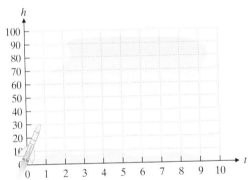

27. The number of mosquitoes in some areas varies with the amount of rain. The number, N, of mosquitoes measured in millions in Hamilton, Massachusetts, in May can be predicted by the equation $N = 9x - x^2$, where x is the number of inches of rain during that month. (*Source*: Massachusetts Environmental Protection Agency.)

(a) Graph the equation.

(b) How many inches of rain produce the maximum number of mosquitoes in May?

(c) What is the maximum number of mosquitoes?

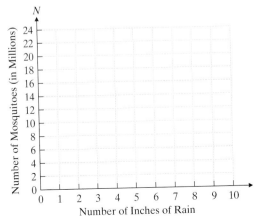

Optional Graphing Calculator Problems

Use your graphing calculator to find the vertex, the y-intercept, and the x-intercepts of the parabolas described by the following equations. Round to the nearest thousandth when necessary.

28. $y = 3x^2 + 7x - 9$

29. $y = x^2 + 10.6x - 212.16$

30. $y = -156x^2 - 289x + 1133$

31. $y = -301x^2 - 167x + 1528$

Cumulative Review Problems

32. If y varies inversely as x, and $y = \dfrac{1}{3}$ when $x = 21$, find y when $x = 7$.

33. If y varies directly as x^2, and $y = 12$ when $x = 2$, find y when $x = 5$.

34. A football player picks up a ball from the field and runs 60 yards. He runs the first half of this distance at 20 feet per second. He runs the second half at 24 feet per second. How long does it take him to run the 60 yards?

Putting Your Skills to Work

Helping the Football Coach Plan the Defense

A football is thrown a distance of 40 yards by a quarterback. The ball is thrown and caught 5 feet above the ground. The football is 15 feet above the ground at the peak of the throw. The ball follows a parabolic path. The vertex of the parabola is the highest point the ball reaches. The height h of the football in feet at a distance of x yards from the quarterback is given by the equation

$$h = -\frac{1}{40}(x - 20)^2 + 15.$$

A graph of the equation is shown in the following figure.

Problems for Individual Investigation

Approximate the answers to the following questions based on the graph.

1. How high is the football above the ground when it is 8 yards from the quarterback?

2. How high is the football above the ground when it is 32 yards from the quarterback?

3. If a member of the opposing team is rushing toward the quarterback and can leap up a distance of 8 feet to intercept or deflect the ball, how close to the quarterback does he have to be to impede the throw of the ball?

4. If a member of the opposing team is covering the receiver and he can leap up a distance of 9 feet to intercept or deflect the ball, how close to the receiver does he need to be to keep the receiver from catching the ball?

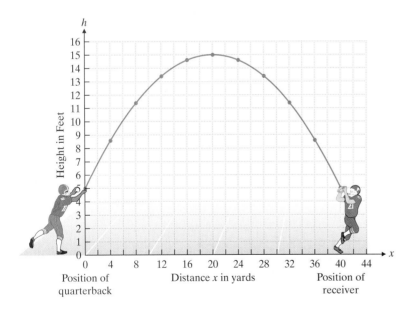

Position of quarterback

Distance x in yards

Position of receiver

Problems for Group Investigation and Analysis

Together with some of your classmates, complete the following.

Suppose that the equation $h = -\dfrac{1}{35}(x - 15)^2 + 12$ describes the most commonly used throw by an opposing quarterback. Graph the equation from $x = 0$ to $x = 32$.

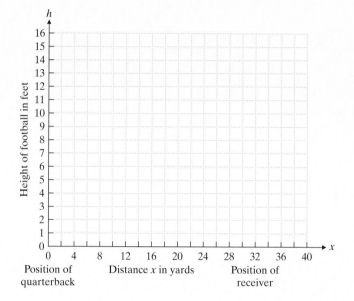

Position of quarterback · Distance x in yards · Position of receiver

Answer the following questions based on your graph.

5. What is the maximum height the football attains?

6. How far will the football have traveled down the field if the receiver catches it at a height of 5 feet above the ground?

7. If a person can leap 7 feet up to intercept or deflect the ball, how close to the quarterback does he have to be to impede the throw of the ball?

8. If a person can leap 9 feet up to intercept or deflect the ball, how close to the receiver does he have to be to keep the receiver from catching the ball?

Note: Using videocameras and computer technology, it is possible to find an equation for the path of a football thrown by a quarterback.

Internet Connections

  Netsite: http://www.prenhall.com/tobey_beginning

Site: Annual Records and Team Accolades (New York Giants) or a related site

9. This site shows the number of games that were won, lost, and tied by the New York Giants each season. Choose any 15-year period of the Giants' history. Tell what time period you chose and give the percent of games won, percent of games lost, and percent of games tied during this period.

10.5 Formulas and Applied Problems

1 Solving Applied Problems Using a Quadratic Equation

We can use what we have learned about quadratic equations to solve word problems. Some word problems will involve geometry.

Recall that the Pythagorean Theorem states that in a right triangle $a^2 + b^2 = c^2$, where a and b are legs and c is the hypotenuse. We will use this theorem to solve word problems involving right triangles.

EXAMPLE 1 The hypotenuse of a right triangle is 25 meters in length. One leg is 17 meters longer than the other. Find the length of each leg.

1. **Understand the problem.**
   Draw a picture.

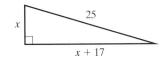

2. **Write an equation.**
   The problem involves a right triangle.
   Use the Pythagorean theorem.

$$a^2 + b^2 = c^2$$
$$x^2 + (x + 17)^2 = 25^2$$

3. **Solve and state the answer.**

$$x^2 + (x + 17)^2 = 25^2$$
$$x^2 + x^2 + 34x + 289 = 625$$
$$2x^2 + 34x - 336 = 0$$
$$x^2 + 17x - 168 = 0$$
$$(x + 24)(x - 7) = 0$$
$$x + 24 = 0 \qquad x - 7 = 0$$
$$x = -24 \qquad x = 7$$

(Note that $x = -24$ is not a valid solution for this particular word problem.)

One leg is 7 meters in length. The other leg is $x + 17 = 7 + 17 = 24$ meters in length.

4. **Check.**
   Check the conditions in the original problem.
   Do the two legs differ by 17 meters?

$$24 - 7 \stackrel{?}{=} 17 \qquad 17 = 17 \ \checkmark$$

Do the sides of the triangle satisfy the Pythagorean theorem?

$$7^2 + 24^2 \stackrel{?}{=} 25^2$$
$$49 + 576 \stackrel{?}{=} 625$$
$$625 = 625 \ \checkmark$$

▲ Practice Problem 1 The hypotenuse of a right triangle is 30 meters in length. One leg is 6 meters shorter than the other leg. Find the length of each leg.

Sometimes it may help to use two variables in the initial part of the problem. Then one variable can be eliminated by substitution.

EXAMPLE 2 The Ski Club is renting a bus to travel to Mount Snow. The members agreed to share the cost of $180 equally. On the day of the ski trip, three members were sick with the flu and could not go. This raised by $10 the share of each person going on the trip. How many people originally planned to attend?

1. *Understand the problem.*

   Let s = the number of students in the ski club

   $\quad$ c = the cost for each student in the original group

2. *Write an equation(s).*

   $$\text{number of students} \times \text{cost per student} = \text{total cost}$$
   $$s \cdot c = 180 \qquad \textbf{(1)}$$

   If three people are sick, the number of students drops by three, but the cost for each increases by $10. The total is still $180. Therefore, we have the following.

   $$(s - 3)(c + 10) = 180 \qquad \textbf{(2)}$$

3. *Solve and state the answer.*

   $$c = \frac{180}{s} \qquad \text{Solve equation (1) for } c.$$

   $$(s - 3)\left(\frac{180}{s} + 10\right) = 180 \qquad \text{Substitute } \frac{180}{s} \text{ for } c \text{ in equation (2).}$$

   $$(s - 3)\left(\frac{180 + 10s}{s}\right) = 180 \qquad \text{Add } \frac{180}{s} + 10. \text{ Use the LCD which is } s.$$

   $$(s - 3)(180 + 10s) = 180s \qquad \text{Multiply both sides of the equation by } s.$$

   $$10s^2 + 150s - 540 = 180s \qquad \text{Use FOIL to multiply the binomials.}$$

   $$10s^2 - 30s - 540 = 0$$

   $$s^2 - 3s - 54 = 0$$

   $$(s - 9)(s + 6) = 0$$

   $$s = 9 \qquad s = -6$$

   The number of students originally in the ski club was 9.

4. *Check.*
   Would the cost increase by $10 if the number of students dropped from nine to six? Nine people in the club would mean that each would pay $20 $(180 \div 9 = 20)$.

   If the number of students was reduced by three, there were six people who took the trip. Their cost was $30 $(180 \div 6 = 30)$. The increase is $10.

   $$20 + 10 \overset{?}{=} 30$$

   $$30 = 30 \quad \checkmark$$

Practice Problem 2 Several friends from the University of New Mexico decided to charter a sport fishing boat for the afternoon while on a vacation at the coast. They agreed to share the rental cost of $240 equally. At the last minute, two of them could not go on the trip. This raised the cost for each person in the group by $4. How many people originally planned to charter the boat?

▲ ◗ **EXAMPLE 3**   Minette is fencing a garden that borders the back of a large barn. She has 120 feet of fencing. She would like a rectangular garden that measures 1350 square feet in area. She wants to use the back of the barn, so she needs to use fencing on three sides only. What dimensions should she use for her garden?

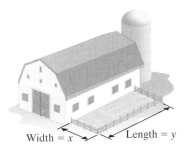

1. *Understand the problem.*
   Draw a picture.
   Let x = the width of the garden in feet and
      y = the length of the garden in feet.

2. *Write an equation(s).*
   Look at the drawing. The 120 feet of fencing will be needed for the width twice (two sides) and the length once (one side).

   $$120 = 2x + y \quad \textbf{(1)}$$

   The area formula is A = (width)(length).

   $$1350 = x(y) \quad \textbf{(2)}$$

3. *Solve and state the answer.*

   $$y = 120 - 2x \qquad \text{Solve for } y \text{ in equation (1)}.$$
   $$1350 = x(120 - 2x) \qquad \text{Substitute for } y \text{ in equation (2)}.$$
   $$1350 = 120x - 2x^2 \qquad \text{Solve for } x.$$
   $$2x^2 - 120x + 1350 = 0$$
   $$x^2 - 60x + 675 = 0$$
   $$(x - 15)(x - 45) = 0$$
   $$x = 15 \qquad x = 45$$

First solution: If the width is 15 feet, the length is 90 feet.

   $$\textit{Check:} \quad 15 + 15 + 90 = 120 \quad ✓$$
   $$(15)(90) = 1350 \quad ✓$$

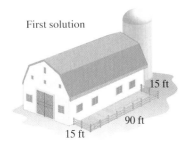

First solution

15 ft
90 ft
15 ft

Second solution: If the width is 45 feet, the length is 30 feet.

   $$\textit{Check:} \quad 45 + 45 + 30 = 120 \quad ✓$$
   $$(45)(30) = 1350 \quad ✓$$

Thus we see that Minette has two choices that satisfy the given requirements (see the figures). She would probably choose the shape that is the most practical for her. (If her barn is not 90 feet long, she could not use the first solution!)

   Remark: Is there another way to do this problem? Yes. Use only the variable x to represent the length and width. If you let x = the width and $120 - 2x$ = the length, you can immediately write equation **(3)**, which is

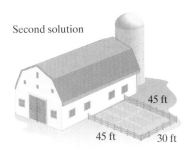

Second solution

45 ft
45 ft   30 ft

   $$1350 = x(120 - 2x). \quad \textbf{(3)}$$

However, many students find that this is difficult to do, and they would prefer to use two variables. Work these problems in the way that is easiest for you.

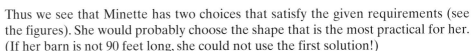

▲ **Practice Problem 3**   An Arizona rancher has a small fenced-in area against a canyon wall. Therefore, he only needs three sides of fencing to make a rectangular holding area for cattle. He has 160 feet of fencing available. He wants to make the rectangular area so that it has an area of 3150 square feet. What dimensions should he use for the rectangular holding area? ◗

Applications

Solve.

▲ **1.** The base of a triangle is 5 cm shorter than the altitude. The area of the triangle is 88 sq cm. Find the lengths of the base and altitude.

▲ **2.** The hypotenuse of a right triangle is 15 yards long. One leg of the triangle is 3 yards shorter than the other leg. Find the length of each leg of the triangle.

▲ **3.** Joe wishes to expand the size of his front garden, which is square. Joe measures his front yard and decides that he can increase the width by 7 feet and the length by 11 feet. If the area of the newly expanded garden is 396 sq ft, what were the dimensions of the old garden?

▲ **4.** Manuel placed a ladder 20 feet long against the house. The distance from the top of the ladder to the bottom of the house is 4 feet greater than the distance from the bottom of the house to the foot of the ladder. How far is the foot of the ladder from the house? At what height does the top of the ladder touch the building?

20 ft

5. To help out Nina and Tom with their new twins, the employees at Tom's office decided to give the new parents a gift certificate worth $420 to furnish the babies' room. The employees planned to share the cost equally. At the last minute, seven people from the executive staff chipped in, lowering the cost per person by $5. How many people originally planned to give the gift?

6. Everyone in the School Street Townhouse Association agreed to equally share the cost of a new $6000 septic system. However, before the system was installed, five additional units were built and the five new families were asked to contribute their share. This lowered the contribution of each family by $200. How many families originally agreed to share the cost of the new septic system?

7. A fraternity was sponsoring a spring fling, for which the $400 DJ cost was to be split evenly among those members attending. If 20 more members than were expected came, reducing the cost per person by $1, find the number of fraternity members that attended.

8. The members of the varsity football team decided to give the coach a gift that cost $80. On the day that they collected money, 10 people were absent. They were in a hurry, so they collected an equal amount from each member and bought the gift. Later the team captain said that if everyone had been present to chip in, each person would have contributed $4 less. How many people actually contributed? How many people in total were on the football team?

▲ **9.** A security fence encloses a rectangular area on one side of Fenway Park in Boston. Three sides of fencing are used, since the fourth side of the area is formed by a building. The enclosed area measures 1250 square feet. Exactly 100 feet of fencing is used to fence in three sides of this rectangle. What are the possible dimensions that could have been used to construct this area?

▲ **10.** The new housing development is planting a cooperative organic garden. The garden is in the shape of a rectangle, and one side rests against the community sports facility, so the co-op needs fencing on only three sides. If they have 140 feet of fencing available and the garden is 2000 square feet in area, what are the possible dimensions for the garden?

11. A pilot is testing a new experimental craft. The cruising speed is classified information. In a test the jet traveled 2400 miles. The pilot revealed that if he had increased his speed 200 mph, the total trip would have taken 1 hour less. Determine the cruising speed of the jet.

12. A gymnasium floor is being covered by square shock-absorbing tiles. The old gym floor required 864 square tiles. The new tiles are 2 inches larger in both length and width than the old tiles. The new flooring will require only 600 tiles. What is the length of a side of one of the new shock-absorbing tiles? (*Hint*: Since the area is the same in each case, write an expression for area with old tiles and one with new tiles. Set the two expressions equal to each other.)

Use the following information to solve exercises 13–18.

Two young college graduates opened a chain of Quick Print Centers in southern California. The chain expanded rapidly during the early 1990s and many new stores were opened. The number of Quick Print Centers in operation during any year from 1990 to 2000 is given by the equation $y = -2.5x^2 + 22.5x + 50$, where x is the number of years since 1990.

13. How many Quick Print Centers were in operation in 1991?

14. How many Quick Print Centers were in operation in 1993?

15. How many more Quick Print Centers were in operation in 1993 than in 1992?

16. How many fewer Quick Print Centers were in operation in 1998 than in 1996?

17. During what two years were there 100 Quick Print Centers in operation?

18. During what two years were there 85 Quick Print Centers in operation?

The Springfield Boot Company has determined that the daily profit of the company can be predicted by the equation $P = -2x^2 + 360x - 14{,}400$, where x is the number of pairs of boots produced each day.

 19. What is the daily profit if 85 pairs of boots are produced each day?

 20. (a) How many pairs of boots need to be produced each day to obtain the maximum daily profit?

(b) What is the maximum daily profit?

To Think About

The sum of all the counting numbers from 1 to 6 is $1 + 2 + 3 + 4 + 5 + 6 = 21$. A shortcut to the answer is $6(7)\left(\dfrac{1}{2}\right) = 21$. In general, the sum of the counting numbers from 1 to n, which we can write as $1 + 2 + \cdots + n$, can be computed by the formula $\dfrac{n(n + 1)}{2} = s$.

21. If the sum of the first *n* counting numbers is 91, find *n*.

22. If the sum of the first *n* counting numbers is 136, find *n*.

Cumulative Review Problems

23. Graph on a number line the solution to $3x + 11 \geq 9x - 4$.

24. Rationalize the denominator. $\dfrac{\sqrt{3}}{\sqrt{5} + 2}$

▲ **25.** A recycling bin that has a length of 3 feet, a width of 2 feet, and a height of 2 feet has three separate sections inside for sorting. Section 1 is for paper, section 2 is for glass and plastic, and section 3 is for metal. Sections 1 and 2 have identical measurements, but section 3 has $\dfrac{2}{3}$ the volume of section 1 or 2. If all three sections have the same width and height, how long is section 3?

Math in the Media

Stopping Distances

A USA TODAY.com Snapshot, Quick Stops, reported that increased speeds on U.S. highways were accompanied by increased stopping distances.

Stopping distances are an important and universal element in highway safety. For this reason, this data is widely collected by many different highway safety agencies throughout the world.

To get some hands-on experience examining the relationship between speed and stopping distance, try answering the questions that follow which are based on data from the Traffic Board of Western Australia.

EXERCISES

While driving, you suddenly see the brake lights of the car in front of you come on. If you assume the car ahead is going to stop and you then apply the brakes, how far will your car travel before it comes to a complete stop? In other words, what is the stopping distance from the moment your brain receives the signal to stop until the car is no longer moving?

The table below contains data for stopping distance in this situation. Note that the total stopping distance is the sum of the reaction distance (the distance traveled from the time you realize that you must brake unitl your foot hits the brake pedal) and the braking distance (distance traveled after the brake pedal is pressed).

V Speed miles/hour	x Speed feet/second	R (reaction time: 0.7 sec) Reaction distance feet	B Braking Distance feet	y Total Stopping Distance feet
55	81	57	219	276
65	95	67	301	368
75	110	77	403	480

Source: Based on data from the Traffic Board of Western Australia. *converted from Metric to US units.

1. Notice the pattern between the second and third columns and find an equation that gives the reaction distance R as a function of the speed x.

2. The model for braking distance B as a function of the speed x is $B = 0.032512x^2 + 0.134975x - 5.24631$. Use this equation and the equation from your answer to question 1, to find an equation for total stopping distance y as a function of the speed x. Write the equation for total stopping distance in simplest form by combining like terms.

616

Chapter 10 Organizer

Topic	Procedure	Examples
Solving quadratic equations by factoring, p. 578.	**1.** Clear the equation of fractions, if necessary. **2.** Write as $ax^2 + bx + c = 0$. **3.** Factor. **4.** Set each factor equal to 0. **5.** Solve the resulting equations.	Solve. $\quad x + \dfrac{1}{2} = \dfrac{3}{2x}$ Multiply each term by LCD $= 2x$. $$2x(x) + 2x\left(\dfrac{1}{2}\right) = 2x\left(\dfrac{3}{2x}\right)$$ $$2x^2 + x = 3$$ $$2x^2 + x - 3 = 0$$ $$(2x + 3)(x - 1) = 0$$ $$2x + 3 = 0 \qquad x - 1 = 0$$ $$2x = -3 \qquad\quad x = 1$$ $$x = -\dfrac{3}{2}$$
Solving quadratic equations: taking the square root of each side of the equation, p. 586.	Begin solving quadratic equations of the form $ax^2 - c = 0$ by solving for x^2. Then use the property that if $x^2 = a$, $x = \pm\sqrt{a}$. This amounts to taking the square root of each side of the equation.	Solve. $\quad 2x^2 + 1 = 99$ $$2x^2 = 99 - 1$$ $$2x^2 = 98$$ $$\dfrac{2x^2}{2} = \dfrac{98}{2}$$ $$x^2 = 49$$ $$x = \pm\sqrt{49} = \pm 7$$
Solving quadratic equations: completing the square, p. 588.	**1.** Put the equation in the form $ax^2 + bx = c$. If $a \neq 1$, divide each term by a. **2.** Square $\dfrac{b}{2}$ and add the result to both sides of the equation. **3.** Factor the left side (a perfect square trinomial). **4.** Use the square root property. **5.** Solve the equations. **6.** Check the solutions in the original equation.	Solve. $\quad 5x^2 - 3x - 7 = 0$ $$5x^2 - 3x = 7$$ $$x^2 - \dfrac{3}{5}x = \dfrac{7}{5}$$ $$x^2 - \dfrac{3}{5}x + \dfrac{9}{100} = \dfrac{7}{5} + \dfrac{9}{100}$$ $$\left(x - \dfrac{3}{10}\right)^2 = \dfrac{149}{100}$$ $$x - \dfrac{3}{10} = \pm\dfrac{\sqrt{149}}{10}$$ $$x = \dfrac{3}{10} \pm \dfrac{\sqrt{149}}{10}$$ $$x = \dfrac{3 \pm \sqrt{149}}{10}$$
Solving quadratic equations: quadratic formula, p. 592.	**1.** Put the equation in standard form: $ax^2 + bx + c = 0$. **2.** Carefully determine the values of a, b, and c. **3.** Substitute these into the formula $$x = \dfrac{-b \pm \sqrt{b^2 - 4ac}}{2a}.$$ **4.** Simplify.	Solve. $\quad 3x^2 - 4x - 8 = 0$ $$a = 3,\ b = -4,\ c = -8$$ $$x = \dfrac{-(-4) \pm \sqrt{(-4)^2 - 4(3)(-8)}}{2(3)}$$ $$= \dfrac{4 \pm \sqrt{16 + 96}}{6}$$ $$= \dfrac{4 \pm \sqrt{112}}{6} = \dfrac{4 \pm \sqrt{16 \cdot 7}}{6}$$ $$= \dfrac{4 \pm 4\sqrt{7}}{6} = \dfrac{2(2 \pm 2\sqrt{7})}{2 \cdot 3} = \dfrac{2 \pm 2\sqrt{7}}{3}$$

Topic	Procedure	Examples
The discriminant, p. 594.	The *discriminant* in the quadratic formula is $b^2 - 4ac$. If it is negative the quadratic equation has no real solutions (since it is not possible to take the square root of a negative number and get a real number).	What kinds of solutions does this equation have? $$3x^2 + 2x + 1 = 0$$ $$b^2 - 4ac = 2^2 - 4 \cdot 3 \cdot 1$$ $$= 4 - 12 = -8$$ This equation has no real solutions.
Properties of parabolas, p. 601.	**1.** The graph of $y = ax^2 + bx + c$ is a parabola. It opens up if $a > 0$ and down if $a < 0$. **2.** The vertex of a parabola is its lowest (if $a > 0$) or highest (if $a < 0$) point. Its x-coordinate is $x = \dfrac{-b}{2a}$.	The vertex of $y = 3x^2 + 4x - 11$ has x-coordinate $$x = \frac{-b}{2a} = \frac{-4}{6} = -\frac{2}{3}.$$ The parabola opens upward.
Graphing parabolas, p. 601.	Graph the vertex of a parabola and several other points to get a good idea of its graph. Usually, we find the y-intercept and the x-intercepts if any exist. The following procedure is helpful. **1.** Determine if $a < 0$ or $a > 0$. **2.** Find the y-intercept. **3.** Find the x-intercepts. **4.** Find the vertex. **5.** Plot one or two extra points if necessary. **6.** Draw the graph.	Graph, $y = x^2 - 2x - 8$ $a = 1, b = -2, c = -8; a > 0$, so the parabola opens upward. Let $x = 0$. $y = 0^2 - 2(0) - 8 = -8$ The y-intercept is $(0, -8)$. $$0 = x^2 - 2x - 8$$ $$0 = (x - 4)(x + 2)$$ $$x = 4, \quad x = -2$$ The x-intercepts are $(4, 0)$ and $(-2, 0)$. The x-coordinate of the vertex is $$x = \frac{-b}{2a} = -\frac{-(-2)}{2} = 1.$$ If $x = 1$, $y = (1)^2 - 2(1) - 8 = -9$. The vertex is at $(1, -9)$.
Formulas and applied problems, p. 610.	Some word problems require the use of a quadratic equation. You may want to follow this four-step procedure for problem solving. **1.** Understand the problem. **2.** Write an equation. **3.** Solve and state the answer. **4.** Check.	The area of a rectangle is 48 square inches. The length is three times the width. Find the dimensions of the rectangle. **1.** Draw a picture. **2.** $(3w)w = 48$ **3.** $3w^2 = 48$ $\quad w^2 = 16$ $\quad w = 4$ inches, $l = 12$ inches **4.** $(4)(12) = 48$

Putting Your Skills to Work

The Mathematics of a Reflecting Pool

When you look at a picture of the Taj Mahal, you are impressed with the beauty of the structure, which is embellished by a carefully designed rectangular reflecting pool in front of the structure. There are actually several reflecting pools near the Taj Mahal. The perimeter of one of the reflecting pools is 640 meters. The area of any rectangle with a perimeter of 640 meters is given by the equation $A = x(320 - x)$, where x is the width of the rectangle.

Problems for Individual Investigation

▲ **1.** Using the preceding equation, find the areas of rectangular reflecting pools that can be constructed if the perimeters are 640 meters and the widths of the rectangles are 10 meters, 20 meters, 40 meters, 80 meters, and 160 meters.

2. Using the values obtained in question 1, graph the equation $A = x(320 - x)$.

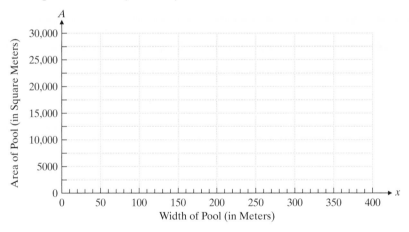

Problems for Group Investigation and Cooperative Learning

▲ **3.** What are the dimensions of a reflecting pool with the maximum possible area and a perimeter of 640 meters?

▲ **4.** The reflecting pool of the Taj Mahal was created to have a small area (to make it easier to keep the surface of the water clean) and a narrow width (so that the eye of an observer standing at the end of the pool is led directly to the structure of the Taj Mahal in the distance). The area of the reflecting pool is 25% of the maximum possible area that can be enclosed by a rectangle with a perimeter of 640 meters. Find the dimensions of the reflecting pool. Round to the nearest tenth.

Internet Connections

  Netsite: http://www.prenhall.com/tobey_beginning

Site: National Parks of Washington, D.C.

▲ **5.** Find the dimensions of the rectangular reflecting pool in front of the Washington Monument in the U.S. capital.

6. Find the maximum area that can be enclosed by a rectangle having the same perimeter as the reflecting pool in Washington, D.C. What percent of this maximum area is the area of the reflecting pool in Washington, D.C.? What percent of the maximum area for the perimeter of this reflecting pool is represented by the actual dimensions of the reflecting pool?

Chapter 10 Review Problems

10.1 *Write in standard form. Determine the values of a, b, and c. Do not solve.*

1. $6x^2 = 5x - 8$

2. $9x^2 + 3x = -5x^2 + 16$

3. $(x - 4)(2x - 1) = x^2 + 7$

4. $x(3x + 1) = -5x + 10$

5. $\dfrac{1}{x^2} - \dfrac{3}{x} + 5 = 0$

6. $\dfrac{x}{x + 2} - \dfrac{3}{x - 2} = 5$

Place in standard form. Solve by factoring.

7. $x^2 + 26x + 25 = 0$

8. $x^2 + 16x + 64 = 0$

9. $x^2 + 6x - 55 = 0$

10. $4x^2 - 16x + 15 = 0$

11. $9x^2 - 24x + 16 = 0$

12. $5x^2 = 6x - 1$

13. $20x^2 = 19x - 3$

14. $3x^2 = -14x - 11$

15. $x^2 + \dfrac{1}{6}x - 2 = 0$

16. $x^2 + \dfrac{7}{6}x - \dfrac{1}{2} = 0$

17. $x^2 + \dfrac{2}{15}x = \dfrac{1}{15}$

18. $1 + \dfrac{13}{12x} - \dfrac{1}{3x^2} = 0$

19. $x^2 + 12x - 3 = 8x - 3$

20. $6x^2 - 13x + 2 = 5x + 2$

21. $1 + \dfrac{2}{3x + 4} = \dfrac{3}{3x + 2}$

22. $2 - \dfrac{5}{x + 1} = \dfrac{3}{x - 1}$

23. $5 + \dfrac{24}{2 - x} = \dfrac{24}{2 + x}$

24. $\dfrac{4}{9} = \dfrac{5x}{3} - x^2$

10.2   *Solve by taking the square root of each side.*

25. $x^2 + 3 = 28$ **26.** $x^2 - 4 = 32$ **27.** $x^2 - 5 = 17$ **28.** $x^2 + 11 = 50$

29. $2x^2 - 1 = 15$ **30.** $3x^2 + 4 = 154$ **31.** $3x^2 + 6 = 60$ **32.** $2x^2 - 5 = 43$

33. $(x - 4)^2 = 7$ **34.** $(x - 2)^2 = 3$ **35.** $(5x + 6)^2 = 20$ **36.** $(3x + 8)^2 = 12$

Solve by completing the square.

37. $x^2 + 10x - 11 = 0$ **38.** $x^2 + 12x + 11 = 0$

39. $2x^2 - 8x - 90 = 0$ **40.** $3x^2 - 36x - 60 = 0$

41. $3x^2 + 6x - 6 = 0$ **42.** $2x^2 + 10x - 3 = 0$

10.3   *Solve using the quadratic formula.*

43. $x^2 + 4x - 6 = 0$ **44.** $x^2 + 4x - 8 = 0$ **45.** $2x^2 - 7x + 4 = 0$

46. $2x^2 + 5x - 6 = 0$ **47.** $2x^2 + 4x - 5 = 0$ **48.** $2x^2 + 6x - 5 = 0$

49. $3x^2 - 5x = 4$ **50.** $4x^2 + 3x = 2$

Solve by any method. If there is no real number solution, so state.

51. $2x^2 - 9x + 10 = 0$

52. $4x^2 - 4x - 3 = 0$

53. $25x^2 + 10x + 1 = 0$

54. $2x^2 - 11x + 12 = 0$

55. $3x^2 - 6x + 2 = 0$

56. $5x^2 - 7x = 8$

57. $4x^2 + 4x = x^2 + 5$

58. $5x^2 + 7x + 1 = 0$

59. $x^2 + 5x = -1 + 2x$

60. $3x^2 = 6 - 7x$

61. $3x^2 + 1 = 73 + x^2$

62. $2x^2 - 1 = 35$

63. $\dfrac{(y - 2)^2}{20} + 3 + y = 0$

64. $\dfrac{(y + 2)^2}{5} + 2y = -9$

65. $3x^2 + 1 = 6 - 8x$

66. $2x^2 + 10x = 2x - 7$

67. $2y - 10 = 10y(y - 2)$

68. $\dfrac{3y - 2}{4} = \dfrac{y^2 - 2}{y}$

69. $\dfrac{y^2 + 5}{2y} = \dfrac{2y - 1}{3}$

70. $\dfrac{5x^2}{2} = x - \dfrac{7x^2}{2}$

10.4 *Graph each quadratic equation. Label the vertex.*

71. $y = 2x^2$

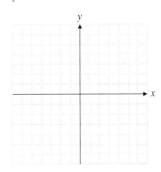

72. $y = x^2 + 4$

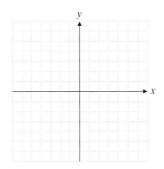

73. $y = x^2 - 3$

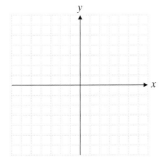

74. $y = -\dfrac{1}{2}x^2$

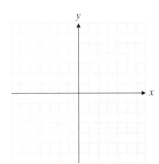

75. $y = x^2 - 3x - 4$

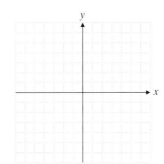

76. $y = \dfrac{1}{2}x^2 - 2$

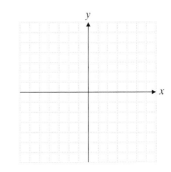

77. $y = -2x^2 + 12x - 17$

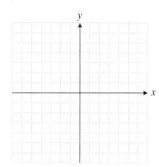

78. $y = -3x^2 - 2x + 4$

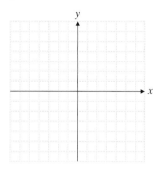

10.5 *Solve.*

▲ **79.** Jon is building a dog pen against the house. He has 11 feet of fencing and will pen in an area of 15 square feet. What are the dimensions of the pen?

▲ **80.** The hypotenuse of a right triangle is 20 centimeters. One leg of the right triangle is 4 centimeters shorter than the other. What is the length of each leg of the right triangle?

81. Last year the Golf Club on campus raised $720 through dues assigned equally to each member. This year there were four fewer members. As a result, the dues for each member went up by $6. How many members were in the Golf Club last year?

82. One positive number is 2 more than twice another positive number. The difference of the squares of these numbers is 119. Find the numbers that satisfy these conditions.

▲ **83.** The length of a room exceeds its width by 4 feet. A rug covers the floor area except for a border 2 feet wide all around it. The area of the border is 68 square feet. What is the area of the rug?

84. Alice drove 90 miles to visit her cousin. Her average speed on the trip home was 15 mph faster than her average speed on the trip there. Her total travel time for both trips was 3.5 hours. What was her average rate of speed on each trip?

Use the following information to solve exercises 85–88.

The Lacorazza family opened a chain of pizza stores in New England in 1980. The number of pizza stores in operation during any year is given by the equation $y = -0.05x^2 + 2.5x + 60$, where x is the number of years since 1980.

85. According to the equation, how many pizza stores will still be in operation in 2030?

86. How many fewer pizza stores will still be in operation in 2040 than in 2030?

87. During what two years will there be exactly 80 pizza stores in operation?

88. During what two years will there be exactly 90 pizza stores in operation?

Solve by any desired method. If there is no real number solution, say so.

1. $5x^2 + 7x = 4$

2. $3x^2 + 13x = 10$

3. $2x^2 = 2x - 5$

4. $6x^2 - 7x - 5 = 0$

5. $12x^2 + 11x = 5$

6. $18x^2 + 32 = 48x$

7. $2x^2 - 11x + 3 = 5x + 3$

8. $2x^2 + 5 = 37$

9. $2x(x - 6) = 6 - x$

10. $x^2 - x = \dfrac{3}{4}$

Graph each quadratic equation. Locate the vertex.

11. $y = 3x^2 - 6x$

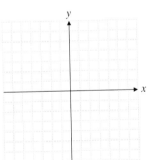

12. $y = -x^2 + 8x - 12$

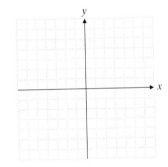

Solve.

▲ 13. The hypotenuse of a right triangle is 15 meters in length. One leg is 3 meters longer than the other leg. Find the length of each leg.

14. When an object is thrown upward, its height (S) in meters is given (approximately) by the quadratic equation

$$S = -5t^2 + vt + h,$$

where v = the initial upward velocity in meters per second,
t = the time of flight in seconds, and
h = the height above level ground from which the object is thrown.
Suppose that a ball is thrown upward with a velocity of 33 meters/second at a height of 14 meters above the ground. How long after it is thrown will it strike the ground?

1. _____

2. _____

3. _____

4. _____

5. _____

6. _____

7. _____

8. _____

9. _____

10. _____

11. _____

12. _____

13. _____

14. _____

Approximately one-half of this test covers Chapters 0–9. The remainder covers the content of Chapter 10.

1. Simplify. $3x\{2y - 3[x + 2(x + 2y)]\}$

2. Evaluate $2x^2 - 3xy + y^2$ when $x = 2$ and $y = -3$.

3. Solve for x. $\dfrac{1}{2}(x - 2) = \dfrac{1}{3}(x + 10) - 2x$

4. Factor completely. $16x^4 - 1$

5. Multiply. $(3x + 1)(2x - 3)(x + 4)$

6. Solve for x. $\dfrac{3x}{x^2 - 4} = \dfrac{2}{x + 2} + \dfrac{4}{2 - x}$

7. Graph the line $y = -\dfrac{3}{4}x + 2$. Identify the slope and the y-intercept.

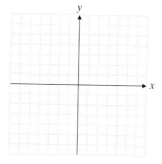

8. Find the equation of a line with slope $= -3$ that passes through the point $(2, -1)$.

9. Solve for (x, y).

$$3x + 2y = 5$$
$$7x + y = 19$$

10. Simplify. $(-3x^3y^2)^4$

1. _____

2. _____

3. _____

4. _____

5. _____

6. _____

7. _____

8. _____

9. _____

10. _____

11. _____

11. Simplify. Assume that all variables are positive. $\sqrt{18x^5y^6z^3}$

12. _____

12. Multiply. $\left(\sqrt{2} + \sqrt{3}\right)\left(2\sqrt{2} - 4\sqrt{3}\right)$

13. _____

13. Rationalize the denominator. $\dfrac{\sqrt{5} - 3}{\sqrt{5} + 2}$

14. _____

14. Solve for r. $A = 3r^2 - 2Ar$

15. _____

15. Solve for b. $H = 25b^2 - 6$

In questions 16–22, solve.

16. _____

16. $2x^2 + 3x = 35$

17. $(2x + 1)^2 = 20$

17. _____

18. $\dfrac{60}{R} = \dfrac{60}{R + 8} + 10$

19. $3x^2 + 11x + 2 = 0$

18. _____

20. $7x^2 + 36x + 5 = 0$

21. $x^2 = 98$

19. _____

22. $3x^2 + 4 = 79$

20. _____

23. Graph $y = x^2 + 6x + 10$. Locate the vertex.

21. _____

22. _____

23. _____

24. _____

▲ **24.** The length of a rectangle is 2 meters less than three times its width. The area is 96 square meters. Find its length and width.

Practice Final Examination

The following questions cover the content of Chapters 0–10. Follow the directions for each question and simplify your answers.

1. Simplify.
$-2x + 3y\{7 - 2[x - (4x + y)]\}$.

2. Evaluate if $x = -2$ and $y = 3$.
$2x^2 - 3xy - 4y$

3. Simplify. $(-3x^2y)(-6x^3y^4)$

4. Combine like terms.
$5x^2y - 6xy + 8xy - 3x^2y - 10xy$

5. Solve for x.
$\dfrac{1}{2}(x + 4) - \dfrac{2}{3}(x - 7) = 4x$

6. Solve for b. $p = 2a + 2b$

7. Solve for x and graph the resulting inequality on a number line.
$5x + 3 - (4x - 2) \le 6x - 8$

8. Multiply. $(2x + y)(x^2 - 3xy + 2y^2)$

9. Factor completely. $4x^2 - 18x - 10$

10. Factor completely. $25x^4 - 25$

11. Combine.
$\dfrac{2}{x - 3} - \dfrac{3}{x^2 - x - 6} + \dfrac{4}{x + 2}$

12. Simplify. $\dfrac{\dfrac{3}{x} + \dfrac{5}{2x}}{1 + \dfrac{2}{x + 2}}$

13. Solve for x. $\dfrac{2}{x + 2} = \dfrac{4}{x - 2} + \dfrac{3x}{x^2 - 4}$

14. Find the slope of the line and then graph the line. $5x - 2y - 3 = 0$

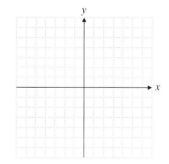

15. Find the equation of the line that has a slope of $-\dfrac{3}{4}$ and passes through the point $(-2, 5)$.

1. _____

2. _____

3. _____

4. _____

5. _____

6. _____

7. _____

8. _____

9. _____

10. _____

11. _____

12. _____

13. _____

14. _____

15. _____

16. Two quarter-circles are attached to a rectangle. The dimensions are indicated in the figure. Find the area of the region. Use 3.14 as an approximation for π.

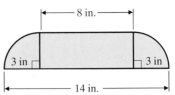

16. _____

17. _____

18. _____

19. _____

20. _____

21. _____

22. _____

23. _____

24. _____

25. _____

26. _____

27. _____

28. _____

29. _____

30. _____

17. Solve for x and y.

$$2x + 3y = 8$$

$$3x - 5y = 31$$

18. Solve for a and b.

$$a - \frac{3}{4}b = \frac{1}{4}$$

$$\frac{3}{2}a + \frac{1}{2}b = -\frac{9}{2}$$

19. Simplify and combine.
$$2x\sqrt{50} + \sqrt{98x^2} - 3x\sqrt{18}$$

20. Multiply and simplify.
$$\sqrt{6}(3\sqrt{2} - 2\sqrt{6} + 4\sqrt{3})$$

21. Simplify by rationalizing the denominator. $\dfrac{\sqrt{3} + \sqrt{7}}{\sqrt{5} - \sqrt{7}}$

22. Solve for x. $3x^2 - 2x - 5 = 0$

23. Solve for y. $2y^2 = 6y - 1$

24. Solve for x. $4x^2 + 3 = 19$

25. In the right triangle shown, sides b and c are given. Find the length of side a.

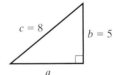

26. A number is tripled and then increased by 6. The result is 21. What is the original number?

27. A rectangular region has a perimeter of 38 meters. The length of the rectangle is 2 meters shorter than double the width. What are the dimensions of the rectangle?

28. Margaret invested $7000 for one year. She placed part of the money in a tax-free bond earning 10% simple interest. She placed the rest in a money market account earning 14% simple interest. At the end of one year she had earned $860 in interest. How much did she invest at each interest rate?

29. A benefit concert was held on campus to raise scholarship money. A total of 360 tickets were sold. Admission prices were $5 for reserved seats and $3 for general admission. Total receipts were $1480. How many reserved-seat tickets were sold? How many general admission tickets were sold?

30. The area of a triangle is 68 square meters. The altitude of the triangle is one meter longer than double the base of the triangle. Find the altitude and base of the triangle.

Glossary

Absolute value of a number (1.1)   The absolute value of a number x is the distance between 0 and the number x on the number line. It is written as $|x|$. $|x| = x$ if $x \geq 0$, but $|x| = -x$ if $x < 0$.

Altitude of a geometric figure (1.8)   The height of the geometric figure. In the three figures shown the altitude is labeled a.

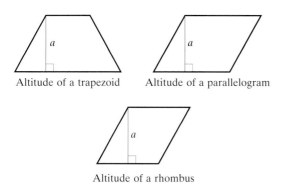

Altitude of a trapezoid   Altitude of a parallelogram

Altitude of a rhombus

Altitude of a triangle (1.8)   The height of any given triangle. In the three triangles shown the altitude is labeled a.

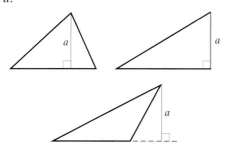

Associative property of addition (1.1)   If a, b, and c are real numbers, then

$$a + (b + c) = (a + b) + c.$$

This property states that if three numbers are added it does not matter *which two numbers* are added first, the result will be the same.

Associative property of multiplication (1.3)   If $a, b,$ and c are real numbers, then

$$a \times (b \times c) = (a \times b) \times c.$$

This property states that if three numbers are multiplied it does not matter *which two numbers* are multiplied first; the result will be the same.

Base (1.4)   The number or variable that is raised to a power. In the expression 2^6, the number 2 is the base.

Base of a triangle (1.8)   The side of a triangle that is perpendicular to the altitude.

Binomial (4.3)   A polynomial of two terms. The expressions $a + 2b$, $6x^3 + 1$, and $5a^3b^2 + 6ab$ are all binomials.

Circumference of a circle (1.8)   The distance around a circle. The circumference of a circle is given by the formula $C = \pi d$ or $C = 2\pi r$, where d is the diameter of the circle and r is the radius of the circle.

Coefficient (4.1)   A coefficient is a factor or a group of factors in a product. In the term $4xy$ the coefficient of y is $4x$, but the coefficient of xy is 4. In the term $-5x^3y$ the coefficient of x^3y is -5.

Commutative property for addition (1.1)   If a and b are any real numbers, then $a + b = b + a$.

Commutative property for multiplication (1.3)   If a and b are any real numbers, then $ab = ba$.

Complex fraction (6.4)   A fraction that contains at least one fraction in the numerator or in the denominator or both. These three fractions are complex fractions:

$$\frac{7 + \dfrac{1}{x}}{x^2 + 2}, \qquad \frac{1 + \dfrac{1}{5}}{2 - \dfrac{1}{7}}, \qquad \text{and} \qquad \frac{\dfrac{1}{3}}{4}$$

Constant (2.3)   Symbol or letter that is used to represent exactly one single quantity during a particular problem or discussion.

Coordinates of a point (7.1)   An ordered pair of numbers (x, y) that specifies the location of a point in a rectangular coordinate system.

Degree of a polynomial (4.3)   The degree of the highest-degree term of a polynomial. The degree of the polynomial $5x^3 + 2x^2 - 6x + 8$ is 3. The degree of the polynomial $5x^2y^2 + 3xy + 8$ is 4.

Degree of a term of a polynomial (4.3)   The sum of the exponents of the variables in the term. The degree of $3x^3$ is 3. The degree of $4x^5y^2$ is 7.

Denominator (0.1) and (6.1)   The bottom number or algebraic expression in a fraction. The denominator of

$$\frac{3x - 2}{x + 4}$$

is $x + 4$. The denominator of $\dfrac{3}{7}$ is 7. The denominator of a fraction may not be zero.

Dependent equations (8.1) Two equations are dependent if every value that satisfies one equation satisfies the other. A system of two dependent equations in two variables will not have a unique solution.

Diagonal of a four-sided figure (1.8) A line connecting two nonadjacent corners of the figure. In each of the figures shown, line *AC* is a diagonal.

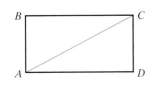

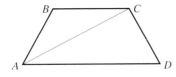

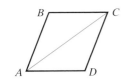

Difference (3.1) The result of subtracting one number or expression from another. The mathematical expression $x - 6$ can be written in words as the difference between *x* and 6.

Difference-of-two-squares polynomial (5.5) A polynomial of the form $a^2 - b^2$ that may be factored by using the formula

$$a^2 - b^2 = (a + b)(a - b).$$

Distributive property (1.6) For all real numbers *a*, *b*, and *c*, $a(b + c) = ab + ac$.

Dividend (0.4) The number that is to be divided by another. In the problem $30 \div 5 = 6$, the three parts are as follows:

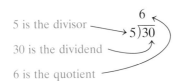

Divisor (0.4) The number you divide into another.

Domain of a relation (7.6) In any relation, the set of values that can be used for the independent variable is called its domain. This is the set of all the first coordinates of the ordered pairs that define the relation.

Equilateral triangle (1.8) A triangle with three sides equal in length and three angles that measure 60°. Triangle *ABC* is an equilateral triangle.

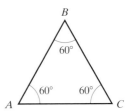

Even integers (1.3) Integers that are exactly divisible by 2, such as …, −4, −2, 0, 2, 4, 6, ….

Exponent (1.4) The number that indicates the power of a base. If the number is a positive integer it indicates how many times the base is multiplied. In the expression 2^6, the exponent is 6.

Expression (4.3) A mathematic expression is any quantity using numbers and variables. Therefore, $2x, 7x + 3$, and $5x^2 + 6x$ are all mathematical expressions.

Extraneous solution (6.5) and (9.6) An obtained solution to an equation that when substituted back into the original equation, does *not* yield an identity. $x = 2$ is an extraneous solution to the equation

$$\frac{x}{x - 2} - 4 = \frac{2}{x - 2}$$

An extraneous solution is also called an extraneous root.

Factor (0.1) and (5.1) When two or more numbers, variables, or algebraic expressions are multiplied, each is called a factor. If we write $3 \cdot 5 \cdot 2$, the factors are 3, 5, and 2. If we write $2xy$, the factors are 2, *x*, and *y*. In the expression $(x - 6)(x + 2)$, the factors are $(x - 6)$ and $(x + 2)$.

Fractions

Algebraic fractions (6.1) The indicated quotient of two algebraic expressions.

$$\frac{x^2 + 3x + 2}{x - 4} \quad \text{and} \quad \frac{y - 6}{y + 8}$$

are algebraic fractions. In these fractions the value of the denominator cannot be zero.

Numerical fractions (0.1) A set of numbers used to describe parts of whole quantities. A numerical fraction can be represented by the quotient of two integers for which the denominator is not zero. The numbers $\frac{1}{5}, -\frac{2}{3}, \frac{8}{2}, -\frac{4}{31}, \frac{8}{1}$, and $-\frac{12}{1}$ are all numerical fractions. The set of rational numbers can be represented by numerical fractions.

Function (7.6) A relation in which no two different ordered pairs have the same first coordinate.

Hypotenuse of a right triangle (9.6) The side opposite the right angle in any right triangle. The hypotenuse is always the longest side of a right triangle. Side AB is the hypotenuse of triangle below.

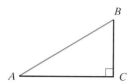

Identity (2.1)   A statement that is always true. The equations $5 = 5, 7 + 4 = 7 + 4,$ and $x + 8 = x + 8$ are examples of identities.

Imaginary number (9.1)   A complex number that is not a real number. Imaginary numbers can be created by taking the square root of a negative real number. The numbers $\sqrt{-9}, \sqrt{-7},$ and $\sqrt{-12}$ are all imaginary numbers.

Improper Fraction (0.1)   A numerical fraction whose numerator is larger than or equal to its denominator. $\frac{8}{3}, \frac{5}{2},$ and $\frac{7}{7}$ are improper fractions.

Inconsistent system of equations (8.1)   A system of equations that does not have a solution.

Independent equations (8.1)   Two equations that are not dependent are said to be independent.

Inequality (2.6), (2.7), and (7.5)   A mathematical relationship between quantities that are not equal. $x \le -3, w > 5,$ and $x < 2y + 1$ are mathematical inequalities.

Integers (1.1)   The set of numbers $\ldots, -5, -4, -3, -2, -1, 0, 1, 2, 3, 4, 5, \ldots$.

Intercepts of an equation (7.2)   The point or points where the graph of the equation crosses the x-axis or the y-axis or both. (*See x-intercept.*)

Irrational number (9.1)   A real number that cannot be expressed in the form $\frac{a}{b}$, where a and b are integers and $b \ne 0.$ $\sqrt{2}, \pi, 5 + 3\sqrt{2},$ and $-4\sqrt{7}$ are irrational numbers.

Isosceles triangle (1.8)   A triangle with two equal sides and two equal angles. Triangle ABC is an isosceles triangle. Angle BAC is equal to angle ACB. Side AB is equal in length to side BC.

Least common denominator of numerical fractions (0.2)   The smallest whole number that is exactly divisible by all denominators of a group of fractions. The least common denominator (LCD) of $\frac{1}{6}, \frac{2}{3},$ and $\frac{3}{5}$ is 30. The least common denominator is also called the lowest common denominator.

Leg of a right triangle (9.6)   One of the two shorter sides of a right triangle. Side AC and side BC are legs of triangle ABC.

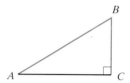

Like terms (1.7)   Terms that have identical variables and exponents. In the expression $5x^3 + 2xy^2 + 6x^2 - 3xy^2$, the term $2xy^2$ and the term $-3xy^2$ are like terms.

Linear equation in two variables (7.2)   An equation of the form $Ax + By = C$, where A, B, and C are real numbers. The graph of a linear equation in two variables is a straight line.

Line of symmetry of a parabola (10.4)   A line that can be drawn through a parabola such that, if the graph were folded on this line, the two halves of the curve would correspond exactly. The line of symmetry through the parabola formed by $y = ax^2 + bx + c$ is given by $x = \frac{-b}{2a}$. The line of symmetry of a parabola always passes through the vertex of the parabola.

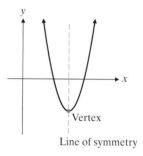

Mixed number (0.1)   A number that consists of an integer written next to a proper fraction. $2\frac{1}{3}, 4\frac{6}{7},$ and $3\frac{3}{8}$ are all mixed numbers. Mixed numbers are sometimes called mixed fractions or mixed numerals.

Monomial (4.3)   A polynomial of one term. The expressions $3xy, 5a^2b^3cd,$ and -6 are all monomials.

Natural numbers (0.1)   The set of numbers 1, 2, 3, 4, 5, ... This set is also called the set of counting numbers.

Numeral (0.1)   The symbol used to describe a number.

Numerator (0.1)   The top number or algebraic expression in a fraction. The numerator of

$$\frac{x + 3}{5x - 2}$$

is $x + 3$. The numerator of $\frac{12}{13}$ is 12.

Numerical coefficient (4.1)   The number that is multiplied by a variable or a group of variables. The numerical coefficient in $5x^3y^2$ is 5. The numerical coefficient in $-6abc$ is -6. The numerical coefficient in x^2y is 1. A numerical coefficient of 1 is not usually written.

Odd integers (1.3)   Integers that are not exactly divisible by 2, such as $-3, -1, 1, 3, 5, 7, 9, \ldots$.

Opposite of a number (1.1)   Two numbers that are the same distance from zero on the number line but lie on different sides of it are considered opposites. The opposite of -6 is 6. The opposite of $\frac{22}{7}$ is $-\frac{22}{7}$.

Ordered pair (7.1)   A pair of numbers presented in a specified order. An ordered pair is often used to specify a location on a graph. Every point in a rectangular coordinate system can be represented by an ordered pair (x, y).

Origin (7.1)   The point $(0, 0)$ in a rectangular coordinate system.

Parabola (10.4)   A curve created by graphing a quadratic equation. The curves shown are all parabolas. The graph of the equation $y = ax^2 + bx + c$, where $a \neq 0$, will always be a parabola.

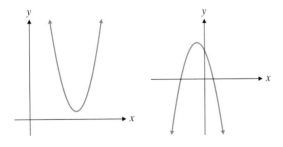

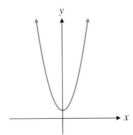

Parallel lines (7.3) and (8.1)   Two straight lines that never intersect. The graph of an inconsistent system of two linear equations in two variables will result in parallel lines.

Parallelogram (1.8)   A four-sided figure with opposite sides parallel. Figure $ABCD$ is a parallelogram.

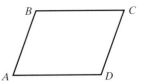

Percent (0.5)   Hundredths or "per one hundred"; indicated by the % symbol. Thirty-seven hundredths $\left(\frac{37}{100}\right) = 37\%$ (thirty-seven percent).

Perfect square number (5.5)   A number that is the square of an integer. The numbers 1, 4, 9, 16, 25, 36, 49, 64, 81, 100, 121, 144, … are perfect square numbers.

Perfect square trinomial (5.5)   A polynomial of the form $a^2 + 2ab + b^2$ or $a^2 - 2ab + b^2$ that may be factored using one of the following formulas:

$$a^2 + 2ab + b^2 = (a + b)^2$$

or

$$a^2 - 2ab + b^2 = (a - b)^2.$$

Perimeter (1.8)   The distance around any plane figure. The perimeter of this triangle is 13. The perimeter of this rectangle is 20

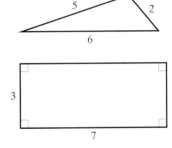

Pi (1.8)   An irrational number, denoted by the symbol π, that is approximately equal to 3.141592654. In most cases 3.14 can be used as a sufficiently accurate approximation for π.

Polynomial (4.3)   Expressions that contain terms with nonnegative integer exponents. The expressions $5ab + 6$, $x^3 + 6x^2 + 3$, -12, and $x + 3y - 2$ are all polynomials. The expressions $x^{-2} + 2x^{-1}$, $2\sqrt{x} + 6$, and $\frac{5}{x} + 2x^2$ are not polynomials.

Prime number (0.1)   Any natural number greater than 1 whose only natural number factors are 1 and itself. The first eight prime numbers are 2, 3, 5, 7, 11, 13, 17, and 19.

Prime polynomial (5.6)   A prime polynomial is a polynomial that cannot be factored by the methods of elementary algebra. $x^2 + x + 1$ is a prime polynomial.

Principal (3.4)   In monetary problems, the principal is the original amount of money invested or borrowed.

Principal square root (9.1)   For any given nonnegative number N, the principal square root of N (written $\sqrt{N}$) is the nonnegative number a if and only if $a^2 = N$. The principal square root of 25 ($\sqrt{25}$) is 5.

Proper fraction (0.1)   A numerical fraction whose numerator is less than its denominator; $\frac{3}{7}, \frac{2}{5},$ and $\frac{8}{9}$ are proper fractions.

Proportion (6.6)   A proportion is an equation stating that two ratios are equal.

$$\frac{a}{b} = \frac{c}{d} \quad \text{where } b, d \neq 0$$

is a proportion.

Pythagorean Theorem (9.6)   In any right triangle, if c is the length of the hypotenuse and a and b are the lengths of the two legs, then $c^2 = a^2 + b^2$.

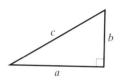

Quadratic equation (5.7)   A quadratic equation is a polynomial equation with one variable that contains at least one term with the variable squared, but no term with the variable raised to a higher power. $5x^2 + 6x - 3 = 0$, $x^2 = 7$, and $5x^2 = 2x$ are all quadratic equations.

Quadratic formula (10.3)   If $ax^2 + bx + c = 0$ and $a \neq 0$, then the solutions to the quadratic equation are given by

$$x = \frac{-b \pm \sqrt{b^2 - 4ac}}{2a}.$$

Quotient (0.4)   The result of dividing one number or expression by another. In the problem $12 \div 4 = 3$, the quotient is 3.

Radical (9.1)   An expression composed of a radical sign and a radicand. The expressions $\sqrt{5x}, \sqrt{\frac{3}{5}}, \sqrt{5x + b},$ and $\sqrt{10}$ are called radicals.

Radical equation (9.6)   An equation that contains one or more radicals. $\sqrt{x + 4} = 12$, $\sqrt{x} = 3$, and $\sqrt{3x} + 4 = x + 2$ are all radical equations.

Radical sign (9.1)   The symbol $\sqrt{\ }$ used to indicate the square root of a number.

Radicand (9.1)   The expression beneath the radical sign. The radicand in $\sqrt{5ab}$ is $5ab$.

Range of a relation (7.6)   In any relation, the set of values that represents the dependent variable is called its range. This is the set of all the second coordinates of the ordered pairs that define the relation.

Ratio (6.6)   The ratio of one number a to another number b is the quotient $a \div b$ or $\frac{a}{b}$.

Rational numbers (1.1) and (9.1)   A number that can be expressed in the form $\frac{a}{b}$, where a and b are integers and $b \neq 0$. $\frac{7}{3}, -\frac{2}{5}, \frac{7}{-8}, \frac{5}{1},$ 1.62, and 2.7156 are rational numbers.

Rationalizing the denominator (9.5)   The process of transforming a fraction that contains a radical in the denominator to an equivalent fraction that does not contain a radical in the denominator.

Original Fraction	Equivalent Fraction with Rationalized Denominator
$\dfrac{3}{\sqrt{2}}$	$\dfrac{3\sqrt{2}}{2}$

Rationalizing the expression (9.5)   The process of transforming a radical that contains a fraction to an equivalent expression that does not contain a fraction inside a radical.

Original Radical	Equivalent Expression after It Has Been Rationalized
$\sqrt{\dfrac{3}{5}}$	$\dfrac{\sqrt{15}}{5}$

Real number (9.1)   Any number that is rational or irrational. $2, 7, \sqrt{5}, \frac{3}{8}, \pi, -\frac{7}{5},$ and $-3\sqrt{5}$ are all real numbers.

Rectangle (1.8)   A four-sided figure with opposite sides parallel and all interior angles measuring 90°. The opposite sides of a rectangle are equal.

Relation (7.6)   A relation is any set of ordered pairs.

Rhombus (1.8)   A parallelogram with four equal sides. Figure $ABCD$ is a rhombus.

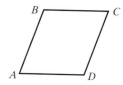

Right angle (1.8)   An angle that measures 90°. Right angles are usually labeled in a sketch by using a small square to indicate that it is a right angle. Here angle ABC is a right angle.

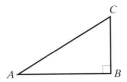

Right triangle (1.8)   A triangle that contains a right angle.

Root of an equation (2.1) and (5.7)   A value of the variable that makes an equation into a true statement. The root of an equation is also called the solution of an equation.

Scientific notation (4.2)   A positive number is written in scientific notation if it is in the form $a \times 10^n$, where $1 \le a < 10$ and n is an integer.

Slope-intercept form (7.3)   The equation of a line that has slope m and the y-intercept at $(0, b)$ is given by $y = mx + b$.

Slope of a line (7.3)   The ratio of change in y over the change in x for any two different points on a nonvertical line. The slope m is determined by

$$m = \frac{y_2 - y_1}{x_2 - x_1},$$

where $x_2 \ne x_1$ for any two points (x_1, y_1) and (x_2, y_2) on a nonvertical line.

Solution of an equation (2.1)   A number that, when substituted into a given equation, yields an identity. The solution of an equation is also called the root of an equation.

Solution of a linear inequality (2.7)   The possible values that make a linear inequality true.

Solution of an inequality in two variables (7.5)   The set of all possible ordered pairs that when substituted into the inequality will yield a true statement.

Solution to a system of two equations in two variables (8.2)   An ordered pair that can be substituted into each equation to obtain an identity in each case.

Square (1.8)   A rectangle with four equal sides.

Square root (9.1)   For any given nonnegative number N, the square root of N is the number a if $a^2 = N$. One square root of 16 is 4 since $(4)^2 = 16$. Another square root of 16 is -4 since $(-4)^2 = 16$. When we write $\sqrt{16}$, we want only the positive root, which is 4.

Standard form of a quadratic equation (5.7)   A quadratic equation that is in the form $ax^2 + bx + c = 0$.

Subscript of a variable (3.2)   A small number or letter written slightly below and to the right of a variable. In the expression $5 = 2(x - x_0)$, the subscript of x is 0. In the expression $t_f = 5(t_a - b)$ the subscript of the first t is f. The subscript of the second t is a. A subscript is used to indicate a different value of the variable.

System of equations (8.1)   A set of two or more equations that must be considered together.

Term (1.7)   A number, a variable, or a product of numbers and variables. For example, in the expression $a^3 - 3a^2b + 4ab^2 + 6b^3 + 8$, there are five terms. They are $a^3, -3a^2b, 4ab^2, 6b^3$, and 8. The terms of a polynomial are separated by plus and minus signs.

Trapezoid (1.8)   A four-sided figure with two sides parallel. The parallel sides are called the bases of the trapezoid. Figure $ABCD$ is a trapezoid.

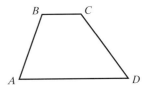

Trinomial (4.3)   A polynomial of three terms. The expressions $x^2 + 6x - 8$ and $a + 2b - 3c$ are trinomials.

Variable (1.4)   A letter that is used to represent a number or a set of numbers.

Variation (9.7)   An equation relating values of one variable to those of other variables. An equation of the form $y = kx$, where k is a constant, indicates *direct variation*. An equation of the form $y = \dfrac{k}{x}$, where k is a constant, indicates *inverse variation*. In both cases, k is called the *constant of variation*.

Vertex of a parabola (10.4)   The lowest point on a parabola opening upward or the highest point on a parabola opening downward. The x-coordinate of the vertex of the parabola formed by the equation $y = ax^2 + bx + c$ is given by $= \dfrac{-b}{2a}$.

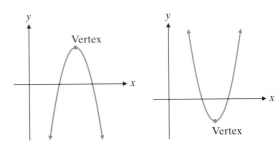

Vertical line test (7.6) If a vertical line can intersect the graph of a relation more than once, the relation is not a function.

Whole numbers (0.1) The set of numbers 0, 1, 2, 3, 4, 5, ...

x-intercept (7.2) The ordered pair $(a, 0)$ is the x-intercept of a line if the line crosses the x-axis at $(a, 0)$. The x-intercept of line l on the following graph is $(0, 4)$.

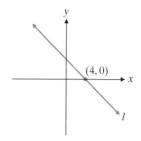

y-intercept (7.2) The ordered pair $(0, b)$ is the y-intercept of a line if the line crosses the y-axis at $(0, b)$. The y-intercept of line p on the following graph is $(0, 3)$.

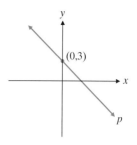

Appendix A
Table of Square Roots

x	$\sqrt{x}$	x	$\sqrt{x}$	x	$\sqrt{x}$	x	$\sqrt{x}$	x	$\sqrt{x}$
1	1.000	41	6.403	81	9.000	121	11.000	161	12.689
2	1.414	42	6.481	82	9.055	122	11.045	162	12.728
3	1.732	43	6.557	83	9.110	123	11.091	163	12.767
4	2.000	44	6.633	84	9.165	124	11.136	164	12.806
5	2.236	45	6.708	85	9.220	125	11.180	165	12.845
6	2.449	46	6.782	86	9.274	126	11.225	166	12.884
7	2.646	47	6.856	87	9.327	127	11.269	167	12.923
8	2.828	48	6.928	88	9.381	128	11.314	168	12.961
9	3.000	49	7.000	89	9.434	129	11.358	169	13.000
10	3.162	50	7.071	90	9.487	130	11.402	170	13.038
11	3.317	51	7.141	91	9.539	131	11.446	171	13.077
12	3.464	52	7.211	92	9.592	132	11.489	172	13.115
13	3.606	53	7.280	93	9.644	133	11.533	173	13.153
14	3.742	54	7.348	94	9.695	134	11.576	174	13.191
15	3.873	55	7.416	95	9.747	135	11.619	175	13.229
16	4.000	56	7.483	96	9.798	136	11.662	176	13.266
17	4.123	57	7.550	97	9.849	137	11.705	177	13.304
18	4.243	58	7.616	98	9.899	138	11.747	178	13.342
19	4.359	59	7.681	99	9.950	139	11.790	179	13.379
20	4.472	60	7.746	100	10.000	140	11.832	180	13.416
21	4.583	61	7.810	101	10.050	141	11.874	181	13.454
22	4.690	62	7.874	102	10.100	142	11.916	182	13.491
23	4.796	63	7.937	103	10.149	143	11.958	183	13.528
24	4.899	64	8.000	104	10.198	144	12.000	184	13.565
25	5.000	65	8.062	105	10.247	145	12.042	185	13.601
26	5.099	66	8.124	106	10.296	146	12.083	186	13.638
27	5.196	67	8.185	107	10.344	147	12.124	187	13.675
28	5.292	68	8.246	108	10.392	148	12.166	188	13.711
29	5.385	69	8.307	109	10.440	149	12.207	189	13.748
30	5.477	70	8.367	110	10.488	150	12.247	190	13.784
31	5.568	71	8.426	111	10.536	151	12.288	191	13.820
32	5.657	72	8.485	112	10.583	152	12.329	192	13.856
33	5.745	73	8.544	113	10.630	153	12.369	193	13.892
34	5.831	74	8.602	114	10.677	154	12.410	194	13.928
35	5.916	75	8.660	115	10.724	155	12.450	195	13.964
36	6.000	76	8.718	116	10.770	156	12.490	196	14.000
37	6.083	77	8.775	117	10.817	157	12.530	197	14.036
38	6.164	78	8.832	118	10.863	158	12.570	198	14.071
39	6.245	79	8.888	119	10.909	159	12.610	199	14.107
40	6.325	80	8.944	120	10.954	160	12.649	200	14.142

Unless the value of $\sqrt{x}$ ends in 000, all values are rounded to the nearest thousandth.

Appendix B
Metric Measurement and Conversion of Units

⬤**1** *Converting from One Metric Unit of Measurement to Another*

Metric measurements are becoming more common in the United States. The metric system is used in many parts of the world and in the sciences. The basic unit of length in the metric system is the meter. For smaller measurements, centimeters are commonly used. There are 100 centimeters in 1 meter. For larger measurements, kilometers are commonly used. There are 1000 meters in 1 kilometer. The following tables give the metric units of measurement for length.

After studying this section, you will be able to:

1 Convert from one metric unit of measurement to another.

2 Convert between metric and U.S. units of measure.

3 Convert from one type of unit of measure to another.

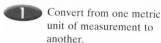

SSM   PH TUTOR CENTER   CD & VIDEO   MATH PRO   WEB

Metric Length

*1 kilometer (km)	=	1000 meters
1 hectometer (hm)	=	100 meters
1 dekameter (dam)	=	10 meters
*1 meter (m)	=	1 meter
1 decimeter (dm)	=	0.1 meter
*1 centimeter (cm)	=	0.01 meter
*1 millimeter (mm)	=	0.001 meter

The four most common units of metric measurement are indicated by an asterisk.

Metric Weight

1 kilogram (kg)	=	1000 grams
1 gram (g)	=	1 gram
1 milligram (mg)	=	0.001 gram

Metric Volume

1 kiloliter (kL)	=	1000 liters
1 liter (L)	=	1 liter
1 milliliter (mL)	=	0.001 liter

One way to convert from one metric unit to another is to multiply by 1. We know mathematically that to multiply by 1 will yield an equivalent expression since $(a)(1) = a$. In our calculations when solving word problems, we treat dimension symbols much as we treat variables in algebra.

EXAMPLE 1 Fred drove a distance of 5.4 kilometers. How many meters is that?

$$5.4 \text{ km} \cdot 1 = 5.4 \text{ km} \cdot \frac{1000 \text{ m}}{1 \text{ km}} = \frac{(5.4)(1000)}{1} \cdot \frac{\text{km}}{\text{km}} \cdot \text{m}$$
$$= 5400 \text{ m}$$

The distance is 5400 meters.

Practice Problem 1 June's room is 3.4 meters wide. How many centimeters is that?

EXAMPLE 2 A chemist measured 67 milliliters of a solution. How many liters is that?

$$67 \text{ mL} \cdot \frac{0.001 \text{ L}}{1 \text{ mL}} = 0.067 \text{ L}$$

The chemist measured 0.067 liters of the solution.

Practice Problem 2 The container has 125 liters of water. How many kiloliters is that?

2 Converting between Metric and U.S. Units of Measure

The most common relationships between the U.S. and metric systems are listed in the following table. Most of these values are approximate.

Metric Conversion Ratios

LENGTH:	1 inch = 2.54 centimeters
	39.37 inches = 1 meter
	1 mile = 1.61 kilometers
	0.62 mile = 1 kilometer
WEIGHT:	1 pound = 454 grams
	2.20 pounds = 1 kilogram
	1 ounce = 28.35 grams
	0.0353 ounce = 1 gram
LIQUID CAPACITY:	1 quart = 946 milliliters
	1.06 quarts = 1 liter
	1 gallon = 3.785 liters

EXAMPLE 3 A box weighs 190 grams. How many ounces is that? Round to the nearest hundredth.

$$190 \text{ g} \cdot \frac{0.0353 \text{ oz}}{1 \text{ g}} = 6.707 \text{ oz}$$
$$\approx 6.71 \text{ oz} \qquad \text{(rounded to the nearest hundredth)}$$

The box weighs approximately 6.71 ounces.

Practice Problem 3 A bag of groceries weighs 5.72 pounds. How many kilograms is that?

EXAMPLE 4 Juanita drives 23 kilometers to work each day. How many miles is the trip? Round to the nearest mile.

$$23 \text{ km} \cdot \frac{0.62 \text{ mi}}{1 \text{ km}} = 14.26 \text{ mi} \approx 14 \text{ mi} \quad \text{(rounded to the nearest mile)}$$

Juanita drives approximately 14 miles to work each day.

Practice Problem 4 Carlos installed an electrical connection that is 8.00 centimeters long. How many inches long is the connection? Round to the nearest hundredth.

EXAMPLE 5 Anita purchased 42.0 gallons of gasoline for her car last month. How many liters did she purchase? Round to the nearest liter.

$$42.0 \text{ gal} \cdot \frac{3.785 \text{ L}}{1 \text{ gal}} = 158.97 \text{ L} \approx 159 \text{ L} \quad \text{(rounded to the nearest liter)}$$

Anita purchased approximately 159 liters of gasoline last month.

Practice Problem 5 Warren purchased a 3-liter bottle of Coca-Cola. How many quarts of Coca-Cola is that? Round to the nearest hundredth of a quart.

3 Converting from One Type of Unit of Measure to Another

Sometimes you will need to convert from one type of unit of measure to another. For example, you may need to convert days to minutes or miles per hour to feet per second. Recall the U.S. units of measure.

Length	*Time*
12 inches = 1 foot	60 seconds = 1 minute
3 feet = 1 yard	60 minutes = 1 hour
5280 feet = 1 mile	24 hours = 1 day
1760 yards = 1 mile	7 days = 1 week

Weight	*Volume*
16 ounces = 1 pound	2 cups = 1 pint
2000 pounds = 1 ton	2 pints = 1 quart
	4 quarts = 1 gallon

EXAMPLE 6 A car was traveling at 50.0 miles per hour. How many feet per second was the car traveling? Round to the nearest tenth of a foot per second.

$$\frac{50 \text{ mi}}{\text{hr}} \cdot \frac{5280 \text{ ft}}{1 \text{ mi}} \cdot \frac{1 \text{ hr}}{60 \text{ min}} \cdot \frac{1 \text{ min}}{60 \text{ sec}} = \frac{(50)(5280) \text{ ft}}{(60)(60) \text{ sec}} = \frac{73.333 \ldots \text{ ft}}{\text{sec}}$$

The car was traveling at 73.3 feet per second, rounded to the nearest tenth of a foot per second.

Practice Problem 6 A speeding car was traveling at 70.0 miles per hour. How many feet per second was the car traveling? Round to the nearest tenth of a foot per second.

B Exercises

1. How many meters are in 34 km?

2. How many meters are in 128 km?

3. How many centimeters are in 57 m?

4. How many centimeters are in 46 m?

5. How many millimeters are in 25 cm?

6. How many millimeters are in 63 cm?

7. How many meters are in 563 mm?

8. How many meters are in 831 mm?

9. How many milligrams are in 29.4 g?

10. How many milligrams are in 75.2 g?

11. How many kilograms are in 98.4 g?

12. How many kilograms are in 62.7 g?

13. How many milliliters are in 7 L?

14. How many milliliters are in 12 L?

15. How many kiloliters are in 4 mL?

16. How many kiloliters are in 3 mL?

Use the table of metric conversion ratios to find each of the following. Round all answers to the nearest tenth.

17. How many inches are in 4.2 cm?

18. How many inches are in 3.8 cm?

19. How many kilometers are in 14 mi?

20. How many kilometers are in 13 mi?

21. How many meters are in 110 in.?

22. How many meters are in 150 in?

23. How many centimeters are in 7 in.?

24. How many centimeters are in 9 in.?

25. A box weighing 2.4 lb would weigh how many grams?

26. A box weighing 1.6 lb would weigh how many grams?

27. A man weighs 78 kg. How many pounds does he weigh?

28. A woman weighs 52 kg. How many pounds does she weigh?

29. Ferrante purchased 3 qt of milk. How many liters is that?

30. Wong Tin purchased 1 gal of milk. How many liters is that?

Answer the following questions. Round all answers to the nearest hundredth.

31. How many inches are in 3050 miles?

32. How many inches are in 4500 miles?

33. A truck traveled at a speed of 40 feet per second. How many miles per hour is that?

34. A car traveled at a speed of 55 feet per second. How many miles per hour is that?

35. How many years are in 3,500,000 seconds? (Use the approximate value that 365 days = 1 year.)

36. How many years are in 2,800,000 seconds? (Use the approximate value that 365 days = 1 year.)

Appendix C
Interpreting Data from Tables, Charts, and Graphs

1 Tables

A table is a device used to organize information into categories. Using it you can readily find details about each category.

After studying this section, you will be able to interpret data from:

1 Tables

2 Charts

3 Pictographs

4 Bar Graphs

5 Line Graphs

6 Pie Graphs and Circle Graphs

SSM PH TUTOR CENTER   CD & VIDEO   MATH PRO   WEB

EXAMPLE 1

Table of Nutritive Values of Certain Popular "Fast Foods"

Type of Sandwich	Calories	Protein (g)	Fat (g)	Cholesterol (g)	Sodium (mg)
Burger King Whopper	630	27	38	90	880
McDonald's Big Mac	500	25	26	100	890
Wendy's Bacon Cheeseburger	440	22	25	65	870
Burger King BK Broiler (chicken)	280	20	10	50	770
McDonald's McChicken	415	19	20	50	830
Wendy's Grilled Chicken	290	24	7	60	670

Source: U.S. Government Agencies and Food Manufacturers

(a) Which food item has the least amount of fat per serving?

(b) Which beef item has the least amount of calories per serving?

(c) How much more protein does a Burger King Whopper have than a McDonald's Big Mac?

(a) The least amount of fat, 7 grams, is in the Wendy's Grilled Chicken Sandwich.

(b) The least amount of calories, 440, for a beef sandwich is the Wendy's Bacon Cheeseburger.

(c) The Burger King Whopper has 2 grams of protein more than the McDonald's Big Mac.

Practice Problem 1

(a) Which sandwich has the lowest level of sodium?

(b) Which sandwich has the highest level of cholesterol?

2 Charts

A chart is a device used to organize information in which not every category is the same. Example 2 illustrates a chart containing different types of data.

EXAMPLE 2 The following chart shows how people in Topsfield indicated they spent their free time.

Survey of Use of Leisure Time

Category	Activity	Hours spent per week
Single men	Gym	6
	Outdoor sports	4
	Dating	7
	Watching pro sports	12
	Reading & TV	3
Single women	Gym	4
	Outdoor sports	2
	Dating	7
	Time with friends	10
	Reading & TV	9
Couples	Time with family	21
	Time as a couple	8
	Time with friends	4
	Reading & TV	9
Children	Watching TV	28
	Playing outside	8
	Reading	1

Use the chart to answer the following questions about people in Topsfield.

(a) What is the average amount of time a couple spends together as a family during the week?

(b) How much more time do children spend watching TV than playing outside?

(c) What activity do single men spend most of their time doing?

(a) The average amount of time a couple spends together as a family is 21 hours per week.

(b) Children spend 20 more hours per week watching TV than playing outside.

(c) Single men spend more time per week watching pro sports (12 hr) than any other activity.

Practice Problem 2

(a) What two categories do single women spend the most time doing?

(b) What do couples spend most of their time doing?

(c) What is the most significant numerical difference in terms of number of hours per week spent by single women versus single men?

3 *Pictographs*

A pictograph uses a visually appropriate symbol to represent an amount of items. A pictograph is used in Example 3.

EXAMPLE 3 Consider the following pictograph.

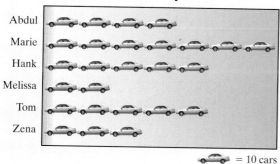

Annual Car Sales in 2000 for Sales Personnel at Oakwood Chrysler

= 10 cars

(a) How many cars did Melissa sell in 2000?

(b) Who sold the greatest number of cars?

(c) How many more cars did Tom sell than Zena?

(a) Melissa sold $2 \times 10 = 20$ cars.

(b) Marie sold the greatest number of cars.

(c) Tom sold $5 \times 10 = 50$ cars. Zena sold $3 \times 10 = 30$ cars. Now $50 - 30 = 20$. Therefore, Tom sold 20 cars more than Zena.

Practice Problem 3

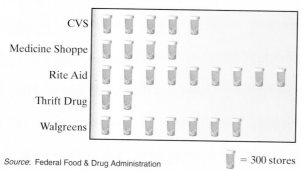

Approximate number of chain pharmacy stores in the United States in 2000

Source: Federal Food & Drug Administration   = 300 stores

(a) Approximately how many stores does Walgreens have?

(b) Approximately how many more stores does Rite Aid have than CVS?

(c) What is the combined number of Thrift Drug and Medicine Shoppe stores?

4 Bar Graphs

A bar graph is helpful for making comparisons and noting changes or trends. A scale is provided so that the height of the bar graph indicates a specific number. A bar graph is displayed in Example 4. A bar graph may be represented horizontally or vertically. In either case the basic concepts of interpreting a bar graph are the same.

EXAMPLE 4   The approximate population of California by year is given in the following bar graph.

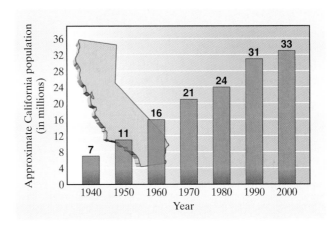

(a) What was the approximate population of California in 2000?

(b) How much greater was the population of California in 1980 than in 1970?

(a) The approximate population of California in 2000 was about 33 million.

(b) In 1980 it was 24 million. In 1970 it was 21 million. The population was approximately 3 million people more in California in 1980 than in 1970.

Practice Problem 4   The following bar graph depicts the number of fatal accidents for U.S. air carriers for scheduled flight service for aircraft with 30 seats or more.

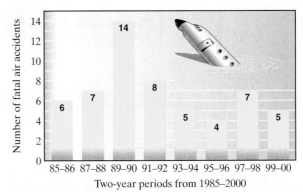

Source: National Transportation Safety Board

(a) What two-year period had the greatest number of fatal accidents?

(b) What was the increase in the number of fatal accidents from the 1987–1988 period to the 1989–1990 period?

(c) What was the decrease in the number of fatal accidents from the 1991–1992 period to the 1993–1994 period?

5 Line Graphs

A line graph is often used to display data when significant changes or trends are present. In a line graph, only a few points are actually plotted from measured values. The points are then connected by straight lines in order to show a trend. The intervening values between points may not exactly lie on the line. A line graph is displayed in Example 5.

EXAMPLE 5 The following line graph shows the number of customers per month coming to a restaurant in a tourist vacation community.

(a) What month had the greatest number of customers?

(b) How many customers came to the restaurant in April?

(c) How many more customers came in July than in August?

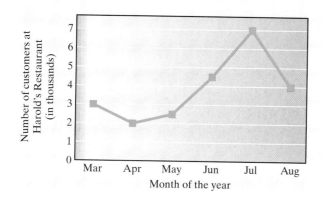

(a) More customers came during the month of July.

(b) Approximately 2000 customers came in April.

(c) In July there were 7000 customers, while in August there were 4000 customers. Thus, there were 3000 more customers in July than in August.

Practice Problem 5 The quality of the air is measured by the Pollutant Standards Index (PSI). To meet the national air quality standards set by the U.S. government, the air in a city cannot have a PSI greater than 100. The following line graph indicates the number of days that the PSI was greater than 100 in the city of Baltimore during an 11-year period.

(a) What was the number of days the PSI exceeded 100 in Baltimore in 1993?

(b) In what year did Baltimore have the fewest days in which the PSI exceeded 100?

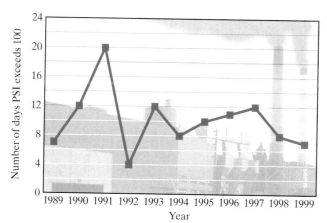

Source: US Environmental Protective Agency

6 *Pie Graphs and Circle Graphs*

A pie graph or a circle graph indicates how a whole quantity is divided into parts. These graphs help you to visualize the size of the relative proportions of parts. Each piece of the pie or circle is called a sector. Example 6 uses a pie graph.

EXAMPLE 6 Together, the Great Lakes from the largest body of fresh water in the world. The total area of these five lakes is about 290,000 square miles, almost all of which is suitable for boating. The percentage of this total area taken up by each of the Great Lakes is shown in the pie graph.

(a) What percentage of the area is taken up by Lake Michigan?

(b) What lake takes up the largest percentage of area?

(c) How many square miles are taken up by Lake Huron and Lake Michigan together?

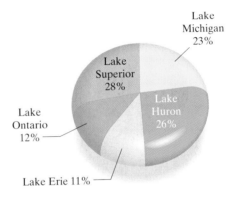

(a) Lake Michigan takes up 23% of the area.

(b) Lake Superior takes up the largest percentage.

(c) If we add 26 + 23, we get 49. Thus Lake Huron and Lake Michigan together take up 49% of the total area. 49% of 290,000 = (0.49)(290,000) = 142,100 square miles.

Practice Problem 6 Seattle receives on average about 37 inches of rain per year. However, the amount of rainfall per month varies significantly. The percent of rainfall that occurs during each quarter of the year is shown by the circle graph.

(a) What percent of the rain in Seattle falls between April and June?

(b) Forty percent of the rainfall occurs in what three-month period?

(c) How many inches of rain fall in Seattle from January to March?

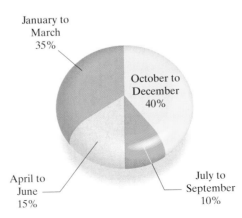

C Exercises

In each case study carefully the appropriate visual display, then answer the questions.

Consider the following table in answering exercises 1–10.

Table of Facts of the Rocky Mountain States

State	Area in Square Miles	Date Admitted To the Union	Estimated 1990 Population	Number of Representatives in U.S. Congress	Popular Name
Colorado	104,247	1876	3,500,000	5	Centennial State
Idaho	83,557	1890	1,100,000	2	Gem State
Montana	147,138	1889	800,000	2	Treasure State
Nevada	110,540	1864	1,100,000	1	Silver State
Utah	84,916	1896	1,700,000	2	Beehive State
Wyoming	97,914	1890	500,000	1	Equality State

1. What is the area of Utah in square miles?

2. What is the area of Montana in square miles?

3. What is the 1990 estimated population of Colorado?

4. What is the 1990 estimated population of Nevada?

5. How many representatives in the U.S. Congress come from Idaho?

6. How many representatives in the U.S. Congress come from Wyoming?

7. What is the popular name for Montana?

8. What is the popular name for Utah?

9. Which of these six states was the first one to be admitted to the Union?

10. In what year did two of these six states both get admitted to the Union?

**Number of new homes built in 2000
in each of five counties**

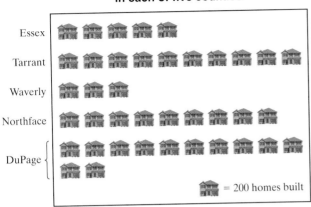

Use this pictograph to answer exercises 11–16.

11. How many homes were built in Tarrant County in the year 2000?

12. How many homes were built in Essex County in the year 2000?

13. In what county were the most homes built?

14. How many more homes were built in Tarrant County than Waverly County?

15. How many homes were built in Essex County and Northface County combined?

16. How many homes were built in DuPage County and Waverly County combined?

Approximate number of apartments in U.S. in 2000

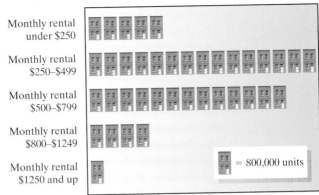

Source: U.S. Bureau of the Census

Use this pictograph to answer exercises 17–20.

17. How many apartment units are rented for under $250 per month?

18. How many apartment units are rented for $800–$1249 per month?

19. How many more apartment units are available in the $500–$799 range than in the $800–$1249 range?

20. How many more apartment units are available in the $800–$1249 range than in the $1250 and up range?

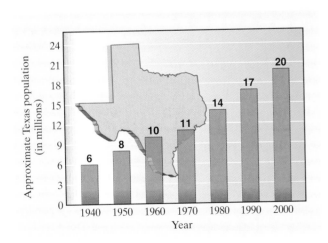

Use this bar graph to answer exercises 21–26.

**Number of people engaged in various activities at
least once in the last twelve months**

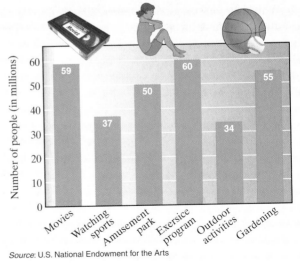

Source: U.S. National Endowment for the Arts

Use this bar graph to answer Questions 27–30.

21. What was the population in Texas in 1950?

22. What was the population in Texas in 1990?

23. Between what two years was the increase in population the greatest?

24. Between what two years was the increase in population the smallest?

25. How many more people lived in Texas in 1970 than in 1950?

26. How many more people lived in Texas in 1980 than in 1960?

27. According to the bar graph, how many people watched a sports event at least once in the last 12 months?

28. According to the bar graph, how many people were involved in gardening at least once in the last 12 months?

29. What two activities were the most common?

30. What two activities were the least common?

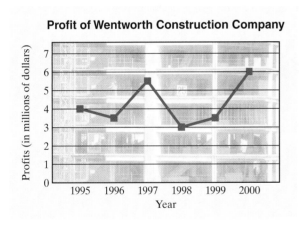

Use this line graph to answer exercises 31–36.

31. What was the profit in 1997?

32. What was the profit in 1996?

33. How much greater was the profit in 2000 than 1999?

34. In what year did the smallest profit occur?

35. Between what two years did the profit decrease the most?

36. Between what two years did the profit increase the most?

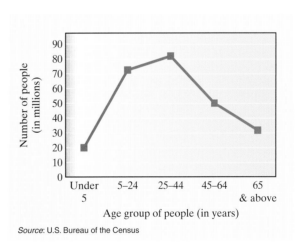

Source: U.S. Bureau of the Census

Use this line graph to answer exercises 37–40.

37. How many people in the U.S. are in the age group of 5 to 24 years?

38. How many people in the U.S. are in the age group of 25 to 44 years?

39. Thirty-three million people are in what age group?

40. Fifty-one million people are in what age group?

Religious faith distribution in the world

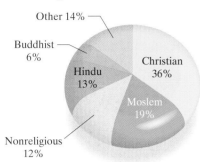

Source: United Nations Statistical Division

Distribution of spending by "average" two-income American family in 2000

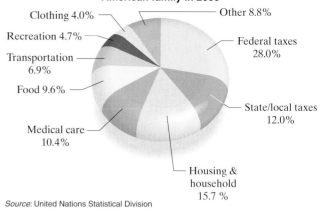

Source: United Nations Statistical Division

Use this circle graph to answer exercises 41–46.

41. What percent of the world's population is either Hindu or Buddhist?

42. What percent of the world's population is either Moslem or nonreligious?

43. What percent of the world's population is *not* Moslem?

44. What percent of the world's population is *not* Christian?

45. If there are approximately 6.5 billion people in the world in 2005, how many of them would we expect to be Hindu?

46. If there are approximately 6.9 billion people in the world in 2010, how many of them would we expect to be Moslem?

Use this circle graph to answer exercises 47–50.

47. What percent of the family income is spent for federal, state, and local taxes?

48. What percent of the family income is spent for food, medical care, and housing?

49. If the average two-income family earns $52,000 per year, how much is spent on transportation?

50. If the average two-income family earns $52,000 per year, how much is spent on recreation?

Appendix D
Inductive and Deductive Reasoning

 Using Inductive Reasoning to Reach a Conclusion

When we reach a conclusion based on specific observations, we are using **inductive reasoning**. Much of our early learning is based on simple cases of inductive reasoning. If a child touches a hot stove or other appliance several times and each time he gets burned, he is likely to conclude, "If I touch something that is hot, I will get burned." This is inductive reasoning. The child has thought about several actions and their outcomes and has made a conclusion or generalization.

The next few examples show how inductive reasoning can be used in mathematics.

After studying this section, you will be able to:

 Use inductive reasoning to reach a conclusion.

 Use deductive reasoning to reach a conclusion.

SSM   PH TUTOR   CD & VIDEO   MATH PRO   WEB
  CENTER

EXAMPLE 1 Find the next number in the sequence 10, 13, 16, 19, 22, 25, 28, …. We observe a pattern that each number is 3 more than the preceding number: $10 + 3 = 13$; $13 + 3 = 16$, and so on. Therefore, if we add 3 to 28, we conclude that the next number in the sequence is 31.

Practice Problem 1 Find the next number in the sequence 24, 31, 38, 45, 52, 59, 66, ….

EXAMPLE 2 Find the next number in the sequence 1, 8, 27, 64, 125, …. The sequence can be written as $1^3, 2^3, 3^3, 4^3, 5^3, …$. Each successive integer is cubed. The next number would be 6^3 or 216.

Practice Problem 2 Find the next number in the sequence.

$$3, 8, 15, 24, 35, 48, 63, 80, ….$$

EXAMPLE 3 Guess the next seven digits in the following irrational number:

$$5.636336333633336…$$

Between 6's there are the digits 3, 33, 333, 3333, and so on. The pattern is that the number of 3's keeps increasing by 1 each time. Thus the next seven digits are 3333363.

Practice Problem 3 Guess the next seven digits in the following irrational number:

$$6.1213314441…$$

EXAMPLE 4 Find the next two figures that would appear in the sequence.

We notice an alternating pattern: square, square, circle, circle, square …. We would next expect a square followed by a circle.

We notice a shading pattern of horizontal, vertical, horizontal, vertical, horizontal …. We would next expect vertical, then horizontal. Thus, the next two figures are

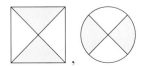

A-17

Practice Problem 4   Find the next two figures that would appear in the sequence.

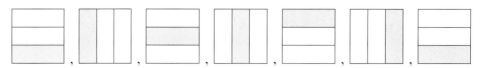

How accurate is inductive reasoning? Do we always come to the right conclusion? Conclusions arrived at by inductive reasoning are always tentative. They may require further investigation. When we use inductive reasoning we are using specific data to reach a general conclusion. However, we may have reached the wrong conclusion. To illustrate:

Take the sequence of numbers 20, 10, 5. What is the next number? You might say 2.5. In the sequence 20, 10, 5, each number appears to be one-half of the preceding number. Thus we would predict 20, 10, 5, 2.5,.... .

But wait! There is another possibility. Maybe the sequence is 20, 10, 5, −5, −10, −20, −25, −35, −40,.... .

To know for sure which answer is correct, we would need more information, such as more numbers in the sequence to verify the pattern. **You should always treat inductive reasoning conclusions as tentative, requiring further verification.**

2   *Using Deductive Reasoning to Reach a Conclusion*

Deductive reasoning requires us to take general facts, postulates, or accepted truths and use them to reach a specific conclusion. Suppose we know the following rules or "facts" of algebra:

For all numbers a, b, and c, and for all numbers $d \neq 0$:

1. *Addition principle of equations:* If $a = b$, then $a + c = b + c$.

2. *Division principle of equations:* If $a = b$, then $\dfrac{a}{d} = \dfrac{b}{d}$.

3. *Multiplication principle of equations:* If $a = b$, then $ac = bc$.

4. *Distributive property:* $a(b + c) = ab + ac$.

EXAMPLE 5   Use deductive reasoning and the four properties listed in the preceding box to justify each step in solving the equation.

$$2(7x - 2) = 38$$

STATEMENT	*REASON*
1. $14x - 4 = 38$	**1.** Distributive property: $a(b + c) = ab + ac$ Here we distributed the 2.
2. $14x = 42$	**2.** Addition principle of equations: If $a = b$ then $a + c = b + c$ Here we added 4 to each side of the equation.
3. $\dfrac{14x}{14} = \dfrac{42}{14}$	**3.** Division principle of equations: If $a = b$ then $\dfrac{a}{d} = \dfrac{b}{d}$
$x = 3$	Here we divided each side by 14.

Practice Problem 5 Use deductive reasoning and the four properties listed in the preceding box to justify each step in solving the equation.

$$\frac{1}{6}x = \frac{1}{3}x + 4$$

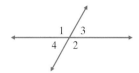

Sometimes we need to make conclusions about angles and lines in geometry. The following properties are useful. We will refer to angles 1, 2, 3, 4.

1. If two lines intersect, the opposite angles are equal.
    Here $\angle 1 = \angle 2$ and $\angle 4 = \angle 3$.

2. If two lines intersect, the adjacent angles are supplementary (they add up to 180°). Here, $\angle 4 + \angle 1 = 180°$, and $\angle 1 + \angle 3 = 180°$, also $\angle 3 + \angle 2 = 180°$, and $\angle 4 + \angle 2 = 180°$.

3. If a transversal (intersecting line) crosses two parallel lines, the alternate interior angles are equal. In the figure at right, line P is a transversal. If line M is parallel to line N, then $\angle 3 = \angle 4$ and $\angle 1 = \angle 2$.

4. If two alternate interior angles on each side of a transversal cutting two straight lines are equal, the two straight lines are parallel. If $\angle 3 = \angle 4$, then line M is parallel to line N.

5. Supplements of equal angles are equal. In the figure at right, if $\angle 1$ and $\angle 2$ are supplementary, and $\angle 3$ and $\angle 4$ are supplementary, then if $\angle 1 = \angle 4$, then $\angle 2 = \angle 3$.

6. Transitive property of equality:

    If $a = b$ and $b = c$, then $a = c$.

We will now use these facts to prove some geometric conclusions.

EXAMPLE 6 Prove that line M is parallel to line N if $\angle 1 = \angle 3$.

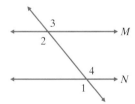

STATEMENT	*REASON*
1. $\angle 3 = \angle 2$ and $\angle 4 = \angle 1$	1. If two lines intersect, the opposite angles are equal.
2. $\angle 2 = \angle 4$	2. Transitive property of equality.
3. Therefore, line M is parallel to line N. (We write this as $M \parallel N$.)	3. If two alternate interior angles on each side of a transversal cutting two straight lines are equal, the straight lines are parallel.

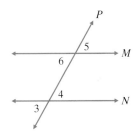

Practice Problem 6   Refer to the diagram in the margin. Line *M* is parallel to line *N*. ∠5 = 26°. Prove∠4 = 26°.

Now let us see if we can use deductive reasoning to solve the following problem.

EXAMPLE 7   For next semester, four professors from the psychology department have expressed the following desires for freshman courses. Each professor will teach only *one* freshman course next semester. The four freshman courses are General Psychology, Social Psychology, Psychology of Adjustment, and Educational Psychology.

1. Professors *A* and *B* don't want to teach General Psychology.
2. Professor *C* wants to teach Social Psychology.
3. Professor *D* will be happy to teach any course.
4. Professor *B* wants to teach Psychology of Adjustment.

Which professor will teach Educational Psychology, if all professors are given the courses they desire?

Let us organize the facts by listing the four freshman courses and each professor: *A*, *B*, *C*, and *D*.

General Psychology	Social Psychology	Psychology of Adjustment	Educational Psychology
A	*A*	*A*	*A*
B	*B*	*B*	*B*
C	*C*	*C*	*C*
D	*D*	*D*	*D*

STEP	*REASON*
1. We cross off Professors *A* and *B* from the General Psychology list.	**1.** Professors *A* and *B* don't want to teach General Psychology.

General Psychology			
~~*A*~~	*A*	*A*	*A*
~~*B*~~	*B*	*B*	*B*
C	*C*	*C*	*C*
D	*D*	*D*	*D*

2. We cross Professor *C* off every list (except Social Psychology) and mark that he will teach it.	**2.** Professor *C* wants to teach Social Psychology.

	Social Psychology		
~~*A*~~			
~~*B*~~			
~~*C*~~	$\boxed{C}$	~~*C*~~	~~*C*~~
D			

STEP	*REASON*
3. Professor *D* is thus the only person who can teach General Psychology. We cross him off every other list and mark that he will teach General Psychology.	**3.** Professor *D* is happy to teach any course.

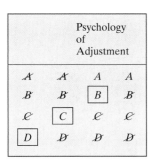

4. We cross out all courses for Professor *B* except Psychology of Adjustment.	**4.** Professor *B* wants to teach Psychology of Adjustment.

5. Professor *A* will teach Educational Psychology.	**5.** He is the only professor left. All others are assigned.

Educational Psychology

A

Practice Problem 7 A Honda, Toyota, Mustang, and Camaro are parked side by side, but not in that order.

1. The Camaro is parked on the right end.
2. The Mustang is between the Honda and the Toyota.
3. The Honda is not next to the Camaro.

Which car is parked on the left end?

D Exercises

Find the next number in the sequence.

1. $2, 4, 6, 8, 10, 12, \ldots$

2. $0, 5, 10, 15, 20, 25, \ldots$

3. $7, 16, 25, 34, 43, \ldots$

4. $12, 25, 38, 51, 64, \ldots$

5. $1, 16, 81, 256, 625, \ldots$

6. $1, 6, 13, 22, 33, 46, \ldots$

7. $-7, 3, -6, 4, -5, 5, -4, 6, \ldots$

8. $2, -4, 8, -16, 32, -64, 128, \ldots$

9. $5x, 6x - 1, 7x - 2, 8x - 3, 9x - 4, \ldots$

10. $60x, 30x, 15x, 7.5x, 3.75x, 1.875x, \ldots$

In exercises 11–16, guess the next seven digits in each irrational number.

11. $8.181181118\ldots$

12. $3.043004300043\ldots$

13. $12.98987987698765\ldots$

14. $7.6574839201102938\ldots$

15. $2.14916253649\ldots$

16. $6.112223333\ldots$

17. Find the next row in this triangular pattern.

```
              1
          1       1
        1     2     1
      1    3     3    1
    1    4    6    4    1
  1    5   10   10    5    1
```

18. Find the next two figures that would appear in the sequence.

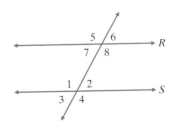

Use deductive reasoning and the four properties discussed in Example 5 to justify each step in solving each equation

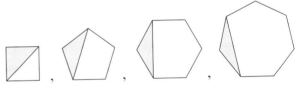

19. $12x - 30 = 6$

20. $3x + 7 = 4$

21. $4x - 3 = 3x - 5$

22. $2x - 9 = 4x + 5$

23. $8x - 3(x - 5) = 30$

24. $3x - 5(x - 1) = -5$

25. $\dfrac{1}{2}x + 6 = \dfrac{3}{2} + 6x$

26. $\dfrac{4}{5}x - 3 = \dfrac{1}{10}x + \dfrac{3}{5}$

In exercises 27–28, use deductive reasoning and the properties of geometry discussed in Example 6 to prove each statement.

27. If $\angle 1 = \angle 5$, prove that line P is parallel to line S.

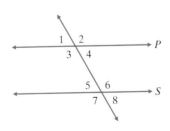

28. If line R is parallel to line S, prove that $\angle 6 = \angle 3$.

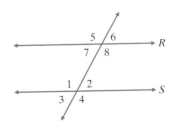

A-22

29. William, Brent, Charlie, and Dave competed in the Olympics. These four divers placed first, second, third, and fourth in the competition.
  1. James ranked between Brent and Dave.
  2. William did better than Brent.
  3. Brent did better than Dave.
  Who finished in each of the four places?

30. The four floors of Hotel Royale are color-coded.
  1. The blue floor is directly below the green floor.
  2. The red floor is next to the yellow floor.
  3. The green floor is above the red floor.
  4. The blue floor is above the yellow floor.
  5. There is no floor below the red floor.
  What is the order of the colors from top to bottom of Hotel Royale?

31. There are four people named Peter, Michael, Linda and Judy. Each of them has one occupation. They are a teacher, a butcher, a baker, and a candlestick maker.
  1. Linda is the baker.
  2. The teacher plans to marry the baker in October.
  3. Judy and Peter were married last year.
  Who is the teacher?

32. A president of a company, a lawyer, a salesman, and a doctor are seated at the head table at a banquet.
  1. The lawyer prosecuted the doctor in a malpractice lawsuit and they should not sit together.
  2. The salesman should be at one end of the table.
  3. The lawyer should be at the far right.
  From left to right how should they be seated?

33. John is considering purchasing a new car when he starts his new full-time job. The car with the best gas mileage is a Honda Civic, the car with the most conveniently located dealer is a Toyota Corolla, the car with the lowest price is the Ford Escort, and the car that has best handling is the Dodge Neon. His next-door neighbor has a Honda and he does not want to copy his neighbor. If he does not buy a Toyota then he will definitely not go to graduate school. He told his best friend that he probably will not purchase a car based on its handling. He has definitely decided to go to graduate school. What car is he most likely to purchase?

34. Detective Smith recently found a stolen Corvette abandoned near Skull Rocks. A small empty boat was anchored 1 mile offshore with no person in the boat. Divers later found the body of a 35-year-old man $\frac{1}{2}$ mile further out to sea than the boat. If the man entered the water after 10:00 A.M. then the current of the outgoing tide was strong enough to take a dead body a distance of $\frac{1}{2}$ mile further out to sea. The coroner determined when the body was found at 4:00 P.M. that the man had been dead for seven hours. The detective determined that he died instantly when his head hit the rocks directly beneath the anchored boat. Could the man have acted alone based on the facts revealed so far?

35. Dr. Parad is deciding whether he should do a root canal in order to save Fred's tooth or just remove the tooth and put in a false tooth. If Dr. Parad works alone on the tooth the procedure will take 2 hours and 15 minutes. Fred wants to save the tooth but he does not want the dental bill for Tuesday's work to be more than $200. Dr. Parad charges $100 per hour for dental procedures. If the bill is under $150 it is because Dr. Parad's efficient dental assistant is helping with the oral surgery. The dental procedure is scheduled for Tuesday, when Dr. Parad does not have any dental assistants available. Will Fred's tooth be saved or will he get a false tooth?

36. If Marcia passes the Bilingual Teacher's exam, she will stay in the Chicago area. If she devotes at least 50 hours of study time to practice her Spanish, she is confident that she can pass the exam. Either Marcia will stay in the Chicago area or she will move back to Massachusetts. If she did not pass the Bilingual Teacher's exam it is because she needed 12 hours of study time per month. Marcia started studying for the exam in April. She will be moving to Massachusetts. When was the Bilingual Teacher's exam given, if we know that Marcia took the exam?

Solutions to Practice Problems

Chapter 0

0.1 Practice Problems

1. (a) $\dfrac{10}{16} = \dfrac{2 \times 5}{2 \times 8} = \dfrac{5}{8}$ **(b)** $\dfrac{24}{36} = \dfrac{12 \times 2}{12 \times 3} = \dfrac{2}{3}$

(c) $\dfrac{42}{36} = \dfrac{6 \times 7}{6 \times 6} = \dfrac{7}{6}$

2. (a) $\dfrac{4}{12} = \dfrac{2 \times 2 \times 1}{2 \times 2 \times 3} = \dfrac{1}{3}$ **(b)** $\dfrac{25}{125} = \dfrac{5 \times 5 \times 1}{5 \times 5 \times 5} = \dfrac{1}{5}$

(c) $\dfrac{73}{146} = \dfrac{73 \times 1}{73 \times 2} = \dfrac{1}{2}$

3. (a) $\dfrac{18}{6} = \dfrac{3 \times 6}{6} = 3$ **(b)** $\dfrac{146}{73} = \dfrac{73 \times 2}{73} = 2$

(c) $\dfrac{28}{7} = \dfrac{7 \times 4}{7} = 4$

4. 56 out of 154 $= \dfrac{56}{154} = \dfrac{2 \times 7 \times 4}{2 \times 7 \times 11} = \dfrac{4}{11}$

5. (a) $\dfrac{12}{7} = 12 \div 7 = 7\overline{)12} = 1\dfrac{5}{7}$

$$\begin{array}{r} 1 \\ 7\overline{)12} \\ \underline{7} \\ 5 \quad \text{Remainder} \end{array}$$

(b) $\dfrac{20}{5} = 20 \div 5 = 5\overline{)20} = 4$

$$\begin{array}{r} 4 \\ 5\overline{)20} \\ \underline{20} \\ 0 \quad \text{Remainder} \end{array}$$

6. (a) $3\dfrac{2}{5} = \dfrac{(3 \times 5) + 2}{5} = \dfrac{15 + 2}{5} = \dfrac{17}{5}$

(b) $1\dfrac{3}{7} = \dfrac{(1 \times 7) + 3}{7} = \dfrac{7 + 3}{7} = \dfrac{10}{7}$

(c) $2\dfrac{6}{11} = \dfrac{(2 \times 11) + 6}{11} = \dfrac{22 + 6}{11} = \dfrac{28}{11}$

(d) $4\dfrac{2}{3} = \dfrac{(4 \times 3) + 2}{3} = \dfrac{12 + 2}{3} = \dfrac{14}{3}$

7. (a) $\dfrac{3}{8} = \dfrac{?}{24}$ Observe $8 \times 3 = 24$ **(b)** $\dfrac{5}{6} = \dfrac{?}{30}$

$\dfrac{3 \times}{8 \times} \dfrac{3}{3} = \dfrac{9}{24}$ $\dfrac{5 \times}{6 \times} \dfrac{5}{5} = \dfrac{25}{30}$

(c) $\dfrac{12}{13} = \dfrac{?}{26}$ **(d)** $\dfrac{2}{7} = \dfrac{?}{56}$

$\dfrac{12 \times}{13 \times} \dfrac{2}{2} = \dfrac{24}{26}$ $\dfrac{2 \times}{7 \times} \dfrac{8}{8} = \dfrac{16}{56}$

(e) $\dfrac{5}{9} = \dfrac{?}{27}$ **(f)** $\dfrac{3}{10} = \dfrac{?}{60}$

$\dfrac{5 \times}{9 \times} \dfrac{3}{3} = \dfrac{15}{27}$ $\dfrac{3 \times}{10 \times} \dfrac{6}{6} = \dfrac{18}{60}$

(g) $\dfrac{3}{4} = \dfrac{?}{28}$ **(h)** $\dfrac{8}{11} = \dfrac{?}{55}$

$\dfrac{3 \times}{4 \times} \dfrac{7}{7} = \dfrac{21}{28}$ $\dfrac{8 \times}{11 \times} \dfrac{5}{5} = \dfrac{40}{55}$

0.2 Practice Problems

1. (a) $\dfrac{3}{6} + \dfrac{2}{6} = \dfrac{3 + 2}{6} = \dfrac{5}{6}$

(b) $\dfrac{3}{11} + \dfrac{2}{11} + \dfrac{6}{11} = \dfrac{3 + 2 + 6}{11} = \dfrac{11}{11} = 1$

(c) $\dfrac{5}{8} + \dfrac{2}{8} + \dfrac{7}{8} = \dfrac{5 + 2 + 7}{8} = \dfrac{14}{8} = 1\dfrac{3}{4}$

---

2. (a) $\dfrac{11}{13} - \dfrac{6}{13} = \dfrac{11 - 6}{13} = \dfrac{5}{13}$

(b) $\dfrac{8}{9} - \dfrac{2}{9} = \dfrac{8 - 2}{9} = \dfrac{6}{9} = \dfrac{2}{3}$

3. Find LCD of $\dfrac{1}{8}$ and $\dfrac{5}{12}$.

$8 = 2 \cdot 2 \cdot 2$
$12 = \downarrow \quad 2 \cdot 2 \cdot 3$
$\quad\quad 2 \cdot 2 \cdot 2 \cdot 3 = 24$
LCD $= 24$

4. Find the LCD using prime factors.

$\dfrac{8}{35}$ and $\dfrac{6}{15}$

$35 = 7 \cdot 5$
$15 = \quad 5 \cdot 3$
$\quad\quad\quad 7 \cdot 5 \cdot 3 \quad$ LCD $= 105$

5. Find LCD of $\dfrac{5}{12}$ and $\dfrac{7}{30}$.

$12 = 3 \cdot 2 \cdot 2$
$30 = 3 \quad 2 \cdot 5$
$\quad\quad 3 \cdot 2 \cdot 2 \cdot 5 \quad$ LCD $= 60$

6. Find LCD of $\dfrac{1}{18}, \dfrac{2}{27}$ and $\dfrac{5}{12}$.

$12 = 2 \cdot 2 \cdot 3$
$18 = 2 \quad 3 \cdot 3$
$27 = \quad\quad 3 \cdot 3 \cdot 3$
$\quad\quad 2 \cdot 2 \cdot 3 \cdot 3 \cdot 3 \quad$ LCD $= 108$

7. Add $\dfrac{1}{8} + \dfrac{5}{12}$

First find the LCD.
$8 = 2 \cdot 2 \cdot 2$
$12 = 2 \cdot 2 \quad \cdot 3$
$\quad\quad 2 \cdot 2 \cdot 2 \cdot 3 \quad$ LCD $= 24$

Then change to equivalent fractions and add.
$\dfrac{1}{8} \times \dfrac{3}{3} + \dfrac{5}{12} \times \dfrac{2}{2} = \dfrac{3}{24} + \dfrac{10}{24} = \dfrac{3 + 10}{24} = \dfrac{13}{24}$

8. $\dfrac{3}{5} + \dfrac{4}{25} + \dfrac{1}{10}$

First find the LCD.
$5 = \quad\quad 5$
$10 = \quad 2 \cdot 5$
$25 = \quad\quad 5 \cdot 5$
$\quad\quad 2 \cdot 5 \cdot 5 \quad$ LCD $= 50$

Then change to equivalent fractions and add.
$\dfrac{3}{5} \times \dfrac{10}{10} + \dfrac{4}{25} \times \dfrac{2}{2} + \dfrac{1}{10} \times \dfrac{5}{5} = \dfrac{30}{50} + \dfrac{8}{50} + \dfrac{5}{50}$

$\quad\quad\quad\quad\quad\quad = \dfrac{30 + 8 + 5}{50} = \dfrac{43}{50}$

9. Add $\dfrac{1}{49} + \dfrac{3}{14}$

First find the LCD.
$14 = 2 \cdot 7$
$49 = \quad 7 \cdot 7$
$\quad\quad 2 \cdot 7 \cdot 7 \quad$ LCD $= 98$

Then change to equivalent fractions and add.

$$\frac{1}{49} \times \frac{2}{2} + \frac{3}{14} \times \frac{7}{7} = \frac{2}{98} + \frac{21}{98} = \frac{2 + 21}{98} = \frac{23}{98}$$

10. $\frac{1}{12} - \frac{1}{30}$

First find the LCD.

$12 = 2 \cdot 2 \cdot 3$
$30 = 2 \cdot 3 \cdot 5$
$ \downarrow \downarrow \downarrow \downarrow$
$ 2 \cdot 2 \cdot 3 \cdot 5 \qquad \text{LCD} = 60$

Then change to equivalent fractions and subtract.

$$\frac{1}{12} \times \frac{5}{5} - \frac{1}{30} \times \frac{2}{2} = \frac{5}{60} - \frac{2}{60} = \frac{5 - 2}{60} = \frac{3}{60} = \frac{1}{20}$$

11. $\frac{2}{3} + \frac{3}{4} - \frac{3}{8}$

First find the LCD.

$3 = 3$
$4 = 2 \cdot 2 \Big|$
$8 = 2 \cdot 2 \cdot 2 \Big\downarrow$
$ 2 \cdot 2 \cdot 2 \cdot 3 \qquad \text{LCD} = 24$

Then change to equivalent fractions and add and subtract.

$$\frac{2}{3} \times \frac{8}{8} + \frac{3}{4} \times \frac{6}{6} - \frac{3}{8} \times \frac{3}{3}$$

$$= \frac{16}{24} + \frac{18}{24} - \frac{9}{24} = \frac{16 + 18 - 9}{24} = \frac{25}{24} = 1\frac{1}{24}$$

12. (a) $1\frac{2}{3} + 2\frac{4}{5} = \frac{5}{3} + \frac{14}{5} = \frac{5}{3} \times \frac{5}{5} + \frac{14}{5} \times \frac{3}{3} = \frac{25}{15} + \frac{42}{15}$

$$= \frac{25 + 42}{15} = \frac{67}{15} = 4\frac{7}{15}$$

(b) $5\frac{1}{4} - 2\frac{2}{3} = \frac{21}{4} - \frac{8}{3} = \frac{21}{4} \times \frac{3}{3} - \frac{8}{3} \times \frac{4}{4} = \frac{63}{12} - \frac{32}{12}$

$$= \frac{63 - 32}{12} = \frac{31}{12} = 2\frac{7}{12}$$

13.

$$\begin{array}{c} 6\frac{1}{2} \\ \fbox{} \\ 6\frac{1}{2} \end{array} \qquad 4\frac{1}{5} \text{ (left)} \quad 4\frac{1}{5} \text{ (right)}$$

$$4\frac{1}{5} + 4\frac{1}{5} + 6\frac{1}{2} + 6\frac{1}{2}$$

$$= \frac{21}{5} + \frac{21}{5} + \frac{13}{2} + \frac{13}{2}$$

$$\text{LCD} = 10$$

$$\frac{21}{5} \times \frac{2}{2} + \frac{21}{5} \times \frac{2}{2} + \frac{13}{2} \times \frac{5}{5} + \frac{13}{2} \times \frac{5}{5}$$

$$= \frac{42}{10} + \frac{42}{10} + \frac{65}{10} + \frac{65}{10} = \frac{42 + 42 + 65 + 65}{10} = \frac{214}{10} = 21\frac{2}{5}$$

The perimeter is $21\frac{2}{5}$ cm.

0.3 Practice Problems

1. (a) $\frac{2}{7} \times \frac{5}{11} = \frac{2 \cdot 5}{7 \cdot 11} = \frac{10}{77}$

(b) $\frac{8}{9} \times \frac{3}{10} = \frac{8 \cdot 3}{9 \cdot 10} = \frac{24}{90} = \frac{4}{15}$

2. (a) $\frac{3}{5} \times \frac{4}{3} = \frac{3 \cdot 4}{5 \cdot 3} = \frac{4}{5}$

(b) $\frac{9}{10} \times \frac{5}{12} = \frac{3 \cdot 3}{2 \cdot 5} \times \frac{5}{2 \cdot 2 \cdot 3} = \frac{3}{8}$

3. (a) $4 \times \frac{2}{7} = \frac{4}{1} \times \frac{2}{7} = \frac{4 \cdot 2}{1 \cdot 7} = \frac{8}{7} = 1\frac{1}{7}$

(b) $12 \times \frac{3}{4} = \frac{12}{1} \times \frac{3}{4} = \frac{2 \cdot 2 \cdot 3}{1} \times \frac{3}{2 \cdot 2} = 9$

4. (a) $2\frac{1}{5} \times \frac{3}{7} = \frac{11}{5} \times \frac{3}{7} = \frac{11 \cdot 3}{5 \cdot 7} = \frac{33}{35}$

(b) $3\frac{1}{3} \times 1\frac{2}{5} = \frac{10}{3} \times \frac{7}{5} = \frac{2 \cdot 5}{3} \times \frac{7}{5} = \frac{14}{3} = 4\frac{2}{3}$

5. $3\frac{1}{2} \times \frac{1}{14} \times 4 = \frac{7}{2} \times \frac{1}{14} \times \frac{4}{1} = \frac{7}{2} \times \frac{1}{2 \cdot 7} \times \frac{2 \cdot 2}{1} = 1$

6. (a) $\frac{2}{5} \div \frac{1}{3} = \frac{2}{5} \times \frac{3}{1} = \frac{6}{5}$

(b) $\frac{12}{13} \div \frac{4}{3} = \frac{2 \cdot 2 \cdot 3}{13} \times \frac{3}{2 \cdot 2} = \frac{9}{13}$

7. (a) $\frac{3}{7} \div 6 = \frac{3}{7} \div \frac{6}{1} = \frac{3}{7} \times \frac{1}{6} = \frac{3}{42} = \frac{1}{14}$

(b) $8 \div \frac{2}{3} = \frac{8}{1} \times \frac{3}{2} = 12$

8. (a) $\dfrac{\frac{3}{11}}{\frac{5}{7}} = \frac{3}{11} \div \frac{5}{7} = \frac{3}{11} \times \frac{7}{5} = \frac{21}{55}$

(b) $\dfrac{\frac{12}{5}}{\frac{8}{15}} = \frac{12}{5} \div \frac{8}{15} = \frac{2 \cdot 2 \cdot 3}{5} \times \frac{3 \cdot 5}{2 \cdot 2 \cdot 2} = \frac{9}{2} = 4\frac{1}{2}$

9. (a) $1\frac{2}{5} \div 2\frac{1}{3} = \frac{7}{5} \div \frac{7}{3} = \frac{7}{5} \times \frac{3}{7} = \frac{3}{5}$

(b) $4\frac{2}{3} \div 7 = \frac{14}{3} \times \frac{1}{7} = \frac{2 \cdot 7}{3} \times \frac{1}{7} = \frac{2}{3}$

(c) $\dfrac{1\frac{1}{5}}{1\frac{2}{7}} = 1\frac{1}{5} \div 1\frac{2}{7} = \frac{6}{5} \div \frac{9}{7} = \frac{6}{5} \times \frac{7}{9} = \frac{2 \cdot 3}{5} \times \frac{7}{3 \cdot 3} = \frac{14}{15}$

10. $64 \div 5\frac{1}{3} = 64 \div \frac{16}{3} = 64 \times \frac{3}{16} = 12 \text{ jars}$

11. $25\frac{1}{2} \times 5\frac{1}{4} = \frac{51}{2} \times \frac{21}{4} = \frac{1071}{8} = 133\frac{7}{8} \text{ miles}$

0.4 Practice Problems

1. (a) $1.371 \qquad 1\frac{371}{1000}$

One and three hundred seventy-one thousandths.

(b) $\frac{9}{100} = \text{Nine hundredths.}$

2. (a) $\frac{3}{8} \qquad \begin{array}{r} 0.375 \\ 8\overline{)3.000} \\ \underline{2\,4} \\ 60 \\ \underline{56} \\ 40 \\ \underline{40} \\ 0 \end{array}$ **(b)** $\frac{7}{200} \qquad \begin{array}{r} 0.035 \\ 200\overline{)7.000} \\ \underline{6\,00} \\ 1\,000 \\ \underline{1\,000} \\ 0 \end{array}$ **(c)** $\frac{33}{20} \qquad \begin{array}{r} 1.65 \\ 20\overline{)33.00} \\ \underline{20} \\ 13\,0 \\ \underline{12\,0} \\ 1\,00 \\ \underline{1\,00} \\ 0 \end{array}$

3. (a) $\frac{1}{6} \qquad \begin{array}{r} 0.166 = 0.1\overline{6} \\ 6\overline{)1.000} \\ \underline{6} \\ 40 \\ \underline{36} \\ 40 \\ \underline{36} \\ 4 \end{array}$ **(b)** $\frac{5}{11} \qquad \begin{array}{r} 0.4545 = .\overline{45} \\ 11\overline{)5.0000} \\ \underline{4\,4} \\ 60 \\ \underline{55} \\ 50 \\ \underline{44} \\ 60 \\ \underline{55} \\ 5 \end{array}$

4. (a) $0.8 \qquad \frac{8}{10} = \frac{2 \cdot 2 \cdot 2}{2 \cdot 5} = \frac{4}{5}$

(b) $0.88 \qquad \frac{88}{100} = \frac{11 \cdot 2 \cdot 2 \cdot 2}{5 \cdot 5 \cdot 2 \cdot 2} = \frac{22}{25}$

(c) $0.45 \qquad \frac{45}{100} = \frac{5 \cdot 3 \cdot 3}{5 \cdot 5 \cdot 2 \cdot 2} = \frac{9}{20}$

(d) $0.148 \qquad \frac{148}{1000} = \frac{2 \cdot 2 \cdot 37}{5 \cdot 5 \cdot 5 \cdot 2 \cdot 2 \cdot 2} = \frac{37}{250}$

(e) $0.612 \qquad \frac{612}{1000} = \frac{17 \cdot 3 \cdot 3 \cdot 2 \cdot 2}{5 \cdot 5 \cdot 5 \cdot 2 \cdot 2 \cdot 2} = \frac{153}{250}$

(f) $0.016 \qquad \frac{16}{1000} = \frac{2 \cdot 2 \cdot 2 \cdot 2}{5 \cdot 5 \cdot 5 \cdot 2 \cdot 2 \cdot 2} = \frac{2}{125}$

5. (a) 3.12
5.08
1.42
9.62

(b) 152.003
− 136.118
15.885

(c) 1.1
3.16
5.123
9.383

(d) 1.0052
− 0.1234
0.8818

6. (a) 0.061
5.0008
1.3
6.3618

(b) 18.000
− 0.126
17.874

7. 0.5
$\times\,0.3$
0.15

8. 0.12
$\times\,0.4$
0.048

9. (a) 1.23
$\times\,0.005$
0.00615
0.02232

(b) 0.00002
$\times$ 0.003
0.00000006

10. $5\overline{)0.11160}$
10
11
10
16
15
10
10
0

11. 30,000.
$.06.\overline{)1800.00.}$

12. $4.9.\overline{)0.0.1764}$
0.0036
$49\overline{)0.1764}$
147
294
294
0

13. (a) $0.0016 \times 100 = 0.16$
Move decimal point 2 places to the right.

(b) $2.34 \times 1000 = 2340$
Move decimal point 3 places to the right.

(c) $56.75 \times 10,000 = 567,500$
Move decimal point 4 places to the right.

14. (a) $\dfrac{5.82}{10}$ (Move decimal point 1 place to the left.) 0.582

(b) $\dfrac{123.4}{1000}$ (Move decimal point 3 places to the left.) 0.1234

(c) $\dfrac{0.00614}{10,000}$ (Move decimal point 4 places to the left.)
0.000000614

0.5 Practice Problems

1. (a) $0.92 = 92\%$ **(b)** $0.418 = 41.8\%$ **(c)** $0.7 = 70\%$
2. (a) $0.0019 = 0.19\%$ **(b)** $0.0736 = 7.36\%$
(c) $0.0003 = 0.03\%$
3. (a) 304%   **(b)** 518.6%   **(c)** 210%
4. (a) 0.07   **(b)** 0.093   **(c)** 0.002
5. (a) 1.31   **(b)** 3.016   **(c)** 0.0004
6. Change % to decimal and multiply.
(a) $0.18 \times 50 = 9$ **(b)** $0.04 \times 64 = 2.56$
(c) $1.56 \times 35 = 54.6$ **(d)** $0.008 \times 60 = 0.48$
(e) $0.013 \times 82 = 1.066$ **(f)** $0.00002 \times 564 = 0.01128$
7. $4.2\% \times 18,000 = 0.042 \times 18,000$
(a) \$756.00   **(b)** $18,000 + 756.00 = \$18,756.00$
8. $\dfrac{37}{148}$ reduces to $\dfrac{37 \cdot 1}{37 \cdot 2 \cdot 2} = \dfrac{1}{4} = 0.25 = 25\%$
9. (a) $\dfrac{24}{48} = \dfrac{2 \cdot 2 \cdot 2 \cdot 3}{2 \cdot 2 \cdot 2 \cdot 2 \cdot 3} = \dfrac{1}{2} = 0.5 = 50\%$
(b) $\dfrac{4}{25} = 0.16 = 16\%$
10. $\dfrac{430}{1256} = \dfrac{215}{628} \approx 0.342 \approx 34\%$

0.6 Practice Problems

1. $100,000 \times 400 = 40,000,000$
2. $12\dfrac{1}{2} \to 10; 9\dfrac{3}{4} \to 10;$ First room $10 \times 10 = 100$ square feet.
$14\dfrac{1}{4} \to 15; 18\dfrac{1}{2} \to 20;$ Second room $15 \times 20 = 300$ square feet.
Both rooms $100 + 300 = 400$ square feet.
3. (a) $422.8 \to 400$ miles $19.3 \to 20$ gallons of gas
$\dfrac{400}{20} = 20$ miles/gallon
(b) $1.69\dfrac{9}{10} \to \$2.00$ per gallon $3862 \to 4000$ miles
$\dfrac{4000 \text{ miles}}{20 \text{ mi/gallon}} = 200$ gallons
$200 \text{ gallons} \times \$2.00 \text{ per gallon} = \$400.00$
4. (a) $\begin{matrix} 3,580,000,000 & \text{estimate} \to & 4,000,000,000 \text{ miles} \\ 43,300 & \text{estimate} \to & 40,000 \text{ mi/hour} \end{matrix}$
100,000 hours
(b) 24 hours/day estimate $\to \dfrac{100,000}{20} = 5000$ days
5. 56.93% estimate $\to 60\% \to 0.6$
$293,567.12$ estimate $\to 300,000$
$0.6 \times 300,000 = \$180,000$

0.7 Practice Problems

1.

MATHEMATICS BLUEPRINT FOR PROBLEM SOLVING			
Gather the Facts	What Am I Solving for?	What Must I Calculate?	Key Points to Remember
Living Room measures $16\frac{1}{2}$ ft $\times$ $10\frac{1}{2}$ ft. The carpet costs $20.00 per square yard.	Area of room in square feet. Area of room in square yards. Cost of the carpet.	Multiply $16\frac{1}{2}$ ft by $10\frac{1}{2}$ ft to get the area in square feet. Divide the number of square feet by 9 to get the number of square yards. Multiply the number of square yards by $20.00.	9 sq feet = 1 square yard

$16\frac{1}{2} \times 10\frac{1}{2} = 173\frac{1}{4}$ square feet
$173\frac{1}{4} \div 9 = 19\frac{1}{4}$ square yards
$19\frac{1}{4} \times 20 = \$385.00$ total cost of carpet
CHECK:
  Estimate area of room $16 \times 10 = 160$ square feet.
  Estimate area in square yards $160 \div 10 = 16$
  Estimate the cost $16 \times 20 = \$320.00$
This is close to our answer of $385.00. Our answer seems reasonable.

2. (a) $\dfrac{10}{55} \approx 0.181 \approx 18\%$ **(b)** $\dfrac{3,660,000}{13,240,000} \approx 0.276 \approx 28\%$

(c) $\dfrac{15}{55} \approx 0.273 \approx 27\%$ **(d)** $\dfrac{3,720,000}{13,240,000} \approx 0.281 \approx 28\%$

(e) We notice that 18% of the company's sales force is located in the Northwest, and they were responsible for 28% of the sales volume. The percent of sales compared to the percent of sales force is about 150%. 27% of the company's sales force is located in the Southwest, and they were responsible for 28% of the sales volume. The percent of sales compared to the percent of sales force is approximately 100%. It would appear that the Northwest sales force is more effective.

Chapter 1

1.1 Practice Problems

1.

	Number	Integer	Rational Number	Irrational Number	Real Number
(a)	$-\frac{2}{5}$		X		X
(b)	1.515151...		X		X
(c)	-8	X	X		X
(d)	π			X	X

2. (a) Population growth of 1,259 is +1,259.
  (b) Depreciation of $763 is −$763.00.
  (c) Wind chill factor of minus 10 is −10.

3. (a) The additive inverse of $+\frac{2}{5}$ is $-\frac{2}{5}$.

  (b) The additive inverse of −1.92 is +1.92.
  (c) The opposite of a loss of 12 yards on a football play is a gain of 12 yards on the play.

4. (a) $|-7.34| = 7.34$

  (b) $\left|\dfrac{5}{8}\right| = \dfrac{5}{8}$ **(c)** $\left|\dfrac{0}{2}\right| = \dfrac{0}{2} = 0$

5. $-23 + (-35)$
   $23 + \quad 35 = 58$
   $-23 + (-35) = -58$

6. $-\dfrac{3}{5} + \left(-\dfrac{4}{7}\right)$

   $-\dfrac{21}{35} + \left(-\dfrac{20}{35}\right)$

   $-\dfrac{21}{35} + \left(-\dfrac{20}{35}\right) = -\dfrac{41}{35}$

7. $-12.7 + (-9.38)$
   $12.7 + \quad 9.38 = 22.08$
   $-12.7 + (-9.38) = -22.08$

8. $-7 + (-11) + (-33)$
   $= -18 + (-33)$
   $= -51$

9. $-9 + 15$
   $15 - 9 = 6$
   $-9 + 15 = 6$

10. $-\dfrac{5}{12} + \dfrac{7}{12} + \left(-\dfrac{11}{12}\right)$

   $= \dfrac{2}{12} + \left(-\dfrac{11}{12}\right) = -\dfrac{9}{12} = -\dfrac{3}{4}$

11. $-6.3 + (-8.0) + 3.5$
   $= -14.3 + 3.5$
   $= -10.8$

12. $-6 \quad\quad + 5 + (-7) + (-2) + 5 + 3$
   $\begin{array}{r} -6 \\ -7 \\ -2 \\ \hline -15 \end{array} \quad \begin{array}{r} +5 \\ +5 \\ +3 \\ \hline 13 \end{array}$
   $-15 + 13 = -2$

13. (a) $-2.9 + (-5.7) = -8.6$ **(b)** $\dfrac{2}{3} + \left(-\dfrac{1}{4}\right)$

  (c) $-10 + (-3) + 15 + 4$
   $= -13 + 15 + 4$ $= \dfrac{8}{12} + \left(-\dfrac{3}{12}\right) = \dfrac{5}{12}$
   $= 2 + 4 = 6$

1.2 Practice Problems

1. $9 - (-3) = 9 + (+3)$
   $= 12$

2. $-12 - (-5)$
   $= -12 + (+5)$
   $= -7$

3. $-\dfrac{1}{5} - \dfrac{1}{4} = -\dfrac{1}{5} + \left(-\dfrac{1}{4}\right) = -\dfrac{4}{20} + \left(-\dfrac{5}{20}\right) = -\dfrac{9}{20}$

4. $-17.3 - (-17.3)$
$= -17.3 + 17.3$
$= 0$

5. (a) $-21 - 9$
$= -21 + (-9)$
$= -30$

  (b) $17 - 36$
$= 17 + (-36)$
$= -19$

6. $350 - (-186)$
$= 350 + 186$
$= 536$ The helicopter is 536 feet from the sunken vessel.

1.3 Practice Problems

1. (a) $(-6)(-2) = 12$

  (b) $(7)(9) = 63$

  (c) $\left(-\dfrac{3}{5}\right)\left(\dfrac{2}{7}\right) = -\dfrac{6}{35}$

  (d) $40(-20) = -800$

2. $(-5)(-2)(-6)$
$= (+10)(-6) = -60$

3. (a) positive; $-2(-3)(-4)(-1)$
$= 6(-4)(-1)$
$= -24(-1)$
$= +24 \text{ or } 24$

  (b) negative; $(-1)(-3)(-2)$
$= 3(-2)$
$= -6$

  (c) positive; $-4\left(-\dfrac{1}{4}\right)(-2)(-6)$
$= 1(-2)(-6)$
$= -2(-6)$
$= +12 \text{ or } 12$

4. (a) $-36 \div (-2) = 18$

  (b) $\dfrac{50}{-10} = -5$

  (c) $-49 \div 7 = -7$

5. (a) $-1.242 \div (-1.8)$

$$1.8_\wedge \overline{)1.2_\wedge 42}\ \ \begin{array}{r}.69\\\hline\end{array}$$
$$\begin{array}{r}10\ 8\\\hline 1\ 62\end{array}$$
$$\begin{array}{r}1\ 62\\\hline 0\end{array}$$

Thus $-1.242 \div (-1.8) = 0.69$

  (b) $0.235 \div (-0.0025)$

$$0.0025_\wedge \overline{)2350_\wedge}\ \ \begin{array}{r}94\ .\\\hline\end{array}$$
$$\begin{array}{r}225\\\hline 100\end{array}$$
$$\begin{array}{r}100\\\hline 0\end{array}$$

Thus $0.235 \div (-0.0025) = -94$

6. $-\dfrac{5}{16} \div \left(-\dfrac{10}{13}\right) = \left(-\dfrac{5}{16}\right)\left(-\dfrac{13}{10}\right) = \left(-\dfrac{\overset{1}{\cancel{5}}}{16}\right)\left(-\dfrac{13}{\underset{2}{\cancel{10}}}\right) = \dfrac{13}{32}$

7. (a) $\dfrac{-12}{-\dfrac{4}{5}} = -12 \div \left(-\dfrac{4}{5}\right) = -12\left(-\dfrac{5}{4}\right) = \left(-\dfrac{\overset{3}{\cancel{12}}}{1}\right)\left(-\dfrac{5}{\underset{1}{\cancel{4}}}\right) = 15$

  (b) $\dfrac{-\dfrac{2}{9}}{\dfrac{8}{13}} = -\dfrac{2}{9} \div \dfrac{8}{13} = -\dfrac{\overset{1}{\cancel{2}}}{9}\left(\dfrac{13}{\underset{4}{\cancel{8}}}\right) = -\dfrac{13}{36}$

8. (a) $6(-10) = -60$ yards

  (b) $7(15) = 105$ yards

  (c) $-60 + 105 = 45$ yards

1.4 Practice Problems

1. (a) $6(6)(6)(6) = 6^4$

  (b) $-2(-2)(-2)(-2)(-2) = (-2)^5$

  (c) $108(108)(108) = 108^3$

  (d) $-11(-11)(-11)(-11)(-11)(-11) = (-11)^6$

  (e) $(w)(w)(w) = w^3$

  (f) $(z)(z)(z)(z) = z^4$

2. (a) $3^5 = (3)(3)(3)(3)(3) = 243$

  (b) $2^2 + 3^3$

  $2^2 = (2)(2) = 4$
  $3^3 = (3)(3)(3) = 27$
  $4 + 27 = 31$

3. (a) $(-3)^3 = -27$

  (b) $(-2)^6 = 64$

  (c) $-2^4 = -(2^4) = -16$

  (d) $-(3^6) = -729$

4. (a) $\left(\dfrac{1}{3}\right)^3 = \left(\dfrac{1}{3}\right)\left(\dfrac{1}{3}\right)\left(\dfrac{1}{3}\right) = \dfrac{1}{27}$

  (b) $(0.3)^4 = (0.3)(0.3)(0.3)(0.3) = 0.0081$

  (c) $\left(\dfrac{3}{2}\right)^4 = \left(\dfrac{3}{2}\right)\left(\dfrac{3}{2}\right)\left(\dfrac{3}{2}\right)\left(\dfrac{3}{2}\right) = \dfrac{81}{16}$

  (d) $(3)^4(4)^2$

  $3^4 = (3)(3)(3)(3) = 81$
  $4^2 = (4)(4) = 16$
  $(81)(16) = 1296$

  (e) $4^2 - 2^4 = 16 - 16 = 0$

1.5 Practice Problems

1. $25 \div 5 \cdot 6 + 2^3$
$= 25 \div 5 \cdot 6 + 8$
$= 5 \cdot 6 + 8$
$= 30 + 8$
$= 38$

2. $(-4)^3 - 2^6$
$= -64 - 64$
$= -128$

3. $6 - (8 - 12)^2 + 8 \div 2$
$= 6 - (-4)^2 + 8 \div 2$
$= 6 - (16) + 8 \div 2$
$= 6 - 16 + 4$
$= -10 + 4$
$= -6$

4. $\left(-\dfrac{1}{7}\right)\left(-\dfrac{14}{5}\right) + \left(-\dfrac{1}{2}\right) \div \left(\dfrac{3}{4}\right)$
$= \left(-\dfrac{1}{7}\right)\left(-\dfrac{14}{5}\right) + \left(-\dfrac{1}{2}\right) \times \left(\dfrac{4}{3}\right)$
$= \dfrac{2}{5} + \left(-\dfrac{2}{3}\right)$
$= \dfrac{2 \cdot 3}{5 \cdot 3} + \left(-\dfrac{2 \cdot 5}{3 \cdot 5}\right)$
$= \dfrac{6}{15} + \left(-\dfrac{10}{15}\right) = -\dfrac{4}{15}$

1.6 Practice Problems

1. (a) $-3(x + 2y) = -3(x) + (-3)(2y) = -3x - 6y$

  (b) $-a(a - 3b) = -a(a) + (-a)(-3b) = -a^2 + 3ab$

2. (a) $-(-3x + y) = (-1)(-3x + y) = (-1)(-3x) + (-1)(y)$
$= 3x - y$

3. (a) $\dfrac{3}{5}(a^2 - 5a + 25) = \left(\dfrac{3}{5}\right)(a^2) + \left(\dfrac{3}{5}\right)(-5a) + \left(\dfrac{3}{5}\right)(25)$
$= \dfrac{3}{5}a^2 - 3a + 15$

  (b) $2.5(x^2 - 3.5x + 1.2)$
$= (2.5)(x^2) + (2.5)(-3.5x) + (2.5)(1.2)$
$= 2.5x^2 - 8.75x + 3$

4. $-4x(x - 2y + 3) = (-4)(x)(x) - (-4)(x)(2)(y) + (-4)(x)(3)$
$= -4x^2 + 8xy - 12x$

5. $(3x^2 - 2x)(-4) = (3x^2)(-4) - (2x)(-4) = -12x^2 + 8x$

6. $400(6x + 9y) = 400(6x) + 400(9y)$
$= 2400x + 3600y$

1.7 Practice Problems

1. (a) $5a$ and $8a$ are like terms.

$2b$ and $-4b$ are like terms.

(b) y^2 and $-7y^2$ are like terms. These are the only like terms.

2. (a) $5a + 7a + 4a = (5 + 7 + 4)a = 16a$

(b) $16y^3 + 9y^3 = (16 + 9)y^3 = 25y^3$

3. $-8y^2 - 9y^2 + 4y^2 = (-8 - 9 + 4)y^2 = -13y^2$

4. (a) $-x + 3a - 9x + 2a = -x - 9x + 3a + 2a = -10x + 5a$

(b) $5ab - 2ab^2 - 3a^2b + 6ab = 5ab + 6ab - 2ab^2 - 3a^2b$
$$= 11ab - 2ab^2 - 3a^2b$$

(c) $7x^2y - 2xy^2 - 3x^2y - 4xy^2 + 5x^2y$
$$= 7x^2y - 3x^2y + 5x^2y - 2xy^2 - 4xy^2 = 9x^2y - 6xy^2$$

5. $5xy - 2x^2y + 6xy^2 - xy - 3xy^2 - 7x^2y$
$$= 5xy - xy - 2x^2y - 7x^2y + 6xy^2 - 3xy^2$$
$$= 4xy - 9x^2y + 3xy^2$$

6. $5a(2 - 3b) - 4(6a + 2ab) = 10a - 15ab - 24a - 8ab$
$$= -14a - 23ab$$

7. $\dfrac{1}{7}a^2 + 2a^2 = \dfrac{1}{7}a^2 + \dfrac{2}{1}a^2 = \dfrac{1}{7}a^2 + \dfrac{2 \cdot 7}{1 \cdot 7}a^2$
$$= \dfrac{1}{7}a^2 + \dfrac{14}{7}a^2 = \dfrac{15}{7}a^2$$

$-\dfrac{5}{12}b - \dfrac{1}{3}b = -\dfrac{5}{12}b - \dfrac{1 \cdot 4}{3 \cdot 4}b = -\dfrac{5}{12}b - \dfrac{4}{12}b$
$$= -\dfrac{9}{12}b = -\dfrac{3}{4}b$$

Thus, our solution is $\dfrac{15}{7}a^2 - \dfrac{3}{4}b$

1.8 Practice Problems

1. $4 - \dfrac{1}{2}x = 4 - \dfrac{1}{2}(-8)$
$$= 4 + 4$$
$$= 8$$

2. (a) $-x^4 = -(-3)^4$
$$= -(81) = -81$$

(b) $(-x)^4 = [-(-3)]^4$
$$= (3)^4 = 81$$

3. $(5x)^3 + 2x = [5(-2)]^3 + 2(-2)$
$$= (-10)^3 + (-4)$$
$$= (-1000) + (-4)$$
$$= -1004$$

4. Area of a triangle is

$A = \frac{1}{2}ba$

altitude = 3 meters (m)

base = 7 meters (m)

$A = \dfrac{1}{2}(7\text{m})(3\text{m})$

$\quad = \dfrac{1}{2}(7)(3)(m)(m)$

$\quad = \left(\dfrac{7}{2}\right)(3)(m)^2$

$\quad = \dfrac{21}{2}(m)^2$

$\quad = 10.5$ square meters

5. Area of a circle is

$A = \pi r^2$

$r = 3$ meters

$A = 3.14(3 \text{ m})^2$

$\quad = 3.14(9)(\text{m})^2$

$\quad = 28.26$ square meters

6. Formula $\quad C = \dfrac{5}{9}(F - 32)$
$$= \dfrac{5}{9}(68 - 32)$$
$$= \dfrac{5}{9}(36)$$
$$= 5(4)$$
$$= 20° \text{ Celsius}$$

7. Use the formula.

$k = 1.61 \ (r)$ Replace r by 35.

$k = 1.61 \ (35)$

$k = 56.35$ The truck is violating the minimum speed limit.

1.9 Practice Problems

1. $5[4x - 3(y - 2)]$
$$= 5[4x - 3y + 6]$$
$$= 20x - 15y + 30$$

2. $3ab - [2ab - (2 - a)]$
$$= 3ab - [2ab - 2 + a]$$
$$= 3ab - 2ab + 2 - a$$
$$= ab + 2 - a$$

3. $3[4x - 2(1 - x)] - [3x + (x - 2)]$
$$= 3[4x - 2 + 2x] - [3x + x - 2]$$
$$= 12x - 6 + 6x - 3x - x + 2$$
$$= 14x - 4$$

4. $-2\{5x - 3x[2x - (x^2 - 4x)]\}$
$$= -2\{5x - 3x[2x - x^2 + 4x]\}$$
$$= -2\{5x - 3x[6x - x^2]\}$$
$$= -2\{5x - 18x^2 + 3x^3\}$$
$$= -10x + 36x^2 - 6x^3$$
$$= -6x^3 + 36x^2 - 10x$$

Chapter 2

2.1 Practice Problems

1. $x + 0.3 = 1.2$

$\underline{\quad -0.3 \quad -0.3 \quad}$ Check. $\quad 0.9 + 0.3 \overset{?}{=} 1.2$

$\qquad x \qquad = 0.9 \qquad\qquad\qquad 1.2 = 1.2$

2. $\quad 17 = x - 5$

$\underline{+ 5 \qquad + 5}$ Check. $\quad 17 \overset{?}{=} 22 - 5$

$\quad 22 = x \qquad\qquad\qquad 17 = 17$

3. $5 - 12 = x - 3$

$\quad -7 = x - 3$

$\underline{+3 \qquad + 3}$ Check. $\quad 5 - 12 \overset{?}{=} -4 - 3$

$\quad -4 = x \qquad\qquad\qquad -7 = -7$

4. $x + 8 = -22 + 6$

$\qquad x = -2$

Check. $\quad -2 + 8 \overset{?}{=} -22 + 6$

$\qquad\qquad\qquad 6 \neq -16 \qquad$ This is not true.

Thus $x = -2$ is not a solution. Solve to find the solution.

$x + 8 = -22 + 6 = -16$

$\qquad x = -16 - 8$

$\qquad x = -24$

5. $\dfrac{1}{20} - \dfrac{1}{2} = x + \dfrac{3}{5}$ Check.

$\dfrac{1}{20} - \dfrac{1 \cdot 10}{2 \cdot 10} = x + \dfrac{3 \cdot 4}{5 \cdot 4} \qquad\qquad \dfrac{1}{20} - \dfrac{1}{2} \overset{?}{=} -1\dfrac{1}{20} + \dfrac{3}{5}$

$\dfrac{1}{20} - \dfrac{10}{20} = x + \dfrac{12}{20} \qquad\qquad \dfrac{1}{20} - \dfrac{10}{20} \overset{?}{=} -\dfrac{21}{20} + \dfrac{12}{20}$

$\qquad -\dfrac{9}{20} = x + \dfrac{12}{20} \qquad\qquad\qquad -\dfrac{9}{20} = -\dfrac{9}{20} \ \checkmark$

$-\dfrac{9}{20} - \dfrac{12}{20} = x + \dfrac{12}{20} - \dfrac{12}{20}$

$\qquad -\dfrac{21}{20} = -1\dfrac{1}{20} = x$

2.2 Practice Problems

1. $(8)\dfrac{1}{8}x = -2(8)$

$x = -16$

2. $\dfrac{\cancel{9}x}{\cancel{9}} = \dfrac{72}{9}$

$x = 8$

3. $\dfrac{\cancel{6}x}{\cancel{6}} = \dfrac{50}{6}$

$x = \dfrac{25}{3}$

4. $\dfrac{-\cancel{27}x}{-\cancel{27}} = \dfrac{54}{-27}$

$x = -2$

5. $\dfrac{-x}{-1} = \dfrac{36}{-1}$

$x = -36$

6. $\dfrac{-51}{-6} = \dfrac{-6x}{-6}$

$\dfrac{17}{2} = x$

7. $\dfrac{21}{4.2} = \dfrac{\cancel{4.2}x}{\cancel{4.2}}$

$5 = x$

2.3 Practice Problems

1. $9x + 2 = 38$

$\dfrac{-2 \quad -2}{}$

$\dfrac{9x}{9} = \dfrac{36}{9}$

$x = 4$

Check. $9(4) + 2 \overset{?}{=} 38$

$36 + 2 \overset{?}{=} 38$

$38 = 38 \quad ✓$

2. $13x = 2x - 66$

$\dfrac{-2x \quad -2x}{}$

$\dfrac{11x}{11} = \dfrac{-66}{11}$

$x = -6$

Check. $13(-6) \overset{?}{=} 2(-6) - 66$

$-78 \overset{?}{=} -12 - 66$

$-78 = -78 \quad ✓$

3. $3x + 2 = 5x + 2$

$\dfrac{-2 \qquad -2}{}$

$3x \quad = 5x$

$\dfrac{-5x \qquad -5x}{}$

$\dfrac{-2x \quad = 0}{}$

$x = 0$

Check. $3(0) + 2 \overset{?}{=} 5(0) + 2$

$2 = 2 \quad ✓$

4. $-z + 8 - z = 3z + 10 - 3$

$-2z + 8 \quad = 3z + 7$

$\dfrac{-3z \qquad -3z}{}$

$-5z + 8 \quad = 7$

$\dfrac{-8 \quad = -8}{}$

$\dfrac{-5z}{-5} = \dfrac{-1}{-5}$

$z = \dfrac{1}{5}$

5. $2x^2 - 6x + 3 = -4x - 7 + 2x^2$

$\dfrac{-2x^2 \qquad\qquad\qquad -2x^2}{}$

$-6x + 3 = -4x - 7$

$\dfrac{+4x \qquad +4x}{}$

$-2x + 3 = \quad -7$

$\dfrac{-3 = \quad -3}{}$

$\dfrac{-2x}{-2} = \dfrac{-10}{-2}$

$x = 5$

6. $4x - (x + 3) = 12 - 3(x - 2)$

$4x - x - 3 = 12 - 3x + 6$

$3x - 3 = \quad -3x + 18$

$\dfrac{+3x \qquad\qquad + 3x}{}$

$6x - 3 = \qquad 18$

$\dfrac{+3 \qquad\qquad +3}{}$

$\dfrac{6x}{6} = \dfrac{21}{6}$

$x = \dfrac{21}{6} = \dfrac{7}{2}$

Check. $4\left(\dfrac{7}{2}\right) - \left(\dfrac{7}{2} + 3\right) \overset{?}{=} 12 - 3\left(\dfrac{7}{2} - 2\right)$

$14 - \dfrac{13}{2} \overset{?}{=} 12 - 3\left(\dfrac{3}{2}\right)$

$\dfrac{28}{2} - \dfrac{13}{2} \overset{?}{=} \dfrac{24}{2} - \dfrac{9}{2}; \qquad \dfrac{15}{2} = \dfrac{15}{2} \quad ✓$

7. $4(-2x - 3) = -5(x - 2) + 2$

$-8x - 12 = -5x + 10 + 2$

$-8x - 12 = -5x + 12$

$\dfrac{+5x \qquad\qquad +5x}{}$

$-3x - 12 = \qquad 12$

$\dfrac{+12 = \qquad +12}{}$

$\dfrac{-3x}{-3} = \dfrac{24}{-3}$

$x = -8$

8. $0.3x - 2(x + 0.1) = 0.4(x - 3) - 1.1$

$0.3x - 2x - 0.2 = 0.4x - 1.2 - 1.1$

$-1.7x - 0.2 = 0.4x - 2.3$

$\dfrac{-0.4x \qquad\qquad -0.4x}{}$

$-2.1x - 0.2 = -2.3$

$\dfrac{+0.2 \qquad +0.2}{}$

$\dfrac{-2.1x}{-2.1} = \dfrac{-2.1}{-2.1}$

$x = 1$

9. $5(2z - 1) + 7 = 7z - 4(z + 3)$

$10z - 5 + 7 = 7z - 4z - 12$

$10z + 2 = \qquad 3z - 12$

$\dfrac{-3z \qquad\qquad - 3z}{}$

$7z + 2 = \qquad\quad - 12$

$\dfrac{-2 \qquad\qquad - 2}{}$

$\dfrac{7z}{7} = \qquad -\dfrac{14}{7}$

$z = -2$

Check. $5[2(-2) - 1] + 7 \overset{?}{=} 7(-2) - 4[-2 + 3]$

$5(-5) + 7 \overset{?}{=} -14 - 4(1)$

$-25 + 7 \overset{?}{=} -18$

$-18 = -18 \quad ✓$

2.4 Practice Problems

1. $\dfrac{3}{8}x - \dfrac{3}{2} = \dfrac{1}{4}x$

$3x - 12 = \quad 2x$

$\dfrac{-2x \qquad\qquad -2x}{}$

$x - 12 = \quad 0$

$\dfrac{+12 \quad +12}{}$

$x = \quad 12$

2. $\dfrac{5x}{4} - 1 = \dfrac{3x}{4} + \dfrac{1}{2}$

$5x - 4 = 3x + 2$

$\dfrac{-3x \qquad -3x}{}$

$2x - 4 = \qquad +2$

$\dfrac{+4 = \qquad +4}{}$

$\dfrac{2x}{2} = \dfrac{6}{2}$

$x = 3$

Check. $\dfrac{5(3)}{4} - 1 \overset{?}{=} \dfrac{3(3)}{4} + \dfrac{1}{2}$

$\dfrac{15}{4} - 1 \overset{?}{=} \dfrac{9}{4} + \dfrac{1}{2}$

$\dfrac{15}{4} - \dfrac{4}{4} \overset{?}{=} \dfrac{9}{4} + \dfrac{2}{4}$

$\dfrac{11}{4} = \dfrac{11}{4} \quad ✓$

3. $\dfrac{5x}{6} - \dfrac{5}{8} = \dfrac{3x}{4} - \dfrac{1}{3}$

$20x - 15 = 18x - 8$

$\underline{-18x \qquad\quad -18x}$

$2x - 15 = \qquad -8$

$\underline{\quad +15 = \qquad +15}$

$\dfrac{2x}{2} = \dfrac{7}{2}$

$x = \dfrac{7}{2}$

4. $\dfrac{1}{3}(x - 2) = \dfrac{1}{4}(x + 5) - \dfrac{5}{3}$ Check. $\dfrac{1}{3}(3 - 2) \stackrel{?}{=} \dfrac{1}{4}(3 + 5) - \dfrac{5}{3}$

$\dfrac{1}{3}x - \dfrac{2}{3} = \dfrac{1}{4}x + \dfrac{5}{4} - \dfrac{5}{3}$ $\dfrac{1}{3}(1) \stackrel{?}{=} \dfrac{1}{4}(8) - \dfrac{5}{3}$

$4x - 8 = 3x + 15 - 20$ $\dfrac{1}{3} \stackrel{?}{=} 2 - \dfrac{5}{3}$

$4x - 8 = 3x - 5$ $\dfrac{1}{3} \stackrel{?}{=} \dfrac{6}{3} - \dfrac{5}{3}$

$\underline{-3x \qquad = -3x}$

$x - 8 = \qquad -5$ $\dfrac{1}{3} = \dfrac{1}{3}$ ✓

$\underline{\quad +8 \qquad\quad +8}$

$x = \qquad 3$

5. $2.8 = 0.3(x - 2) + 2(0.1x - 0.3)$

$2.8 = 0.3x - 0.6 + 0.2x - 0.6$

$10(2.8) = 10(0.3x) - 10(0.6) + 10(0.2x) - 10(0.6)$

$28 = 3x - 6 + 2x - 6$

$28 = 5x - 12$

$\underline{+12 \qquad\quad + 12}$

$40 = 5x$

$8 = x$

2.5 Practice Problems

1. $d = rt$

$3525 = r(2.5)$

$\dfrac{3525}{2.5} = r$

$1410 \text{ mph} = r$

2. Solve for m. $\dfrac{E}{c^2} = \dfrac{mc^2}{c^2}$ $\dfrac{E}{c^2} = m$

3. Solve for y. $8 - 2y + 3x = 0$

$\underline{\qquad\qquad -3x = \quad - 3x}$

$8 - 2y = - 3x$

$\underline{-8 \qquad\qquad\quad - 8}$

$\dfrac{-2y}{-2} = \dfrac{-3x - 8}{-2}$

$y = \dfrac{3x + 8}{2} \text{ or } y = \dfrac{3}{2}x + 4$

4. Solve for d. $C = \pi d$

$\dfrac{C}{\pi} = \dfrac{\pi d}{\pi}$

$\dfrac{C}{\pi} = d$

2.6 Practice Problems

1. (a) $7 > 2$ **(b)** $-3 > -4$ **(c)** $-1 < 2$

(d) $-8 < -5$ **(e)** $0 > -2$ **(f)** $\dfrac{2}{5} > \dfrac{3}{8}$

2. (a) $x > 5$; x is greater than 5

(b) $x \le -2$; x is less than or equal to -2

(c) $3 > x$; 3 is greater than x (or x is less than 3)

(d) $x \ge -\dfrac{3}{2}$; x is greater than or equal to $-\dfrac{3}{2}$

3. (a) $t \le 180$ **(b)** $d < 15{,}000$

2.7 Practice Problems

1. (a) $9 + 4 > 6 + 4$

$13 > 10$

(b) $-2 - 3 < 5 - 3$

$-5 < 2$

(c) $1(2) > -3(2)$

$2 > -6$

(d) $\dfrac{10}{5} < \dfrac{15}{5}$

$2 < 3$

2. (a) $7 > 2$ **(b)** $-3 < -1$

$-14 < -4$ $3 > 1$

(c) $-10 \ge -20$ **(d)** $-15 \le -5$

$1 \le 2$ $3 \ge 1$

3. $8x - 2 < 3$

$\underline{\quad +2 \quad +2}$

$\dfrac{8x}{8} < \dfrac{5}{8}$

$x < \dfrac{5}{8}$

4. $4 - 5x > 7$

$\underline{-4 \qquad\quad -4}$

$\dfrac{-5x}{-5} < \dfrac{3}{-5}$

$x < \dfrac{-3}{5}$

5. $\dfrac{1}{2}x + 3 < \dfrac{2}{3}x$

$3x + 18 < 4x$

$\underline{-4x \qquad\quad -4x}$

$-x + 18 < 0$

$\underline{\quad -18 \quad -18}$

$\dfrac{-x}{-1} > \dfrac{-18}{-1}$

$x > 18$

6. $\dfrac{1}{2}(3 - x) \le 2x + 5$

$\dfrac{3}{2} - \dfrac{1}{2}x \le 2x + 5$

$3 - x \le 4x + 10$

$-5x \le 7$

$x \ge -\dfrac{7}{5}$

7. $2000n - 700{,}000 \ge 2{,}500{,}000$

$2{,}000n \ge 3{,}200{,}000$

$n \ge 1{,}600$

Chapter 3

3.1 Practice Problems

1. (a) $x + 4$ **(b)** $3x$

  (c) $x - 8$ **(d)** $\dfrac{1}{4}x$

2. (a) $3x + 8$ **(b)** $3(x + 8)$

3. (a) $3x - 4$ **(b)** $\dfrac{2}{3}(x + 5)$

4. Let a = Ann's hours per week.
Then $a - 17$ = Marie's hours per week.

5. width $= w$
length $= 2w + 5$

6. 1st angle $= s - 16$
2nd angle $= s$
3rd angle $= 2s$

7. Let x = the number of students in the fall.
$\dfrac{2}{3}x$ = the number of students in the spring.
$\dfrac{1}{5}x$ = the number of students in the summer.

3.2 Practice Problems

1. $\dfrac{3}{4}x = -81$

$x = -108$

2. $3x - 2 = 49$
$3x = 51$
$x = 17$

3. $x + (3x - 12) = 24$
$4x - 12 = 24$
$4x = 36$
$x = 9$

First number is 9. Second number $= 3(9) - 12 = 15$.

4. Let precipitation in British Columbia, Canada $= x$. Then precipitation in Texas, US $= x + 24$.
$x + x + 24 = 62$
$2x + 24 = 62$
$2x = 38$
$x = 19$
B.C., Canada had 19 inches of rain.
Texas, US had $19 + 24 = 43$ inches of rain.

5. (a) $\dfrac{220}{4} = 55$ mph   **(b)** $\dfrac{225}{4.5} = 50$ mph

  (c) The trip leaving the city by 5 mph.

6. $\dfrac{x + x + 78 + 80 + 100 + 96}{6} = 90$

$354 + 2x = 540$
$2x = 186$
$x = 93$

She needs a 93 on the final exam.

3.3 Practice Problems

1. short piece $= x$
long piece $= x + 17$
$x + (x + 17) = 89$
$2x + 17 = 89$
$2x = 72$
$x = 36$ feet

Therefore   Short piece   36 feet
           Long piece $36 + 17 = 53$ feet

2. Family 1 $= x + 360$
Family 2 $= x$
Family 3 $= 2x - 200$
$(x) + (x + 360) + (2x - 200) = 3960$
$4x + 160 = 3960$
$x = 950$

Therefore   Family #2 = \$950.00
           Family #3 $= 2(950) - 200 = \$1700.00$
           Family #1 $= 950 + 360 = \$1310.00$

3. Let the second side $= x$ Therefore the
     first side $= x - 30$ second side $= 300$ meters
     third side $= \frac{1}{2}x$ first side $= 270$ meters
$x + x - 30 + \frac{1}{2}x = 720$ third side $= 150$ meters
$5x - 60 = 1440$
$5x = 1500$
$x = 300$

3.4 Practice Problems

1. $3(25) + (0.20)m = 350$
$75 + 0.20m = 350$
$0.20m = 275$
$m = 1375$ miles

2. $0.38x = 4560$
$x = \$12,000.00$

3. $x + 0.07x = 13,910$
$1.07x = 13,910$
$x = \$13,000.00$

4. $x = 7000(0.12)(1)$
$x = \$840.00$

5. $0.09x + 0.07(8000 - x) = 630$
$0.09x + 560 - 0.07x = 630$
$0.02x + 560 = 630$
$0.02x = 70$
$x = \$3500$

Therefore, she invested \$3500.00 at 9%.
$(8000 - 3500) = \$4500.00$ at 7%

6. x = Dimes
$x + 5$ = Quarters
$0.10x + 0.25(x + 5) = 5.10$
$0.10x + 0.25x + 1.25 = 5.10$
$0.35x + 1.25 = 5.10$
$0.35x = 3.85$
$x = 11$

Therefore, she has 11 dimes and 16 quarters.

7.   Nickels $= 2x$
     Dimes $= x$
     Quarters $= x + 4$
$0.05(2x) + 0.10(x) + 0.25(x + 4) = 2.35$
$0.1x + 0.10x + 0.25x + 1 = 2.35$
$0.45x + 1 = 2.35$
$0.45x = 1.35$
$x = 3$

Therefore, the boy has 3 dimes, 6 nickels, and 7 quarters.

3.5 Practice Problems

1. $A = \dfrac{1}{2}ab$

$= \dfrac{1}{2}(20)(14)$

$= (10)(14)$

$= 140$ square inches

2. $A = lw$
$120 = l(8)$
$\dfrac{120}{8} = l$
$l = 15$ yards

3. $A = \dfrac{1}{2}a(b_1 + b_2)$

$256 = \dfrac{1}{2}a(12 + 20)$

$256 = \dfrac{1}{2}(a)(32)$

$256 = 16a$

$\dfrac{256}{16} = a$

$a = 16$ feet

4. $C = 2\pi r$

$C = 2(3.14)(15)$

$C = 94.2$

$C \approx 94$ meters

5. $P = a + b + c$

$P = 15 + 15 + 15$

$P = 45$ cm

6.

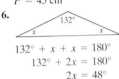

$132° + x + x = 180°$

$132° + 2x = 180°$

$2x = 48°$

$x = 24°$

7. Surface Area $= 4\pi r^2$

$\qquad = 4(3.14)(5)^2$

$\qquad = 314$ m^2

8. $V = \pi r^2 h$

$V = 3.14(3)^2(4) \approx 113$ ft^3

9. Calculate the area of the pool.

$A = lw$ $A = lw$

$A = 12 \times 8 = 96$ sq. ft. $A = 18 \times 14 = 252$ sq. ft.

Now add 6 feet to the length   Now subtract the areas.

and 6 feet to the width of the pool $252 - 96 = 156$ sq. ft.

and calculate the area.   at \$12 a square foot.

     $156 \times 12 = \$1872$

10. $V = lwh$

$V = (6)(5)(8) = 240$ ft^3

Weight $= 240$ ft$^3 \times \dfrac{62.4 \text{ lb}}{1 \text{ ft}^3} = 14{,}976$ lb $\approx 15{,}000$ lb

3.6 Practice Problems

1. (a) height ≤ 6 **(b)** speed > 65

    (c) area ≥ 560 **(d)** profit margin ≥ 50

2. $\dfrac{1050 + 1250 + 950 + x}{4} \ge 1100$

$\qquad\qquad \dfrac{x + 3250}{4} \ge 1100$

$\qquad\qquad x + 3250 \ge 4400$

$\qquad\qquad\qquad x \ge 1150$

The rope must hold at least 1150 pounds.

3. $1400 + 0.02x > 2200$

$\qquad 0.02x > 800$

$\qquad\qquad x > 40{,}000$

Rita must sell more than \$40,000 worth of products each month.

Chapter 4

4.1 Practice Problems

1. (a) $a^7 \cdot a^5 = a^{7+5} = a^{12}$ **(b)** $w^{10} \cdot w = w^{10+1} = w^{11}$

2. (a) $x^3 \cdot x^9 = x^{3+9} = x^{12}$ **(b)** $3^7 \cdot 3^4 = 3^{7+4} = 3^{11}$

    (c) $a^3 \cdot b^2 = a^3 \cdot b^2$ (cannot be simplified)

3. (a) $(-a^8)(a^4) = (-1 \cdot 1)(a^8 \cdot a^4)$

$\qquad\qquad\qquad = -1(a^8 \cdot a^4)$

$\qquad\qquad\qquad = -1a^{12} = -a^{12}$

    (b) $(3y^2)(-2y^3) = (3)(-2)(y^2 \cdot y^3) = -6y^5$

    (c) $(-4x^3)(-5x^2) = (-4)(-5)(x^3 \cdot x^2) = 20x^5$

4. $(2xy)\left(-\dfrac{1}{4}x^2y\right)(6xy^3) = (2)\left(-\dfrac{1}{4}\right)(6)(x \cdot x^2 \cdot x)(y \cdot y \cdot y^3)$

$\qquad\qquad\qquad\qquad\qquad = -3x^4y^5$

5. (a) $\dfrac{10^{13}}{10^7} = 10^{13-7} = 10^6$ **(b)** $\dfrac{x^{11}}{x} = x^{11-1} = x^{10}$

6. (a) $\dfrac{c^3}{c^4} = \dfrac{1}{c^{4-3}} = \dfrac{1}{c}$ **(b)** $\dfrac{10^{31}}{10^{56}} = \dfrac{1}{10^{56-31}} = \dfrac{1}{10^{25}}$

7. (a) $\dfrac{-7x^7}{-21x^9} = \dfrac{1}{3x^{9-7}} = \dfrac{1}{3x^2}$ **(b)** $\dfrac{15x^{11}}{-3x^4} = -5x^{11-4} = -5x^7$

8. (a) $\dfrac{x^7y^9}{y^{10}} = \dfrac{x^7}{y}$ **(b)** $\dfrac{12x^5y^6}{-24x^3y^8} = -\dfrac{x^2}{2y^2}$

9. (a) $\dfrac{10^7}{10^7} = 1$

    (b) $\dfrac{12a^4}{15a^4} = \dfrac{4}{5}\left(\dfrac{a^4}{a^4}\right) = \dfrac{4}{5}(1) = \dfrac{4}{5}$

10. $\dfrac{-20a^3b^8c^4}{28a^3b^7c^5} = -\dfrac{5a^0b}{7c} = -\dfrac{5(1)b}{7c} = -\dfrac{5b}{7c}$

11. $\dfrac{(-6ab^5)(3a^2b^4)}{16a^5b^7} = \dfrac{-18a^3b^9}{16a^5b^7} = -\dfrac{9b^2}{8a^2}$

12. (a) $(a^4)^3 = a^{4\cdot3} = a^{12}$

    (b) $(10^5)^2 = 10^{5\cdot2} = 10^{10}$ **(c)** $(-1)^{15} = -1$

13. (a) $(3xy)^3 = (3)^3x^3y^3 = 27x^3y^3$ **(b)** $(yz)^{37} = y^{37}z^{37}$

14. $\left(\dfrac{4a}{b}\right)^6 = \dfrac{4^6a^6}{b^6} = \dfrac{4096a^6}{b^6}$

15. (a) $\left(\dfrac{-2x^3y^0z}{4xz^2}\right)^5 = \left(\dfrac{-x^2}{2z}\right)^5 = \dfrac{(-1)^5(x^2)^5}{2^5z^5} = -\dfrac{x^{10}}{32z^5}$

4.2 Practice Problems

1. (a) $x^{-12} = \dfrac{1}{x^{12}}$ **(b)** $w^{-5} = \dfrac{1}{w^5}$ **(c)** $z^{-2} = \dfrac{1}{z^2}$

2. $4^{-3} = \dfrac{1}{4^3} = \dfrac{1}{64}$

3. (a) $\dfrac{3}{w^{-4}} = 3w^4$ **(b)** $\dfrac{x^{-6}y^4}{z^{-2}} = \dfrac{y^4z^2}{x^6}$ **(c)** $x^{-6}y^{-5} = \dfrac{1}{x^6y^5}$

4. (a) $(2x^4y^{-5})^{-2} = 2^{-2}x^{-8}y^{10} = \dfrac{y^{10}}{2^2x^8} = \dfrac{y^{10}}{4x^8}$

    (b) $\dfrac{y^{-3}z^{-4}}{y^2z^{-6}} = \dfrac{z^6}{y^2y^3z^4} = \dfrac{z^6}{y^5z^4} = \dfrac{z^2}{y^5}$

5. (a) $78{,}200 = 7.82 \times 10{,}000 = 7.82 \times 10^4$

    (b) $4{,}786{,}000 = 4.786 \times 1{,}000{,}000 = 4.786 \times 10^6$

6. (a) $0.98 = 9.8 \times 10^{-1}$ **(b)** $0.000092 = 9.2 \times 10^{-5}$

7. (a) $2.96 \times 10^3 = 2.96 \times 1000 = 2960.0$

    (b) $1.93 \times 10^6 = 1.93 \times 1{,}000{,}000 = 1{,}930{,}000.0$

    (c) $5.43 \times 10^{-2} = 5.43 \times \dfrac{1}{100} = 0.0543$

    (d) $8.562 \times 10^{-5} = 8.562 \times \dfrac{1}{100{,}000} = 0.00008562$

8. $30{,}900{,}000{,}000{,}000{,}000 = 3.09 \times 10^{16}$ meters

9. (a) $(56{,}000)(1{,}400{,}000{,}000) = (5.6 \times 10^4)(1.4 \times 10^9)$

$\qquad\qquad\qquad\qquad\qquad\qquad = (5.6)(1.4)(10^4)(10^9)$

$\qquad\qquad\qquad\qquad\qquad\qquad = 7.84 \times 10^{13}$

    (b) $\dfrac{0.000111}{0.00000037} = \dfrac{1.11 \times 10^{-4}}{3.7 \times 10^{-7}} = \dfrac{1.11}{3.7} \times \dfrac{10^{-4}}{10^{-7}}$

$\qquad\qquad\qquad\qquad = \dfrac{1.11}{3.7} \times \dfrac{10^7}{10^4} = 0.3 \times 10^3 = 3.0 \times 10^2$

10. $159 \text{ parsecs} = (159 \text{ parsecs}) \dfrac{(3.09 \times 10^{13} \text{ kilometers})}{1 \text{ parsec}}$

$$= 491.31 \times 10^{13} \text{ kilometers}$$

$d = r \times t$

$$491.31 \times 10^{13} \text{ km} = \frac{50{,}000 \text{ km}}{1 \text{ hr}} \times t$$

$$4.9131 \times 10^{15} \text{ km} = \frac{5 \times 10^4 \text{ km}}{1 \text{ hr}} \times t$$

$$\frac{4.9131 \times 10^{15} \text{ km}}{\dfrac{5 \times 10^4 \text{ km}}{1 \text{ hr}}} = t$$

$$\frac{4.9131 \times 10^{15} \text{ km} \,(1 \text{ hr})}{5.0 \times 10^4 \text{ km}} = t$$

$$0.98262 \times 10^{11} \text{ hr} = t$$

$$9.83 \times 10^{10} \text{ hours}$$

4.3 Practice Problems

1. (a) This polynomial is of degree 5. It has two terms, so it is a binomial.

   (b) This polynomial is of degree 7, since the sum of the exponents is $3 + 4 = 7$. It has one term, so it is a monomial.

   (c) This polynomial is of degree 3. It has three terms, so it is a trinomial.

2. $(-8x^3 + 3x^2 + 6) + (2x^3 - 7x^2 - 3)$

$$= [-8x^3 + 2x^3] + [3x^2 - 7x^2] + [6 - 3]$$

$$= [(-8 + 2)x^3] + [(3 - 7)x^2] + [6 - 3]$$

$$= -6x^3 - 4x^2 + 3$$

3. $\left(-\dfrac{1}{3}x^2 - 6x - \dfrac{1}{12}\right) + \left(\dfrac{1}{4}x^2 + 5x - \dfrac{1}{3}\right)$

$$= \left[-\frac{1}{3}x^2 + \frac{1}{4}x^2\right] + [-6x + 5x] + \left[-\frac{1}{12} - \frac{1}{3}\right]$$

$$= \left[\left(-\frac{1}{3} + \frac{1}{4}\right)x^2\right] + [(-6 + 5)x] + \left[-\frac{1}{12} - \frac{1}{3}\right]$$

$$= \left[\left(-\frac{4}{12} + \frac{3}{12}\right)x^2\right] + [-x] + \left[-\frac{1}{12} - \frac{4}{12}\right]$$

$$= -\frac{1}{12}x^2 - x - \frac{5}{12}$$

4. $(3.5x^3 - 0.02x^2 + 1.56x - 3.5) + (-0.08x^2 - 1.98x + 4)$

$$= 3.5x^3 + [-0.02 - 0.08]x^2 + [1.56 - 1.98]x + [-3.5 + 4]$$

$$= 3.5x^3 - 0.1x^2 - 0.42x + 0.5$$

5. $(5x^3 - 15x^2 + 6x - 3) - (-4x^3 - 10x^2 + 5x + 13)$

$$= (5x^3 - 15x^2 + 6x - 3) + (4x^3 + 10x^2 - 5x - 13)$$

$$= [5 + 4]x^3 + [-15 + 10]x^2 + [6 - 5]x + [-3 - 13]$$

$$= 9x^3 - 5x^2 + x - 16$$

6. $(x^3 - 7x^2y + 3xy^2 - 2y^3) - (2x^3 + 4xy - 6y^3)$

$$= [1 - 2]x^3 - 7x^2y + 3xy^2 - 4xy + [-2 + 6]y^3$$

$$= -x^3 - 7x^2y + 3xy^2 - 4xy + 4y^3$$

7. (a) 1974 is 4 years later than 1970, so $x = 4$.

$$0.03(4) + 5.4 = 0.12 + 5.4$$

$$= 5.52$$

We estimate that the average truck in 1974 obtained 5.52 miles per gallon.

   (b) 2006 is 36 years later than 1970, so $x = 36$.

$$0.03(36) + 5.4 = 1.08 + 5.4$$

$$= 6.48$$

We predict that the average truck in 2006 will obtain 6.48 miles per gallon.

4.4 Practice Problems

1. $4x^3(-2x^2 + 3x) = 4x^3(-2x^2) + 4x^3(3x)$

$$= 4(-2)(x^3)(x^2) + 4(3)(x^3)(x)$$

$$= -8x^5 + 12x^4$$

2. (a) $-3x(x^2 + 2x - 4) = -3x^3 - 6x^2 + 12x$

   (b) $6xy(x^3 + 2x^2y - y^2) = 6x^4y + 12x^3y^2 - 6xy^3$

3. $(-6x^3 + 4x^2 - 2x)(-3xy) = 18x^4y - 12x^3y + 6x^2y$

4. $(5x - 1)(x - 2) = 5x^2 - 10x - x + 2 = 5x^2 - 11x + 2$

5. $(8a - 5b)(3a - b) = 24a^2 - 8ab - 15ab + 5b^2$

$$= 24a^2 - 23ab + 5b^2$$

6. $(3a + 2b)(2a - 3c) = 6a^2 - 9ac + 4ab - 6bc$

7. $(3x - 2y)(3x - 2y) = 9x^2 - 6xy - 6xy + 4y^2$

$$= 9x^2 - 12xy + 4y^2$$

8. $(2x^2 + 3y^2)(5x^2 + 6y^2) = 10x^4 + 12x^2y^2 + 15x^2y^2 + 18y^4$

$$= 10x^4 + 27x^2y^2 + 18y^4$$

9. $A = (\text{length})(\text{width}) = (7x + 3)(2x - 1)$

$$= 14x^2 - 7x + 6x - 3$$

$$= 14x^2 - x - 3$$

There are $(14x^2 - x - 3)$ square feet in the room.

4.5 Practice Problems

1. $(6x + 7)(6x - 7) = (6x)^2 - (7)^2 = 36x^2 - 49$

2. $(3x + 5y)(3x - 5y) = (3x)^2 - (5y)^2 = 9x^2 - 25y^2$

3. (a) $(5x + 4)^2 = (5x)^2 + 2(5x)(4) + (4)^2 = 25x^2 + 40x + 16$

   (b) $(4a - 9b)^2 = (4a)^2 - 2(4a)(9b) + (9b)^2$

$$= 16a^2 - 72ab + 81b^2$$

4.

$$
\begin{array}{r}
3x^2 - 2xy + 4y^2 \\
x - 2y \\
\hline
-6x^2y + 4xy^2 - 8y^3 \\
+\,3x^3 - 2x^2y + 4xy^2 \\
\hline
3x^3 - 8x^2y + 8xy^2 - 8y^3
\end{array}
$$

5. $(2x^2 - 5x + 3)(x^2 + 3x - 4)$

$$= 2x^2(x^2 + 3x - 4) - 5x(x^2 + 3x - 4) + 3(x^2 + 3x - 4)$$

$$= 2x^4 + 6x^3 - 8x^2 - 5x^3 - 15x^2 + 20x + 3x^2 + 9x - 12$$

$$= 2x^4 + x^3 - 20x^2 + 29x - 12$$

6. $(3x - 2)(2x + 3)(3x + 2) = (3x - 2)(3x + 2)(2x + 3)$

$$= [(3x)^2 - 2^2](2x + 3)$$

$$= (9x^2 - 4)(2x + 3)$$

$$
\begin{array}{r}
9x^2 - 4 \\
2x + 3 \\
\hline
27x^2 + 0x - 12 \\
18x^3 + 0x^2 - 8x \\
\hline
18x^3 + 27x^2 - 8x - 12
\end{array}
$$

Thus we have

$$(3x - 2)(2x + 3)(3x + 2) = 18x^3 + 27x^2 - 8x - 12.$$

4.6 Practice Problems

1. $\dfrac{15y^4 - 27y^3 - 21y^2}{3y^2} = \dfrac{15y^4}{3y^2} - \dfrac{27y^3}{3y^2} - \dfrac{21y^2}{3y^2} = 5y^2 - 9y - 7$

2.

$$
\begin{array}{r}
x^2 + 6x + 7 \\
x + 4\,\overline{)\,x^3 + 10x^2 + 31x + 35} \\
\underline{x^3 + 4x^2} \\
6x^2 + 31x \\
\underline{6x^2 + 24x} \\
7x + 35 \\
\underline{7x + 28} \\
7
\end{array}
$$

Ans. $x^2 + 6x + 7 + \dfrac{7}{x + 4}$

3.

$$
\begin{array}{r}
2x^2 + x + 1 \\
x - 1 \overline{\smash{\big)}\ 2x^3 - x^2 + 0x + 1} \\
\underline{2x^3 - 2x^2} \\
x^2 + 0x \\
\underline{x^2 - x} \\
x + 1 \\
\underline{x - 1} \\
2
\end{array}
$$

Ans. $2x^2 + x + 1 + \dfrac{2}{x - 1}$

4.

$$
\begin{array}{r}
5x^2 + x - 2 \\
4x - 3 \overline{\smash{\big)}\ 20x^3 - 11x^2 - 11x + 6} \\
\underline{20x^3 - 15x^2} \\
4x^2 - 11x \\
\underline{4x^2 - 3x} \\
-8x + 6 \\
\underline{-8x + 6} \\
0
\end{array}
$$

Ans. $5x^2 + x - 2$

Check. $(4x - 3)(5x^2 + x - 2)$
$= 20x^3 + 4x^2 - 8x - 15x^2 - 3x + 6$
$= 20x^3 - 11x^2 - 11x + 6$

Chapter 5

5.1 Practice Problems

1. (a) $7(3a - b)$ **(b)** $p(1 + rt)$
2. $4a(3a + 4b^2 - 3ab)$
3. (a) $8(2a^3 - 3b^3)$ **(b)** $r^3(s^2 - 4rs + 7r^2)$
4. $9ab^2c(2a^2 - 3bc - 5ac)$
5. $6xy^2(5x^2 - 4x + 1) = 30x^3y^2 - 24x^2y^2 + 6xy^2$
6. $(a + b)(3 + x)$ **7.** $(8y - 1)(9y^2 - 2)$
8. $\pi b^2 - \pi a^2 = \pi(b^2 - a^2)$

5.2 Practice Problems

1. $(2x - 7)(3y - 8)$
2. $3x(2x - 5) + 2(2x - 5)$
$= (2x - 5)(3x + 2)$
3. $a(x + 2) + 4b(x + 2)$
$= (x + 2)(a + 4b)$
4. $6a^2 + 3ac + 10ab + 5bc$
$= 3a(2a + c) + 5b(2a + c)$
$= (2a + c)(3a + 5b)$
5. $6xy + 14x - 15y - 35$
$= 2x(3y + 7) - 5(3y + 7)$
$= (3y + 7)(2x - 5)$
6. $3x + 6y - 5ax - 10ay$
$= 3(x + 2y) - 5a(x + 2y)$
$= (3 - 5a)(x + 2y)$
7. $10ad + 27bc - 6bd - 45ac$
$= 10ad - 6bd - 45ac + 27bc$
$= 2d(5a - 3b) - 9c(5a - 3b)$
$= (2d - 9c)(5a - 3b)$
Check: $10ad - 6bd - 45ac + 27bc$

5.3 Practice Problems

1. $(x + 6)(x + 2)$ **2.** $(x + 2)(x + 15)$
3. $(x - 9)(x - 2)$ **4.** $(y - 3)(y - 8)$
5. $(a + 3)(a - 8)$
6. $(x + 20)(x - 3)$
Check: $x^2 - 3x + 20x - 60$
$= x^2 + 17x - 60$
7. $(x + 5)(x - 12)$ **8.** $(a^2 + 7)(a^2 - 6)$
9. $3x^2 + 45x + 150$
$= 3(x^2 + 15x + 50)$
$= 3(x + 5)(x + 10)$

10. $4x^2 - 8x - 140$
$= 4(x^2 - 2x - 35)$
$= 4(x + 5)(x - 7)$
11. $x(x + 1) - 4(5)$
$= x^2 + x - 20$
$= (x + 5)(x - 4)$

5.4 Practice Problems

1. $(2x - 5)(x - 1)$ **2.** $(9x - 1)(x - 7)$
3. $(3x - 7)(x + 2)$
4. $2x^2 - 2x - 5x + 5$
$= 2x(x - 1) - 5(x - 1)$
$= (2x - 5)(x - 1)$
5. $9x^2 - 63x - x + 7$
$= 9x(x - 7) - 1(x - 7)$
$= (x - 7)(9x - 1)$
6. $3x^2 + 6x - 2x - 4$
$= 3x(x + 2) - 2(x + 2)$
$= (x + 2)(3x - 2)$
7. $8x^2 - 8x - 6$
$= 2(4x^2 - 4x - 3)$
$= 2(2x + 1)(2x - 3)$
8. $24x^2 - 38x + 10$
$= 2(12x^2 - 19x + 5)$
$= 2(12x^2 - 4x - 15x + 5)$
$= 2[4x(3x - 1) - 5(3x - 1)]$
$= 2(4x - 5)(3x - 1)$

5.5 Practice Problems

1. $(1 + 8x)(1 - 8x)$ **2.** $(6x + 7)(6x - 7)$
3. $(10x + 9y)(10x - 9y)$
4. $(x^4 + 1)(x^4 - 1)$
$= (x^4 + 1)(x^2 + 1)(x^2 - 1)$
$= (x^4 + 1)(x^2 + 1)(x + 1)(x - 1)$
5. $(4x + 1)(4x + 1) = (4x + 1)^2$
6. $(5x - 3)(5x - 3) = (5x - 3)^2$
7. (a) $(5x - 6y)(5x - 6y) = (5x - 6y)^2$
     (b) $(8x^3 - 3)(8x^3 - 3) = (8x^3 - 3)^2$
8. $9x^2 - 12x - 3x + 4$
$= 3x(3x - 4) - 1(3x - 4)$
$= (3x - 4)(3x - 1)$
9. $20x^2 - 45$
$= 5(4x^2 - 9)$
$= 5(2x - 3)(2x + 3)$
10. $3(25x^2 - 20x + 4)$
$= 3(5x - 2)(5x - 2)$
$= 3(5x - 2)^2$

5.6 Practice Problems

1. (a) $9y^2(x^4 - 1)$
$= 9y^2(x^2 + 1)(x^2 - 1)$
$= 9y^2(x^2 + 1)(x + 1)(x - 1)$
     (b) $-(4x^2 - 12x + 9)$
$= -(2x - 3)(2x - 3)$
$= -(2x - 3)^2$
     (c) $3(x^2 - 12x + 36)$
$= 3(x - 6)(x - 6)$
$= 3(x - 6)^2$
     (d) $5x(x^2 - 3xy + 2x - 6y)$
$= 5x[x(x - 3y) + 2(x - 3y)]$
$= 5x(x + 2)(x - 3y)$

2. $x^2 - 9x - 8$

The factors of -8 are

$$(-2)(4) = -8$$
$$(2)(-4) = -8$$
$$(-8)(1) = -8$$
$$(-1)(8) = -8$$

None of these pairs will add up to be the coefficient of the middle term. Thus the polynomial cannot be factored. It is prime.

3. $25x^2 + 82x + 4$

Check to see if this is a perfect square trinomial.

$$2[(5)(2)] = 2(10) = 20$$

This is not the coefficient of the middle term. The grouping number equals 100. No factors add to 82. It is prime.

5.7 Practice Problems

1. $10x^2 - x - 2 = 0$

$$(5x + 2)(2x - 1) = 0 \qquad 5x + 2 = 0 \qquad 2x - 1 = 0$$
$$5x = -2 \qquad 2x = 1$$
$$x = -\frac{2}{5} \qquad x = \frac{1}{2}$$

Check. $10\left(-\frac{2}{5}\right)^2 - \left(-\frac{2}{5}\right) - 2 \stackrel{?}{=} 0 \qquad 10\left(\frac{1}{2}\right)^2 - \frac{1}{2} - 2 \stackrel{?}{=} 0$

$$10\left(\frac{4}{25}\right) + \frac{2}{5} - 2 \stackrel{?}{=} 0 \qquad 10\left(\frac{1}{4}\right) - \frac{1}{2} - 2 \stackrel{?}{=} 0$$
$$\frac{8}{5} + \frac{2}{5} - 2 \stackrel{?}{=} 0 \qquad \frac{5}{2} - \frac{1}{2} - 2 \stackrel{?}{=} 0$$
$$\frac{10}{5} - \frac{10}{5} \stackrel{?}{=} 0 \qquad \frac{4}{2} - 2 \stackrel{?}{=} 0$$
$$0 = 0 \qquad 2 - 2 \stackrel{?}{=} 0$$
$$0 = 0$$

Thus $-\frac{2}{5}$ and $\frac{1}{2}$ are both roots for the equation.

2. $3x^2 - 5x + 2 = 0 \qquad 3x - 2 = 0 \qquad x - 1 = 0$

$$(3x - 2)(x - 1) = 0 \qquad 3x = 2 \qquad x = 1$$
$$x = \frac{2}{3}$$

3. $7x^2 + 11x = 0 \qquad x = 0 \qquad 7x + 11 = 0$

$$x(7x + 11) = 0 \qquad 7x = -11$$
$$x = \frac{-11}{7}$$

4. $x^2 - 6x + 4 = -8 + x \qquad x - 3 = 0 \qquad x - 4 = 0$

$$x^2 - 7x + 12 = 0 \qquad x = 3 \qquad x = 4$$
$$(x - 3)(x - 4) = 0$$

5. $\frac{2x^2 - 7x}{3} = 5$

$$2x^2 - 7x = 15 \qquad 2x + 3 = 0 \qquad x - 5 = 0$$
$$2x^2 - 7x - 15 = 0 \qquad x = -\frac{3}{2} \qquad x = 5$$
$$(2x + 3)(x - 5) = 0$$

6. Let $w =$ width, then

$3w + 2 =$ length.

$$(3w + 2)w = 85 \qquad 3w + 17 = 0 \qquad w - 5 = 0$$
$$3w^2 + 2w = 85 \qquad w = -\frac{17}{3} \qquad w = 5$$
$$3w^2 + 2w - 85 = 0$$
$$(3w + 17)(w - 5) = 0$$

The only valid answer is width $= 5$ meters.

$$\text{length} = 3(5) + 2 = 17 \text{ meters}$$

7. Let $b =$ base.

$b - 3 =$ altitude.

$$\frac{b(b - 3)}{2} = 35$$
$$b^2 - 3b = 70$$
$$b^2 - 3b - 70 = 0$$
$$(b + 7)(b - 10) = 0$$

$$b + 7 = 0 \qquad b - 10 = 0$$
$$b = -7 \qquad b = 10$$

This is not a valid answer. Thus the base $= 10$ centimeters.

altitude $= 10 - 3 = 7$ centimeters

8. $-5t^2 + 45 = 0$

$$-5(t^2 - 9) = 0$$
$$t^2 - 9 = 0$$
$$(t + 3)(t - 3) = 0$$
$$t = 3 \qquad t = -3 \qquad t = -3 \text{ is not a valid answer}$$

Thus it will be 3 seconds before he breaks the water's surface.

Chapter 6

6.1 Practice Problems

1. (a) $\frac{7}{3} = \frac{?}{18}$

$$\frac{7 \cdot 6}{3 \cdot 6} = \frac{42}{18}$$

(b) $\frac{28}{63} = \frac{7 \times 2 \times 2}{7 \times 3 \times 3} = \frac{4}{9}$

2. $\frac{12x - 6}{14x - 7} = \frac{6(2x - 1)}{7(2x - 1)} = \frac{6}{7}$

3. $\frac{4x - 6}{2x^2 - x - 3} = \frac{2(2x - 3)}{(2x - 3)(x + 1)} = \frac{2}{x + 1}$

4. $\frac{x^3 + 11x^2 + 30x}{3x^3 + 17x^2 - 6x} = \frac{x(x^2 + 11x + 30)}{x(3x^2 + 17x - 6)} = \frac{(x + 5)(x + 6)}{(3x - 1)(x + 6)}$

$$= \frac{x + 5}{3x - 1}$$

5. $\frac{2x - 5}{5 - 2x} = \frac{-1(-2x + 5)}{(5 - 2x)} = \frac{-1(5 - 2x)}{(5 - 2x)} = -1$

6. $\frac{4x^2 + 3x - 10}{25 - 16x^2} = \frac{(4x - 5)(x + 2)}{(5 + 4x)(5 - 4x)} = \frac{(4x - 5)(x + 2)}{-1(4x - 5)(5 + 4x)}$

$$= \frac{x + 2}{-1(5 + 4x)} = -\frac{x + 2}{5 + 4x}$$

7. $\frac{4x^2 - 9y^2}{4x^2 + 12xy + 9y^2} = \frac{(2x + 3y)(2x - 3y)}{(2x + 3y)(2x + 3y)} = \frac{2x - 3y}{2x + 3y}$

8. $\frac{12x^3 - 48x}{6x - 3x^2} = \frac{12x(x^2 - 4)}{3x(2 - x)} = \frac{12x(x + 2)(x - 2)}{3x(2 - x)} = -4(x + 2)$

6.2 Practice Problems

1. $\frac{10x}{x^2 - 7x + 10} \cdot \frac{x^2 + 3x - 10}{25x}$

$$= \frac{10x}{(x - 5)(x - 2)} \cdot \frac{(x - 2)(x + 5)}{25x}$$
$$= \frac{2(x + 5)}{5(x - 5)}$$

2. $\frac{2y^2 - 6y - 8}{y^2 - y - 2} \cdot \frac{y^2 - 5y + 6}{2y^2 - 4y - 6}$

$$= \frac{2(y + 1)(y - 4)}{(y + 1)(y - 2)} \cdot \frac{(y - 2)(y - 3)}{2(y + 1)(y - 3)} = \frac{y - 4}{y + 1}$$

3. $\frac{x^2 + 5x + 6}{x^2 + 8x} \div \frac{2x^2 + 5x + 2}{2x^2 + x}$

$$= \frac{(x + 2)(x + 3)}{x(x + 8)} \cdot \frac{x(2x + 1)}{(2x + 1)(x + 2)} = \frac{x + 3}{x + 8}$$

4. $\frac{x + 3}{x - 3} \div (9 - x^2) = \frac{x + 3}{x - 3} \cdot \frac{1}{(3 + x)(3 - x)}$

$$= \frac{1}{(x - 3)(3 - x)}$$

6.3 Practice Problems

1. $\dfrac{2s+t}{2s-t} + \dfrac{s-t}{2s-t} = \dfrac{2s+t+s-t}{2s-t} = \dfrac{3s}{2s-t}$

2. $\dfrac{b}{(a-2b)(a+b)} - \dfrac{2b}{(a-2b)(a+b)} = \dfrac{b-2b}{(a-2b)(a+b)}$

$= \dfrac{-b}{(a-2b)(a+b)}$

3. $\dfrac{7}{6x+21}, \dfrac{13}{10x+35}$ $\qquad 6x+21 = 3(2x+7)$
$\qquad\qquad\qquad\qquad\qquad\quad 10x+35 = 5(2x+7)$
$\qquad\qquad\qquad\qquad\quad \text{LCD} = 3\cdot 5\cdot(2x+7) = 15(2x+7)$

4. (a) $\dfrac{3}{50xy^2z}, \dfrac{19}{40x^3yz}$ $\qquad 50xy^2z = 2\cdot 5^2\cdot x\cdot y^2\cdot z$
$\qquad\qquad\qquad\qquad\qquad\quad 40x^3yz = 2^3\cdot 5\cdot x^3\cdot y\cdot z$
$\qquad\qquad\qquad\qquad\qquad \text{LCD} = 2^3\cdot 5^2\cdot x^3\cdot y^2\cdot z$
$\qquad\qquad\qquad\qquad\qquad \text{LCD} = 200x^3y^2z$

(b) $\dfrac{2}{x^2+5x+6}, \dfrac{6}{3x^2+5x-2}$
$\qquad x^2+5x+6 = (x+3)(x+2)$
$\qquad 3x^2+5x-2 = (3x-1)(x+2)$
$\qquad\qquad \text{LCD} = (x+2)(x+3)(3x-1)$

5. $\dfrac{7}{a} + \dfrac{3}{abc} = \dfrac{7bc+3}{abc} \qquad \text{LCD} = abc$

6. $\dfrac{2a-b}{(a+2b)(a-2b)} + \dfrac{2}{(a+2b)} \qquad \text{LCD} = (a+2b)(a-2b)$

$= \dfrac{2a-b}{(a+2b)(a-2b)} + \dfrac{2(a-2b)}{(a+2b)(a-2b)}$

$= \dfrac{2a-b+2a-4b}{(a+2b)(a-2b)} = \dfrac{4a-5b}{(a+2b)(a-2b)}$

7. $\dfrac{7a}{a^2+2ab+b^2} + \dfrac{4}{a^2+ab}$

$= \dfrac{7a}{(a+b)(a+b)} + \dfrac{4}{a(a+b)} \qquad \text{LCD} = a(a+b)^2$

$= \dfrac{7a^2}{a(a+b)(a+b)} + \dfrac{4(a+b)}{a(a+b)(a+b)} = \dfrac{7a^2+4a+4b}{a(a+b)^2}$

8. $\dfrac{x+7}{3x-9} - \dfrac{x-6}{x-3} = \dfrac{x+7}{3(x-3)} - \dfrac{x-6}{x-3} \qquad \text{LCD} = 3(x-3)$

$= \dfrac{x+7-3(x-6)}{3(x-3)} = \dfrac{x+7-3x+18}{3(x-3)} = \dfrac{-2x+25}{3(x-3)}$

9. $\dfrac{x-2}{x^2-4} - \dfrac{x+1}{2x^2+4x} = \dfrac{x-2}{(x+2)(x-2)} - \dfrac{x+1}{2x(x+2)}$

$\text{LCD} = 2x(x+2)(x-2)$

$= \dfrac{2x(x-2)}{2x(x+2)(x-2)} - \dfrac{(x-2)(x+1)}{2x(x+2)(x-2)}$

$= \dfrac{2x^2-4x-(x^2-x-2)}{2x(x+2)(x-2)} = \dfrac{2x^2-4x-x^2+x+2}{2x(x+2)(x-2)}$

$= \dfrac{x^2-3x+2}{2x(x+2)(x-2)} = \dfrac{(x-2)(x-1)}{2x(x+2)(x-2)} = \dfrac{x-1}{2x(x+2)}$

6.4 Practice Problems

1. $\dfrac{\frac{1}{a}+\frac{1}{b}}{\frac{2}{ab^2}} = \dfrac{\frac{b+a}{ab}}{\frac{2}{ab^2}}$

$= \dfrac{b+a}{ab} \div \dfrac{2}{ab^2} = \dfrac{b+a}{ab} \cdot \dfrac{ab^2}{2} = \dfrac{ab^2(a+b)}{2ab} = \dfrac{b(a+b)}{2}$

2. $\dfrac{\frac{1}{a}+\frac{1}{b}}{\frac{1}{a}-\frac{1}{b}} = \dfrac{\frac{b+a}{ab}}{\frac{b-a}{ab}} = \dfrac{b+a}{ab}\cdot\dfrac{ab}{b-a} = \dfrac{b+a}{b-a}$

3. $\dfrac{\frac{x}{x^2+4x+3}+\frac{2}{x+1}}{x+1} = \dfrac{\frac{x}{(x+1)(x+3)}+\frac{2}{x+1}}{x+1}$

$= \dfrac{\frac{x+2(x+3)}{(x+1)(x+3)}}{x+1} = \dfrac{\frac{x+2x+6}{(x+1)(x+3)}}{(x+1)}$

$= \dfrac{3x+6}{(x+1)(x+3)} \cdot \dfrac{1}{(x+1)}$

$= \dfrac{3(x+2)}{(x+1)^2(x+3)}$

4. $\dfrac{\frac{6}{x^2-y^2}}{\frac{1}{x-y}+\frac{3}{x+y}} = \dfrac{\frac{6}{(x+y)(x-y)}}{\frac{(x+y)+3(x-y)}{(x+y)(x-y)}} = \dfrac{6}{(x+y)(x-y)}$

$\div \dfrac{x+y+3x-3y}{(x+y)(x-y)} = \dfrac{6}{(x+y)(x-y)} \cdot \dfrac{(x+y)(x-y)}{4x-2y}$

$= \dfrac{6}{(x+y)(x-y)} \cdot \dfrac{(x+y)(x-y)}{2(2x-y)} = \dfrac{3}{2x-y}$

5. $\dfrac{\frac{2}{3x^2}-\frac{3}{y}}{\frac{5}{xy}-4} \qquad \text{LCD} = 3x^2y$

$\dfrac{3x^2y\left(\frac{2}{3x^2}\right) - 3x^2y\left(\frac{3}{y}\right)}{3x^2y\left(\frac{5}{xy}\right) - 3x^2y(4)} = \dfrac{2y-3x^2(3)}{3x(5)-12x^2y} = \dfrac{2y-9x^2}{15x-12x^2y}$

6. $\dfrac{\frac{6}{x^2-y^2}}{\frac{7}{x-y}+\frac{3}{x+y}} = \dfrac{\frac{6}{(x+y)(x-y)}}{\frac{7}{x-y}+\frac{3}{x+y}}$

$\text{LCD} = (x+y)(x-y)$

$\dfrac{(x+y)(x-y)\left(\frac{6}{(x+y)(x-y)}\right)}{(x+y)(x-y)\left(\frac{7}{x-y}\right) + (x+y)(x-y)\left(\frac{3}{x+y}\right)}$

$= \dfrac{6}{7(x+y)+3(x-y)} = \dfrac{6}{7x+7y+3x-3y} = \dfrac{6}{10x+4y}$

$= \dfrac{3}{5x+2y}$

6.5 Practice Problems

1. $\dfrac{2}{x}+\dfrac{4}{x} = 3 - \dfrac{1}{x} - \dfrac{17}{8}$

$\qquad \text{LCD} = 8x$

$\quad 16+32 = 24x-8-17x \qquad Check. \quad \dfrac{2}{8}+\dfrac{4}{8}\overset{?}{=}3-\dfrac{1}{8}-\dfrac{17}{8}$

$\qquad\qquad 48 = 7x-8 \qquad\qquad\qquad\qquad \dfrac{6}{8}\overset{?}{=}3-\dfrac{18}{8}$

$\qquad\qquad 56 = 7x \qquad\qquad\qquad\qquad\quad \dfrac{6}{8}\overset{?}{=}\dfrac{24}{8}-\dfrac{18}{8}$

$\qquad\qquad\quad x = 8 \qquad\qquad\qquad\qquad\quad \dfrac{6}{8}=\dfrac{6}{8} \checkmark$

2. $\dfrac{4}{2x+1} = \dfrac{6}{2x-1} \qquad \text{LCD} = (2x+1)(2x-1)$

$(2x+1)(2x-1)\left[\dfrac{4}{2x+1}\right] = (2x+1)(2x-1)\left[\dfrac{6}{2x-1}\right]$

$\qquad\qquad 4(2x-1) = 6(2x+1)$

$\qquad\qquad 8x-4 = 12x+6$

$\qquad\qquad -4x = 10$

$\qquad\qquad\quad x = -\dfrac{5}{2}$

Check. $\dfrac{4}{2\left(-\dfrac{5}{2}\right)+1} \overset{?}{=} \dfrac{6}{2\left(-\dfrac{5}{2}\right)-1}$

$$\dfrac{4}{-5+1} \overset{?}{=} \dfrac{6}{-5-1}$$

$$\dfrac{4}{-4} \overset{?}{=} \dfrac{6}{-6}$$

$$-1 = -1 \ \checkmark$$

3. $\dfrac{x-1}{x^2-4} = \dfrac{2}{x+2} + \dfrac{4}{x-2}$

$\dfrac{x-1}{(x+2)(x-2)} = \dfrac{2}{x+2} + \dfrac{4}{x-2}$ LCD $= (x+2)(x-2)$

$(x+2)(x-2)\left[\dfrac{x-1}{(x+2)(x-2)}\right]$

$\qquad = (x+2)(x-2)\left[\dfrac{2}{x+2}\right] + (x+2)(x-2)\left[\dfrac{4}{x-2}\right]$

$x-1 = 2(x-2) + 4(x+2)$

$x-1 = 2x-4 + 4x+8$

$x-1 = 6x+4$

$-5x = 5$

$x = -1$

Check. $\dfrac{-1-1}{(-1)^2-4} \overset{?}{=} \dfrac{2}{-1+2} + \dfrac{4}{-1-2}$

$$\dfrac{-2}{-3} \overset{?}{=} \dfrac{2}{1} + \dfrac{4}{-3}$$

$$\dfrac{2}{3} \overset{?}{=} \dfrac{6}{3} - \dfrac{4}{3}$$

$$\dfrac{2}{3} = \dfrac{2}{3} \ \checkmark$$

4. $\dfrac{2x}{x+1} = \dfrac{-2}{x+1} + 1$ LCD $= (x+1)$

$(x+1)\left[\dfrac{2x}{x+1}\right] = (x+1)\left[\dfrac{-2}{x+1}\right] + (x+1)[1]$

$2x = -2 + x + 1$

$2x = x - 1$

$x = -1$

Check. $\dfrac{2(-1)}{-1+1} \overset{?}{=} \dfrac{-2}{-1+1} + 1$

$$\dfrac{-2}{0} \overset{?}{=} \dfrac{-2}{0} + 1$$

These expressions are not defined; therefore, there is no solution to this problem.

6.6 Practice Problems

1. $\dfrac{8}{420} = \dfrac{x}{315}$

$8(315) = 420x$

$2520 = 420x$

$x = 6$

It would take Brenda 6 hours to drive 315 miles.

2. $\dfrac{\dfrac{5}{8}}{30} = \dfrac{2\dfrac{1}{2}}{x}$

$\dfrac{5}{8}x = 30\left(2\dfrac{1}{2}\right)$

$\dfrac{5}{8}x = 75$

$x = 120$

Therefore $2\dfrac{1}{2}$ inches would represent 120 miles.

3. $\dfrac{13}{x} = \dfrac{16}{18}$

$13(18) = 16x$

$234 = 16x$

$x = 14\dfrac{5}{8}$ cm

4. $\dfrac{6}{7} = \dfrac{x}{38.5}$

$6(38.5) = 7x$

$231 = 7x$

$x = 33$ feet

5. Train A time $= \dfrac{180}{x+10}$ Train B time $= \dfrac{150}{x}$

$\dfrac{180}{x+10} = \dfrac{150}{x}$

$180x = 150(x+10)$

$180x = 150x + 1500$

$30x = 1500$

$x = 50$

Train B travels 50 kilometers per hour. Train A travels $50 + 10 = 60$ kilometers per hour.

6.

	Number of Hours	Part of the Job Done in One Hour
John	6 hours	$\dfrac{1}{6}$
Dave	7 hours	$\dfrac{1}{7}$
John & Dave Together	x	$\dfrac{1}{x}$

$\dfrac{1}{6} + \dfrac{1}{7} = \dfrac{1}{x}$ LCD $= 42x$

$7x + 6x = 42$

$13x = 42$

$x = 3\dfrac{3}{13}$ $\dfrac{3}{13}$ hour $\times \dfrac{60 \text{ min}}{1 \text{ hour}} = \dfrac{180}{13}$ min $= 13.846$ min

Thus, doing the job together will take 3 hours and 14 minutes.

Chapter 7

7.1 Practice Problems

1. Point B is 3 units to the right on the x-axis and 4 units up from the point where we stopped on the x-axis.

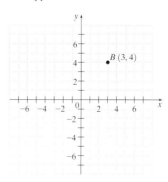

2. (a) Begin by counting 2 squares to the left starting at the origin. Since the y-coordinate is negative, count 4 units down from the point where we stopped on the x-axis. Label the point I.

(b) Begin by counting 4 squares to the left of the origin. Then count 5 units up because the y-coordinate is positive. Label the point J.

(c) Begin by counting 4 units to the right of the origin. Then count 2 units down because the y-coordinate is negative. Label the point K.

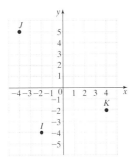

3. The points are plotted in the figure.

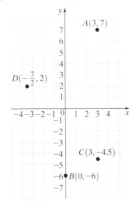

4. Move along the x-axis to get as close as possible to B. We end up at 2. Thus the first number of the ordered pair is 2. Then count 7 units upward on a line parallel to the y-axis to reach B. So the second number of the ordered pair is 7. Thus, $B = (2, 7)$.

5. $A = (-2, -1)$; $B = (-1, 3)$; $C = (0, 0)$; $D = (2, -1)$; $E = (3, 1)$

6. (a) Replace x by 0 in the equation.
$$3(0) - 4y = 12$$
$$0 - 4y = 12$$
$$y = -3 \qquad (0, -3)$$

(b) Replace the variable y by 3.
$$3x - 4(3) = 12$$
$$3x - 12 = 12$$
$$3x = 24$$
$$x = 8 \qquad (8, 3)$$

(c) Replace the variable y by -6.
$$3x - 4(-6) = 12$$
$$3x + 24 = 12$$
$$3x = -12$$
$$x = -4 \qquad (-4, -6)$$

7. (a)

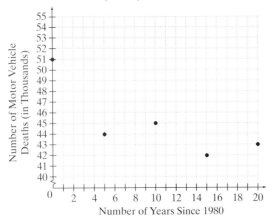

Number of Motor Vehicle Deaths (in Thousands) vs Number of Years Since 1980

(b) 1980 was a significant high in motor vehicle deaths. During 1985–2000, the number of motor vehicle deaths was relatively stable.

7.2 Practice Problems

1. Graph $x + y = 10$.
Let $x = 0$.
$$0 + y = 10$$
$$y = 10$$

Let $x = 5$.
$$5 + y = 10$$
$$y = 5$$

Let $x = 2$.
$$2 + y = 10$$
$$y = 8$$

Plot the ordered pairs.
$(0, 10), (5, 5),$ and $(2, 8)$

2.
$$7x + 3 = -2y + 3$$
$$7x + 3 - 3 = -2y + 3 - 3$$
$$7x = -2y$$
$$7x + 2y = -2y + 2y$$
$$7x + 2y = 0$$

Let $x = 0$.
$$7(0) + 2y = 0$$
$$2y = 0$$
$$y = 0$$

Let $x = -2$.
$$7(-2) + 2y = 0$$
$$-14 + 2y = 0$$
$$2y = 14$$
$$y = 7$$

Let $x = 2$.
$$7(2) + 2y = 0$$
$$14 + 2y = 0$$
$$2y = -14$$
$$y = -7$$

Graph the ordered pairs. $(0, 0), (2, -7),$ and $(-2, 7)$

3. $2y - x = 6$
Find the two intercepts.

Let $x = 0$.	Let $y = 0$.
$2y - 0 = 6$	$2(0) - x = 6$
$2y = 6$	$-x = 6$
$y = 3$	$x = -6$

Find the third point.
Let $y = 1$.
$$2(1) - x = 6$$
$$2 - x = 6$$
$$-x = 4$$
$$x = -4$$

Graph the ordered pairs.
$(0, 3), (-6, 0),$ and $(-4, 1)$

4. $2y - 3 = 0$
Solve for y.
$$2y = 3$$
$$y = \frac{3}{2}$$

This line is parallel to the x-axis. It is a horizontal line $1\frac{1}{2}$ units above the x-axis.

5. $x + 3 = 0$
Solve for x.
$$x = -3$$

This line is parallel to the y-axis. It is a vertical line 3 units to the left of the y-axis.

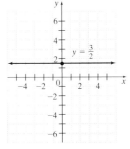

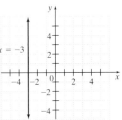

7.3 Practice Problems

1. $m = \dfrac{y_2 - y_1}{x_2 - x_1} = \dfrac{-1 - 1}{-4 - 6} = \dfrac{-2}{-10} = \dfrac{1}{5}$

2. $m = \dfrac{y_2 - y_1}{x_2 - x_1} = \dfrac{1 - 0}{-1 - 2} = \dfrac{1}{-3} = -\dfrac{1}{3}$

3. (a) $m = \dfrac{4 - 0}{-4 - (-4)} = \dfrac{4}{0}$

$\dfrac{4}{0}$ is undefined. Therefore there is no slope and the line is a vertical line through $x = -4$.

(b) $m = \dfrac{-11 - (-11)}{3 - (-7)} = \dfrac{0}{10} = 0$

$m = 0$. The line is a horizontal line through $y = -11$.

4. Solve for y.

$4x - 2y = -5$

$-2y = -4x - 5$

$y = \dfrac{-4x - 5}{-2}$

$y = 2x + \dfrac{5}{2} \qquad$ Slope $= 2 \qquad$ y-intercept $= \left(0, \dfrac{5}{2}\right)$

5. (a) $y = mx + b$

$m = -\dfrac{3}{7} \qquad$ y-intercept $= \left(0, \dfrac{2}{7}\right)$

$y = -\dfrac{3}{7}x + \dfrac{2}{7}$

(b) $y = -\dfrac{3}{7}x + \dfrac{2}{7}$

$7(y) = 7\left(-\dfrac{3}{7}x\right) + 7\left(\dfrac{2}{7}\right)$

$7y = -3x + 2$

$3x + 7y = 2$

6. y-intercept $= (0, -1)$. Thus the coordinates of the y-intercept for this line are $(0, -1)$. Plot the point. Slope is $\dfrac{\text{rise}}{\text{run}}$. Since the slope for this line is $\dfrac{3}{4}$, we will go up (rise) 3 units and go over (run) 4 units to the right from the point $(0, -1)$. This is the point $(4, 2)$.

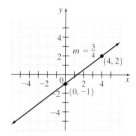

7. $y = -\dfrac{2}{3}x + 5$

The y-intercept is $(0, 5)$ since $b = 5$. Plot the point $(0, 5)$. The slope is $-\dfrac{2}{3} = \dfrac{-2}{3}$. Begin at $(0, 5)$, go down 2 units and to the right 3 units. This is the point $(3, 3)$. Draw a line that connects the points $(0, 5)$ and $(3, 3)$.

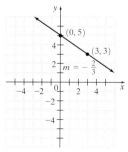

8. (a) Parallel lines have the same slope. Line j has a slope of $\dfrac{1}{4}$.

(b) Perpendicular lines have slopes whose product is -1.

$m_1 m_2 = -1$

$\dfrac{1}{4}m_2 = -1$

$4\left(\dfrac{1}{4}\right)m_2 = -1(4)$

$m_2 = -4$

Thus line k has a slope of -4.

9. (a) The slope of line n is $\dfrac{1}{4}$. The slope of a line that is parallel to line n is $\dfrac{1}{4}$.

(b) $m_1 m_2 = -1$

$\dfrac{1}{4}m_2 = -1$

$m_2 = -4$

The slope of a line that is perpendicular to n is -4.

7.4 Practice Problems

1. $y = mx + b$

$12 = -\dfrac{3}{4}(-8) + b$

$12 = 6 + b$

$6 = b$

The equation of the line is $y = -\dfrac{3}{4}x + 6$.

2. Find the slope.

$m = \dfrac{y_2 - y_1}{x_2 - x_1} = \dfrac{1 - 5}{-1 - 3} = \dfrac{-4}{-4} = 1$

Using either of the two points given, substitute x and y values into the equation $y = mx + b$.

$m = 1 \quad x = 3 \quad$ and $\quad y = 5$.

$y = mx + b$

$5 = 1(3) + b$

$5 = 3 + b$

$2 = b$

The equation of the line is $y = x + 2$.

3. The y-intercept is $(0, 1)$. Thus $b = 1$. Look for another point in the line. We choose $(6, 2)$. Count the number of vertical units from 1 to 2 (rise). Count the number of horizontal units from 0 to 6 (run).

$m = \dfrac{1}{6} \quad$ Now we can write the equation of the line.

$y = mx + b$

$y = \dfrac{1}{6}x + 1$

7.5 Practice Problems

1. Graph $x - y \geq -10$.

Begin by graphing the line $x - y = -10$. You may use any method discussed previously. Since there is an equal sign in the inequality, we will draw a solid line to indicate that the line is part of the solution set. The easiest test point is $(0, 0)$. Substitute $x = 0$, $y = 0$ in the inequality.

$x - y \geq -10$

$0 - 0 \geq -10$

$0 \geq -10 \quad$ True

Therefore shade the side of the line that includes the point $(0, 0)$.

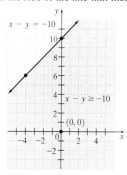

2. Step 1 Graph $y = \dfrac{1}{2}x$. Since $>$ is used, the line should be a dashed line.

Step 2 We see the line passes through $(0, 0)$.

Step 3 Choose another test point. We will choose $(-1, 1)$.

$$y > \frac{1}{2}x$$

$$1 > \frac{1}{2}(-1)$$

$$1 > -\frac{1}{2} \quad \text{true}$$

Shade the region that includes $(-1, 1)$, that is, the region above the line.

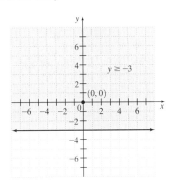

3. Step 1 Graph $y = -3$. Since $\geq$ is used, the line should be solid.

Step 2 Test $(0, 0)$ in the inequality.

$$y \geq -3$$
$$0 \geq -3 \quad \text{true}$$

Shade the region that includes $(0, 0)$, that is, the region above the line $y = -3$.

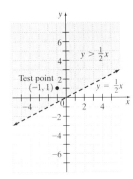

7.6 Practice Problems

1. The domain is $\{-3, 3, 0, 20\}$. The range is $\{-5, 5\}$.

2. (a) Look at the ordered pairs. No two ordered pairs have the same first coordinate. Thus this set of ordered pairs defines a function.

(b) Look at the ordered pairs. Two different ordered pairs, $(60, 30)$ and $(60, 120)$, have the same first coordinate. Thus this relation is not a function.

3. (a) Looking at the table, we see that no two different ordered pairs have the same first coordinate. The cost of gasoline is a function of the distance traveled.

Note that cost depends on distance. Thus distance is the independent variable. Since a negative distance does not make sense, the domain is {all nonnegative real numbers}.

The range is {all nonnegative real numbers}.

(b) Looking at the table, we see two ordered pairs, $(5, 20)$ and $(5, 30)$, have the same first coordinate. Thus this relation is not a function.

4. Construct a table, plot the ordered pairs and connect the points.

x	$y = x^2 - 2$	y
-2	$y = (-2)^2 - 2 = 2$	2
-1	$y = (-1)^2 - 2 = -1$	-1
0	$y = 0 - 2 = -2$	-2
1	$y = (1)^2 - 2 = -1$	-1
2	$y = (2)^2 - 2 = 2$	2

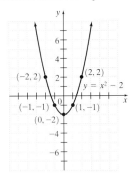

5. Select values of y and then substitute it into the equation to obtain x.

y	$x = y^2 - 1$	x	y
-2	$x = (-2)^2 - 1 = 3$	3	-2
-1	$x = (-1)^2 - 1 = 0$	0	-1
0	$x = (0)^2 - 1 = -1$	-1	0
1	$x = (1)^2 - 1 = 0$	0	1
2	$x = (2)^2 - 1 = 3$	3	2

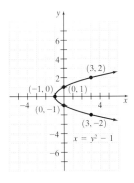

6. $y = \dfrac{6}{x}$

x	$y = \dfrac{6}{x}$	y
-3	$y = \dfrac{6}{-3} = -2$	-2
-2	$y = \dfrac{6}{-2} = -3$	-3
-1	$y = \dfrac{6}{-1} = -6$	-6
0	We cannot divide by 0.	
1	$y = \dfrac{6}{1} = 6$	6
2	$y = \dfrac{6}{2} = 3$	3
3	$y = \dfrac{6}{3} = 2$	2

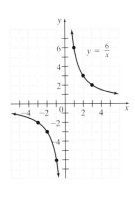

7. (a) The graph of a vertical line is not a function.

(b) This curve is a function. Any vertical line will cross the curve in only one location.

(c) This curve is not the graph of a function. There exist vertical lines that will cross the curve in more than one place.

8. $f(x) = -2x^2 + 3x - 8$

(a) $f(2) = -2(2)^2 + 3(2) - 8$
$f(2) = -2(4) + 3(2) - 8$
$f(2) = -8 + 6 - 8$
$f(2) = -10$

(b) $f(-3) = -2(-3)^2 + 3(-3) - 8$
$f(-3) = -2(9) + 3(-3) - 8$
$f(-3) = -18 - 9 - 8$
$f(-3) = -35$

(c) $f(0) = -2(0)^2 + 3(0) - 8$
$f(0) = -2(0) + 3(0) - 8$
$f(0) = 0 + 0 - 8$
$f(0) = -8$

Chapter 8

8.1 Practice Problems

1. $x + y = 12$
$-x + y = 4$
Graph the equations on the same coordinate plane.
The lines intersect at the point $x = 4$, $y = 8$. Thus the solution to the system of equations is $(4, 8)$.

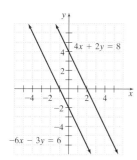

2. $4x + 2y = 8$
$-6x - 3y = 6$
Graph both equations on the same coordinate plane.
These lines are parallel. They do not intersect. Hence there is no solution to this system of equations.

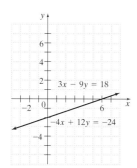

3. $3x - 9y = 18$
$-4x + 12y = -24$
Graph both equations on the same coordinate plane.
Notice both equations represent the same line. Thus there is an infinite number of solutions to this system.

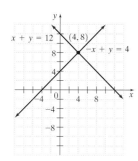

4. (a) Bill Tupper's Electrical Service charges $100 for a house call and $30 per hour. Thus we obtain the first equation,

$y = 100 + 30x$. Wire for Hire charges $50 for a house call and $40 per hour. Thus we obtain the second equation, $y = 50 + 40x$.

(b) **Bill Tupper's Electrical Service** $y = 100 + 30x$
Let $x = 0$ $y = 100 + 30(0) = 100$
Let $x = 4$ $y = 100 + 30(4) = 220$
Let $x = 8$ $y = 100 + 30(8) = 340$

x	y
0	100
4	220
8	340

Wire for Hire $y = 50 + 40x$
Let $x = 0$ $y = 50 + 40(0) = 50$
Let $x = 4$ $y = 50 + 40(4) = 210$
Let $x = 8$ $y = 50 + 40(8) = 370$

x	y
0	50
4	210
8	370

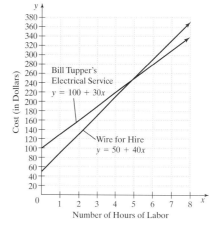

(c) We see that the graphs of the two lines intersect at $(5, 250)$. Thus the two companies will charge the same if 5 hours of electrical repairs are required.

(d) We draw a dashed line at $x = 6$. We see the line representing Wire for Hire is higher than the line representing Bill Tupper's Electrical Service after 5 hours. Thus the cost would be less if we use Bill Tupper's Electrical Service for 6 hours of work.

8.2 Practice Problems

1. $5x + 3y = 19$
$2x - y = 12$
Step 1 Solve the second equation for y.
$2x - y = 12$
$-y = -2x + 12$
$y = 2x - 12$
Step 2 Now substitute $y = 2x - 12$ into the first equation.
$5x + 3y = 19$
$5x + 3(2x - 12) = 19$
Step 3 Solve this equation.
$5x + 6x - 36 = 19$
$11x - 36 = 19$
$11x = 55$
$x = 5$
Step 4 Now obtain the value for the second variable.
$2x - y = 12$
$2(5) - y = 12$
$10 - y = 12$
$-y = 2$
$y = -2$
The solution is $(5, -2)$.
Step 5 Check.

$$5x + 3y = 19 \qquad\qquad 2x - y = 12$$
$$5(5) + 3(-2) \stackrel{?}{=} 19 \qquad 2(5) - (-2) \stackrel{?}{=} 12$$
$$25 - 6 \stackrel{?}{=} 19 \qquad\qquad 10 + 2 \stackrel{?}{=} 12$$
$$19 = 19 \checkmark \qquad\qquad 12 = 12 \checkmark$$

2. $\dfrac{x}{3} - \dfrac{y}{2} = 1$

$x + 4y = -8$

Clear the first equation of fractions. We observe that the LCD is 6.

$$6\left(\dfrac{x}{3}\right) - 6\left(\dfrac{y}{2}\right) = 6(1)$$

$$2x - 3y = 6$$

Step 1 Solve the second equation for x.

$$x + 4y = -8$$

$$x = -4y - 8$$

Step 2 Substitute $x = -4y - 8$ into the equation $2x - 3y = 6$.

$$2(-4y - 8) - 3y = 6$$

Step 3 Solve this equation.

$$-8y - 16 - 3y = 6$$

$$-11y - 16 = 6$$

$$-11y = 22$$

$$y = -2$$

Step 4 Obtain the value of the second variable.

$$x + 4y = -8$$

$$x + 4(-2) = -8$$

$$x - 8 = -8$$

$$x = 0$$

The solution is $(0, -2)$.

Step 5 Check.

$$\dfrac{x}{3} - \dfrac{y}{2} = 1 \qquad\qquad x + 4y = -8$$

$$\dfrac{0}{3} - \dfrac{(-2)}{2} \overset{?}{=} 1 \qquad 0 + 4(-2) \overset{?}{=} -8$$

$$0 + 1 \overset{?}{=} 1 \qquad\qquad -8 = -8 \;\checkmark$$

$$1 = 1 \;\checkmark$$

8.3 Practice Problems

1. $3x + y = 7$

$5x - 2y = 8$

Multiply the first equation by 2.

$$2(3x) + 2(y) = 2(7)$$

$$\begin{array}{r} 6x + 2y = 14 \\ + \; 5x - 2y = \; 8 \\ \hline 11x \qquad = 22 \\ x \qquad = 2 \end{array}$$

Substitute $x = 2$ into one of the original equations.

$$5(2) - 2y = 8$$

$$10 - 2y = 8$$

$$-2y = -2$$

$$y = 1$$

The solution is $(2, 1)$.

2. $4x + 5y = 17$

$3x + 7y = 12$

Multiply the first equation by -3 and the second equation by 4.

$$-3(4x) + (-3)(5y) = (-3)(17)$$

$$4(3x) + 4(7y) = 4(12)$$

$$\begin{array}{r} -12x - 15y = -51 \\ 12x + 28y = \; 48 \\ \hline 13y = -3 \end{array}$$

$$y = -\dfrac{3}{13}$$

Substitute $y = -\dfrac{3}{13}$ into one of the original equations.

$$4x + 5\left(-\dfrac{3}{13}\right) = 17$$

$$4x - \dfrac{15}{13} = 17$$

$$13(4x) - 13\left(\dfrac{15}{13}\right) = 13(17)$$

$$52x - 15 = 221$$

$$52x = 236$$

$$x = \dfrac{59}{13}$$

The solution is $\left(\dfrac{59}{13}, -\dfrac{3}{13}\right)$.

3. Multiply the first equation by 12.

$$12\left(\dfrac{2}{3}x\right) - 12\left(\dfrac{3}{4}y\right) = 12(3)$$

$$8x - 9y = 36$$

We now have an equivalent system without fractions.

$$8x - 9y = 36$$

$$-2x + y = 6$$

Eliminate the x-variable.

$$4(-2x) + 4(y) = 4(6)$$

$$\begin{array}{r} -8x + 4y = 24 \\ 8x - 9y = 36 \\ \hline -5y = \; 60 \\ y = -12 \end{array}$$

Substitute $y = -12$ into one of the original equations.

$$-2x + (-12) = 6$$

$$-2x - 12 = 6$$

$$-2x = 18$$

$$x = -9$$

The solution is $(-9, -12)$.

4. Multiply each term of each equation by 10.

$$10(0.2x) + 10(0.3y) = 10(-0.1)$$

$$10(0.5x) - 10(0.1y) = 10(-1.1)$$

$$2x + 3y = -1$$

$$5x - y = -11$$

Eliminate the y-variable.

$$\begin{array}{r} 2x + 3y = -1 \\ 15x - 3y = -33 \\ \hline 17x \qquad = -34 \\ x = -2 \end{array}$$

Substitute $x = -2$ into one of the equations.

$$2(-2) + 3y = -1$$

$$-4 + 3y = -1$$

$$3y = 3$$

$$y = 1$$

The solution is $(-2, 1)$.

8.4 Practice Problems

1. (a) $3x + 5y = 1485$

$x + 2y = 564$

Solve for x in the second equation and solve using the substitution method.

$$x = -2y + 564$$

$$3(-2y + 564) + 5y = 1485$$

$$-6y + 1692 + 5y = 1485$$

$$-y = -207$$

$$y = 207$$

Substitute $y = 207$ into the second equation and solve for x.

$$x + 2(207) = 564$$

$$x + 414 = 564$$

$$x = 150$$

The solution is $(150, 207)$.

(b) $7x + 6y = 45$
$6x - 5y = -2$
Using the addition method, multiply the first equation by -6 and the second equation by 7.
$-6(7x) + (-6)(6y) = (-6)(45)$
$7(6x) - 7(5y) = 7(-2)$

$-42x - 36y = -270$
$\underline{42x - 35y = -14}$
$-71y = -284$
$y = 4$

Substitute $y = 4$ into one of the original equations and solve for x.
$7x + 6(4) = 45$
$7x + 24 = 45$
$7x = 21$
$x = 3$
The solution is $(3, 4)$.

2. $4x + 2y = 2$
$-6x - 3y = 6$
Use the addition method. Multiply the first equation by 3 and the second equation by 2.
$12x + 6y = 6$
$\underline{-12x - 6y = 12}$
$0 = 18$
The statement $0 = 18$ is inconsistent. There is no solution to this system of equations.

3. $3x - 9y = 18$
$-4x + 12y = -24$
Use the addition method. Multiply the first equation by 4 and the second equation by 3.
$12x - 36y = 72$
$\underline{-12x + 36y = -72}$
$0 = 0$
The statement $0 = 0$ is true and can identify the equations as dependent. There are an infinite number of solutions to the system.

8.5 Practice Problems

1. Let x = number of gallons/hour pumped by the smaller pump and y = number of gallons/hour pumped by the larger pump.
1st day $\qquad 8x + 5y = 49,000$
2nd day $\qquad 5x + 3y = 30,000$
Use the addition method.
Multiply the first equation by 5 and the second equation by -8.
$40x + 25y = 245,000$
$\underline{-40x - 24y = -240,000}$
$y = 5,000$
Substitute $y = 5,000$ into one of the original equations and solve for x.
$8x + 5(5000) = 49,000$
$8x + 25,000 = 49,000$
$8x = 24,000$
$x = 3,000$
The smaller pump removes 3000 gallons per hour.
The larger pump removes 5000 gallons per hour.

2. Let x = cost/gallon of unleaded premium gasoline and y = cost/gallon of unleaded regular gasoline.
Last week's purchase $\qquad 7x + 8y = 27.22$
This week's purchase $\qquad 8x + 4y = 22.16$
Use the addition method. Multiply the first equation by 8 and the second equation by -7.
$56x + 64y = 217.76$
$\underline{-56x - 28y = -155.12}$
$36y = 62.64$
$y = 1.74$
Substitute $y = 1.74$ into one of the original equations and solve for x.

$7x + 8(1.74) = 27.22$
$7x + 13.92 = 27.22$
$7x = 13.30$
$x = 1.90$
Unleaded premium gasoline costs \$1.90/gallon.
Unleaded regular gasoline costs \$1.74/gallon.

3. Let x = the amount of 20% solution and y = the amount of 80% solution.
Total amount of 65% solution
$x + y = 4000$
Amount of acid in 65% solution
$0.2x + 0.8y = 0.65(4000)$
Use the substitution method. Solve for x in the first equation.
$x = 4000 - y$
Substitute $x = 4000 - y$ into the second equation.
$0.2(4000 - y) + 0.8y = 0.65(4000)$
$800 - 0.2y + 0.8y = 2600$
$800 + 0.6y = 2600$
$0.6y = 1800$
$y = 3000$
Now substitute $y = 3000$ into the first equation.
$x + y = 4000$
$x + 3000 = 4000$
$x = 1000$
Therefore he will need 1000 liters of 20% solution and 3000 liters of 80% solution.

4. To go downstream with the current, we have as the rate of travel
$$r = \frac{d}{t} = \frac{72}{3} = 24 \text{ miles/hour}$$
To go upstream against the current, we have as the rate of travel
$$r = \frac{d}{t} = \frac{72}{4} = 18 \text{ miles/hour}$$
Let x = the speed of the boat in miles/hour and y = the speed of the current in miles/hour.
Thus we have the equations $x + y = 24$ and $x - y = 18$. Use the addition method to solve.
$x + y = 24$
$\underline{x - y = 18}$
$2x = 42$
$x = 21$
Substitute $x = 21$ into one of the original equations and solve for y.
$21 + y = 24$
$y = 3$
Thus the boat speed is 21 miles/hour and the speed of the current is 3 miles/hour.

Chapter 9

9.1 Practice Problems

1. (a) $\sqrt{64} = 8$ because $8^2 = 64$.
(b) $-\sqrt{121} = -11$ because $(11)^2 = 121$.

2. (a) $\sqrt{\dfrac{9}{25}} = \dfrac{\sqrt{9}}{\sqrt{25}} = \dfrac{3}{5}$ **(b)** $-\sqrt{\dfrac{121}{169}} = \dfrac{-\sqrt{121}}{\sqrt{169}} = -\dfrac{11}{13}$

3. (a) $-\sqrt{0.0036} = -0.06$ **(b)** $\sqrt{2500} = 50$

4. (a) 1.732 **(b)** 5.916 **(c)** 11.269

9.2 Practice Problems

1. (a) 6
(b) 13^2
(c) 18^6
2. (a) y^9
(b) x^{15}
3. (a) $25y^2$
(b) $x^8 y^{11}$
(c) $11x^6 y^3$

4. (a) $\sqrt{49} \cdot \sqrt{2} = 7\sqrt{2}$

(b) $\sqrt{4} \cdot \sqrt{3} = 2\sqrt{3}$

(c) $\sqrt{25} \cdot \sqrt{3} = 5\sqrt{3}$

5. (a) $\sqrt{x^{11}} = \sqrt{x^{10}} \cdot \sqrt{x} = x^5\sqrt{x}$

(b) $\sqrt{x^5 y^3} = \sqrt{x^4 y^2} \cdot \sqrt{xy} = x^2 y\sqrt{xy}$

6. (a) $\sqrt{48x^{11}} = \sqrt{16x^{10}} \cdot \sqrt{3x} = 4x^5\sqrt{3x}$

(b) $\sqrt{121x^6 y^7 z^8} = \sqrt{121x^6 y^6 z^8} \cdot \sqrt{y} = 11x^3 y^3 z^4\sqrt{y}$

9.3 Practice Problems

1. (a) $11\sqrt{11}$ **(b)** $\sqrt{t}$

2. $3\sqrt{x} - 2\sqrt{xy} + 7\sqrt{xy} - 5\sqrt{y}$
$= 3\sqrt{x} + 5\sqrt{xy} - 5\sqrt{y}$

3. (a) $\sqrt{50} - \sqrt{18} + \sqrt{98}$
$= \sqrt{25} \cdot \sqrt{2} - \sqrt{9} \cdot \sqrt{2} + \sqrt{49} \cdot \sqrt{2}$
$= 5\sqrt{2} - 3\sqrt{2} + 7\sqrt{2} = 2\sqrt{2} + 7\sqrt{2} = 9\sqrt{2}$

(b) $\sqrt{12} + \sqrt{18} - \sqrt{50} + \sqrt{27}$
$= \sqrt{4} \cdot \sqrt{3} + \sqrt{9} \cdot \sqrt{2} - \sqrt{25} \cdot \sqrt{2} + \sqrt{9} \cdot \sqrt{3}$
$= 2\sqrt{3} + 3\sqrt{3} + 3\sqrt{2} - 5\sqrt{2} = 5\sqrt{3} - 2\sqrt{2}$

4. $\sqrt{9x} + \sqrt{8x} - \sqrt{4x} + \sqrt{50x}$
$= 3\sqrt{x} + 2\sqrt{2x} - 2\sqrt{x} + 5\sqrt{2x}$
$= \sqrt{x} + 7\sqrt{2x}$

5. $\sqrt{27} - 4\sqrt{3} + 2\sqrt{75}$
$= \sqrt{9}\sqrt{3} - 4\sqrt{3} + 2\sqrt{25}\sqrt{3}$
$= 3\sqrt{3} - 4\sqrt{3} + 10\sqrt{3} = 9\sqrt{3}$

6. $2\sqrt{12x} - 3\sqrt{45x} - 3\sqrt{27x} + \sqrt{20}$
$= 2\sqrt{4} \cdot \sqrt{3x} - 3\sqrt{9} \cdot \sqrt{5x} - 3\sqrt{9} \cdot \sqrt{3x} + \sqrt{4}\sqrt{5}$
$= 2(2)\sqrt{3x} - 3(3)\sqrt{5x} - 3(3)\sqrt{3x} + 2\sqrt{5}$
$= 4\sqrt{3x} - 9\sqrt{3x} - 9\sqrt{5x} + 2\sqrt{5}$
$= -5\sqrt{3x} - 9\sqrt{5x} + 2\sqrt{5}$

9.4 Practice Problems

1. $\sqrt{3a}\sqrt{6a} = \sqrt{18a^2} = 3a\sqrt{2}$

2. (a) $(2\sqrt{3})(5\sqrt{5}) = 10\sqrt{15}$

(b) $(4\sqrt{3x})(2x\sqrt{6x}) = 8x\sqrt{18x^2} = 24x^2\sqrt{2}$

3. $(\sqrt{180})(\sqrt{150})$
$= (6\sqrt{5})(5\sqrt{6})$
$= 30\sqrt{30}$ square millimeters

4. $2\sqrt{3}(\sqrt{3} + \sqrt{5} - \sqrt{12})$
$= 2\sqrt{9} + 2\sqrt{15} - 2\sqrt{36}$
$= 2(3) + 2\sqrt{15} - 2(6) = 6 + 2\sqrt{15} - 12 = -6 + 2\sqrt{15}$

5. $2\sqrt{x}(4\sqrt{x} - x\sqrt{2})$
$= 8\sqrt{x^2} - 2x\sqrt{2x}$
$= 8x - 2x\sqrt{2x}$

6. $(\sqrt{2} + \sqrt{6})(2\sqrt{2} - \sqrt{6})$
$= 2\sqrt{4} - \sqrt{12} + 2\sqrt{12} - \sqrt{36}$
$= 2(2) + \sqrt{12} - 6 = 4 + 2\sqrt{3} - 6 = -2 + 2\sqrt{3}$

7. $(\sqrt{6} + \sqrt{5})(\sqrt{2} + 2\sqrt{5})$
$= \sqrt{12} + 2\sqrt{30} + \sqrt{10} + 2\sqrt{25}$
$= 2\sqrt{3} + 2\sqrt{30} + \sqrt{10} + 2(5)$
$= 2\sqrt{3} + 2\sqrt{30} + \sqrt{10} + 10$

8. $(3\sqrt{5} - \sqrt{10})^2$
$= (3\sqrt{5} - \sqrt{10})(3\sqrt{5} - \sqrt{10})$
$= 9\sqrt{25} - 3\sqrt{50} - 3\sqrt{50} + \sqrt{100}$
$= 45 - 6\sqrt{25}\sqrt{2} + 10 = 55 - 30\sqrt{2}$

9.5 Practice Problems

1. (a) $\sqrt{\dfrac{x^3}{x}} = \sqrt{x^2} = x$ **(b)** $\sqrt{\dfrac{49}{x^2}} = \dfrac{7}{x}$

2. $\sqrt{\dfrac{50}{a^4}} = \dfrac{5\sqrt{2}}{a^2}$

3. $\dfrac{9}{\sqrt{7}} \cdot \dfrac{\sqrt{7}}{\sqrt{7}} = \dfrac{9\sqrt{7}}{7}$

4. (a) $\dfrac{\sqrt{2}}{\sqrt{12}} \cdot \dfrac{\sqrt{3}}{\sqrt{3}} = \dfrac{\sqrt{6}}{\sqrt{36}} = \dfrac{\sqrt{6}}{6}$

(b) $\dfrac{6a}{\sqrt{a^7}} \cdot \dfrac{\sqrt{a}}{\sqrt{a}} = \dfrac{6a\sqrt{a}}{\sqrt{a^8}} = \dfrac{6a\sqrt{a}}{a^4} = \dfrac{6\sqrt{a}}{a^3}$

5. $\dfrac{\sqrt{2x}}{\sqrt{8x}} = \dfrac{\sqrt{2x}}{2\sqrt{2x}} \cdot \dfrac{\sqrt{2x}}{\sqrt{2x}} = \dfrac{\sqrt{4x^2}}{2\sqrt{4x^2}} = \dfrac{2x}{4x} = \dfrac{1}{2}$

6. (a) $\dfrac{4}{\sqrt{3} + \sqrt{5}} \cdot \dfrac{\sqrt{3} - \sqrt{5}}{\sqrt{3} - \sqrt{5}}$

$= \dfrac{4(\sqrt{3} - \sqrt{5})}{3 - 5} = \dfrac{4(\sqrt{3} - \sqrt{5})}{-2}$

$= \dfrac{2(\sqrt{3} - \sqrt{5})}{-1} = -2(\sqrt{3} - \sqrt{5})$

(b) $\dfrac{\sqrt{a}}{(\sqrt{10} - 3)} \cdot \dfrac{(\sqrt{10} + 3)}{(\sqrt{10} + 3)}$

$= \dfrac{\sqrt{a}(\sqrt{10} + 3)}{\sqrt{100} + 3\sqrt{10} - 3\sqrt{10} - 9}$

$= \dfrac{\sqrt{a}(\sqrt{10} + 3)}{10 - 9} = \sqrt{a}(\sqrt{10} + 3)$

7. $\dfrac{\sqrt{7} - \sqrt{x}}{\sqrt{7} + \sqrt{x}} \cdot \dfrac{\sqrt{7} - \sqrt{x}}{\sqrt{7} - \sqrt{x}}$

$= \dfrac{7 - 2\sqrt{7x} + x}{7 - x} = \dfrac{7 + x - 2\sqrt{7x}}{7 - x}$

9.6 Practice Problems

1. $c^2 = a^2 + b^2$
$c^2 = 9^2 + 12^2$
$c^2 = 81 + 144$
$c^2 = 225$
$c = \pm\sqrt{225}$
$c = \pm 15$
$c = +15$ centimeters because length is not negative.

2. $(\sqrt{17})^2 = (1)^2 + b^2$
$17 = 1 + b^2$
$16 = b^2$
$\sqrt{16} = b$
$4 = b$
The leg is 4 meters long.

3. $c^2 = 3^2 + 8^2$
$c^2 = 9 + 64$
$c^2 = 73$
$c = \sqrt{73}$
$c \approx 8.5$ meters

4. $\sqrt{3x - 2} - 7 = 0$ *Check.*
$\sqrt{3x - 2} = 7$ $\sqrt{3(17) - 2} - 7 \overset{?}{=} 0$
$(\sqrt{3x - 2})^2 = (7)^2$ $\sqrt{51 - 2} - 7 \overset{?}{=} 0$
$3x - 2 = 49$ $\sqrt{49} - 7 \overset{?}{=} 0$
$3x = 51$ $7 - 7 \overset{?}{=} 0$
$x = 17$ $0 = 0$ ✓

5. $\sqrt{5x + 4} + 2 = 0$ *Check.*
$\sqrt{5x + 4} = -2$ $\sqrt{5(0) + 4} + 2 \overset{?}{=} 0$
$(\sqrt{5x + 4})^2 = (-2)^2$ $\sqrt{4} + 2 \overset{?}{=} 0$
$5x + 4 = 4$ $2 + 2 \overset{?}{=} 0$
$5x = 0$ $4 \neq 0$
$x = 0$
No, this does not check. There is no solution.

6. $-2 + \sqrt{6x - 1} = 3x - 2$

$\sqrt{6x - 1} = 3x$

$(\sqrt{6x - 1})^2 = (3x)^2$

$6x - 1 = 9x^2$

$0 = 9x^2 - 6x + 1$ Factor.

$0 = (3x - 1)^2$

$0 = 3x - 1$

$x = \dfrac{1}{3}$ is a double root.

Check.

$-2 + \sqrt{6\left(\dfrac{1}{3}\right) - 1} \stackrel{?}{=} 3\left(\dfrac{1}{3}\right) - 2$

$-2 + \sqrt{1} \stackrel{?}{=} 1 - 2$

$-2 + 1 \stackrel{?}{=} 1 - 2$

$-1 = -1$ ✓

7. $2 - x + \sqrt{x + 4} = 0$

$\sqrt{x + 4} = x - 2$

$(\sqrt{x + 4})^2 = (x - 2)^2$

$x + 4 = x^2 - 4x + 4$

$0 = x^2 - 5x$

$0 = x(x - 5)$

$x = 0$ $x - 5 = 0$

$x = 5$

Check.

$2 - 0 + \sqrt{0 + 4} \stackrel{?}{=} 0$

$2 + \sqrt{4} \stackrel{?}{=} 0$

$2 + 2 \stackrel{?}{=} 0$

$4 \neq 0$

$x = 0$ does not check.

$2 - 5 + \sqrt{5 + 4} \stackrel{?}{=} 0$

$-3 + \sqrt{9} \stackrel{?}{=} 0$

$-3 + 3 \stackrel{?}{=} 0$

$0 = 0$ ✓

The only solution is $x = 5$.

8. $\sqrt{2x + 1} = \sqrt{x - 10}$

$(\sqrt{2x + 1})^2 = (\sqrt{x - 10})^2$

$2x + 1 = x - 10$

$x = -11$

$x = -11$ produces a negative radicand and thus is extraneous. There is no solution.

9.7 Practice Problems

1. Change of $100°C$ = change of $180°F$.

Let C = the change in C,

f = the change in F, and

k = the constant of variation.

$$C = kf$$
$$100 = k \cdot 180$$
$$k = \frac{5}{9}$$

if $C = kf$ then

$$C = \frac{5}{9}f$$

if $f = -20$ then

$$C = \frac{5}{9}(-20)$$

$$C = -11\frac{1}{9}$$

The temperature drops $11\dfrac{1}{9}$ degrees Celsius.

2. $y = kx$

To find k we substitute $y = 18$ and $x = 5$.

$$18 = k(5)$$
$$\frac{18}{5} = k$$

We now write the variation equation with k replaced by $\dfrac{18}{5}$.

$$y = \frac{18}{5}x$$

Replace x by $\dfrac{20}{23}$ and solve for y.

$$y = \frac{18}{5} \cdot \frac{20}{23}$$

$$y = \frac{72}{23}$$

Thus $y = \dfrac{72}{23}$ when $x = \dfrac{20}{23}$.

3. Let d = the distance to stop the car,

s = the speed of the car, and

k = the constant of variation.

Since the distance varies directly as the square of the speed, we have

$$d = ks^2$$

To evaluate k we substitute the known distance and speed.

$$60 = k(20)^2$$
$$60 = k(400)$$
$$k = \frac{3}{20}$$

Now write the variation equation with the constant evaluated.

$$d = \frac{3}{20}s^2$$

$$d = \frac{3}{20}(40)^2 \quad \text{Substitute } s = 40.$$

$$d = \frac{3}{20}(1600)$$

$$d = 240$$

The distance to stop the car going 40 mph on an ice-covered road is 240 feet.

4. $y = \dfrac{k}{x}$

$8 = \dfrac{k}{15}$ Substitute known values of x and y to find k.

$120 = k$

$y = \dfrac{120}{x}$ Write the variation equation with k replaced by 120.

$y = \dfrac{120}{\dfrac{3}{5}}$ Find y when $x = \dfrac{3}{5}$.

$y = 200$ Thus $y = 200$ when $x = \dfrac{3}{5}$.

5. Let V = the volume of sales,

p = the price of the calculator, and

k = the constant of variation.

Since volume varies inversely as the price, then

$$V = \frac{k}{p} \qquad \text{Evaluate } k \text{ by substituting known values for } V \text{ and } p.$$

$$120{,}000 = \frac{k}{30}$$

$$3{,}600{,}000 = k$$

$$V = \frac{3{,}600{,}000}{24} \qquad \text{Write the variation equation with the evaluated constant and substitute } p = 24.$$

$$V = 150{,}000$$

Thus the volume increased to 150,000 calculators sold per year when the price was reduced to $24 per calculator.

6. Let R = the resistance in the circuit,

i = the amount of current, and

k = the constant of variation.

Since the resistance in the circuit varies inversely as the square of the amount of current,

$$R = \frac{k}{i^2}.$$

To evaluate k, substitute the known values for R and i.

$$800 = \frac{k}{(0.01)^2}$$
$$k = 0.08$$
$$R = \frac{0.08}{(0.02)^2} \quad \text{Write the variation equation with the evaluated constant of } k = 0.08 \text{ and substitute } i = 0.02.$$
$$R = \frac{0.08}{0.0004}$$
$$R = 200$$

Thus the resistance is 200 ohms if the amount of current is increased to 0.02 ampere.

Chapter 10

10.1 Practice Problems

1. (a) $2x^2 + 12x - 9 = 0$
$a = 2, b = 12, c = -9$

(b) $7x^2 = 6x - 8$
$7x^2 - 6x + 8 = 0$
$a = 7, b = -6, c = 8$

(c) $-x^2 - 6x + 3 = 0$
$x^2 + 6x - 3 = 0$
$a = 1, b = 6, c = -3$

(d) $10x^2 - 12x = 0$
$a = 10, b = -12, c = 0$

2. $2x^2 - 7x - 6 = 4x - 6$
$2x^2 - 11x = 0$
$x(2x - 11) = 0$
$x = 0 \qquad 2x - 11 = 0$
$$x = \frac{11}{2}$$

The two roots are 0 and $\frac{11}{2}$.

3. $3x + 10 - \dfrac{8}{x} = 0$
$3x^2 + 10x - 8 = 0$
$(3x - 2)(x + 4) = 0$
$3x - 2 = 0 \qquad x + 4 = 0$
$$x = \frac{2}{3} \qquad\qquad x = -4$$

Check. $3\left(\dfrac{2}{3}\right) + 10 - \dfrac{8}{\frac{2}{3}} \stackrel{?}{=} 0 \qquad 3(-4) + 10 - \dfrac{8}{-4} \stackrel{?}{=} 0$
$2 + 10 - 12 \stackrel{?}{=} 0 \qquad\qquad -12 + 10 + 2 \stackrel{?}{=} 0$
$2 - 2 \stackrel{?}{=} 0 \qquad\qquad\qquad -2 + 2 \stackrel{?}{=} 0$
$0 = 0 \; ✓ \qquad\qquad\qquad 0 = 0 \; ✓$

Both roots check, so $x = \dfrac{2}{3}$ and $x = -4$ are the two roots that satisfy the equation $3x + 10 - \dfrac{8}{x} = 0$.

4. $2x^2 + 9x - 18 = 0$
$(2x - 3)(x + 6) = 0$
$2x - 3 = 0 \qquad x + 6 = 0$
$$x = \frac{3}{2} \qquad\qquad x = -6$$

Check.
$2\left(\dfrac{3}{2}\right)^2 + 9\left(\dfrac{3}{2}\right) - 18 \stackrel{?}{=} 0 \qquad 2(-6)^2 + 9(-6) - 18 \stackrel{?}{=} 0$
$2\left(\dfrac{9}{4}\right) + \dfrac{27}{2} - 18 \stackrel{?}{=} 0 \qquad 2(36) - 54 - 18 \stackrel{?}{=} 0$
$\dfrac{9}{2} + \dfrac{27}{2} - 18 \stackrel{?}{=} 0 \qquad\qquad 72 - 54 - 18 \stackrel{?}{=} 0$
$\dfrac{36}{2} - 18 \stackrel{?}{=} 0 \qquad\qquad\qquad 18 - 18 \stackrel{?}{=} 0$
$18 - 18 \stackrel{?}{=} 0 \qquad\qquad\qquad\qquad 0 = 0 \; ✓$
$0 = 0 \; ✓$

Both roots check, so $x = \dfrac{3}{2}$ and $x = -6$ are the two roots that satisfy the equation $2x^2 + 9x - 18 = 0$.

5. $-4x + (x - 5)(x + 1) = 5(1 - x)$
$-4x + x^2 - 4x - 5 = 5 - 5x$
$x^2 - 8x - 5 = 5 - 5x$
$x^2 - 3x - 10 = 0$
$(x + 2)(x - 5) = 0$
$x + 2 = 0 \qquad x - 5 = 0$
$x = -2 \qquad\qquad x = 5$

Check. $-4(-2) + (-2 - 5)(-2 + 1) \stackrel{?}{=} 5(1 + 2)$
$8 + (-7)(-1) \stackrel{?}{=} 5(3)$
$8 + 7 \stackrel{?}{=} 15$
$15 = 15 \; ✓$

$-4(5) + (5 - 5)(5 + 1) \stackrel{?}{=} 5(-4)$
$-20 + 0 \stackrel{?}{=} -20$
$-20 = -20 \; ✓$

Both roots check, so $x = -2$ and $x = 5$ are the two roots that satisfy the equation $-4x + (x - 5)(x + 1) = 5(1 - x)$.

6. $\dfrac{10x + 18}{x^2 + x - 2} = \dfrac{3x}{x + 2} + \dfrac{2}{x - 1}$

$\dfrac{10x + 18}{(x - 1)(x + 2)} = \dfrac{3x}{x + 2} + \dfrac{2}{x - 1} \quad$ Multiply by the LCD.

$10x + 18 = 3x(x - 1) + 2(x + 2)$
$10x + 18 = 3x^2 - 3x + 2x + 4$
$0 = 3x^2 - 11x - 14$
$0 = (3x - 14)(x + 1)$
$3x - 14 = 0 \qquad x + 1 = 0$
$$x = \frac{14}{3} \qquad\qquad x = -1$$

7. t = the number of truck routes
n = the number of cities they can service
$$t = \frac{n^2 - n}{2}$$
$$28 = \frac{n^2 - n}{2}$$
$56 = n^2 - n$
$0 = n^2 - n - 56$
$0 = (n + 7)(n - 8)$
$n + 7 = 0 \qquad n - 8 = 0$
$n = -7 \qquad\qquad n = 8$

We reject $n = -7$. There cannot be a negative number of cities. Thus our answer is 8 cities.

10.2 Practice Problems

1. (a) $x^2 = 1$
$x = \pm\sqrt{1}$
$x = \pm 1$

(b) $x^2 - 50 = 0$
$x^2 = 50$
$x = \pm\sqrt{50}$
$x = \pm 5\sqrt{2}$

(c) $3x^2 = 81$
$$x^2 = 27$$
$$x = \pm\sqrt{27} = \pm 3\sqrt{3}$$

2. $4x^2 - 5 = 319$
$$4x^2 = 324$$
$$x^2 = 81$$
$$x = \pm\sqrt{81} = \pm 9$$

3. $(2x - 3)^2 = 12$
$$\sqrt{(2x - 3)^2} = \pm\sqrt{12}$$
$$2x - 3 = \pm 2\sqrt{3}$$
$$2x = 3 \pm 2\sqrt{3}$$
$$x = \frac{3 \pm 2\sqrt{3}}{2}$$

4. $x^2 + 10x = 3$
$$x^2 + 10x + (5)^2 = 3 + 25$$
$$(x + 5)^2 = 28$$
$$\sqrt{(x + 5)^2} = \pm\sqrt{28}$$
$$x + 5 = \pm\sqrt{28}$$
$$x = -5 \pm 2\sqrt{7}$$

5. $x^2 - 5x = 7$
$$x^2 - 5x + \left(\frac{5}{2}\right)^2 = 7 + \left(\frac{5}{2}\right)^2$$
$$\left(x - \frac{5}{2}\right)^2 = 7 + \frac{25}{4}$$
$$\left(x - \frac{5}{2}\right)^2 = \frac{28}{4} + \frac{25}{4}$$
$$\left(x - \frac{5}{2}\right)^2 = \frac{53}{4}$$
$$x - \frac{5}{2} = \pm\sqrt{\frac{53}{4}}$$
$$x = \frac{5}{2} \pm \frac{\sqrt{53}}{2}$$
$$x = \frac{5 \pm \sqrt{53}}{2}$$

6. $4y^2 + 4y - 3 = 0$
$$(2y + 3)(2y - 1) = 0$$
$$2y + 3 = 0 \qquad 2y - 1 = 0$$
$$y = -\frac{3}{2} \qquad y = \frac{1}{2}$$

10.3 Practice Problems

1. $x^2 - 7x + 6 = 0$
$$x = \frac{-(-7) \pm \sqrt{(-7)^2 - 4(1)(6)}}{2(1)}$$
$$x = \frac{7 \pm \sqrt{49 - 24}}{2}$$
$$x = \frac{7 \pm \sqrt{25}}{2}$$
$$x = \frac{7 \pm 5}{2}$$
$$x = \frac{7 + 5}{2} \qquad x = \frac{7 - 5}{2}$$
$$x = 6 \qquad x = 1$$

2. $3x^2 - 8x + 3 = 0$
$$x = \frac{-(-8) \pm \sqrt{(-8)^2 - 4(3)(3)}}{2(3)}$$
$$x = \frac{8 \pm \sqrt{64 - 36}}{6}$$

$$x = \frac{8 \pm \sqrt{28}}{6}$$
$$x = \frac{8 \pm 2\sqrt{7}}{6}$$
$$x = \frac{2(4 \pm \sqrt{7})}{6} = \frac{4 \pm \sqrt{7}}{3}$$

3. $\qquad x^2 = 7 - \frac{3}{5}x$
$$5x^2 = 35 - 3x$$
$$5x^2 + 3x - 35 = 0$$
$$x = \frac{-3 \pm \sqrt{(3)^2 - 4(5)(-35)}}{2(5)}$$
$$x = \frac{-3 \pm \sqrt{9 + 700}}{10}$$
$$x = \frac{-3 \pm \sqrt{709}}{10}$$

4. $2x^2 = 13x + 5$
$$2x^2 - 13x - 5 = 0$$
$$x = \frac{-(-13) \pm \sqrt{(-13)^2 - 4(2)(-5)}}{2(2)}$$
$$x = \frac{13 \pm \sqrt{169 + 40}}{4}$$
$$x = \frac{13 \pm \sqrt{209}}{4}$$
$$x = \frac{13 + \sqrt{209}}{4} \qquad x = \frac{13 - \sqrt{209}}{4}$$
$$x \approx 6.864 \quad \text{and} \quad x \approx -0.364$$

5. $5x^2 + 2x = -3$
$$5x^2 + 2x + 3 = 0$$
$$x = \frac{-2 \pm \sqrt{(2)^2 - 4(5)(3)}}{2(5)}$$
$$x = \frac{-2 \pm \sqrt{4 - 60}}{10}$$
$$x = \frac{-2 \pm \sqrt{-56}}{10}$$
There is no real number that is $\sqrt{-56}$.
There is no solution to this problem.

6. (a) $5x^2 - 3x - 2 = 0$
$$b^2 - 4ac = (-3)^2 - 4(5)(-2)$$
$$= 9 + 40 = 49$$
Because $49 > 0$ there will be two real roots.

(b) $2x^2 - 4x - 5 = 0$
$$b^2 - 4ac = (-4)^2 - 4(2)(-5)$$
$$= 16 + 40 = 56$$
Because $56 > 0$ there will be two real roots.

10.4 Practice Problems

1. (a) $y = 2x^2$

x	y
-2	8
-1	2
0	0
1	2
2	8

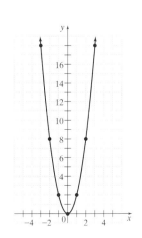

(b) $y = -2x^2$

x	y
-2	-8
-1	-2
0	0
1	-2
2	-8

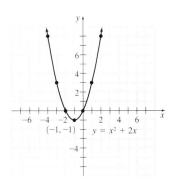

2. (a) $y = x^2 + 2x$

x	y	
-4	8	
-3	3	
-2	0	
-1	-1	vertex
0	0	
1	3	
2	8	

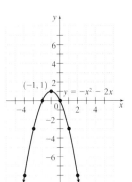

(b) $y = -x^2 - 2x$

x	y	
-4	-8	
-3	-3	
-2	0	
-1	1	vertex
0	0	
1	-3	
2	-8	

3. $y = x^2 - 4x - 5$
$$x = \frac{-b}{2a} = \frac{-(-4)}{2(1)} = 2$$
If $x = 2$, then
$y = (2)^2 - 4(2) - 5$
$y = 4 - 8 - 5$
$y = -9$
The point $(2, -9)$ is the vertex. x-intercepts:
$x^2 - 4x - 5 = 0$
$(x + 1)(x - 5) = 0$
$x + 1 = 0 \qquad x - 5 = 0$
$x = -1 \qquad\quad x = 5$
The x-intercepts are the points $(-1, 0)$ and $(5, 0)$.

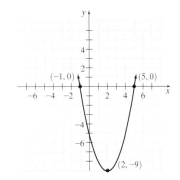

4. $h = -16t^2 + 32t + 10$

(a) Since the equation is a quadratic equation, the graph is a parabola. a is negative, therefore the parabola opens downward and the vertex is the highest point.

$$\text{Vertex} = t = \frac{-b}{2a} = \frac{-32}{2(-16)} = \frac{-32}{-32} = 1$$

If $t = 1$,   then $h = -16(1)^2 + 32(1) + 10$
$h = -16 + 32 + 10$
$h = 26$

The vertex (t, h) is $(1, 26)$.
Now find the h-intercept. The point where $t = 0$.
$$h = -16t^2 + 32t + 10$$
$$h = 10$$
The h-intercept is at $(0, 10)$.

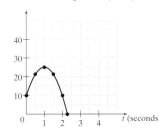

Table Values

t	h
0	10
0.5	22
1	26
1.5	22
2	10
2.275	0

(b) The ball is 26 feet above the ground one second after it is thrown. This is the greatest height of the ball.

(c) The ball hits the ground when $h = 0$.
$$0 = -16t^2 + 32t + 10$$
$$t = \frac{-32 \pm \sqrt{(32)^2 - 4(-16)(10)}}{2(-16)}$$
$$t = \frac{-32 \pm \sqrt{1024 + 640}}{-32}$$
$$t = \frac{-32 \pm \sqrt{1664}}{-32}$$
$$t = \frac{4 \pm \sqrt{26}}{4}$$
$$t \approx \frac{4 - 5.099}{4} \quad \text{and} \quad t \approx \frac{4 + 5.099}{4}$$
$$t \approx -0.275 \qquad\qquad t \approx 2.275$$
In this situation it is not useful to find points for negative values of t. Therefore $t \approx 2.3$ seconds. The ball will hit the ground approximately 2.3 seconds after it is thrown.

10.5 Practice Problems

1. $a^2 + b^2 = c^2$ \qquad Let $a = x$
$\qquad\qquad\qquad\qquad\qquad b = x - 6$
$x^2 + (x - 6)^2 = 30^2$
$x^2 + x^2 - 12x + 36 = 900$
$2x^2 - 12x + 36 = 900$
$2x^2 - 12x - 864 = 0$
$x^2 - 6x - 432 = 0$
$(x + 18)(x - 24) = 0$
$x + 18 = 0 \qquad x - 24 = 0$
$x = -18 \qquad\quad x = 24$
One leg is 24 meters, the other leg is $x - 6$ or $24 - 6 = 18$ meters.

2. Let c = cost for each student in the original group
s = number of students
Number of students $\times$ cost per student = total cost
$$s \times c = 240$$
If 2 people cannot go, then the number of students drops by 2 but the cost of each increases by $4.
$$(s - 2)(c + 4) = 240$$

If $s \times c = 240$ then

$$c = \frac{240}{s}$$

$(s - 2)\left(\dfrac{240}{s} + 4\right) = 240$ Multiply by the LCD which is s.

$(s - 2)(240 + 4s) = 240s$

$4s^2 + 232s - 480 = 240s$

$4s^2 - 8s - 480 = 0$

$s^2 - 2s - 120 = 0$

$(s - 12)(s + 10) = 0$

$\qquad s - 12 = 0 \qquad\qquad s + 10 = 0$

$\qquad\qquad s = 12 \qquad\qquad\quad s = -10$

The number of students originally going on the trip was 12.

Check.   If 12 students originally planned the trip then

$\qquad 12 \times c = 240.$

$\qquad\quad c = \$20$ per student

If 2 students dropped out, does this mean they must increase the cost of the trip by \$4?

$10 \times c = 240$

$c = \$24.00$ Yes.

3. Let $x =$ width

$\qquad y =$ length

$160 = 2x + y$

The area of a rectangle is (width)(length).

$\qquad\qquad A = x(y)$ Substitute $y = 160 - 2x$.

$\qquad 3150 = x(160 - 2x)$

$\qquad 3150 = 160x - 2x^2$

$2x^2 - 160x + 3150 = 0$

$\quad x^2 - 80x + 1575 = 0$

$\quad (x - 45)(x - 35) = 0$

$\qquad\qquad x = 45 \qquad x = 35$

First solution:

If the width is 45 feet then the length is

$\qquad\qquad 160 = 2(45) + y$

$\qquad\qquad 160 = 90 + y$

$\qquad\qquad\quad y = 70$ feet

Check. $\qquad 45 + 45 + 70 \overset{?}{=} 160$

$\qquad\qquad\qquad 90 + 70 \overset{?}{=} 160$

$\qquad\qquad\qquad\quad 160 = 160$ ✓

Second solution:

If the width is 35 feet then the length is

$\qquad\qquad 160 = 2(35) + y$

$\qquad\qquad 160 = 70 + y$

$\qquad\qquad\quad y = 90$ feet

Check. $\qquad 35 + 35 + 90 \overset{?}{=} 160$

$\qquad\qquad\qquad 70 + 90 \overset{?}{=} 160$

$\qquad\qquad\qquad\quad 160 = 160$ ✓

Answers to Selected Exercises

Chapter 0

Pretest Chapter 0

1. $\frac{3}{11}$ **2.** $\frac{2}{5}$ **3.** $3\frac{3}{4}$ **4.** $\frac{33}{7}$ **5.** 6 **6.** 35 **7.** 120 **8.** $\frac{5}{7}$ **9.** $\frac{19}{42}$ **10.** $8\frac{5}{12}$ **11.** $1\frac{33}{40}$ **12.** $\frac{10}{9}$ or $1\frac{1}{9}$ **13.** $\frac{21}{2}$ or $10\frac{1}{2}$
14. $\frac{7}{2}$ or $3\frac{1}{2}$ **15.** $\frac{28}{39}$ **16.** 0.875 **17.** $0.\bar{5}$ **18.** 19.651 **19.** 2.0664 **20.** 9.715 **21.** 2.246 **22.** 90% **23.** 0.7% **24.** 3.27
25. 0.02 **26.** 407.5 **27.** 7.2% **28.** 10,000 **29.** 21,000,000 **30.** 20 miles per gallon **31.** 28.4 miles per gallon **32.** 44%

0.1 Exercises

1. 12 **3.** When two or more numbers are multiplied, each number that is multiplied is called a factor. In 2×3, 2 and 3 are factors.
5. 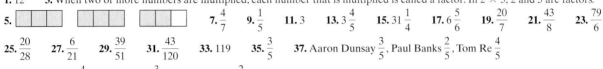 **7.** $\frac{4}{7}$ **9.** $\frac{1}{5}$ **11.** 3 **13.** $3\frac{4}{5}$ **15.** $31\frac{1}{4}$ **17.** $6\frac{5}{6}$ **19.** $\frac{20}{7}$ **21.** $\frac{43}{8}$ **23.** $\frac{79}{6}$
25. $\frac{20}{28}$ **27.** $\frac{6}{21}$ **29.** $\frac{39}{51}$ **31.** $\frac{43}{120}$ **33.** 119 **35.** $\frac{3}{5}$ **37.** Aaron Dunsay $\frac{3}{5}$, Paul Banks $\frac{2}{5}$, Tom Re $\frac{4}{5}$
39. Juliet Martin $\frac{4}{7}$, Letitia Stone $\frac{3}{4}$, Maureen Robb $\frac{2}{3}$

0.2 Exercises

1. Answers may vary. A sample answer is: 8 is exactly divisible by 4.
2. You must change the fractions to equivalent fractions with a common denominator.
3. 105 **5.** 20 **7.** 63 **9.** 90 **11.** $\frac{23}{24}$ **13.** $\frac{3}{28}$ **15.** $\frac{19}{18}$ or $1\frac{1}{18}$ **17.** $\frac{31}{63}$ **19.** $\frac{41}{24}$ or $1\frac{17}{24}$ **21.** $\frac{17}{36}$ **23.** $\frac{3}{2}$ or $1\frac{1}{2}$
25. $\frac{31}{60}$ **27.** $\frac{11}{36}$ **29.** $5\frac{7}{24}$ **31.** $7\frac{5}{12}$ **33.** $3\frac{13}{20}$ **35.** $2\frac{3}{4}$ **37.** $7\frac{10}{21}$ **39.** $6\frac{3}{5}$ **41.** $10\frac{7}{24}$ miles **43.** $4\frac{1}{12}$ hours
45. $A = 12$ inches, $B = 15\frac{7}{8}$ inches **47.** $1\frac{5}{8}$ inches **49.** $\frac{9}{11}$ **50.** $\frac{133}{5}$

Putting Your Skills to Work

1. $10\frac{1}{8}$ in. **2.** from 1952 to 1972

0.3 Exercises

1. First, change each number to an improper fraction. Look for a common factor in the numerator and denominator to divide by, and, if one is found, peform the division. Multiply the numerators. Multiply the denominators. **2.** Invert the second fraction. Multiply the result by the first fraction.
3. 6 **5.** $\frac{18}{5}$ or $3\frac{3}{5}$ **7.** $\frac{3}{5}$ **9.** $\frac{1}{7}$ **11.** 14 **13.** $\frac{15}{16}$ **15.** $\frac{8}{35}$ **17.** $\frac{7}{12}$ **19.** 1 **21.** $\frac{5}{2}$ or $2\frac{1}{2}$ **23.** $\frac{219}{4}$ or $54\frac{3}{4}$
25. $\frac{17}{2}$ or $8\frac{1}{2}$ **27.** 28 **29.** $\frac{3}{16}$ **31.** $71\frac{1}{2}$ yds **33.** $2\frac{4}{5}$ mi **35.** $\frac{29}{31}$ **36.** $\frac{3}{7}$

0.4 Exercises

1. 10; 100; 10,000; and so on **3.** three, left **5.** 0.15 **7.** $0.\bar{6}$ **9.** 0.8 **11.** $\frac{3}{20}$ **13.** $\frac{2}{25}$ **15.** $\frac{9}{8}$ **17.** 0.00153 **19.** 9.38 **21.** 208.791
23. 44.48 **25.** 18.512 **27.** 0.02834 **29.** 77.28 **31.** 3.18 **33.** 6.2 **35.** 6.3 **37.** 0.0076 **39.** 3450 **41.** 243 **43.** 0.73892
45. 0.007296 **47.** 52.73 **49.** 25,712 **51.** 0.003472 **53.** 666.9 miles **55.** Yes, it will yield 13.494 pounds of meat.
57. Yes, he exceeded his goal by $11.50. **58.** $\frac{2}{3}$ **59.** $\frac{1}{6}$ **60.** $\frac{93}{100}$ **61.** $\frac{11}{10}$ or $1\frac{1}{10}$

Putting Your Skills to Work

1. 563,000,000 people **2.** 10,000,000 people

0.5 Exercises

1. Answers may vary. Sample answers are below. 19% means 19 out of 100 parts. Percent means per 100. 19% is really a fraction with a denominator of 100. In this case it would be $\frac{19}{100}$. **3.** 56.8% **5.** 40.9% **7.** 239% **9.** 0.03 **11.** 0.004 **13.** 2.5 **15.** 358 **17.** 0.52 **19.** 72.8
21. 150% **23.** 5% **25.** 137.5% **27.** 85% **29.** $4.92 tip $37.72 new total **31.** Approximately 21% **33.** 540 **35.** 77%
37. (a) 12,090 mi **(b)** $3747.90 **39.** $2.97 **41.** $538.13 **42.** $6220.50 **43.** 22 miles per gallon **44.** 4.0 in.

0.6 Exercises

1. 210,000   **3.** 180,000,000   **5.** 240   **7.** 30,000   **9.** 0.1   **11.** $4000.00   **13.** 4500 square feet   **15.** $6400   **17.** 25 miles per gallon
19. $480.00   **21.** $1,600,000,000   **23.** $240   **25.** 2.1   **26.** 56.25%   **27.** $6.73   **28.** 270 tires

0.7 Exercises

1. $269.17   **3.** 95 cubic yards   **5.** Jog $2\frac{2}{3}$ miles; walk $3\frac{1}{9}$ miles; rest $4\frac{4}{9}$ minutes; walk $1\frac{7}{9}$ miles   **7.** Betty; Melinda increases each activity by
$\frac{2}{3}$ by day 3 and Betty increases each activity by $\frac{7}{9}$ by day 3.   **9.** $4\frac{1}{2}$ miles   **11.** $98,969.00   **13.** 49%: 61%; business travelers; Answers may
vary: A sample is: Men 21 or older spend more of their travel time on business than for pleasure.   **15.** 18%   **17.** 69%

Chapter 0 Review Problems

1. $\frac{3}{4}$ **2.** $\frac{3}{10}$ **3.** $\frac{18}{41}$ **4.** $\frac{3}{5}$ **5.** $\frac{19}{7}$ **6.** $6\frac{4}{5}$ **7.** $6\frac{3}{4}$ **8.** 15   **9.** 5   **10.** 12   **11.** 22   **12.** $\frac{17}{20}$ **13.** $\frac{29}{24}$ or $1\frac{5}{24}$ **14.** $\frac{4}{15}$ **15.** $\frac{17}{48}$
16. $\frac{173}{30}$ or $5\frac{23}{30}$ **17.** $\frac{79}{20}$ or $3\frac{19}{20}$ **18.** $\frac{8}{15}$ **19.** $\frac{23}{12}$ or $1\frac{11}{12}$ **20.** $\frac{30}{11}$ or $2\frac{8}{11}$ **21.** $\frac{21}{2}$ or $10\frac{1}{2}$ **22.** $\frac{19}{8}$ or $2\frac{3}{8}$ **23.** $\frac{20}{7}$ or $2\frac{6}{7}$
24. $\frac{3}{10}$ **25.** $\frac{10}{3}$ or $3\frac{1}{3}$ **26.** $\frac{3}{20}$ **27.** 6   **28.** 7.201   **29.** 7.737   **30.** 21.572   **31.** 4.436   **32.** 0.03745   **33.** 362,341
34. 0.07956   **35.** 10.368   **36.** 0.00186   **37.** 0.07132   **38.** 1.3075   **39.** 90   **40.** 0.07   **41.** 0.5   **42.** 37.5%
43. $\frac{6}{25}$ **44.** 0.014   **45.** 0.361   **46.** 0.0002   **47.** 1.253   **48.** 510   **49.** 90   **50.** 18   **51.** 12.5%   **52.** 60%   **53.** 98,280,000   **54.** 75%
55. 400,000,000,000   **56.** 2500   **57.** 600,000   **58.** 19   **59.** $12,000   **60.** 30   **61.** $640   **62.** $300   **63.** $349.07   **64.** 8%
65. 1840 miles; 1472 miles   **66.** 1500 miles; 1050 miles   **67.** $2\frac{2}{3}$ cups of sugar   **68.** $3\frac{2}{3}$ miles   **69.** $515\frac{5}{8}$ pounds   **70.** 46 packages
71. $1585.50   **72.** $371.20

Chapter 0 Test

1. $\frac{8}{9}$ **2.** $\frac{7}{8}$ **3.** $\frac{45}{7}$ **4.** $3\frac{3}{11}$ **5.** $\frac{17}{12}$ or $1\frac{5}{12}$ **6.** $\frac{39}{8}$ or $4\frac{7}{8}$ **7.** $\frac{5}{6}$ **8.** $\frac{4}{3}$ or $1\frac{1}{3}$ **9.** $\frac{5}{24}$ **10.** $\frac{7}{2}$ or $3\frac{1}{2}$ **11.** $\frac{65}{8}$ or $8\frac{1}{8}$
12. $\frac{159}{74}$ or $2\frac{11}{74}$ **13.** 14.64   **14.** 1806.62   **15.** 1.312   **16.** 16.32   **17.** 230   **18.** 263,259   **19.** 7.3%   **20.** 1.965   **21.** 63
22. 0.336   **23.** 5%   **24.** 30%   **25.** 18   **26.** 100   **27.** 100   **28.** 65%   **29.** 60 tiles

Chapter 1

Pretest Chapter 1

1. 3   **2.** -15 **3.** -18 **4.** 20   **5.** -42 **6.** $\frac{8}{3}$ **7.** 9.2   **8.** -4 **9.** 16   **10.** 64   **11.** $-\frac{8}{27}$ **12.** -256 **13.** 28   **14.** 197
15. $\frac{107}{12}$ **16.** 6.37   **17.** $-6x^2 + 4x^2y - 2xz$ **18.** $-3x + 4y + 12$ **19.** $5x^2 - 6x^2y - 11xy$ **20.** $7x - 2y + 7$ **21.** $5x - 7y$
22. $-x^2y + 17y^2$ **23.** 18   **24.** 29   **25.** 25°C   **26.** $-9x - 3y$ **27.** $-x^2 + 6xy$

1.1 Exercises

1. Whole number, rational number, real number   **3.** Irrational number, real number   **5.** Rational number, real number   **7.** Rational number, real number   **9.** Irrational number, real number   **11.** $-20,000$ **13.** $-2\frac{3}{8}$ **15.** $+7$ **17.** $-\frac{3}{4}$ **19.** 2.73   **21.** 1.3   **23.** $\frac{5}{6}$ **25.** -11
27. $\frac{1}{3}$ **29.** -31 **31.** $-\frac{7}{13}$ **33.** -3.8 **35.** 0.4   **37.** -25 **39.** $\frac{1}{35}$ **41.** -6 **43.** 3   **45.** 17   **47.** -28 **49.** $\frac{1}{15}$ **51.** $-\frac{11}{12}$
53. -3.4 **55.** 12   **57.** 1   **59.** 16.39   **61.** $-5°F$ **63.** -265 feet   **65.** no; $16.65 short   **67.** $140   **69.** $22,000,000   **71.** 28
73. 41%; 120 more kittens   **74.** $\frac{2}{3}$ **75.** $\frac{8}{27}$ **76.** $\frac{1}{12}$ **77.** $\frac{25}{34}$

1.2 Exercises

1. First change subtracting -3 to adding a positive three. Then use the rules for addition of two real numbers with different signs. Thus
$-8 - (-3) = -8 + 3 = -5$.   **3.** -26 **5.** -11 **7.** 8   **9.** 5   **11.** -5 **13.** 0   **15.** -3 **17.** -0.9 **19.** 4.47   **21.** 3.7   **23.** $\frac{27}{20}$
25. $-\frac{19}{12}$ **27.** -53 **29.** -99 **31.** 7.1   **33.** $\frac{35}{4}$ **35.** $\frac{43}{5}$ **37.** $-\frac{21}{20}$ **39.** -8.5 **41.** 0.0499   **43.** $-\frac{29}{5}$ **45.** -6.0023 **47.** 7
49. -48 **51.** -2 **53.** 11   **55.** -62 **57.** 7   **59.** 38   **61.** $149   **63.** $-16°F$ **65.** -21 **66.** -51 **67.** -19 **68.** $-8°C$
69. $6\frac{2}{3}$ miles were snow-covered.

1.3 Exercises

1. To multiply two real numbers, multiply the absolute values. The sign of the result is positive if both numbers have the same sign, but negative if the two numbers have opposite signs. **3.** -24 **5.** 0 **7.** 24 **9.** 0.264 **11.** -1.75 **13.** -24 **15.** $\frac{4}{15}$ **17.** -3 **19.** 0 **21.** -5
23. -16 **25.** -2 **27.** -9 **29.** -0.9 **31.** $-\frac{3}{10}$ **33.** $-\frac{20}{3}$ **35.** -30 **37.** $\frac{9}{16}$ **39.** $-\frac{7}{9}$ **41.** -24 **43.** 24 **45.** -16
47. -18 **49.** $-\frac{8}{35}$ **51.** $\frac{2}{27}$ **53.** -4 **55.** -2 **57.** 17 **59.** -72 **61.** -1 **63.** \$30 **65.** \$328.50 **67.** 20 yards. **69.** 70 yards
71. The Panthers gained 20 yards. **73.** The Panthers would have gained 105 fewer yards. **74.** -6.69 **75.** -72 **76.** 4.63 **77.** -88
78. 266 square yards

1.4 Exercises

1. The base is 4 and the exponent is 4. Thus you multiply $(4)(4)(4)(4) = 256$. **3.** The answer is negative. When you raise a negative number to an odd power the result is always negative. **5.** If you have parentheses surrounding the -2, then the base is -2 and the exponent is 4. The result is 16. If you do not have parentheses, then the base is 2. You evaluate to obtain 16 and then take the negative of 16, which is -16. Thus $(-2)^4 = 16$ but $-2^4 = -16$. **7.** 27 **9.** 81 **11.** 343 **13.** -27 **15.** 64 **17.** -125 **19.** $\frac{1}{16}$ **21.** $\frac{8}{125}$ **23.** 0.81 **25.** 0.0016
27. 256 **29.** -256 **31.** 6^5 **33.** w^2 **35.** x^4 **37.** $(3q)^3$ **39.** 161 **41.** -91 **43.** 152 **45.** -576 **47.** -512 **49.** 16,777,216
51. 2 **53.** -19 **54.** $-\frac{5}{3}$ **55.** -8 **56.** 2.52 **57.** \$1672

1.5 Exercises

1. $3(4) + 6(5)$ **3. (a)** 90 **(b)** 42 **5.** 12 **7.** -29 **9.** 14 **11.** 13 **13.** -6 **15.** 42 **17.** $\frac{9}{4}$ **19.** 0.848 **21.** $\frac{3}{10}$ **23.** 7.56
25. $\frac{1}{4}$ **27.** 0.125 **28.** $-\frac{19}{12}$ **29.** -1 **30.** $\frac{72}{125}$ **31.** 45 ounces

1.6 Exercises

1. variable **3.** Here we are multiplying 4 by x by x. Since we know from the definition of exponents that x multiplied by x is x^2, this gives us an answer of $4x^2$.
5. Yes, $a(b - c)$ can be written as $a[b + (-c)]$
$$3(10 - 2) = (3 \times 10) - (3 \times 2)$$
$$3 \times 8 = 30 - 6$$
$$24 = 24$$

7. $-x - 4y$ **9.** $-6a + 15b$ **11.** $8x + 2y - 4$ **13.** $6x^2 - 9xy + 3xz$ **15.** $8x^2 - 2xy - 12x$ **17.** $-15x - 45 + 35y$ **19.** $\frac{x^2}{4} + \frac{x}{2} - 2$
21. $-18a^4 + 6a^2 - 14$ **23.** $y^2 - \frac{4xy}{3} - 2y$ **25.** $6a^2 + 3ab - 3ac - 12a$ **27.** $-20x - 4$ **29.** $12x^2 - 15xy - 18x$
31. $-4a^2b + 2ab^2 + ab$ **33.** $-8x^2y + 4xy^2 - 12xy$ **35.** $3.75a^2 - 8.75a + 5$ **37.** $-1.89q^2 + 0.18qr + 0.72qs$
39. $8400x + 5600y$ square feet **41.** $4500x - 12xy$ square feet **43.** -16 **44.** 64 **45.** 14 **46.** 56% **47.** 14 days

1.7 Exercises

1. A term is a number, a variable, or a product of numbers and variables. **3.** The two terms $5x$ and $-8x$ are like terms because they both have the variable x with the exponent of one. **5.** The only like terms are $7xy$ and $-14xy$ because the other two have different exponents even though they have the same variables. **7.** $-25b^2$ **9.** $18x^4 + 7x^2$ **11.** $-4ab - 7$ **13.** $7.1x - 3.5y$ **15.** $-2x - 8.7y$ **17.** $\frac{3}{4}x^2 - \frac{10}{3}y$
19. $-\frac{1}{15}x - \frac{2}{21}y$ **21.** $5p + q - 18$ **23.** $5x^2y - 10xy^6 - xy^2$ **25.** $5bc - 6ac$ **27.** $x^2 - 10x + 3$ **29.** $-10y^2 - 16y + 12$
31. $5a - 3ab - 8b$ **33.** $-12x + 13y$ **35.** $14x^2 + 2xy$ **37.** $-17xy + 27y^2$ **39.** $71x - 27$ **41.** $20x + 4$ meters **43.** $32a + 10$ feet
44. $-\frac{2}{15}$ **45.** $-\frac{5}{6}$ **46.** $\frac{23}{50}$ **47.** $-\frac{15}{98}$ **48.** 0.2 liter

1.8 Exercises

1. -5 **3.** -11 **5.** $\frac{25}{2}$ **7.** -26 **9.** 10 **11.** 3 **13.** -24 **15.** $\frac{25}{4}$ **17.** 9 **19.** 39 **21.** -2 **23.** 15 **25.** -9 **27.** 29
29. 42 **31.** 29 **33.** 4968 square feet **35.** 129 square millimeters **37.** 166.5 square inches **39.** 133 square feet **41.** 78.5 square meters
43. 14°F **45.** \$300.26 **47.** $-76°F$ to $-22°F$ **49.** 12.4 miles **50.** 16 **51.** $-x^2 + 2x - 4y$ **52.** 6.2 minutes/song

Putting Your Skills to Work

1. 21.3 **2.** yes; 25.6

1.9 Exercises

1. $-(3x + 2y)$ **3.** distributive **5.** $3x + 6y$ **7.** $5a + 3b$ **9.** $-245 + 25x$ **11.** $8x^3 - 4x^2 + 12x$ **13.** $4x - 6y - 3$ **15.** $15a - 60ab$
17. $-x^3 + 2x^2 - 3x - 12$ **19.** $3a^2 + 16b + 12b^2$ **21.** $-7a + 8b$ **23.** $12a^2 - 8b$ **25.** 219 successful attempts; 8541 unsuccessful attempts
27. 97.52°F **28.** 453,416 square feet **29.** 300,000 square feet; \$16,500,000 **30.** 11.375 square feet; \$1387.75

Putting Your Skills to Work

1. $335.20 **2.** $22,050.00 **3.** $598.00 **4.** $507.00

Chapter 1 Review Problems

1. -8 **2.** -4.2 **3.** -9 **4.** 1.9 **5.** $-\dfrac{1}{3}$ **6.** $-\dfrac{7}{22}$ **7.** $\dfrac{1}{6}$ **8.** $\dfrac{22}{15}$ **9.** 8 **10.** 13 **11.** -33 **12.** 9.2 **13.** $-\dfrac{13}{8}$ **14.** $\dfrac{1}{2}$

15. -22.7 **16.** -88 **17.** -4 **18.** 16 **19.** -29 **20.** 1 **21.** -3 **22.** 18 **23.** 32 **24.** $-\dfrac{2}{3}$ **25.** $-\dfrac{25}{7}$ **26.** -72 **27.** 30

28. -30 **29.** $-\dfrac{1}{2}$ **30.** $-\dfrac{4}{7}$ **31.** -30 **32.** -5 **33.** -9.1 **34.** 0.9 **35.** 10.1 **36.** -1.2 **37.** 1.9 **38.** -1.3 **39.** 24 yards

40. $-22°$F **41.** 7363 feet **42.** $2\dfrac{1}{4}$ point loss **43.** -243 **44.** -128 **45.** 625 **46.** $\dfrac{8}{27}$ **47.** -81 **48.** 0.36 **49.** $\dfrac{25}{36}$ **50.** $\dfrac{27}{64}$

51. -44 **52.** 30 **53.** 1 **54.** $15x - 35y$ **55.** $6x^2 - 14xy + 8x$ **56.** $-7x^2 + 3x - 11$ **57.** $-6xy^2 - 3xy + 3y^2$ **58.** $-5a^2b + 3bc$

59. $-3x - 4y$ **60.** $-5x^2 - 35x - 9$ **61.** $10x^2 - 8x - \dfrac{1}{2}$ **62.** -55 **63.** 1 **64.** -4 **65.** -15 **66.** 10 **67.** -16 **68.** $\dfrac{32}{5}$

69. $810 **70.** $86°$F **71.** $2119.50 **72.** $8580.00 **73.** $100,000$ square feet; $200,000 **74.** 10.45 square feet; $689.70 **75.** $-2x + 42$
76. $-17x - 18$ **77.** $-2 + 10x$ **78.** $-12x^2 + 63x$ **79.** $5xy^3 - 6x^3y - 13x^2y^2 - 6x^2y$ **80.** $x - 10y + 35 - 15xy$ **81.** $10x - 22y - 36$
82. $-10a + 25ab - 15b^2 - 10ab^2$ **83.** $-3x - 9xy + 18y^2$ **84.** $10x + 8xy - 32y$

Chapter 1 Test

1. 4 **2.** 0.2 **3.** 96 **4.** -70 **5.** 4 **6.** -3 **7.** -64 **8.** 1.69 **9.** $\dfrac{16}{81}$ **10.** 8 **11.** -25 **12.** $-5x^2 - 10xy + 35x$

13. $6a^2b^2 + 4ab^3 - 14a^2b^3$ **14.** $2a^2b + \dfrac{15}{2}ab$ **15.** $8a^2 + 20ab$ **16.** $5a + 30$ **17.** $14x - 16y$ **18.** 122 **19.** 37 **20.** $\dfrac{13}{6}$

21. 96 kilometers/hour **22.** $22,800$ square feet **23.** $23.12 **24.** 452.16 square feet **25.** $-3a - 9ab + 3b^2 - 3ab^2$ **26.** $-69x + 90y - 51$

Chapter 2

Pretest Chapter 2

1. $x = -16$ **2.** $x = 48$ **3.** $x = -4$ **4.** $x = 4$ **5.** $x = -\dfrac{7}{2}$ **6.** $x = -\dfrac{1}{7}$ **7.** $x = 1$ **8.** $x = -4$ **9.** $x = \dfrac{5}{2}$ **10.** $x = \dfrac{17}{10}$

11. $x = -\dfrac{2}{3}$ **12.** $x = 17$ **13. (a)** $F = \dfrac{9C + 160}{5}$ **(b)** $F = 5°$ **14. (a)** $r = \dfrac{I}{Pt}$ **(b)** 6% **15.** $<$ **16.** $>$ **17.** $>$ **18.** $<$

19. **20.**

2.1 Exercises

1. equals, equal **3.** solution **5.** Answers may vary. A sample answer is to isolate the variable. **7.** $x = 6$ **9.** $6 = x$ **11.** $x = 16$
13. $x = 7$ **15.** $x = 1$ **17.** $x = 0$ **19.** $x = -2$ **21.** $x = 66$ **23.** $x = 27$ **25.** $x = -11$ **27.** no; $x = 26$ **29.** yes **31.** no; $x = 8$

33. yes **35.** 4 **37.** 22 **39.** 3.5 **41.** $x = -2.9$ **43.** $\dfrac{1}{3}$ **45.** $x = -\dfrac{1}{5}$ **47.** $\dfrac{7}{6}$ or $1\dfrac{1}{6}$ **49.** $27\dfrac{1}{8}$ **51.** 7.2 **53.** $x = -3.783$

54. $-2x - 4y$ **55.** $-2y^2 - 4y + 4$ **56.** 76.5% **57.** $14.92 **58.** 117 feet

2.2 Exercises

1. 6 **3.** 7 **5.** $x = 35$ **7.** $x = -27$ **9.** $x = 80$ **11.** $x = -15$ **13.** $x = 4$ **15.** $x = -\dfrac{8}{3}$ **17.** $x = 50$ **19.** $x = 15$ **21.** $x = -7$

23. $x = 0.2$ or $\dfrac{1}{5}$ **25.** no, $x = -7$ **27.** yes **29.** $y = -0.8$ **31.** $t = \dfrac{8}{3}$ **33.** $y = -0.7$ **35.** $x = 3$ **37.** $x = -4$ **39.** $x = \dfrac{7}{9}$ **41.** $x = \dfrac{7}{6}$

43. $x = -5.26$ **45.** To solve an equation, we are performing steps to get an equivalent equation that has the same solution. Now $a = b$ and $a(0) = b(0)$ are not equivalent equations because they do not have the same solution. So we must have the requirement that when we multiply both sides of the equation by c, it is absolutely essential that c is nonzero. **47.** 42 **48.** -37 **49.** 21

50. $93\dfrac{1}{3}\%$ is acceptable. **51.** $632 **52.** 104 calves

2.3 Exercises

1. 2 **3.** 6 **5.** -17.5 **7.** 13 **9.** 15 **11.** 40 **13.** -36 **15.** 8 **17.** 3 **19.** 7 **21.** -9 **23.** yes

25. no; $x = 12$ **27.** $7; 7$ **29.** 0 **31.** 5 **33.** -4 **35.** 6 **37.** 5 **39.** $\dfrac{1}{2}$ **41.** 1 **43.** 3.1 **45.** -4 **47.** 0 **49.** 2

51. 1.2 **53.** -1.2 **55.** 2 **57.** $-\dfrac{3}{5}$ **59.** 1 **61.** 1 **63.** -7 **65.** 42 **67.** -4 **69.** 2.78 **70.** $14x^2 = 14xy$ **71.** $-10x - 60$

72. $1215.70 **73. (a)** $91.00 **(b)** $94.50

2.4 Exercises

1. 2 **3.** $2\dfrac{1}{2}$ or $\dfrac{5}{2}$ **5.** 1 **7.** 24 **9.** 20 **11.** 7 **13.** 18 **15.** -3.5 **17.** yes **19.** no **21.** 1 **23.** 0 **25.** 2 **27.** -3 **29.** 4

31. −22   **33.** 2   **35.** −12   **37.** $-\dfrac{13}{12}$ or $-1\dfrac{1}{12}$ **39.** $\dfrac{4}{113}$ **41.** $\dfrac{18}{17}$ **43.** no solution   **45.** infinite number of solutions

47. $\dfrac{27}{14}$ or $1\dfrac{13}{14}$ **48.** $\dfrac{19}{20}$ **49.** $\dfrac{539}{40}$ or $13\dfrac{19}{40}$ **50.** $\dfrac{22}{5}$ or $4\dfrac{2}{5}$ **51.** 264 pairs   **52.** $108   **53.** $226.08   **54.** $273.60

2.5 Exercises

1. Multiply each term by 5. Then add −160 to each side. Then divide each side by 9. We would obtain $\dfrac{5F - 160}{9} = C$.

3. (a) 10 meters   **(b)** 16 meters   **5. (a)** $y = \dfrac{3}{5}x - 3$ or $\dfrac{15 - 3x}{-5}$ **(b)** −6   **7.** $b = \dfrac{2A}{h}$ **9.** $P = \dfrac{I}{rt}$ **11.** $m = \dfrac{y - b}{x}$ **13.** $t = \dfrac{A - P}{Pr}$

15. $y = \dfrac{5}{6}x - 1$ **17.** $x = -\dfrac{4}{3}y + 12$ **19.** $y = \dfrac{c - ax}{b}$ **21.** $r^2 = \dfrac{A}{\pi}$ **23.** $g = \dfrac{2S}{t^2}$ **25.** $h = \dfrac{S - 2\pi r^2}{2\pi r}$ **27.** $h = \dfrac{V}{\pi r^2}$ **29.** $L = \dfrac{V}{WH}$

31. $r^2 = \dfrac{3V}{\pi h}$ **33.** $W = \dfrac{P - 2L}{2}$ **35.** $a^2 = c^2 - b^2$ **37.** $C = \dfrac{5(F - 32)}{9}$ or $C = \dfrac{5F - 160}{9}$ **39.** $T = \dfrac{PV}{k}$ **41.** $S = \dfrac{360A}{\pi r^2}$

43. The width is 0.8 miles.   **45.** The length is 30 feet.   **47. (a)** $\dfrac{V - 7050}{1100} = x$ **(b)** 2002   **49.** I doubles.   **51.** A increases four times.

53. 31.2   **54.** 0.096   **55.** 40%   **56.** 2.5%   **57.** 39,000 square feet   **58.** $10\dfrac{7}{12}$ hours

Putting Your Skills to Work

1. 3.3 million tons   **2.** 1.9 million tons

2.6 Exercises

1. Yes, both statements imply 5 is to the right of −6 on the number line.   **3.** >   **5.** <   **7.** >   **9.** <   **11.** <   **13.** <   **15.** 0.0247

27. $x \geq -\dfrac{2}{3}$ **29.** $x < -20$ **31.** $x > 6.5$ **33.** $V > 580$ **35.** $h \geq 37$ **37.** $h \geq 48$ **38.** $C \leq 18$

41. 6.08   **42.** 16.25   **43.** 2%   **44.** 37.5%

45. 9947 attempts were successful. 14,200,053 attempts were unsuccessful.   **46.** 495 feet

2.7 Exercises

1. $x \leq -3$ **3.** $x > -9$ **5.** $x \geq 8$

7. $x < \dfrac{7}{2}$ **9.** $x > -\dfrac{5}{4}$ **11.** $x > -6$

13. $x > \dfrac{1}{3}$

15. $3 > 1$; Adding any number to both sides of an inequality does not reverse the direction.   **17.** $x > 2$ **19.** $x \geq 4$ **21.** $x < -1$ **23.** $x \leq 14$

25. $x < -3$ **27.** $x > -\dfrac{4}{11}$ **29.** $x > -1.46$ **31.** 76 or greater   **33.** 8 days or more   **35.** 260 feet   **36.** 4.5 inches

37. 63.6 square inches   **38.** 15,600 square feet

Chapter 2 Review Problems

1. −10   **2.** −1   **3.** 2   **4.** −3   **5.** 2.75   **6.** −13   **7.** 40.4   **8.** −7   **9.** −2   **10.** −64   **11.** −11   **12.** $\dfrac{22}{3}$ or $7\dfrac{1}{3}$ **13.** −3   **14.** −1

15. 4   **16.** 3   **17.** 3   **18.** −7   **19.** 3   **20.** $-\dfrac{7}{2}$ or $-3\dfrac{1}{2}$ or −3.5   **21.** −3   **22.** 2.1   **23.** $-\dfrac{7}{3}$ or $-2\dfrac{1}{3}$ **24.** $-\dfrac{2}{7}$ **25.** 5   **26.** 0

27. 1   **28.** 20   **29.** $\dfrac{2}{3}$ **30.** 5   **31.** $\dfrac{35}{11}$ **32.** 4   **33.** −17   **34.** $\dfrac{2}{5}$ or 0.4   **35.** 32   **36.** $\dfrac{26}{7}$ **37.** −1   **38.** $-\dfrac{5}{2}$ or $-2\dfrac{1}{2}$ or −2.5   **39.** 4

40. −17   **41.** 4   **42.** 0   **43.** 4   **44.** −5   **45.** 23   **46.** −32   **47.** $-\dfrac{17}{5}$ or −3.4   **48.** 9   **49.** 0   **50.** 9.75 or $9\dfrac{3}{4}$ or $\dfrac{39}{4}$

51. $y = 3x - 10$ **52.** $y = \dfrac{-5x - 7}{2}$ **53.** $r = \dfrac{A - P}{Pt}$ **54.** $h = \dfrac{A - 4\pi r^2}{2\pi r}$ **55.** $p = \dfrac{3H - a - 3}{2}$ **56.** $y = \dfrac{c - ax}{b}$ **57.** $d = \dfrac{6c - H}{5}$

58. $b = \dfrac{4H - 3c}{2}$ **59. (a)** $T = \dfrac{1000C}{WR}$ **(b)** $T = 6000$ **60. (a)** $y = \dfrac{5}{3}x - 4$ **(b)** $y = 11$ **61. (a)** $R = \dfrac{E}{I}$ **(b)** $R = 5$

62. $x \leq 2$ **63.** $x \geq 1$ **64.** $x < -4$ **65.** $x > -3$

66. $x \geq 3$

67. $x < 2$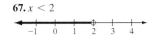

68. $x \leq -8$

69. $x \geq 3$

70. $x < 10$

71. $x > -3$

72. $x > \dfrac{17}{2}$

73. $x \geq \dfrac{19}{7}$

74. $x \geq -15$

75. $x > \dfrac{7}{5}$

76. $n \leq 46$ **77.** $n \leq 17$

Chapter 2 Test

1. 2   **2.** $\dfrac{1}{3}$ **3.** $-\dfrac{7}{2}$ or $-3\dfrac{1}{2}$ or -3.5 **4.** 8.4 or $8\dfrac{2}{5}$ or $\dfrac{42}{5}$ **5.** 8   **6.** -1.2 **7.** 7   **8.** $\dfrac{7}{3}$ **9.** 13   **10.** 125   **11.** 10   **12.** -4 **13.** 12

14. $-\dfrac{1}{5}$ or -0.2 **15.** 3   **16.** 2   **17.** 18   **18.** $w = \dfrac{A - 2P}{3}$ **19.** $w = \dfrac{6 - 3x}{4}$ **20.** $a = \dfrac{2A - hb}{h}$ **21.** $y = \dfrac{10ax - 5}{8ax}$

22. $\dfrac{P - 2L}{2} = W$ **23.** 18 feet

24. $x \leq -\dfrac{1}{2}$

25. $x > -\dfrac{5}{4}$

26. $x < 2$

27. $x \geq \dfrac{1}{2}$

Cumulative Test for Chapters 0–2

1. $\dfrac{4}{21}$ **2.** $\dfrac{79}{20}$ or $3\dfrac{19}{20}$ **3.** $\dfrac{32}{15}$ or $2\dfrac{2}{15}$ **4.** 0.6888   **5.** 0.12   **6.** 117   **7.** 30   **8.** $12ab - 28ab^2$ **9.** $25x^2$ **10.** $2x + 6y - 4xy + 16xy^2$

11. $\dfrac{40}{11}$ or $3\dfrac{7}{11}$ **12.** 4   **13.** 1   **14.** $y = \dfrac{3x + 2}{7}$ **15.** $b = \dfrac{3H - 8a}{2}$ **16.** $t = \dfrac{I}{Pr}$ **17.** $a = \dfrac{2A}{h} - b$

18. $x < 5$ **19.** $x \leq 3$ **20.** $x > -15$ **21.** $x \leq -1$ **22.** $x \geq -12$

23. 92

Chapter 3

Pretest Chapter 3

1. $2x - 30$ **2.** $0.40x + 150$ **3.** 5   **4.** 1st number = 18; 2nd number = 9; 3rd number = 29   **5.** One side = 18 feet; second side = 7 feet; third side = 13 feet   **6.** 1st package = 7.5 pounds; 2nd package = 4 pounds; 3rd package = 5.5 pounds   **7.** width = 11 feet; length = 33 feet   **8.** $400 at 12%; $600 at 9%   **9.** 22,000 people   **10.** 3 nickels; 7 dimes; 10 quarters   **11.** 904.32 cubic inches   **12.** 339.12 cubic meters   **13.** 12.86 square inches   **14.** $6859.80   **15.** $w \leq 28$ **16.** $f \geq 250$ **17.** $s < 23$ **18.** $t > 18,000$ **19.** more than $15,000 per week   **20.** length ≤ 16 feet

3.1 Exercises

1. $x + 5$ **3.** $x - 6$ **5.** $\dfrac{x}{8}$ **7.** $x - 7$ **9.** $\dfrac{2}{x + 8}$ **11.** $6(x + 8)$ **13.** $\dfrac{1}{2}x - 4$ **15.** $4x + \dfrac{1}{2}x$ **17.** $7x - \dfrac{1}{3}x$

19. x = income from retirement fund; $x - 833$ = income from mutual fund   **21.** $3w + 3$ = length of rectangle; w = width of rectangle   **23.** s = attendance on Saturday; $s + 1600$ = attendance on Friday; $s - 783$ = attendance on Thursday   **25.** 1st angle = $t + 19$; 2nd angle = $3t$; 3rd angle = t **27.** m = number of medals Carl Lewis won; $m + 2$ = number of medals Mark Spitz won   **29.** x = the cost of his history book in dollars; $x + 13$ = the cost of his biology book in dollars; $x - 27$ = the cost of his English book in dollars   **31.** p = number of 8th graders; $50 + p$ = number of 7th graders; $\frac{3}{4}p$ = number of 6th graders   **33.** x = the atomic weight of chromium; $3x + 51.2$ = the atomic weight of lead   **35.** x = income of women aged 55–64 years $x - 159$ = income of women aged 16–24 years $x - 12$ = income of women aged 25–34 years $x + 24$ = income of women aged 35–44 years $x + 68$ = income of women aged 45–54 years $x - 99$ = income of women aged 65 years or older   **36.** $x = 17$ **37.** $x = 10$ **38.** $x = 12$ **39.** $w = 7$

3.2 Exercises

1. 1261   **3.** 2368   **5.** 182   **7.** 43   **9.** 51   **11.** -5 **13.** 12   **15.** 15 used bikes   **17.** 40 months   **19.** 6 CDs   **21.** 4 French fries   **23.** 550 lb   **25.** 120 minutes or 2 hours   **27.** 5 miles   **29.** He traveled 52 mph on the mountain road. It was 12 mph faster on the highway.

31. 96   **33.** 9 mpg   **35.** (a) $F - 40 = \dfrac{x}{4}$ (b) 200 chirps   (c) 77°F   **36.** $10x^3 - 30x^2 - 15x$ **37.** $-2a^2b + 6ab - 10a^2$

38. $-5x - 6y$ **39.** $-4x^2y - 7xy^2 - 8xy$ **40.** 3000 apples   **41.** 60 people

3.3 Exercises

1. Long piece is 32 m long; short piece is 15 m long   **3.** David worked 30 hours; Sarah worked 45 hours; Kate worked 25 hours   **5.** Mount McKinley is 20,320 feet high; Mount Whitney is 14,494 feet high; Mount Oxford is 14,153 feet high   **7.** width is 100 meters; length is 250 meters   **9.** width = 32.5 cm; length = 62.5 cm   **11.** The cheetah can run at 70 miles per hour. The jackal can run at 35 miles per hour. The elk can run at 45 miles per hour.   **13.** longest side = 18 in.; shortest side = 13 in.; third side = 15 in.   **15. (a)** 38 miles per hour   **(b)** 57 miles per hour **(c)** 19 miles per hour   **17.** Original square was 11 m × 11 m.   **19.** $-8x^3 + 12x^2 - 32x$ **20.** $5a^2b + 30ab - 10a^2$ **21.** $-19x + 2y - 2$ **22.** $9x^2y - 6xy^2 + 7xy$ **23.** 82%   **24.** 20%

3.4 Exercises

1. 11 bags   **3.** 7 hours   **5.** $360   **7.** Walter made $4500. Jim made $5040.   **9.** $6000   **11.** They invested $3000 at 7% and $2000 at 5%. **13.** They invested $250,000 in the conservative fund and $150,000 in the growth fund.   **15.** $12,000   **17.** 13 quarters; 9 nickels   **19.** 18 nickels; 6 dimes; 9 quarters   **21.** eight $10 bills; sixteen $20 bills; eleven $100 bills   **23.** first angle measures 70°; second angle measures 35°; third angle measures 75°   **25.** $1328 at 7%; $1641 at 11%   **27.** more than 225 miles   **29.** 12   **30.** 15   **31.** −28   **32.** −25   **33.** $95.20 **34.** $19,040

Putting Your Skills to Work

1. Monthly payments will be $248.90. He will pay $1947.20 in interest.   **2.** Monthly payments will be $141.21. She will pay $389.04 in interest.

3.5 Exercises

1. distance around   **3.** surface   **5.** 180°   **7.** 98 in.^{2} **9.** 10 feet   **11.** 125.6 inches   **13.** 28.26 m^2 **15.** 24 inches   **17. (a)** 24 cm **(b)** 55.7 cm   **19.** 9 inches   **21.** length should be 32 inches; width should be 20 inches   **23.** 13°   **25.** first angle = 100°; second angle = 50°; third angle = 30°   **27.** They each measure 17°.   **29.** first angle is 63°; second angle is 21°; third angle is 96°   **31.** The altitude is 20 feet. **33.** 3 feet   **35. (a)** 8 inches   **(b)** 552 square inches   **37. (a)** 113.04 cm^3 **(b)** 113.04 cm^2 **39.** approximately 7 batches   **41. (a)** 285.74 sq. in. **(b)** $2143.05   **43. (a)** 36 cubic yd   **(b)** $126   **45. (a)** 113.04 in.^{2} **(b)** 31.4 in.^{2}; the sphere   **47.** $3x^2 + 8x - 24$ **48.** $-11x - 17$ **49.** $-4x - 9$ **50.** $22x - 84$

Putting Your Skills to Work

1. July   **2.** January   **3.** 1.8 inches   **4.** 2.59 inches

3.6 Exercises

1. $x > 67{,}000$ **3.** $x \le 120$ **5.** $x < 34$ **7.** $x \ge 93$ **9.** It must be less than or equal to 140 cm.   **11.** 41 or more customers **13.** Depth must be less than or equal to 7.5 feet.   **15.** $x \le 384.6$ miles   **17.** $F < 230°$.   **19.** The budget will exceed $307.1 billion after 2005. **21.** He must purchase 12 chairs or fewer.   **23.** 455 or more discs must be manufactured and sold.   **25.** $x < -4$ **26.** $x \le 3$ **27.** $x \ge 4\frac{2}{3}$ **28.** $x > -4$ **29. (a)** 2920 gallons   **(b)** $24.82

Chapter 3 Review Problems

1. $x + 19$ **2.** $\frac{2}{3}x$ **3.** $x - 56$ **4.** $3x$ **5.** $2x + 7$ **6.** $2x - 3$ **7.** r = the number of retired people; $4r$ = the number of working people; $0.5r$ = the number of unemployed people   **8.** $3w + 5$ = the length; w = the width   **9.** b = the number of degrees in angle B; $2b$ = the number of degrees in angle A; $b - 17$ = the number of degrees in angle C **10.** a = the number of students in algebra; $a + 29$ = the number of students in biology; $0.5a$ = the number of students in geology   **11.** 18   **12.** 12   **13.** $40   **14.** 16 years old   **15.** 6.6 hours; 6 hours   **16.** 90 **17.** 1st side = 8 yd; 2nd side = 15 yd; 3rd side = 17 yd   **18.** 1st angle = 32°; 2nd angle = 96°; 3rd angle = 52°   **19.** 31.25 yd; 18.75 yd **20.** Jon = $30,000; Lauren = $18,000   **21.** 310 kilowatt-hours   **22.** 280 miles   **23.** $22,500   **24.** $200   **25.** $7000 at 12%; $2000 at 8% **26.** $2000 at 4.5%; $3000 at 6%   **27.** 18 nickels; 6 dimes; 9 quarters   **28.** 7 nickels; 8 dimes; 10 quarters   **29.** 8 m   **30.** $16\frac{1}{2}$ in.   **31.** 71° **32.** 42 sq. mi.   **33.** 1920 ft^3 **34.** 113.04 cm^3 **35.** 254.34 cm^2 **36.** 12 in.   **37.** $3440   **38.** $23,864   **39.** 200 miles or less   **40.** more than $300,000   **41.** 3 hours or more   **42.** less than $16.15   **43.** $24\frac{4}{5}$ min   **44.** 4 ounces   **45.** more than 3 years   **46.** more than 60 months   **47.** The height is 4 feet. The width is 10 feet.   **48.** It will take 10 years.   **49.** They made 27 field goals and 9 free throws.   **50.** The width is 16 feet and the length is 22 feet.   **51.** Michael should pay $6 and Scotty should pay $10.   **52.** They invested $9000 in the high tech funds, and $3000 in bond funds.   **53.** first angle measures 60°; second angle measures 40°; third angle measures 80°;   **54.** $1224   **55.** 94   **56.** No, she traveled 9000 miles for business. Her correct deduction was $2970.   **57.** It can travel 640 miles in one hour. It would take approximately 169 minutes.   **58.** 12 seconds   **59.** 5 Explorers and 10 Caravans

Putting Your Skills to Work

1. d_B(for boat) $= 5t$; d_H(for helicopter) $= 120\left(t - \dfrac{1}{12}\right)$ **2.** $5t + 120\left(t - \dfrac{1}{12}\right) = 100$; $t = \dfrac{22}{25}$. Time is 52.8 minutes. Yes.

Chapter 3 Test

1. 35   **2.** 36   **3.** −12   **4.** first side = 20 m; second side = 30 m; third side = 16 m   **5.** width = 20 m; length = 47 m **6.** 1st pollutant = 8 ppm; 2nd pollutant = 4 ppm; 3rd pollutant = 3 ppm   **7.** 15 months   **8.** $8500   **9.** $1400 at 14%; $2600 at 11% **10.** 16 nickels; 7 dimes; 8 quarters   **11.** 213.52 in.   **12.** 192 square inches   **13.** 4187 in.^{3} **14.** 96 cm^2 **15.** $450   **16.** at least an 82 **17.** more than $110,000   **18.** 600 hours

Cumulative Test for Chapters 0–3

1. 3.69 **2.** $\dfrac{31}{24}$ **3.** $-23y + 18$ **4.** 21 **5.** $\dfrac{2H - 3a}{5} = b$ **6.** $x \geq -3$; **7.** 40

8. Psychology students $= 50$; World Hist. students $= 84$ **9.** width $= 7$ cm; length $= 32$ cm **10.** \$135,000 **11.** \$3000 at 15%; \$4000 at 7% **12.** 7 nickels; 12 dimes; 4 quarters **13.** $A = 162.5$ m^2; \$731.25 **14.** $V = 113.04$ cubic inches; 169.56 pounds **15.** 25 days

Chapter 4

Pretest Chapter 4

1. 3^{16} **2.** $-10x^7$ **3.** $-24a^5b^4$ **4.** x^{23} **5.** $-\dfrac{2}{x^2y^2}$ **6.** $\dfrac{5bc}{3a}$ **7.** x^{50} **8.** $-8x^6y^3$ **9.** $\dfrac{64a^9b^3}{c^6}$ **10.** $\dfrac{3x^2}{y^3z^4}$ **11.** $-\dfrac{2b^2}{a^3c^4}$ **12.** $\dfrac{1}{9}$
13. 6.38×10^{-4} **14.** $1,894,000,000,000$ **15.** $-2x^2 + 7x - 29$ **16.** $x^3 - 5x^2 + 11x - 12$ **17.** $-6x + 21x^2$ **18.** $2x^4 + 8x^3 - 2x^2$
19. $5x^4y^2 - 15x^2y^3 + 5xy^4$ **20.** $x^2 + 14x + 45$ **21.** $12x^2 + xy - 6y^2$ **22.** $2x^4 - 7x^2y^2 + 3y^4$ **23.** $64x^2 - 121y^2$
24. $25x^2 - 30xy + 9y^2$ **25.** $25a^2b^2 - 60ab + 36$ **26.** $4x^3 - 13x^2 + 11x - 2$ **27.** $16x^4 - 81$ **28.** $7x^3 - 6x^2 + 11x$

29. $5x^2 - 6x + 2 - \dfrac{3}{3x - 2}$

4.1 Exercises

1. When you multiply exponential expressions, keep the base the same and add the exponents. **3.** A sample example is $\dfrac{2^2}{2^3} = \dfrac{\cancel{2} \cdot \cancel{2}}{\cancel{2} \cdot \cancel{2} \cdot 2} = \dfrac{1}{2} = \dfrac{1}{2^{3-2}}$

5. 6; x, y; 11 and 1 **7.** 2^2a^3b **9.** $-3a^2b^2c^3$ **11.** 3^{15} **13.** 5^{26} **15.** $3^5 \cdot 8^2$ **17.** $-54x^5$ **19.** $12x^5$ **21.** $-12x^{11}$ **23.** $\dfrac{3}{4}x^7y^5$

25. $-8.05wx^5y^4$ **27.** 0 **29.** $80x^3y^7$ **31.** $-28a^5b^6$ **33.** 0 **35.** 0 **37.** $-24w^5xyz^6$ **39.** $\dfrac{1}{y^3}$ **41.** y^7 **43.** $\dfrac{1}{13^{10}}$ **45.** 3^4

47. $\dfrac{a^8}{4}$ **49.** $\dfrac{x^7}{y^9}$ **51.** $\dfrac{x^4}{2}$ **53.** 1 **55.** $-\dfrac{2x^4}{y^5}$ **57.** $\dfrac{t^2}{20s^3}$ **59.** $6x^2$ **61.** $\dfrac{y^2}{16x^3}$ **63.** $\dfrac{3a}{4}$ **65.** $\dfrac{5x^6}{7y^8}$ **67.** $3x^2$ **69.** $-\dfrac{2}{3a^3}$

71. $-7ab^5$ **73.** Answers will vary. **75.** x^{3a} **77.** c^{y+2} **79.** w^{40} **81.** $a^{12}b^4$ **83.** $m^{15}n^{10}p^5$ **85.** $3^8x^4y^8$ **87.** $16a^{20}$

89. $\dfrac{12^5x^5}{y^{10}}$ **91.** $\dfrac{16a^{16}}{81b^{12}}$ **93.** $-32a^{25}b^{10}c^5$ **95.** $-64x^3z^{12}$ **97.** $\dfrac{a}{4b^4}$ **99.** $-8a^7b^{11}$ **101.** $\dfrac{64}{x^{18}}$ **103.** $\dfrac{a^{15}b^5}{c^{25}d^5}$ **105.** $16x^9y^3$

107. $\pm 2x^5y^4z^7$ **108.** -11 **109.** -46 **110.** $-\dfrac{7}{4}$ **111.** 5 **112.** 1500 people **113.** 8700 people **114.** 4050 people

4.2 Exercises

1. $\dfrac{3}{x^2}$ **3.** $\dfrac{1}{16x^4y^2}$ **5.** $\dfrac{3xz^3}{y^2}$ **7.** $3x$ **9.** $\dfrac{wy^3}{x^5z^2}$ **11.** $\dfrac{1}{8}$ **13.** $\dfrac{z^8}{9x^2y^4}$ **15.** $\dfrac{1}{x^6y}$ **17.** 1.2378×10^5 **19.** 7.42×10^{-4} **21.** 7.652×10^9
23. $302,000$ **25.** 0.000033 **27.** $983,000$ **29.** 1.67×10^{-27} kilogram **31.** 0.000007 meter **33.** 1.0×10^1 **35.** 3.2×10^{-19}
37. 4.5×10^5 **39.** 2.03×10^4 dollars **41.** 1.90×10^{11} hours **43.** 1.15×10^{11} hours **45.** 4.717×10^{32} joules

47. 46.3% **49.** -0.8 **50.** -1 **51.** $-\dfrac{1}{28}$ **52.** candidate #1 shook 2524 hands; candidate #2 shook 1016 hands

53. Mario earns \$36,094; Alfonso earns \$27,352; Gina earns \$48,554

Putting Your Skills to Work

1. 3.02×10^8 acres **2.** 2.203×10^9 acres

4.3 Exercises

1. A polynomial in x is the sum of a finite number of terms of the form ax^n, where a is any real number and n is a whole number. An example is $3x^2 - 5x - 9$. **3.** The degree of a polynomial in x is the largest exponent of x in any of the terms of the polynomial. **5.** degree 4; monomial

7. degree 5; trinomial **9.** degree 5; binomial **11.** $5x - 28$ **13.** $-3x^2 - 10x + 10$ **15.** $\dfrac{5}{6}x^2 + \dfrac{1}{2}x - 9$ **17.** $1.7x^3 - 3.4x^2 - 13.2x - 5.4$

19. $5x - 24$ **21.** $-\dfrac{1}{6}x^2 + 7x - \dfrac{3}{10}$ **23.** $-2x^3 + 5x - 10$ **25.** $-4.7x^4 - 0.7x^2 - 1.6x + 0.4$ **27.** $10x + 10$ **29.** $-3x^2y + 6xy^2 - 4$

31. $x^4 - 3x^3 - 4x^2 - 24$ **33.** 5.4 **35.** 2020 **37.** 727,000 **39.** 434,400 **41.** $3x^2 + 12x$ **43.** $x = \dfrac{3y - 2}{8}$ **44.** $B = \dfrac{3A}{2CD}$

45. $b = \dfrac{2A}{h} - c$ or $b = \dfrac{2A - ch}{h}$ **46.** $d = \dfrac{5xy}{B}$ **47.** approximately 696.8 billion dollars

4.4 Exercises

1. $-12x^4 + 2x^2$ **3.** $-15x^3 + 10x^2 - 25x$ **5.** $6x^7 - 4x^6 + 10x^4 - 2x^3$ **7.** $x + \dfrac{3}{2}x^2 + \dfrac{5}{2}x^3$ **9.** $-15x^4y^2 + 6x^3y^2 - 18x^2y^2$

11. $4b^4 + 6b^3 - 8b^2$ **13.** $3x^4 - 9x^3 + 15x^2 - 6x$ **15.** $-2x^3y^3 + 12x^2y^2 - 16xy$ **17.** $-28x^5y + 12x^4y + 8x^3y - 4x^2y$
19. $-6c^2d^5 + 8c^2d^3 - 12c^2d$ **21.** $12x^7 - 6x^5 + 18x^4 + 54x^3$ **23.** $-24x^7 + 12x^5 - 8x^3$ **25.** $x^2 + 13x + 30$ **27.** $x^2 + 8x + 12$
29. $x^2 - 3x - 18$ **31.** $x^2 - 11x + 30$ **33.** $-14x^2 - 23x - 3$ **35.** $3x^2 + 4xy - 27x - 36y$ **37.** $12y^2 - y - 6$ **39.** $20y^2 - 22y + 6$
41. The last term is incorrect. The result should be $-3x + 7$. **43.** $25x^2 - 1$; it is the difference of two perfect squares
45. $6b^4 - 31b^2c + 35c^2$ **47.** $64x^2 - 32x + 4$ **49.** $25a^4 - 30a^2b^2 + 9b^4$ **51.** $0.8x^2 + 11.94x - 0.9$ **53.** $30x^2 + 43xy - 8y^2$

55. $\frac{1}{9}x^2 - \frac{1}{10}x - \frac{1}{10}$ **57.** $5bx - 10b^2 - 7cx + 14bc$ **59.** $10x^2 - 11x - 6$ **61.** $x = -10$ **62.** $w = -\frac{25}{7}$ **63.** 8 dimes; 11 quarters
64. -5 **65.** 9 twenties; 8 tens; 23 fives **66.** 19 hours **67.** 22 hours **68.** 31 hours **69.** 32.5 hours

4.5 Exercises

1. binomial **3.** The middle term is missing. The answer should be $16x^2 - 56x + 49$. **5.** $y^2 - 49$ **7.** $x^2 - 81$ **9.** $64x^2 - 9$ **11.** $4x^2 - 49$
13. $25x^2 - 9y^2$ **15.** $0.36x^2 - 9$ **17.** $9y^2 + 6y + 1$ **19.** $25x^2 - 40x + 16$ **21.** $81x^2 + 90x + 25$ **23.** $9x^2 - 42x + 49$
25. $\frac{4}{9}x^2 + \frac{1}{3}x + \frac{1}{16}$ **27.** $36w^2 + 60wz + 25z^2$ **29.** $49x^2 - 9y^2$ **31.** $49c^6 - 84c^3d + 36d^2$ **33.** $x^3 - 11x + 6$
35. $4x^4 - 7x^3 + 2x^2 - 3x - 1$ **37.** $3x^3 - 2x^2 - 25x + 24$ **39.** $3x^3 - 13x^2 - 6x + 40$ **41.** $2x^3 - 7x^2 - 32x + 112$
43. $a^4 + a^3 - 13a^2 + 17a - 6$ **45.** $24x^3 + 14x^2 - 11x - 6$ **47.** \$11,000 at 7%; \$7,000 at 11% **48.** width is 7 meters; length is 10 meters
49. $-28.12°$F **50.** 2.0625×10^5 meters

Putting Your Skills to Work

1. 1.53×10^6 nanometers **2.** 1.53 millimeters

4.6 Exercises

1. $5x^3 - 3x + 4$ **3.** $2y^2 - 3y - 1$ **5.** $7x^4 - 3x^2 + 8$ **7.** $8x^4 - 9x + 6$ **9.** $3x + 5$ **11.** $x - 2 - \frac{20}{x - 7}$ **13.** $3x^2 - 4x + 8 - \frac{10}{x + 1}$
15. $2x^2 - 3x - 2 - \frac{5}{2x + 5}$ **17.** $2x^2 + x - 2$ **19.** $6y^2 + 3y - 8 + \frac{7}{2y - 3}$ **21.** $y^2 - 4y - 1 - \frac{9}{y + 3}$
23. $y^3 + 2y^2 - 5y - 10 - \frac{25}{y - 2}$. **25.** $2y^2 + \frac{1}{2}y + \frac{7}{8} - \frac{49}{32y - 8}$ **27.** 77,000 gallons **28.** approximately 3.4 million
29. 200 cats **30.** 519 and 520

Putting Your Skills to Work

1. 4.1 million **2.** 700,000

Chapter 4 Review Problems

1. $-18a^7$ **2.** 5^{23} **3.** $6x^4y^6$ **4.** 8^{17} **5.** $\frac{1}{7^{12}}$ **6.** $\frac{1}{x^5}$ **7.** y^{14} **8.** $\frac{x^4}{3}$ **9.** $-\frac{3}{5x^5y^4}$ **10.** $-\frac{2a}{3b^6}$ **11.** x^{24} **12.** $125x^3y^6$ **13.** $9a^6b^4$
14. $\frac{2x^4}{3y^2}$ **15.** $\frac{25a^2b^4}{c^6}$ **16.** $\frac{y^9}{64w^{15}z^6}$ **17.** $\frac{1}{x^3}$ **18.** $\frac{1}{x^5y^{11}}$ **19.** $\frac{2y^3}{x^6}$ **20.** $\frac{x^5}{2y^6}$ **21.** $\frac{1}{4x^6}$ **22.** $\frac{3y^2}{x^3}$ **23.** $\frac{4w^2}{x^5y^6z^8}$ **24.** $\frac{b^5c^3d^4}{27a^2}$
25. 1.563402×10^{11} **26.** 1.79632×10^5 **27.** 7.8×10^{-3} **28.** 6.173×10^{-5} **29.** 120,000 **30.** 83,670,000,000 **31.** 3,000,000 **32.** 0.25
33. 0.00000005708 **34.** 0.000000006 **35.** 2.0×10^{13} **36.** 9.36×10^{19} **37.** 9.6×10^{-10} **38.** 7.8×10^{-11} **39.** 3.504×10^8 kilometers
40. 7.94×10^{14} cycles **41.** 6×10^9 **42.** $-5x^2 - 11x - 18$ **43.** $6.7x^2 - 11x + 3$ **44.** $-x^3 + 2x^2 - x + 8$ **45.** $7x^3 - 3x^2 - 6x + 4$
46. $5x^3y^3 - 2x^2y^2 + 10xy - 4$ **47.** $\frac{1}{4}x^2 - \frac{1}{4}x + \frac{1}{10}$ **48.** $-x^2 - 2x + 10$ **49.** $-6x^2 - 6x - 3$ **50.** $15x^2 + 2x - 1$
51. $28x^2 - 29x + 6$ **52.** $20x^2 + 48x + 27$ **53.** $10x^3 - 30x^2 + 15x$ **54.** $-6x^3y^3 + 10x^2y^2 - 12xy$ **55.** $4x^4 - 12x^3 + 20x^2 - 8x$
56. $5a^2 - 8ab - 21b^2$ **57.** $8x^4 - 10x^2y - 12x^2 + 15y$ **58.** $-15x^6y^2 - 9x^4y + 6x^2y$ **59.** $9x^2 - 12x + 4$ **60.** $25x^2 - 9$
61. $49x^2 - 36y^2$ **62.** $25a^2 - 20ab + 4b^2$ **63.** $64x^2 + 144xy + 81y^2$ **64.** $4x^3 + 27x^2 + 5x - 3$ **65.** $2x^3 - 7x^2 - 42x + 72$
66. $2y^2 + 3y + 4$ **67.** $6x^3 + 7x^2 - 18x$ **68.** $4x^2y - 6x + 8y$ **69.** $53x^3 - 12x^2 + 19x + 13$ **70.** $4x + 1$ **71.** $3x + 7$
72. $3x^2 + 2x + 4 + \frac{9}{2x - 1}$ **73.** $2x^2 - 5x + 13 - \frac{27}{x + 2}$ **74.** $4x + 1$ **75.** $4x - 5 + \frac{11}{2x + 1}$ **76.** $x^2 + 3x + 8$
77. $2x^2 + 4x + 5 + \frac{11}{x - 2}$ **79.** 1.147×10^9 people **81.** 3.3696×10^{31} joules **83.** $2x^2 - 4y^2$

Chapter 4 Test

1. 3^{34} **2.** 3^{28} **3.** 8^{24} **4.** $12x^4y^{10}$ **5.** $-\frac{7x^3}{5}$ **6.** $-125x^3y^{18}$ **7.** $\frac{49a^{14}b^4}{9}$ **8.** $\frac{1}{4a^5b^6}$ **9.** $\frac{1}{64}$ **10.** $\frac{6c^5}{a^4b^3}$ **11.** $\frac{2w^6}{x^3y^4z^8}$
12. 5.482×10^{-4} **13.** 582,000,000 **14.** 2.4×10^{-6} **15.** $-2x^2 + 5x$ **16.** $3x^2 + 3xy - 6y - 7y^2$ **17.** $-21x^5 + 28x^4 - 42x^3 + 14x^2$
18. $15x^4y^3 - 18x^3y^2 + 6x^2y$ **19.** $10a^2 + 7ab - 12b^2$ **20.** $6x^3 - 11x^2 - 19x - 6$ **21.** $49x^4 + 28x^2y^2 + 4y^4$ **22.** $81x^2 - 4y^2$
23. $12x^4 - 14x^3 + 25x^2 - 29x + 10$ **24.** $3x^4 + 4x^3y - 15x^2y^2$ **25.** $3x^3 - x + 5$ **26.** $2x^2 - 7x + 4$ **27.** $2x^2 + 6x + 12$
28. 7.85×10^1 people **29.** 4.18×10^6 miles

Cumulative Test for Chapter 0–4

1. $-\frac{11}{24}$ **2.** -0.74 **3.** $-\frac{6}{7}$ **4.** \$8.96 **5.** $2x^2 - 13x$ **6.** 35 **7.** $x = -\frac{9}{2}$ **8.** $x = 44$ **9.** $x > -1$
10. $f = \frac{2B}{3a} - \frac{c}{3}$ or $f = \frac{2B - ac}{3a}$ **11.** 12,400 employees **12.** \$199.20 **13.** $15x^2 - 47x + 28$ **14.** $9x^2 - 30x + 25$
15. $6x^3 - 17x^2 - 26x - 8$ **16.** $-20x^5y^8$ **17.** $-\frac{2x^3}{3y^9}$ **18.** $-27x^3y^{12}z^6$ **19.** $\frac{9z^8}{w^2x^3y^4}$ **20.** 1.36×10^{15} **21.** 5.6×10^{-4} **22.** 4.0×10^{-35}
23. $5x^3 + 7x^2 - 6x + 50$ **24.** $-36x^3y^2 + 18x^2y^3 - 48xy^4$ **25.** $2x^4 - 15x^3 + 28x^2 - 33x + 12$ **26.** $x + 5 + \frac{3}{x - 3}$

Chapter 5

Pretest Chapter 5

1. $2x(x - 3y + 6y^2)$ **2.** $(3x + 4y)(a - 2b)$ **3.** $18ab(2b - 1)$ **4.** $(a - 2b)(5 - 3x)$ **5.** $(3x - 4)(x + y)$ **6.** $(7x - 3)(3x - 2)$
7. $(x - 24)(x + 2)$ **8.** $(x - 5)(x - 3)$ **9.** $(x + 8)(x + 1)$ **10.** $2(x + 6)(x - 2)$ **11.** $3(x - 9)(x + 7)$ **12.** $(5x - 2)(3x - 2)$
13. $(3y - 2z)(2y + 3z)$ **14.** $4(3x + 5)(x + 2)$ **15.** $(9x^2 + 4)(3x + 2)(3x - 2)$ **16.** $(7x - 2y)^2$ **17.** $(5x + 8)^2$
18. $3x(2x - 1)(x + 3)$ **19.** $2y^2(4x - 3)^2$ **20.** cannot be factored **21.** $x = 1, x = -\dfrac{3}{2}$ **22.** $x = 5, x = 6$ **23.** $x = -\dfrac{2}{3}, x = 3$

24. altitude $= 8$ cm, base $= 15$ cm

5.1 Exercises

1. factors **3.** no; $6a^3 + 3a^2 - 9a$ has a common factor of 3a **5.** $2c(c + 1)$ **7.** $9wz(2 - 3w)$ **9.** $2x(4x^2 - 5x - 7)$
11. $6(2xy - 3yz - 6xz)$ **13.** $b^2(2ab + 3x - 5b^2 + 2)$ **15.** $2x^5(3x^4 - 4x^2 + 2)$ **17.** $9ab(ab - 4)$ **19.** $8a(5a - 2b - 3)$
21. $8y(6x - 3y + 5)$ **23.** $(3a + b)(6 - z)$ **25.** $(x - 7)(5x + 3)$ **27.** $(3y + 5z)(7x - 6t)$ **29.** $(bc - 1)(5a + b + c)$
31. $(bc - 3a)(3c - 2 - 6b)$ **33.** $(x - 2y)(3x^2 - 1)$ **35.** $(5x - 3)(d - 1)$ **37.** $C = 29.95(a + b + c + d)$ **38.** 17, 19, 21
39. \$26,000 **40.** 2.24 ft/min **41.** 208 people **42.** 345 people

5.2 Exercises

1. $(a + 4)(b - 3)$ **3.** $(a + 3b)(2x - y)$ **5.** $(x - 4)(x^2 + 3)$ **7.** $(3a + b)(x - 2)$ **9.** $(a + 2b)(5 + 6c)$ **11.** $(a - b)(5 - 2x)$
13. $(y - 2)(y - 3)$ **15.** $(7 + y)(2 - y)$ **17.** $(3x + y)(2a - 1)$ **19.** $(2x - 3)(x + 4)$ **21.** $(4x + 3w)(7x + 2y^2)$ **23.** We must rearrange the terms in a different order so that the expression in the parentheses is the same in each case. We use the order $6a^2 - 8ad + 9ab - 12bd$
to factor $2a(3a - 4d) + 3b(3a - 4d) = (3a - 4d)(2a + 3b)$. **24.** 8 seconds **25.** \$420 **26.** \$126.9 million **27.** \$170.4 million

5.3 Exercises

1. product; sum **3.** $(x + 1)^2$ **5.** $(x + 7)(x + 5)$ **7.** $(x - 3)(x - 1)$ **9.** $(x - 7)(x - 4)$ **11.** $(x + 4)(x - 3)$
13. $(x - 14)(x + 1)$ **15.** $(x + 7)(x - 5)$ **17.** $(x - 6)(x + 4)$ **19.** $(x + 7)(x - 2)$ **21.** $(x - 6)(x - 4)$ **23.** $(x + 3)(x + 10)$
25. $(y - 5)(y + 1)$ **27.** $(a + 8)(a - 2)$ **29.** $(x - 4)(x - 8)$ **31.** $(x + 7)(x - 3)$ **33.** $(x + 5)(x + 8)$ **35.** $(x + 2)(x - 23)$
37. $(x + 12)(x - 3)$ **39.** $(x + 3y)(x - 5y)$ **41.** $(x - 7y)(x - 9y)$ **43.** $2(x - 2)(x - 4)$ **45.** $3(x + 4)(x - 6)$
47. $4(x + 5)(x + 1)$ **49.** $7(x + 5)(x - 2)$ **51.** $6(x + 1)(x + 2)$ **53.** $3(x - 1)(x - 5)$ **55.** $4(\pi x^2 - 9)$ **57.** $t = \dfrac{A - P}{Pr}$

58. $x \geq -\dfrac{5}{3}$
59. 120 miles **60.** \$130,000 **61.** 1:30 A.M. **62.** 9:00 P.M. **63.** 25°C **64.** June

5.4 Exercises

1. $(4x + 1)(x + 3)$ **3.** $(2x - 1)(x - 2)$ **5.** $(3x - 7)(x + 1)$ **7.** $(2x + 1)(x - 3)$ **9.** $(5x - 2)(x + 1)$ **11.** $(3x - 5)(5x - 3)$
13. $(2x - 5)(x + 4)$ **15.** $(3x + 1)(3x + 2)$ **17.** $(3x + 2)(2x - 3)$ **19.** $(3x - 2)(2x - 5)$ **21.** $(x - 2)(7x + 9)$
23. $(9y - 4)(y - 1)$ **25.** $(5a + 2)(a - 3)$ **27.** $(6x - 1)(2x - 3)$ **29.** $(5x - 2)(3x + 2)$ **31.** $(6x + 5)(2x + 3)$
33. $(6x + 1)(2x - 3)$ **35.** $(2x^2 - 1)(x^2 + 8)$ **37.** $(2x - y)(2x + 5y)$ **39.** $(5x - 4y)(x + 4y)$ **41.** $2(5x + 6)(x + 1)$
43. $3(2x - 1)(2x - 3)$ **45.** $5(2x + 1)(x - 3)$ **47.** $2x(3x + 1)(x - 3)$ **49.** $(2x + 5)(6x - 7)$ **51.** $(4x - 3)(5x - 3)$ **53.** $x = \dfrac{1}{10}$
54. 18.8 million children **55.** 36.7 million children

5.5 Exercises

1. $(9x - 4)(9x + 4)$ **3.** $(4 - 3x)(4 + 3x)$ **5.** $(3x + 5)(3x - 5)$ **7.** $(2x - 5)(2x + 5)$ **9.** $(6x - 5)(6x + 5)$ **11.** $(1 - 7x)(1 + 7x)$
13. $(4x - 7y)(4x + 7y)$ **15.** $(5 + 11x)(5 - 11x)$ **17.** $(9x + 10y)(9x - 10y)$ **19.** $(5a + 7)(5a - 7)$ **21.** $(3x + 1)^2$ **23.** $(y - 3)^2$
25. $(3x - 4)^2$ **27.** $(7x + 2)^2$ **29.** $(x + 7)^2$ **31.** $(5x - 4)^2$ **33.** $(9x + 2y)^2$ **35.** $(5x - 3y)^2$ **37.** $(4a + 9b)^2$ **39.** $(3x^2 - y)^2$
41. $(7x + 1)(7x + 9)$ **43.** $(4x^2 + 1)(2x + 1)(2x - 1)$ **45.** $(x^5 + 6y^5)(x^5 - 6y^5)$ **47.** $(3x^5 - 2)^2$ **49.** Because no matter what combination you try, you cannot multiply two binomials to equal $9x^2 + 1$. **51.** 49; one answer **53.** $4(2x - 3)(2x + 3)$ **55.** $3(7x - y)(7x + y)$
57. $3(2x - 3)^2$ **59.** $2(7x + 3)^2$ **61.** $(x - 2)(x - 7)$ **63.** $(2x - 1)(x + 3)$ **65.** $(4x - 11)(4x + 11)$ **67.** $(3x + 7)^2$
69. $3(x + 5)(x - 3)$ **71.** $5(x - 4)(x + 4)$ **73.** $5(x + 2)^2$ **75.** $2(x - 9)(x - 7)$
77. $x^2 + 3x + 4 + \dfrac{-3}{x - 2}$ **78.** $2x^2 + x - 5$ **79.** 1.2 ounces of greens, 1.05 ounces of bulk vegetables, 0.75 ounce of fruit
80. 1.44 ounces of greens, 1.26 ounces of bulk vegetables, 0.9 ounce of fruit **81.** 3838 ft above sea level **82.** 5 miles

5.6 Exercises

1. $a(6a + 2b - 3)$ **3.** $9(2x - y)(2x + y)$ **5.** $(3x - 2y)^2$ **7.** $(x + 5)(x + 3)$ **9.** $(3x + 2)(5x - 1)$ **11.** $(x - 3y)(a + 2b)$
13. $3(x^2 + 2)(x^2 - 2)$ **15.** $(2x - 3)^2$ **17.** $(2x - 3)(x - 4)$ **19.** $(x - 10y)(x + 7y)$ **21.** $(a + 3)(x - 5)$ **23.** $5x(3 - x)(3 + x)$
25. $5xy^3(x - 1)^2$ **27.** $3xy(3z + 2)(3z - 2)$ **29.** $3(x + 7)(x - 5)$ **31.** $5(x - 2)(x - 4)$ **33.** $-1(2x^2 + 1)(x + 2)(x - 2)$ **35.** prime
37. $5(x^2 + 2xy - 6y)$ **39.** $3x(2x + y)(5x - 2y)$ **41.** $4(2x - 1)(x + 4)$ **43.** prime **45.** \$28,000 **46.** 372 live strains
47. 57 hardcover books, 94 softcover books, 47 magazines **48.** 55 hardcover books, 92 softcover books, 46 magazines

5.7 Exercises

1. $-3, 7$ **3.** $-\dfrac{1}{2}, 3$ **5.** $\dfrac{3}{2}, 2$ **7.** $\dfrac{2}{3}, \dfrac{3}{2}$ **9.** $0, -13$ **11.** $3, -3$ **13.** $0, 1$ **15.** $-3, 2$ **17.** $-\dfrac{1}{2}$ **19.** $0, -2$ **21.** $-3, -4$ **23.** $-\dfrac{5}{3}, 3$
25. You can always factor out x.   **27.** $L = 14$ m; $W = 10$ m   **29.** 182 games   **31.** 15 teams   **33.** 12 meters above ground after 2 seconds
35. 5 additional helicopters   **37.** 2415 telephone calls   **38.** 3160 telephone calls   **39.** 18 people   **41.** $-12x^3y^7$ **42.** $12a^{10}b^{13}$
43. $-\dfrac{3a^4}{2b^2}$ **44.** $\dfrac{1}{3x^5y^4}$

Putting Your Skills to Work

1. \$765,000   **2.** \$3,160,000

Chapter 5 Review Problems

1. $3x^2y(5x - 3y)$ **2.** $8x(5x - 4)$ **3.** $7xy(x - 2y - 3x^2y^2)$ **4.** $25a^4b^4(2b - 1 + 3ab)$ **5.** $9x^2(3x - 1)$ **6.** $2(x - 2y + 3z + 6)$
7. $(a + 3b)(2a - 5)$ **8.** $3xy(5x^2 + 2y + 1)$ **9.** $(3x - 7)(a - 2)$ **10.** $(a + 5b)(a - 4)$ **11.** $(x^2 + 3)(y - 2)$ **12.** $(4a - 5)(d + 5c)$
13. $(5x - 1)(3x + 2)$ **14.** $(5w - 3)(6w + z)$ **15.** $(x - 7)(x + 5)$ **16.** $(x - 4)(x - 6)$ **17.** $(x + 6)(x + 8)$
18. $(x + 3y)(x + 5y)$ **19.** $(x^2 + 7)(x^2 + 6)$ **20.** $(x - 17)(x + 3)$ **21.** $5(x + 1)(x + 3)$ **22.** $3(x + 1)(x + 12)$
23. $2(x - 6)(x - 8)$ **24.** $4(x - 5)(x - 6)$ **25.** $(4x - 5)(x + 3)$ **26.** $(3x - 1)(4x + 5)$ **27.** $(5x + 4)(3x - 1)$
28. $(3x - 2)(2x - 3)$ **29.** $(2x - 3)(x + 1)$ **30.** $(3x - 4)(x + 2)$ **31.** $(10x - 1)(2x + 5)$ **32.** $(5x - 1)(4x + 5)$
33. $(4a + 1)(a - 3)$ **34.** $(4a + 3)(a - 1)$ **35.** $2(x - 1)(3x + 5)$ **36.** $2(x + 1)(3x - 5)$ **37.** $2(2x - 3)(x - 5)$
38. $4(x - 9)(x + 4)$ **39.** $(3x - 1)(4x + 3)$ **40.** $2(x - 1)(8x + 15)$ **41.** $(3x - 2y)(2x - 5y)$ **42.** $2(3x - y)(x - 5y)$
43. $(7x + y)(7x - y)$ **44.** $4(2x - 3y)(2x + 3y)$ **45.** $(3x - 2)^2$ **46.** $(8x + 1)(8x - 1)$ **47.** $(5x - 6)(5x + 6)$
48. $(10x - 3)(10x + 3)$ **49.** $(1 - 7x)(1 + 7x)$ **50.** $(2 - 7x)(2 + 7x)$ **51.** $(6x + 1)^2$ **52.** $(5x - 2)^2$ **53.** $(4x - 3y)^2$
54. $(7x - 2y)^2$ **55.** $2(x - 3)(x + 3)$ **56.** $3(x - 5)(x + 5)$ **57.** $2(2x + 5)^2$ **58.** $2(5x - 6)^2$ **59.** $(2x + 3y)(2x - 3y)$
60. $(x + 3)^2$ **61.** $(x - 3)(x - 6)$ **62.** $(x + 15)(x - 2)$ **63.** $(x - 1)(6x + 7)$ **64.** $(5x - 2)(2x + 1)$ **65.** $4(3x + 4)$
66. $4xy(2xy - 1)$ **67.** $10x^2y^2(5x + 2)$ **68.** $13ab(2a^2 - b^2 + 4ab^3)$ **69.** $x(x - 8)^2$ **70.** $2(x + 10)^2$ **71.** $3(x - 3)^2$
72. $x(5x - 6)^2$ **73.** $(7x + 5)(x - 2)$ **74.** $(4x + 3)(x - 4)$ **75.** $xy(3x + 2y)(3x - 2y)$ **76.** $x^3a(3a + 4x)(a - 5x)$
77. $2(3a + 5b)(2a - b)$ **78.** $(4a - 5b)^2$ **79.** $(a - 1)(7 - b)$ **80.** $(3d - 4)(1 - c)$ **81.** $(2x - 1)(1 + b)$ **82.** $(b - 7)(5x + 4y)$
83. $x(2a - 1)(a - 7)$ **84.** $x(x + 4)(x - 4)(x + 1)(x - 1)$ **85.** $(x^2 + 9y^6)(x + 3y^3)(x - 3y^3)$ **86.** $(3x^2 - 5)(2x^2 + 3)$
87. $yz(14 - x)(2 - x)$ **88.** $x(3x + 2)(4x + 3)$ **89.** $(2w + 1)(8w - 5)$ **90.** $3(2w - 1)^2$ **91.** $2y(2y - 1)(y + 3)$
92. $(5y - 1)(2y + 7)$ **93.** $8y^8(y^2 - 2)$ **94.** $49(x^2 + 1)(x + 1)(x - 1)$ **95.** prime   **96.** prime   **97.** $4y(2y^2 - 5)(y^2 + 3)$
98. $3x(3y + 7)(y - 2)$ **99.** $(4x^2y - 7)^2$ **100.** $2xy(8x + 1)(8x - 1)$ **101.** $(2x + 5)(a - 2b)$ **102.** $(2x + 1)(x + 3)(x - 3)$
103. $-3, 6$ **104.** $3, -9$ **105.** $0, \dfrac{1}{6}$ **106.** $0, -\dfrac{11}{6}$ **107.** $-5, \dfrac{1}{2}$ **108.** $-8, -3$ **109.** $-5, -9$ **110.** $-\dfrac{3}{5}, 2$ **111.** -3 **112.** $-3, \dfrac{3}{4}$
113. $\dfrac{1}{5}, 2$ **114.** base $= 10$ cm, altitude $= 7$ cm   **115.** width $= 7$ feet, length $= 15$ feet   **116.** 6 seconds   **117.** 8 amperes, 12 amperes

Putting Your Skills to Work

1. 3.0 times greater   **2.** 5.0 times greater

Chapter 5 Test

1. $(x + 14)(x - 2)$ **2.** $(5x + 7y)(5x - 7y)$ **3.** $(5x + 1)(2x + 5)$ **4.** $(3a - 5b)^2$ **5.** $x(7 - 9x + 14y)$ **6.** $(x + 2y)(3x - 2w)$
7. $2x(3x - 4)(x - 2)$ **8.** $c(5a - b)(a - 2b)$ **9.** $4(5x^2 + 2y^2)(5x^2 - 2y^2)$ **10.** $(3x - y)(3x - 4y)$ **11.** $7x(x - 6)$ **12.** prime
13. prime   **14.** $-5y(2x - 3y)^2$ **15.** $(9x + 1)(9x - 1)$ **16.** $(x^8 + 1)(x^4 + 1)(x^2 + 1)(x + 1)(x - 1)$ **17.** $(x + 3)(2a - 5)$
18. $(a + 2b)(w + 2)(w - 2)$ **19.** $3(x - 6)(x + 5)$ **20.** $x(2x + 5)(x - 3)$ **21.** $-5, -9$ **22.** $-\dfrac{7}{3}, -2$ **23.** $-\dfrac{5}{2}, 2$
24. width $= 7$ miles, length $= 13$ miles

Cumulative Test for Chapters 0–5

1. 15%   **2.** 0.494   **3.** -10.36 **4.** $8x^4y^{10}$ **5.** 81   **6.** $27x^2 + 6x - 8$ **7.** $2x^3 - 12x^2 + 19x - 3$ **8.** $x \le -3$ **9.** 2   **10.** -15
11. $t = \dfrac{2s - 2a}{3}$ **12.** $(3x - 1)(2x - 1)$ **13.** $(3x + 4)(2x - 1)$ **14.** $(3x + 2)(3x - 1)$ **15.** $(11x + 8y)(11x - 8y)$
16. $-4(5x + 6)(4x - 5)$ **17.** prime   **18.** $x(4x + 5)^2$ **19.** $(9x^2 + 4b^2)(3x + 2b)(3x - 2b)$ **20.** $(2x + 3)(a - 2b)$
21. $(x^2 + 5)(x^2 + 3)$ **22.** $-8, 3$ **23.** $\dfrac{5}{3}, 2$ **24.** base $= 6$ miles, altitude $= 19$ miles

Chapter 6

Pretest Chapter 6

1. -2 **2.** $\dfrac{x - 3}{2x + 3}$ **3.** $\dfrac{b}{2a - b}$ **4.** $\dfrac{2a - b}{6}$ **5.** $\dfrac{x}{x + 5}$ **6.** 1   **7.** $\dfrac{y}{x - 1}$ **8.** 3   **9.** $\dfrac{-2y - 7}{(y - 1)(2y + 3)}$ **10.** $\dfrac{1}{x + 5}$
11. $\dfrac{5x^2 + 6x - 9}{3x(x - 3)^2}$ **12.** $\dfrac{2a - 3}{5a^2 + a}$ **13.** $\dfrac{a^2 - 2a - 2}{3a^2(a + 1)}$ **14.** $\dfrac{-x^2 - y^2}{x^2 + 2xy - y^2}$ **15.** $x = -5$ **16.** $x = 7$ **17.** 9.3   **18.** \$263.50

6.1 Exercises

1. 3 **3.** $\dfrac{6}{x}$ **5.** $\dfrac{2}{x-4}$ **7.** $\dfrac{x+y^2}{xy}$ **9.** $\dfrac{x+2}{x}$ **11.** $\dfrac{x-5}{3x-1}$ **13.** $\dfrac{x(x-4)}{x+6}$ **15.** $\dfrac{3x-2}{x+4}$ **17.** $\dfrac{3x-5}{4x-1}$ **19.** $\dfrac{x-5}{x+1}$ **21.** $\dfrac{-3}{2}$

23. $\dfrac{-2x-3}{x+5}$ **25.** $\dfrac{4x+5}{2x-1}$ **27.** $\dfrac{2x-3}{-x+5}$ **29.** $\dfrac{a-b}{2a-b}$ **31.** $\dfrac{3x-2y}{3x+2y}$ **33.** $\dfrac{x(x-2)}{2(x+3)}$ **35.** $9x^2-42x+49$ **36.** $100x^2-81y^2$

37. $2x^3-9x^2-2x+24$ **38.** $2x^4-5x^3-10x^2-14x-15$ **39.** $1\dfrac{5}{8}$ acre **40.** 6 hours, 25 minutes **41.** 885,000,000 people

42. 332,000,000 people

6.2 Exercises

1. factor the numerator and denominator completely and divide out any common factors
3. $\dfrac{2(x+4)}{(x-4)}$ **5.** $\dfrac{x^2}{2(x-2)}$ **7.** $\dfrac{x-2}{x+3}$ **9.** $x+2$ **11.** $\dfrac{3(x+2y)}{4(x+3y)}$ **13.** $\dfrac{-3(x-2)}{x+5}$ **15.** $\dfrac{(x+5)(x-2)}{3x-1}$ **17.** 1
19. By definition, the denominator of a rational expression cannot have the value zero. So the original expression cannot have a replacement value of 2 (or the first denominator would be zero) or 6 (or the second denominator would be zero). If we multiply the first fraction by the reciprocal of the second fraction, then the denominator $x+7$ cannot be zero. Thus x also cannot have a replacement value of -7.

21. $\dfrac{3}{2}$ **22.** $7x^3-22x^2+2x+3$ **23.** 5 milligrams **24.** $\$713,989.5x$ **25.** $\$2.38$

6.3 Exercises

1. The LCD would be a product that contains each factor. However, any repeated factor in any one denominator must be repeated the greatest number of times it occurs in any one denominator. So the LCD would be $(x+5)(x+3)^2$.
3. $\dfrac{3x+1}{x+5}$ **5.** $\dfrac{-1}{x-4}$ **7.** $\dfrac{2x-7}{5x+7}$ **9.** $5a^3$ **11.** $112x^2y^4$ **13.** x^2-16 **15.** $(x+2)(4x-1)^2$ **17.** $\dfrac{3x+2}{(x-1)(x-1)}$

19. $\dfrac{4y}{(y+1)(y-1)}$ **21.** $\dfrac{10a+5b+2ab}{2ab(2a+b)}$ **23.** $\dfrac{x^2-3x+24}{4x^2}$ **25.** $\dfrac{5x+21}{(x-4)(x+3)(x+5)}$ **27.** $\dfrac{2x}{(x+1)(x-1)}$ **29.** $\dfrac{a+5}{6}$

31. $\dfrac{9x+2}{(3x-4)(4x-3)}$ **33.** $\dfrac{-4x-2}{x(x-2)^2}$ **35.** $\dfrac{x^2-5x-2}{(x+2)(x+3)(x-1)}$ **37.** $\dfrac{9y}{y-3}$ **39.** $\dfrac{3}{y+4}$ **41.** $\dfrac{x^2+9x-10}{(x-4)(x+4)}$

43. $\dfrac{(4y+1)(y-1)}{2y(3y+1)(y-3)}$ **45.** $\dfrac{1}{10(3x+2)}$ **46.** $x=-7$ **47.** $y=\dfrac{5ax+6bc}{2a}$ **48.** $x<\dfrac{3}{2}$ **49.** $81x^{12}y^{16}$ **50.** at least 17 days

51. 566,100 more people **52.** **(a)** 900,000 fish **(b)** three more months **53.** approximately 7.8%

Putting Your Skills to Work

1. $\$883$ billion **2.** $\$423$ billion

6.4 Exercises

1. $\dfrac{5}{7x}$ **3.** $y+x$ **5.** $\dfrac{y}{x}$ **7.** $\dfrac{x-3}{x}$ **9.** $\dfrac{-2x}{3(x+1)}$ **11.** $\dfrac{3a^2+9}{a^2+2}$ **13.** $\dfrac{(2y+1)(y-2)}{2y}$ **15.** $\dfrac{x^2-2x}{4+5x}$ **17.** $\dfrac{2x-5}{3(x+3)}$ **19.** $\dfrac{y+1}{-y+1}$

21. No expression in any denominator can be allowed to be zero, since division by zero is undefined. So -3, 5, and 0 are not allowable replacements

for the variable x. **23.** $w=\dfrac{P-2l}{2}$ **24.** $x>-1$ **25.** $\$1875$ **26.** approximately 57%

6.5 Exercises

1. 3 **3.** -8 **5.** $-\dfrac{3}{4}$ **7.** -2 **9.** 4 **11.** $\dfrac{-14}{3}$ **13.** There is no solution. **15.** -3 **17.** -5 **19.** 12 **21.** There is no solution.

23. There is no solution. **25.** -11 **27.** 2, 4 **29.** $(3x+4)(2x-3)$ **30.** $\dfrac{-10}{3}$ **31.** width $=7$ m, length $=20$ m **32.** $\$946.45$

33. 21.9% **34.** 90.9% **35.** 10.3 million people **36.** 3.5 million people

6.6 Exercises

1. 18 **3.** $40\dfrac{4}{5}$ **5.** $\dfrac{40}{3}$ **7.** 22.75 **9.** 110 miles **11.** **(a)** 522.88 British pounds **(b)** $\$214.20$ **13.** 56 miles per hour **15.** 29 miles

17. $18\dfrac{17}{20}$ inches **19.** $9\dfrac{1}{3}$ in. **21.** $38\dfrac{2}{5}$ m **23.** 48 inches **25.** 86 meters **27.** 61.5 miles per hour **29.** commuter airline, 250

kilometers per hour; helicopter, 210 kilometers per hour **31.** **(a)** $\$22.06$ **(b)** $\$24.71$ **(c)** $\$27$ **33.** $2\dfrac{2}{9}$ hours or 2 hours, 13 minutes

35. $3\dfrac{3}{7}$ hours or 3 hours, 26 minutes **37.** 6.316×10^{-7} **38.** 582,000,000 **39.** $\dfrac{w^8}{x^3y^2z^4}$ **40.** $\dfrac{27}{8}$ or $3\dfrac{3}{8}$

Putting Your Skills to Work

1. 462 days **2.** 21 orbits

Chapter 6 Review Problems

1. $-\dfrac{4}{5}$ **2.** $\dfrac{x}{x-y}$ **3.** $\dfrac{x+3}{x-4}$ **4.** $\dfrac{x+2}{x+4}$ **5.** $\dfrac{x+3}{x-7}$ **6.** $\dfrac{2(x+4)}{3}$ **7.** $\dfrac{2x+3}{2x}$ **8.** $\dfrac{x}{x-5}$ **9.** $\dfrac{2(x-4y)}{2x-y}$ **10.** $\dfrac{2-y}{3y-1}$

11. $\dfrac{x-2}{(5x+6)(x-1)}$ **12.** $4x+2y$ **13.** $\dfrac{x-25}{x+5}$ **14.** $\dfrac{2(y-3)}{3y}$ **15.** $\dfrac{(y+2)(2y+1)(y-2)}{(2y-1)(y+1)^2}$ **16.** $\dfrac{4y(3y-1)}{(3y+1)(2y+5)}$

17. $\dfrac{3y(4x+3)}{2(x-5)}$ **18.** $\dfrac{x+2}{2}$ **19.** $\dfrac{3}{16}$ **20.** $\dfrac{2(x+3y)}{x-4y}$ **21.** $\dfrac{9x+2}{x(x+1)}$ **22.** $\dfrac{5x^2+7x+1}{x(x+1)}$ **23.** $\dfrac{(x-1)(x-2)}{(x+3)(x-3)}$

24. $\dfrac{10x-22}{(x+2)(x-4)}$ **25.** $\dfrac{2xy+4x+5y+6}{2y(y+2)}$ **26.** $\dfrac{(2a+b)(a+4b)}{ab(a+b)}$ **27.** $\dfrac{3x-2}{3x}$ **28.** $\dfrac{2x^2+7x-2}{2x(x+2)}$ **29.** $\dfrac{1-2x-x^2}{(x+5)(x+2)}$

30. $\dfrac{3}{2(x-9)}$ **31.** $\dfrac{1}{11}$ **32.** $\dfrac{5}{3x^2}$ **33.** $w-2$ **34.** 1 **35.** $-\dfrac{y^2}{2}$ **36.** $\dfrac{x+2y}{y(x+y+2)}$ **37.** $\dfrac{-1}{a(a+b)}$ **38.** $\dfrac{-3a-b}{b}$ **39.** $\dfrac{-3y}{2(x+2y)}$

40. $\dfrac{5y(x+5y)^2}{x(x-6y)}$ **41.** 15 **42.** 2 **43.** $\dfrac{2}{7}$ **44.** -8 **45.** -4 **46.** -2 **47.** $\dfrac{1}{2}$ **48.** $\dfrac{1}{5}$ **49.** no solution **50.** $-4, 6$ **51.** 2 **52.** 6

53. -2 **54.** no solution **55.** 9 **56.** 0 **57.** 1.3 **58.** 2.8 **59.** 26.4 **60.** 2 **61.** 16 **62.** 9.1 **63.** 8.3 gallons

64. 167 cookies **65.** 46 gallons **66.** 91.5 miles **67.** train, 60 miles per hour; car, 40 miles per hour **68.** 3 hours, 5 minutes **69.** 1200 feet

70. 182 feet **71.** $3\dfrac{1}{3}$ hours or 3 hours, 20 minutes **72.** 12 hours

Chapter 6 Test

1. $\dfrac{2}{3a}$ **2.** $\dfrac{2x^2(2-y)}{(y+2)}$ **3.** $\dfrac{5}{12}$ **4.** $\dfrac{1}{3y(x-y)}$ **5.** $\dfrac{(a-2)(4a+7)}{(a+2)^2}$ **6.** $\dfrac{4x+18}{(x+5)(x-2)(x+3)}$ **7.** $\dfrac{x-a}{ax}$ **8.** $-\dfrac{x+2}{x+3}$ **9.** $\dfrac{x}{4}$

10. $\dfrac{x+1}{x(x+3)^2}$ **11.** $\dfrac{2x-3y}{4x+y}$ **12.** $\dfrac{x}{(x+2)(x+4)}$ **13.** $x=3$ **14.** $x=4$ **15.** no solution **16.** $x=\dfrac{47}{6}$ **17.** $x=\dfrac{45}{13}$

18. $x=37.2$ **19.** 6 hours, 25 minutes **20.** \$368 **21.** 102 feet

Cumulative Test for Chapters 0–6

1. 0.006 cm **2.** \$375 **3.** \$92.50 **4.** $x=6$ **5.** $h=\dfrac{A}{\pi r^2}$ **6.** $x>1.25$ 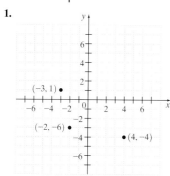 **7.** $x>5$

8. $(a+b)(3x-2y)$ **9.** $2a(4a+b)(a-5b)$ **10.** $-12x^5y^7$ **11.** $\dfrac{2x+5}{x+7}$ **12.** $\dfrac{(x-2)(3x+1)}{3x(x+5)}$ **13.** $\dfrac{1}{2}$ **14.** $\dfrac{11x-3}{2(x+2)(x-3)}$

15. $\dfrac{x^2+2x+6}{(x+4)(x-3)(x+2)}$ **16.** $x=-2$ **17.** $x=-\dfrac{9}{2}$ **18.** $\dfrac{x+8}{6x(x-3)}$ **19.** $\dfrac{3ab^2+2a^2b}{5b^2-2a^2}$ **20.** $\dfrac{98}{5}$ or $19\dfrac{3}{5}$ **21.** 208 miles

22. 484 phone calls

Chapter 7

Pretest Chapter 7

1.

2. $A\,(-1,-2)$; $B\,(3,-2)$; $C\,(-1,2)$
3. (a) $(-1,10)$ **(b)** $(2,1)$

4.

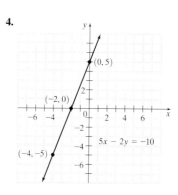

5.

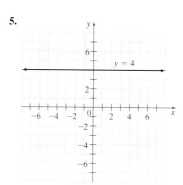

6. $m=-1$ **7.** $m=\dfrac{4}{3}$ **8. (a)** $y=4x-5$ **(b)** $4x-y=5$ **9.**

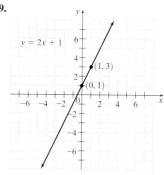

10. $y = -\dfrac{3}{2}x + 2$ **11. (a)** $m = \dfrac{3}{5}$ **(b)** $m = -\dfrac{5}{3}$ **12.** $y = \dfrac{2}{3}x - 1$ **13.** $y = -x + 5$

14.

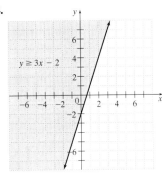

15.

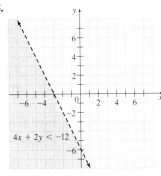

16. function **17.** not a function
18. (a) -13 **(b)** 14

19. (a) 72 **(b)** $\dfrac{1}{2}$

7.1 Exercises

1.

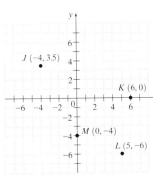

3. $R: (-3, -5)$; $S: \left(-4\dfrac{1}{2}, 0\right)$; $X: (3, -5)$, $Y: \left(2\dfrac{1}{2}, 6\right)$ **5.** 0

7. The order in which you write the numbers matters. The graph of $(5, 1)$ is not the same as the graph of $(1, 5)$.

9. $(-5, -3)$, $(-4, -4)$, $(-4, -5)$, $(-3, -3)$, $(-2, -1)$, $(-2, -2)$, $(-1, -3)$, $(0, -4)$, $(0, -5)$, and $(1, -3)$

11. (a) $(0, 8)$ **(b)** $(4, 20)$ **13. (a)** $(-6, 15)$ **(b)** $(3, -3)$ **15. (a)** $(7, 13)$ **(b)** $(-1, -7)$

17. (a) $(10, 4)$ **(b)** $(-5, -8)$ **19. (a)** $(3, -12)$ **(b)** $\left(\dfrac{3}{2}, 6\right)$ **21.** D5 **23.** D1 **25.** C3

27. (a)

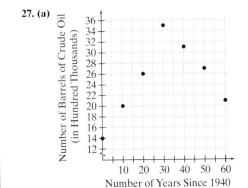

(b) The numbers of barrels of crude oil increased significantly from 1940 to 1970. From 1970 to 2000, it has decreased significantly

28. 8 **29.** $\dfrac{x - y}{x - 3y}$

30. 1133.54 square yards **31.** 922 **32.** 95.7%
33. (a) $\$55$ per square foot **(b)** $\$6160$

7.2 Exercises

1. no; replacing x by -2 and y by 5 in the equation does not result in a true statement **3.** x-axis

5. $(0, 1)$, $(-2, 5)$, $(1, -1)$

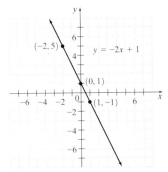

7. $(0, -4)$, $(2, -2)$, $(4, 0)$

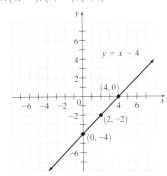

9.

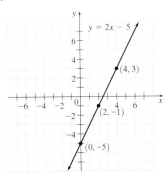

11.
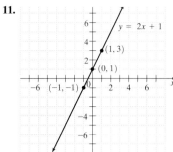
$y = 2x + 1$
(1, 3)
(0, 1)
(−1, −1)

13.
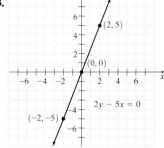
(2, 5)
(0, 0)
$2y - 5x = 0$
(−2, −5)

15.
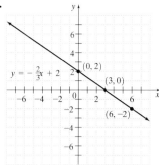
(0, 2)
$y = -\frac{2}{3}x + 2$
(3, 0)
(6, −2)

17.
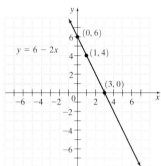
(0, 6)
$y = 6 - 2x$
(1, 4)
(3, 0)

19.
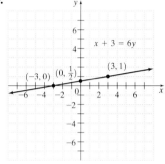
$x + 3 = 6y$
(3, 1)
(−3, 0) $\left(0, \frac{1}{2}\right)$

21.
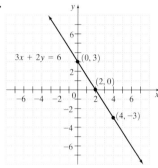
$3x + 2y = 6$
(0, 3)
(2, 0)
(4, −3)

23.
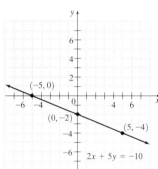
(−5, 0)
(0, −2)
(5, −4)
$2x + 5y = -10$

25.
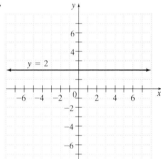
$y = 2$

27. $C = 0$; $C = 120$; $C = 240$; $C = 360$; $C = 480$; $C = 600$

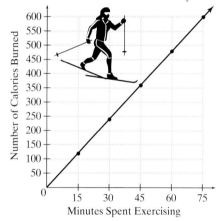

29.
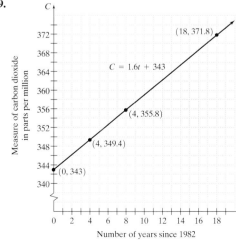

31. $x = 28$

32. $x \geq -\dfrac{14}{3}$;

33. $L = 17$ meters; $W = 9.5$ meters

34. 212   **35.** $1.24   **36.** 76 milligrams

7.3 Exercises

1. -1 **3.** $\dfrac{3}{5}$ **5.** $-\dfrac{2}{5}$ **7.** $-\dfrac{4}{3}$ **9.** $-\dfrac{16}{5}$ **11.** no; division by zero is impossible, so the slope is undefined **13.** $m = 8$; $(0, 9)$

15. $m = -3$; $(0, 4)$ **17.** $m = \dfrac{5}{6}$; $\left(0, -\dfrac{2}{9}\right)$ **19.** $m = -6$; $(0, 0)$ **21.** $m = -6$; $\left(0, \dfrac{4}{5}\right)$ **23.** $m = -\dfrac{5}{2}$; $\left(0, \dfrac{3}{2}\right)$ **25.** $m = \dfrac{7}{3}$; $\left(0, -\dfrac{4}{3}\right)$

27. (a) $y = \dfrac{3}{4}x + 2$ **(b)** $3x - 4y = -8$ **29. (a)** $y = 6x - 3$ **(b)** $6x - y = 3$ **31. (a)** $y = -\dfrac{5}{4}x - \dfrac{3}{4}$ **(b)** $5x + 4y = -3$

33.

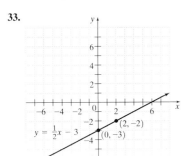

35.

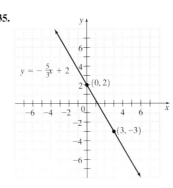

37.

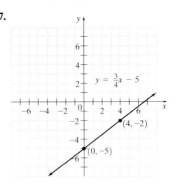

39.

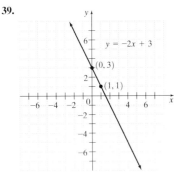

41.

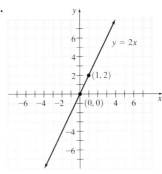

43. (a) $m = \dfrac{13}{5}$ **(b)** $m = -\dfrac{5}{13}$ **45. (a)** $m = -\dfrac{5}{8}$ **(b)** $m = \dfrac{8}{5}$ **47.** yes; $2x - 3y = 18$

49. (a) $y = 2.2x + 83$ **(b)** $m = 2.2$; $(0, 83)$. **(c)** the slope is the increase in the number of civilians employed in the United States in millions for each year during the period from 1970 to 1990; the slope indicates how fast the number of civilians employed is growing

50. $x > 4$; **51.** $x < \dfrac{12}{5}$; **52.** $x \leq 24$;

53. $x \leq -22$; **54.** 148 miles **55.** 23 **56.** 3.5%

Putting Your Skills to Work

1. 25 pounds per square inch **2.** 44 feet **3.** $\dfrac{5}{11}$ **4.**

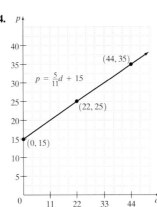

5. $\dfrac{5}{11}$; yes; you can determine the slope of a line using the coordinates of two points, from the equation in slope–intercept form, and from the graph by counting

6.

d	0	11	22	33	44
p	15	20	25	30	35

(a) 5 **(b)** 11 **(c)** the slope is the ratio of these two numbers; slope is the difference between the p values over the difference between the d values

7.4 Exercises

1. $y = 4x + 12$ **3.** $y = -2x + 11$ **5.** $y = -3x + \dfrac{7}{2}$ **7.** $y = -\dfrac{2}{5}x - 1$ **9.** $y = -2x - 6$ **11.** $y = -3x + 3$ **13.** $y = 5x - 10$

15. $y = \dfrac{1}{3}x + \dfrac{1}{2}$ **17.** $y = -\dfrac{2}{3}x + 1$ **19.** $y = \dfrac{2}{3}x - 4$ **21.** $y = 3x$ **23.** $y = -2$ **25.** $y = -2$ **27.** $y = 2$ **29.** $y = \dfrac{3}{4}x - 4$

31. $y = -\dfrac{1}{2}x + 4$ **33.** $y = 2.4x + 227$ **35.** $t = \dfrac{24}{5}$ **36.** $\dfrac{(2x + 5)(x - 3)}{(2x - 5)^2}$ **37.** \$61.20 **38.** 290 minutes **39.** 4.95% **40.** \$12,640

7.5 Exercises

1. No, all points in one region will be solutions to the inequality while all points in the other region will not be solutions. Thus testing any point will give the same result, as long as the point is not on the boundary line.

3.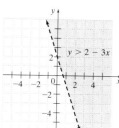

5.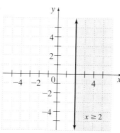

7.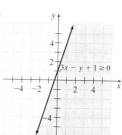

9.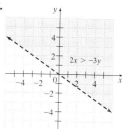

11.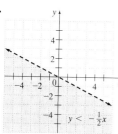

13.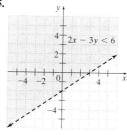

15.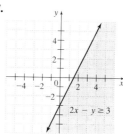

17.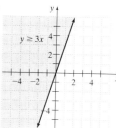

19.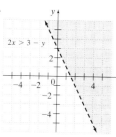

21.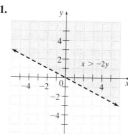

23. (a) $x < 3$ **(b)** $x < 3$ **(c)** $3x + 3$ will be greater than $5x - 3$ for those values of x where the graph of $y = 3x + 3$ lies *above* the graph of $y = 5x - 3$. **24.** $x = -\dfrac{3}{2}, x = -2$ **25.** $y = -\dfrac{7}{3}x + 7$ **26.** 9200 miles **27.** 16

7.6 Exercises

1. You can describe a function using a table of values, an algebraic equation, or a graph. **3.** Possible values; independent **5.** If a vertical line can intersect the graph more than once, the relation is not a function. If no such line exists, then the relation is a function.

7. (a) Domain $= \left\{\dfrac{1}{4}, \dfrac{1}{2}, \dfrac{3}{4}\right\}$; Range $= \{5, 6, 10\}$ **(b)** not a function **9. (a)** Domain $= \{0, 2, 7.3\}$; Range $= \{1, 8\}$ **(b)** function

11. (a) Domain $= \{5, 5.6, 5.8, 6\}$; Range $= \{5.8, 6, 8\}$ **(b)** function **13. (a)** Domain $= \{85, 95, 110\}$; Range $= \{3, 11, 15, 20\}$ **(b)** not a function

15.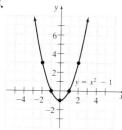

17.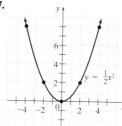

19.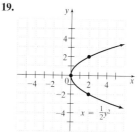

21.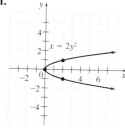

23.

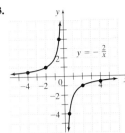

$y = -\dfrac{2}{x}$

25.

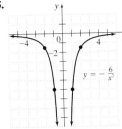

$y = -\dfrac{6}{x^2}$

27.

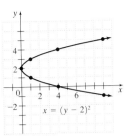

$x = (y - 2)^2$

29.

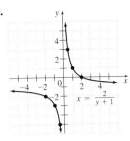

$x = \dfrac{2}{y + 1}$

31. function   **33.** not a function   **35.** function   **37.** not a function   **39. (a)** 8   **(b)** −2   **(c)** −7   **41. (a)** −4   **(b)** −3   **(c)** 12   **43. (a)** −8
(b) 7   **(c)** 4   **45.** 30%   **47.** between 1985 and 1990

49.

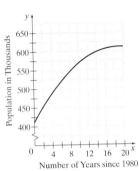

Population in Thousands

Number of Years since 1980

$f(0) = 411; f(4) = 489.4; f(10) = 571; f(20) = 611;$
The curve slopes upward more steeply for smaller values of x.
Growth rate is decreasing as x gets larger.

50. $-2x^2 - 14x + 6$ **51.** $-3x^2 - 7x + 11$ **52.** $4x^3 + 6x^2 - x + 6$ **53.** $2x + 4 + \dfrac{-1}{3x - 1}$

Putting Your Skills to Work

1. 213,408 acres; 78,000,000 acres   **2.** 4337 million acres or 4,337,000,000 acres   **3.** 4181 million acres; 3947 million acres

Chapter 7 Review Problems

1.

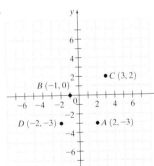

2.

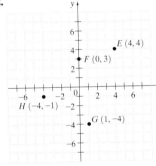

3. (a) $(0, -5)$ **(b)** $(3, 4)$ **4. (a)** $(1, 2)$ **(b)** $(-4, 4)$
5. (a) $(6, -1)$ **(b)** $(6, 3)$

6.

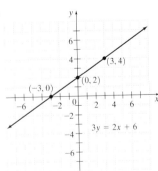

$3y = 2x + 6$

7.

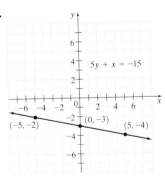

$5y + x = -15$

8.

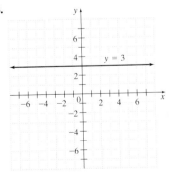

$y = 3$

9. $m = -\dfrac{5}{6}$ **10.** $m = \dfrac{9}{11}; \left(0, \dfrac{5}{11}\right)$ **11.** $y = -\dfrac{1}{2}x + 3$ **12.** $\dfrac{3}{2}$

13.
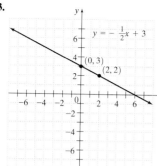

14.
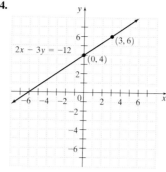

15.
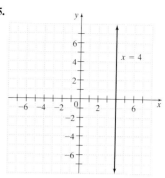

16. $y = 2x - 4$ **17.** $y = -6x + 14$ **18.** $y = \dfrac{1}{3}x - 5$ **19.** $y = 7$ **20.** $y = \dfrac{2}{3}x - 3$ **21.** $y = -3x + 1$ **22.** $x = -5$

23.
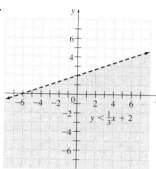

24.
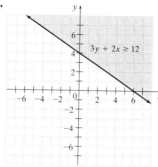

25.
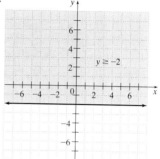

26. Domain: $\{-6, -5, 5\}$; Range: $\{-6, 5\}$; not a function   **27.** Domain: $\{-7, -3, 3, 7\}$; Range: $\{-7, -3, 3, 7\}$; function   **28.** function
29. not a function   **30.** function

31.
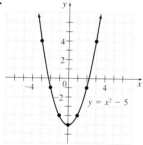

32.
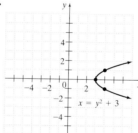

33.
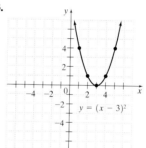

34. (a) 7 **(b)** 31   **35. (a)** -1 **(b)** -5 **36. (a)** 1 **(b)** -6 **37. (a)** 1 **(b)** $\dfrac{1}{5}$ **38. (a)** $-\dfrac{6}{5}$ **(b)** $-\dfrac{4}{7}$ **39.** $300   **40.** $600
41. $y = 0.15x + 150$; 0.15   **42.** The cost of the trip increases $0.15 for each mile.   **43.** 2600 miles   **44.** 4000 miles   **45.** $210   **46.** $174
47. $y = 0.09x + 30$ (0, 30); it tells us that if Russ and Norma use no electricity, the minimum cost is $30   **48.** $m = 0.09$; the electric bill increases
$0.09 for each kilowatt-hour of use   **49.** 1300 kilowatt-hours   **50.** 2400 kilowatt-hours

Chapter 7 Test

1.
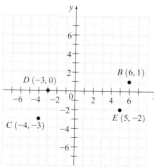

2.
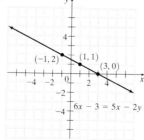

3.
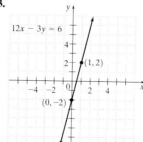

4.
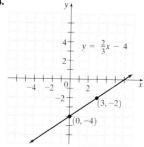

5. $m = -\dfrac{3}{2}; \left(0, \dfrac{5}{2}\right)$ **6.** $m = 1$ **7.** $y = \dfrac{1}{2}x - 4$ **8.** $m = 0$ **9.** $y = -\dfrac{1}{3}x + \dfrac{17}{3}$ **10.** $x = 2$, undefined

11.

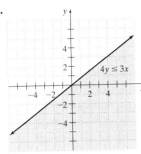

12.

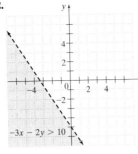

13. yes; no two different ordered pairs have the same first coordinate

14. yes; It passes the vertical line test.

15.

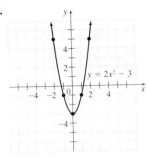

16. (a) -4 **(b)** -2 **17. (a)** -3 **(b)** $-\dfrac{1}{5}$

Cumulative Test for Chapters 0–7

1. $-27x^9y^{12}z^3$ **2.** $-\dfrac{3a^2}{4b^2}$ **3.** $x = 7$ **4.** $R = 40$ **5.** $x = 5$ **6.** $w = \dfrac{A - lh}{l}$ or $w = \dfrac{A}{l} - h$ **7.** $2(5a + 7b)(5a - 7b)$

8. $(3x + 7)(x - 3)$ **9.** $2(x - 1)$ **10.** $y = 3x - 10$ **11.** $x = 7$ **12.** $y = \dfrac{1}{3}x + \dfrac{11}{3}$ **13.** $m = 0$ **14.** $m = \dfrac{3}{7}$

15.

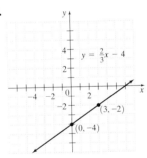

16.

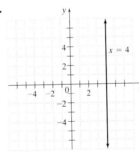

17.

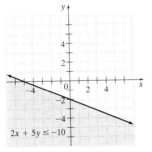

18. no

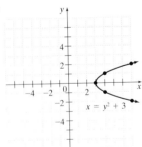

19. yes   **20. (a)** -39 **(b)** 3

Chapter 8

Pretest Chapter 8

1. $(-3, 2)$

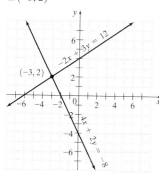

2. no solution

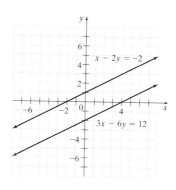

3. $(-3, -2)$ **4.** $\left(\dfrac{3}{5}, \dfrac{3}{5}\right)$ **5.** $(2, 0)$

6. $(6, 2)$ **7.** $(2, 3)$ **8.** intersect

9. are parallel **10.** coincide **11.** $\left(3, \dfrac{1}{2}\right)$

12. $(0, -5)$ **13.** $(2.5, 7.5)$

14. 290 letters/hour; 690 letters/hour

8.1 Exercises

1.

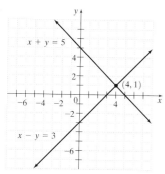

3.

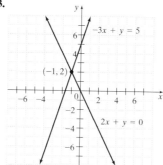

5.

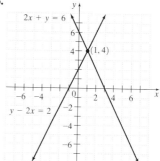

7.

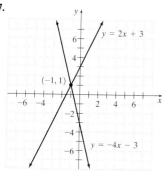

9.

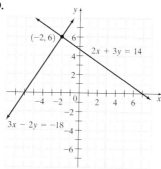

11.

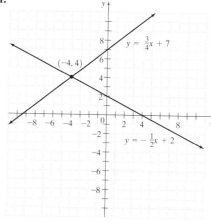

13.

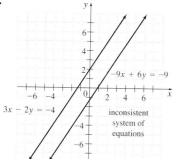

15.

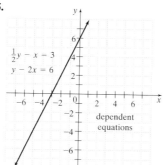

17. The lines are parallel. One line intersects the y-axis at -5. The other line intersects the y-axis at 6. There is no solution. The system is inconsistent.

19. intersect; one

21. The lines intersect at the y-intercept. The y-intercept is a solution of the system.

23. (a) Camp Property Care: $y = 100 + 60x$; Manchester Landscape Designs: $y = 200 + 40x$

(b)

x	$y = 100 + 60x$
0	100
5	400
10	700

x	$y = 200 + 40x$
0	200
5	400
10	600

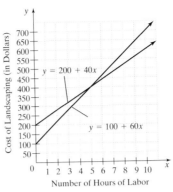

(c) 5 hours   **(d)** Manchester Landscape Designs.   **25.** $(21, 27)$ **27.** $(-29.61, -71.52)$ **28.** $\frac{1}{8} \leq x$ **29.** $d = \frac{5y + 16}{x}$ **30.** $9x^2 - 42x + 49$

31. $(5x - 6)^2$ **32.** 90,000 people   **33.** 300,000 people   **34.** 547,500 people   **35.** 38.9%

8.2 Exercises

1. $(3, -1)$ **3.** $(1, 2)$ **5.** $(3, -5)$ **7.** $(0, 4)$ **9.** $(9, 5)$ **11.** $(2, 0)$ **13.** $\left(\frac{2}{3}, \frac{5}{3}\right)$ **15.** $(2, -3)$ **17.** $(-3, 6)$ **19.** $(3, -2)$

21. $(7, -2)$ **23.** $(8, 1)$ **25.** 3; 7; an extra equation is needed for each additional variable to reduce the system to one equation with one unknown   **27.** both   **29.** no solution; parallel lines have no points in common   **31. (a)** $y = 5000 + 2000x$; $y = 3000 + 2200x$ **(b)** 10 months; $25,000   **(c)** 40 months; Sunset Acres

32. $m = -\frac{7}{11}$; $\left(0, \frac{19}{11}\right)$ **33.**

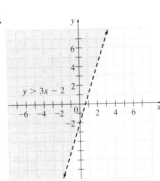

$y > 3x - 2$

34. 398 meters/minute   **35.** $3,914,000
36. 8736 nights in rooms
37. $162

8.3 Exercises

1. (a) Eliminate y by multiplying the second equation by 2, then adding the equations.   **(b)** Eliminate x by multiplying the first equation by -1, then adding the equations.   **(c)** Eliminate y by multiplying the first equation by 4, multiplying the second equation by 3, then adding the equations.

3. $(1, 3)$ **5.** $(1, 2)$ **7.** $(0, 5)$ **9.** $\left(\frac{1}{4}, -3\right)$ **11.** $(4, 2)$ **13.** $\left(-\frac{9}{4}, 1\right)$ **15.** $(2, 2)$ **17.** $(1, -2)$ **19.** $(3, 7)$ **21.** $\left(2, \frac{1}{5}\right)$

23. $(3, -2)$ **25.** $\left(-2, \frac{4}{3}\right)$ **27.** $(-4, -2)$ **29. (a)** multiply equation **(1)** by 10; $5x - 3y = 1$, $5x + 3y = 6$ **(b)** multiply equation **(1)** by 100, multiply equation **(2)** by 10; $8x + 100y = 5$, $20x - y = 30$ **(c)** multiply equation **(1)** by 10, multiply equation **(2)** by 100; $40x + 5y = 90$, $20x - 5y = 100$ **31.** $(1, 0)$ **33.** $(-4, -7)$ **35.** $(8, -10)$ **37.** $\left(\frac{1}{5}, -\frac{1}{3}\right)$ **39.** $(2, 5)$ **40.** between 550 and 770 airplanes

41. 8 minutes   **42.** 99 miles/hour or greater   **43.** 36 games   **44.** $3(5x - 1)(5x + 1)$ **45.** $(2x - 5)(3x - 5)$

Putting Your Skills to Work

1. Aero 8, $81,500; plus 8, $52,600   **2.** Aero 8, $83,000; plus 8, $54,000

8.4 Exercises

1. Parallel lines; inconsistent   **3.** intersect; independent; consistent   **5.** $(3, -7)$ **7.** no solution   **9.** $\left(\frac{9}{31}, -\frac{109}{31}\right)$

11. $(50, 40)$ **13.** $(3, 1)$ **15.** $\left(\frac{1}{2}, \frac{2}{3}\right)$ **17.** $(-4, -3)$ **19.** $x = 2$ **20.** $x = 25$ **21.** $21,500   **22.** $38,500

23. 1261 tickets   **24.** approximately 52.4 cents/mile

8.5 Exercises

1. A loaf from Meilleur Pain is $0.42. A loaf from Corner Loaf is $0.30.   **3.** five etchings at $35 each; seven etchings at $40 each   **5.** width 225 feet; length 300 feet   **7.** 8 quarts of 50% antifreeze; 8 quarts of 80% antifreeze.   **9.** 30 pounds of nuts; 20 pounds of raisins   **11.** The wind speed is 50 kilometers per hour. The plane's speed in still air is 550 kilometers per hour.   **13.** 2040; 115,000 employees   **15.** 2010; 5,680,000

17. $29,000 for the heart kits; $36,000 for the kidney kits   **19.** $\dfrac{2x^2 - 12}{x^2 - 9}$ **20.** 35   **21.** $-64x^9y^6$ **22.** 35%

Putting Your Skills to Work

1. $P = 36.5 + 0.2x$ **2.** $P = 30.9 + 0.3x$

Chapter 8 Review Problems

1.

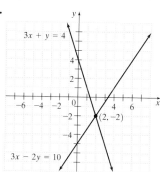

2.

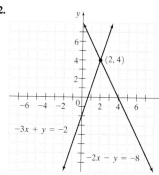

3.

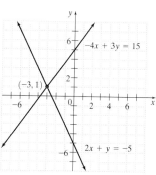

4.

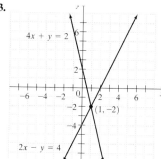

5. $(3, 3)$ **6.** $(3, 5)$ **7.** $(14, 66)$ **8.** $(2, -3)$ **9.** $(2, 1)$

10. $(-1, 0)$ **11.** $(3, 1)$ **12.** $(-4, -2)$ **13.** $(4, -4)$ **14.** $(10, 2)$

15. $\left(\dfrac{4}{5}, -\dfrac{6}{5}\right)$ **16.** $\left(\dfrac{33}{13}, -\dfrac{28}{13}\right)$ **17.** $(-3, -1)$ **18.** $(2, -3)$ **19.** no solution, inconsistent system

20. $(9, 4)$ **21.** $\left(-3, -\dfrac{2}{3}\right)$ **22.** infinite number of solutions, dependent equations   **23.** $(-2, -2)$

24. $(-1, -2)$ **25.** $(4, -3)$ **26.** $\left(-\dfrac{24}{41}, \dfrac{2}{41}\right)$ **27.** infinite number of solutions, dependent equations

28. $(1960, 1040)$ **29.** $(8, -5)$ **30.** $(-4, -7)$ **31.** $\left(\dfrac{29}{28}, \dfrac{1}{7}\right)$ **32.** $(4, 10)$ **33.** $(10, 8)$

34. $(-12, -15)$ **35.** $\left(\dfrac{1}{2}, -\dfrac{2}{3}\right)$ **36.** $\left(\dfrac{15}{13}, \dfrac{36}{13}\right)$

37. infinite number of solutions, dependent equations   **38.** infinite number of solutions, dependent equations   **39.** $(-5, 4)$ **40.** $(9, -8)$
41. $(2, 4)$ **42.** $(3, 1)$ **43.** $(3, 2)$ **44.** $(12, -12)$ **45.** 120 reserved seats; 640 general admission seats   **46.** 7 pounds of apples; 5 pounds
of oranges   **47.** 275 miles/hour in still air; 25 miles/hour wind speed   **48.** 32 liters   **49.** 44 regular-sized cars; 42 compact cars   **50.** 8 tons
of 15% salt; 16 tons of 30% salt   **51.** speed of the boat, 19 kilometers per hour; speed of the current, 4 kilometers per hour   **52.** $3 for a gram
of copper; $4 hourly labor rate   **53.** 14 outfield grandstand seats; 9 right field roof seats   **54.** 13 upper bleacher seats; 14 lower bleacher seats

Putting Your Skills to Work

1. loss; it costs $100 more to produce   **2.** $325 profit

Chapter 8 Test

1. $(-3, -4)$ **2.** $(-3, 4)$ **3.**

   4. $(6, 2)$ **5.** $(4, 3)$ **6.** $(5, 1)$ **7.** $(3, 0)$
8. no solution, inconsistent system
9. infinite number of solutions, dependent equations
10. $(-2, 3)$ **11.** $(-5, 3)$ **12.** $(-3.5, -4.5)$ **13.** $5, -3$
14. 16,000 of the $12 tickets; 14,500 of the $8 tickets
15. A shirt costs $20 and a pair of slacks costs $24.
16. 500 per hour by 1st machine; 400 per hour by 2nd machine
17. 50 kilometers/hour; 450 kilometers/hour

Cumulative Test for Chapters 0–8

1. $4.09   **2.** -8 **3.** $15x^3 - 19x^2 + 11x - 3$ **4.** $\dfrac{1}{(x + 3)(x - 4)}$ **5.** $x \le \dfrac{9}{5}$ **6.** $x = \dfrac{1}{2}, -\dfrac{2}{3}$ **7.** 3000   **8.** $x = \dfrac{5}{3}$

9. $5a(a - 4)(a + 3)$ **10.** $x = 3$ **11.** $(8, -8)$ **12.** $(24, 8)$ **13.** infinite number of solutions; dependent equations
14. $(5, 1)$ **15.** inconsistent system; no solution   **16.** 4000 liters of 8%; 8000 liters of 20%   **17.** 1 mile/hour   **18.** the printer that broke prints
1200 labels/hour; the other printer prints 1800 labels/hour

Chapter 9

Pretest Chapter 9

1. 9 **2.** $\frac{5}{8}$ **3.** 12 **4.** -10 **5.** 2.236 **6.** not real **7.** x^2 **8.** $5x^2y^3$ **9.** $2\sqrt{10}$ **10.** $x^2\sqrt{x}$ **11.** $6x\sqrt{x}$ **12.** $2a^2b\sqrt{2b}$
13. $9\sqrt{7}$ **14.** $15\sqrt{2}$ **15.** $4\sqrt{2}$ **16.** $13\sqrt{7} - 7$ **17.** $11\sqrt{2} - 18\sqrt{3}$ **18.** $30\sqrt{2}$ **19.** $2\sqrt{3} + 6$ **20.** $23 + 8\sqrt{7}$ **21.** 1 **22.** -1
23. $17 - 4\sqrt{15}$ **24.** $\frac{6}{x}$ **25.** $\frac{5\sqrt{3}}{3}$ **26.** $3\sqrt{5} - 6$ **27.** $13 - 2\sqrt{42}$ **28.** $2\sqrt{13}$ **29.** $\sqrt{11}$ **30.** 7 **31.** $\frac{35}{3}$ **32.** $\frac{1}{2}$ **33.** 7

9.1 Exercises

1. The principal square root of N, where $N \geq 0$, is a nonnegative number a that has the property $a^2 = N$. **3.** no; $(0.3)(0.3) = 0.09$ **5.** ± 3
7. ± 7 **9.** 6 **11.** 9 **13.** -7 **15.** 0.9 **17.** $\frac{6}{11}$ **19.** $\frac{7}{8}$ **21.** 30 **23.** -100 **25.** 12 **27.** $-\frac{1}{8}$ **29.** $\frac{3}{4}$ **31.** 0.07 **33.** 130
35. 13 feet **37.** 6.481 **39.** 8.602 **41.** -11.533 **43.** -13.964 **45.** 7.5 seconds **47.** 9.5 seconds **49.** 3 **51.** -4 **53.** 3 **55.** 3
57. no; a fourth power of a real number must be nonnegative **59.** $x = 2$; $y = 1$ **60.** no solution, inconsistent system of equations **61.** Each
snowboard is $250. Each pair of goggles is $50. **62.** 470 mph **63.** approximately 11.4% **64.** approximately 20.5%

9.2 Exercises

1. 8 **3.** 10^2 **5.** 9^3 **7.** 33^4 **9.** 5^{70} **11.** x^7 **13.** t^9 **15.** y^{13} **17.** $6x^4$ **19.** $12x$ **21.** x^3y^2 **23.** $4xy^{10}$ **25.** $10x^6y^4$
27. $2\sqrt{3}$ **29.** $5\sqrt{3}$ **31.** $4\sqrt{2}$ **33.** $2\sqrt{15}$ **35.** $5\sqrt{5}$ **37.** $3\sqrt{6}$ **39.** $2y^2\sqrt{3y}$ **41.** $5x^2\sqrt{x}$ **43.** $7x^2\sqrt{x}$ **45.** $5x^3\sqrt{2xy}$
47. $2xy\sqrt{3y}$ **49.** $3a^4y^3\sqrt{3y}$ **51.** $6a^2c\sqrt{5ab}$ **53.** $3a^2b^3c\sqrt{7c}$ **55.** $9x^6y^5w^2\sqrt{yw}$ **57.** $x + 2$ **59.** $2x + 5$ **61.** $2(2x + 1)$
63. $7\sqrt{3}$ square millimeters **65. (a)** $35\sqrt{3}$ square millimeters **(b)** five times larger **66.** $x = 4$; $y = -1$ **67.** $x = -4$
68. 810 million tons **69.** approximately 31.7 quadrillion Btu **70.** approximately 71.4%

9.3 Exercises

1. 1. simplify each radical term. 2. combine like radicals. **3.** $6\sqrt{5}$ **5.** $5\sqrt{2} + 3\sqrt{3}$ **7.** $4\sqrt{5}$ **9.** $17\sqrt{2}$ **11.** $\sqrt{2}$ **13.** $9\sqrt{2}$
15. $4\sqrt{3} + 2\sqrt{5} + 6$ **17.** $8\sqrt{3} - \sqrt{2}$ **19.** $13\sqrt{2x}$ **21.** $0.2\sqrt{3x}$ **23.** $7\sqrt{5y}$ **25.** $5y\sqrt{y}$ **27.** $6\sqrt{7x} - 15x\sqrt{7x}$ **29.** $-5x\sqrt{2x}$
31. $x\sqrt{3}$ **33.** $2y\sqrt{6y} - 6y\sqrt{6}$ **35.** $-110x\sqrt{2x}$ **37.** $(4 + \sqrt{14})$ miles **38.** $x \geq -\frac{4}{3}$

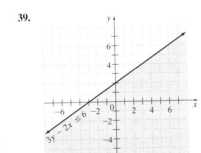

40. 3 CDs at $7.99 and 7 CDs at $11.99 **41.** $1600 at 7% and $6400 at 5%
42. 47 levels above the lobby, a total of 48 levels

39.

9.4 Exercises

1. $\sqrt{30}$ **3.** $2a\sqrt{3}$ **5.** $6\sqrt{30}$ **7.** $8a^3\sqrt{b}$ **9.** $-6b\sqrt{a}$ **11.** $\sqrt{21} + 15$ **13.** $6\sqrt{2} + 15\sqrt{5}$ **15.** $2x - 16\sqrt{5x}$
17. $2\sqrt{3} - 18 + 4\sqrt{15}$ **19.** $6\sqrt{ab} + 2a\sqrt{b} - 4a$ **21.** $-6 - 9\sqrt{2}$ **23.** $21 + 14\sqrt{2}$ **25.** $5 - \sqrt{21}$ **27.** $-6 + 9\sqrt{2}$
29. $2\sqrt{14} + 34$ **31.** $29 - 12\sqrt{5}$ **33.** $53 + 10\sqrt{6}$ **35.** $48 - 6\sqrt{15}$ **37.** $9x^2y - 5$ **39. (a)** It is not true when a is negative.
(b) In order to deal with real numbers. **41.** 12 square feet **43.** $(6x + 7y)(6x - 7y)$ **44.** $(2x + 5)(3x - 2)$
45. $m = \left(\frac{1519}{1320}\right)k$, 40.3 miles per hour **46.** $8\frac{1}{3}$ years

9.5 Exercises

1. 2 **3.** $\frac{1}{2}$ **5.** $\frac{\sqrt{6}}{x^2}$ **7.** 5 **9.** $\frac{3\sqrt{2}}{a^2}$ **11.** $\frac{3\sqrt{7}}{7}$ **13.** $\frac{x\sqrt{2x}}{2}$ **15.** $\frac{2\sqrt{2x}}{x}$ **17.** $\frac{\sqrt{3}}{2}$ **19.** $\frac{7\sqrt{a}}{a^2}$ **21.** $\frac{\sqrt{2x}}{2x^2}$ **23.** $\frac{3\sqrt{x}}{x^2}$ **25.** $\frac{\sqrt{15}}{5}$
27. $\frac{9\sqrt{2x}}{8x}$ **29.** $2\sqrt{3} + 2$ **31.** $\sqrt{5} - \sqrt{2}$ **33.** $2 + \sqrt{2}$ **35.** $x(2\sqrt{2} + \sqrt{5})$ **37.** $2\sqrt{14} - 7$ **39.** $\frac{7 - 2\sqrt{10}}{3}$ **41.** $3\sqrt{2} + 4$
43. $5\sqrt{6} + 8\sqrt{2}$ **45.** $\sqrt{x} - 5$ **47. (a)** $x^2 - 2$ **(b)** $(x + \sqrt{2})(x - \sqrt{2})$ **(c)** $(x + 2\sqrt{3})(x - 2\sqrt{3})$ **49.** 1.183215957. The values are the
same. They are equivalent expressions. **51.** $(3\sqrt{5} - 6)$ meters **53.** length, 4.236 meters; width, 0.708 meter **55.** $x = 7.38$ **56.** 448
57. 71 mg **58.** more than 15 doctor visits

9.6 Exercises

1. 13 **3.** 25 **5.** $\sqrt{74}$ **7.** $3\sqrt{7}$ **9.** $2\sqrt{19}$ **11.** $\sqrt{57}$ **13.** 9.91 **15.** 9.9 ft **17.** 98 ft **19.** top cable is 145.40 ft long; bottom cable is 140.77 ft **21.** $x = 7$ **23.** $x = -\dfrac{2}{3}$ **25.** $x = 7$ **27.** $x = \dfrac{81}{2}$ **29.** $x = 5$ **31.** $y = 0, y = 3$ **33.** $x = 1$ **35.** $y = 0, y = 1$

37. $y = \dfrac{1}{2}$ **39.** $x = \dfrac{2}{9}$ **41.** $x = 5$ **42.** $25x^2 - 70x + 49$ **43.** 20 **44.** relation **45.** 450 birds **46.** approximately \$37.21

9.7 Exercises

1. $y = 72$ **3.** $y = \dfrac{1029}{2}$ **5.** $y = 1296$ **7.** 169 lb/sq in. **9.** 56 min **11.** $y = \dfrac{48}{7}$ **13.** $y = 2$ **15.** $y = \dfrac{40}{27}$ **17.** 3.45 minutes

19. $444\dfrac{4}{9}$ pounds **21. (a)** 15 km, 8.2 km, 2.7 km **(b)**

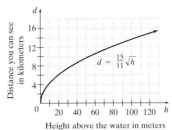

23. 5000 Btu/hr **25.** $a = 160$ **26.**

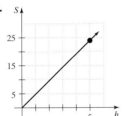

27. $4(3x + 4)(x - 3)$ **28.** $x = 7$

Chapter 9 Review Problems

1. 11 **2.** 12 **3.** 13 **4.** 14 **5.** not a real number **6.** not a real number **7.** 7 **8.** 16 **9.** 15 **10.** 17 **11.** 0.9 **12.** 0.6
13. $\dfrac{1}{5}$ **14.** $\dfrac{6}{7}$ **15.** 10.247 **16.** 14.071 **17.** 8.775 **18.** 9.381 **19.** $5\sqrt{2}$ **20.** $4\sqrt{3}$ **21.** $7\sqrt{2}$ **22.** $6\sqrt{2}$ **23.** $2\sqrt{10}$ **24.** $4\sqrt{5}$
25. x^4 **26.** y^5 **27.** $x^2y^3\sqrt{x}$ **28.** $ab^2\sqrt{a}$ **29.** $4xy^2\sqrt{xy}$ **30.** $7x^2y^3\sqrt{2}$ **31.** $2x^2\sqrt{3x}$ **32.** $3x^3\sqrt{3x}$ **33.** $5x^5\sqrt{3}$ **34.** $5x^6\sqrt{5}$
35. $2ab^2c^2\sqrt{30ac}$ **36.** $11a^3b^2\sqrt{c}$ **37.** $2x^3y^4\sqrt{14xy}$ **38.** $3x^6y^3\sqrt{11xy}$ **39.** $4\sqrt{2}$ **40.** $11\sqrt{5}$ **41.** $7x\sqrt{3}$ **42.** $8a\sqrt{2} + 2a\sqrt{3}$
43. $-7\sqrt{5} + 2\sqrt{10}$ **44.** $9\sqrt{6} - 15\sqrt{2}$ **45.** $6x^2$ **46.** $-10a\sqrt{b}$ **47.** $4ab\sqrt{a}$ **48.** $-15x^4$ **49.** 5 **50.** -18 **51.** $\sqrt{10} - \sqrt{6} - 4$
52. $\sqrt{30} - 10 + 5\sqrt{2}$ **53.** $20 + 3\sqrt{11}$ **54.** $27 + 8\sqrt{10}$ **55.** $8 - 4\sqrt{3} + 12\sqrt{6} - 18\sqrt{2}$ **56.** $15 - 3\sqrt{2} - 10\sqrt{3} + 2\sqrt{6}$
57. $66 + 36\sqrt{2}$ **58.** $74 - 40\sqrt{3}$ **59.** $\dfrac{\sqrt{3x}}{3x}$ **60.** $\dfrac{2y\sqrt{5}}{5}$ **61.** $\dfrac{x^2y\sqrt{2}}{4}$ **62.** $\dfrac{3a\sqrt{2b}}{2}$ **63.** $\dfrac{\sqrt{21}}{7}$ **64.** $\dfrac{\sqrt{2}}{3}$ **65.** $\dfrac{a^2\sqrt{2}}{2}$ **66.** $\dfrac{x\sqrt{3}}{3}$
67. $\sqrt{5} - \sqrt{2}$ **68.** $\dfrac{2(\sqrt{6} + \sqrt{3})}{3}$ **69.** $-7 + 3\sqrt{5}$ **70.** $\dfrac{3 - 2\sqrt{3}}{3}$ **71.** $c = \sqrt{89}$ **72.** $a = \sqrt{2}$ **73.** $b = \dfrac{\sqrt{51}}{2}$ **74.** $c = 3.124$
75. 30 meters **76.** 45.8 miles **77.** 7.1 inches **78.** $x = 13$ **79.** $x = 42$ **80.** $x = -1$ **81.** $x = 6$ **82.** $x = 4$ **83.** $x = 7$
84. $x = 2$ **85.** $x = 7$ **86.** $y = 77$ **87.** $y = \dfrac{2}{15}$ **88.** $y = \dfrac{1}{2}$ **89.** 400 lb **90.** approximately 134 ft **91.** it will be 8 times as much

Putting Your Skills to Work

1.

x	0	64	100	144	256	400	784
t	0	2	2.5	3	4	5	7

2.

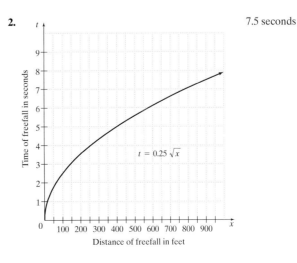

7.5 seconds

Chapter 9 Test

1. 11   **2.** $\dfrac{3}{10}$ **3.** $4xy^3\sqrt{3y}$ **4.** $10xz^2\sqrt{xy}$ **5.** $\sqrt{5}$ **6.** $8\sqrt{a}+5\sqrt{2a}$ **7.** $12ab$ **8.** $3\sqrt{2}-\sqrt{6}+45$ **9.** $10-4\sqrt{6}$

10. $19+\sqrt{10}$ **11.** $\dfrac{\sqrt{5x}}{5}$ **12.** $\dfrac{\sqrt{3}}{2}$ **13.** $\dfrac{3\sqrt{2}-16}{34}$ **14.** $a(\sqrt{5}-\sqrt{2})$ **15.** 12.49   **16.** $x=\sqrt{133}$ **17.** $x=3\sqrt{3}$

18. $x=\dfrac{35}{2}$ **19.** $x=9$ **20.** 16 inches   **21.** \$912.05   **22.** $A=21.2\ \text{cm}^2$

Cumulative Test for Chapters 0–9

1. $\dfrac{4}{5}$ **2.** $8\dfrac{11}{12}$ **3.** $\dfrac{15}{16}$ **4.** 0.0007   **5.** 3   **6.** 31   **7.** $-2x^3yz^2+4xyz$ **8.** $27x^3-54x^2+36x-8$ **9.** $16x^2-40x+25$

10. $9x^2-121$ **11.** $\dfrac{4x+4}{x+2}$ **12.** $\dfrac{(x-3)(x+1)}{(x-6)(x+2)}$ **13.** $7x^2y^3\sqrt{2x}$ **14.** $14\sqrt{3}$ **15.** $9-6\sqrt{2}$ **16.** $5+2\sqrt{6}$ **17.** $10+22\sqrt{3}$

18. $\dfrac{3+\sqrt{2}}{7}$ **19.** -4 **20.** not a real number   **21.** $c=\sqrt{23}$ **22.** $b=4\sqrt{5}$ **23.** $x=5$ **24.** $y=6$ **25.** $y=10$ **26.** 78.5 m^2

Chapter 10

Pretest Chapter 10

1. $x=16,\ x=-3$ **2.** $x=0,\ x=\dfrac{7}{5}$ **3.** $x=\dfrac{2}{5},\ x=4$ **4.** $x=\dfrac{1}{4},\ x=-\dfrac{1}{2}$ **5.** $x=\pm3\sqrt{2}$ **6.** $x=\pm5$ **7.** $x=-4\pm\sqrt{11}$

8. $x=\dfrac{-3\pm\sqrt{65}}{4}$ **9.** $x=\dfrac{-2\pm\sqrt{14}}{2}$ **10.** $x=\dfrac{7\pm\sqrt{17}}{4}$ **11.** $x=\dfrac{-4\pm\sqrt{13}}{3}$ **12.** no real solution

13. 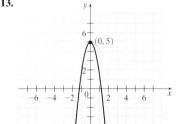   **14.** 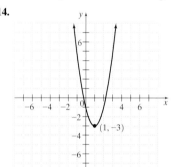   **15.** base $=6$ cm; altitude $=13$ cm

10.1 Exercises

1. $a=1,\ b=8,\ c=7$ **3.** $a=8,\ b=-11,\ c=0$ **5.** $a=1,\ b=3,\ c=-15$ **7.** $x=0,\ x=\dfrac{1}{3}$ **9.** $x=0,\ x=\dfrac{8}{7}$ **11.** $x=0,\ x=\dfrac{3}{2}$

13. $x=7,\ x=-4$ **15.** $x=-\dfrac{1}{2},\ x=8$ **17.** $x=\dfrac{1}{3},\ x=5$ **19.** $x=6,\ x=-3$ **21.** $x=-\dfrac{2}{5},\ x=-\dfrac{5}{3}$ **23.** $m=-5,\ m=-6$

25. $y=\dfrac{1}{4},\ y=\dfrac{3}{2}$ **27.** $x=\dfrac{6}{5}$ **29.** $x=0,\ x=3$ **31.** $x=0,\ x=-\dfrac{14}{3}$ **33.** $n=-3,\ n=6$ **35.** $x=3,\ x=7$ **37.** $x=1$

39. $x=\dfrac{5}{2},\ x=-4$ **41.** $x=-\dfrac{5}{2},\ x=6$ **43.** $x=-7,\ x=4$ **45.** $x=4,\ x=2$ **47.** $x=-\dfrac{8}{3},\ x=5$ **49.** You can always factor out x.

51. $a=\dfrac{14}{9},\ c=-14$ **53.** 190 truck routes   **55.** 7 cities   **57.** 11 cities   **59.** 140 or 160 machines

61. The average is 150 machines. The cost of producing 150 machines is \$4500, which is less than the cost of producing 140 or 160 machines.
63. The cost of producing 148 machines or 152 machines is \$4500.80. This result is less than the result of exercise 59 but more than the result of exercise 61. It appears that the minimum cost of \$4500 occurs when 150 machines are manufactured.
64. $\dfrac{(2x-1)(x+2)}{2x}$ **65.** $\dfrac{2x^2-4x+2}{3x^2-11x+10}$ **66.** approximately 4.32 mg   **67.** approximately 130 mg

10.2 Exercises

1. $x=\pm8$ **3.** $x=\pm7\sqrt{2}$ **5.** $x=\pm2\sqrt{10}$ **7.** $x=\pm3$ **9.** $x=\pm2\sqrt{5}$ **11.** $x=\pm2\sqrt{7}$ **13.** $x=\pm2\sqrt{3}$ **15.** $x=\pm\sqrt{5}$

17. $x=3\pm\sqrt{5}$ **19.** $x=-4\pm\sqrt{6}$ **21.** $x=\dfrac{-5\pm\sqrt{2}}{2}$ **23.** $x=\dfrac{1\pm\sqrt{7}}{3}$ **25.** $x=\dfrac{-2\pm2\sqrt{3}}{7}$ **27.** $x=\dfrac{3\pm5\sqrt{3}}{5}$

29. $x=3\pm2\sqrt{5}$ **31.** $x=-3\pm\sqrt{2}$ **33.** $x=6\pm\sqrt{41}$ **35.** $x=0,\ x=-3$ **37.** $x=\dfrac{3}{2},\ x=\dfrac{1}{2}$ **39.** $x=\dfrac{9}{2},\ x=-1$

41. $x=\dfrac{-b\pm\sqrt{b^2-4c}}{2}$ **42.** $a=-4,\ b=-4$ **43.** $x=1,\ x=5$ **44.** 200 pounds per square inch

45. A total of 16 tires were defective. 6 had both kinds of defects, 7 had only defects in workmanship, and 3 had only defects in materials.

10.3 Exercises

1. $x = \dfrac{-3 \pm \sqrt{29}}{2}$ **3.** $x = \dfrac{3 \pm \sqrt{41}}{2}$ **5.** $x = 2, x = -\dfrac{3}{4}$ **7.** $x = \dfrac{5 \pm \sqrt{17}}{4}$ **9.** $x = \dfrac{3 \pm \sqrt{33}}{12}$ **11.** $x = \dfrac{1 \pm \sqrt{19}}{6}$

13. no real solution   **15.** $x = 6, x = 1$ **17.** $y = \dfrac{-7 \pm \sqrt{109}}{10}$ **19.** $d = \dfrac{4}{3}, d = -3$ **21.** $x = 0.372, x = -5.372$

23. $x = 0.686, x = -2.186$ **25.** $x = -1.894, x = -0.106$ **27.** $t = \dfrac{3}{2}, t = \dfrac{2}{3}$ **29.** $s = 3, s = \dfrac{1}{3}$ **31.** $t = -1 \pm \sqrt{23}$ **33.** $y = \dfrac{1 \pm \sqrt{51}}{5}$

35. $x = \pm\dfrac{\sqrt{55}}{5}$ **37.** $x = 1 \pm 2\sqrt{2}$ **39.** 2 feet   **41.** $\dfrac{3(x + 5)}{x + 2}$ **42. (a)** $y = -\dfrac{5x}{13} - \dfrac{14}{13}$ **(b)** $5x + 13y = -14$

43. $x^2 + 5x + 2$ **44.** $\dfrac{9p - 20}{3f} = g$ **45.** 96.76 cubic feet   **46.** 138.16 square inches

10.4 Exercises

1.
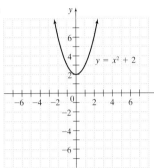

3.
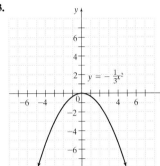

5.
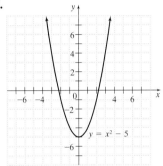

7.
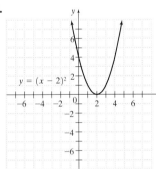

9.
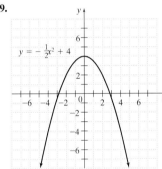

11.
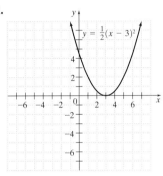

13.
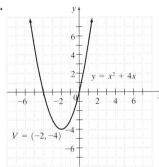

15.
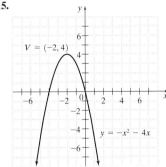

17.
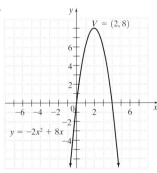

19.
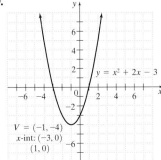

21.
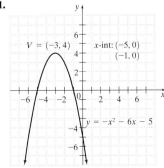

23.
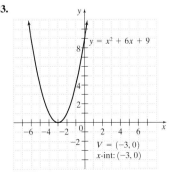

25. (a)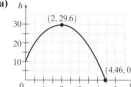

(b) 24.7 meters **(c)** 29.6 meters **27. (a)** 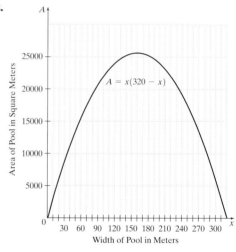 **(b)** 4.5 inches **(c)** 20.25 million
(d) about 4.46 seconds

29. vertex $(5.3, -240.25)$, y-intercept $(0, -212.6)$, x-intercepts $(10.2, 0)$ and $(-20.8, 0)$ **31.** vertex $(-0.277, 1551.164)$, y-intercept $(0, 1528)$,
x-intercepts $(1.993, 0)$ and $(-2.548, 0)$ **32.** $y = 1$ **33.** $y = 75$ **34.** 8.25 seconds

Putting Your Skills to Work

1. 11.4 feet **2.** 11.4 feet **3.** 3 yards **4.** 5 yards

10.5 Exercises

1. altitude is 16 centimeters, base is 11 centimeters **3.** 11 ft by 11 ft **5.** 21 people **7.** 80 members were expected. 100 members attended.
9. The width is 25 feet. The length along the building is 50 feet. **11.** 600 mph **13.** 70 **15.** 10 **17.** 1994 and 1995 **19.** $1750 **21.** 13
23.   **24.** $\sqrt{15} - 2\sqrt{3}$ **25.** 0.75 feet

Putting Your Skills to Work

1.

x	10	20	40	80	160
A	3100	6000	11,200	19,200	25,600

2.

Chapter 10 Review Problems

1. $a = 6, b = -5, c = 8$ **2.** $a = 14, b = 3, c = -16$ **3.** $a = 1, b = -9, c = -3$ **4.** $a = 3, b = 6, c = -10$ **5.** $a = 5, b = -3, c = 1$

6. $a = 4, b = 5, c = -14$ **7.** $x = -25, x = -1$ **8.** $x = -8$ **9.** $x = 5, x = -11$ **10.** $x = \dfrac{3}{2}, x = \dfrac{5}{2}$ **11.** $x = \dfrac{4}{3}$ **12.** $x = \dfrac{1}{5}, x = 1$

13. $x = \dfrac{1}{5}, x = \dfrac{3}{4}$ **14.** $x = -\dfrac{11}{3}, x = -1$ **15.** $x = -\dfrac{3}{2}, x = \dfrac{4}{3}$ **16.** $\dfrac{1}{3}, -\dfrac{3}{2}$ **17.** $\dfrac{1}{5}, -\dfrac{1}{3}$ **18.** $\dfrac{1}{4}, -\dfrac{4}{3}$ **19.** $0, -4$ **20.** $0, 3$

21. $0, -\dfrac{5}{3}$ **22.** $0, 4$ **23.** $-\dfrac{2}{5}, 10$ **24.** $\dfrac{1}{3}, \dfrac{4}{3}$ **25.** ± 5 **26.** ± 6 **27.** $\pm\sqrt{22}$ **28.** $\pm\sqrt{39}$ **29.** $\pm 2\sqrt{2}$ **30.** $x = \pm 5\sqrt{2}$ **31.** $\pm 3\sqrt{2}$

32. $\pm 2\sqrt{6}$ **33.** $4 \pm \sqrt{7}$ **34.** $2 \pm \sqrt{3}$ **35.** $\dfrac{-6 \pm 2\sqrt{5}}{5}$ **36.** $\dfrac{-8 \pm 2\sqrt{3}}{3}$ **37.** $x = 1, x = -11$ **38.** $x = -1, x = -11$ **39.** $9, -5$

40. $6 \pm 2\sqrt{14}$ **41.** $-1 \pm \sqrt{3}$ **42.** $\dfrac{-5 \pm \sqrt{31}}{2}$ **43.** $-2 \pm \sqrt{10}$ **44.** $-2 \pm 2\sqrt{3}$ **45.** $\dfrac{7 \pm \sqrt{17}}{4}$ **46.** $\dfrac{-5 \pm \sqrt{73}}{4}$ **47.** $\dfrac{-2 \pm \sqrt{14}}{2}$

48. $\dfrac{-3 \pm \sqrt{19}}{2}$ **49.** $\dfrac{5 \pm \sqrt{73}}{6}$ **50.** $\dfrac{-3 \pm \sqrt{41}}{8}$ **51.** $\dfrac{5}{2}, 2$ **52.** $\dfrac{3}{2}, -\dfrac{1}{2}$ **53.** $-\dfrac{1}{5}$ **54.** $4, \dfrac{3}{2}$ **55.** $\dfrac{3 \pm \sqrt{3}}{3}$ **56.** $\dfrac{7 \pm \sqrt{209}}{10}$

57. $\dfrac{-2 \pm \sqrt{19}}{3}$ **58.** $\dfrac{-7 \pm \sqrt{29}}{10}$ **59.** $\dfrac{-3 \pm \sqrt{5}}{2}$ **60.** $-3, \dfrac{2}{3}$ **61.** ± 6 **62.** $\pm 3\sqrt{2}$ **63.** -8 only **64.** -7 only **65.** $\dfrac{-4 \pm \sqrt{31}}{3}$

66. $\dfrac{-4 \pm \sqrt{2}}{2}$ **67.** $\dfrac{11 \pm \sqrt{21}}{10}$ **68.** $-4, 2$ **69.** $-3, 5$ **70.** $0, \dfrac{1}{6}$

71.

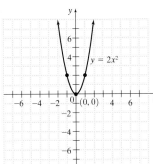

72.

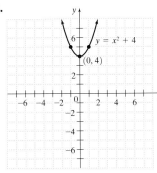

73.

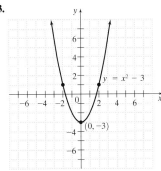

74.

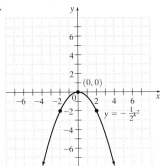

75.

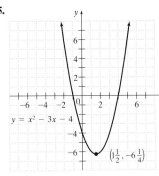

76.

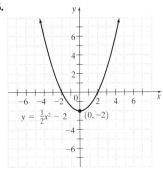

77.

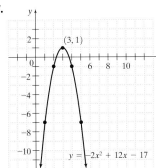

78.

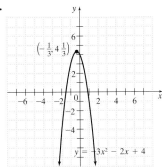

79. 6 feet by 2.5 feet **80.** 16 centimeters, 12 centimeters **81.** 24 members **82.** 5 and 12 **83.** 38.25 ft^2
84. going 45 mph, returning 60 mph **85.** 60 pizza stores **86.** 30 fewer pizza stores **87.** 1990 and 2020 **88.** 2000 and 2010

Chapter 10 Test

1. $\dfrac{-7 \pm \sqrt{129}}{10}$ **2.** $-5, \dfrac{2}{3}$ **3.** no real solution **4.** $-\dfrac{1}{2}, \dfrac{5}{3}$ **5.** $-\dfrac{5}{4}, \dfrac{1}{3}$ **6.** $\dfrac{4}{3}$ **7.** $0, 8$ **8.** ± 4 **9.** $-\dfrac{1}{2}, 6$ **10.** $\dfrac{3}{2}, -\dfrac{1}{2}$

11.

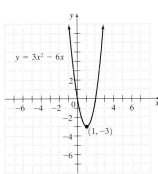

12.

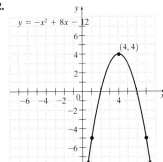

13. 9 meters, 12 meters **14.** 7 seconds

Cumulative Test for Chapters 1–10

1. $-27x^2 - 30xy$ **2.** 35 **3.** 2 **4.** $(2x - 1)(2x + 1)(4x^2 + 1)$ **5.** $6x^3 + 17x^2 - 31x - 12$ **6.** $-\dfrac{12}{5}$

7.

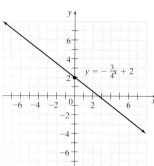

slope $-\dfrac{3}{4}$; y-intercept = $(0, 2)$

$y = -\dfrac{3}{4}x + 2$

8. $y = -3x + 5$ **9.** $(3, -2)$ **10.** $81x^{12}y^8$ **11.** $3x^2y^3z\sqrt{2xz}$ **12.** $-2\sqrt{6} - 8$ **13.** $-5\sqrt{5} + 11$ **14.** $r = \dfrac{A \pm \sqrt{A^2 + 3A}}{3}$

15. $b = \pm\dfrac{\sqrt{H + 6}}{5}$ **16.** $\dfrac{7}{2}, -5$ **17.** $\dfrac{-1 \pm 2\sqrt{5}}{2}$ **18.** $-12, 4$ **19.** $\dfrac{-11 \pm \sqrt{97}}{6}$ **20.** $\dfrac{-1}{7}, -5$ **21.** $\pm 7\sqrt{2}$ **22.** ± 5

23.

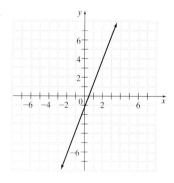

$y = x^2 + 6x + 10$

$(-3, 1)$

24. length 16 meters, width 6 meters

Practice Final Examination

1. $-2x + 21y + 18xy + 6y^2$ **2.** 14 **3.** $18x^5y^5$ **4.** $2x^2y - 8xy$ **5.** $x = \dfrac{8}{5}$ **6.** $b = \dfrac{p - 2a}{2}$ **7.** $x \geq 2.6$

8. $2x^3 - 5x^2y + xy^2 + 2y^3$ **9.** $2(2x + 1)(x - 5)$ **10.** $25(x^2 + 1)(x + 1)(x - 1)$ **11.** $\dfrac{6x - 11}{(x + 2)(x - 3)}$ **12.** $\dfrac{11(x + 2)}{2x(x + 4)}$ **13.** $-\dfrac{12}{5}$

14. slope $= \dfrac{5}{2}$

15. $3x + 4y = 14$ **16.** 38.13 sq in. **17.** $x = 7, y = -2$ **18.** $a = -2, b = -3$ **19.** $8x\sqrt{2}$ **20.** $6\sqrt{3} - 12 + 12\sqrt{2}$

21. $-\dfrac{\sqrt{15} + \sqrt{35} + \sqrt{21} + 7}{2}$ **22.** $\dfrac{5}{3}, -1$ **23.** $\dfrac{3 \pm \sqrt{7}}{2}$ **24.** ± 2 **25.** $\sqrt{39}$ **26.** 5 **27.** width 7 meters, length 12 meters

28. \$3000 at 10%, \$4000 at 14% **29.** 200 reserved-seat tickets, 160 general admission tickets **30.** base 8 meters, altitude 17 meters

Appendix B Exercises

1. 34,000 m   **3.** 5700 cm   **5.** 250 mm   **7.** 0.563 m   **9.** 29,400 mg   **11.** 0.0984 kg   **13.** 7000 mL   **15.** 0.000004 kL   **17.** 1.7 in.
19. 22.5 km   **21.** 2.8 m   **23.** 17.8 cm   **25.** 1089.6 g   **27.** 171.6 lb   **29.** 2.8 L   **31.** 193,248,000 in.   **33.** 27.27 miles per hour   **35.** 0.11 yr

Appendix C Exercises

1. 84,916 square miles   **3.** 3,500,000   **5.** 2   **7.** Treasure State   **9.** Nevada   **11.** 2000 homes   **13.** Dupage County   **15.** 2800 homes
17. 4,000,000 apartment units   **19.** 7,200,000 more apartment units   **21.** 8,000,000   **23.** It is a three-way tie. The increase was 3 million in three cases. It occurred between 1970 and 1980, and again between 1980 and 1990, and finally again between 1990 and 2000.   **25.** 3 million more
27. 37 million   **29.** Movies and an Exercise Program   **31.** $5.6 million   **33.** $2.5 million greater   **35.** Between 1993 and 1994
37. 74 million   **39.** an age group of 65 years and above   **41.** 19%   **43.** 83%   **45.** 754 million people   **47.** 40%   **49.** $3588

Appendix D Exercises

1. 14   **3.** 52   **5.** 1296   **7.** −3   **9.** $10x - 5$ **11.** 1111811   **13.** 9876549   **15.** 6481100   **17.** 1   6   15   20   15   6   1

19. $x = 3$, steps will vary   **21.** $x = -2$, steps will vary   **23.** $x = 3$, steps will vary   **25.** $x = \dfrac{9}{11}$, steps will vary   **27.** answers will vary

29. William was first, Brent was second, James was third, Dave was fourth   **31.** The teacher is Michael   **33.** Toyota Corolla   **35.** Unless Fred is willing to pay over $200 for Tuesday's procedure, he will probably get a false tooth.

Applications Index

Index

Photo Credits

CHAPTER 0   CO Corbis   **p. 31** Daemmrich/Stock Boston   **p. 39** Adam Jones/Photo Researchers, Inc.   **p. 43** Peter Hvizdak/The Image Works

CHAPTER 1   CO Charles Gupton/Stock Boston   **p. 68** Peter Skinner/Photo Researchers, Inc.   **p. 111** John Coletti/Stock Boston   **p. 114** Wojnarowicz/The Image Works

CHAPTER 2   CO Mark Richards/PhotoEdit   **p. 141** Jerry Wachter/Photo Researchers, Inc.   **p. 171** Jeffrey Dunn/Stock Boston   **p. 174** Stephen J. Krasemann/Photo Researchers, Inc.

CHAPTER 3   CO Larry Kolvoord/The Image Works   **p. 199** NASA/Science Source/Photo Researchers, Inc.   **p. 200** Townsend P. Dickinson/The Image Works
p. 201 Bob Daemmrich/Stock Boston   **p. 205** Dave Bartruff/Corbis
p. 217 Dana White/PhotoEdit   **p. 231** George Ranalli/Photo Researchers, Inc.
p. 240 Neil Rabinowitz/Corbis

CHAPTER 4   CO D. Young-Wolff/PhotoEdit   **p. 283** Bsip Laurent/Photo Researchers, Inc.

CHAPTER 5   CO Joseph Nettis/Stock Boston   **p. 305** Malcolm Fielding, Johnson Matthey PLC/Science Photo Library/Photo Researchers, Inc.   **p. 329** Alok Kavan/Photo Researchers, Inc.   **p.342** Laima Druskis/Stock Boston   **p. 348** Alan Schein/Corbis Stock Market

CHAPTER 6   CO Owen Franken/Stock Boston

CHAPTER 7   CO Kaz Mori/The Image Bank   **p. 438** Rondi/Tani/Science Source/Photo Researchers, Inc.   **p. 461** Jacques Jangoux/Stone

CHAPTER 8   CO Topham/The Image Works   **p. 487** Rob Crandall/The Image Works

CHAPTER 9   CO Jump Run Productions/The Image Bank
p. 537 World Perspectives/Stone   **p. 556** Juan Silva Production/The Image Bank
p. 571 Joseph McBride/Stone

CHAPTER 10   CO Trevor Wood/Stone   **p. 591** David Muench/Stone
p. 608 Bettmann/Corbis